Tenei te rūrū te koukou mai nei
Kīhai mahitihiti
Kīhai mārakaraka
Te upoko nui o te rūrū
TEREKOU!

He pō, he pō
He ao, he ao
Ka awatea!

THE NATURAL HISTORY OF SOUTHERN NEW ZEALAND

Edited by **John Darby, R. Ewan Fordyce, Alan Mark, Keith Probert, Colin Townsend**

University of Otago Press
PO Box 56/56 Union Street West
Dunedin, New Zealand
Ph: 64 3 479 8807. Fax: 64 3 479 8385
Email: university.press@otago.ac.nz

First published 2003

Published in association with the Otago Museum
ISBN 877133 51 5

Illustrated by Allan Kynaston
Printed in Hong Kong through Condor Production Ltd

Front cover: Moeraki Boulders, North Otago. *David Wall*
Back cover: Karearea, New Zealand falcon. *Rod Morris*
Title page: Sea tulip (see page 273). *Elizabeth Batham*

CONTENTS

EDITORS AND CONTRIBUTORS

Editors

John Darby, Assistant Director, Otago Museum, P.O. Box 6202, Dunedin (Coordinating Editor).

Associate Professor Ewan Fordyce, Department of Geology, University of Otago, P.O. Box 56, Dunedin.

Professor Alan Mark, Department of Botany, University of Otago, P.O. Box 56, Dunedin.

Dr Keith Probert, Department of Marine Science, University of Otago, P.O. Box 56, Dunedin.

Professor Colin Townsend, Department of Zoology, University of Otago, P.O. Box 56, Dunedin.

Contributors

Dr Ralph Allen, Landcare Research, P.O. Box 1930, Dunedin (now Wildland Consultants, c/o Landcare Research, P.O. Box 1930, Dunedin).

Dr Richard Allibone, Biodiversity Recovery Unit, Department of Conservation, P.O. Box 10420, Wellington.

Associate Professor Mike Barker, Department of Marine Science, University of Otago, P.O. Box 56, Dunedin.

Dr Barbara Barratt, AgResearch Ltd, Private Bag 50034, Mosgiel.

Professor Carolyn Burns, Department of Zoology, University of Otago, Box 56, Dunedin.

Tony Brett, Ministry of Fisheries, Private Bag 1926, Dunedin.

Dr Lionel Carter, National Institute Water and Atmospheric Research Ltd, P.O. Box 14901, Wellington.

Professor R.L. Carter, School of Earth Sciences, James Cook University, Townsville, Queensland 4811, Australia.

Jeff Connell, Otago Conservator, Department of Conservation, P.O. Box 5244, Dunedin.

Associate Professor David Craw, Department of Geology, University of Otago, P.O. Box 56, Dunedin.

Dr Alison Cree, Department of Zoology, University of Otago, P.O. Box 56, Dunedin.

John Darby, Assistant Director, Otago Museum, P.O. Box 6202, Dunedin.

Associate Professor Lloyd Davis, Department of Zoology, University of Otago, P.O. Box 56, Dunedin.

Dr Norm Davis, Department of Zoology, University of Otago, P.O. Box 56, Dunedin.

Dr Stephen Dawson. Department of Marine Science, University of Otago, P.O. Box 56, Dunedin.

Dr Jenifer Dugan, Marine Science Institute, University of California, Santa Barbara, CA 93106, U.S.A.

Eric Edwards, Department of Conservation, Southland Conservancy, P.O. Box 743, Invercargill.

Barry Fahey, Landcare Research, P.O. Box 69, Lincoln. Canterbury.

Professor Blair Fitzharris, Department of Geography, University of Otago, P.O. Box 56, Dunedin.

Associate Professor Ewan Fordyce, Department of Geology, University of Otago, P.O. Box 56, Dunedin.

Dr Lyn Forster, Otago Museum, P.O. Box 6202, Dunedin.

Dr Ray Forster, Director, Otago Museum, P.O. Box 6202, Dunedin. (Deceased).

Anthony Harris, Otago Museum, P.O. Box 6202, Dunedin.

Dr Jill Hamel, 42 Ann Street, Roslyn, Dunedin.

Dr Michael Heads, Department of Biological Sciences, University of the South Pacific, Suva, Fiji.

Dr Alan Hewitt, Landcare Research, P.O. Box 1930, Dunedin.

Dr David Hubbard, Museum of Systematics and Ecology, Department of Ecology, Evolution and Marine Biology, University of California, Santa Barbara, CA 93106, U.S.A.

Associate Professor Alex Huryn, Department of Biological Sciences, University of Maine, 5722 Deering Hall, Orono, Maine 04469-5722, U.S.A.

Associate Professor John Jillett, Department of Marine Science, University of Otago, P.O. Box 56, Dunedin.

Dr Peter Johnson, Landcare Research, P.O. Box 1930, Dunedin.

Dr William Lee, Landcare Research, PO Box 1930, Dunedin.

Professor Alan Mark, Department of Botany, University of Otago, P.O. Box 56, Dunedin.

Dr David Galloway, Landcare Research, P.O. Box 1930, Dunedin.

Dr Matt McGlone, Landcare Research, P.O. Box 69, Lincoln.

Professor Philip Mladenov, Department of Marine Science, University of Otago, P.O. Box 56, Dunedin.

Dr Rick McGovern-Wilson, Senior Archaeologist, Historic Places Trust, P.O. Box 2629, Wellington.

Professor Richard Norris, Department of Geology, University of Otago, P.O. Box 56, Dunedin.

Brian Patrick, Manager Collections and Research, Otago Museum, P.O. Box 6202, Dunedin.

Dr Barrie Peake, Department of Chemistry, University of Otago, P.O. Box 56, Dunedin.

Dr Keith Probert, Department of Marine Science, University of Otago, P.O. Box 56, Dunedin.

Dr Abigail Smith, Department of Marine Science, University of Otago, P.O. Box 56, Dunedin.

Dr Tony Reay, Department of Geology, University of Otago, P.O. Box 56, Dunedin.

Associate Professor Donald Scott, Department of Zoology, University of Otago, P.O. Box 56, Dunedin.

Dr Ian Smith, Department of Anthropology, University of Otago, P.O. Box 56, Dunedin.

Associate Professor Hamish Spencer, Department of Zoology, University of Otago, P.O. Box 56, Dunedin.

Susan Walker, Landcare Research, P.O. Box 1930, Dunedin.

Associate Professor Graham Wallis, Department of Zoology, University of Otago, P.O. Box 56, Dunedin.

Dr Peter Wardle, Landcare Research, P.O. Box 69, Lincoln.

Dr Jon Waters, Department of Zoology, University of Otago, P.O. Box 56, Dunedin.

Dr John B Ward, Canterbury Museum, Rolleston Avenue, Christchurch.

Dr Susan Walker, Landcare Research, Private Bag 1930, Dunedin

Associate Professor David Wharton, Department of Zoology, University of Otago, P.O. Box 56, Dunedin.

Trevor Worthy, Paleofaunal Surveys, 2A Willow Park Drive, Masterton.

ACKNOWLEDGEMENTS

The Editors would like to thank the many writers, photographers and illustrators who have contributed to this book: their patience and skills have been much appreciated. In particular, we thank Edward Ellison and Neville Peat for their opening pieces. Allan Kynaston has made a wonderful job of the graphics and we are grateful for his attention to detail. Particular thanks are due to Mrs Joy Soper who made the photographic collection of her late husband Dr Mike Soper freely available. Likewise, Rod Morris has donated his photographic expertise to this work. Peter Batson, Bill Brockie, Jim Fyfe, Brian Stewart, Laurel Tierney and Barbara Williams gave advice on chapters 11 and 12.

John Darby would like to thank Professor Carolyn Burns and staff of the Department of Zoology, University of Otago, for their welcome to the department and for providing facilities for work on the final stages of this book.

The Managing Editor of the University of Otago Press, Wendy Harrex, produced this book. The Editors are grateful for her unswerving commitment to this project while coping with inadequate facilities. Thanks also to Elaine Ross, Jenny Cooper, Fiona Moffat and Dr Susan Hallas for their assistance with typesetting, design and copy-editing.

MIHI

Nō muri rā anō mātou tēnei taiao i tiaki mai i kā tāpuhipuhi mauka tae atu ki te aumoana, ā, whaiatia toutia mō te turoataka o tēnei ao.

First, I speak on behalf of the Kāi Tahu people. We have long acted as guardians of this natural world, from mountain summits to ocean depths, and we continue to seek allies in our struggle for sustainability.

Haere tou ōku mihi ki a koutou, nāhana ō koutou huatau kāmehameha i waiho.

Second, I wish to acknowledge the many authors and contributors who have seen fit to offer their own thoughts and expertise to this valuable work.

Ka mutu, maioha tou atu tēnei ki tēra i whakarauika mai. Kāhore e kore, ka tipu tō tātou māramataka, ka tika.

Furthermore, it is important to pay tribute to the intent of this collection. Our understanding of the natural world and human interaction with it will be continually enhanced.

Heoti anō, koutou i aka mai ki tēnei wānaka-ā-pukapuka. Naia te manuka i whakatakoto ai, kia tiki rānei koe. I te mea tēnei kaupapa he whakaharahara ki te ao pāhekeheke, ā, mo kā uri ā muri ake nei.

In conclusion, I invite you the reader to grow in your knowledge and understanding of the many issues and challenges identified in the following pages. It is through your enhanced knowledge and understanding that we can achieve sustainable outcomes for the present and future generations.

EDWARD ELLISON

Aerial view to the east across the southern Hunter Mountains in southeastern Fiordland showing the 27 km^3 Green Lake Landslide, one of the world's largest. Extending some nine kilometres, from the far shores of Green Lake (right) along the Hunter Mountains to Mt Burns (1645 m) and the road across Borland Saddle (980 m) just below the treeline (left), this slide is about one kilometre deep and extends three kilometres down to the Grebe Valley (foreground). The slide probably resulted from a large (magnitude 7 or greater) earthquake towards the end of the Last Glacial, about 13,000 years ago. Debris from the slide has choked the Grebe Valley, which is now occupied by an extensive wetland. Evergreen beech forest occupies most of the undulating surface of the landslide, apart from several lakes and tarns, and irregular patches of copper tussock grassland in other depressions. The Takitimu Mountains show in the distance. *Lloyd Homer, photo supplied by IGNS. Crown Copyright.*

FOREWORD

Southern New Zealand projects to the rest of New Zealand – and to the world – images of diverse habitats, unique biota, wild places and remoteness. Here is a region of mountains, lakes, rivers and fiords, of breathtakingly beautiful scenery and sometimes weird wildlife, of landscapes and climatic conditions so contrasting you would swear they were in different countries instead of next door to each other.

From the Mount Aspiring region all the way south to Stewart Island/Rakiura, superlatives leap out of the landscape . . . one of New Zealand's tallest, most glaciated mountains, its deepest lakes, largest river, widest stretches of wilderness. Climatic diversity chimes in with the coldest, driest and windiest localities in New Zealand, and some of the hottest and wettest places.

As you move across the region, the habitats change with astonishing rapidity. An east-west transect along latitude 45 degrees south, which cuts through North Otago, Central Otago and Fiordland, touches grassland, herbfield, alpine cushionfield and fellfield, saltpans, gravel river terraces, an ultramafic zone, lakeshore, rainforest, fiord and a wave-pounded coastline. In keeping with habitat diversity of this order, southern New Zealand lives and breathes biological diversity. Many species of flora and fauna are unique to the region. They evolved here and survive nowhere else. Some have but a critical toehold. Southern New Zealand is the last refuge for the world's heaviest parrot (kakapo), largest rail (takahe) and most endangered kiwis (Haast tokoeko and Okarito brown kiwi). It provides a breeding platform for uncommon and threatened marine life, such as sea lions, yellow-eyed penguins and royal albatrosses. Numerous lizards, fish and invertebrate animals are specifically southern, and in the world of plants, the endemics are spread from near sea level to the highest places plants can grow in this region.

The first humans here must have gasped at the strangeness of the landforms, plants and animals. The human experience is relatively short, for New Zealand was the last temperate landmass to be settled by people, and its southern region no doubt the last bit to be explored, from about 800 years ago.

Just as it was the last frontier and southern outpost of Polynesia, it remains a frontier of world significance for the life and earth sciences. In recognition of its outstanding wildlife and dynamic geology, a vast tract of southern New Zealand has had World Heritage status conferred on it, under the formal title, Te Wahipounamu Southwest New Zealand. The area included in this book covers three national parks – Fiordland, Mount Aspiring and Rakiura – and a good deal of intervening land.

Actually, Stewart Island's Rakiura has only recently become the country's fourteenth national park and is the only one outside the two main islands. Stewart Island is a treasure island – an emerald isle where rainforest descends to the shoreline and granite domes shine greyly through a squally climate, looking like something out of Middle Earth.

Natural history is more than a catalogue of terrestrial, freshwater and marine species and their habitats. It starts from the origins of the rocks and works upwards and outwards, gathering complexity. It tries to describe not only the species, but also how they interact with each other and with their environments.

To understand the existing landforms, flora and fauna you need to delve into the past. Acknowledging this, the five editors of this book have commissioned chapters on geological history, the fossil record, biogeography, climates past and present, environmental change since the last glaciation, and how humans and their other camp followers, the mammalian predators in particular, have impacted on the indigenous life. Nothing stays still for long – even the rocks, as earthquakes remind us – and issues to do with climate change, conservation and the possibility of new arrivals on the scene are never far from the natural history arena. There are new frontiers to explore, especially in the ocean. Ecology is a journey just begun.

Much has been written about the nature of southern New Zealand in scientific journals and popular publications. This work takes the published knowledge to a new and unprecedented level of richness and detail. Fifty-three authors, most from scientific disciplines and leaders in their specialist fields, describe the nature of the region in thirteen chapters.

At the same time, you can never know everything about species, about how they evolved, and how their relationships are panning out today. Some things are bound to remain mysterious.

Southern New Zealand natural history generates so many powerful images and countless intriguing stories. The front cover features an image of the Moeraki boulders, which are possibly New Zealand's best-known set of rocks. They are symbols of the dynamic nature of the land and of culture coming to terms with nature, as through the Araiteuru tradition that explains the presence of these spheres on a North Otago beach. Although the stories here are essentially scientific, they leave the door open to further enquiry and interpretation. Understanding nature is the ultimate challenge.

NEVILLE PEAT

CHAPTER 1

GEOLOGY

TONY REAY

Macraes in Otago is the largest gold mine in New Zealand. Mining commenced in 1990 and on 20 January 2002 produced its 1,000,000th ounce of gold. *J. Youngson*

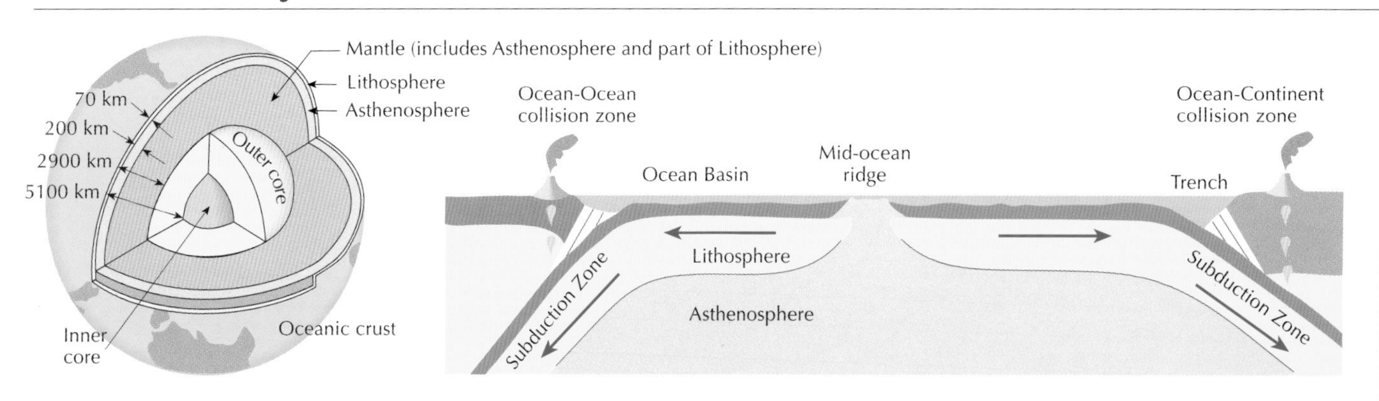

Fig. 1. Cross-section of the earth, showing the major subdivisions.

Fig. 2. Schematic representation of plate tectonics. New thin oceanic crust produced at the mid-ocean ridges moves towards the trenches, where it is being destroyed in subduction zones.

1 PLATE COLLISION AND THE ALPINE FAULT

The Alpine Fault is one of New Zealand's major geological features. It is a major fracture in the earth's crust running from Fiordland in the south to Arthurs Pass, where it splits into several north-east trending faults. It was first recognised by Harold Wellman of the New Zealand Geological Survey in 1953.

The fault is the present on-land contact between the Pacific Plate and the Australian Plate and links two subduction zones (Fig. 3). The southern subduction zone underlies southern Fiordland, where the Australian Plate is being dragged down in an easterly direction under the Pacific Plate. The Pacific Plate is being subducted westward under the Australian Plate at the northern subduction zone.

The Alpine Fault is an example of a transform fault and has been described as one of the great faults of the world. It first formed about 25 million years ago. Since then, the rocks on the north-west (Nelson) side of the fault have moved northeast relative to those on the south-east (Otago) side by a distance of 480 km (Fig. 3). This equates to an average rate of movement of 19.2 mm a year, although this is not continuous. Geological research on the Alpine Fault north of Milford Sound has demonstrated that it moves by a series of discrete jumps which amount to about 8 m every approximately 200 years.

Movement on the Alpine Fault is the result of oblique collision between the two plates. Currently, this oblique collision is a consequence of the movement of the Pacific Plate relative to the Australian Plate. This results in a displacement of the Pacific Plate of 39 mm per year in a south-west direction. As a result, Christchurch will move to the present position of Milford Sound, 400 km away, in just over ten million years.

The collision between the plates causes the rapid uplift of the Southern Alps. Rates of uplift can be calculated from the presence of dated marine fossils now above sea level on the West Coast and have been shown to be fastest in the Mt Cook area. Uplift rates are 5mm to 10 mm/year (5000 to 10,000 m/ million years). However, the height of the Southern Alps (Aoraki/Mt Cook is 3764 m) is a function of dynamic interaction between uplift and erosion. As uplift takes place, the prevailing westerly winds are forced to pass over higher and higher ranges, generating increasingly heavy rainfall. Rainfall leads to increased slope and stream erosion and this in turn lowers the mountain chain.

Uplift along the Alpine Fault also means that rocks are uplifted faster than they can cool, so that hot rock lies close to the surface. Water percolating through the rock becomes heated, rises, and emerges as hot springs, for example in the Copland Valley. This hot water also deposits minerals; some of the gold veins in the Alps, such as those in the Callery River, may form in this way. Gold may be depositing from hot springs at the present time.

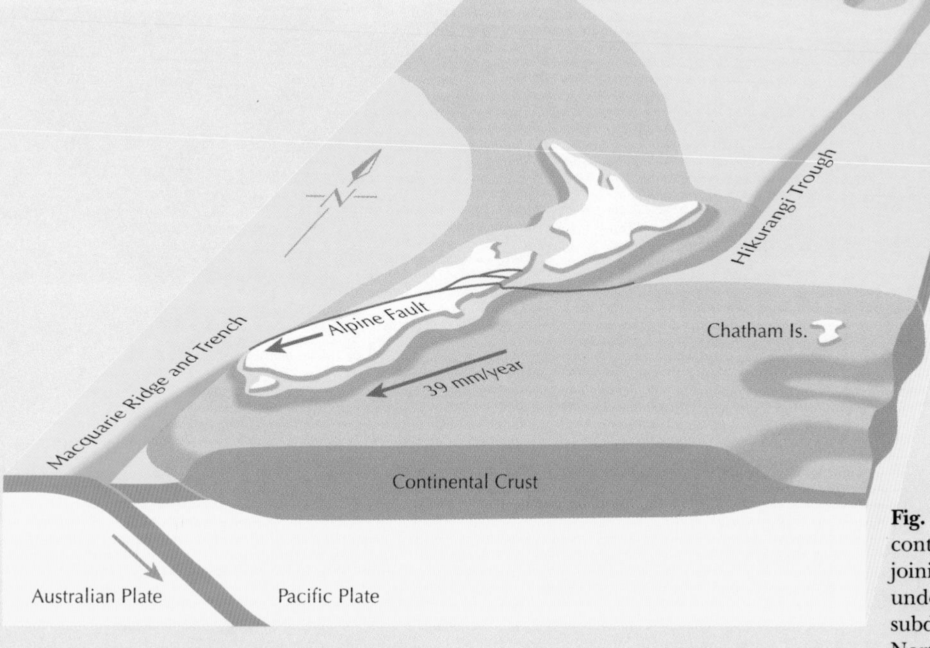

Fig. 3. Tectonic setting of New Zealand. The continental crust is cut by the Alpine Fault joining the east-dipping subduction zone under Fiordland, with the west-dipping subduction zone under the east coast of the North Island. (After Stevens, 1980)

GEOLOGY

New Zealand is known colloquially as the 'shaky isles': a reference to its large number of earthquakes. They are one of many processes that are happening within a very active geological environment. Earthquakes are the result of movement within the outer few hundred kilometres of the earth, and are an indication that the earth is an active body. The major expression of this activity is the movement of large sections of the crust, or 'plates' (about 100 km thick), over the outer part of the earth. This movement started soon after the earth was formed 4.56 billion years ago and continues today, with Otago moving south-west at a rate of 39 mm per year (Box 1).

An examination of an outline map of the world leads to speculation on the complementary shapes of the west coast of Africa and the east coast of South America, an observation made in 1620 by Francis Bacon. Despite numerous attempts to explain this close match, geologists had to wait until the 1960s for an adequate theory to account for the close fit of these two major continents. Since then, this theory, called Plate Tectonics, has revolutionised the way in which we think about the earth.

The crust of the earth is of two types: thick and usually old 'continental crust' (Fig. 1) forming the continents, and thin young 'oceanic crust' flooring the oceans. New oceanic crust is produced at the ocean ridge systems, which lie within and often central to the ocean basins, e.g. the mid Atlantic Ridge (Fig. 2). As more and more new crust is formed by the injection of hot liquid magma from below the mid-ocean ridge, the previously formed older oceanic crust is moving sideways. In other words, the oceanic crust is moving. Rates vary from 1–5 cm a year for the mid Atlantic Ridge, to 18 cm a year for the east Pacific sea floor. These rates of movement are about the speed at which fingernails grow. In the context of the geological time scale, a rate of 1 mm per year equals 1 km in 1 million years. The continents also move over the surface of the earth at similar variable velocities; in recent years, precise measurement using satellites has con-firmed these rates of plate movement.

The surface of the earth is covered by six large plates, each of which usually includes one of the major continents. In addition, many small plates exist. The plates are rigid, and are able to slide over a weak layer in the earth's interior called the asthenosphere (Fig.1). The movement is generated by processes internal to the earth. Since the plates are moving at different rates, they inevitably collide with one another; the junctions between plates are scenes of intense geological activity. The contact between two plates is known as the plate boundary, an area of very active geological processes such as earthquakes, volcanic eruptions, and mountain building. Fig. 2 illustrates the major types of plate boundary.

Although plate tectonic theory has led to a better under-standing of the way in which the earth operates, this insight has been aided by a second geological revolution: the use of radio-active isotopes to date rocks and minerals. We can now date most rocks with a high degree of precision, and thus provide the time frame necessary to understand geological processes. In an example of the use of the technique, Dr Trevor Ireland, born in Oamaru and educated at University of Otago, has dated meteorites at 4.56 billion years, providing the best estimate of the time of origin of the earth. A time frame has been developed for southern New Zealand and is given in Fig. 4, along with the geological scale of time.

The South Island of today sits astride a plate boundary separating the Australian Plate to the west from the Pacific Plate (Fig. 3). Originally part of south-eastern Australia, which itself was a section of the supercontinent Gondwana, the southern half of the South Island is made up of at least six different ancient plates. These collided and stuck together on the edge of Gondwana over a period of tens of millions of years. Such plates, or fragments of plates, each with its own geological history, are known as 'terranes'.

Gondwana started breaking up into several plates about 160 million years ago. At about 85 million years ago, New Zealand separated and drifted southeast from Australia, creating new oceanic crust and forming the Tasman Sea. Since that time, geological activity within the southern part of the South Island has been more sedate, although it has included the development of lakes in the Central Otago area, flooding by sea of most of eastern Otago and much of Southland, and the widespread formation of volcanoes over North and East Otago. This chapter looks at the geological development of southern New Zealand during the past 450 million years, working through from the oldest geological events to the present day.

THE BASEMENT OF OTAGO/SOUTHLAND

The oldest rocks in an area are often referred to as 'the basement', because they are the foundation for most subsequent geological activity. Within the South Island, the oldest rocks – which occur in Fiordland and north-west Nelson – were laid down under the sea about 500 million years ago. The oldest rocks in Otago/Southland are fragments of plates which were stitched together by about 110 million years ago. The terrane map of the South Island (Fig. 5) shows the present distribution

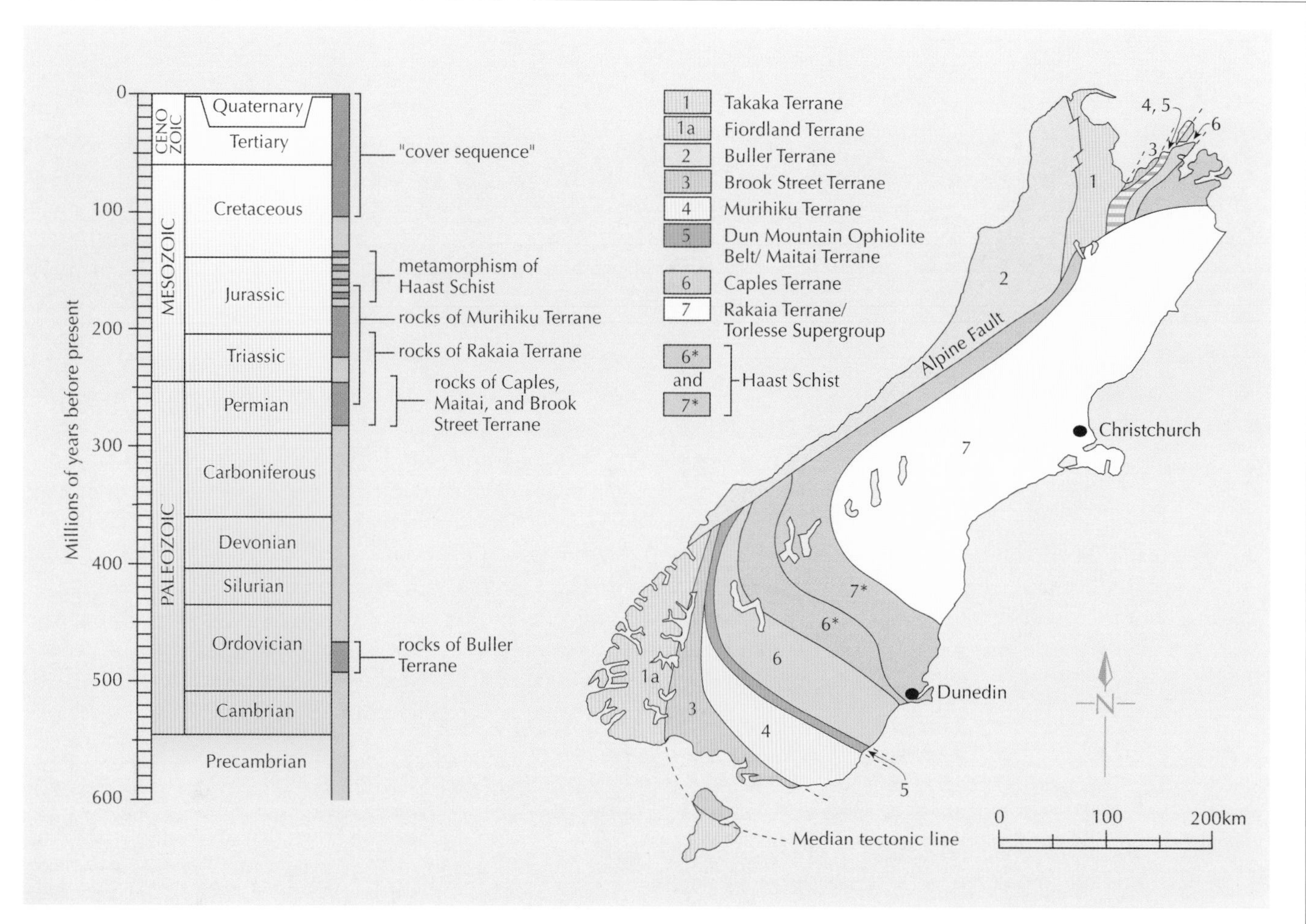

Fig. 4. Geological time scale and sedimentary rocks of the South Island.

Fig. 5 (top right). The South Island, showing distribution of the major terranes, remnants of ancient plates. *After Bishop et al., 1985*

Fig. 6 (below). Skippers Range, east of Lake McKerrow. Brook Street volcanics appear in the foreground with Fiordland Terrane, Mt Tutoko, in the background. *A. Reay*

of these rocks. A terrane is a large body of rock which is internally continuous and is bounded by faults. It is usually named with reference to a key locality. The Fiordland Terrane is a fragment of the continental crust of Gondwana, on to which the other five terranes have been accreted. These five terranes consist essentially of material from island arc volcanoes and sediments deposited over a period of 180 million years. Each terrane will be described in order, from west to east.

The Fiordland Terrane

Most westerly of the terranes is the Fiordland Terrane, which lies to the west of the Waiau, Eglinton and Hollyford Valleys. Geologically, this is the least known area in New Zealand; a combination of remoteness, steep country, thick bush, heavy rain, sandflies and geological complexity has meant that we are just beginning to piece together its geological history. North-west Nelson is a fragment of the same terrane, which has been displaced 480 km north-east, along the Alpine Fault. (Box 1)

The rocks now exposed in the north-west Nelson region were formed at a shallower depth; they have suffered significantly less physical and chemical change and subsequently experienced less erosion than those exposed in Fiordland. As a result, the geological history of the northwest Nelson area is much better understood. Deeper burial of the Fiordland rocks has obscured some of the evidence as to their origin. However, recent isotopic dating studies have given a significantly better view of geological events in Fiordland. Even so, the geology is still constrained, with ideas on the region's geological history being modified at frequent intervals. Fiordland may contain more than one terrane, but it will be treated here as a single unit.

A major clue to the origin of the Fiordland Terrane is provided by the presence of fossils in the Preservation Inlet area. They are the oldest rocks, consisting of deep-water marine sediments about 450 million years old. Both the fossils and the rocks containing them are almost identical to those in south-eastern Victoria, in Australia. They were once part of Gondwana. Since about 370 million years ago (Devonian period), Fiordland has been undergoing complex geological activity, the details of which are still being worked out. This activity has included a cycle of deposition of sands and muds, deep burial, injection of magmas and subsequent uplift, which may have been repeated three times. The largest single geological unit is the Western Fiordland Orthogneiss, which was formed by the injection of molten rock at 120 million years (Early Cretaceous) ago. This unit, which now forms some of the highest mountains in the area (Fig. 6), stretches from just south of Milford Sound to Doubtful Sound, a distance of about 110 km. In eastern Fiordland, many granites were injected into the crust at the same time, and today are seen as peaks such as Titiroa, just south of Lake Manapouri, and the related granite along the Borland Road (Fig. 7).

An intriguing feature of the area is the evenness of the height of the summits of the mountains, which increases northwards from the south coast to reach a maximum of 2400 m in the Milford area. The simplest explanation is that this concordance represents an old erosion surface, possibly of marine origin, and hence once at sea level, which has been uplifted within the

Fig. 7. Tors of 120 million-year-old granite intruded along the eastern edge of the Fiordland Terrane, Borland Road. *A. Reay*

last 1.6 million years. If so, this represents an uplift rate of 1.5 mm a year (1.5 km in one million years).

The western boundary of the Fiordland Terrane is the Alpine Fault, forming the boundary between the Australian and Pacific Plates in this area and one of the best documented major fault boundaries in the world. The line of the Alpine Fault can be seen just north of the entrance to Milford Sound.

The contact between the Fiordland Terrane and the five terranes to the east is complex, and much of it is hidden beneath the gravels deposited by the rivers bounding eastern Fiordland. Where the contact is exposed, the interpretation is difficult and much effort is being expended at the present time in an attempt to characterise the nature of the boundary. This boundary, the Median Tectonic Line or Zone, separates the continental crust of the Fiordland Terrane from the five terranes to the east, most of which were originally laid down in an oceanic environment. Subsequent collision with the edge of Gondwana has increased their thickness and converted them to continental crust.

The Brook Street Terrane

This terrane is named from a locality in Nelson, and is present in western Southland as a result of displacement by the Alpine Fault. Outcrops on either side of the fault are now 480 km apart. The terrane is well displayed in Otago in the Skippers Range east of Lake McKerrow (Fig. 6). It continues southward through the Eglinton Valley, the Takitimu Mountains and the Longwood Range to Bluff. The Brook Street rocks are very well dated as

Permian, about 265 million years old. The rocks are a distinctive collection of volcanic lavas and ashes, and rocks produced by the erosion of these materials. They have a distinctive chemical signature that implies they formed part of a string of island arc volcanoes, where one oceanic plate was dragged below or 'subducted' beneath another (Fig. 2). A modern example would be the volcanoes developed along the Tonga/Kermadec arc.

Part of the Brook Street Terrane exposed along the south coast of Southland, between Colac Bay and Orepuki, differs from the rock suites to the north in containing mineral grains which measure a few millimetres across. This implies they have cooled and thus crystallised much more slowly than the faster cooling surface volcanic rocks, whose mineral grains are generally less than one millimetre across. These coarse-grained rocks have cooled at a depth of 6–10 km and have the general term plutonic rocks, named after the Greek god of the underworld, Pluto. It is thought that these rocks are the remnants of arc magma chambers that stored and supplied the lavas extruded at the surface.

Within the Brook Street Terrane, the Takitimu Mountains rise to a height of 1700 m. They consist primarily of volcanic and associated sedimentary rocks, originally deposited under the sea as sub-horizontal beds. The beds are now vertical, with a measured thickness of 14 km. An east-to-west walk across the mountains will take you from the top of the volcanic pile to somewhere near its base although we never see the base exposed at the surface. Weathering and erosion are currently producing spectacular scree slopes; presumably this process has been going on ever since the rocks were pushed up to form mountains. If we examine the rocks that were deposited in the adjacent trough of the Waiau basin (Box 2), between Fiordland and the Takitimus, we find that deposition has been going on for at least 65 million years. However, debris derived from the Takitimu Mountains is found only in the most recent deposits in the Waiau basin, dated at less than 5 million years old (Late Miocene and younger), indicating that these, like the Southern Alps, are very young mountains.

The Murihiku Terrane

This terrane extends from the coast a short way south of the Clutha River mouth to Fortrose, and is best seen as the rocks forming the hills south of the Waimea Plains. The rocks are mainly volcanic ashes deposited in the sea between about 250 and 170 million years ago (Triassic–Middle Jurassic). They have a total thickness of about 10 km and have become world famous as the type locality of rocks of the Zeolite facies. These are rocks which have been formed at the surface and then recrystallised to a new set of minerals following burial to a depth of a few kilometres. The volcanoes that produced the Murihiku rocks were formed at the collision zone between an oceanic plate and a continent, as is seen today on the west coast of South America. Although the majority of the Murihiku rocks were deposited under marine conditions, some contain leaves and stems of trees and occasional forests, such as that at Curio Bay, indicating that at various times land existed in this area.

A major geological feature of the Southland rocks is the Southland Syncline. The Murihiku rocks, which were laid down as a sequence of essentially horizontal layers, have been compressed into a U shape (syncline) which is apparent when viewed from the east. The northern limb of the U is vertical, whereas the southern limb is much more gently dipping.

The Dun Mountain Ophiolite Belt/ Maitai Terrane

This terrane extends as a narrow, discontinuous belt of rocks extending from the Alpine Fault in north-west Otago 335 km south-east. This thin, distinctive group of rocks, dated as being 280 million years old (Early Permian), is a section of the top of the mantle, the overlying sea floor and up to 6 km of sedimentary rocks. The sequence was thrust to the surface as a result of collision between two major plates. The Red Mountain area, one of the most striking mountain scenes in New Zealand, is underlain by this terrane. The chemistry of the rocks is such that little grows on the soils developed from them and most of the rock is exposed as a dun-coloured outcrop. In Nelson, Dun Mountain is part of the same unit and has given its name to the distinctive rock dunite.

2 OIL EXPLORATION

News of oil exploration in the region hits the headlines. Natural gas has been reported from the Lillburn Valley in western Southland and at Madagascar Beach in north-west Otago. The source rocks for oil and gas production are fine-grained sediments rich in organic (principally plant) material and buried quickly in oxygen-poor environments. The organic material is converted to oil and gas at between 60°C and 150°C, as the sediments are buried and heated. A promising area for hydrocarbons is the valley of the Waiau River, a major site of deposition of marine sediments from Cretaceous to Pliocene times. Here up to 8000 m of sediment was deposited in topographic lows within the basement. The edges of this deep and narrow basin are defined by major faults, that link with the Moonlight Fault of Central Otago. While conditions for the production of oil and gas have been attained within the valley, no commercial finds have been made. The two recent oil exploration wells near Solander Island in Foveaux Strait were dry. The only oil produced commercially (c. 1900) in the area is that distilled from Oligocene oil shales in the Orepuki area.

Of much greater significance to the region is the presence of oil in large structures in the Great South Basin, east of Stewart Island. This basin contains up to 8000 m of sediment deposited over a 80 million-year period, from mid Cretaceous to essentially the present day. Wells drilled by Hunt Petroleum in the 1970s revealed the presence of both oil and gas in significant quantities. The rough seas and ocean depth, at 1000 m, pose major problems of extraction.

Caples Terrane

This terrane is not well understood, and even its age is not known with any certainty, since fossils are scarce and little dating has taken place. Like the Brook Street and Murihiku Terranes, its rocks are mainly sands eroded from island arc volcanics. These were produced in a collision-zone environment. Most of the area west of the northern arm of Lake Wakatipu and the Thompson Mountains is made up of this unit. The Caples rocks are the source of some of the South Island's greenstone (Box 3), which is present as thin pods in the Greenstone, Routeburn and Caples Valleys. The pods may represent another piece of uplifted sea floor similar to the Dun Mountain Ophiolite Belt.

3 GREENSTONE

Probably the best known of New Zealand's rocks and minerals is pounamu/greenstone. It was used extensively by the Maori for a wide variety of purposes including tools, weapons and ornaments, with some artefacts dating back to the early parts of the second millennium.

Within the Otago area, sources of greenstone are located in three main places: Jackson River, Wakatipu and Wanaka. A related material, bowenite, is found at Anita Bay.

Greenstone is composed of the mineral tremolite, a calcium magnesium silicate which forms a mass of interlocking crystals resulting in a very tough rock. The source of most of New Zealand's greenstone is part of the upper mantle that has been squeezed to the surface at a subduction zone (Fig. 2). The Wakatipu and Wanaka sources are related to the Pounamu ultramafics, which represent old sea floor and upper mantle now exposed in the schist of the western part of the Rakaia Terrane. The bowenite is made up of a much softer mineral, serpentine, and is of little use for tools.

Rakaia Terrane

The basement rocks to the north-east of the Caples Terrane were deposited 250 to 130 million years ago, with the age decreasing in a north-easterly direction. The rocks are predominantly sandstone and finer-grained sediments derived by the weathering and erosion of a piece of continental crust such as that exposed in Fiordland today. The source may have been Marie Byrd Land in Antarctica. This terrane is the major rock unit in the South Island in terms of surface exposure, as it dominates outcrops north of the Dunedin area and forms much of the distinctive tor topography of eastern and Central Otago. The naming of these rocks is complicated by several factors, including the long time of deposition, the very extensive surface exposure, and a convoluted contact with the Caples Terrane, which has suffered deep burial.

During deposition, early rocks were buried by later sediments. As the rocks were buried deeper with time, temperature and pressure increased and the minerals present in the rocks changed to new minerals, a process known as metamorphism. This process peaked 160 to 180 million years ago, when rocks now at the surface in Central Otago were buried to a depth of about 15 km. One result is the development of minerals known as micas, that crystallise in flat sheets lying parallel to each other. The rocks break along these sheets, forming the characteristic schist slabs of the Central Otago landscape. Where the rocks are schistose, the basement is known as Haast Schist and includes parts of both the Caples Terrane and the Rakaia Terrane. Moving northward from the schist of Central and eastern Otago, the rocks of the remainder of the terrane have been buried to shallower levels and have undergone less metamorphism (lower temperatures and pressures), so that in North Otago the mica minerals have not formed and the rocks lack the ability to split into thin slabs.

Another result of the metamorphism was the development of mineral veins after the metamorphic climax. These veins have been worked as a source of gold (Box 4), tungsten (from the mineral scheelite), and antimony (derived from the mineral stibnite). After the peak of metamorphism, the rocks were uplifted and gradually eroded over a period of 50 million years.

Although most of the Rakaia Terrane is a very thick sequence of gravels, sands, and muds deposited as a very large submarine delta, other rock types are present. Submarine lava flows or volcanic ashes are now present in the Haast Schist as greenschist horizons, often associated with manganese minerals. These include the very distinctive mineral piemontite, which colours the pink Central Otago schist used as a decorative building stone (Fig. 8). The manganese leached out of the volcanic rocks when they came in contact with sea water.

Fig. 8. Like many buildings in Dunedin, the Registry building of the University of Otago utilises rocks from the Otago region: Port Chalmers Breccia for the base, Bluestone of Leith Valley trachy andesite for the exterior and Oamaru stone for window frames. *J. Darby*

Macraes mine gold-processing plant under construction, 1990. Otago peneplain surface in the middle distance, with two smooth-outline volcanic cones of Miocene age on the skyline. *J. Youngson*

4 GOLD

No account of the geology of this region would be complete without reference to the goldfields. The gold rush of the nineteenth century, following the discovery of gold in Gabriels Gully in 1861, laid the foundation for Otago's prosperity. Total production to date in the region is approaching 300 tonnes (10.7 million ounces), with a gross value in excess of NZ$5 billion. Alluvial mining at Island Block and Nokomai, and hard rock mining at Macraes (above), produced 3.73 tonnes of gold in 1994.

Gold occurs in Otago in two ways. Primary vein gold (hard rock gold) was deposited from hot water solutions in the Rakaia Terrane rocks at intervals over the past 140 million years. The gold was originally dissolved from much deeper rocks by water under high temperature and pressure. At the same time, other elements such as silicon, tungsten, iron and arsenic were also dissolved. As these solutions migrated to the surface, the temperature dropped and minerals such as gold, scheelite (calcium tungstate), pyrite (iron sulphide) and arsenopyrite (iron arsenic sulphide) crystallised, along with the dominant quartz (silicon dioxide). Much of the gold is present as minute inclusions in the pyrite and arsenopyrite. Most of the gold deposition took place soon after the metamorphic climax (140 million years ago) and before the deposition of the cover strata, although gold is probably still being deposited in the Southern Alps.

Only about four per cent of the gold mined comes from hard rock mining. Most of Otago's gold is derived from river and stream gravels in modern or old river beds (below). Following its original deposition in veins, most of the hard rock gold has been released by weathering and erosion. One distinctive property of gold, its density (19.3 times greater than water), means that it is very difficult to transport in the river systems. As a result, most of the gold naturally weathered from the veins is probably still fairly close to source, its principal movement being vertical as the mountains have been eroded. Moving down the Clutha River away from the source of the primary gold, the gold becomes much smaller in grain size. It remains in the river systems, where a little may dissolve and reprecipitate around existing gold grains, thus increasing their size and forming nuggets. No large nuggets have been found in Otago, probably because the process of solution and redeposition is very slow and there hasn't been enough time for large nuggets to form.

Auriferous gravels sluiced for gold in old river terrace, sitting on Haast Schist, Shotover River. *A. Reay*

MARINE SEDIMENT COVER OF SOUTHERN SOUTHLAND AND THE EASTERN OTAGO REGION

A long period of erosion of the mountains in the Rakaia and Caples Terranes led to the development of a very extensive low relief surface (a peneplain) over southern New Zealand. However, during this time considerable relief on the peneplain occurred in a few places, probably as the result of faulting. Massive scree deposits were laid down following rapid erosion near these faults. These localised deposits are well developed in the Naseby and Henley areas.

By about 65 million years ago most of the surface of the Otago area was a peneplain, a flat surface with a gentle slope to the north-east and crossed by a few mature meandering rivers with swampy backwaters. As the very resistant mineral, quartz, was the major mineral to survive the long period of erosion, the surface was covered in places by white quartz sands and gravels, the last remnants of what was once a mountain chain. A few small areas of clay are also present, represented today by the deposits at Hyde. Both the quartz sands and the clays are the products of chemical and physical weathering of the schist. The clay results from the chemical weathering of the mineral feldspar, a major component of the schist. Southland was also predominantly a low-lying area, with slow meandering rivers flanked by forests covering the area to the south of the Hokonui hills. Subsequent burial of some of this organic matter produced many of the coal seams found in the Gore/Mataura area. However, west of Stewart Island and in the Waiau Valley marine sediments were being deposited; this continued within recent times (Box 2).

Also by about 65 million years ago (end of the Cretaceous) the sea began to flood over the land surface (Fig. 9). It is not certain whether the land was sinking or the sea rising, but the net result was that the sea began flooding over the Otago peneplain from east to west, reaching its maximum extent about 34 million years ago. The furthest west in Otago that marine sediments deposited from this shallow sea have been found is at Naseby. There is no evidence that the sea continued through to Central Otago (Box 5). At 20 million years, the sea began to recede and by 13 million the Otago coastline was close to its present position. Whether the sea encroached along only pre-existing river valleys or whether it covered all of the peneplain surface is uncertain.

Fig. 9. Vertical schistosity in Haast Schist (lower half of outcrop) overlain by fossiliferous horizontal marine sediments, Raupo Creek, north Otago. The contact between the two rock types represents a period of about 30 to 40 million years. *A. Reay*

During the 50 million year period that it transgressed over the land surface, much of Otago was submerged and a series of sediments were laid down on the sea floor. However, the incursion in the Southland area was much less extensive, beginning much later, and resulting in the deposition of extensive Oligocene limestone deposits (Box 6). The sequence of marine deposits changes from Southland through the Dunedin area to North Otago. The Dunedin sequence is well documented and will be used as a basis for discussion.

5 MARINE SEDIMENTS OF WAKATIPU

A surprising feature of the geology of Central Otago is the presence of marine sediments, including limestones, at Bob's Cove on the northern shores of Lake Wakatipu. The rocks contain fossils deposited in marine conditions about 25 million years ago. Slivers of these rocks are preserved from the upper Shotover through Bob's Cove to Mt Nicholas, a distance of about 40 km. From what is known of the geology of the area south-west of the Queenstown district, these sediments were deposited in a narrow inlet that extended south-west to join the sea in the Te Anau region. It marked the position of a major fault in the earth's crust, the Moonlight Fault system, branches of which extend from the south coast as far north as the Wanaka area. This arm of the sea was probably in existence for about 10 million years, originally following the fault. After the regression of the sea, the Bob's Cove sediments were subjected to major earth movement, which resulted in tilting and folding.

6 LIMESTONE

Limestone is an important raw material in southern New Zealand. A principal use is agricultural lime, derived from major quarries in the Oamaru, Milburn and Winton areas. However, it is also used extensively as a building stone (Oamaru, see Fig. 10), as a raw material for cement production and as a chemical added to the gold extraction line at Macraes (Dunback). A more recent application is as a stabiliser in the base course of roads.

The major limestones are all marine, with the Oamaru limestones being deposited around 35 million years ago and the Winton and Milburn (below) limestones 22 million years ago during the major marine inflooding over southern New Zealand.

Tertiary limestone. Deeply eroded surface of the Milburn limestone north of Milton.
Don Weston

Fig. 10. Ototara limestone (Oamaru stone) being cut for use as building stone, Parkside Quarry, Weston. *Don Weston*

The limestone at Creamery Road, Brighton, is the oldest marine deposit known in the area, being laid down 65 million years ago. It contains a suite of fossils, including some of the world's last belemnites (a relative of the squid), which became extinct along with many other animals at the end of the Cretaceous, after a massive impact by a meteorite in Mexico. As the sea rose gradually over the next 30 million years and the shoreline moved inland, the sea water at any one point, for example at Fairfield, became deeper. As a result, the type of sediment being deposited was changing. Although the general sequence of rock units throughout Otago is similar, rocks deposited at the same time in different areas have been given different names.

The major geological units in the Dunedin area are shown in the stratigraphic column (Fig. 11), with the oldest marine units, the Fairfield Greensand and the Brighton Limestone, at the base. The Fairfield Greensand, as seen at the Fairfield quarry (Fig. 12), contains the green mineral glauconite, which forms only in marine conditions, through a slow reaction between muds on the ocean floor and sea water. The Fairfield Greensand was being deposited at the time of the great mass extinction of the dinosaurs, although there is no obvious evidence in the rocks of southern New Zealand of any catastrophic event that may have been responsible for the extinction.

The succeeding Abbotsford Formation, with a maximum thickness of 300 m, was deposited over 20 million years (latest Paleocene to Early Eocene). This equates to an average annual rate of deposition of less than two hundredths of a millimetre per year. This unit is dominated by clay minerals, including montmorillonite, that has negligible structural strength and whose presence was one of the factors responsible for the Abbotsford slip of 1979 (Box 7). Overlying the Abbotsford Formation is the Green Island Loose Sand, which was probably an offshore sandbar.

The maximum sea depth in the Dunedin district is marked by the presence of the Burnside Mudstone, a clay-calcite (calcium carbonate) rock also known as a marl, which was the raw material for the cement works formerly located at Burnside. Topping this mudstone is a boundary representing a long period of non-deposition of sediment. At this time 34 million years ago, major changes were taking place in the orientation of the plates in the South Pacific area. In particular, when Australia separated from Antarctica, new ocean current systems formed and swept up the east coast of the South Island, eroding soft sediment and preventing deposition of sediment for several million years. Following on this break is the Concord Greensand, another glauconite-bearing unit, well known locally for its shark tooth content. The sea then gradually receded off the Otago peneplain, marine conditions became quiet and a thick layer of calcareous sandstone was laid down, known as the Caversham Sandstone, which is dramatically exposed in the cliffs north and south of Tunnel Beach (Fig. 13). The incoming of quartz in these sandstones is an indication of geological change, with uplift in Central Otago rejuvenating the river systems, leading to increased erosion of the schist. In a few places, notably north of Karitane and at a couple of spots around the Dunedin harbour, limestones were deposited on top of the calcareous sandstones.

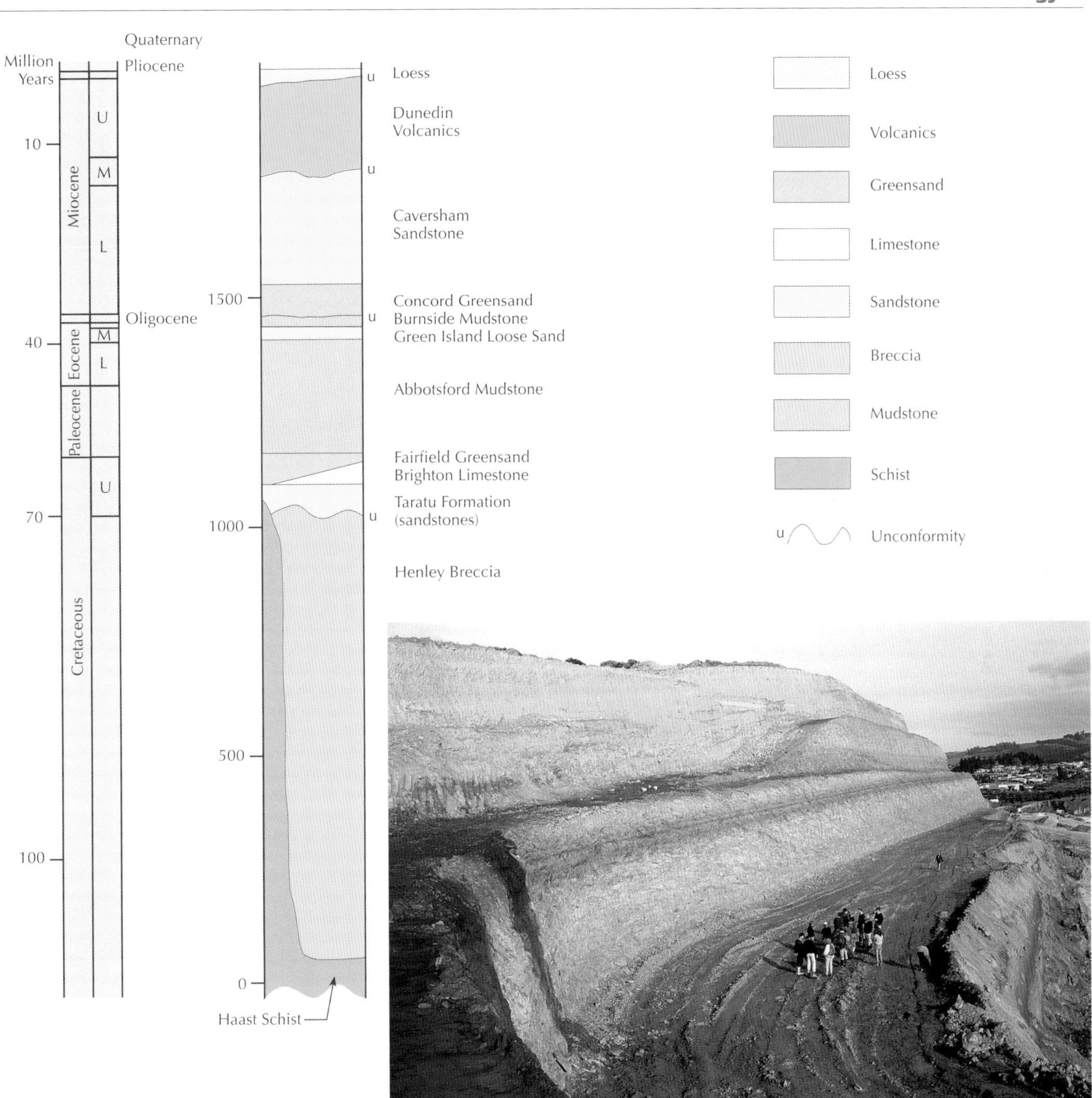

Fig. 11. Stratigraphic column showing major geological units of the Dunedin area.

Fig. 12. Buff weathering Fairfield Greensand overlying fresh greensand, Fairfield Quarry. The Cretaceous Tertiary boundary is located within the steep face to the right and below the figures. *A. Reay*

Fig.13. Cliffs in Caversham Sandstone on coast south of Dunedin, with Blackhead, a basaltic intrusion, in the background. Tunnel Beach is situated in front of the prominent peninsula (centre). *A. Reay*

Housing destroyed in the extensional area at the head of the Abbotsford landslide.
Geology Dept, University of Otago

7 THE EAST ABBOTSFORD LANDSLIDE

The landslide in the Dunedin suburb of Abbotsford on 8 August 1979 was a geological event which captured world headlines, being seen on television worldwide and appearing on the cover of *Time* magazine. The Abbotsford Mudstone is a geological unit notorious for its ability to move downslope at the slightest provocation. Professor Benson wrote a paper in 1940 describing a fast landslide in Abbotsford mudstone on the south-eastern slopes of Saddle Hill and suggesting that building on such a slip-prone geological unit was foolhardy. Throughout the Dunedin area, wherever the Abbotsford Mudstone is at the surface, major slips occur and are identified by their hummocky topography (e.g. Saddle Hill, Abbotsford, Green Island, the Kilmog section of State Highway 1 and Seacliff). These slips may stop moving over time, but are often reactivated by heavy rainfall.

Major movement of the East Abbotsford block was first noted in late May and early June 1979 when cracks started opening at the head of the slip. (With hindsight, the initial movements on the slip lane may have started as early as 1969.) Movement accelerated, reaching a maximum of 300 mm/day prior to the catastrophic failure, when the block moved 50 m in half an hour. A total of 62 houses on the slip were destroyed or removed.

Probably no one condition was responsible for the landslide but contributory factors were: the very weak nature of the mudstone due to its content of montmorillonite, a clay with very little mechanical strength; lubrication of the slip plane by water, possibly from precipitation or from a leaking water pipe uphill; and the removal of the base of the hill (quarried away for sand), which removed some support.

8 MOERAKI BOULDERS

The Moeraki Boulders are probably the highest profile geological phenomenon in Otago/Southland. They were reported scientifically in 1850 and figure prominently in pre-European Maori legends.

The boulders are weathering out of a dark grey mudstone which was deposited on the ocean floor about 60 million years ago (Paleocene). These large – some more than 2 m in diameter – spherical boulders are known as concretions, and are composed predominantly of the calcium carbonate mineral calcite. Research at University of Otago has demonstrated that once the mudstone was buried, the concretions started to grow as calcium carbonate migrated towards a centre. Equal rates of growth away from the centre led to a spherical shape. Concretions of 95 cm in diameter are estimated to have grown over several million years.

Carbonate concretions, the Moeraki Boulders have been exposed by the weathering of Eocene mudstones. The village of Moeraki in the background is underlain by the mudstone and associated volcanic ashes and basaltic rocks. *Rod Morris*

LAKE MANUHERIKIA

About 16–18 million years ago a major freshwater lake developed in the Central Otago area. Lake Manuherikia probably extended from Mt Pisa to the Maniototo, and from the Nevis Valley to Roxburgh. Sediments deposited in the lake are present in a number of separate areas as a result of later earth movement. Some of the sands and silts contain plant fossils that indicate that the climate at the time was warmer than that today. The accumulation of plant material in the lake subsequently led to the development of low-grade coal, which has been mined in several places in Otago, including Roxburgh, the Home Hills and the Ida Valley.

THE MIOCENE VOLCANIC EPISODE

The Dunedin volcano is the largest volcano in the region and is the latest major product of volcanism. (The much smaller Solander Island in Foveaux Strait was an island arc volcano active about one to two million years ago, formed when oceanic crust of the Australian Plate subducted under the Pacific Plate.)

During the later period of the marine transgression, plate movement was such that the continental crust of the Otago region was slowly stretched. Geological evidence from the south-west of the South Island suggests that minor stretching or extension of the region may have begun about 35 million years ago (Late Eocene), with the rate of extension in the Otago area speeding up in the Miocene period about 23 million years ago. Extension of continental plates usually gives rise to faulting, and often to volcanic activity, the volcanoes of the east African rift system (which includes Mt Kilimanjaro) being a prime example. The faults act as conduits, allowing magma to reach the surface. In a continental extensional setting, the lavas are typically dark grey basalts. Indeed, this is the dominant lava type of the Otago Miocene volcanic rocks.

The earliest volcanism related to this period of stretching is a suite of volcanics squeezed into the upper crust, in the Wanaka/Haast Pass area. Radiometric dates for these are about 24 million years. East of this area, and north and west of Dunedin, there are about two hundred small volcanic cones (Fig. 14). These erupted between about 22 and 10 million years: an eruption rate of about 1 every 50,000 years. The cones are often of volcanic ash, but may include lava flows. Most appear to have been very short-lived, probably lasting from only a few days to a few years. The biggest volume of lava flows can be seen in the hills south of Waipiata. Many of these contain rock fragments from the upper mantle, enabling us to work out how fast the lava has risen from its source at about 80 km depth in the upper mantle. This lava has often travelled up major faults at speeds of as much as 36 km a day, taking two or three days to reach the surface. An intriguing volcanic ash deposit in a road cutting at Longlands Station, on the edge of the Maniototo, contains small spherical bodies of ash formed when ash and rain water fell together.

Towards the end of this volcanic episode, a major volcano started to develop in the Dunedin area. Again, the movement of the molten rock appears to have been controlled by major faults, since the Dunedin volcano sits astride the two big faults that define either edge of the Taieri plain. The first evidence of the eruption of the Dunedin volcano is the presence of thin ash layers in the marine limestone deposited 13 million years ago. The volcano began as a submarine eruption, but very quickly built up a sub-aerial cone centred on the Port Chalmers/Portobello area. Volcanic activity continued for three million years, with the youngest rock dated being one from Mt Cargill

9 COAL

Southern New Zealand has been a major producer of coal, and huge reserves remain. Coal has been worked in North Otago, Shag Point, Central Otago, Green Island, Kaitangata, Pomohaka, Mataura, Forest Hill, Orepuki and Ohai, with minor workings widespread throughout the region. A 1988 survey identified the largest reserves as being in the Kapuka (Ashers/Waituna) field, with just over one billion tonnes available at a depth of less than 150 m. The total reserves in the Southland plains is estimated at 6760 million tonnes.

Coal is formed from peat, vegetation that has been preserved in oxygen-poor environments such as swamps. Subsequent burial, with consequential rise in temperature and pressure, alters the peat both physically and chemically, converting it first to lignite, then to sub-bituminous and bituminous coals and in extreme cases to anthracite.

The highest quality coals near Dunedin are bituminous coals at Shag Point, although this deposit is exhausted. The Ohai coal field yields sub-bituminous coals, whereas all the other fields are of lignite. Although these deposits were formed by the same process, they are of differing ages in the region. The eastern and southern Otago fields are developed within the Taratu sands and gravels of Late Cretaceous age resting on the Otago peneplain. The deposits of eastern Southland were formed on a delta that was advancing across a shallow sea of Oligocene-Early Miocene age which had flooded the Southland plains. In Central Otago, the coals were associated with peat-forming areas in deltas, which developed on the edge of the freshwater Lake Manuherikia.

Thick lignite seams in an open-cast coal mine, Kaitangata. *J. Lindqvist*

Fig. 14. Looking south along the Otago peneplain surface near Dunback, with isolated peaks of small volcanic cones in the distance. *A. Reay*

at ten million years ago. The Dunedin volcano differs from the outlying volcanic vents in being much bigger and much longer-lived. Geophysical surveys tell us that about 600 km^3 of molten rock was present in a number of shallow magma chambers below the Port Chalmers area. Thus, instead of coming straight from the upper mantle, the lava for the Dunedin volcano rose up into the crust, where it ponded in magma chambers for some time before being erupted at the surface.

Over the three million-year history of the volcano, there were times when parts of it were active whilst other parts were quiet. Individual lava flows are well exposed at Second Beach (Fig. 15), where red horizons mark the tops of flows. The red colour is due to oxidation, resulting from iron in ash or soil on top of the lava flow being heated by the overlying flow at 1100°C. Magnificent examples of thin lava flows are exposed in the cliffs just north of Aramoana, where as many as fifty successive lava flows separated by thin ash or rubble horizons are visible.

The geological history of the Dunedin volcano has been the subject of much research at the University of Otago, leading to the publication of the Benson geological map of the volcano in 1968. The volcano extends from Taiaroa Head to the Chain Hills and from St Clair to Waitati, and covers an area of about 300 square kilometres. The total volume of extruded lava is estimated at about 100–150 km^3.

The contact between the older marine rocks and the overlying volcanics can be seen in several places, the most spectacular being below Cargill's Castle, where the lavas can be seen abutting into a hill in the Caversham Sandstone and gradually rising up the hill before overtopping it and extending further west. The contact can also be seen in Stone Street, Kaikorai, where the lower part of the roadside is yellow/grey Caversham Sandstone and the upper sections are basalt. The last phase of volcanism in the Dunedin volcano was the injection of a number of domes of very viscous lava which are now exposed as some of the high points on the Mt Cargill/Mihiwaka ridge (Fig. 16).

Earlier volcanism includes 105 million-year-old rocks in the Shag Valley, which are rhyolites similar to those produced by the volcanoes that have erupted from Taupo in historic times. In addition, there is scattered volcanism in the area north of Moeraki, most of which erupted into shallow seas.

Fig. 16. Lava domes of the Mt Cargill ridge, final activity of the Dunedin volcano. *A. Reay*

THE LAST TEN MILLION YEARS

Following the cessation of volcanic activity, the Dunedin volcano suffered erosion, helped perhaps by the presence of fault(s). Two major stream systems developed along the line of Otago harbour, with one flowing north-east and the other south-west from the Port Chalmers/Portobello area. Subsequently, a rise in sea level flooded the two stream valleys and the Otago Peninsula became an island. At this time the offshore Southland current was moving sand northwards along the coast, and much was deposited in the area between St Clair and Waverley, to form the south Dunedin flat. Thus the peninsula is a classic tombola, a former island now joined to the mainland by a sandbar.

At 10 million years ago the Australian and Pacific Plates started moving towards one another. Compression was a major feature of the area and volcanism ceased. Results of the present 39 mm per year south-west movement of the Pacific Plate can be seen throughout Otago. Apart from the Southern Alps, perhaps the most spectacular result of this compression is that the rocks of Central Otago are being deformed to make a distinctive basin-and-range topography. Rough Ridge, on the west side of the Maniototo, is rising and deflecting stream courses at its northern end. Further east, earthquakes along the projected line of the Akatore Fault are another expression of compression.

Fig. 15 (left). Old quarry at Second Beach, Dunedin, exposing basalt lava flows from the Dunedin volcano. Horizontal features are contacts between flows. The red colour is due to an overriding lava flow baking the top of a previous flow (possibly of ash). *A. Reay*

In the south, the exposure of rock on the Southland plains is limited by a cover of gravels deposited during Pleistocene–Holocene times. These gravels are an indication of limited geological activity, but to the west both margins of the Fiordland area are very active, with earthquakes a common occurrence, reflecting the proximity to the plate boundary.

A worldwide phenomenon over the past ten million years has been a general global cooling, culminating in the Pleistocene glaciation which may have contained more than twenty glacial episodes. Evidence of glaciation is shown by the presence of extensive glaciated surfaces around Lake Wakatipu, the U-shaped valleys of Fiordland, and the glacial outwash deposits in the Clutha Valley. Glaciation results in the over-steepening of valley sides; as the ice retreats, some of these valley walls collapse as landslides. A major landslide with typically hummocky topography occurred in the Grebe valley at the Borland Saddle and is one of the biggest of its type recorded. Recent glaciation locked up much of the earth's surface water as ice: this had the effect of lowering river levels and sea level as much as 100 m, exposing much of the muds of the river channels and the shallow submarine shelf. This mud, rock dust of glacial origin, was dispersed by wind over large areas of Otago and Southland, where it accumulated as a mantle of yellow loess, the clay-like curse of local gardeners. Thick deposits of the loess can be seen in many coastal road cuts around the region and are particularly thick near Cape Wanbrow, Oamaru (Fig. 17).

In summary, over the past 450 million years the southern

Exfoliating granite domes, Gog and Magog, in southern Stewart Island, part of the Fiordland Terrane. *A. Reay*

10 STEWART ISLAND

This island is composed almost entirely of igneous rocks that have formed by the cooling of molten magma, most of which has cooled at depth to produce coarse-grained plutonic rocks. Much of the island is made up of the common plutonic rock granite, and may be related to the granites of south-west Fiordland.

Spectacular granite outcrops, the result of long continued weathering, are exposed in southern Stewart Island. In the latter part of the nineteenth century mineral veins south of Mount Allen, the highest point in the island, were worked for tin. These veins occur in a large fragment of sedimentary rock that has been enveloped by the granite as it was squeezed into the upper levels of the crust.

The eastern slopes of Stewart Island form a very planar feature when viewed from a distance and this feature is probably related to the Otago peneplain.

half of New Zealand has experienced a great deal of geological activity. A brief schematic history based on the Rakaia Terrane is given in Fig. 18. Much of that activity continues today, with the result that the area attracts many world-class geologists to examine this natural laboratory of geological processes.

Million years before present	
today	earthquakes, landslides, north-east trending ridges continue to form, all geological processes operating
2–1.8	widespread glaciation starts, loess deposition over most of Otago
2?	continental arc volcano, Solander Island erupts in Foveaux Strait
3–0	formation of the Southern Alps
10–0	formation of north-east trending ridge and valley system,
10	Otago/Southland area under compression, Dunedin volcano becomes extinct
13	major volcano, Dunedin starts to erupt, first eruption submarine
24	minor volcanism starts in West Otago
25	Alpine Fault initiated? seas start retreating off peneplain
34	major current sweeps east coast, break in deposition
40–34	minor, mainly submarine volcanism in North Otago
45	boundary between Australian and Pacific Plates forms through New Zealand
55	Australia separates from Antarctica
65–30	sea transgresses westwards across Otago, deposition of marine muds, sands and limestones
85	New Zealand separates from Gondwanaland, Tasman Sea forms
105	volcanism in the Shag Valley
110–80?	large scree deposits from uprising fault scarps
120–65	uplift and massive erosion, formation of Otago peneplain with residual clays, quartz gravels and coal
140–150	Macraes gold deposited
160	metamorphism
250–130	marine muds, sands and gravels derived from weathering and erosion of a continental source, little volcanics

Fig. 17. Submarine basaltic lava (35+ milliion years old) is overlain by thick yellow loess deposits at Boatmans Harbour, Oamaru. *A. Reay*

Fig. 18 (left). Schematic history of the Rakaia Terrane and its overlying sequences.

CHAPTER 2

LANDFORMS

D. CRAW AND R.J. NORRIS

A typical feature of the Central Otago landscape – tor, Old Man Range (see page 20). *D. Craw*

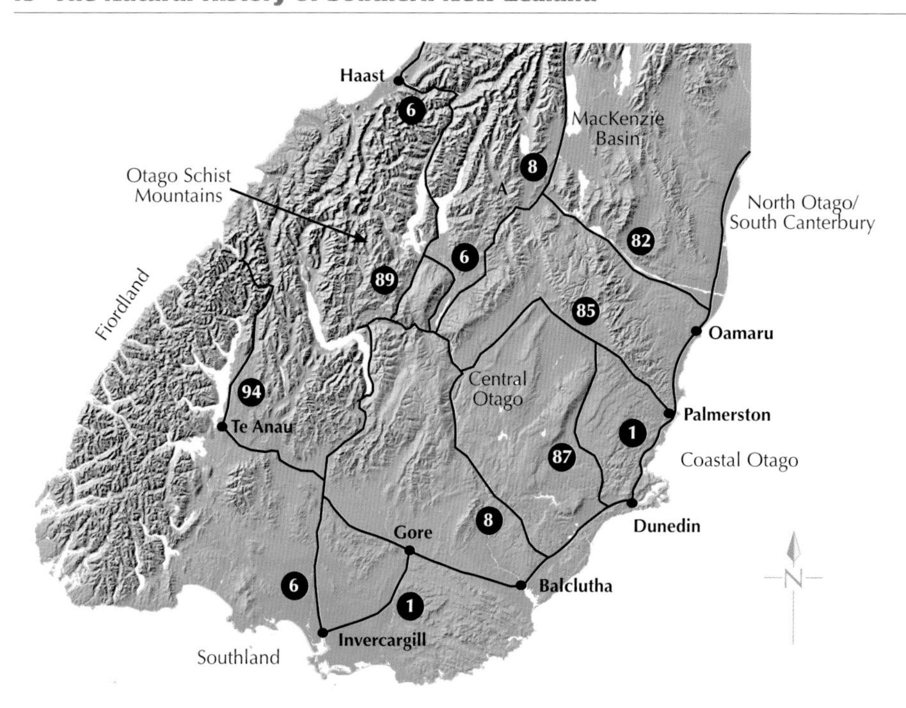

Fig. 1. Digital elevation model of the southern South Island, compiled from 1:250,000 topographic data by GeographX (www.geographx.co.nz). The model shows the principal regional scale landforms and the geographic subregions described in the text. Principal state highways are indicated, following many important geological structures (e.g. see Fig. 2). A = Ahuriri Valley (Fig. 3).

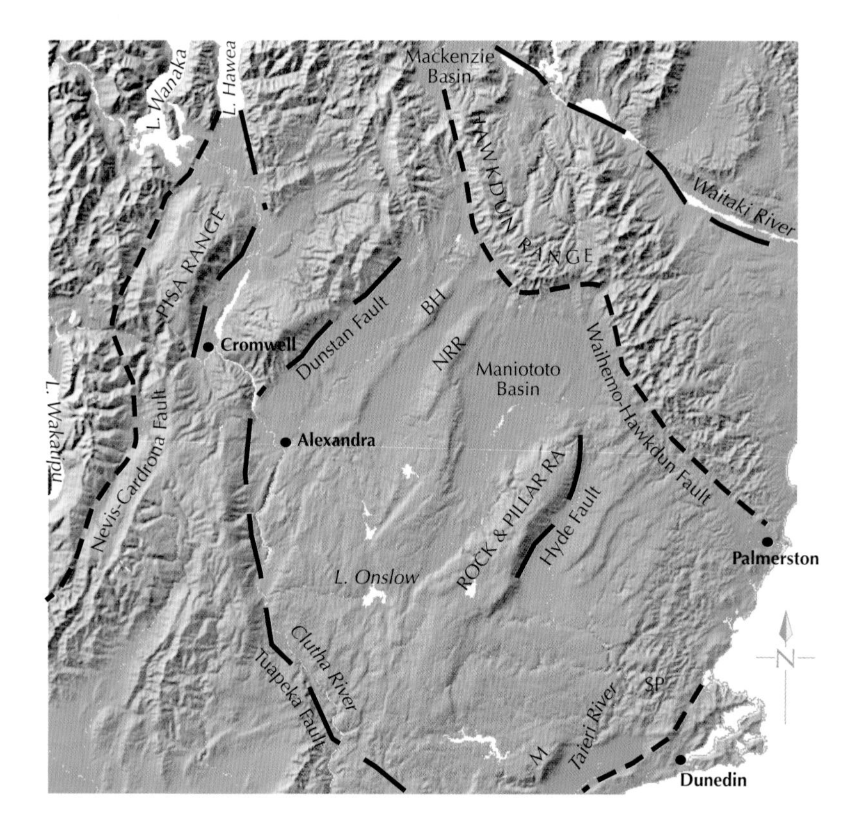

Fig. 2. Enlarged view of the digital elevation model in Fig. 1, showing principal landforms and active faults in Otago. Initialled features are mentioned in the text: SP = Silver Peaks; M = Maungatua; R & P = Rock and Pillar Range; NRR = North Rough Ridge; BH = Blackstone Hill.

Our thanks to Peter Koons and Chuck Landis for discussions on tectonic geomorphology.

LANDFORMS

The dynamic geological setting of southern New Zealand (Chapter 1) means that almost all landforms reflect underlying processes of active earth deformation. As a result, observation of landforms out the car window gives immediate insight into the geological processes that are going on around us. Every landform feature has a story to tell. The landscape may appear to be stationary and permanent, but it is not. Hills and valleys are moving relative to each other, and the shapes of these hills and valleys are changing constantly. Most of the time, this will be unnoticed by individual humans, but over time small changes add up to major reorganisation of the shape of our environment.

The underlying driving forces of these changes are the tectonic plates (plate tectonics, Chapter 1), moving at about the rate of fingernail growth: not visible while waiting for a bus, but remarkably persistent as the years roll by. But changes to landforms do not necessarily occur at a constant rate. A long period of essentially no change may be terminated with a catastrophic event – such as flood erosion, landslide, or earthquake – which completely alters the shape of the land. These events are a normal part of our environment, and we are learning to live with them. Examination of the landforms around us allows us to gain some insight into the types of processes to be expected in different parts of the southern region.

Landforms are also governed by the underlying bedrock (Chapter 1). Southern New Zealand can be divided into broad subregions with distinctive landforms that reflect the combination of bedrock geology and active plate tectonic processes (Fig. 1). These subregions are described below. Further landform development has occurred as a result of glaciation in mountainous areas, adding an additional set of distinctive features.

FAULTS AND ROAD CORRIDORS

Active faults cut many parts of southern New Zealand; these faults are defined at the surface by zones of soft crushed rocks that can be seen along many of the main roads (Fig. 1). The soft rocks are readily eroded by streams and landslides, resulting in characteristic landforms: long linear zones of low relief, controlling river valleys and low passes through high ridges. These zones were exploited as travelling routes by early settlers of the region, both Maori and Pakeha, and eventually became the routes of most of the major highways.

State Highway 1 follows a fault from Waitati to Dunedin through Leith Saddle, then goes parallel to a related fault (Titri Fault) south from Mosgiel through Milton almost to Balclutha. The highway then follows a major fault zone (including the Dun Mountain Ophiolite Belt, Chapter 1) immediately to the north of the Murihiku Escarpment, to the Southland plains at Gore. Highway 94 continues from Gore along this same fault zone parallel to the Murihiku Escarpment as far as Mossburn. The fault zone swings northwards, and Highway 94 rejoins it in the Eglinton Valley, where several active faults have converged to make a low pass (The Divide) to the Hollyford Valley.

Highway 85 (Pigroot) follows several different branches of the Waihemo-Hawkdun Fault (Fig. 2) from Palmerston to the Manuherikia River near St Bathans, before heading south-west along the edge of the growing folds of the Raggedy Range to Alexandra. Highway 8 tracks the Tuapeka Fault from Milton to Beaumont, and then the Old Man Fault and related structures through to Alexandra. The entrance to the Cromwell Gorge at Clyde is fault-controlled, and the Clyde Dam is built on an active fault. From Cromwell, Highway 6 runs along the Pisa Fault and related faults to Hawea, then crosses to a parallel fault that controls the Makarora Valley and Haast Pass. The spectacular wall of mountains that is the Southern Alps ensures that Highway 6 follows the Alpine Fault north from Haast. Highway 89 is a diversion from Highway 6 over the Crown Range between Arrowtown and Hawea, along the Nevis-Cardrona Fault.

The Waitaki Valley is incised into a set of active faults, and State Highway 82 follows on or close to them (Fig. 1, 2), past hydro-electric dams built across branches of the faults. These faults are linked to Central Otago faults by a large fault parallel to the road through Danseys Pass, and a primitive road along the continuation of the Waihemo-Hawkdun Fault through Omarama Saddle to Omarama and the Mackenzie Basin. Highway 8 takes advantage of low topography formed by a family of active faults through the Lindis Pass to the Mackenzie Basin, then continues along the Ostler Fault to Twizel. Highway 87 follows the Hyde Fault along the eastern edge of the Rock and Pillar Range to the Maniototo basin, where it joins Highway 85 and the Waihemo-Hawkdun Fault.

All the faults described here are active to some degree, and movement along them is inevitable in the future. Most of the faults have surface ruptures made during earthquakes hundreds or thousands of years apart. Earthquake damage to roads has happened many times in New Zealand's human history. This damage may appear as steps or gaps in a road, the complete offset of one portion relative to another, or landslides from neighbouring hillsides covering the road. This is just a normal part of the active environment that we live in.

CLIMATE AND LANDFORMS

In low rainfall areas, rivers follow geological depressions and cause little erosion. High rainfall encourages rapid erosion, and rivers cut deeply into bedrock. The Southern Alps on the western side of the South Island (Fig. 1) form a long narrow mountain barrier rising into a belt of prevailing westerly winds in an oceanic environment. This mountain barrier causes the moisture-laden winds to rise and release large amounts of rain on the western side, and dry winds to sweep from the mountains on the eastern side (see Chapter 4). The dramatic differences in rainfall on the eastern and western sides of the mountains are reflected in the landforms. High rainfall ensures that uplifted rocks are rapidly eroded by streams and rivers. These streams and rivers cut deep steep-sided gorges into the bedrock. River courses are little affected by weaknesses in the underlying bedrock, such as layering and faults, and cut relatively direct paths to the sea. Thin slow-moving landslides are rare, because regular floods can remove loose rock rapidly, preventing these types of landslide developing.

In contrast, rivers in the rain-shadow area east of the mountains have less erosive power; they merely modify the rising mountains and valleys. Many valleys in the rainshadow have not been cut by rivers: they are areas which have been uplifted less than the surrounding mountains. This is especially true in Central Otago and South Canterbury (Figs 1, 2), where the large basins which host the main rivers were formed when ridges on either side were uplifted at a faster rate than the basin floors. Rivers and streams can cause minor incision into mountain slopes and basin floors, but the general shape of the topography on a large scale is unaffected by water erosion. Downslope movement of rock material via landslides and screes is an important process of erosion and landform modification. Soft rocks in fault zones become sites of localised stream and landslide erosion.

NORTH OTAGO AND SOUTH CANTERBURY

The North Otago sub-region (Fig. 1, 2) consists mainly of mountains of schist and greywacke with the Waipounamu Erosion Surface (Box 1) sloping gently north-east. The erosion surface has been steeply dissected by rivers, but broad upland plateaux are preserved over large areas. The greywacke mountains have blocky outcrops separated by scree-covered hillsides, and no tors (Fig. 3). Active screes are unvegetated, and older screes have become variably revegetated. The lower slopes of the mountains west of Oamaru have a veneer of Tertiary sediments and volcanic rocks which dominate the landforms out to the coast. Limestone layers form steep white scarps along river valleys, and intervening hills are flat-topped, capped with limestone or hard volcanic flows.

The Waitaki River valley follows a set of active faults which have uplifted greywacke ranges on the northern side of the valley (Figs 1, 2). This major river has built a large gravel delta at the coast north of Oamaru (Fig. 1) during times of glacial advance. The large inland lakes of the Mackenzie Basin now trap much of the material eroded from the mountains, and the delta is currently inactive. Consequently, the delta coastline is eroding back at about 1 m per year (horizontal).

North of the Waitaki River (Fig. 1), the greywacke mountain ranges are being formed by uplift along faults, and intervening basins have undergone less uplift. The ranges have remnants of the Waipounamu Erosion Surface on their summits and lower slopes. Preservation of the erosion surface is progressively poorer as the ranges grow higher towards the west. The erosion surface is preserved beneath the large depression of the Mackenzie Basin, where the bedrock has been folded into a trough (synform) whose base is below sea level. This depression is partially filled by several generations of alluvial fans and gravel-bed rivers (Box 2) that have been pouring into the basin for over two million years. Older fan gravels are being uplifted by active faults to form lines of low hills within the basin. The

1 WAIPOUNAMU EROSION SURFACE (OR 'OTAGO PENEPLAIN')

The flat surface on top of the schist and greywacke bedrock is one of the most significant landforms in southern New Zealand. This surface was cut mainly by erosion as the sea came over the South Island between 65 and 30 million years ago, as described in Chapter 1. The broad flat topography that dominates the skyline over much of Otago (e.g. Lake Onslow area, Fig. 2) is made up of remnants of this erosion surface. The surface, called the Waipounamu Erosion Surface by W. LeMasurier and C.A. Landis in 1996, is also recognisable in north-west Nelson and Antarctica.

Uplift over the past 30 million years has encouraged removal of most of the overlying thin layer of sediments and volcanic rocks (Chapter 1), and exposed the erosion surface again. Groundwater in the sediment layers had penetrated into the bedrock, causing clay alteration, so the bedrock at the erosion surface was soft and friable, and easily eroded as well. Harder patches of schist resisted this alteration and erosion, and are left behind as tors on the erosion surface (right). The tors have been accentuated by cold climate erosion (by wind and ice) over the past million years.

The erosion surface has been disrupted by many active faults and folds in southern New Zealand. These later structures have constructed many of the large-scale landforms that we see: the mountain ranges and intervening valleys (Fig. 2). Small amounts of uplift allow preservation of the erosion surface as broad flat areas between deeply incised streams. As uplift progresses, the streams cut deeper and wider channels into the bedrock, and the erosion surface is progressively reduced to isolated remnants. After more than about 2 km of uplift, the erosion surface is completely destroyed (e.g. west of the Nevis-Cardrona Fault, Fig. 2).

Schist tor at the Waipounamu Erosion Surface west of Alexandra. Soft clay-altered rocks at the erosion surface have been removed, leaving this hard rock remnant which has been accentuated by cold climate erosion. The original erosion surface level was immediately above the top of the tor.

Fig. 3. Greywacke mountains of the Ahuriri Valley, with blocky outcrops and extensive scree development on most slopes above the alluvial fans at valley level. A small lake (tarn) in the foreground fills a high basin (cirque) left after melting of glacial ice.

Waipounamu Erosion Surface is not preserved anywhere west of the Mackenzie Basin, where the greywacke ranges are steep and scree-covered (Fig. 3), with blocky and friable ridge crests. The mountain ranges trend northwards, controlled by faults on their flanks, and intersect the Main Divide at their northern ends (Fig. 1).

Waihemo-Hawkdun Fault Zone

This zone of active faults is traceable northwest from the Otago coast near Palmerston to the Mackenzie Basin through northern Central Otago (Fig. 2). The faults have been slowly uplifting the northeastern side for the past five million years, creating a prominent fault scarp that forms the kilometre-high steep southern faces of the Horse, Kakanui, Ida and Hawkdun Ranges (Figs 4, 5). The Waihemo-Hawkdun Fault Zone is the boundary between screes and blocky outcrops of greywacke to the north-east, and smoother Otago Schist slopes with scattered tors to the south-west.

Fig. 4. Low relief surfaces on the crest of the Hawkdun Range (greywacke) are remnants of the Waipounamu Erosion Surface that have been uplifted nearly 2000 m along the Waihemo-Hawkdun Fault (Fig. 2). South-western slopes (left) are deeply incised by glacial and stream erosion of the fault scarp. The view is towards the west, and the high peaks of the Mt Aspiring area are visible on the horizon.

Unless otherwise indicated, all photographs in this chapter are by the authors. Thanks to Steve Read for help with the digital elevations.

2 ALLUVIAL FANS AND GRAVEL BED RIVERS

Alluvial fans are fan-shaped deposits of gravel and sand formed at the base of steep slopes throughout southern New Zealand (above). They range from 100 m to over 1 km in width. The fans are formed as streams emanate from steep gullies in bedrock mountains and deposit material eroded from the mountains. The streams are small for most of the time, and commonly disappear underground in dry periods. However, the streams rise rapidly during heavy rainfall, and can carry 1 m boulders at times of flood. The positions of the streams vary across the fans with time, progressively building up the fan structures. Debris flows, or slurries of boulders, mud and water, accompany flooded stream flow and further build up the fans. Inactive parts of the fans become vegetated as the active channels migrate away and soil develops, but renewed activity quickly erodes this surficial cover. Reactivation is inevitable in all parts of the fans, but timing of this reactivation is not predictable and fans may remain apparently stable for hundreds or thousands of years. Fans are ultimately uplifted and streams become incised into the fan gravels, forming new fans downstream (opposite). Reworking of fan gravels results in concentration of gold, and many such deposits have been mined over the past 140 years (below).

Streams on alluvial fans merge into lower-gradient gravel-bed rivers (opposite), which drain most of the mountain areas of southern New Zealand. These rivers carry little sediment most of the time, but can transport large volumes of coarse debris during floods. Wide valleys permit gravel-bed rivers to become braided, with several channels. Steep-sided narrow gorges are cut through bedrock by the high erosive power of these rivers. As the river gradient decreases towards the sea, the rivers transport mainly sand and silt.

Top: Aerial view of the eastern slopes of the Rock and Pillar Range (lower centre) and other ranges to the west (Fig. 2). The flat top of the range (snow covered) is the uplifted Waipounamu Erosion Surface. This surface has been partly eroded on the east face by deeply incised streams which are forming alluvial fans where they emanate from the range (bottom of picture). The Hyde Fault (Fig. 2) is at the boundary between the eroded schist range and the alluvial fans.

Right: Gravel cliffs of an uplifted and eroded alluvial fan at the base of the Dunstan Range (Fig. 2, and opposite). The cliffs were formed by nineteenth-century goldminers sluicing for gold in enriched layers in the gravels, which have been cemented by ground water activity, and are now hard and resistant to erosion. Similar exposures are found throughout Central Otago and northern Southland.

Opposite: Satellite image of the Waikerikeri alluvial fan (lower left) at the eastern base of the Dunstan Range (Fig. 2). The fan developed as the Dunstan Mountains (top left) rose along the Dunstan Fault over the past 500,000 years. The fan has itself been uplifted and is being eroded by streams emanating from the Dunstan Range. The lower reaches of the fan have been eroded by the Clutha River (bottom left) and Manuherikia River (centre). The fan partially fills a downwarp (synform) occupied by the Manuherikia River between upwarped (antiform) schist ridges (top left and lower right). Remnants of the Waipounamu Erosion Surface underlain by schist form the broad low-relief but irregular landform on the lower right.

N
1 km
DUNSTAN MOUNTAINS
Dunstan Fault
Waikerikeri Fan
Manuherikia River
Clutha River
Alexandra

Fig. 5. Crest of Blackstone Hill looking north-east to the flat-topped Hawkdun Range on the horizon (Figs 2, 4). The Waipounamu Erosion Surface is folded at Blackstone Hill, and slopes down to the right and left. Schist tors stud the surface in the foreground and middle distance, but lateral slopes are dominated by shallow landslides. In contrast, the greywacke of the distant Hawkdun Range has steep slopes with screes and few landslides.

The Waihemo-Hawkdun Fault scarp is being eroded by numerous small steep streams (Fig. 4) with alluvial fans (Box 2) at their bases. Large alluvial fans formed in the early stages of uplift on the fault zone have themselves been uplifted by further fault movement and are being eroded near Kyeburn, Naseby and St Bathans. Erosion of these old alluvial fans by rivers and gold miners has formed the yellow gravel cliffs that are characteristic of these historic gold mining areas (Box 2).

CENTRAL OTAGO

Central Otago landforms are dominated by folds of the schist basement and the Waipounamu Erosion Surface, deformed like a wrinkled carpet (Fig. 2) as the South Island is compressed by plate boundary processes. Broad upfolds (antiforms) of schist basement and the Waipounamu Erosion Surface form north-east trending mountain ranges separated by wide basins (Figs 5, 6). The ranges are asymmetrical in profile, with a steeper south-eastern side rising from a faulted margin. The basins are downfolds of the schist basement (synforms), whose bedrock floors are at or near sea level. The basins contain remnants of sediments resting on the Waipounamu Erosion Surface, as well as alluvial fan gravels deposited as the mountain ranges rose.

Fig. 6 (opposite). Satellite image of Raggedy Range (lower left) and Blackstone Hill (upper right; Fig. 2), with flanking Ida Valley (centre right), Poolburn Gorge (lower centre) and Manuherikia Valley (left). North Rough Ridge (Fig. 2) lies on the lower right. The now-continuous Raggedy Range-Blackstone Hill ridge has developed from several small domes which have merged. The Poolburn Gorge cuts through this ridge in the centre of the picture. Before the ranges merged, the Poolburn and Ida rivers flowed separately to the Manuherikia Valley. Growth of the ranges forced these two rivers closer together until they combined and cut the Poolburn Gorge. An old abandoned and uplifted channel of the Poolburn is visible to the south-west of the gorge. Range growth and associated river diversion occurred over about one million years and is still active.

Schist ridges are smooth where the Waipounamu Erosion Surface has been exposed, and higher ridges are tor-studded (Fig. 5). Schist hillsides do not have scree, and commonly have a mantle of vegetated hummocky landslides (Box 3) instead. Streams in schist are less steep than in greywacke, but alluvial fans occur at the edges of the higher schist ridges (Box 2) where streams have become more deeply incised (Fig. 6). Rivers flow down the basins (Fig. 6) but cause little erosion of the basin floors. River erosion occurs principally within rising mountain ranges (e.g. the Clutha River is cutting through the Dunstan Range between Cromwell and Alexandra, Fig. 2).

The Nevis-Cardrona Fault is a major active feature which forms the western boundary to the Central Otago folds. It marks a change in landforms, from the smooth topography associated with the Waipounamu Erosion Surface, to the rough rugged mountains to the west (Fig. 2). Mountain uplift rates are over 1 mm per year on the western side of the fault, and less than 1 mm

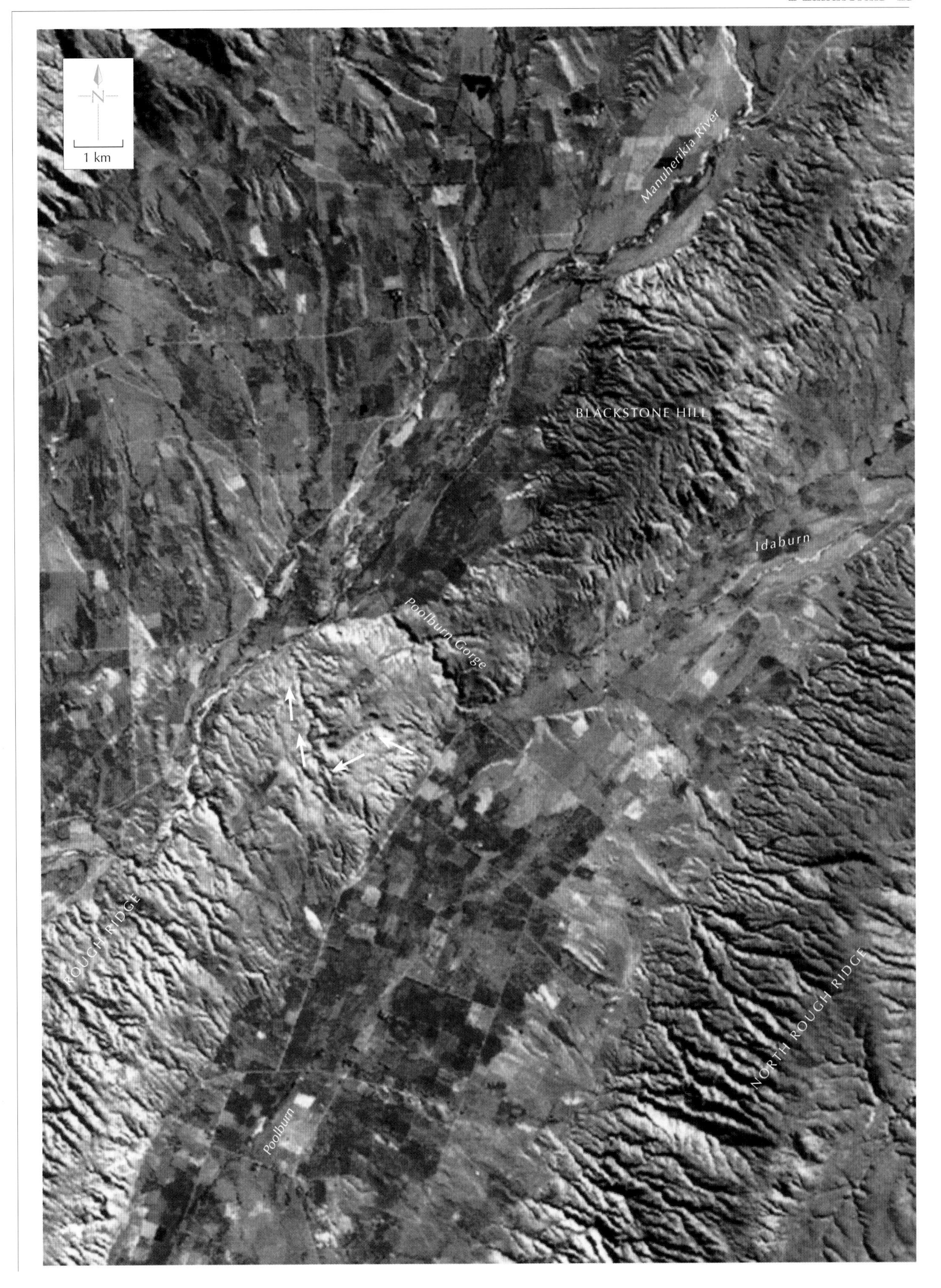
N
1 km
Manuherikia River
BLACKSTONE HILL
Idaburn
Poolburn Gorge
ROUGH RIDGE
NORTH ROUGH RIDGE
Poolburn

3 LANDSLIDES

Slow-moving landslides blanket large areas of southern New Zealand, particularly in the Otago Schist (Chapter1) to the east of the Main Divide. In these areas, the mountains are becoming higher due to slow uplift, and the rivers stay at the same level due to erosion of their beds. Hence, the slopes are getting steeper and landslides are the principal form of erosion in this situation. Hillsides are normally delicately balanced, with landslides poised to move faster in response to some external stimulus such as high rainfall or earthquakes, and catastrophic failure can follow.

Slow-moving landslides are characterised by rounded hummocky tussock-covered slopes devoid of tors. The most famous are those of the Cromwell Gorge, where extensive (and expensive) earthworks were required to stabilise landslides before Lake Dunstan was filled behind the Clyde Dam (opposite top). These landslides begin at least 1500 m above the river, are tens of metres thick, and can be kilometres wide. Highway 6 near the head of Lake Wanaka is built on similar landslides.

Landslides are common along the east coast of Otago, where clay-rich mudstones occur (Chapter 1). Hummocky slopes on coastal hills, beneath volcanic caps, are characteristic of landsliding mudstones. Landslides are less common in the greywacke mountains of north Otago, where screes transfer eroded rock from hills to valleys (Fig. 3). High rainfall west of the Main Divide helps to remove eroded debris, and landslides are generally smaller. However, sloping schist slabs encourage landslide movement, and a landslide down such slabs at the Gates of Haast Bridge is a perennial engineering problem. Large earthquakes on the nearby Alpine Fault every 250–300 years can cause sudden catastrophic landslide movements as well.

Melting glaciers over the past 12,000 years have exposed steep glaciated rock walls in the mountains. These steep slopes can collapse progressively or via catastrophic rockfalls and landslides. Retreat of glaciers over the past fifty years has resulted in some dramatic rockfalls in the high mountains, including the sudden collapse of the top of Aoraki/Mt Cook, and two sudden rockfall events around Mt Aspiring which displaced glacial lake waters, devastating valleys (and huts) downstream. Ancient examples of catastrophic landslides include the Lochnagar slide in the Shotover Valley, the Acheron and Snowdon slides in the Upukerora Valley, and the Green Lake slide in eastern Fiordland (opposite bottom). This last has an area of 45 square kilometres and is reputed to be one of the largest landslides on Earth. These large catastrophic landslides may be triggered by earthquake events.

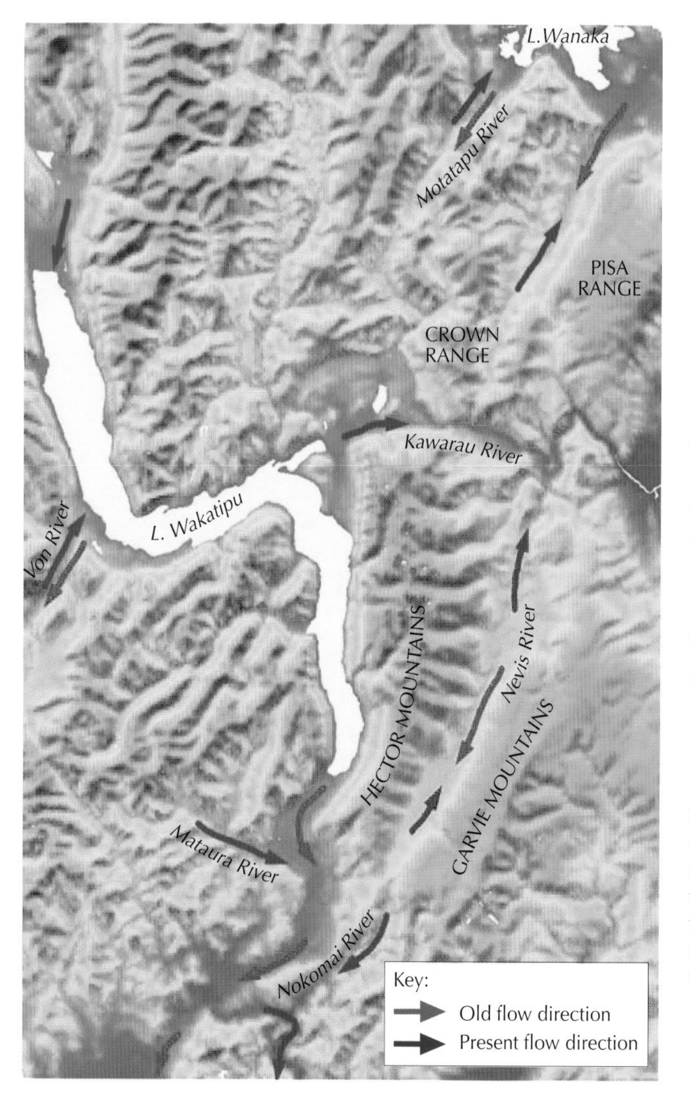

per year to the east. Continued uplift of fold mountain ranges in Central Otago can cause changes in river drainage patterns (river capture) if a river does not have sufficient erosive power to cut through rising bedrock (Fig. 6). Farther west, the Cardrona River used to flow southwards to the Kawarau River, but simultaneous uplift of the Crown and Pisa Ranges has caused these ranges to meet, pinching off the south-flowing river and reversal of its drainage (Fig. 7). Before this reversal, rivers in the Lake Wanaka catchment flowed south via the Motatapu Valley to the Kawarau River (Fig. 7). Likewise, the Nevis River flowed south via the Nokomai Valley to the Mataura River until the Garvie and Hector Ranges grew together and pinched off the valley (Fig. 7). This caused a low divide to be breached, and a gorge was cut down to the Kawarau River. Similar river capture events have occurred elsewhere as the fold ranges of Central Otago have developed over the past million years. The resulting complex history of drainage evolution has affected biota such as galaxiid speciation (Chapters 5 and 10). Only large rivers such as the Kawarau, Clutha and Taieri have been able to maintain their courses as ranges grew up across their paths, and deep gorges have formed.

Fig. 7. Digital elevation model of the Lake Wakatipu area of Central Otago and northern Southland, constructed from LINZ topographic data by S. Read. Present flow directions of major rivers are shown with blue arrows. Old flow directions of several rivers are indicated with red arrows, as described in the text. Changes in river flow directions are caused by growth of mountain ranges, and glacial advances and retreat.

Slow-moving landslides above Lake Dunstan in the Cromwell Gorge between Cromwell and Alexandra (Fig. 2). The gorge is being cut by the Clutha River as it crosses an actively rising schist ridge (including the Dunstan Range, Box 2). Lake Dunstan was formed by damming the river for hydro-electricity at the lower end of the gorge. As this dam and nearby towns were potentially threatened by catastrophic landsliding into the lake, extensive earthworks have been undertaken to reduce the risk. Landslide activity is enhanced by water within and beneath the landslide. Hence,the most active landslide in this picture has been covered with an impervious coating to prevent rain penetration. In addition, extensive underground drainage structures are removing ground water from the landslide. Smooth tussock slopes around the engineered portion are also underlain by landslides, but these are less active. Only the skyline ridges and the blocky lakeshore outcrops are free of landslide activity.

Green Lake landslide area, Fiordland (Fig. 10), reputedly one of the largest catastrophic landslides in the world. Post-glacial collapse of a glaciated hillside (foreground) resulted in infilling of the adjacent valley with hummocky piles of debris over several square kilometres (centre of picture). Green Lake (centre left) is impounded in a hollow in the landslide debris. Ridges of approximately even height in the background are typical of the sea-cut erosion surface that is partially preserved across most of Fiordland. *A.F. Mark*

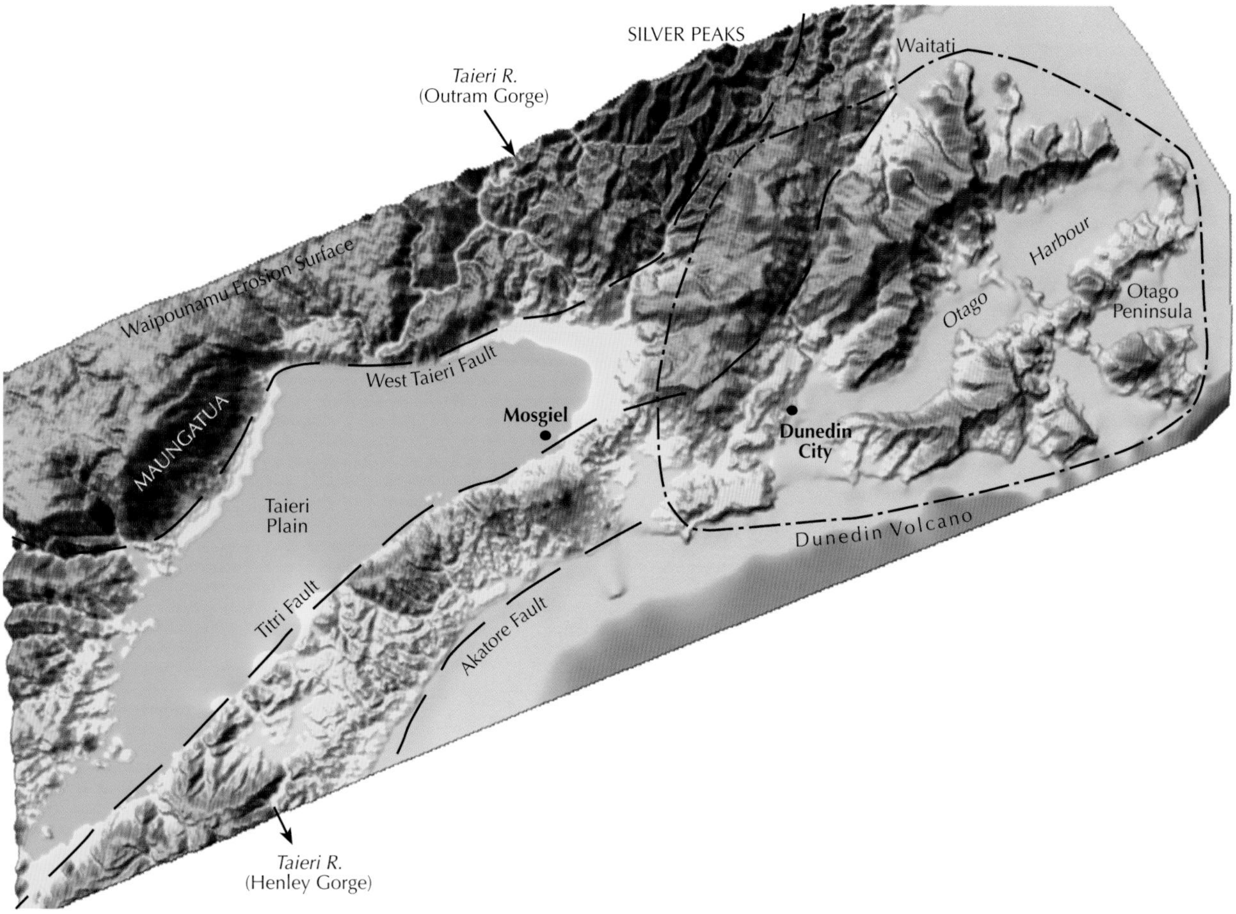

COASTAL OTAGO

The broad uplands of the uplifted Waipounamu Erosion Surface in Central Otago slope gently towards the east coast (Fig. 2), where the erosion surface is capped by remnants of Tertiary sediments and volcanic rocks (Chapter 1). The erosion surface is also disrupted by faults with a north-east trend. These faults have uplifted schist bedrock to form high ridges, such as the Silver Peaks and Maungatua (Fig. 8). The Taieri and Waikouaiti Rivers are cutting steep-sided gorges through these uplifted blocks. The sides of the Taieri River (Outram Gorge, Fig. 8) are collapsing via deep-seated landslides, so that bridges have had to be shortened by up to several metres over the past century.

The Taieri Plain is a depression between two faults (Fig. 8), and an extension of this depression continues south to Balclutha. The base of the Taieri depression has been lowered below sea level, while the Taieri River deposits progressively more sediment at sea level. The river is cutting a gorge (Henley Gorge, Fig. 8) through rising coastal hills to the sea. The active Akatore Fault parallels the coastline east of the Taieri Plain (Fig. 8), and goes onshore to the south. An offshore reef was formed by two metres of uplift in a single earthquake on the fault about 1000 years ago. The lower reaches of the Taieri Plain are currently tidal; the sea extended across most of the plain about 8000 years ago. The extent of tidal incursion in the future will depend on balances between sea-level rise, fault movement and sedimentation.

Remnants of the Dunedin volcano (Chapter 1) form a prominent feature at the northeast end of the Taieri Plain and Akatore Fault (Fig. 1). Consequently, landforms around Dunedin city are starkly different from those of the surrounding schist bedrock. The city skyline is dominated by eroded remnants of volcanic domes (Mt Cargill, Swampy Summit, Flagstaff and Highcliff Hill) perched on the rim of the old volcanic edifice. The Otago Harbour occupies the volcanic crater depression, which was long and narrow. Hillsides are steep, and coastal erosion is making them steeper. Erosion occurs mainly via thin slow-moving landslides and bouldery mudflows, which are particularly active in wet weather.

Fig. 8 (above). Digital elevation model of the Dunedin area, constructed from LINZ topographic data by S. Read, showing the principal landforms. Dunedin city and immediate environs are built on the remnants of the Dunedin volcano (dashed circle). The centre of the volcano was in the middle of the Otago Harbour, and the harbour formed when the crater was flooded by the sea. Small but prominent hills such as Saddle Hill are remnants of outlying volcanic centres. The Taieri Plain is a fault-bounded depression in schist bedrock. Continuation of the Titri Fault forms the prominent linear feature followed by the northern motorway to Waitati. That of the West Taieri Fault has resulted in uplift of the high Silver Peaks. The coastline south of Dunedin is controlled by the active Akatore Fault, whose continuation into the city area is uncertain.

Fig. 9 (right). Enlarged view of the digital elevation model in Fig. 1, showing principal landforms and active faults in Southland. The most prominent feature is the Murihiku Escarpment, a locally active fault. To the south of this escarpment, linear features are ridges of hard sandstones separated by more easily eroded mudstones. The smooth topography of the Central Otago schist (top centre and right) changes southwards to more rugged greywacke ranges.

SOUTHLAND

Landforms of Southland show the strong contrast of fertile alluvial plains with steep low mountains rising from the plains (Fig. 9). The most prominent feature is the Murihiku Escarpment and associated ranges of hills underlain by greywackes of the Southland Syncline (Fig. 9; Chapter 1). The greywackes show prominent layering, which is tilted almost vertical over large areas; these layers are traceable for several kilometres laterally (Fig. 9). The Murihiku Escarpment runs all the way from the east coast (Nuggets) to Mossburn. It is breached by the Mataura River at Gore, where faults cut across it, and by the Oreti River at Lumsden. To the north of the Murihiku Escarpment, alluvial fans of varying age emerge from rising greywacke and schist mountains on the southern edge of Central Otago. Older alluvial fans are being uplifted and eroded, exposing yellow cliffs as in Central Otago (Box 2). These old fans have been exploited for gold in the Waikaia and Wakaka catchments.

Water from Lake Wakatipu currently flows down the Kawarau River to the Clutha River, as Lake Wakatipu is dammed by a small glacial moraine at its southern end. However, Wakatipu catchment water has discharged southwards into the Mataura River several times in the past (Fig. 7), and it would take a rise in lake level of only fifty metres for this to happen again. Likewise, Lake Wakatipu has discharged at times from its western side to the Oreti River via the Von Valley (Fig. 7). Further, there is a pile of alluvial fan gravels only fifty metres high preventing the Mataura River joining the Oreti River north of Lumsden (Fig. 9); discharge here has also occurred in the past. These diversions are controlled mainly by deposits formed during glacial advances and retreats over the past 100,000 years, and represent radical reorganisation of the largest rivers in southern New Zealand. The Manapouri hydro-electric scheme had a similar effect, by diverting Waiau River water to Doubtful Sound (Fig. 10), thereby moving the Main Divide fifty kilometres to the east.

Western Southland (Waiau Valley, Fig. 9) is a complexly faulted sedimentary basin underlain mainly by mudstones, which form low relief landforms. Sandstones uplifted along faults such as the Moonlight Fault form prominent ridges, like the Haycocks in the Mararoa Valley (Fig. 10). The Takitimu Mountains are being uplifted from beneath these sediments, and form a spectacular isolated massif, of mainly greywacke and volcanic rocks, which forms large screes from ridge-top to valley floor. Related rocks form ranges of hills to the south, culminating in the crystalline rocks of the Longwood Range. These rocks do not form screes, and are resistant to erosion, so they have persisted as rounded hills for millions of years. Likewise, Bluff Hill (Fig. 9) has remained steadfast as an island or peninsula, variably surrounded by sea and low alluvial plains, for 30 million years.

Greywacke mountains rise to the north of the Takitimu Mountains, extending as far as Lake Wakatipu (Fig. 10). Valleys follow ancient and modern faults, to give a north to north-east

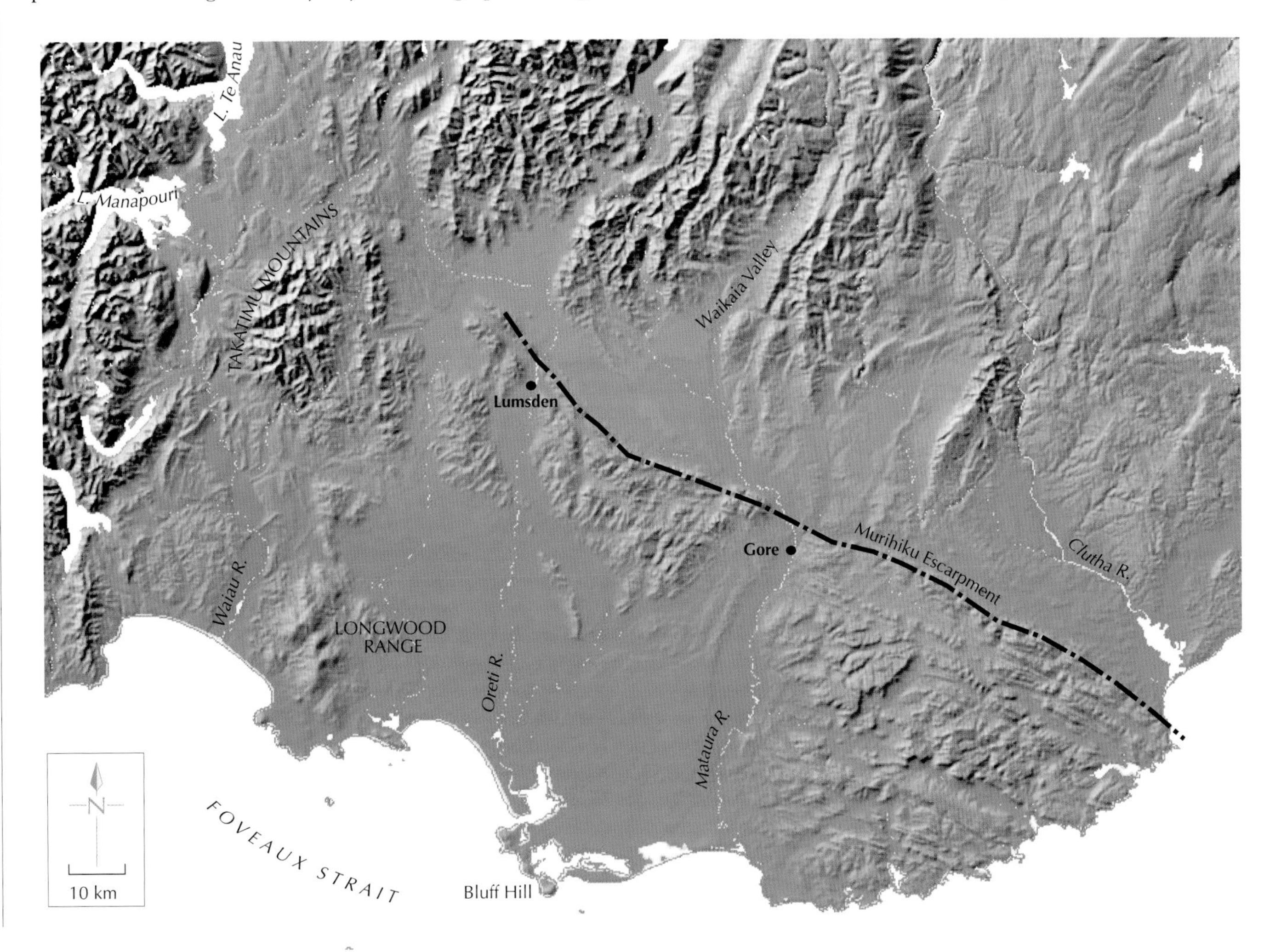

grain to the topography (Fig. 11). The mountains are clothed in scree on their upper slopes, and broad alluvial fans extend from steep gullies into wide trunk valleys (Fig. 11). Serpentine and related rocks (Dun Mountain Ophiolite Belt, Chapter 1) are largely barren of vegetation in the centre of this area (Fig. 12). Blocks of hard crystalline rocks, from metres to hundreds of metres across, protrude from soft serpentine hillsides like tors.

Stewart Island is made up of two distinct topographic massifs (Fig. 13). The northern portion, which includes Mt Anglem, is geologically related to Bluff Hill and the Longwood Range to the north, and has similar topography. The southern portion of the island consists of crystalline gneisses with prominent granite knobs (Chapter 1: Box 10, Fig. 23), and is geologically related to southern Fiordland. The two massifs are separated by the swamps of the fault-controlled Freshwater Valley, which are near sea level across the island. The Waipounamu Erosion Surface is preserved as gently sloping bedrock surfaces on the southeast part of the island (Fig. 13). The island has been little affected by glaciation apart from minor ice-sculpting at Mt Anglem. Foveaux Strait was a broad coastal plain, probably similar to today's Freshwater catchment, when sea level was lower during glaciation. This plain was inundated by rising sea level about 10,000 years ago, apart from scattered rocky hills which are now islands.

FIORDLAND

A set of active faults (Fiordland Boundary Faults, Fig. 10) separates the low relief of Southland from the rugged mountains of Fiordland, a large and relatively uniform block of hard crystalline rocks, including granites and gabbros (Chapter 1). These rocks have few weaknesses, such as faults and fractures, so they are capable of maintaining steep slopes. The area is a complex jumble of peaks, ridges and valleys with little regular topographic grain, in contrast to the greywacke ranges to the east (Fig. 10). Peaks and ridges are generally even in height (Box 3) and are remnants of the Waipounamu Erosion Surface (Box 1) and of younger sea-cut surfaces that have been deeply dissected by glaciated valleys. The erosion surface has been uplifted and dissected most in northern Fiordland, where peak heights are typically above 2000 m, including Mt Tutoko which is the steepest mountain in New Zealand (2723 m, less than 8 km from the sea). At the other end of the relict erosion surface, remnants of sea-cut surfaces are preserved near the coast of south-west Fiordland. A spectacular set of such surfaces, ranging up to 1000 m above sea level, occurs in the Waitutu Forest (Fig. 10).

The resistant nature of the Fiordland rock, combined with high rainfall, means that slow-moving landslides are rare, and clean glaciated bedrock is preserved over large areas. Catastrophic landslides occur sporadically.

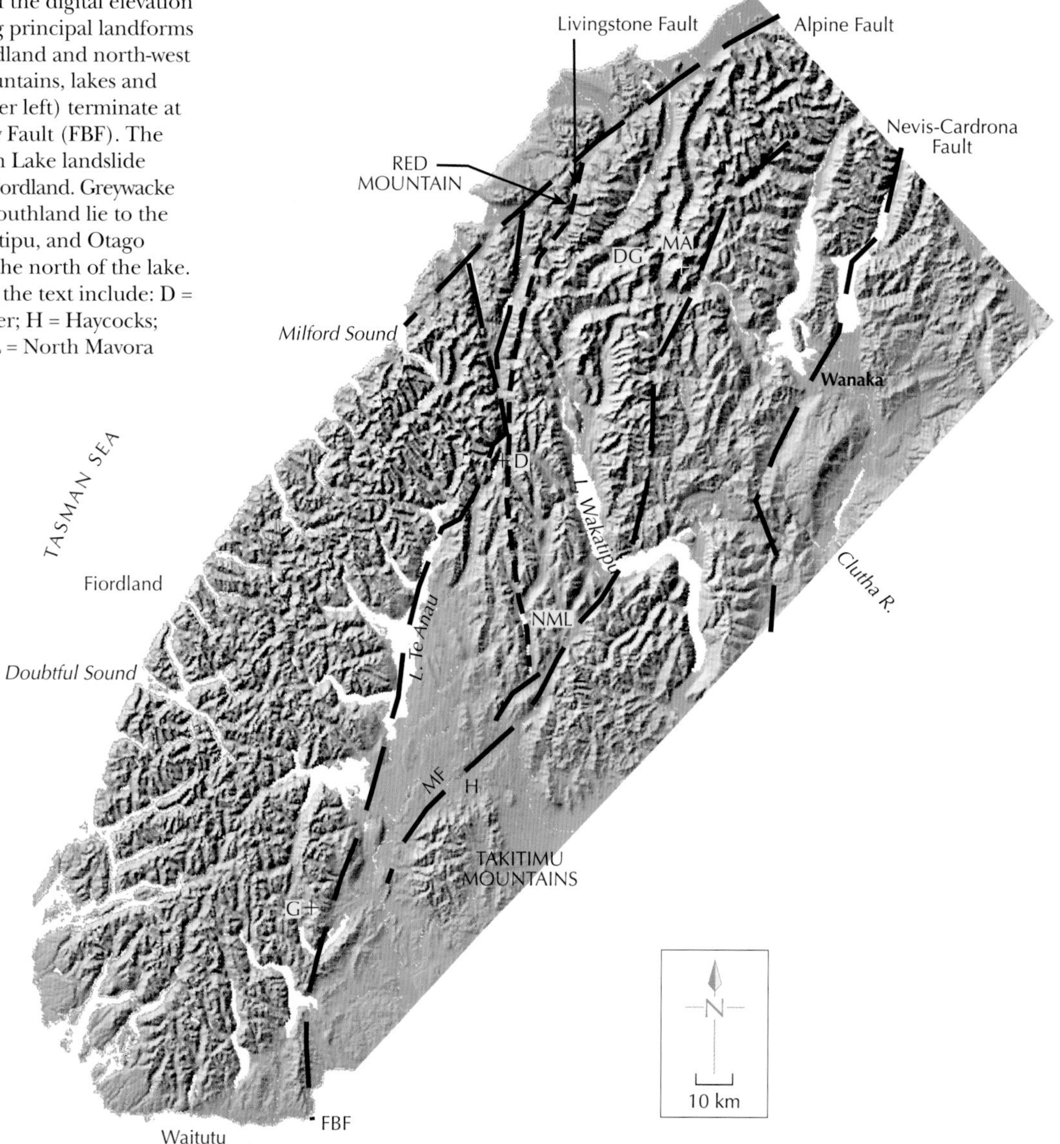

Fig. 10. Enlarged view of the digital elevation model in Fig. 1, showing principal landforms and active faults in Fiordland and north-west Otago. The rugged mountains, lakes and fiords of Fiordland (lower left) terminate at the Fiordland Boundary Fault (FBF). The large catastrophic Green Lake landslide (Box 3) is at point G in Fiordland. Greywacke mountains of northern Southland lie to the southwest of Lake Wakatipu, and Otago Schist mountains lie to the north of the lake. Localities mentioned in the text include: D = Divide; DG = Dart Glacier; H = Haycocks; MA = Mt Aspiring; NML = North Mavora Lake.

Fig. 11. North Mavora Lake (Fig. 10) and the upper Mararoa River (right centre) are controlled by faults in the greywacke mountains of northern Southland. These valleys have been extensively modified by glaciers which reached several hundred metres in thickness. Mt Earnslaw, a schist mountain at the head of Lake Wakatipu, is visible on the skyline (middle distance).

Fig. 12 (left). View, looking south, of greywacke ridges and associated Dun Mountain Ophiolite Belt hills of northern Southland, to the west of North Mavora Lake (Fig. 10, 11). Barren serpentine screes of the ophiolite belt are in the immediate foreground and left middle distance. The Acheron Lakes (centre) are trapped behind a large (kilometre-scale) landslide. Another large landslide has removed a strip of forest from Mt Snowdon (top right distance).

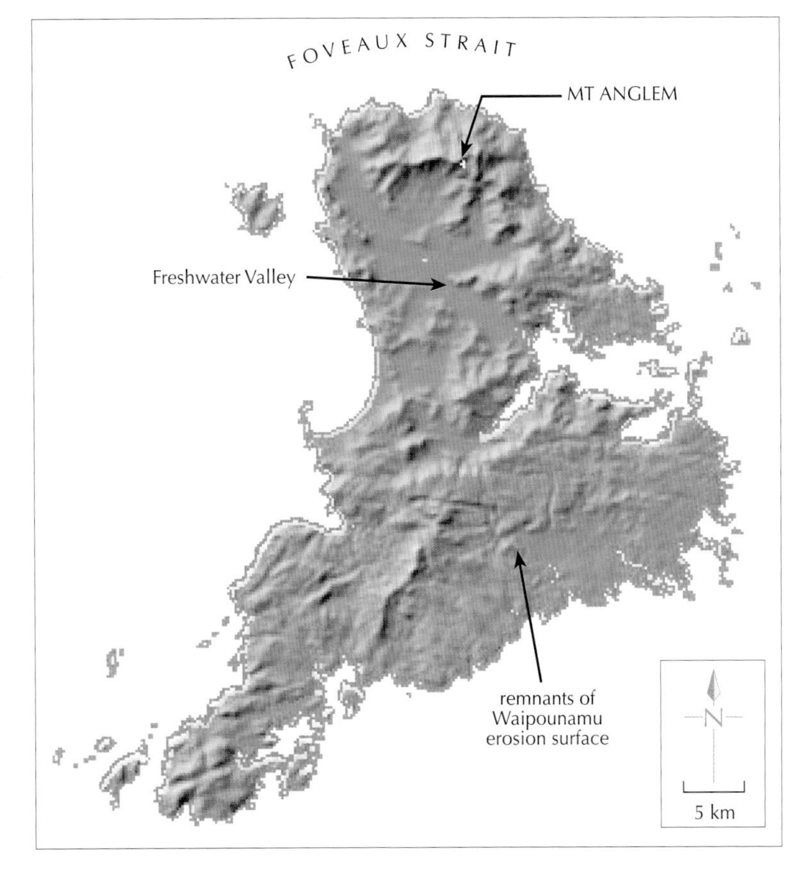

Fig. 13. Enlarged view of the digital elevation model described in Fig. 1 caption, showing principal landforms of Stewart Island, as described in the text.

OTAGO SCHIST MOUNTAINS

The rugged schist mountains of north-west Otago lie west of the Nevis-Cardrona Fault and its continuation up Lake Hawea (Fig. 2). The slabby nature of schist, imparted by aligned mica grains, dominates landforms in these mountains. Slabs slope steeply to moderately east or west (Fig. 14) and trend northwards or north-eastwards throughout the mountains. Slabs control the orientations of many ridges and valleys, imparting a north to north-east grain to the topography (Fig. 10). Rivers have eroded deep into the schist as it was uplifted, forming steep-sided narrow tortuous slabby gorges (Fig. 15). The steepest, narrowest gorges are those cut into resistant layers of greenschist (originally submarine lava flows, Chapter 1). Many mountains are asymmetrical in shape, with a smooth side dominated by slabby faces and a steep blocky side where erosion has cut across the slabs (Fig. 14).

The schist mountains have been uplifted several kilometres over the past five million years, and the Waipounamu Erosion Surface has been destroyed. Small remnants of sediments, and underlying erosion surface, are preserved near Queenstown, and these features create distinctive low-relief saddles in ridges. One of these is above Gibbston, on the east flank of the Remarkables, and the other is in a narrow belt between Bobs Cove (on Lake Wakatipu) to Mt Aurum (Shotover River catchment) (Chapter 1, Box 6). This latter feature is along the Moonlight Fault, an important north-east trending structure in the schist mountains. The high mountains of the Mt Aspiring area are on the west (uplifted) side of the Moonlight Fault, and the fault scarp is a series of high rock walls immediately west of the east branch of the Matukituki River (Fig. 16). Related faults to the north have less impressive topographic changes across them, but they form a series of low passes between valleys from the Wilkin catchment to the Haast Pass area.

The slabby slopes of schist mountains readily form slow-moving landslides (Box 3) that coat most mountain slopes from near ridge crests down to valley level. These landslides form smooth hummocky slopes in tussock country and in the beech forests below.

The schist mountains extend as far west as Red Mountain (Fig. 10), where they abut against the Dun Mountain Ophiolite Belt (Chapter 1). The ultramafic rocks of the ophiolite belt support only limited vegetation, and consist of barren ridges with blocky screes unlike any other southern landforms (Fig. 17). The boundary between the ophiolite belt and the schist mountains is the Livingstone Fault. This fault is visible as a prominent linear zone, with numerous low passes, from the Alpine Fault to the Moonlight Fault, through the remote western mountains (Fig. 10).

Fig. 14 (opposite). Schist slabs in the Otago Schist mountains slope down to the left (west) in the right foreground (by person), and in the mountains in the distance. Mt Aspiring (top left), the highest mountain in Otago (3027 m) is most commonly climbed via the left (North-west) ridge which is controlled by the west-dipping schist slabs. Likewise, Mt Avalanche (top right) is generally climbed via the slabs on its left flank.

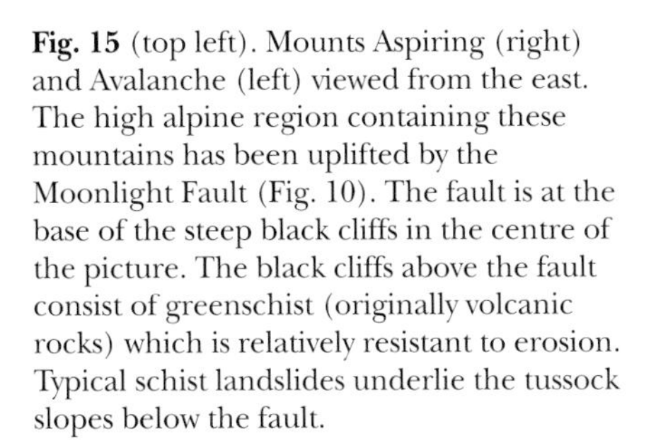

Fig. 15 (top left). Mounts Aspiring (right) and Avalanche (left) viewed from the east. The high alpine region containing these mountains has been uplifted by the Moonlight Fault (Fig. 10). The fault is at the base of the steep black cliffs in the centre of the picture. The black cliffs above the fault consist of greenschist (originally volcanic rocks) which is relatively resistant to erosion. Typical schist landslides underlie the tussock slopes below the fault.

Fig. 16 (top right). Red Mountain (Fig. 10), viewed from the north-west, the most spectacular landform in the Dun Mountain Ophiolite Belt. Blocky outcrops and screes of ultramafic rock are largely devoid of vegetation. The foreground grey rocks are gabbro, also part of the ophiolite belt (Chapter 1). *J.M. Sinton*

Fig. 17 (right). Typical deeply incised stream in schist slab country north of Lake Wakatipu (Fig. 10). Hillsides in the distance are smoother as they are underlain by landslides. This combination of narrow gorges separated by smooth landslide slopes dominates the landforms of this area.

GLACIATION

Repeated advances of glaciers over the past million years have modified some of the topography of the western areas. Valley glaciers followed pre-existing river valleys, deepening the valleys and steepening the valley walls, to a characteristic U-shaped profile (Fig. 18). This U profile is preserved in Fiordland, where the firm hard rock resists landsliding. The U profiles have been modified by landslides and screes in the schist and greywacke mountains, and slopes have become gentler (Box 3). The valley glaciers were commonly over 1000 m thick and reached up to the present beech bushline. A prominent topographic shoulder at about bushline records the ancient ice level throughout the mountains. Steep-sided amphitheatres (cirques), in side-valleys and the main river headwaters, were cut by glacial ice. Some of these cirques are still occupied by glaciers, although these are melting rapidly. The Dart Glacier (Fig. 10), one of the larger glaciers in the Lake Wakatipu catchment, is currently 6 km long; its terminus has retreated 5 km over the past 150 years. At various times this glacier, joined by numerous others in adjacent valleys, has extended beyond the end of Lake Wakatipu (Figs 7, 10).

Cirques from which all the ice has melted typically contain small lakes, nestled amongst glacially scoured rounded rock outcrops (Fig. 3). Lake Quill is one of the more spectacular examples, discharging water to form the Sutherland Falls (with a 580 m drop) beside the Milford Track. Ice-sculptured rocks are evident in the lower reaches of mountain valleys as well, particularly as rounded hills with smooth upstream slopes and steeper down-stream scarps (roches moutonnées), such as Mt Iron, near Wanaka and Mt Alfred at the head of Lake Wakatipu.

Glaciers carry large volumes of rock debris (till), and deposit these as terminal moraines at the farthest extent of advance. Terminal moraines from the last major glacial advance (15,000 years ago) dam all of the large southern lakes (Fig. 1). Remnants of moraines from older glacial advances are recognisable between Cromwell and Wanaka. The 'sounds' of Fiordland (Fig. 10) are in fact fiords: river valleys deepened by ice when the sea level was lower. Submarine moraine ridges were deposited at their Tasman Sea ends. Large boulders (glacial erratics) accompany most terminal moraine deposits. Huge volumes of gravel were transported from melting glaciers into downstream river systems, then sculpted into numerous sets of high river terraces by changing river courses and lowering river levels. These terraces are best displayed in the Clutha catchment, especially around Cromwell and Alexandra. Alluvial fans (Box 2) from various glacial periods spread out from mountain ranges and merge with these terraces.

Fig. 18. (above) A U-shaped glaciated side valley meets Milford Sound, which is a U-shaped glaciated valley (fiord) flooded by the sea. The hard crystalline rock which underlies the area does not succumb to landslides, so glaciated landforms are preserved. Landslides in schist mountains rapidly obscure these glacial landforms as the glaciers retreat.

CHAPTER 3

FOSSILS AND THE HISTORY OF LIFE

R. EWAN FORDYCE

Curio Bay in the south Catlins area of Southland is the site of a spectacular fossil forest of Jurassic age, exposed on a tidal platform. The site contains many logs and stumps. *Neville Peat*

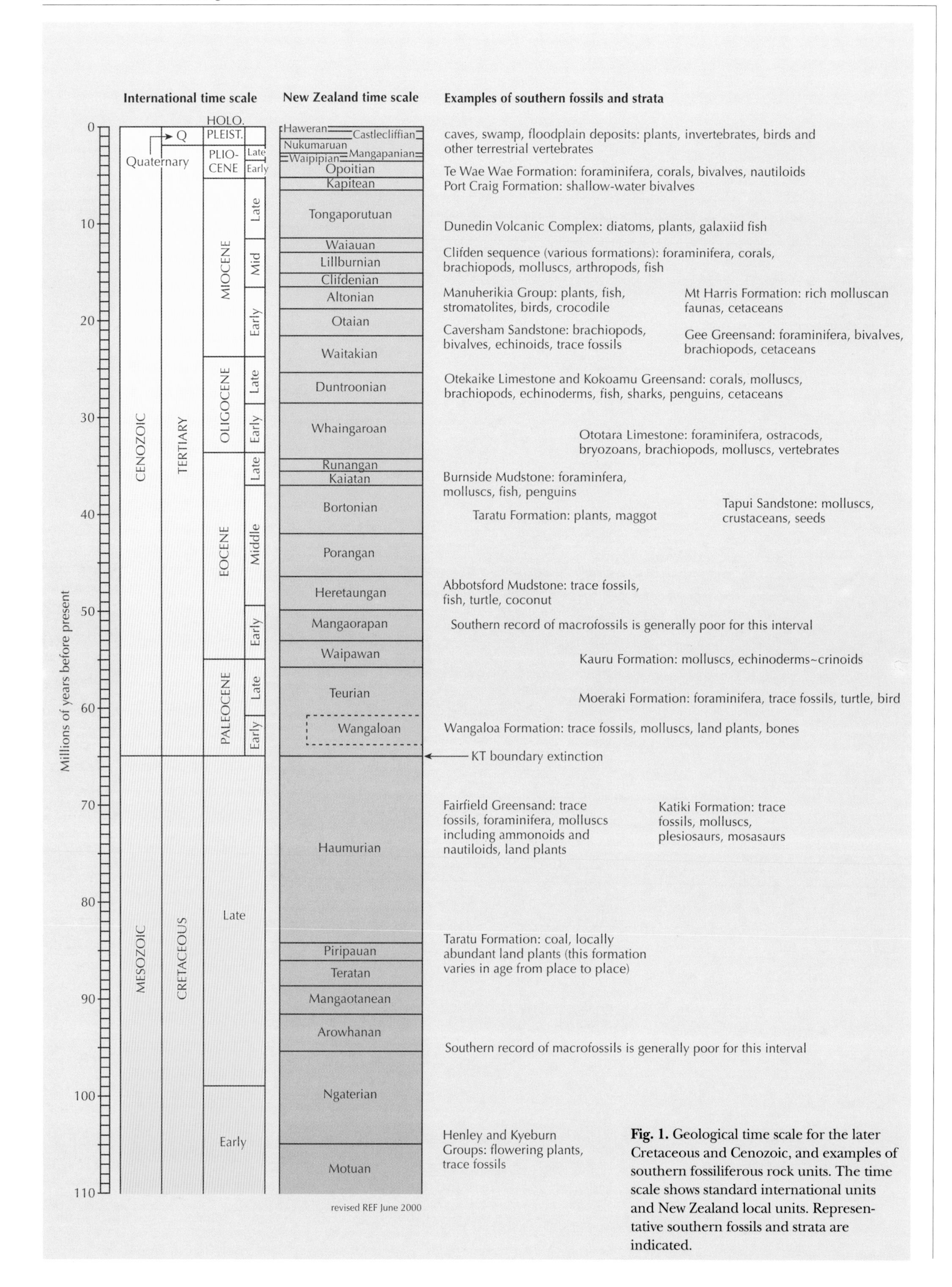

Fig. 1. Geological time scale for the later Cretaceous and Cenozoic, and examples of southern fossiliferous rock units. The time scale shows standard international units and New Zealand local units. Representative southern fossils and strata are indicated.

FOSSILS AND THE HISTORY OF LIFE

New Zealand has a complicated geological heritage spanning more than 500 million years. Fossils are a key to understanding this complex geological past. For much of its history, proto-New Zealand lay on the margin of Gondwanaland – the ancient supercontinent which eventually broke apart to form Antarctica, India, Australia, South America and Africa. This Gondwanan position produced a distinctive southern signal in our fossil record and that signal is still apparent in our modern biota. After Gondwanaland broke apart, proto-New Zealand drifted away into the Pacific, from about 80 to 55 Ma (Ma = Mega-annum, 10^6 years, or million years ago). The landmass was isolated, forming a sort of 'Noah's ark' on which plants and animals evolved in isolation, leading to a suite of endemic – or distinctively local – components of fauna and flora. The fossil record of southern New Zealand reveals much of this history, while missing details are provided by fossil-bearing rocks elsewhere in the country.

Rocks, time and environments

Fossils occur in sedimentary rocks which, in southern New Zealand, are well known through more than 130 years of geological mapping. Initially, southern fossils were used mainly as tools to determine the age of rocks. In this role, they have been critical in producing geological maps of southern New Zealand (page 4). Fossils have also helped to determine ancient environments, indicating, for example, whether sedimentary rocks were deposited in marine conditions (e.g. continental shelf or continental slope) or non-marine settings (e.g. lake or river). Paleontology – the study of fossils – has broadened its outlook in recent decades, using fossils to unravel ancient global geography, the origins of the New Zealand biota and the evolutionary history and adaptations of diverse plant and animal groups.

Time scale

Because fossils are a part of geology, and geology deals with the ordering of earth events, a time scale is fundamental in paleontology. Fossils are usually placed in the international geological time scale, which has a standard sequence of time units. Intervals range from the Cambrian, effectively the period of oldest organised multicellular life, through to the Quaternary, the geologically recent ice ages and modern day (Fig. 1). These named intervals are relative units, that is, they are identified in terms of 'older than' or 'younger than.' An analogy is seen in the easily remembered generations of our own family histories. Absolute ages, in millions of years before present, sometimes are used in discussing geological time, but with rather less confidence. Again, there is a parallel with human history, where we must link the well-known generations with actual dates of birth and death. As with many other parts of the world, New Zealand also has its own distinctive local time units, used because of problems in correlating with the international geological time scale.

Diversity of fossils

What fossils are known from the south? The record is restricted to mainland South Island, with no significant finds reported from Stewart Island. Most of the fossils have some sort of skeleton or mineralised support, whether bone, shell or limy framework, and most major skeleton-bearing groups occur, particularly in marine rocks. The fossils are small and large, animal and plant, and marine and non-marine. Important microfossils (Chapter 6: Box 1) include the single-celled animal groups foraminifera and radiolaria, and unicellular plants such as diatoms and coccoliths; all have mineralised skeletons formed of calcium carbonate or silica. For worms, there are fossils of skeleton-forming annelids and many fossilised traces of activity which have disturbed sediment without preserving the original soft-bodied animal. Corals include single and, less commonly, colonial forms. Fossils of larger shelled invertebrates are dominated by molluscs – the gastropods (snails), bivalves (e.g. pipi, scallop) and cephalopods – *Nautilus* and its shelled relatives. The double-valved (that is, double-shelled) brachiopods, which are inconspicuous today, are overwhelmingly important at some times. Bryozoans or moss animals are inconspicuous, yet are major rock formers. Crabs and other crustaceans are rare but often dramatic. Echinoderms – the spiny-skinned animals – include sea urchins and rarer crinoids (sea lilies) and relatives.

NAMES

Most of the fossils discussed in this chapter are named formally. Unlike living organisms, none have common names, so that the two-part generic and specific names are normally used by paleontologists. This review emphasises the genus, used as a single name, e.g. *Waipatia*. Occasionally, fossils are placed into a genus for convenience, until relationships are determined confidently. In such cases, names of convenience are given in quotes, e.g. '*Rhynchonella*'.

To detail the diversity of all these organisms, their occurrences and their implications, is beyond this overview. Rather, individual case-studies of particular fossils and settings are used to reveal the strengths of the southern record.

The extinct graptolites, preserved as little more than smears on shale, are important in some of the oldest rocks. Vertebrate fossils are conspicuous but quite rare; they include sharks, bony fish, penguins, turtles and whales and dolphins. Plant fossils sometimes also occur in marine strata. Mostly, though, plants are the main fossils in non-marine rocks. Ferns, gymnosperms (podocarps and relatives) and angiosperms (flowering plants) locally are abundant. Rare animal remains are known also from non-marine rocks.

THE SUCCESSION OF BIOTAS

Paleozoic

The southern record includes fossils from the three major Eras or subdivisions of geological time: Paleozoic, Mesozoic and Cenozoic. For the oldest interval, Early Paleozoic life in the south is reported only from Preservation Inlet in Fiordland. Here, graptolites in slates of the Buller Terrane represent the Ordovician Period, some 460–470 Ma. Otherwise, there are no reports of older (Cambrian) fossils. (The term terrane is used widely for a large body of rock which is internally continuous and is bounded by faults. Terranes and other rock units, such as formations, are usually named with reference to a key locality; thus, Buller Terrane takes its name from the district near Westport, far to the north.) Younger Paleozoic time, spanning the Silurian to Carboniferous Periods, is not represented in the south, although this interval is known from North-west Nelson.

The close of Paleozoic time is marked by Permian invertebrates and plants from two different southern sequences. First, rocks in the Productus Creek region of the Wairaki Hills near Ohai form part of a long narrow sliver, the Brook Street Terrane, which is faulted against neighbouring sequences (see pages 5–6). Second, sparse Permian fossils occur in the narrow and discontinuously exposed Dun Mountain/Maitai Terrane (pages 6–7), forming a strip north-west of Balclutha. Both the Brook Street and Dun Mountain/Maitai Terranes are known better from the Nelson district. In summary, most of the Paleozoic is not represented by fossiliferous rocks in the south. There is no record of the early rapid evolution of life in the sea and on land – no early trilobites, extensive reefs nor early land plants and arthropods. Finally, the Permian/Triassic boundary, which saw one of the earth's dramatic mass extinctions at about 250 Ma, has not been identified clearly in the southern record.

Mesozoic

Mesozoic fossils are represented well in a kilometres-thick sequence of marine shelf and non-marine rocks of the Murihiku Supergroup (also known as the Murihiku Terrane, page 6). These rocks, which form a huge rather simple fold termed the Southland Syncline, span South Otago and northern Southland, and they extend north-west from Nugget Point towards the Southern Alps. Rarer Mesozoic fossils also occur in the structurally more complex rocks of the Torlesse Supergroup (Rakaia Terrane, page 7) in North Otago.

The oldest Mesozoic fossils are rare invertebrates of uncertain Early Triassic age. Apparently, the very start of Triassic time is not represented by southern fossils, so that we have little local understanding of biological events immediately following the mass extinction at the end of the Permian. Well-known later Triassic and earlier Jurassic biotas include microfossils such as foraminifera, radiolaria, and the tooth-like conodonts, and larger invertebrates such as coelenterates, worms, brachiopods, gastropods, bivalves, cephalopods, scaphopods, crustaceans, crinoids and echinoids. Vertebrates include fish teeth, ichthyosaurs (dolphin-like marine reptiles) and unidentified fragments. Most of the Murihiku fossils were buried in shallow-water sediments rich in volcanic debris, while those from Torlesse rocks typically occur in granite-derived sediments which were deposited in shallow or deep waters. The later Triassic marine fossils from Murihiku rocks help to define time intervals (stages) which are used throughout New Zealand.

Mesozoic plants deserve mention. Triassic floras occur in the Torlesse rocks of the Waitaki Valley, while Early to Middle Jurassic plant beds are present at Curio Bay and other southern localities. Later Jurassic and earlier Cretaceous plants are not reported, but non-marine beds of roughly middle Cretaceous age (Kyeburn Group, Central Otago and Henley Breccia, East Otago) are a source of occasional plant remains. Dinosaurs have not been reported yet.

Later Cretaceous and Cenozoic cover sequence

A thick continuous sequence of fossiliferous rocks represents the end of the Mesozoic (Late Cretaceous) and the succeeding Cenozoic Era. Rocks and fossils identify diverse ancient environments: alluvial fans and plains, swamps, lakes and rivers, lagoons, estuaries and sandy shores and shallow to deep marine settings. These are the 'cover' strata which span about the last 80 to 100+ million years (Fig. 1). The cover strata normally overlap older rocks at a major unconformity or time gap, usually marked by a prominent physical break, seen over much of New Zealand. The time spanned by the unconformity or break in sequence varies from place to place, but in the south it spans roughly Late Jurassic to Early Cretaceous. Generally, the underlying Jurassic and older rocks are more steeply dipping, folded, faulted and more metamorphosed than the overlying Cretaceous and younger cover strata. The unconformity is thought widely to mark a time of tectonic activity – burial and metamorphism, mountain-building and uplift – followed by extensive erosion before cover strata were finally deposited.

Broadly, the cover strata preserve what is termed a transgressive sequence – one associated with long-term submergence of land, or a rise in sea level, or both. In this sequence, the oldest rocks at the base are non-marine strata which are overlain by shallow marine and then deeper marine sediments. This transgressive sequence roughly marks the time during which proto-New Zealand separated from the Gondwana margin, starting perhaps beyond 100 Ma, and moved north-east into the Pacific from 80+ to 55 Ma. It marks erosion of the land, increasing submergence, and the evolutionary divergence of the biota from that of other Gondwanan landmasses.

Rocks in about the middle of the cover strata are dominated by limestone and greensand, sediments which hint that land was low-lying and producing little rock debris. It is useful to view this as a quiet or quiescent phase in New Zealand's history.

Shallow extensive seaways were present. Some geologists and paleontologists propose that New Zealand was fully submerged at this time, with no land exposed at all; geological evidence for complete submergence is debatable, and there are compelling biological reasons to think that some land remained . Whatever happened, this quiescent phase in New Zealand history spans much of the Oligocene (34–23 Ma). Higher in the sequence, though, above the transgressive and quiescent strata, shallowing marine sediments are followed by non-marine rocks which mark a regression, or withdrawal of the sea. The regression reflects a combination of falling sea levels and uplift of a no-longer-drifting New Zealand from about the Early Miocene onwards (23 Ma to present). Thus, our later Cretaceous and Cenozoic cover sequence reflects transgression, quiescence and regression.

The transgressive/regressive cover strata include widespread coal measures, which are an important source of fossil plants. Marine rocks produce enormously diverse fossils including microfossil groups such as single-celled diatoms, coccoliths, foraminifera, and radiolarians, and multicellular forms such as the tiny double-shelled ostracods. Amongst invertebrates sponges, corals, brachiopods, molluscs, annelid worms, bryozoans, crustaceans and other arthropods, and echinoderms are represented. Vertebrates include bony fish, sharks, turtles and other marine and terrestrial reptiles, birds and marine mammals.

Widespread Quaternary non-marine sediments (younger than about 2 Ma) are a major source of pollens, and of bones of flightless birds. Although these are studied using basic principles of paleontology, they are much younger than fossils mentioned above, and usually command the attention of biologists rather than paleontologists. In the south, the widely studied Quaternary fossils are mostly only a few thousand years old.

Within the cover strata, also, the Cretaceous/Cenozoic (or Cretaceous/Tertiary – 'KT' – or Cretaceous/Paleocene) boundary is marked in the south by the extinction of some foraminifera and cephalopods, and of marine reptiles such as plesiosaurs and mosasaurs. Other biotic changes during the Cenozoic help us understand the changing geography and climate of southern New Zealand, and provide powerful evidence of extinction and evolution as natural processes.

WINDOWS ON THE PAST
Ordovician graptolites in Fiordland

The oldest southern fossils are from Preservation Inlet, south-west Fiordland. Here, graptolites occur as light-coloured elongate streaks in dark slaty mudstones of Early Ordovician age (perhaps 460–470 Ma) (Fig. 2). Better-preserved specimens from other parts of the world show that graptolites are extinct marine colonial animals which may reach more than 100 mm long. Graptolites typically preserve as squashed siliceous impressions which represent an originally three-dimensional organic skeleton. The original skeleton was probably formed of tough organic material like fingernail; it became mineralised with complex silicates after burial. A graptolite colony will have usually one, two or four branches or some multiple of this. Every branch is made of successive tiny cup-like structures (thecae), and each theca apparently held one tiny soft-bodied filter-feeding individual. Fiordland graptolites mainly represent planktonic forms which probably floated in surface waters, filtering food from the water. Upon death, the graptolite colony sank into quiet deep waters, to become buried in dark mudstone. Some stemmed forms are known, such as *Dictyonema*, which lived in a bush-like colony anchored on the sea floor.

Graptolites are used worldwide to correlate Ordovician and Silurian rocks. The Preservation Inlet fossils represent forms such as *Dictyonema*, *Loganograptus* and *Clonograptus*, which are also known from the Early Ordovician of Victoria, Australia. It is the globally widespread occurrence of many different sorts of graptolites which points to a floating or planktonic habit. Fiordland graptolites also can be correlated with those of North-west Nelson, where Ordovician rocks – originally contiguous with those of Fiordland – occur on the north-western side of the Alpine Fault. Both the Fiordland and Nelson graptolite-bearing rocks belong to the Buller Terrane. Our understanding of the Fiordland graptolite-bearing sequences rests largely on the field work of W.N. Benson (University of Otago).

Fig. 2. Ordovician graptolites from Preservation Inlet, Fiordland.
(a) *Loganograptus*; smallest division on scale is 1 mm.
(b) *Dictyonema*, scale bar 100 mm. *All photographs, R. Ewan Fordyce*

Permian: Productus Creek and other localities

Amongst the best known Permian rocks of New Zealand are those at Productus Creek, near Ohai in Southland. Productus Creek takes its name from some of the brachiopods found there in shallow-water marine sedimentary rocks. In particular, the brachiopod *Terrakea,* from the Letham and Mangarewa Formations, shows the characteristic shape of the extinct *Productus*-group of brachiopods: its lower shell (the ventral, or pedicle, valve) has quite a rounded to convex profile and is spiny, while the upper shell or valve is more or less flat (Fig. 3). *Terrakea* and relatives lacked a prominent pedicle, which is the fleshy stalk-like attaching organ found in living brachiopods. Instead, *Terrakea* probably rested on its convex ventral valve, anchored partly by its long spines. The Productus Creek rocks provide a range of other marine fossils, such as the large brachiopods *Trigonotreta* and *Tomiopsis.* Molluscs are present, represented by gastropods and by bivalves such as the large '*Atomodesma*' (more properly known as *Mytilidesmatella*). Corals and layers of bryozoans growing on top of each other form small reef-like accumulations.

The Productus Creek rocks are part of the Brook Street Terrane (pages 5–6). Other fossiliferous rocks of the Brook Street Terrane occur at Mokomoko Inlet, near Bluff (Fig. 3). The Mokomoko sequence is dominated by fine-grained volcanic sediments which originated in shallow waters then moved, as a sediment-water slurry, down-slope into deeper waters. Specimens of the large thin-shelled mussel-like bivalved mollusc *Mytilidesmatella* can be found, and limestones within the sequence are probably formed from broken shells of this bivalve. More obvious, however, are trace fossils, such as *Chondrites,* which represent the activity of organisms that burrowed into the soft sediment.

In the south, Permian fossils also occur in patchy outcrops of the Dun Mountain/Maitai Terrane, immediately north of the Murihiku rocks. Dun Mountain/Maitai rocks form a strip which runs north-west from Balclutha to beyond Mossburn. Notable localities near Clinton, Waipahi and Arthurton produce abundant specimens of *Mytilidesmatella* (Fig. 3). Further north-west, near Mossburn, pebbly conglomerate of the Countess Formation has a Late Permian fauna which includes some of the youngest trilobites known from New Zealand (Fig. 3). Also present are peculiar extinct molluscan groups, the rostroconchs

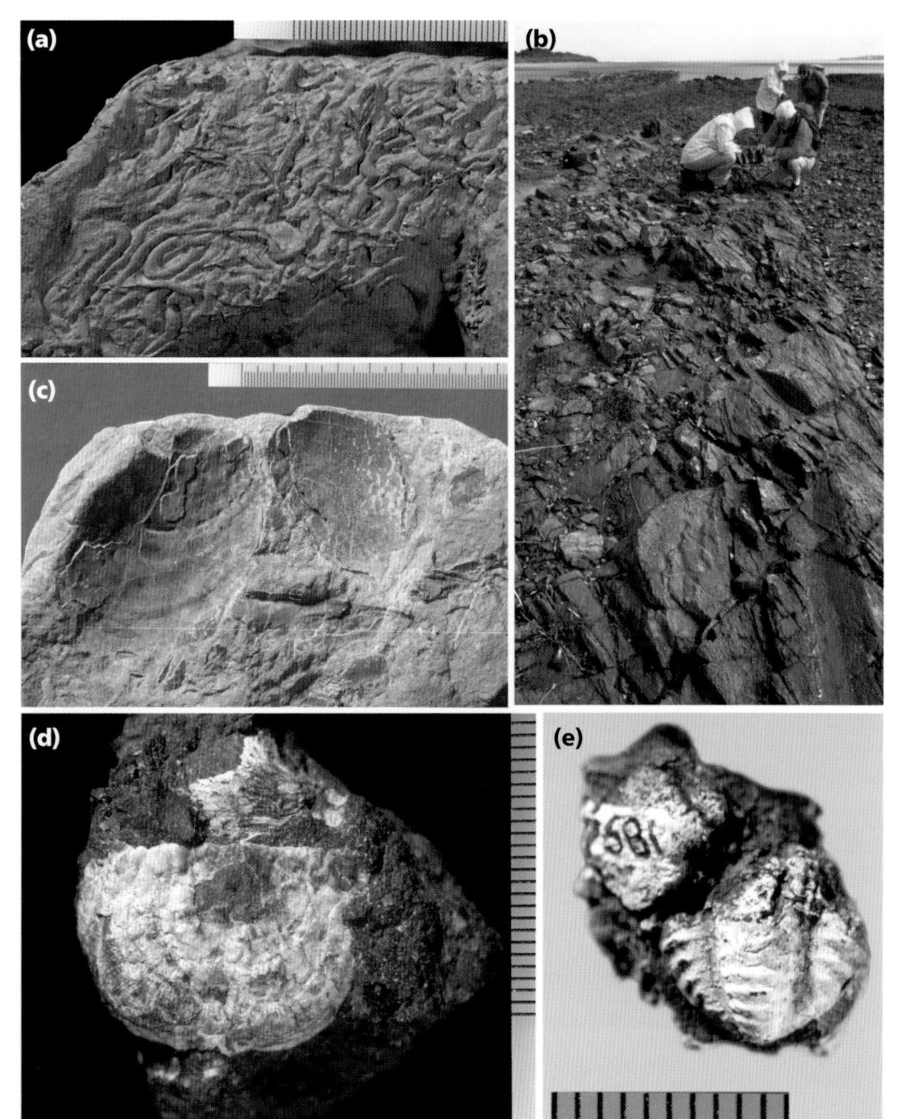

Fig. 3. Permian fossils and strata. Smallest division on scale is 1 mm.
(a) Trace fossils naturally exposed by weathering, on a bedding plane of volcanic-derived sediments from Mokomoko Inlet near Bluff; the finest tubes are Chondrites.
(b) Dipping volcanic-derived sediments at Mokomoko Inlet near Bluff; this sequence produces abundant trace fossils and rare molluscs.
(c) Large smooth-shelled bivalve *Mytilidesmatella,* from near Arthurton.
(d) Brachiopod *Terrakea,* showing some remnants of fine spines, from Productus Creek region of Wairaki Hills.
(e) Tail region of unidentified species of trilobite, family Phillipsiidae, from near Mossburn.

and bellerophontids, as well as productid and spiriferid brachiopods. These fossils represent ecosystems which thrived just before the dramatic mass extinction at the end of the Permian, about 250 Ma.

The Gondwanan seed-fern *Glossopteris*

Tongue-shaped fossil leaves named *Glossopteris* have been known for more than a century from Permian rocks in India, Africa, South America, Australia and Antarctica (Fig. 4). The Antarctic finds were made by Scott's ill-fated expedition of 1911–12. Plants from the *Glossopteris* flora of the Transantarctic Mountains were considered important enough to retain even in dire straits; the fossils were with the party when all the explorers perished. *Glossopteris* was important in helping to develop concepts of continental drift and rearrangement of global geography. Before about the 1960s, the distribution of *Glossopteris* on southern continents and India was difficult to explain in light of modern geography. Clearly, the fossil localities are now separated by wide oceans, yet the large seed-fern which produced the *Glossopteris* leaves probably had poor dispersal capabilities and could not spread across such oceans. Only a few southern hemisphere paleontologists and geologists seriously considered the then-unfashionable idea of rearranging global geography to explain such fossil distributions. In the 1960s, though, with the recognition of plate tectonics as a mechanism for continental drift, the distribution of fossil *Glossopteris* became widely cited as evidence supporting a former southern supercontinent, Gondwanaland.

In the mid 1960s, *Glossopteris* was identified conclusively from New Zealand, based on a leaf from the shallow-water marine Mangarewa Formation at Productus Creek. Other specimens, also associated with shallow water marine invertebrates, have been found since then (Fig. 4). There is no reason to think that the leaves might have travelled far, and a nearby source, perhaps close to a shoreline, is likely. Studies elsewhere have also pointed to a shoreline habitat for some glossopterids. It seems that *Glossopteris* was deciduous, although it is not known if the tree lost all of its leaves in winter. The plant is regarded commonly as an indicator of cool conditions. Perhaps glossopterids persisted into the Triassic, represented by leaves such as *Linguifolium* (known from the Triassic of southern New Zealand; see below).

Fig. 4. *Glossopteris*, a Permian seed fern from Gondwanaland. Shows (a) inset, specimen from marine rocks at Productus Creek, Wairaki Hills, scale 10 mm, and (b) glossopterids from the Ohio Range, Transantarctic Mountains, Antarctica; smallest division on scale is 1 mm.

Triassic – Jurassic fossils from the Murihiku Supergroup, Southland Syncline

Rocks of the Murihiku Supergroup, particularly on the north limb of the great trough-like fold of the Southland Syncline, provide an excellent glimpse of Middle Triassic to Early Jurassic marine life. This interval spans roughly 180–220+ Ma (Fig. 5). The fossils are abundant in shellbeds, some of which can be traced with little change in composition for many kilometres. Other shellbeds are discontinuous, pointing to local unconformities or time gaps. For example, beds of the thin-shelled bivalved mollusc *Monotis* are prominent in some parts of the sequence but not in others; *Monotis* is characteristic of the Warepan Stage, later Late Triassic. Generally, also, fossils are more sporadic in massive fine-grained sediments which presumably formed in deeper water than coarse sandstones. Thicker sequences of sparsely fossiliferous and usually coarser-grained rock often alternate with finer sediments bearing marine fossils; the coarser beds may represent phases of terrestrial deposition. Non-marine rocks with sporadic plant fossils dominate the Jurassic part of the Murihiku sequence, and are well exposed on the south limb of the Southland Syncline and in the axis of the syncline.

Murihiku fossils have been critically important in establishing the sequence of New Zealand time units, or stages, which form an important tool for subdividing local Triassic rocks. These stages are named the Malakovian, Etalian, Kaihikuan, Oretian, Otamitan, Warepan and Otapirian; they form a sequence which spans the Middle and Late Triassic, and possibly include the Triassic/Jurassic boundary at about 210 Ma (Fig. 5). The stages were first used in modern form by J. Marwick (N.Z. Geological Survey), and they have been refined by later geologists and paleontologists, particularly J.D. Campbell (University of Otago). Because the stages can be recognised easily in the field using fossils, they can provide a practical field nomenclature; thus, geologists can readily comment that they are 'working in the Warepan' or some other stage. In contrast, the formational names that are usually applied to sedimentary rocks by field geologists are used less commonly for Murihiku rocks. Within the Murihiku rocks, representative fossiliferous marine sequences include those of Roaring Bay and Otamita Valley, given as examples below.

Roaring Bay fossils and rocks

The South Otago coast from Kaka Point to Nugget Point and beyond (Fig. 6) provides important reference sections for Triassic fossils. South of Nugget Point, fossiliferous rocks of the Oretian Stage are exposed on the track to Roaring Bay and for a few hundred metres southward beyond the boulder beach. Sometimes found here are isolated specimens of the marine bivalve *Maoritrigonia nuggetensis*, which occur as distinctively ribbed large cavities, or natural moulds of the shells, in loose boulders of sandstone. *Maoritrigonia* (Fig. 6) belongs in the family Trigoniidae, which was once diverse but is now represented by only one living species in Australia. Cephalopods including ammonites and the nautiloid *Proclydonautilus* occur, usually with the original shell leached away to reveal the internal chambers

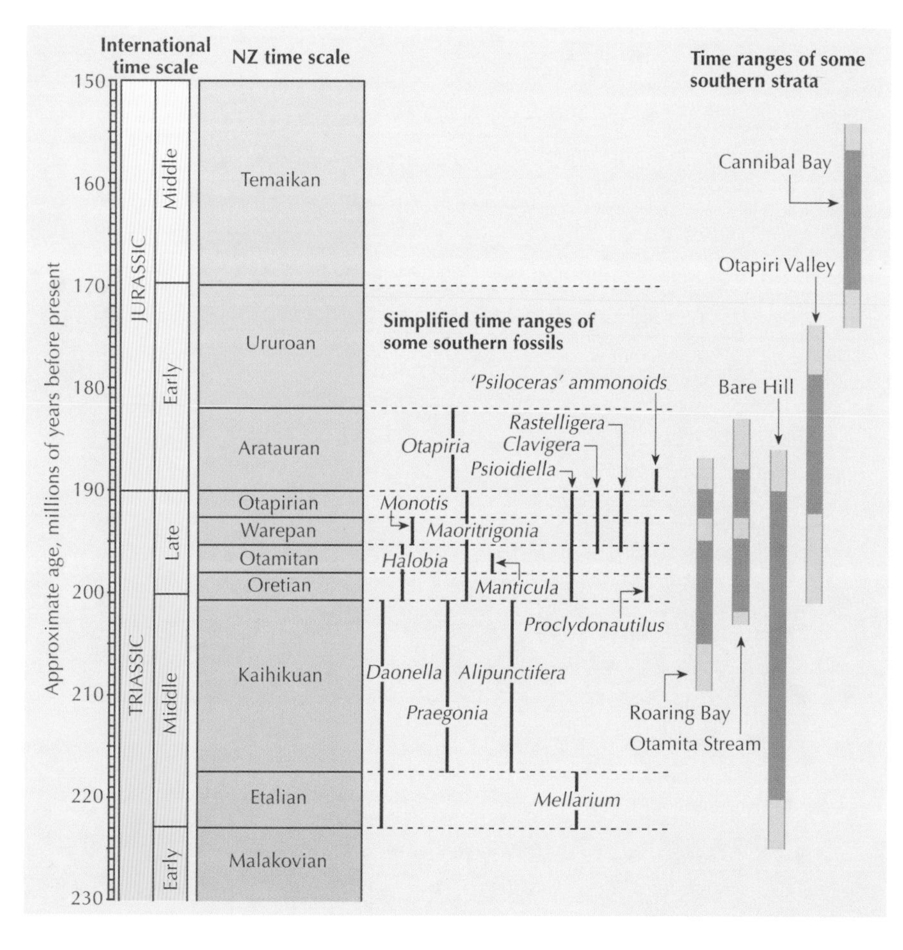

Fig. 5. Geological time scale for the Triassic and part of the Jurassic, showing time ranges of representative invertebrate fossils and time ranges for some southern fossiliferous rock sequences.

Fig. 6 (opposite). Triassic fossils and strata. (a) View south showing Roaring Bay, with oldest (Oretian) rocks on the right, and youngest (Otapirian) rocks on the left; sediments are dipping steeply to the left. (b) Problematica shellbed, formed by millions of specimens of the thin-shelled bivalve *Manticula problematica*, exposed part-way along section at Roaring Bay; bedding dips steeply to the left. (c) Specimens of the large brachiopod *Clavigera*, Otapirian Stage, Roaring Bay; 10 cent coin upper right, for scale, is 24 mm diameter. (d) Isolated vertebra from an ichthyosaur in pebbly sandstone of the Otamitan Stage, Roaring Bay. (e) Detail of a specimen of the bivalve *Manticula*, in the Problematica Shellbed, Roaring Bay. (f) Natural mould of the bivalve *Maoritrigonia*, from a loose boulder of presumed Oretian age, Roaring Bay; scale divisions in mm.

of the shell. Trace fossils are present, giving fossil evidence of activity within or on sediment, without preserving any original mineralised skeleton. One distinctive trace is *Zoophycos*; this has a thin radial structure which lies roughly parallel to bedding and may exceed 1 m in diameter. It was formed by an unknown soft-bodied organism.

In younger rocks, further to the south in Roaring Bay, the base of the Otamitan Stage is marked by several metres of siltstone dominated by thousands of densely packed thin shells. These fossils represent a rather featureless large mussel-like bivalve, *Manticula problematica* (Fig. 6): the name 'problematica' refers to past uncertainty about the exact identity of the fossil. Curiously, the bivalve gives its name to the Problematica siltstone, also known as Problematica shellbed, or Manticula shellbed (Fig. 6). The shellbed may be traced far to the north-west and also can be identified in the Nelson sequence of Triassic rocks on the other side of the Alpine Fault. Other Otamitan bivalves of note here and higher in the sequence include the large *Hokonuia* and smaller radially-ribbed *Halobia*. Cephalopods – with large coiled shells, unlike living squid and octopus – include the ammonoid *Rhacophyllites* and distinctively sculptured large nautiloid *Proclydonautilus*. Brachiopods, such as *Psioidiella* and *Oxycolpella*, are present. The Problematica siltstone also is the source of some of the few Triassic foraminifera described from New Zealand.

Further southwards, that is, moving into younger rocks further up the rock sequence, one finds medium to very coarse sandstone and granule conglomerate, with plant fragments. The rock type and fossils point to a shallow-water setting, and rare bones of ichthyosaurs – dolphin-like marine reptiles (Fig. 6) – confirm that the environment was marine. There are no significant marine invertebrates to date this part of the sequence. At the south end of Roaring Bay, a sheer near-vertical bedding plane is dominated by well-preserved shells of the large spiriferid brachiopod *Clavigera* (Fig. 6). This marks the Otapirian Stage, at the very end of Triassic time. Associated invertebrates are the dramatically winged spiriferid brachiopod *Rastelligera* and the bivalves *Maoritrigonia* and *Otapiria*; all indicate marine conditions. Missing from the Roaring Bay section is one of the most distinct local Triassic invertebrates, the bivalve *Monotis*. Thus, the Warepan Stage is not known at this locality. *Monotis* does occur, however, a few kilometres to the north-west near Glenomaru.

Hokonui Hills

Catchments of the Otamita, Lora and Otapiri Streams within the Hokonui Hills, west of Gore, preserve a more complete fossiliferous sequence than that seen at Roaring Bay. Marine fossils of the Etalian to Ururoan Stages are known in this North Limb of the Southland Syncline (Fig. 5). Because the rocks are less continuously exposed than on the coast, the fossiliferous sequence must be pieced together from scattered localities.

On the slopes of Bare Hill, for example, a sparse Etalian fauna of bivalves, cephalopods, brachiopods and trace fossils represents some of the oldest rocks on the North Limb of the Syncline. The large brachiopod *Alipunctifera kaihikuana* occurs at the peak of Bare Hill, indicating an age of Kaihikuan. To the south, on

the slopes of Bare Hill, we can find abundant specimens of the bivalve *Monotis* (Warepan Stage) and the winged brachiopod *Rastelligera* (Otapirian Stage). Further south-west, ribbed and many-whorled psiloceratid ammonoids – '*Psiloceras*' of older literature – are found on the banks of Taylors Stream (Aratauran Stage, Early Jurassic). Nearby are brachiopods and bivalves which represent the Ururoan Stage, forming the youngest marine rocks of this part of the Southland Syncline.

To the east, within Otamita Valley, the massive grey Trechmann Siltstone is a particularly productive horizon in the Otamitan Stage (Fig. 7). This unit has been collected thoroughly for several decades by staff and students of the Department of Geology at University of Otago, so that fossil assemblages are known well. The sediments and fossils are consistent with a quiet setting, probably in deep water on the continental shelf. Abundant invertebrates (Fig. 7) include many articulate brachiopods such as '*Rhynchonella*', *Oxycolpella* and *Psioidiella*. The range of brachiopods here and in other Murihiku rocks contrasts with the low diversity of brachiopods in most modern seas. Also in the Trechmann Siltstone are molluscs – gastropods (*Raha*, '*Pleurotomaria*'), bivalves (abundant *Halobia*, *Palaeoneilo*, *Maoritrigonia*, *Triaphorus*) and cephalopods (*Cenoceras*, *Proclydonautilus*, *Arcestes*). Less common are echinoderms (short sections of crinoid stems), and wood and leaves. Some of these fossils, such as the large nautiloid *Proclydonautilus* (Fig. 7), also occur in the overlying Mandeville Sandstone. In turn, the Mandeville Sandstone is probably the source horizon for the articulated toothed jaws of an indeterminate small ichthyosaur described by J.D. Campbell.

The Trechmann Siltstone assemblage can be put into sequence easily, for both older and younger fossiliferous rocks are exposed nearby. To the north is the slightly older and underlying Problematica shellbed (also Otamitan), with specimens of *Manticula* preserved mainly as decalcified moulds; that is, the original limy shell is lost, leaving only a rock replica. Further to the south, *Monotis*-bearing beds have not been found, so that the Warepan Stage is unknown, but other outcrops are known to produce the brachiopod *Rastelligera* which here is diagnostic of the Otapirian Stage (latest Triassic). A little younger are species of the finely-ribbed bivalve *Otapiria* which indicate Aratauran Stage (earliest Jurassic).

Is there a Maorian province?

By 1900, New Zealand's Triassic fossils were quite well known. In those early days of paleontology, it seemed that many of the local Triassic fossils were endemic, that is, they had no close relatives elsewhere. This apparent endemism indicated to many paleontologists that New Zealand must have had a long interval of geographic isolation, allowing little if any exchange of organisms with other regions. Eventually, in 1916, the concept of a Maorian paleobiogeographic province was established to formalise the apparent endemism and geographic isolation of proto-New Zealand. (To explain terms here, biogeography deals with the distribution of organisms on earth, while paleo-biogeography considers ancient distributions of organisms as revealed by fossils.) The idea of a Maorian province persisted

Fig. 7. Trechmann Siltstone and representative fossils, Maling Bluff, Otamita Stream. Scale divisions in mm unless specified.
(a) Outcrop of siltstone, dipping to the left.
(b) Gastropod, *Raha* (whitened with magnesium oxide to show detail).
(c) Nautiloid cephalopod, *Cenoceras*, as found in the field; most original shell is lost to reveal a natural mould.
(d) Ammonite, *Cladiscites*.
(e) Thin-shelled bivalve, *Halobia*, on bedding planes of Trechmann Siltstone.
(f) Nautiloid cephalopod, *Proclydonautilus*.
(g) Smaller invertebrates including three specimens of '*Pleurotomaria*' gastropod upper left, the brachiopod *Oxycolpella* in lower left, the elongate and nearly-cylindrical small nautiloid *Michelinoceras* upper right, and two specimens of the small bivalve *Paleoneilo* lower right.
(h) and (i) Brachiopod, *Oxycolpella*.

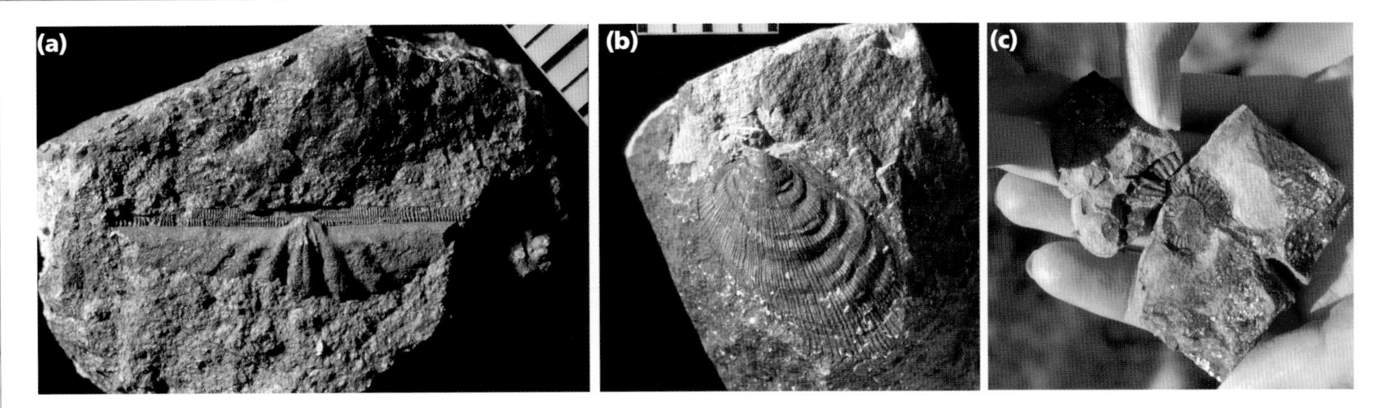

until recently, and some paleontologists and stratigraphers may still regard it as a useful concept.

Different sorts of fossils carry different geographic signals. Whenever paleobiogeographic provinces are considered, the cosmopolitan elements of the biota – those that occur widely around the world – tell little about geographic relationships. Amongst New Zealand's Triassic fossils, cosmopolitan elements include ammonoids, certain groups of bivalves (*Halobia*, *Daonella*, *Monotis* and relatives) and conodonts. Such organisms are found almost everywhere, hence their identity as cosmopolitan elements. Conversely, many groups of brachiopods, gastropods, conulariids, bryozoans, barnacles and echinoids – commonly regarded as organisms that might not disperse easily – include most of the apparently endemic fossils. For the New Zealand Triassic biota, these supposed Maorian elements apparently persisted until early in the Jurassic, when faunas became more 'Tethyan' – that is, there were more obvious similarities with warm water marine faunas from northern latitudes near southern Eurasia. However, recent discoveries of Triassic rocks around the Pacific margin (e.g. Papua New Guinea, Mexico, Peru, Chile, and Antarctica) show that many of our supposedly endemic taxa occur beyond New Zealand. This is particularly the case with brachiopods, such as *Clavigera* (Fig. 6), *Rastelligera* (Fig. 8) and *Oxycolpella* (Fig. 7), on which the concept of the Maorian province was largely founded. So, the endemism is perceived rather than real. As paleontologists take more of a global view of their groups, and assess taxonomic relationships more critically, the evidence for a Maorian marine benthic province is gradually disappearing. Clearly, claims of endemism rely on not finding or recognising a particular taxon elsewhere, and are quite sensitive to biases in collecting (has the region been studied fully?) and in taxonomy (are the fossils identified properly?).

Fig. 8 (above). Representative Triassic-Jurassic invertebrates from the Otamita Valley, north limb of Southland Syncline. Smallest scale divisions 5 mm.
(a) Natural mould of the wide-hinged brachiopod, *Rastelligera*; Otapirian Stage; prominent fine teeth and sockets are visible along the horizontal hinge line towards the top.
(b) Natural mould of the bivalve *Otapiria*, Aratauran Stage.
(c) Ammonoid similar to *Psiloceras*, as found in the field, Aratauran Stage.

Fig. 9. *Monotis*, a thin-shelled bivalve from the Warepan Stage, north limb of Southland Syncline.
(a) Abundant natural moulds of *Monotis* on a bedding plane, Glenomaru.
(b) Natural moulds of *Monotis* from Taringatura Hills, Southland; smallest scale divisions 5 mm.

Triassic bivalves of the *Monotis* group

Monotis is a short-lived genus of marine bivalved mollusc (Fig. 9) which, along with some related bivalves, was abundant in the later Triassic in many parts of the world. Extreme abundance implies that *Monotis* was probably important in marine ecology. Further, a short geological range and almost cosmopolitan distribution makes *Monotis* and relatives important in both local and long-distance stratigraphic correlation. Accordingly, the *Monotis* group has attracted much attention from paleontologists in New Zealand and overseas. *Monotis* was first recognised in New Zealand in 1859, by Austrian geologist F. von Hochstetter, who collected specimens near Nelson. Hochstetter was impressed with the similarity between the New Zealand fossils and those from the Austrian alps, and used *Monotis* to infer a Triassic age for the Nelson strata.

Near its hinge region, the very thin shell of *Monotis* has an anterior ear or projection, and an indentation termed the byssal notch. Through the notch, hair-like threads of the byssus emerged to attach onto firm surfaces like rock, other shells, and wood. The shell has radial ribs which may be quite prominent. In Warepan rocks of the Southland Syncline, as elsewhere, *Monotis* may occur in immense numbers with few or no other fossils apparent. From place to place, *Monotis* shells vary markedly in shape, but there is not yet a full consensus on how to interpret this variation. If we accept that there are only a few species of *Monotis* in New Zealand, we would also have to accept wider variation in structure for *Monotis* than is usual amongst living bivalves. Alternatively, to accept more than a dozen species, as proposed by J.A. Grant-Mackie (University of Auckland), is difficult to explain in terms of the ecology of living bivalves. To accept many species, one must accept also that they were many sympatric, that is, found literally next to one another and, by implication, living together. However, for living organisms, it is a widely used ecological principle that closely related species tend not to have overlapping occurrences.

Monotis, and its relatives such as *Halobia* (Fig. 7), *Daonella* and *Otapiria* (Fig. 8), occur widely beyond New Zealand, in such scattered regions as the European alps, Siberia, Alaska and the Himalayas. Perhaps this wide distribution means that these bivalves were able to disperse easily around and across oceans. Alternatively, perhaps the present distribution resulted when fossiliferous terranes, already including the fossil bivalves, were fragmented by tectonic activity and scattered, eventually to collide and merge with other landmasses including proto-New Zealand.

Curio Bay plant beds

A shore platform at Curio Bay, in coastal Southland, is a classic locality where many fossil tree stumps may be seen in growth position (Fig. 10). The setting is clearly non-marine, and is of Middle Jurassic (Temaikan) age. Details have been published by M.S. Pole (formerly University of Otago). At Curio Bay, fossilised stumps, which are protected by reserve status, lie some metres apart. Amongst the stumps is a scatter of fallen logs which are straight, unbranched, and up to 20 m long. Stumps and logs show obvious growth rings which indicate a forest many decades old, growing in a markedly seasonal climate. Climatic seasonality is consistent with the likely paleolatitude for New Zealand in the middle of the Jurassic – it was probably at high latitudes, possibly even polar, with long summer days and cold dark winters.

Wood from the Curio Bay trees represents two major gymnosperm groups (*Mesembrioxylon* and *Dadoxylon*) which are similar to modern araucarians such as kauri, and podocarps such as rimu. Less common stumps represent tree ferns, some similar to living species in having trunks with matted fibrous rootlets. Also rare but distinctive are stumps of the strange multiple-branched gymnosperm *Pentoxylon*, which occurs elsewhere in India and Australia. Cones, fern fronds, horse-tails and abundant pollens are present. At this phase of plant history, flowering plants are unknown.

The Curio Bay forest is preserved in volcanic-derived sediment which perhaps accumulated as ash falls or as water-borne floods of volcanic debris. No terrestrial animal remains have been reported, but the sort of environment represented by the Curio Bay beds might well be source horizons for dinosaur footprints or bones. Jurassic non-marine rocks with plant fossils occur widely elsewhere in the south limb of the Southland Syncline, but they have not attracted the attention accorded to Curio Bay.

Marine and non-marine Triassic rocks of the Torlesse Supergroup, North Otago

Fossiliferous Triassic rocks occur in the north, bordering the Waitaki River, as well as in the Southland Syncline, but the origin of the Torlesse Supergroup is quite different from the Murihiku rocks. Generally, Torlesse rocks include a varied mix of sediments dominated by quartz and feldspars, but also with a significant component of rock fragments. Such materials probably arose from weathering of a granitic landmass. There are localised deposits of carbonates and occasional basalt-derived rocks. Sequences are thick, but commonly are broken by faulting, so that few individual beds can be traced far. Also, fossils are rare, so that occurrences such as those in North Otago are rather important. The North Otago Torlesse rocks probably represent the Rakaia Terrane, which had a different origin to the Murihiku rocks in the South.

The shallow marine to terrestrial rocks of the Corbies Creek Group (Fig. 11), near Otematata in the Waitaki Valley, provide a valuable close association between land plants and marine invertebrates. The sequence and fossils were detailed by R.J. Ryburn (formerly University of Otago) and G.J. Retallack. Invertebrates include age-diagnostic fossils such as the bivalves *Daonella* and *Praegonia*, the large limpet *Patella* and other gastropods and brachiopods including the large *Alipunctifera kaihikuana*. Tiny coiled encrusting tube-worms and fragmentary crinoids also may be found. In the upper parts of the Corbies Creek Group, marine fossils become scarce then absent, while plant remains become increasingly common. This part of the sequence marks a shift from intertidal to fully non-marine settings. Distinctive trace fossils are present. Terrestrial rocks of the Long Gully Formation have abundant and well-preserved plant macrofossils (leaves, logs; Chapter 6: Box 1), coal accumulations, and roots, in places associated with paleosols or ancient soil horizons. According to G.J. Retallack, two main plant associations occur. One, with gymnosperm-like plants (e.g. *Linguifolium*) represents a swamp-woodland setting some distance from the shore, while the other, a fern-like association (e.g. *Pachydermophyllum*), may represent a coastal mangrove-like setting. Comparable plant associations are found in other parts of Gondwana. For North Otago, the low diversity of the flora and rather small size of plants point to a seasonal, humid, cool temperate climate. In turn, this setting is consistent with a southern position in Gondwanaland. Despite the view that Torlesse and Murihiku rocks represent quite different terranes, floras of the Murihiku Supergroup of South Otago and Southland are not radically different from those of the Corbies Creek Group.

Middle Cretaceous fossils and the breakup of Gondwana

'Breakup' biotas of the Southern Hemisphere – those that were living when parts of Gondwanaland fragmented in the Cretaceous – are of great interest to paleontologists worldwide. By about 80–85 Ma, New Zealand had separated enough from the Gondwanan margin to allow the Tasman Sea to open between Australia and New Zealand. However, some rifting must have occurred earlier, judging from the presence of localised wedges of sparsely fossiliferous Early Cretaceous rocks (about 110 Ma) which occur in a few places in the South Island. In Otago, the key units are the Kyeburn Formation of Central Otago and the Henley Breccia of South Otago. These non-marine rocks

Fig. 10 (above). Jurassic plants from the Southland Syncline. Left: Petrified forest on shore platform at Curio Bay; a log trends obliquely away from mid-view, and many slightly elevated stumps are scattered across the outcrop. Right: Incomplete fern frond, *Cladophlebis*, Temaikan, from Brothers Point-Long Beach, Catlins; scale divisions in mm.

Fig. 11. Triassic fossils and strata of the Torlesse Supergroup at Corbies Creek, North Otago. Smallest division on scale is 1 mm.
(a) Searching for fossils at an outcrop of marine sandstone.
(b) Natural moulds of gastropods, possibly a species of Pleurotomariidae; specimen about 92 mm across.
(c) Natural moulds of outer surface, bivalve *Praegonia*.
(d) Natural mould of inner surface of brachiopod, *Alipunctifera*.

probably accumulated in rivers, lakes and fans associated with fault-bounded depressions and nearby hills or mountains. More broadly, the setting is one of large fault-bounded rift valleys which formed as New Zealand pulled away from the Australian/Antarctic margin of Gondwanaland. In Victoria, Australia, rocks from similar settings or of comparable age have attracted great attention, because of important finds of dinosaurs and mammals.

Amongst the few macrofossils from this interval found in Otago, leaves from the Henley Breccia represent some of the oldest records of flowering plants (angiosperms). Specimens are too poorly preserved to be identified accurately; thus, they are of limited value in understanding the early history of angiosperms. Rare plant fossils also occur in the Kyeburn Formation near Naseby (Fig. 12), roughly the same age as the Henley sequence. Here the fossils include coalified stems in upright position, fragmented leaves and small seeds. Invertebrate trace fossils – trails – have been recognised in fine-grained sediments which probably formed in a lake setting. However, no fully terrestrial vertebrates or other animal fossils have been reported. It need hardly be stated that a single amphibian, reptile, bird or mammal from this interval could have the most profound biogeographic implications.

Fig. 12. Cretaceous fossils and strata of the Kyeburn Formation near Naseby, Central Otago. (a) Boulders of conglomerate, probably deposited originally in a river system; plant macro-fossils occur rarely in this rock. (b) Gently dipping siltstone, probably deposited originally in a lake. Inset: Cross section through a small seed, from calcareous siltstone; scale 10 mm.

The end of the Mesozoic: latest Cretaceous

As New Zealand drifted away from the other elements of Gondwanaland and submerged slowly in the later part of the Cretaceous and Paleocene, a blanket of 'cover strata' started to accumulate (Fig. 1). The cover sequence, richly fossiliferous in places, provides information on two important biological events. First, it reveals a record of long-term biological changes resulting from the breakup of Gondwana, particularly the evolution of organisms now regarded as characteristic of New Zealand. Second, the sequence also preserves evidence of dramatic events associated with the catastrophic global mass extinction at the end of the Cretaceous. The extinction is widely quoted as the 'K/T,' or Cretaceous/Tertiary, boundary event, but it also marks the Cretaceous/Cenozoic and Cretaceous/Paleocene boundaries. It seems that a large meteorite collided with the earth at this time. The impact generated a crater, widespread fires, tsunami (gigantic waves) and much atmospheric dust, leading to major short-term global climate change. Many food chains probably collapsed, and many once-common animal groups disappeared. Amongst the survivors were groups that underwent rapid evolution in the Paleocene or later. Amongst those that disappeared are dinosaurs, marine reptiles and a range of invertebrates including planktic foraminifera, some bivalves, ammonoids and belemnites. Southern New Zealand provides important insights into the biotas that existed just before and just after the end-Cretaceous event. Two Late Cretaceous biotas considered here are from the Katiki Formation of East Otago, and the Fairfield Greensand near Dunedin.

Fig. 13 (opposite). Cretaceous fossils and strata of the Katiki Formation, Haumurian Stage or Late Cretaceous, Shag Point, North Otago. (a) Naturally exposed cluster of poorly preserved gastropods; smallest division on scale is 1 mm. (b) Belemnite, or internal skeleton of an extinct squid, *Dimitobelus*. (c) Cluster of vertebrae of a small unidentified species of mosasaur; block is about 220 mm across. (d) Excavating natural concretions which encase the skeleton of a large plesiosaur, shore platform at Shag Point. (e) Partly prepared skeleton of Shag Point cryptoclidid plesiosaur removed from the concretion of Figure 13d, viewed from the tail end with the neck in the far distance; bones are preserved as natural moulds; person in background for scale. (f) Lateral view of right side of skull of Shag Point cryptoclidid plesiosaur; narrow lower jaw and teeth are apparent below, with large orbit or eye-socket in mid-right, and temporal fossa partly filled with rock in upper left; smallest division on scale is 5 mm. (g) Reconstruction of two swimming individuals of Shag Point cryptoclidid plesiosaur; body length about 6.5 m.
Art by C. Gaskin held in Geology Museum, University of Otago

Shag Point plesiosaur and other fossils from the Katiki Formation

One of New Zealand's most dramatic fossils is an articulated plesiosaur – a large marine reptile – discovered in the 1980s, by an amateur fossil collector, near Shag Point in coastal East Otago (Fig. 13). The plesiosaur includes the skull, jaws, vertebrae and ribs almost in life position, with a length of more than 6.5 m. One reasonably complete limb and parts of others are preserved. This fossil is one of the most complete plesiosaurs known from New Zealand, and is one of few for which skull and limb elements are clearly associated. Elsewhere in the world, complete specimens show that plesiosaurs had a long neck, a large body with four flipper-like limbs, and a short tail. Their structure shows clearly that they are unrelated to dinosaurs.

Skull features indicate that the animal is a new species in the family Cryptoclididae, a 'long-necked, short-headed' group of plesiosaurs which lived from the Middle Jurassic to Late Cretaceous. Previously, cryptoclidids have not been reported from New Zealand. Features of the skull (Fig. 13) include the large forward-looking orbit which hints that partial binocular vision was possible, and a very large temporal fossa (skull opening) behind the orbit. The size and shape of the fossa suggest large fast-acting and/or powerful jaw muscles. Scores of teeth are slim, small and roughly uniform in shape and size, indicating a diet of medium-sized soft-bodied prey. The neck is long, but probably was not particularly flexible. Proportions of the hind flipper are consistent with fast swimming capabilities. Perhaps the plesiosaur fed at depth, where the large eyes would have been at an advantage.

When compared with other species of cryptoclidid plesiosaurs, the Shag Point fossil seems to lie closer to some Northern Hemisphere Jurassic forms, rather than with the two other southern Late Cretaceous cryptoclidids (*Morturneria* of Seymour Island, and *Aristonectes* of Patagonia). It is intriguing that the Cryptoclididae is represented in the north by old fossils, of Middle to Late Jurassic age (145–175 Ma), but in the south by much younger fossils, of later Cretaceous age (65–100 Ma). Further, the southern fossils occur in settings which were probably close to the Cretaceous pole. Where are the cryptoclidids of intermediate age, and in localities between the far south and mid-northern occurrences?

The Shag Point plesiosaur is from the massive dark grey silty sandstone of the Katiki Formation. The lack of calcareous foraminifera and ammonites suggests a setting with restricted access to the open sea, perhaps a sheltered embayment or an estuary floored with a soft, soupy, oxygen-poor muds. Trace fossils, such as the branched *Thalassinoides* (shrimp burrow), and clusters of agglutinated foraminifera are abundant. Wood occurs frequently. Rare molluscs from sandier horizons include often-decalcified gastropods and bivalves and a few examples of the age-diagnostic belemnite *Dimitobelis* (Fig. 13). Belemnites are internal bullet-shaped skeletons from extinct fossil squid. These invertebrates indicate an age of Haumurian Stage, or latest Cretaceous. Other vertebrates from the Katiki Formation include mosasaurs, which are marine lizards related closely to living snakes and lizards.

Fairfield Greensand

At Fairfield, near Dunedin, the marine Fairfield Greensand (Fig. 14) is about the same age as the Katiki Formation. Grains of the green mineral glauconite dominate, in a muddy sand matrix; the formation is thin, and bedding is largely obliterated by extensive burrowing. Thus, the formation is described as extensively bioturbated. At first glance, the greensand seems sparsely fossiliferous, but over the years concretions from within the greensand have produced a diverse assemblage of macrofossils. Most dramatic of these is the ammonoid *Kossmaticeras bensoni*, which has been found in clusters of up to five individuals sometimes associated with fossil wood. The shell of *Kossmaticeras* (Fig. 14) is somewhat inflated, rather than compressed, from side to side. It is also moderately evolute; that is, successive turns, or whorls, partly overlap earlier whorls. There are prominent simple ribs associated with small raised nodules. Incomplete specimens of another ammonoid, *Maorites*, also occur. Another cephalopod is the rarer smooth-shelled nautiloid *Eutrephoceras* (Fig. 14). Intriguingly, ammonoids became extinct at the Cretaceous/Tertiary boundary, while nautiloids persisted through to modern days. At Fairfield, both groups of cephalopods are preserved as dark carbonised and three-dimensional structures from which all the original limy shell has been leached away.

Other Fairfield invertebrates also are carbonised rather than shelly, making identification difficult. Rare gastropods, such as *Protodolium* and limpet-like forms, are present. Bivalves are more common, including the large *Pacitrigonia* (Trigoniidae; now extinct in New Zealand) and *Lahillia* (Cardiidae). Smaller bivalves include *Leionucula* (Nuculidae) and *Nuculana* (Nuculanidae). Plant macrofossils, including leaves, kauri-like scales and bored wood, are abundant in places, but the flora has not yet been formally described. The only common foraminiferan is *Bathysiphon*, in which each individual occupies a long narrow white tube about 1 mm in diameter. The tube, or test, is typically constructed of tiny grains of quartz selected by the foraminiferan and glued together to form its dwelling. Abundant *Bathysiphon* suggests low oxygen levels in the original depositional setting. Some absences are noteworthy: belemnites and marine reptiles have not been found yet.

Fig. 14. Cretaceous fossils of the Fairfield Greensand, Haumurian Stage or Late Cretaceous, Fairfield, Dunedin district. Smallest division on scale is 1 mm unless specified.
(a) Outcrop of Fairfield Greensand, overlain by light purple-brown Abbotsford Formation and underlain by light yellow Taratu Formation on which people are standing; fossils are mainly from concretions from the lower 2–3 m of the greensand.
(b) Characteristic preservation of fossils from Fairfield Greensand, with original carbonate shells leached away to leave dark carbonaceous moulds; specimens include leaves, moulds of small bivalves, and narrow small white tubes of the foraminiferan *Bathysiphon*.
(c) Carbonised natural mould of bivalve *Pacitrigonia*.
(d) Carbonised natural mould of ammonite *Kossmaticeras*, with aperture opening at lower left; surface details indicate the outer surface of the shell but do not show internal partitions; smallest scale divisions 5 mm.
(e) Carbonised natural mould of nautiloid cephalopod *Eutrephoceras*, with aperture opening at lower right; surface details show simple suture lines which mark the positions of septa or internal partitions.

Wangaloa Formation: Paleocene biotas after the crisis

There is great interest worldwide in the 'recovery' biotas which diversified after the extinction at the end of the Cretaceous. It seems that there were rapid evolutionary radiations in many groups, which quickly occupied niches earlier emptied through the mass extinction. In detail, however, much is uncertain about the oldest 'recovery' biotas, their taxonomic composition and ecological partitioning.

One biota which is of key importance in the Southern Hemisphere is that of Wangaloa, in South Otago. Here is preserved the diverse Wangaloan molluscan fauna. A shallow-water inner-shelf setting is indicated by the nature of sediments, which, according to J.K. Lindqvist, include storm-influenced deposits. In some beds, turret shells are clearly aligned, probably by current activity (Fig. 15). Abundant *Teredo*-bored araucariacean (kauri-like) logs, trace fossils such as the shrimp-burrow *Ophiomorpha*, and the range of invertebrate species in the Wangaloa fauna, also are consistent with an inner-shelf setting.

The Wangaloan molluscan fauna (Fig. 15) has an intriguing mix of elements of both Cretaceous and Cenozoic (Tertiary or Paleocene) aspect. Older elements include the gastropods *Conchothyra* (Struthiolariidae) and *Struthioptera* (Aporrhaidae), and the cardiid bivalve *Lahillia*, while molluscs of younger Tertiary aspect include the gastropods *Leptocolpus* (Turritellidae) and *Magnatica* (Naticidae), and the bivalve *Purpurocardia* (?) (Carditidae). These and other molluscs have been reviewed in many publications and also in a thesis by J.D. Stilwell (formerly University of Otago). There are no key Mesozoic indicators such as reptiles, belemnites, or ammonoids. Bone fragments are known, but none is preserved well enough to identify.

Amongst microfossils, foraminifera are rare, probably because the original environment prevented good preservation. However, dinoflagellates, which are a form of unicellular algae, have been used successfully to date the Wangaloa assemblage. The fossils are of early to middle Paleocene age, equivalent to the middle of the local Teurian Stage (perhaps 58–62 Ma). This is clearly a few million years younger than the end-Cretaceous extinction.

The Wangaloa assemblage includes species, such as the early struthiolariid gastropod *Conchothyra australis*, which are direct descendants of known Cretaceous forms. This demonstrates

Fig. 15. Paleocene fossils and strata of the Wangaloa Formation, Wangaloan Stage or about middle Early Paleocene, Wangaloa, South Otago. Smallest division on scale is 1 mm.
(a) Natural outcrop of sandstone exposed on beach, showing shellbeds dominated by large shallow-marine molluscs.
(b) Molluscs from the Wangaloa Formation, including *Globisinum* (gastropod) in upper left, *Heteroterma* (gastropod) in upper middle, *Conchothyra* (gastropod) in upper right, *Lahillia* (bivalve) in lower left; *Glycymerita* (bivalve) in lower middle, and *Polinices* (gastropod) in lower right.
(c) Eroded slab of Wangaloa Formation dominated by the turret shell *Leptocolpus*, which were aligned probably by current activity before burial.

clearly that different lineages of organisms continued across the Cretaceous/Cenozoic boundary, in spite of the effects of global mass extinction. Finally, whereas New Zealand's Cretaceous molluscan faunas have strong geographic links with other parts of Gondwana, the Wangaloan fauna has a marked endemic (distinctively New Zealand) component which points to increasing physical isolation from other southern landmasses.

Beyond Wangaloa: the Paleocene and Eocene

The south has produced few major macrofossils of later Paleocene and earlier Eocene age; fossils from the Middle to Late Eocene are well-known from a few localities detailed below. To give an overview, coal measures and overlying marine sediments in western Southland (e.g. Beaumont Coal Measures, Orauea Mudstone) are a source of well-documented pollens, and of plant macrofossils, which have received less attention; there are negligible reports of significant animal fossils. Around and north of Dunedin, widespread outcrops of the thick Abbotsford Mudstone produce reasonable microfaunas. However, there is a disappointing range of macrofossils, and most of these have lost their original limy skeletons to become preserved as natural moulds. The fossils include bivalve and gastropod molluscs, corals, rare fish and turtle bones, and plant debris including a coconut (Fig. 16). Equivalent units to the Abbotsford Formation occur along the North Otago coast, near Moeraki and Hampden; they include the famous Moeraki Boulders. Macrofossils encompass mainly rare vertebrates: an incomplete leg bone of a large presumed penguin, fragments of sea turtles and the skull of a large bony fish.

The thick, massive and calcareous Burnside Mudstone of Dunedin has produced rare but finely preserved macrofossils including occasional molluscs, sea pens, brachiopods, shark vertebrae, bony fish and penguins (Fig. 16). Foraminifera and other microfossils are richly abundant, and indicate an age of Late Eocene. According to D. Hamilton, the mudstone was deposited in deep water, probably beyond 200 m on the upper continental slope. Burnside Mudstone also occurs near Hampden, in North Otago, where macrofossils are again rare but well preserved. The Hampden section is a significant source of nautiloid cephalopods and the coral Madrepora (Fig. 16).

Fig. 16. Early and Middle Eocene fossils and strata of Otago. Smallest division on scale is 1 mm.
(a) Humerus of extinct leatherback turtle, Abbotsford Mudstone, Boulder Hill.
(b) Fossil coconut, probably *Cocos*, from Abbotsford Mudstone, Boulder Hill; upper view shows apex, lower view shows lateral face.
(c) Upper surface of crab, *Portunites*, in a concretion from Snowdrift Quarry near Milburn.
(d) Outcrop of Burnside Mudstone at disused Burnside Quarry near Dunedin.
(e) Sea pen, *Bensonularia*, from Burnside Quarry; coin for scale.
(f) Lateral view of left side of skull of unnamed large extinct penguin from Burnside Quarry; skull faces the left and is slightly rotated to show oblique view of underside; lower jaws extend both downwards and upwards from the right side; maximum preserved length of skull is 195 mm.
(g) Coral, *Madrepora*, Burnside Mudstone, Hampden coast, North Otago.

North Otago's Bortonian strata

Rocks in North Otago are important in revealing the biota from the Bortonian Stage, in Middle Eocene times about 40 Ma. In the Waitaki Valley at Bortons, between Duntroon and Pukeuri, are outcrops of shallow marine sandstone, while a little to the west, near Livingstone, one can see non-marine sediments which formed at about the same time. Marine fossils are scattered in the Tapui Sandstone at Bortons (Fig. 17); microfaunas are poor, but macrofossils include shallow-water bivalves (*Cucullaea, Duplipecten, Hedecardium*), gastropods (*Monalaria, Mauira, Speightia*) and crustaceans. R.M. Feldmann (Kent State University) has described the crustaceans from the Bortons sequence and others of similar age around Otago including Snowdrift Quarry near Milburn, South Otago. The crustaceans include surface-dwelling and burrowing forms, such as species of *Lyreidus, Callianassa* and *Portunites.*

At Livingstone, in the Maerewhenua Valley, non-marine mudstones and sandstones of Bortonian age have produced a rich plant megaflora (Fig. 17) detailed by M.S. Pole. A surprising associated fossil is the single larva or maggot of a bibionid fly, described by A.C. Harris (Otago Museum).

Marine fossils from Eocene/Oligocene shallow-water volcanic shoals

The coarse-grained and pure Ototara Limestone (Fig. 18) is an important source of marine fossils in coastal North Otago. The limestone – the Oamaru stone of common use – is associated with a coastal strip of Deborah and Waiareka Volcanics. It seems that the limestone accumulated in shallow settings on the flanks of, and between, active small marine volcanic cones. Tuffs, or water-lain fine volcanic sediments, occur within the limestones, and in places there are coarse-grained volcanic rocks and lavas. Foraminifera and, less commonly, macroinvertebrates from the limestone and associated tuffs indicate an age of Kaiatan, Runangan and lower Whaingaroan local stages, or Late Eocene and Early Oligocene, close to 34 Ma. Details of these sequences were given in monographs by M. Gage and by A.R. Edwards (both formerly N.Z. Geological Survey), and in multi-authored papers by D.S. Coombs (University of Otago) and by R.A.F. Cas (Monash University).

Two fossil groups dominate the Ototara Limestone, the bryozoans and brachiopods. Relatively pure assemblages of each are known, but both groups may occur together. Bryozoan-rich

Fig. 17. Middle Eocene fossils and strata of the Bortonian Stage, North Otago. Smallest division on scale is 1 mm.
(a) Poorly exposed outcrop of brown marine sediment, the Tapui Sandstone. at the classic outcrop of Bortons.
(b) Gastropods from the Tapui Sandstone at Bortons, with 5 specimens of the volute *Mauira* on the left and 3 specimens of *Speightia* on the right.
(c) Concentration of solitary corals, *Flabellum,* in Tapui Sandstone probably from Maerewhenua Valley.
(d), (e) Leaves of flowering plants from fine-grained strata of the non-marine Taratu Formation, near Livingstone, North Otago; original carbonaceous material is gone, with leaves preserved as moulds.

limestone often comprises well-sorted broken fragments, and in places this is pure and thick enough to be quarried as a building stone. Parkside Quarry, at Weston, exposes more than 50 m of limestone (Fig. 18). Near Oamaru Cape, a sequence at Boatmans Harbour shows clearly the relationship between limestone and associated volcanic rocks. To the south, the now-disused McDonald's Quarry near Kakanui preserves some metres of bryozoan-brachiopod limestone (Fig. 18) which accumulated in deep quiet waters. Here, the most common brachiopods are the rhynchonellides *Aetheia* and *Tegulorhynchia*, and terebratulides including *Terebratulina*, *Liothyrella* and *Stethothyris*; all have been studied extensively by D.E. Lee (University of Otago). Nearby at the disused Everett's Quarry, the Ototara Limestone is a brachiopod shell limestone, again with *Aetheia*, *Tegulorhynchia* and *Terebratulina*. Amongst other fossil groups, macrofossils generally are rare in the limestone, but more-common in tuffs. Bivalves are known from large prismatic shells ('horse mussel', *Pinna*?) and occasional pectinids – scallops. Large, subspherical echinoids known as cidaroids occur sporadically, both as skeletons and as isolated long spines.

Vertebrates are quite uncommon in the Ototara Limestone, but are important both taxonomically and in the history of development of New Zealand paleontology. Three bones from one bird represent the only described specimen of *Pachydyptes ponderosus*, which is the largest reported fossil penguin. Another species, *Palaeeudyptes antarcticus*, which was named in 1859 by British scientist T.H. Huxley, is the first fossil penguin described; this species is known from a single bone uncertainly from the Ototara Limestone at Kakanui. Other undescribed but more

Fig. 18. Early Oligocene fossils and strata of the Whaingaroan Stage, North Otago. Smallest division on scale is 1 mm. (a) Thick exposures of Ototara Limestone at Parkside Quarry, Weston; limestone here comprises coarse sandstone formed of bryozoan fragments. (b) Volcanic sedimentary rocks at Boatmans Harbour, Oamaru; the sequence dips to the right, so that oldest rocks (bedded fine-grained volcanic sediments) are at lower left and youngest (coarse volcanic agglomerate) at upper right; one person is examining a thin bed of yellow-brown fossiliferous Ototara Limestone. (c) Cluster of brachiopods, mostly *Liothyrella*, Everetts Quarry, Kakanui; in mid right is the lacy structure of a bryozoan; scale bar 20 mm. (d) Partial skeleton of a large cidaroid echinoid, probably *Phyllacanthus*, in bryozoan limestone from Everetts Quarry, Kakanui.

complete penguins are known. One fragmentary whale represented by teeth and parts of a skull is a very primitive mysticete (proto-baleen) whale, and carapace plates, limb bones and skull fragments from a marine turtle are known.

Microfossils are important and revealing about both age and paleoenvironment. A few localities near Weston are renowned as sources of the world-famous Oamaru Diatomite (Runangan Stage, Late Eocene); A.R. Edwards usefully reviewed localities and previous research on this unit. Foraminifera include benthic or bottom-dwelling forms characteristic of warm shallow waters, and also warm-water planktonic or floating forms such as the extinct elegantly-spined *Hantkenina.* These fossils lived during the latest phases of Eocene warmth, before global climates became cooler and more varied because of oceanic circulation changes and expansion of the Antarctic ice cap from the Oligocene to Recent.

Late Oligocene fossils from the Waitaki Valley

Late Oligocene shelf sediments – the Kokoamu Greensand and overlying Otekaike Limestone (Fig. 19) – from the Waitaki Valley region preserve diverse assemblages of vertebrates, particularly whales and dolphins, but also penguins, sharks and bony fish (Fig. 20). These vertebrates provide early records for many groups. On a global scale, the whales and dolphins help to understand the rapid and explosive Oligocene radiation of the living cetacean groups Odontoceti (toothed whales and dolphins) and Mysticeti (baleen whales). Furthermore, the vertebrates are associated with diverse assemblages of invertebrates and microfossils which, on a national scale, are important in their own right. The Waitaki Valley sequences are significant because of their history: they were buried to only moderate depth before being uplifted and exposed, so that fossils were not obliterated but are preserved well. Elsewhere, limestone and greensand of similar age are widespread around New Zealand, including, in the south, the Te Anau and Fiordland regions. However, most of the sequences were buried more deeply than were the Waitaki Valley sequences, so that the rocks

Fig. 19. Late Oligocene fossils and strata of the Kokoamu Greensand and Otekaike Limestone, Duntroonian and Waitakian Stages, North Otago. Smallest division on scale is 1 mm.
(a) Natural outcrop at the 'Earthquakes,' near Duntroon; about 5 m of Kokoamu Greensand is exposed at the level of the shrubs, with nearly 20 m of Otekaike Limestone overlying.
(b) Cross section through a natural exposure of the 'main' shellbed in the upper Maerewhenua Member of Otekaike Limestone, Trig Z, Otiake; most of the larger invertebrates are molluscs.
(c) Bedding plane of the 'main' shellbed in the upper Maerewhenua Member of Otekaike Limestone, Trig Z, Otiake; smooth-shelled scallops, Lentipecten, reach 55 mm across.
(d, inset) Tiny sea urchin or echinoid, *Fibularia*; Otekaike Limestone, Hakataramea Valley; largest specimens are about 7 mm across.
(e) Mosaic of invertebrates from the Kokoamu Greensand – bivalves (*Spissatella, Notocorbula*), gastropods (*Cirsotrema, Austrofusus*), a scaphopod (*Fissidentalium*), a solitary coral (Stephanocyathus), a barnacle, brachiopods (small *Aetheia* and larger *Waiparia*), spines from a cidaroid echinoid, and the heart-shaped echinoid *Lovenia.*
(f) Brachiopod *Waiparia*, Kokoamu Greensand near Duntroon; largest specimens are about 40 mm maximum dimensions.
(g) Large heart urchin, *Pericosmus*, Otekaike Limestone near Duntroon.

are rather cemented and the fossils are not as well preserved.

In terms of stages, or local time divisions, the Kokoamu Greensand is Whaingaroan in the base and Duntroonian in the upper part; thus it spans about 26–29 million years. The overlying Otekaike Limestone is Duntroonian in the lower parts, passing up into the Waitakian Stage; at the top it represents the earliest Miocene, with an age range of about 23–26 million years. Microfossils from these rocks indicate quiet mid- to outer-shelf depths in a sheltered setting, consistent with the depositional environment inferred from sediment character. Benthic foraminifera abound, but age-diagnostic adult planktic foraminifera are less common. Ostracods and coccoliths are other microfossil groups important in biostratigraphy.

Macroinvertebrates are diverse, generally more so in the finer-grained Otekaike Limestone than in the coarser and sometimes leached Kokoamu Greensand (Fig. 19). Corals include common *Flabellum* and less common *Stephanocyathus* and tiny *Peponocyathus*; all are solitary scleractinian corals (hexacorals) which did not form reef-like structures. Brachiopods are particularly abundant in some parts of the Kokoamu Greensand, and are scattered throughout both the greensand and limestone. Indeed, brachiopods from the Kokoamu Greensand and Otekaike Limestone have had a long history of study, most recently by D.E. Lee, D.I. MacKinnon (University of Canterbury) and colleagues. Perhaps the most characteristic brachiopods are the articulates *Aetheia*, *Tegulorhynchia* (both rhynchonellides) and the terebratulides *Rhizothyris* and *Waiparia*. Strangely, the supposedly key indicator brachiopod for the Duntroonian Stage, '*Liothyrella*' *landonensis*, is very rare. Molluscs, which have been researched particularly by P.A. Maxwell (formerly N.Z. Geological Survey), include the conspicuous bivalves *Limopsis*, *Lentipecten*, '*Chlamys*,' *Cucullaea* and *Spissatella*. Gastropods often are represented only by the calcite-shelled *Cirsotrema*, but may include turritellids (turret shells – species of *Zeacolpus* and *Maoricolpus*) and predators such as *Magnatica*, *Austrofusus*, *Proximitra*, *Eoturris* and *Parasyrinx*. Large scaphopods (tusk shells, e.g. *Fissidentalium*) are sometimes conspicuous. Echinoids or sea urchins locally are common in the Otekaike Limestone. Mostly these are large heart urchins (Spatangoidea) like *Taimanawa* and *Pericosmus*, which lived within the sediment. Occasionally, regular echinoids, such as the long-spined cidaroid *Histocidaris*, are found; these probably dwelt on the surface. One tiny echinoid, *Fibularia*, is sometimes extremely abundant (Fig. 19). It is interesting to note that *Fibularia* is dimorphic, with two different sizes which probably represent male and female. Bryozoans or moss-animals are less readily identified in the field, but broken fragments are a significant component of the limestone. Larger arthropods other than barnacle plates are rare.

Research since 1982 has greatly expanded the previously-described fauna of fossil whales and dolphins (Cetacea) to some scores of species, and has significantly revised previous identifications. A key factor in these advances has been the use of power tools to collect taxonomically-important skulls and associated elements, fossils which previously were difficult to recover.

Of the three different groups of cetaceans, the most primitive whales, archaeocetes, are known only poorly. Several fragmentary specimens including a partial skull have been described from the Waihao Greensand of South Canterbury. Mysticete whales, which today are baleen-bearing and toothless, are reasonably diverse in the southern Oligocene. The most tantalising are fragmentary archaic toothed forms such as *Mammalodon*, which are structurally intermediate between archaeocetes and true toothless baleen-bearing whales. The latter have an excellent record which includes some of the oldest global records reported (Fig. 20). For the New Zealand Oligocene as a whole, mysticetes are unusually diverse in structure, and contrast with their living relatives which encompass fewer species with more conservative structure. In North Otago the Oligocene mysticetes include *Mauicetus*-like species which are widely quoted in the literature because of the research efforts of B.J. Marples (formerly University of Otago). For some decades, species of *Mauicetus* were placed in the extinct family Cetotheriidae along with other primitive baleen whales; now, however, it seems that some may be distant ancestors of living rorquals while others probably belong in a new family.

Odontocetes – the echolocating toothed whales and dolphins – mostly belong in the group Platanistoidea. This group includes a diversity of Oligocene marine species, but is now represented by only one or two endangered living freshwater species restricted to India and Pakistan. Platanistoid dolphins from the Kokoamu Greensand and Otekaike Limestone include species of the extinct families Squalodontidae (shark-toothed dolphins), Waipatiidae (e.g. *Waipatia*), Squalodelphinidae and Dalpiazinidae (Fig. 20). Archaic true dolphins, representing the widely-distributed Kentriodontidae and also some problematic specimens, are rare, and there is only one fragmentary record of a possible sperm whale.

All of the fossil Cetacea from the Waitaki region are extinct species. Although some living families appeared in Oligocene times, no species has persisted to the present. Larger mysticetes dominate the cetacean assemblages, with small odontocetes also conspicuous, but large odontocetes have not been reported. The shallow broad seaway in which the Kokoamu Greensand and Otekaike Limestone was deposited could have been a breeding ground for mysticetes. Further, it is possible that the abundance of mysticetes – which first appeared early in the Oligocene – signals a recently developed Southern Ocean ecosystem south of New Zealand. Why platanistoids are common but delphinoids and sperm whales are rare is not clear. Perhaps the latter taxa were mainly oceanic or deep-diving.

Waipatia, a Late Oligocene platanistoid dolphin

Waipatia maerewhenua is an extinct dolphin known only from a single specimen from the Otekaike Limestone (24–25 Ma) near Duntroon, North Otago (Fig. 20). The skull, jaws, earbones, teeth, some vertebrae and ribs are preserved. Judging from skull size, body length was 4–5 m, comparable to a medium-sized living species like the common dolphin *Delphinus delphis*. In *Waipatia*, the advanced structure of facial bones is similar to that of most living dolphins, in which the dished bones of the face carry muscles used to produce echolocation sounds in the soft tissues around the blowhole. Features of the skull base and ear bones also are consistent with the reception of high-frequency sound. *Waipatia*, therefore, is thought to have been an active echolocator, producing and receiving high-frequency sound to navigate and to detect prey. In both the upper and lower jaws, many of the teeth have multiple cutting projections or denticles, in contrast to the simple conical teeth of modern dolphins. The anterior-most teeth, which were found loose by the skull, are narrow and projecting, and seem too delicate to have functioned in feeding. Rather, perhaps, they were for display.

Previously, archaic dolphins like this were placed in the widely-distributed family Squalodontidae. A new family, Waipatiidae, was proposed for *Waipatia maerewhenua* because it differs from squalodontids in its small size, smaller and simpler teeth, slightly asymmetrical skull, and rather different earbones. Other *Waipatia*-like fossils have been collected from around the Waitaki Valley, and similar specimens are known from Eurasia, North America, and Sakhalin. Whether the family Waipatiidae became extinct because of climate change or because of competition from other odontocete groups is uncertain.

An Oligocene penguin, *Platydyptes amiesi*

The extinct robust penguin *Platydyptes*, named by B.J. Marples, is known from many specimens, including one almost complete skeleton (Fig. 21) of *Platydyptes amiesi*. The latter is from the Otekaike Limestone (Waitakian; about 24 million years old) near Duntroon in North Otago. This specimen is taxonomically valuable, since it links together many elements previously known from isolated or unassociated bones. Elements include forelimb bones (pectoral girdle and wing – sternum, scapula, coracoid, humerus, radius), pelvis and leg bones (including femur, tibiotarsus, tarso-metatarsus), vertebrae and ribs. The association of bones from one specimen helps resolve long-standing problems: which bones go with which, and what species are involved.

The skeleton will also be useful for functional studies. Of note, the pectoral girdle is robust and of unusual construction. The sternum (breast bone) is more robust than in other fossil penguins, and has strong struts (coracoid bones) that run outwards to the shoulder. The proportions suggest a body that was deeper from front to back than in modern species. Further, the relative lengths of the limbs vary; the leg is similar in size to that of the yellow-eyed penguin, while the wing is larger, comparable with the emperor penguin. Thus, the bird had big

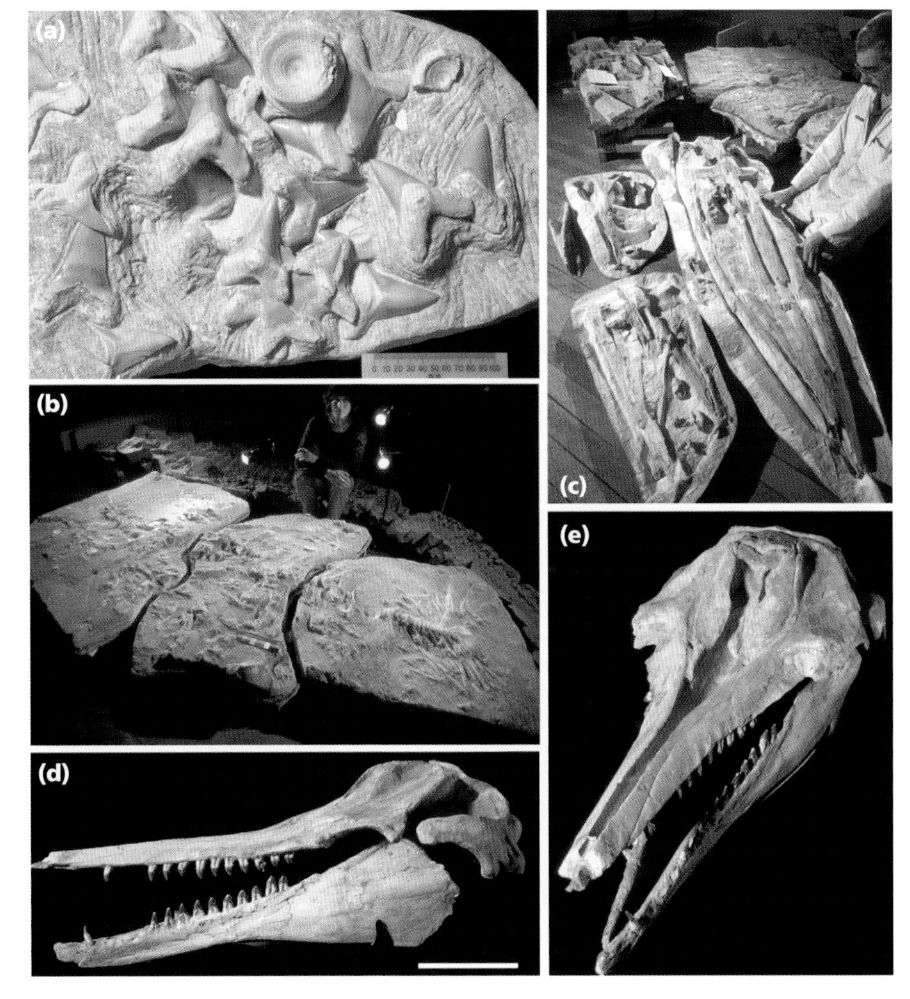

Fig. 20. Late Oligocene fossil vertebrates from the Kokoamu Greensand and Otekaike Limestone, Duntroonian and Waitakian Stages, North Otago.
(a) Teeth and a vertebra of extinct white shark, *Carcharodon angustidens*, near Tokarahi; scale 100 mm long.
(b) Partly articulated skeleton of un-named large bony fish, probably related to living moon-fish (*Lampris*), near Tokarahi; tail is on the right, with fin-rays visible, while vertebrae, which indicate the spine, extend from mid right towards upper left.
(c) Partial and complete skulls of four archaic *Mauicetus*-like baleen whales, from near Duntroon and Hakataramea Valley; the largest and most complete specimen, on the right by the person, shows both the cylindrical lower jaws either side of the flat upper jaw.
(d) and (e) Skull and lower jaws of the archaic dolphin *Waipatia maerewhenua*, from lower Otekaike Limestone near Duntroon; (d) lateral view; (e) oblique view; scale bar is 100 mm.

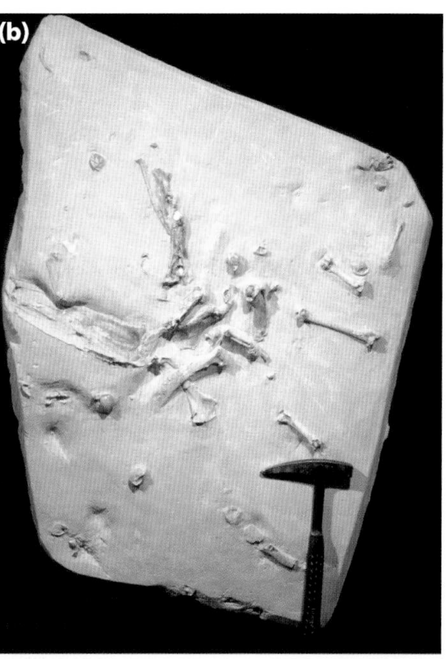

Fig. 21. Extinct penguin *Platydyptes*, from the Otekaike Limestone near Duntroon, North Otago.
(a) Field extraction of skeleton of *Platydyptes*.
(b) Skeleton when partly prepared, showing naturally-associated bones of *Platydyptes* scattered across a bedding plane; hammer is 330 mm long.
(c) Reconstruction of two individuals of *Platydyptes* coming ashore. *Art by C. Gaskin held in Geology Museum, University of Otago*

wings and short legs. Given these unusual proportions, and the lack of a fully-articulated skeleton, it is difficult to judge overall size, but perhaps *Platydyptes amiesi* was intermediate between the latter two living species. The bird was probably 90 cm from tip of bill to end of tail, and stood about 55 cm high at the shoulder (Fig. 21).

Platydyptes includes three species of fossil penguins, another six fossil penguin species that have been named on the basis of specimens from around the Waitaki Valley and even more species that have been recognised but have yet to be named formally. Furthermore, many of the specimens, like the skeleton discussed above, include many bones from single individuals. This abundance of material makes the North Otago/South Canterbury region one of the major sources of fossil penguins known worldwide.

Molluscan assemblages from the Mount Harris Formation

During the Early Miocene, 16–22 Ma, fossiliferous marine sediments accumulated along much of what is now coastal Otago including Dunedin district, and over most of Southland between Fiordland and the Catlins. The Southland strata include richly fossiliferous bryozoan limestones which probably formed in quite shallow waters, but in coastal North Otago we can find massive grey mudstones which accumulated at depth near the edge of the continental shelf. In these mudstones of the Mount Harris Formation, macrofossils are generally rare but well preserved; shellbeds of worn molluscs occur in places. The molluscs, which have been studied closely by P.A. Maxwell, are particularly diverse, with more than 300 species recorded. Many are small, and most are well preserved (Fig. 22). Bivalves include *Neilo*, *Cucullaea*,

Lima, *Limopsis*, '*Cyclocardia*' and *Spissatella*, while the more-common gastropods are *Amplicolpus*, *Typhis*, *Penion*, *Austrotoma*, and *Comitas*. In places, the Mount Harris Formation yields particularly spectacular volutes (Gastropoda: Volutidae), such as *Spinomelon*, *Alcithoe* and *Metamelon*. Other invertebrates are represented by barnacles, crabs, corals, worms, rare brachiopods, bryozoans and echinoids. Vertebrates are rare but significant; otoliths, or ear stones, and fragmentary skeletons are known for fish. Important but mostly unprepared cetaceans include small dolphins and small to moderate-sized baleen whales. Previously, the Mount Harris Formation produced one of the first Tertiary vertebrates to be described from New Zealand: the dolphin *Phocaenopsis*, which was named by T.H. Huxley in 1859. The nature of the sediment, the preservation of the fossils and the composition of the molluscan assemblages indicate deposition in deep quiet conditions near the continental shelf edge, perhaps at depths of 150–200 m. Fossils indicate an age of, mainly, Altonian Stage, or later Early Miocene.

Non-marine fossils from the Manuherikia Group, Central Otago

During the later part of Early Miocene time (Altonian Stage, about 16–18 Ma), an extensive shallow lake complex, Lake Manuherikia, covered much of what is now Central Otago. Lake and river settings are represented in two rock units which comprise the Manuherikia Group (Figs 23, 24). Of these, the Dunstan Formation includes diverse bodies of conglomerate, sandstone, and mudstone, with carbonaceous to coaly rocks. Likely environments include braided river, meandering river, fluvial plain, swamps, levees, distributary channels and delta plain. A lake setting is represented by the thicker overlying Bannockburn Formation, which generally is finer-grained and more massive. Fossils of the Manuherikia Group span a wide range of non-marine organisms. Plants – both pollen and plant macrofossils – are the best-known elements of the biota (Fig. 23), but algal stromatolites and rarer animal remains are present (Fig. 24).

Many details are known about the flora of the Manuherikia Group, particularly through the detailed studies of M.S. Pole and also J.D. Campbell and associates. Fossil plants include ferns, assorted conifers (Podocarpaceae, Araucariaceae – related to kauri and Norfolk Island pine), and she-oaks (*Casuarina*; now extant in Australia but extinct in New Zealand). Palms, preserved as fronds and nuts, are related to living nikau, *Rhopalostylus*. The climber supplejack *Rhipogonum* is present. Tantalising fossils of a legume and possible divaricate plant, with tiny leaves and a perpendicular branching pattern, are known. Of larger trees, myrtaceans include *Eucalyptus*, which is now extinct in New Zealand, represented by characteristic sickle-shaped leaves (Fig. 23). *Metrosideros*, which today includes rata and pohutukawa, is present. Southern beeches – genus *Nothofagus* – include the extinct species *N. novaezealandiae* and *N. azureus*, the latter first

Fig. 22. Early Miocene fossils and mudstone of the Mount Harris Formation, Altonian Stage, near Awamoa, North Otago.
(a) Beach outcrop of Mount Harris Formation.
(b) Molluscs collected from the Mount Harris Formation by P.A. Maxwell; scale bar is 100 mm; top row, left to right, *Penion*, *Magnatica*, *Typhis*; middle row, left to right, *Galeodea*, *Echinophoria*, *Coluzea*, *Zeamacies*, *Spinomelon*, *Spinomelon*, *Penion*; bottom row, left to right, *Lima*, *Neilo*, *Spissatella*, *Cucullaea*, *Comitas*, *Cirsotrema*, *Amplicolpus*. Specimens in collection of P.A. Maxwell.

recognised from the Manuherikia Group.

According to M.S. Pole's research, the older of the Manuherikia floras represents a forest dominated by *Nothofagus*. *Casuarina* occurred at forest margins, in fire-disturbed sites, or both. Younger floras are dominated by araucarians and eucalypts, possibly marking drier and/or warmer conditions. Elsewhere in New Zealand, there is strong evidence that this was a time of near-tropical warmth.

Rough-surfaced and subspherical fossil algal stromatolites (Fig. 24) occur in the Bannockburn Formation at a few localities north-east of Alexandra; they were documented in detail by J.K. Lindqvist (University of Otago). Stromatolites are formed by layer upon microscopic layer of calcium carbonate and mud, trapped by sticky films of blue-green bacteria in shallow lake waters within the photic zone – the zone of significant light penetration. Most of the stromatolites are 50–100 mm in diameter, but larger specimens are known. In places, the stromatolites are common enough to form an extensive 'carpet.' Fish, not yet formally named, are known from massive clays and from laminated mudstone. Material from the clays comprises mostly disarticulated bones including abundant otoliths, while specimens from laminated mudstone may be articulated. *Hyridella*, or freshwater mussel, occurs both as fragmented shell (in massive mudstone) and decalcified moulds (in carbonaceous laminated mudstone). A fossil freshwater crayfish which represents an extinct new species in the still-living genus *Paranephrops* is one of the most exquisite fossils from the Manuherikia Group (Fig. 24). Rare bird bones and broken eggshell are known from massive clays near St Bathans. The fossil birds are disarticulated, hampering identification, but enough is preserved to indicate a small species, probably duck, and a large species, perhaps goose or swan. A crocodile (species uncertain) is the most recent freshwater vertebrate identified (by R. Molnar, Queensland Museum), raising hopes that other reptiles and amphibians eventually may be found.

Clifden, Southland

A fossiliferous succession on the banks of the Waiau River near Clifden, Southland (Fig. 25), provides another glimpse of the biota of the southern Early and Middle Miocene (local Altonian to Waiauan Stages, perhaps 10–18 Ma). Over 400 m of gently-dipping massive and bedded marine mudstone, sandstone and limestone were deposited in a shallow marine setting. The basal limestone is dominated by bryozoans – 'moss animals' – and thus is similar to limestone of the Southland Plains, but the overlying sediments, with a higher proportion of land-derived debris, provide diverse molluscan assemblages not well represented elsewhere in the south. Fossils are both scattered and in beds, and there is a significant molluscan microfauna as well as larger fossils. Diverse bivalved molluscs include pectinids (e.g. *Serripecten*, *Hinnites*, '*Chlamys*'), oysters (*Crenostrea*), the large cucullaeids and glycymeridids (dog-cockles), carditids and

Fig. 23. Early Miocene fossils and strata of the Manuherikia Group, Altonian Stage, Central Otago. Fossils other than (a) are from Bannockburn Formation near Bannockburn.
(a) Stem, branches and small leaves of a proposed divaricate plant, Nevis Valley.
(b) J.D. Campbell collecting plant fossils from outcrop of Bannockburn Formation near Bannockburn. (c) Leaf of *Eucalyptus*. (d) Frond of *Casuarina*, running obliquely from lower right to upper left. (e) Detail of cone of *Casuarina*.

Fig. 25 (opposite). Early to Middle Miocene fossils and strata of the 'Clifden sequence', Waiau River, Clifden, Southland. Smallest division on scale is 1 mm.
(a) Clifden Limestone outcrop (Altonian, Early Miocene) below old swing bridge; younger rocks are hidden from view upstream of the bridge.
(b) Large terebratulide brachiopod *Erihadrosia*, Park Bluff Sandstone, Lillburnian Stage.
(c) Well-preserved specimen of nautiloid cephalopod, *Aturia*, showing septa or internal partitions; Nga Pari Formation, Altonian Stage.
(d) Three species in lineage of the gastropod *Neocola*, from ancestral *Neocola alpha* in the base to *Neocola beta* in the middle and *Neocola gamma* at the top; Nga Pari Formation, Altonian-Clifdenian Stages.

venerids. Gastropods also are diverse, including struthiolariids (*Struthiolaria*), turret shells (*Zeacolpus*), volutes (*Alcithoe*), turrids, conids (cone shells, *Conus*) and rare epitoniids (*Cirsotrema*). Clifden has a particularly important succession of fossil whelks (family Buccinidae) in the subgenus *Neocola*, which in turn is placed in the genus *Austrofusus*. There is an evolutionary lineage showing gradual change from the ancestral species, *Neocola alpha*, to descendant species *Neocola beta* and, later, *Neocola gamma* (Fig. 25). Also notable amongst the molluscs are the nautiloid cephalopod *Aturia*, scattered scaphopods or tusk shells, and occasional fossil chitons. Brachiopods are abundant in places; they include the inarticulate *Lingula* and, amongst articulates, *Rhizothyris* and huge species of the terebratulide '*Stethothyris*' – now known as *Erihadrosia* (Fig. 25). Solitary corals including *Flabellum* and *Notocyathus* are reported. Beyond these familiar macroinvertebrate groups, barnacles and annelids (polychaetes) also occur. Clifden is a locality for many fish otoliths, including type specimens for some species, but otherwise vertebrates have received little mention.

Amongst microfossils from Clifden, the evolutionary succession of planktic or free-floating foraminifera in the so-called *Orbulina* bioseries is particularly important. N. de B. Hornibrook (N.Z. Geological Survey) documented this lineage, in which the shell or test of successive species becomes increasingly spherical over time. The oldest or basal species is *Globigerinoides bisphericus*, which appears in the Altonian Stage. It is succeeded by *Praeorbulina glomerosa*, followed by, in the Lillburnian Stage, *Orbulina suturalis* and *Orbulina universa*. The latter species, *O. universa*, is still living, and has thus persisted for about 16 million years. This succession of planktic foraminifera is important in subdividing the Middle Miocene in New Zealand.

Many of the Clifden macrofossils, including the cone shell *Conus*, the large bivalve *Cucullaea*, and the inarticulate brachiopod *Lingula*, are now extinct in the south or, in some cases, all of New Zealand. However, many have living relatives in

Fig. 24 (above). Early Miocene fossils and strata of the Bannockburn Formation, Manuherikia Group, Altonian Stage, Central Otago. Smallest division on scale is 1 mm unless specified.
(a) Cluster of disarticulated fish bones and scales, near St Bathans.
(b) Bird bones, presumed to represent a duck and small goose; white cardboard microscope slide mid left shows fragmented eggshell found with bones; near St Bathans.
(c) Outcrop of Manuherikia Group near St Bathans; river-deposited quartz gravels of Dunstan Formation at lower left are overlain by lake deposits of the Bannockburn Formation including green clays, brown mudstone, and stromatolite limestone; pick and spade in mid picture about 1 m long.
(d) Extinct species of freshwater crayfish, *Paranephrops*; smallest division on scale is 5 mm.
(e) Cross section of algal stromatolite, from near Lauder.

warm waters elsewhere. The fossils thus point to substantially warmer conditions – warm temperate to subtropical – than at present.

Forests and ponds on the flanks of Dunedin volcano

A flora including the fossil southern beech species *Nothofagus novaezealandiae* occurs in freshwater sediments within the thick Dunedin Volcanic Complex, revealing a flora that is a few million years younger than that of the Manuherikia Group. These plants have been studied variously by J.D. Campbell and M.S. Pole. The source unit, the Kaikorai Valley Leaf Beds (Fig. 26), is associated with the so-called Older Floodplain Conglomerate, at a break between two volcanic phases. The leaf beds themselves are undated but the Dunedin Volcanic Complex is no older than Waiauan Stage (late Middle Miocene); radiometric dates for these rocks range around 10–13 Ma, or mainly Late Miocene.

According to J.D. Campbell, localised occurrences of leaf-beds in Kaikorai Valley may reflect original ponding on the flanks of the Dunedin volcano. Here, some metres of fine-grained leaf beds (Fig. 26) are dominated by diatoms, indicating deposition over many years in still, clear freshwater. The dominance of diatoms also implies low input of land-derived sediment which otherwise would have overwhelmed the diatom component. Plant fossils indicate a flora at least partly similar to the older floras of the Manuherikia Group discussed above. Abundant beech pollen of the *Nothofagus brassii*-group which occurs in the leaf beds probably represents the extinct southern beech species *N. novaezealandiae*, for the latter is known from leaves from the same locality. Other plant fossils include cones of the 'she-oak' *Casuarina*, assorted nuts which have not been identified to species and leaves of proteas, Proteaceae. The proteaceans are possibly related to the living trees *Knightia* (rewarewa) and *Toronia* (toronia) of warmer parts of modern New Zealand, and clearly are extinct in the Otago region. The well preserved leaves point to transport over only a short distance before burial, so it is possible that some trees flanked the pond or lake.

In a few localities of Kaikorai Valley Leaf Beds near Dunedin, fossil galaxiid ('whitebait') fish have been collected (Fig. 26). The specimens have been identified as representing the genus *Galaxias*, but they lack features that allow the particular species to be identified.

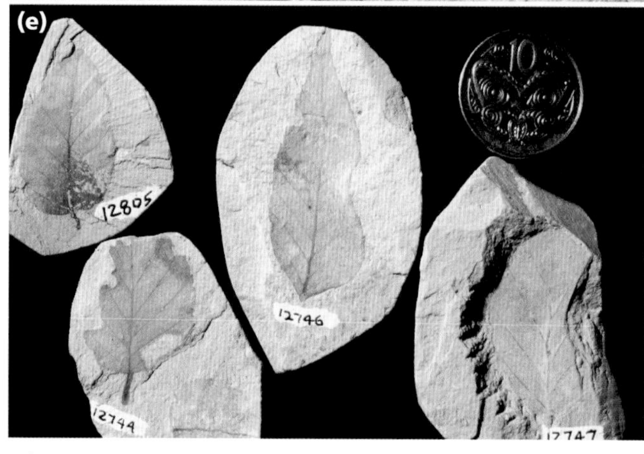

Fig. 26. Late Miocene fossils and strata of the Kaikorai Valley Leaf Beds, Dunedin Volcanic Complex, Dunedin district.
(a) Kaikorai Valley Leaf Beds, Frasers Gully, beside Kaikorai Stream.
(b) Detail of Kaikorai Valley Leaf Beds at same locality; lens cap lower middle is 50 mm diameter.
(c) Bedded volcanic-derived sediments, probably deposited in a lake setting, near Taiaroa Head, Otago Peninsula; upper beds here have produced *Nothofagus* (beech) fossils; hammer is 420 mm long.
(d) Fossil galaxiid fish, 'whitebait'; Kilmog, north of Dunedin; fish is about 30 mm long.
(e) *Nothofagus* (beech) leaves from Kaikorai Valley Leaf Beds, Frasers Gully; coin is 24 mm diameter.

Fig. 27 (opposite). Late Miocene fossils and strata of the Port Craig and Te Waewae Formations, Te Waewae Bay, western Southland.
(a) Mollusc-rich muddy grit of lower Port Craig Formation, Port Craig; Kapitean Stage, Late Miocene; hammer is 420 mm long.
(b) Whale vertebra in Port Craig Formation near Port Craig; flat articular surface of vertebra is visible towards top of photograph; Kapitean Stage, Late Miocene; knife is 85 mm long.
(c) Cross section through siltstone of the Te Waewae Formation, showing abundant articulated bivalves (mainly *Limopsis*); probably upper Kapitean, latest Miocene; western Te Waewae Bay.
(d) *Flabellum*-like colonial coral, siltstone of the Te Waewae Formation; probably upper Kapitean, latest Miocene; western Te Waewae Bay.

Te Waewae Bay macrofaunas

By the end of Miocene times, a little more than 5 million years ago, much of New Zealand was emerging as the proto-Southern Alps rose. Marine sediments and their contained fossils are, therefore, more localised than for earlier times. In southern New Zealand, perhaps the best-known fossil assemblages are those of the Port Craig Formation (Kapitean, Late Miocene) and overlying Te Waewae Formation (Kapitean–Opoitian, Late Miocene–Early Pliocene) at Te Waewae Bay. Here, on the Southland coast, gritty, sandy and muddy sediments, cemented calcareous sandstone and siltstone represent a sequence which ranges from shallow to deep water.

Basal gritty calcareous sandstone within the Port Craig Formation (Fig. 27) has a distinctive fauna dominated by bivalves – the 'dog cockle' *Glycymeris*, ribbed scallop *Mesopeplum* and oyster '*Ostrea*'. These bivalves represent a shallow water assemblage which probably accumulated on a fairly firm substrate. Higher in the sequence, passing into the Te Waewae Formation, the rocks become more massive and finer grained, and are presumed to indicate a deeper-water setting. The fossil fauna in these finer-grained rocks also differs; there are abundant and well-preserved specimens of the small bivalve *Limopsis* (Fig. 27) and a scatter of other bivalves and gastropods such as the volute *Alcithoe*. An uncommon but important element in the fauna is the small nautiloid *Aturia*; this extinct relative of the living *Nautilus* indicates warm waters and, incidentally, provides the youngest New Zealand record for *Aturia*. Te Waewae Bay rocks yield the only specimens of a bizarre pseudo-colonial form of scleractinian coral apparently related to *Flabellum* (Fig. 27). Rare isolated bones are from whales and dolphins, but no complete specimens have been reported yet.

The youngest parts of the southern record

Southern marine rocks and fossils younger than Late Miocene are known poorly. In the north-east, the only significant macrofossil locality is at Oamaru, where Opoitian (Early Pliocene) fossils such as the distinctive gastropod struthiolariid *Struthiolaria* (*Callusaria*) *obesa* wash up on the shore. These fossils are presumed to come from submarine occurrences near to shore. In the far south, the Birchs Mill Shellbed on the shores of Te Waewae Bay is a densely packed lens of shallow marine molluscs, of Opoitian or Early Pliocene age. Beyond these

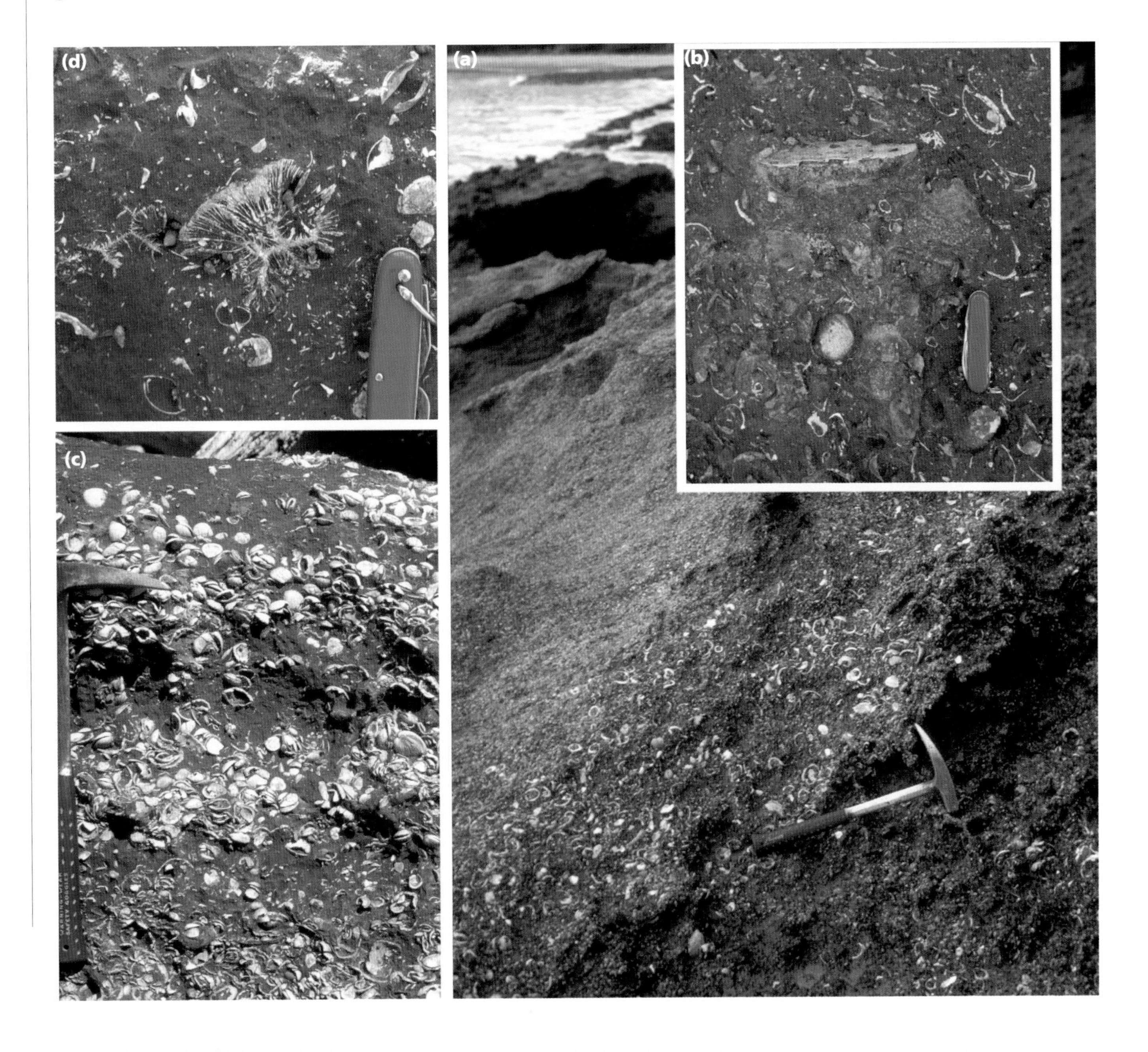

examples, the Pliocene is recognised mostly by pollens from non-marine sediments.

One other unusual southern occurrence involves invertebrates, especially crabs, found in hard concretions of mudstone which wash up on beaches around the entrance to Otago Harbour. Far from being ancient fossils, these are living species which have been entombed in harbour muds perhaps within the last few thousand years.

The south has a good historical record for the terrestrial late Quaternary, perhaps covering only the last few tens of thousands of years, but the material can hardly be called fossil and indeed is rarely the subject of study by geologists. Most of the remains – species represented by pollen, for example – are still living. Further, the preservation of extinct Quaternary species (such as moa) is so good that we can study the specimens essentially as if they are the remains of still-living organisms. Moa are known from abundant material, perhaps thousands more specimens per species than for older vertebrates; mummified soft tissues are known, and stomach contents have been recovered from moa elsewhere in New Zealand. Southern occurrences have been documented widely, most recently by T.H. Worthy (see Chapter 6). Finally, the overlap between most Quaternary species and humans takes these young fossil remains away from paleontology and into the realms of zoology and anthropology.

WHAT THE SOUTHERN RECORD SHOWS

Fossils from southern New Zealand provide many insights. While living organisms allow us to infer historical events, it is only fossils that give the direct evidence for the nature of past life – what was present when and where. And, these basic sorts of information are, in turn, the foundation of many broader contributions.

To move beyond the intrinsic patchiness of the record, southern fossils clearly show broad successions over time. In some cases, such as the gastropod *Neocola* and the planktonic foraminiferan *Orbulina*, they reveal direct ancestor-to-descendant lineages – which, for the *Orbulina* bioseries, involves living species. Such examples are amongst the best documented evidence of macroevolution, the evolution of new species. Paleontology, however, can contribute little about the microevolutionary processes – those genetic processes operating at the individual to population level and over short intervals of time – which presumably underlie macroevolution. Generally, also, it is difficult to demonstrate that descendant species are clearly better adapted than ancestral species, but judging from their abundance and persistence, past organisms were sophisticated and successful. Some had unexpected body forms which indicate different lifestyles from living relatives. So, we cannot argue that the record always provides clear evidence of ongoing 'improvement' over time.

Southern fossils provide important information about the minimum age of some living groups, particularly for some recently evolved vertebrates such as whales and dolphins. Such age information is increasingly important in helping to calibrate 'molecular clock' studies of living organisms. In other cases, southern fossils include some of the youngest representatives of particular taxa. Noteworthy among such 'last' records are those of Late Permian trilobites and of Late Cretaceous plesiosaurs and mosasaurs. Such examples raise the topic of extinction, a process inextricably linked with evolution.

The southern record shows that, as elsewhere, there has been continual piecemeal change of biotas, with species evolving or going extinct in seemingly random pattern. But, as well as piecemeal change, the fossil record also shows some horizons of mass extinction – such as the 'K/T' or Cretaceous/Tertiary boundary. Such extinction events, in turn, commonly foretell rapid adaptive radiations. In the south, adaptive radiations are reflected in records of many molluscs, foraminifera, mammals and birds in the early Cenozoic, following the end-Cretaceous extinction.

Finally, paleontology demonstrates full well the chanciness of survival through extinction events. For that reason, a paleontologist might not wish to predict what sort of biota will be present in New Zealand in some thousands of years, as the current human-induced extinctions presumably decline in rate and magnitude.

RESOURCES BEYOND THIS REVIEW

To take paleontology further, one may collect in the field, look at individual fossils held in collections or read the literature. Field sites are mentioned above. The southern fossils come from sedimentary rocks, which form a major topic for study in their own right. Chapter 1 of this book reviews southern geology in general, and some important sedimentary rocks are mentioned in the last paragraph below.

Fossils from southern New Zealand may be seen in the Geology Museum of the University of Otago, and in the Otago Museum. Some major collections are held elsewhere, particularly in Wellington at the Institute of Geological and Nuclear Sciences.

Many publications provide more details on these fossiliferous sedimentary rocks, including a two-volume set on the geology of New Zealand edited by R.P. Suggate and others, an old but detailed stratigraphic lexicon edited by C.A. Fleming, and 1: 250,000 geological maps and reports produced by the New Zealand Geological Survey and by the Institute of Geological and Nuclear Sciences. *The Field Guide to New Zealand Geology*, by J. Thornton, also provides a useful introduction.

CHAPTER 4

CLIMATE

BLAIR FITZHARRIS

Weather in southern New Zealand is always changeable – snow disrupts Dunedin hill traffic.

Otago Daily Times

Fig. 1. Extreme cold has caused the Arrow River, Central Otago, to freeze, 10 July 1986. *Otago Daily Times*

Fig. 2. Drought has dried up this stream in North Otago to the extent that the fishing is poor, January 1988. *Otago Daily Times*

CLIMATE DIVERSITY

No other region of New Zealand displays the diversity of climates that are to be found in the southern portion of the country. Climates range from some of the wettest to some of the driest, from maritime to semi-continental, from lowland to upland. The decisive influence of topography on weather systems and the complicated nature of the terrain create a surprisingly wide variety of very distinctive climatic types within a relatively small area. Another hallmark of the region is the large day-to-day variability of the weather, yet on a longer time scale the climate is considered to be moderate and well watered.

CLIMATE

MAIN FACTORS CONTROLLING THE CLIMATE

The larger setting

The most important factor that controls the climate of southern New Zealand is its location at the crossroads of several weather systems. To understand the climate, it is necessary to see where southern New Zealand lies within the larger hemispheric perspective. The most relevant features of the hemisphere's wind belts and atmospheric circulation are the belt of very strong westerlies (the Roaring Forties) and the zone of seemingly endless fine weather in the subtropics that is the favourite haunt of sun lovers (Fig. 3). Southern New Zealand lies largely within the westerlies, which circle the hemisphere over the southern oceans and about Antarctica. More formally, these are called the circumpolar disturbed westerlies. On the other hand, the northern part of the country is on the southern fringe of the subtropical high pressure belt, a region of persistent anticyclones that causes the deserts of Australia, southern Africa and northern Chile.

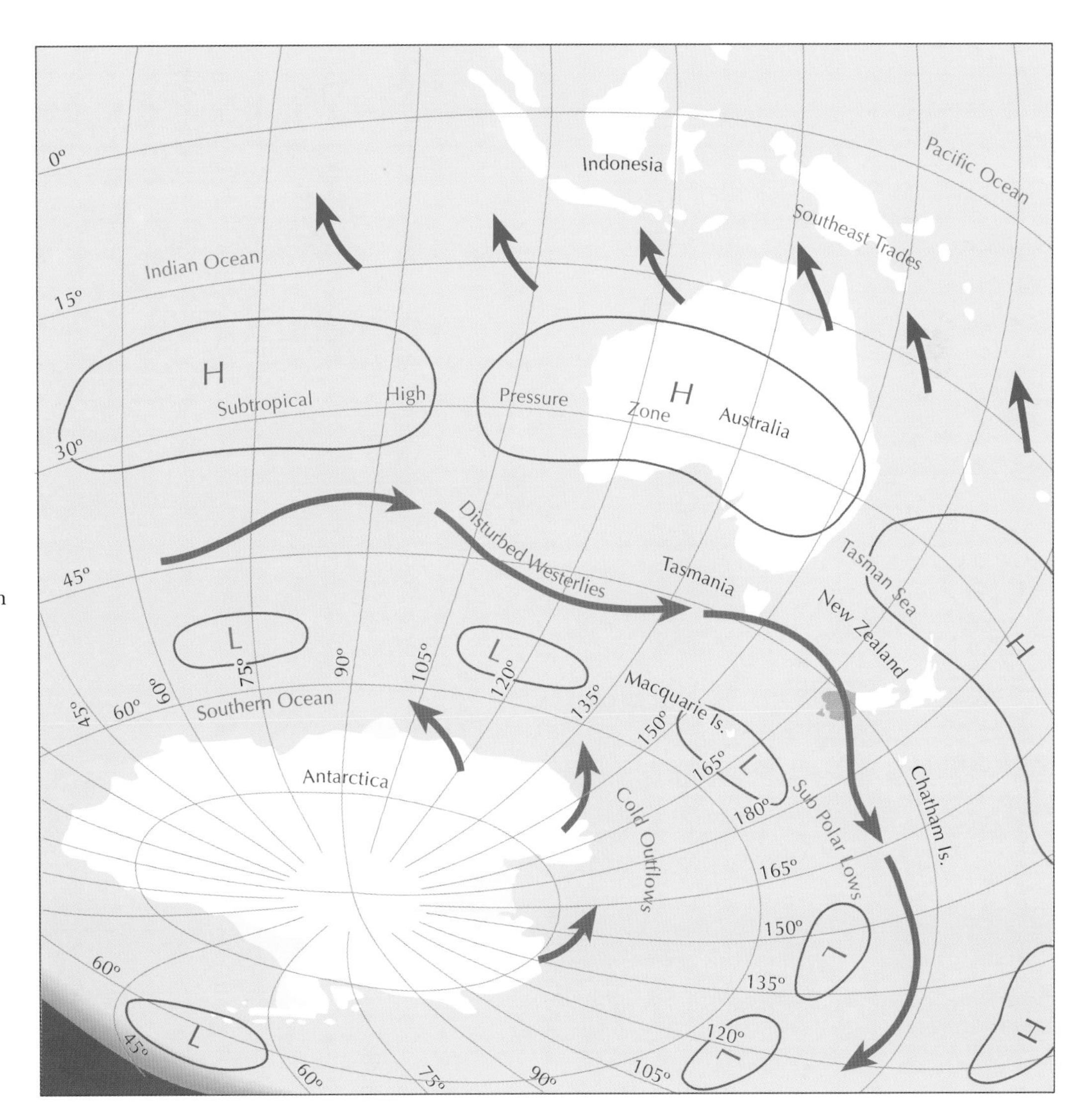

Fig. 3. Location of southern New Zealand with respect to the main features of south-west Pacific atmospheric and oceanic circulation patterns.

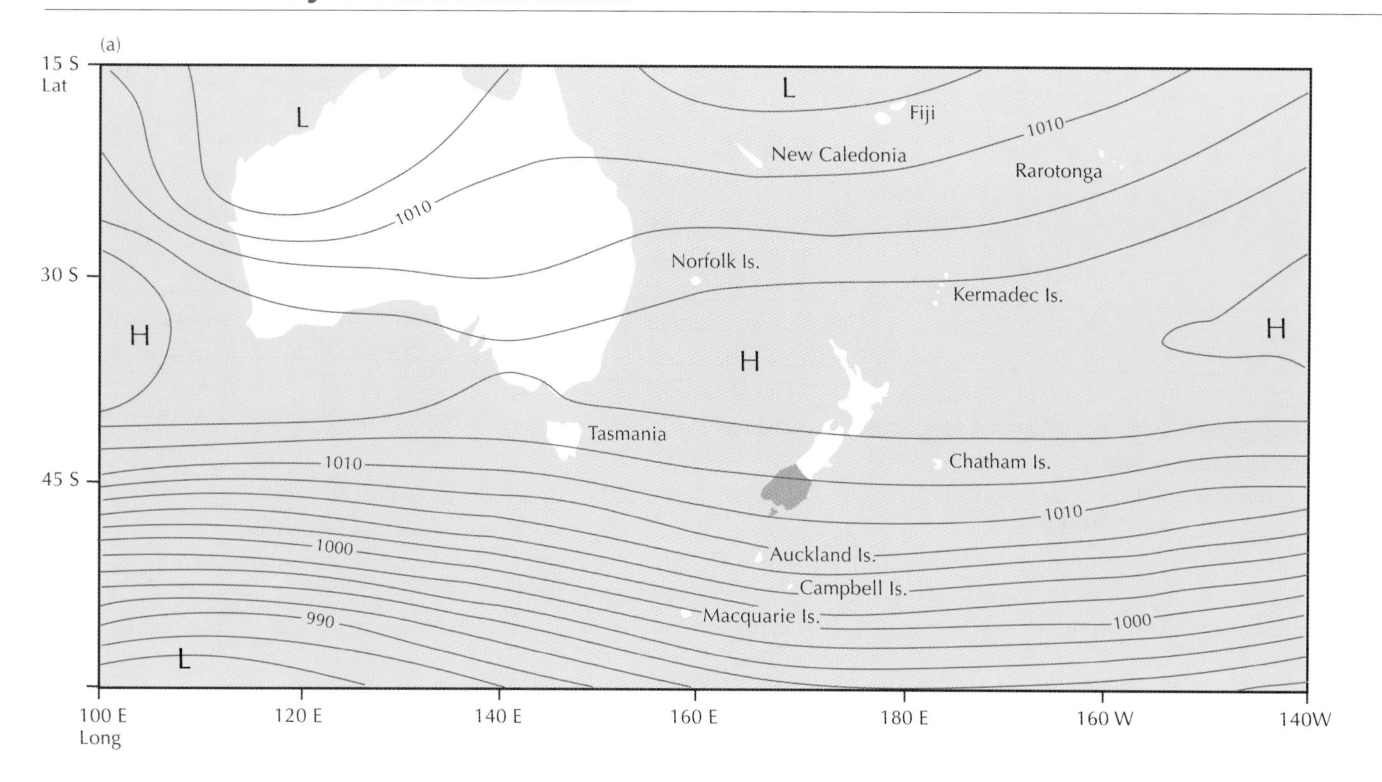

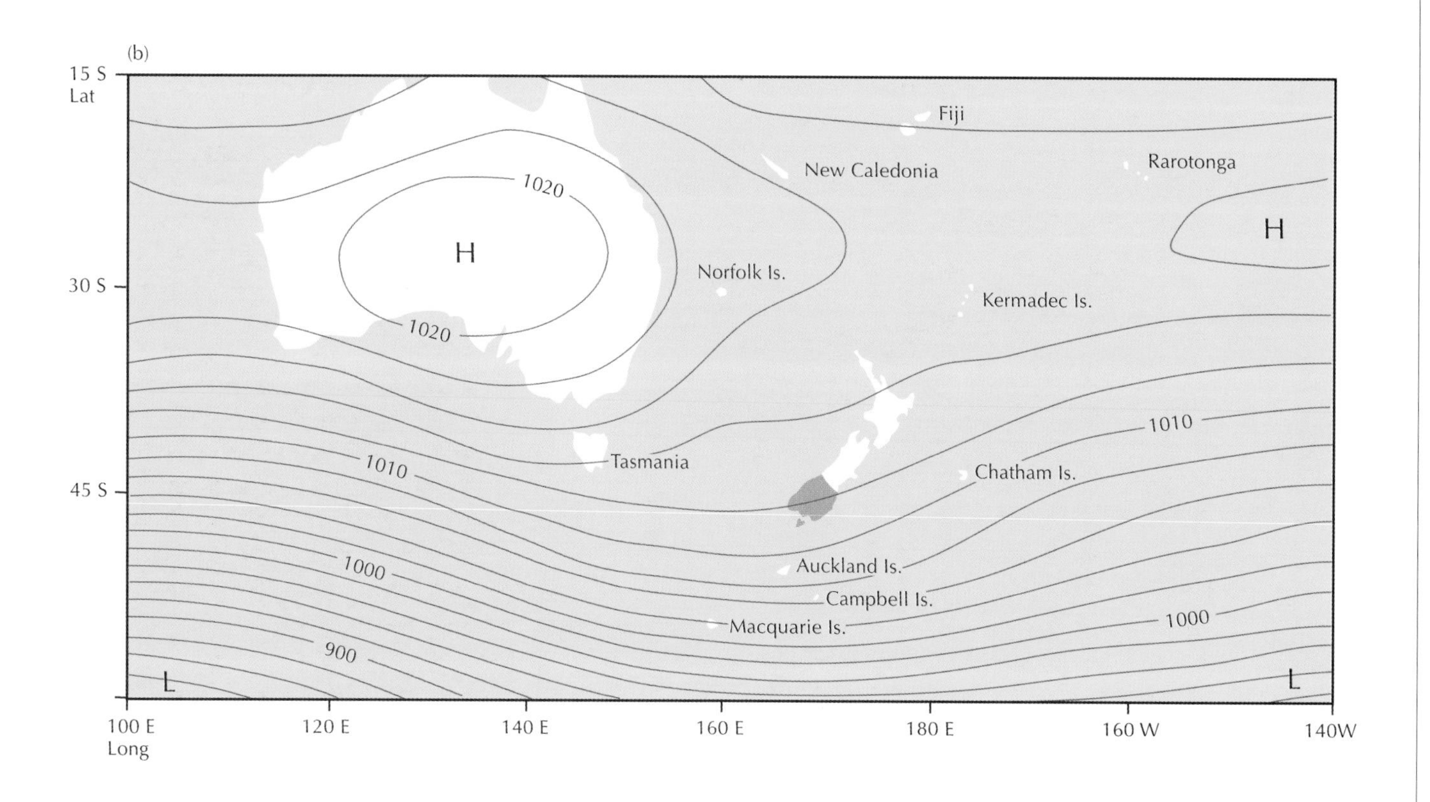

Fig. 4. Mean sea level pressure patterns over the south-west Pacific in winter (a) and summer (b). Isobars are in hPa.

Seasonal changes

These circulation features shift towards the South Pole in the summer and towards the equator in the winter, so that southern New Zealand is influenced by one or the other depending upon the season. However, the influence of the Australian continent complicates this general picture. It is such a large landmass that it distorts the pattern of wind belts, and generates its own pressure features. During summer it becomes much warmer than the surrounding oceans. This heated air rises, so that a thermal low pressure area is created at the surface in the interior of the continent. This tends to push the subtropical anticyclone south into the region of the Great Australian Bight, and from here a ridge often extends eastward over the Tasman Sea and the North Island (Fig. 4). The consequent rearrangement of the pressure field has the effect of strengthening the westerly winds over southern New Zealand.

Something quite different happens in winter. It is so dry and cloud-free over central Australia, that much radiation escapes to space, causing the surface to cool dramatically, especially at night. Consequently an extensive, dense layer of cold air develops in the lower atmosphere to form a thermal high pressure region. The thermal low of summer is replaced by an anticyclone over Australia. Down stream, over southern New Zealand, the air flow is forced to become more south-westerly and wind speeds tend to slacken. Thus, while the westerlies are further north in winter than in summer, they are often weaker.

Other influences

Southern New Zealand is surrounded by a vast expanse of ocean. Australia is 2000 km and Antarctica 2500 km away. As a result, all air masses that reach the region must travel several days over water. This greatly moderates their temperature. It is not surprising, therefore, that winter or summer, temperatures of coastal locations do not vary a lot. The region's location in the middle of a vast ocean also means that most air masses that arrive are usually heavily moisture-laden. When coupled with winds that blow against the mountains, so producing orographic uplift, the situation is ideal for producing frequent and sometimes heavy rainfall.

One important difference between the New Zealand climate and those of places in the Northern Hemisphere, say Great Britain, is the pervasive influence of the great ice mass of Antarctica. It generates enormous quantities of cold air that spill out over the surrounding Southern Ocean, regardless of season. Thus, any time the atmospheric circulation causes air to flow north from there, a cold air outbreak occurs. These frequently reach southern New Zealand as cold fronts. Since Antarctic ice melts little in summer, this advection of cold air from the south can occur in even the warmest part of the year, so that the mountains can receive fresh snow and frosts may occur at any time. Thus southern New Zealand locations tend to be cooler in summer than their counterparts at the same latitude in the Northern Hemisphere. On the other hand, its maritime location means that winters are milder.

A further factor that must be recognised is the role played by the collar of sea ice that encircles Antarctica. This expands substantially in winter, to lie within 1500 km of southern New Zealand. The time that cold air is heated by its passage over the oceans is very much reduced. Thus a sustained southerly change in winter can be very cold indeed, dropping temperatures close to freezing. Antarctica helps maintain a very strong latitudinal temperature gradient between it and Australia, the hot continent basking under the subtropical high pressure belt. The speed of the westerlies is directly related to this gradient. It is strongest in spring and early summer, when Australia is very warm and yet Antarctica remains cold. At these times of the year the westerlies strengthen to give equinoctial gales and enhanced rainfall west of the mountains. On the other hand, the latitudinal gradient is weaker in winter when the Australian continent cools, and therefore the westerlies tend to weaken. Thus many places in the west and mountains have a rainfall minimum at this time of year. Winter is often quoted as having more settled weather, and it is the best time to visit scenic gems such as Milford Sound and Mount Aspiring.

Day-to-day weather in southern New Zealand is controlled by the type of air mass. If the air mass has had its origins in the subtropical Pacific Ocean and is advected over the region from the north, then the air tends to be much warmer and moister than normal. Such an air mass produces wetter conditions, except in the lee of mountains, and increases atmospheric humidity. Contrast this with an air mass that has its origins over the polar ice regions. Initially very dry and cold, such air will be modified from below as it passes over an increasingly warmer ocean. When reaching southern New Zealand, it tends to be colder than usual, giving a 'raw' feel to the weather. It may have picked up some moisture, but on the whole does not usually produce near the rainfall of subtropical air masses, and may sometimes bring snow. Occasionally an air mass arrives directly over the Tasman Sea from the desert interior of Australia, and may be laden with dust or smoke from bush fires there, so that it leaves a reddish brown tinge to the snow fields. Such air is warmer and drier than normal.

The role of topography

The terrain and relief of southern New Zealand exerts a major influence on weather and local climates and helps to explain the tremendous variety in land use and landscape. The orientation of the mountain ranges, more or less athwart the westerlies, is crucial for the protection they offer both to eastern parts (for south-west to north airflows), and to the west (for south to north-east air flows). The interaction of the mountains with the predominant airflow direction means that the western part of the region tends to be much wetter than the eastern part. However, this pattern can be reversed with easterly weather. Gaps in the main axial ranges have important climatic implications because they funnel the wind. Alpine valleys, mountain passes, the broad block-faulted mountain crests of Central Otago and areas near to Foveaux Strait are all extremely windy. As the air descends in the lee of mountain ranges it warms and dries. The classic foehn wind, the nor'wester, is the result (Fig. 3). It is famed for its irritating gustiness, low humidity and extremely warm temperatures. At the same time, the nor'wester creates fantastic cloud forms known as lenticular clouds or 'hog's backs' and arches of cloud in the western sky. Because the nor'wester is a frequent phenomenon, it creates a marked west to east climatic gradient across the region, as suggested in Fig. 5. It is very wet in the west, and dry in the interior. Areas of regular subsidence in the lee of mountains form sunshine pockets, such as Cromwell, Alexandra, Queenstown and Omarama.

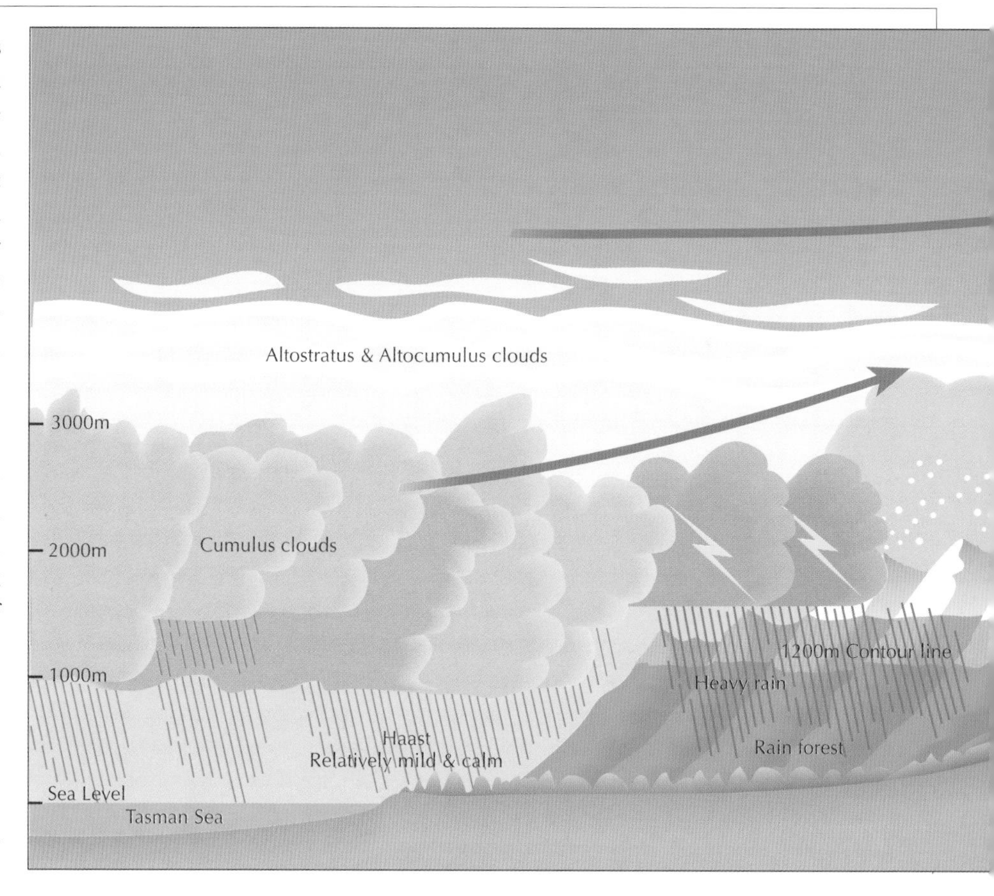

	Coastline	Roaring Billy	Western Mountains
Typical Storm Precipitation (mm)	140	200	440
Temperature (°C) Summer storm	16	14	7
Temperature (°C) Winter storm	8	6	-1
Wind Velocity (km/h)	10	20	50

Fig. 5. Cross-section of a typical Nor'wester storm.
Based on publications of Mt Aspiring National Park, Department of Conservation

Snow and ice

Perennial snow embraces less than five per cent of the region. Despite this small area, the glaciers of southern New Zealand are of wider global interest: they represent one of the few places in the vast area of the southern latitudes where their termini and end-of-summer snowlines are regularly monitored. The main glaciers are about Mount Aspiring and the Olivine Ice Plateau, although small ones are to be found in Fiordland. To the west of the Main Divide, the glaciers descend to lower levels because of the greater precipitation. Seasonal snow is highly variable, but the snow line is in the elevational range 1000–1500 m in winter, rising to 1500–2200 m in summer, depending on aspect and latitude. In winter, seasonal snow covers about 40 per cent of southern New Zealand. Normally, all of this melts each year, contributing 10–30 per cent to the annual flow of the larger rivers. Snow-covered areas generate much cold air on winter nights, which drains downslope across lowlands and plains, ponding in topographic depressions and becoming dammed behind even small ridges. Frost is frequent in such places, so that the winter and night climates of the plains and inter-montane basins are much cooler than would be the case if southern New Zealand were not mountainous.

Fig. 6. Spring landscape in mountains of Central Otago, showing melting snow patches and typical southwesterly sky. Scattered cumulus clouds have well-defined bases at about 2000 m elevation. *Blair Fitzharris*

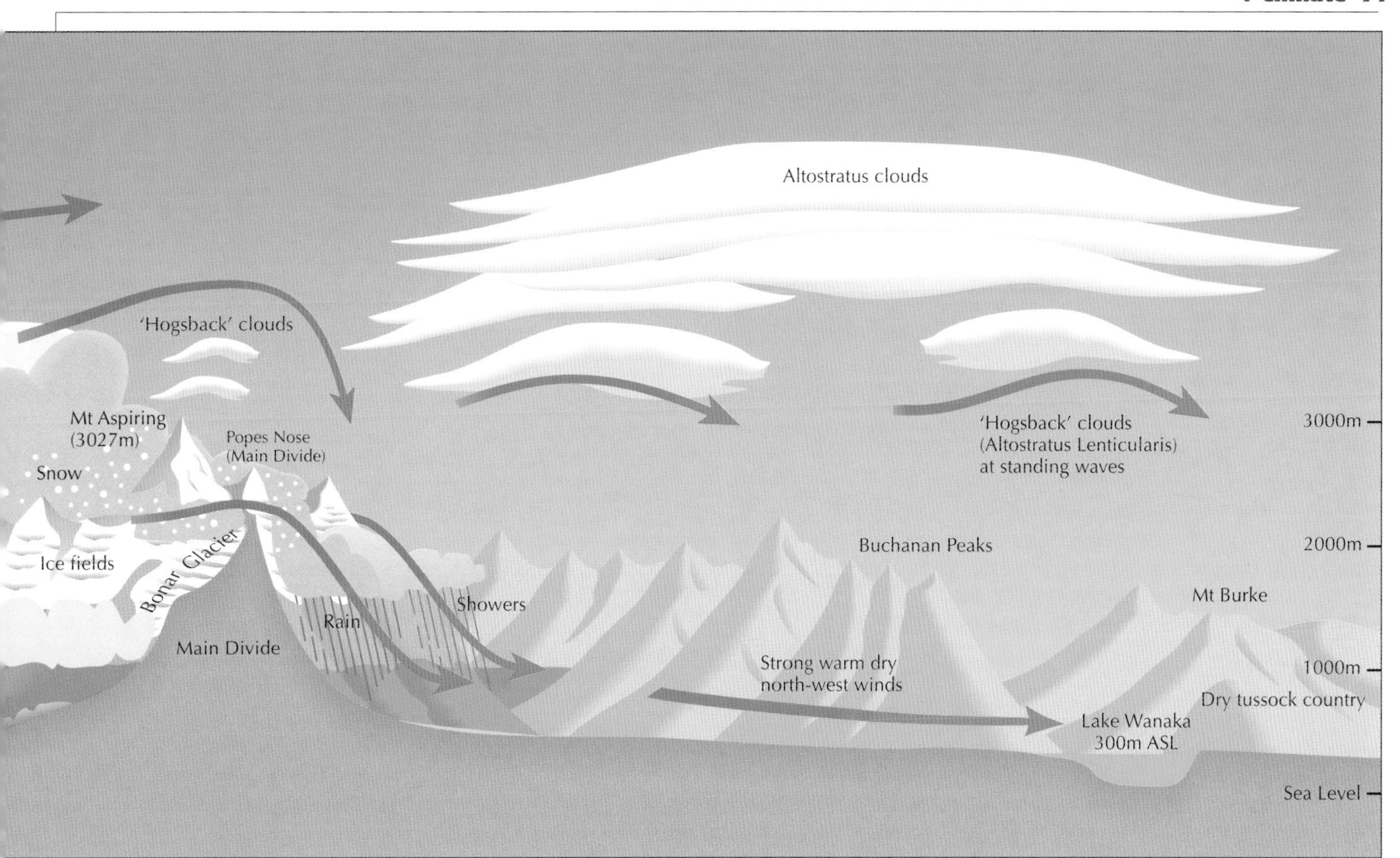

	Divide	Makarora	Wanaka
	320	80	20
	-2	18	24
	-8	6	16
	100+	45	30

Fig. 7. Typical 'sky scape', with strong westerly flow over mountains. Visible on the right are stacked lenticular clouds, while areas of blue sky occur where air is subsiding in the lee of the mountains. Obelisk Road, Old Man Range. *Blair Fitzharris*

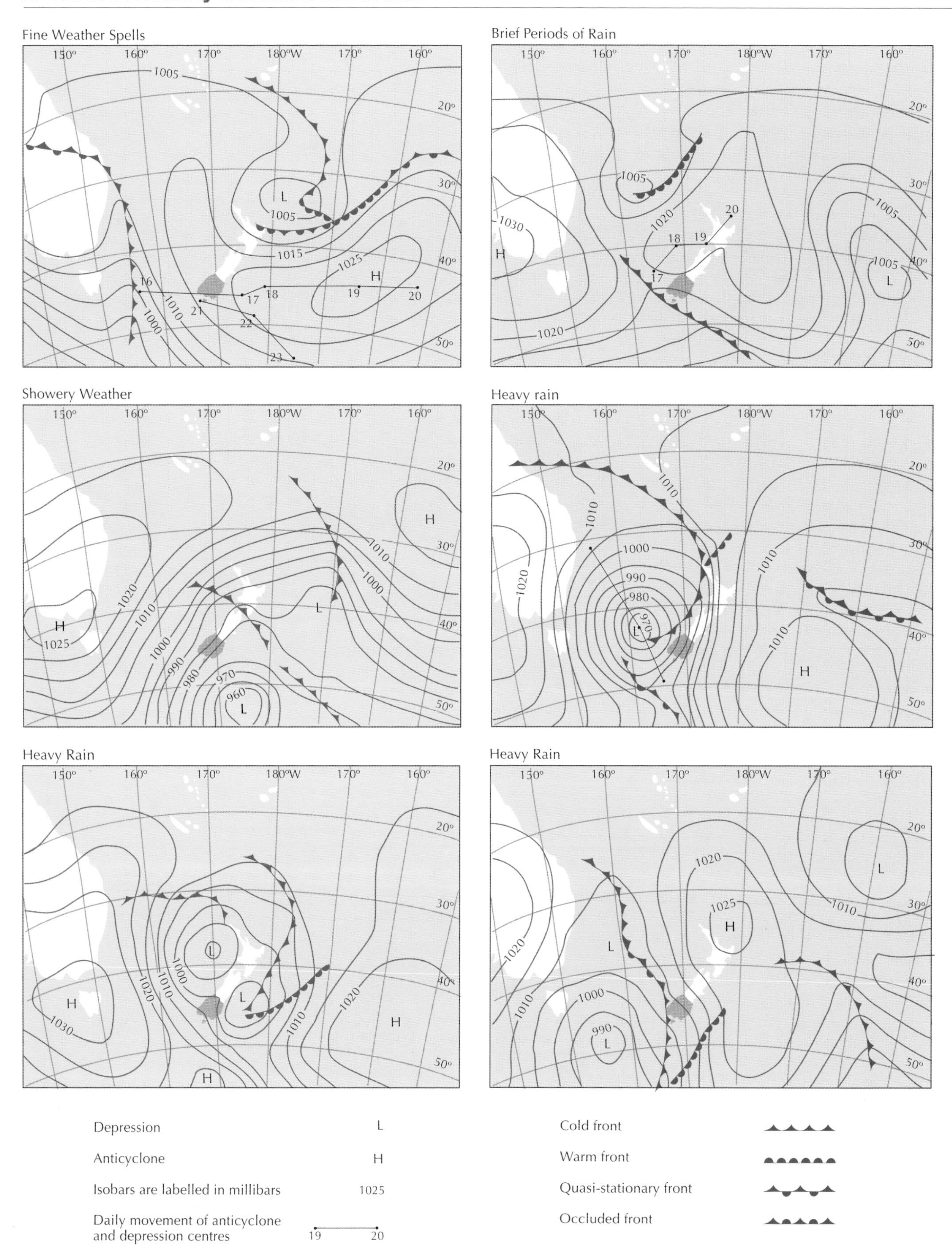

Fig. 8. Meteorological situations producing commonly occurring weather conditions in southern New Zealand.

Based on a publication of the New Zealand Meteorological Service

WEATHER SYSTEMS

On a day-to-day basis, the weather of southern New Zealand is determined by the relative positions of high and low pressure systems, their movement and whether they are in a process of development or decay. The combination determines our climate. At average intervals of six to seven days, the region is traversed by a succession of eastward-moving anticyclones, or ridges. The path followed by the centre of anticyclones is usually to the north of the region, but in summer and autumn it can sometimes be over the region. In a typical sequence, an anticyclone is followed by a trough of low pressure, often with one or more fronts. As a front passes over the region, it gives a period of increasing cloud and rain in the west. The front is preceded by fresh north to north-west winds, and followed by winds from a westerly or southerly quarter, and a lowering of temperatures. Substantial falls of rain occur, even in central Otago, when depressions lie over the district, or when fronts are especially vigorous. Certain commonly occurring weather conditions are produced by four basic types of weather pattern (Fig. 8). The weather conditions are:

Fine weather spells

Prolonged spells of fine weather of six days or more are usually associated with an anticyclone moving slowly eastward over the South Island. A ridge of high pressure may still extend over the region even when the anticyclone is centred far to the east or south-east. Fine sunny weather prevails, and winds are light, except for coastal sea breezes. Temperatures are several degrees above average in the daytime, with frosts expected in any month outside of summer. Frosts may be severe in winter.

Brief periods of rain

Cold fronts between anticyclones often cause an increase in cloud and a period of rain in the west. A typical case is shown in Fig. 8. North-west winds preceding a front are sometimes gusty and the temperatures rise. Rain may spread east from time to time, especially as the front comes closer and cloud thickens. Winds behind the front are usually from the west to south. Weather becomes cooler, with clearing skies in the west and inland areas, and there is risk of frost the following night. If post-frontal airflow is from a southerly quarter, the east is no longer protected by the mountains, so that it suffers cloud and showers.

Showery weather

The weather map in Fig. 8 shows a common type of situation giving showery weather over southern New Zealand. Inland, the showers are usually confined to the surrounding hills, where they may occur as snow. This weather pattern is usually caused when an anticyclone is almost stationary near Tasmania and pressures remain low to the south and south-east of New Zealand. A disturbed southwesterly airstream flows over the region. At intervals of about two days, cold fronts in the air stream pass over the region, freshening the winds and lowering temperatures further.

Heavy rain

Several distinctive synoptic situations give heavy rain to the region, as shown in Fig. 8.

Fiordland can get heavy rain with any well-developed west or north-west airflow. For heavy rain to fall in other areas, a depression passing over the region is necessary. Complex frontal systems are often involved. The weather remains cloudy for more than a day with steady rain. Fiordland may receive in excess of 400 mm from such a situation. Note, however, that rainfalls are still only moderate in Central Otago, often less than 50 mm.

SPATIAL PATTERNS OF MAIN CLIMATIC PARAMETERS

Solar radiation and sunshine

The distribution of mean annual solar radiation over southern New Zealand is shown in Fig. 9 for January, close to the time of its maximum. The west and southern coastal areas are amongst the cloudiest in the whole country. Along the eastern coast, solar radiation levels are reduced because of incursions of marine stratus associated with the frequent north-easter. Highest solar radiation levels are in the intermontane basins. The patterns very generally relate to sunshine hours experienced. In the far west, along the south coast and about Dunedin, annual totals of bright sunshine are less than 1700 hours. Inland they rise to about 2000 hours. Coastal North Otago and Haast receive approximately 1900 hours.

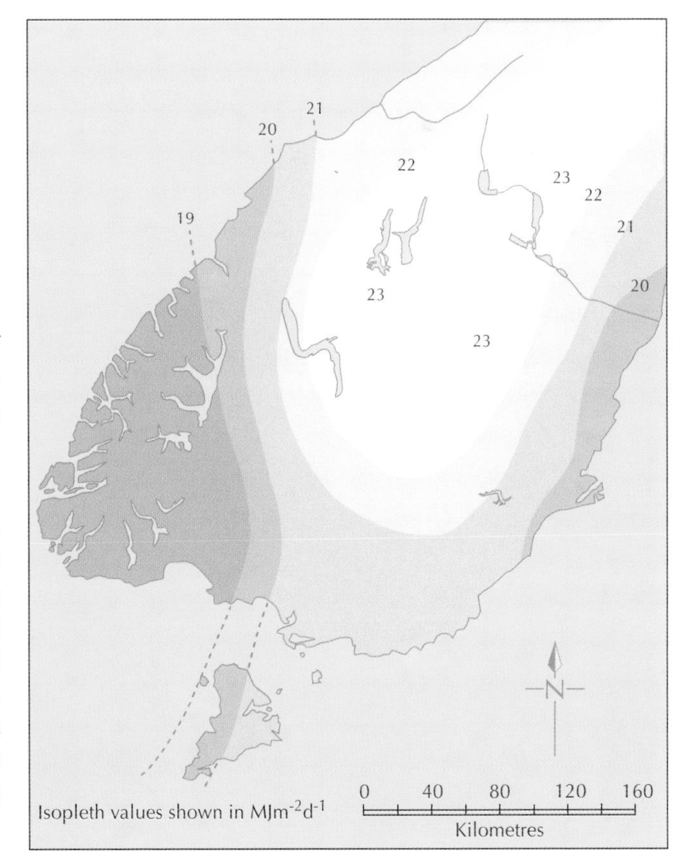

Fig. 9. Mean annual solar radiation patterns over southern New Zealand.

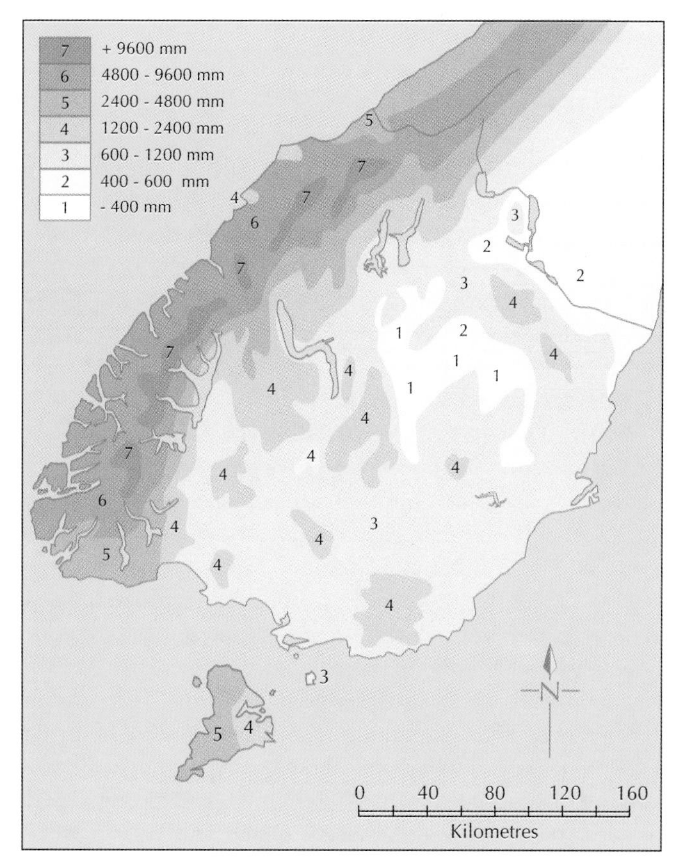

Fig. 10. Mean annual precipitation patterns over southern New Zealand.

Fig. 11. Mean annual rain day patterns over southern New Zealand.

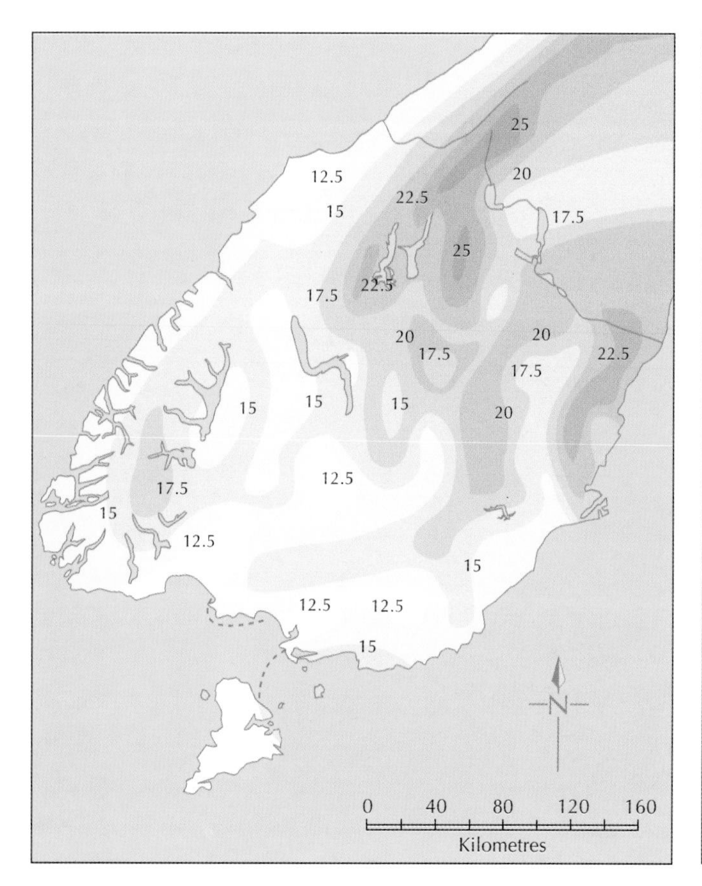

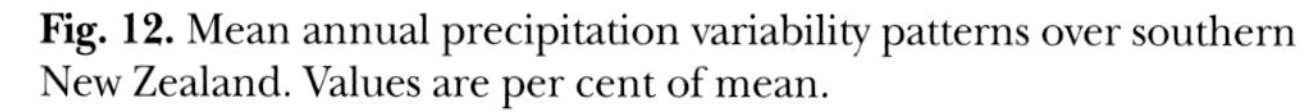

Fig. 12. Mean annual precipitation variability patterns over southern New Zealand. Values are per cent of mean.

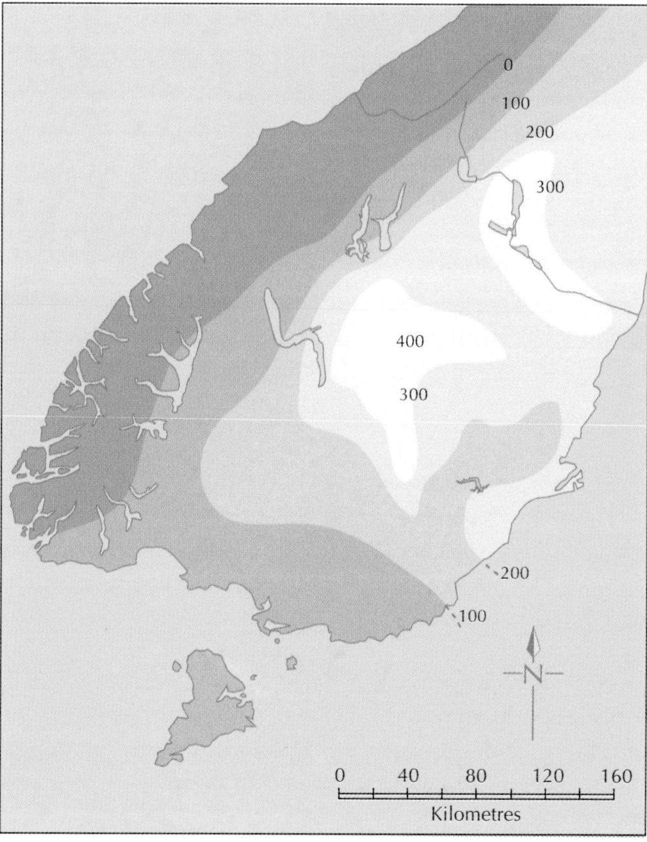

Fig. 13. Mean annual water deficit patterns over southern New Zealand. Values are mm per annum.

Fig. 14. The result of a heavy precipitation event in the western mountains of Otago: major flooding of Queenstown, November 1999. *Otago Daily Times*

Precipitation

Mean annual precipitation over southern New Zealand is shown in Fig. 10. Values show a strong west to east gradient. Lowest precipitation is in Central Otago (less than 400 mm). Highest is in the western mountains, where it exceeds 9600 mm. Extreme heavy precipitation events in these mountains can result in large inflows into Lakes Wakatipu and Wanaka, producing flooding of downstream towns (Fig. 14). For much of the east and south, annual rainfall is 600 mm to 1200 mm.

The mean annual number of rain days (days with at least 0.1 mm) show broadly similar patterns, as shown in Fig. 11. There are over 200 days with rainfall in the west of Fiordland and Stewart Island. Rain falls on less than 60 days in parts of the interior of Otago. Along the south coast, equivalent values are typically 160 days. Eastern coastal districts have 160 rain days in the south, declining to about 120 days near Dunedin and to 80 days at the Waitaki River.

Precipitation variability can be expressed as the coefficient of variability, which is the standard deviation of annual values expressed as a percentage of their means. As Fig. 12 shows, it is most variable about and east of Lakes Wanaka and Hawea, and in North Otago. Precipitation is least variable in the west, Stewart Island and the Catlins.

Water deficit

Annual water deficit can be calculated from a simple water balance model and daily precipitation data. Evaporation is assumed to take place according to the Penman potential rate until 75 mm of soil water storage is used. There is no annual deficit in Fiordland and western Otago (Fig. 13), so that soils here are almost always saturated. Near the lakes and along the south coast deficits are 100–200 mm. Largest deficits are in Central Otago, the Maniototo and the Waitaki Valley, where they exceed 300 mm. These are the largest to be found anywhere in the country, and show that growth of plants is severely constrained without irrigation.

Temperature

Given the large variability in other parameters, the pattern of mean annual temperature across the region is surprisingly uniform (Fig. 15). Warmest areas are coastal North Otago and about Dunedin. Coolest areas are in the mountains, although it should be noted that Fig. 15 is based on data from lowland climate stations. Hence, it is not a true reflection of temperatures for higher elevations above 1000 m. The biggest factor is elevation, so that many of the higher mountains have mean annual temperatures that are at least 7°C cooler than the valleys and coasts.

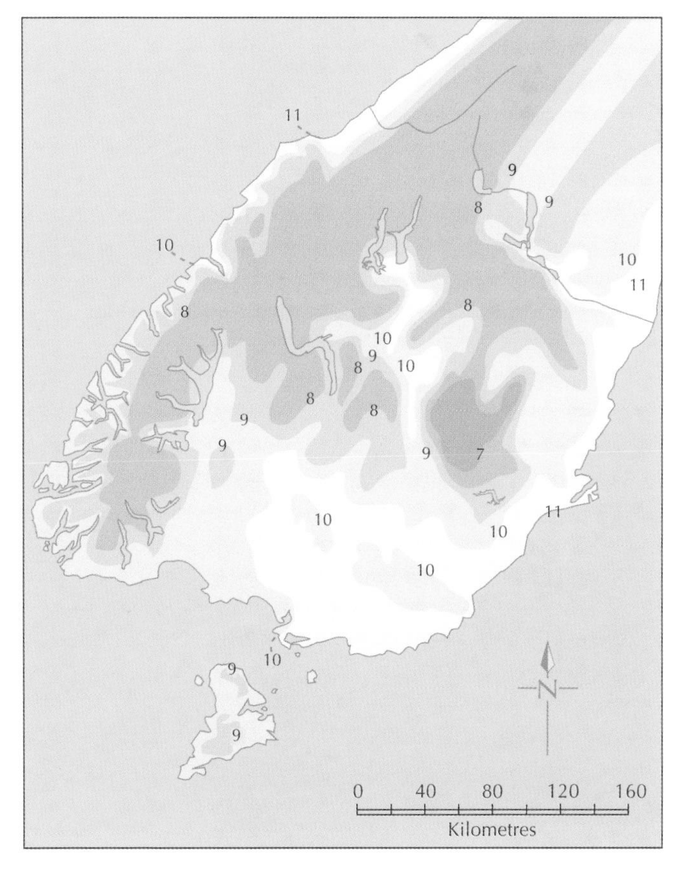

Fig. 15. Mean annual temperature patterns in °C.

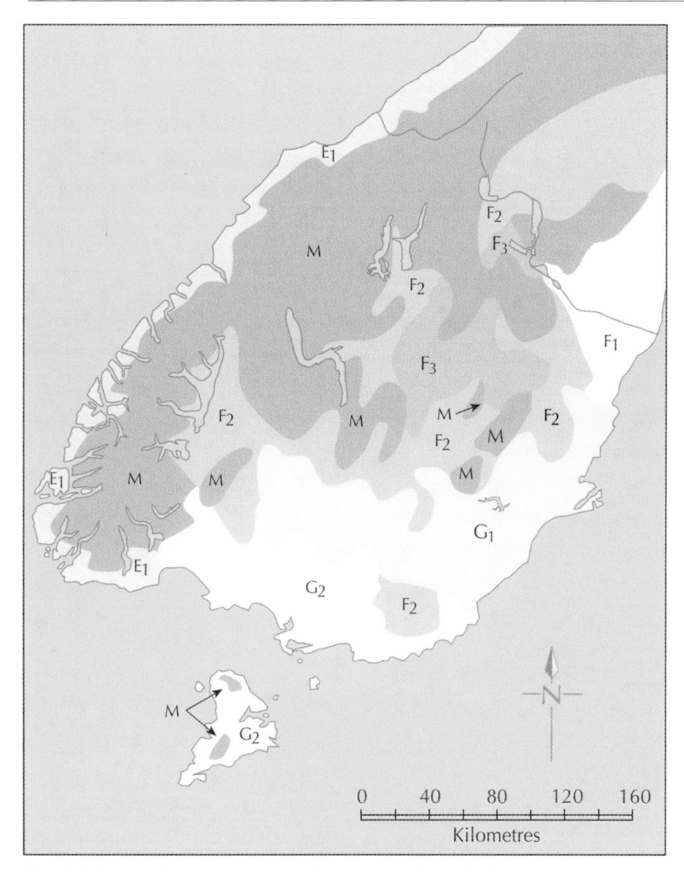

Fig. 16. Main climate regions of southern New Zealand.
Based on a map from the NZ Meteorological Service

Region and Station	90%ile	Mean	10%ile	Max/ day	Days
E1 Milford Sound	7643	6267	5067	520	182
F1 Oamaru Airport	716	537	425	88	76
F2 Queenstown	990	805	609	122	92
F3 Cromwell	477	401	319	64	68
G1 Dunedin,	963	784	619	229	120
G2 Invercargill	1201	1037	875	73	157
M Mountains	13,400	11,000	8,900	*800	200

* estimate

Table 1. Mean annual precipitation data (mm or rain days) for selected stations of the main climate regions.

Region and Station	January mean	July mean	Annual mean	Highest recorded	Lowest recorded
E1 Milford Sound	14.4	5.4	10.3	28.3	–4.9
F1 Oamaru Airport	15.1	5.5	10.6	37.7	–5.3
F2 Queenstown	15.8	3.7	10.2	34.1	–7.8
F3 Cromwell	17.7	2.9	10.8	36.6	–9.6
G1 Dunedin	15.0	6.4	11.0	34.5	–8.0
G2 Invercargill	13.7	5.1	9.7	32.2	–7.4
M Mountains*	6.0	-7.5	0.5	22	–20

* estimated values that will vary depending on elevation and distance from the coast.

Table 2. Mean annual air temperature data (°C) for selected stations of the main climate regions.

CLIMATE REGIONS

The complexity of the climates of southern New Zealand is illustrated by Fig. 16. Seven distinct regions are shown (these are modified from a map from the former Meteorological Service). A key climate station of each region has been selected and their long-term mean climate statistics are given in Tables 1–3. The main characteristics of these climate regions are as follows:

E1 Western Coastal Climate

The temperature range is small. High annual rainfall of greater than 5000 mm at the coast increases even more with elevation. Lowest rainfall is in winter. Rather cloudy with about 200 rain days. Prevailing winds are from the south-west, but gales are infrequent in spite of the exposed coastline.

F1 North Otago Climate

Annual rainfall is low, ranging from 500 mm to 800 mm. There tends to be more in winter than in other seasons. There are fewer than 100 rain days and severe droughts can occur. Summers are warm, with occasional hot north-westerlies giving temperatures above 30°C. Cool winters with frequent frosts and occasional snow. Prevailing wind is north-easterly.

F2 Hill Climate

These areas are cooler, cloudier and wetter than F1. Rainfalls average 800 mm to 1500 mm annually. North-westerlies predominate, with occasional very strong gales, especially along river valleys. Snow may lie for weeks in winter.

F3 Central Otago Climate

A semi-arid, semi-continental climate. Annual rainfall is 300 mm to 500 mm, with fewer than 80 rain days. Drought is endemic. Very hot and sunny summers and cold frosty winters. Foggy in autumn and early winter.

G1 Eastern Otago Climate

Moderate to warm summers and cool winters. Rainfall is 500 mm to 900 mm and evenly distributed throughout the year, but with a slight winter minimum. Rather cloudy. Winds tend to be from the south-west, or from the north-east along the coast.

G2 Southland Coastal Lowland Climate

Wetter than G1, with annual rainfall of 900 mm to 1300 mm. Generally windier, with frequent showers in coastal districts. Rather cloudy.

M Mountain Climate

The climate varies substantially, depending on elevation and proximity to coasts. Eastern and central mountains have annual precipitation of at least 1200 mm. In the west, precipitation is much higher, sometimes above 12,000 mm. Much of winter precipitation falls as snow and may lie on the ground for many months. Temperatures cool by about 0.7°C for every 100 m increase in elevation .

1 DROUGHT

The driest places in the country are in southern New Zealand. In numerous places, periods of 60 days without rain have been recorded. In Central Otago it is so dry that potential evaporation always exceeds precipitation, except for a few months in winter. Drought is almost continuous, and irrigation is required for productive agriculture. In North Otago, potential evaporation and precipitation are on average similar in magnitude. Small year-to-year variations can mean either a water surplus or drought (right). A chronology of drought for North Otago is shown in Fig. 17, based on 130 years of records.

A sustained major drought struck at the end of the nineteenth century, although droughts had been common until then. Drought was also severe every decade from 1910 to 1930, with each period separated by wetter than average spells. However, from 1935 to 1965, dry conditions were rarer and less severe. These are sometimes called the 'Green Years'. Fortunately they coincided with the return of soldiers after the Second World War, with subsequent settlement of hill country and rapidly expanding agriculture helping to fuel growth in the economy at that time. Then came the frequent and severe droughts from about 1965 onwards.

With climatological hindsight, we can see that the recent droughts represent a return to conditions typical of the past, but they came as a great shock to many farmers in North Otago (and to power planners and city water engineers elsewhere). Their perception of the rainfall climate was based on the Green Years, and was unrealistic. The record from the past shows that drought is endemic to North Otago: it will undoubtedly return in the future, and may even be increasing in its intensity.

Drought is not confined to the interior or the east. It can affect even the west of southern New Zealand, where conditions are usually wet. For example, in the first three months of 1965, Milford Sound recorded but 10 mm of rainfall, compared with its average of 1748 mm! Such dry periods have led to low levels in hydro-electric storage lakes and electricity crises, as occurred in the winter of 1992.

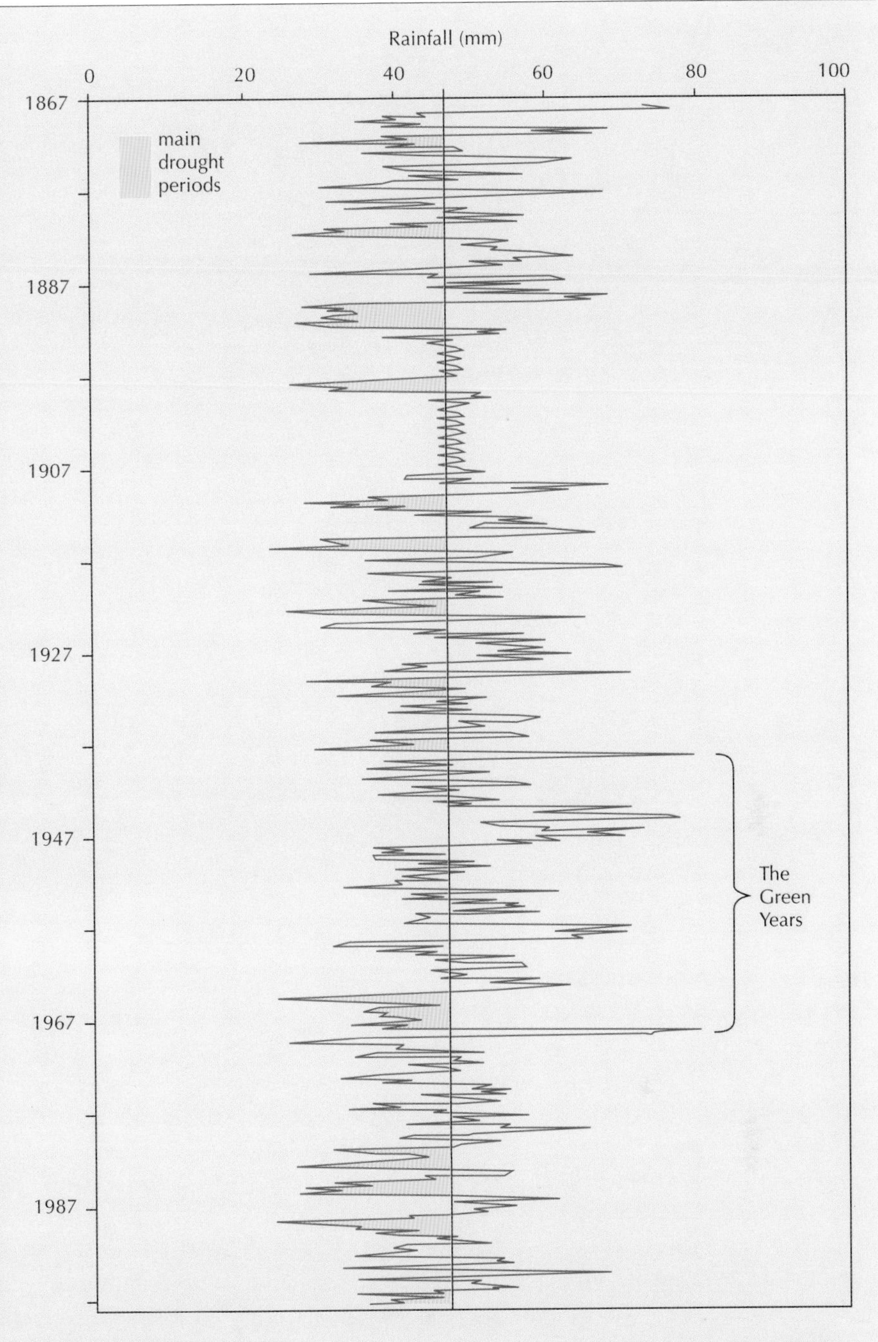

Incidence of drought as shown by rainfall variations in North Otago for the period 1867–1996. The vertical line is the long-term monthly mean rainfall. The curve is a 10-month moving average. Drought is indicated by sustained dips in this curve below the vertical line.

Region and Station (days of)	Ground frost	Air frost	Gales	Fog	Rel. humidity at 9 am (%)	Sunshine (hours/year)
E1 Milford Sound	56.1	28.5	5.7	2.2	87	1600
F1 Oamaru Airport	72.2	31.1	1.4	19.1	76	1857
F2 Queenstown	140.7	50.3	2.8	0.9	73	1921
F3 Cromwell	174.3	87.6	0.2	23.7	74	2060*
G1 Dunedin	77.7	9.7	2.9	5.7	74	1685
G2 Invercargill	111.1	46.3	10.3	50.0	83	1621
M Mountains*	—	292	50	200	92	1200

*estimate

Table 3. Mean annual climate data (number of days) for selected stations of the main climate regions.

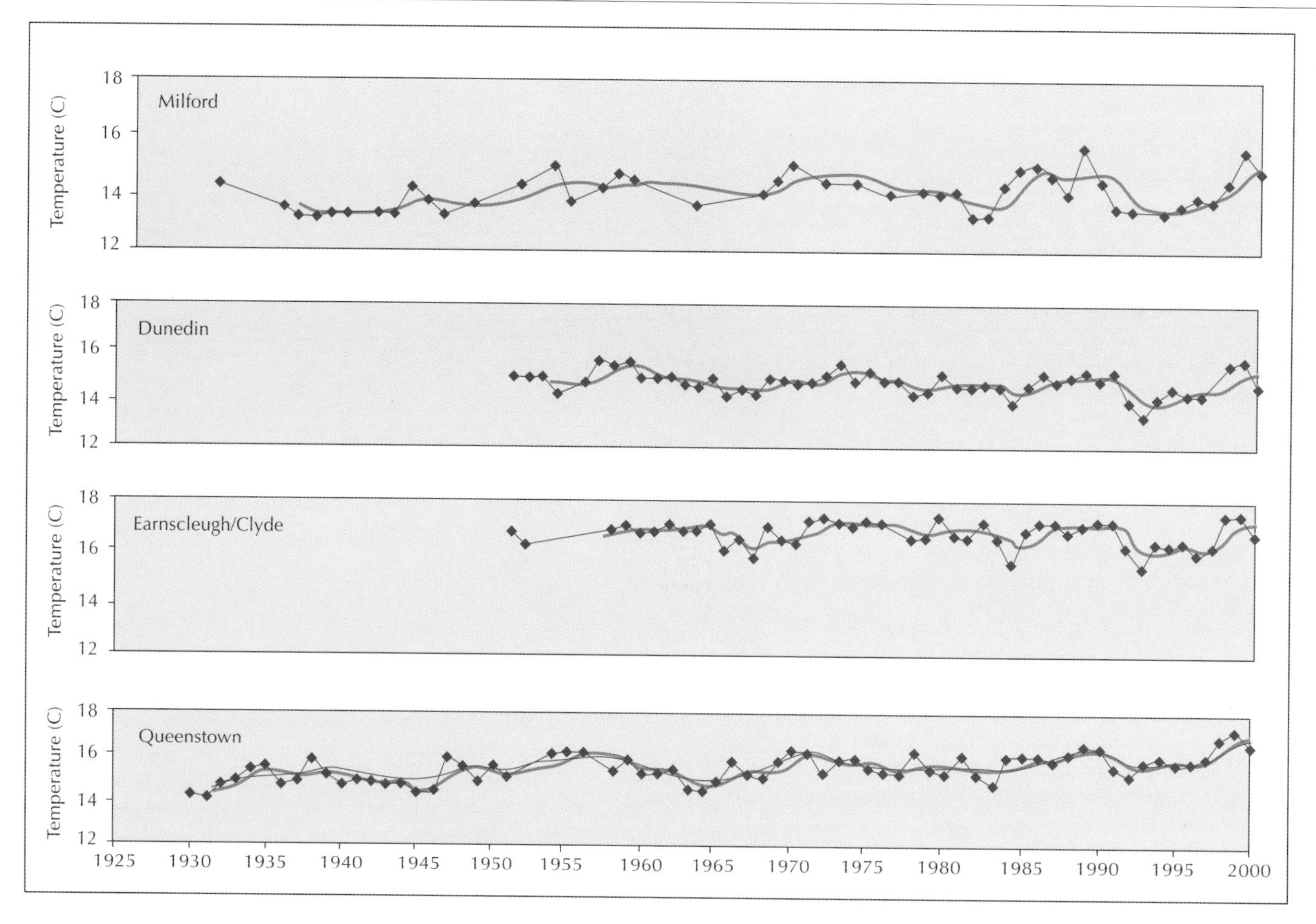

CLIMATE VARIABILITY

PAST VARIATIONS AND TRENDS

Figures 17–20 show variations in key climate variables over a long period of time. Information is presented for climate stations with long records and to reflect the broad climate gradient from west to east. The data have not been corrected for site changes, so discontinuities sometimes occur in the graphs. For example data are for Earnscleugh up until 1983, when that station closed, and thereafter for Clyde Dam.

Annual values of mean daily maximum (day) and minimum (night) temperature show that there is only small variation from year to year (Fig. 17). Overall, there is a long-term trend of warming, although at Dunedin and Earnscleugh/Clyde, the days appear to be getting cooler and the nights warmer. At Queenstown, both day and night-time temperatures have warmed.

There is a clear trend of fewer days with air frost at Queenstown, Earnscleugh/Clyde and Dunedin (Fig. 18). These declines in frost frequency are substantial and are probably associated with changes in atmospheric circulation patterns and night-time cloudiness.

There is no long-term trend in annual rainfall (Fig. 19), although Central Otago stations have become wetter since about 1980, probably due to an increase in the westerlies and El Niño. While not cyclical, the moving averages show that there are quasi-periodicities in the rainfall record, with groups of years tending to be wetter or drier than usual.

Number of fog days is highly dependent on the vigilance of the observer, so trends over time must be examined cautiously. However, there does seem to be an increase in fog days at Earnscleugh/Clyde since about 1965, and this trend may be associated with the creation of hydro lakes and irrigation (Fig. 20). However, fog has also increased at Dunedin. Milford was foggier than usual about 1940 and in the 1950s and 1960s.

Ozone depletion

Over the last part of the twentieth century the stratospheric ozone layer was significantly depleted due to human activity. Ozone monitoring shows a decline in ozone concentration since the 1980s (Fig. 21). In only two of the 120 months between 1985 and 1995 did ozone levels exceed the average of 1970s levels. If continued, these changes will increase UV-B radiation from the sun, and could have impacts on the biota of both natural and managed ecosystems. A drop of ten per cent in total ozone concentrations increases UV-B radiation at the surface by an estimated 20 per cent.

Possible causes of climate variability

A marked feature of the above analyses is that the climate varies considerably from year to year and even from decade to decade. For example, the last quarter of the twentieth century saw periods with persistent drought followed by calamitous floods. During 1988, eastern areas suffered dry conditions, sheep lost condition and had to be tracked to wetter areas, and river flow dwindled. One year later, farmers prayed for the rain to stop. Are such changes part of the natural climate, or are they freak occurrences? Research shows that sequences of weather

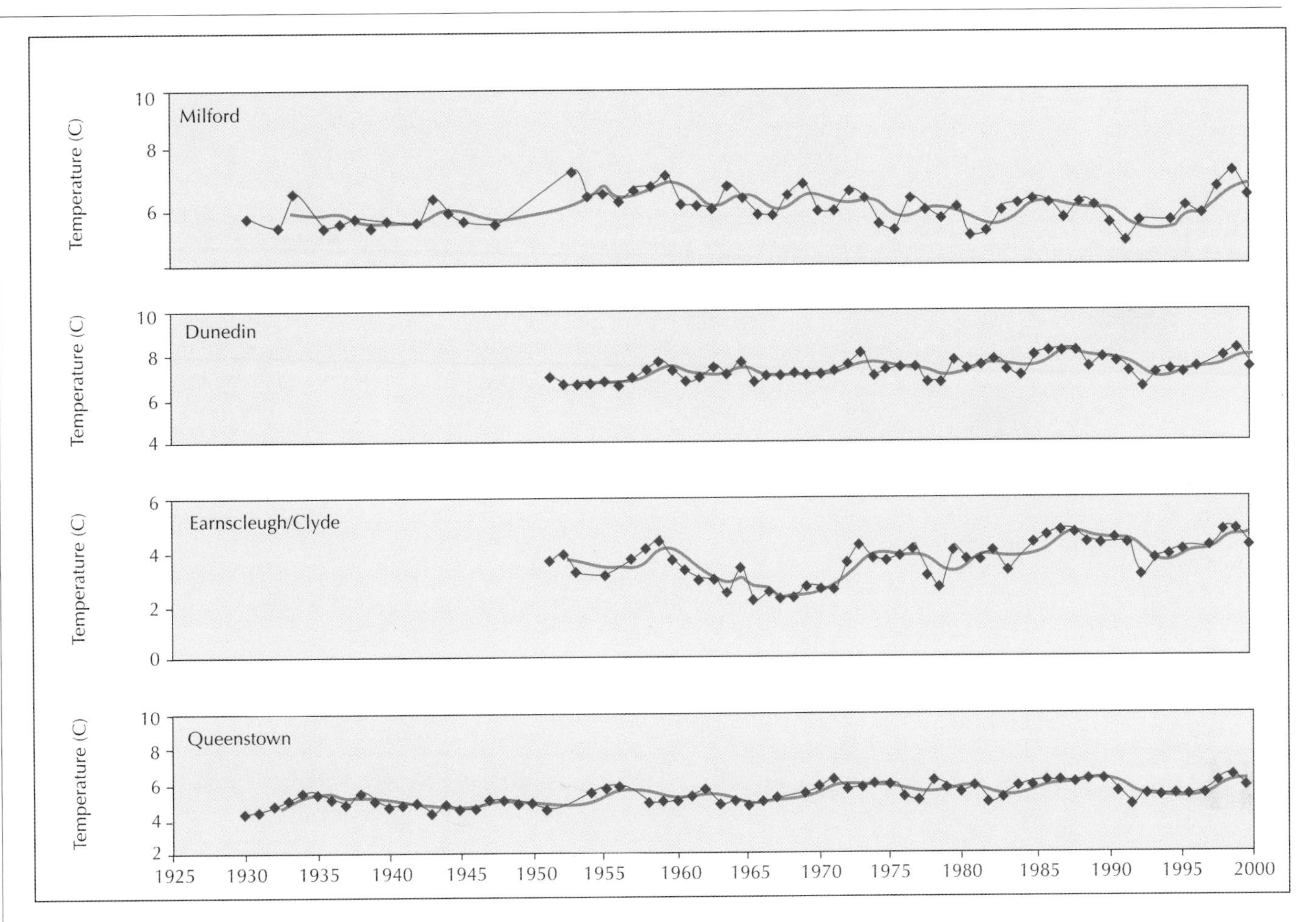

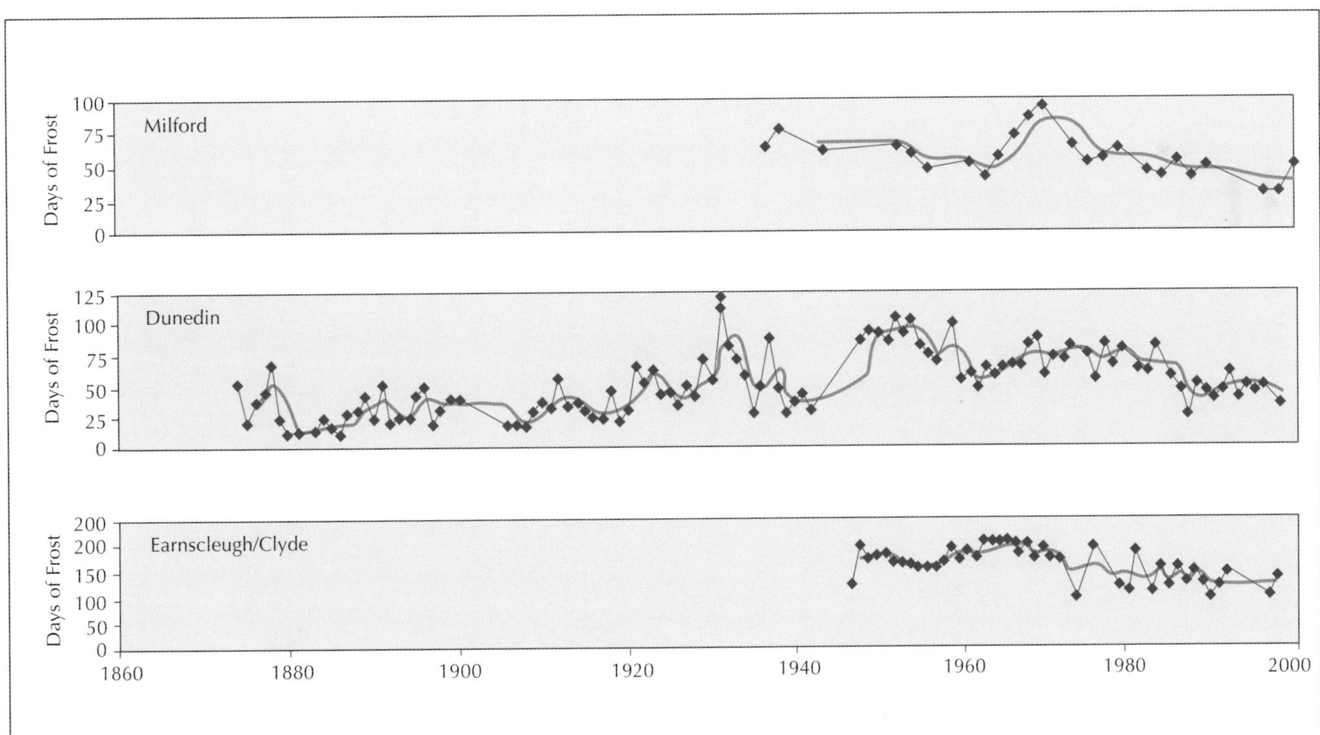

Fig. 17 (top left and right). Long-term variations in annual mean maximum, or day-time, temperatures (top left) and minimum daily temperatures, or night-time, temperatures (top right). Values are shown for each year and as a five-year running mean. *Data from NIWA*

Fig. 18 (above). Long-term variations in annual number of days of ground frost and five-year running mean. *Data from NIWA*

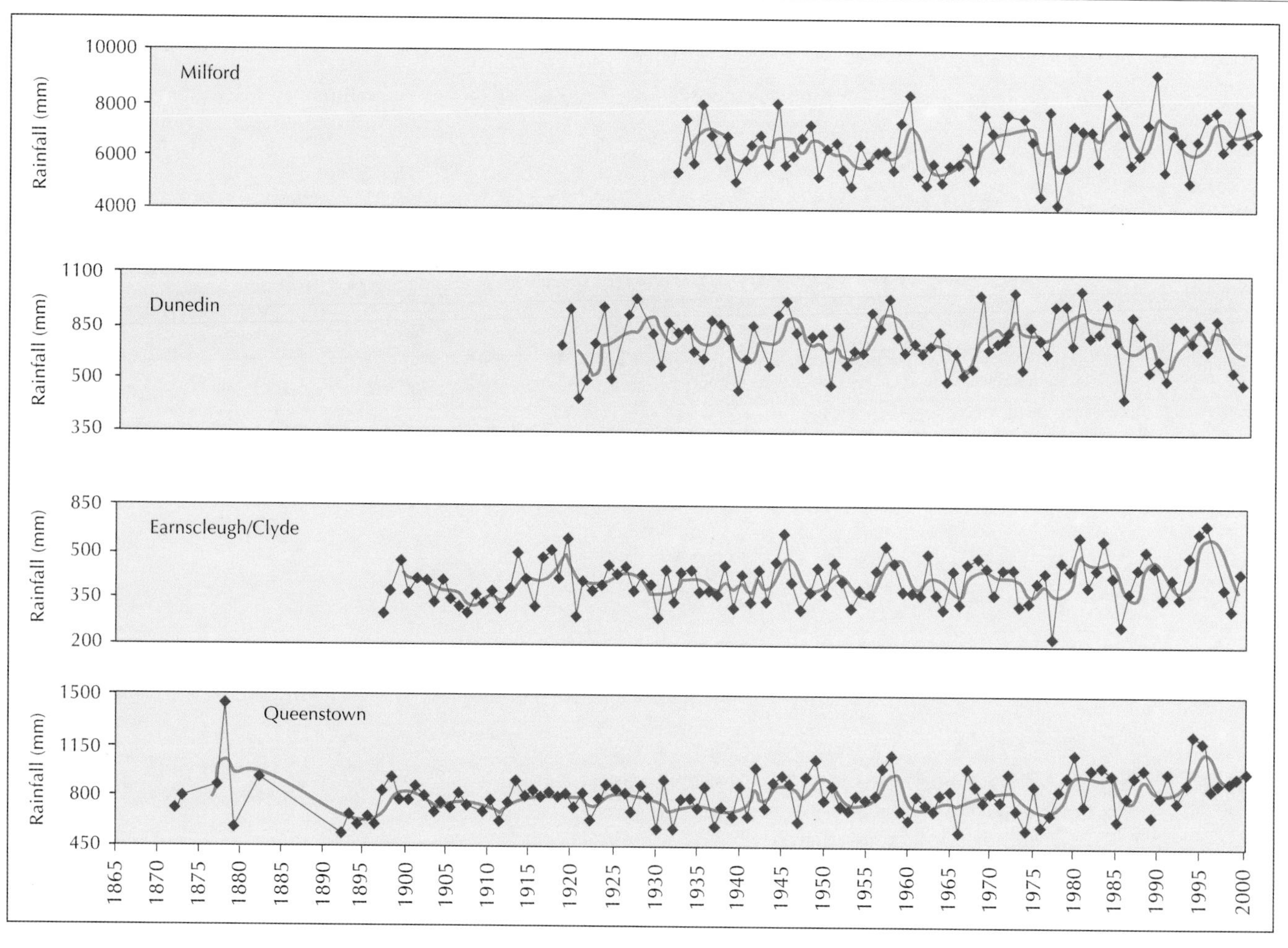

Milford
Dunedin
Earnscleugh/Clyde
Queenstown
Rainfall (mm)
10000
8000
6000
4000
1100
850
500
350
850
500
350
200
1500
1150
800
450
1865
1870
1875
1880
1885
1890
1895
1900
1905
1910
1915
1920
1925
1930
1935
1940
1945
1950
1955
1960
1965
1970
1975
1980
1985
1990
1995
2000

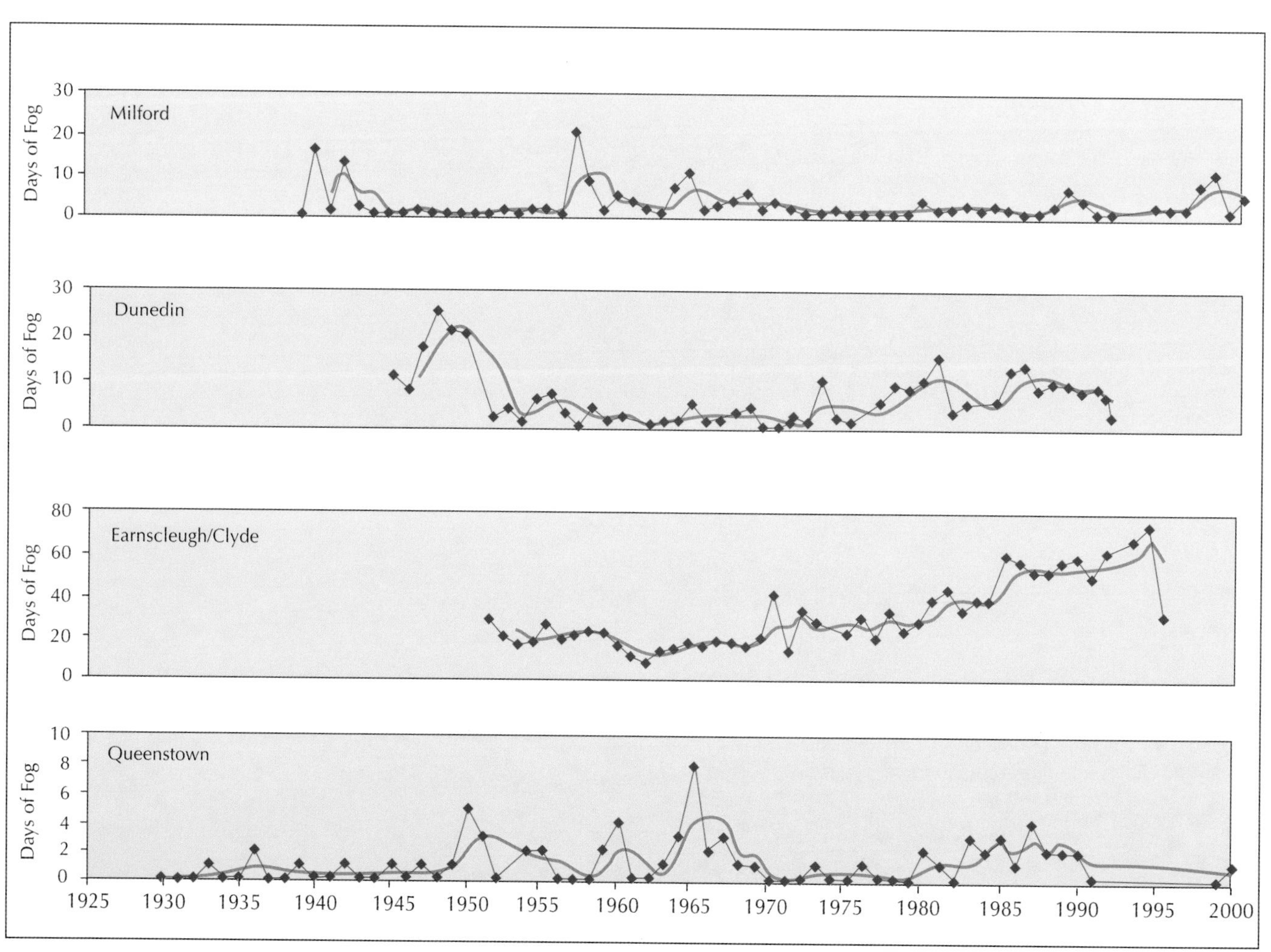

Milford
Dunedin
Earnscleugh/Clyde
Queenstown
Days of Fog
30
20
10
0
80
60
40
20
10
8
6
4
2
1925
1930
1935
1940
1945
1950
1955
1960
1965
1970
1975
1980
1985
1990
1995
2000

Fig. 19 (opposite top). Long-term variations in annual rainfall and five-year running mean. *Data from NIWA*

Fig. 20 (opposite bottom). Long-term variations in annual number of days with fog and five-year running mean. *Data from NIWA*

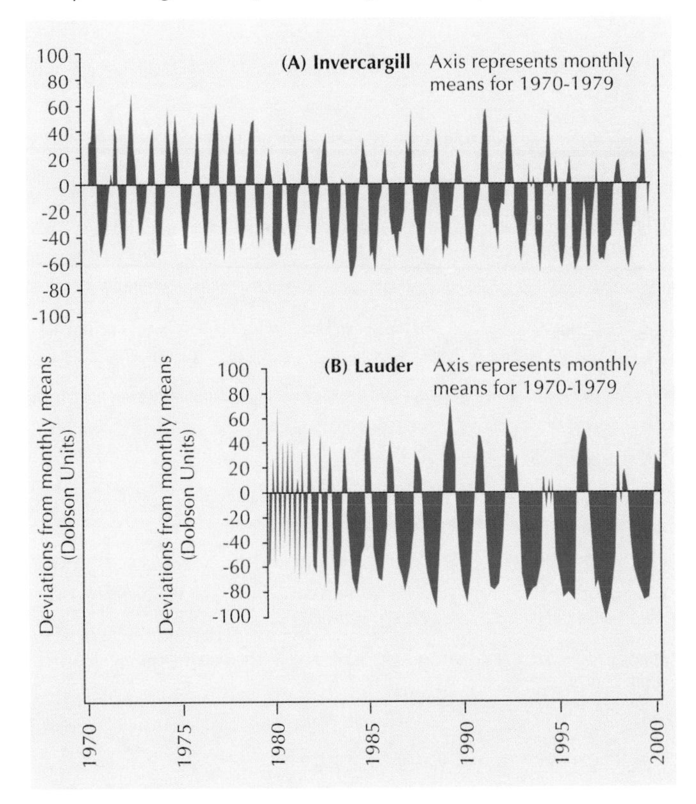

Fig. 21. Deviations in monthly mean ozone for 1970–2000, Invercargill and Lauder records combined. *Data from NIWA*

can persist for many years, then suddenly operate with a different frequency. The climate of southern New Zealand is influenced by a series of wider global influences, many of which operate as 'quasi-periodicities'. These are not cycles, because they do not occur in exactly regular form, but they cause considerable variability over time. Some of these influences are not fully understood, but those known to be of consequence follow.

The Southern Oscillation

In recent years, climatologists have discovered that an important modulating effect on weather and climate comes from the influence of the Southern Oscillation, and its two extreme conditions known as 'El Niño' and 'La Niña'. The Southern Oscillation swings from one extreme to the other, as can be seen in the index of its status based on the standardised pressure difference (Darwin minus Tahiti; Fig. 22). This is the SOI. When the SOI is strongly negative an El Niño event occurs, and when it is strongly positive a La Niña is indicated. However, the periodicity of the changes in the index vary, as does its amplitude.

The spatial correlation pattern of sea level pressure with the SOI shows a remarkable linkage across wide areas of the globe, pointing to what is known as teleconnections (Fig. 23). When the pressure is anomalously low over the equatorial central and east Pacific, it is high over a large area of Australia and south-east Asia. Pressures are anomalously low in the area to the south and east of New Zealand. This is the El Niño phase of the SOI. The pressure patterns of Fig. 23 reverse during La Niña events. That is, lower than usual pressures then occur over northern Australia and south-east Asia. At the same time higher pressures are found in the other mentioned areas.

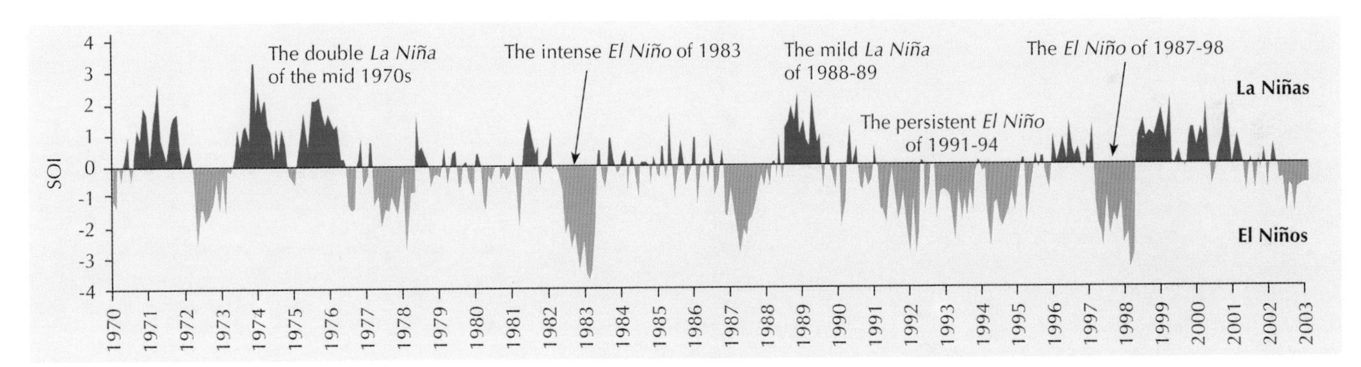

Fig. 22 (above). Variations of the Southern Oscillation Index (SOI).

Fig. 23. General pressure anomaly patterns during a typical El Niño event. Units are hPa.

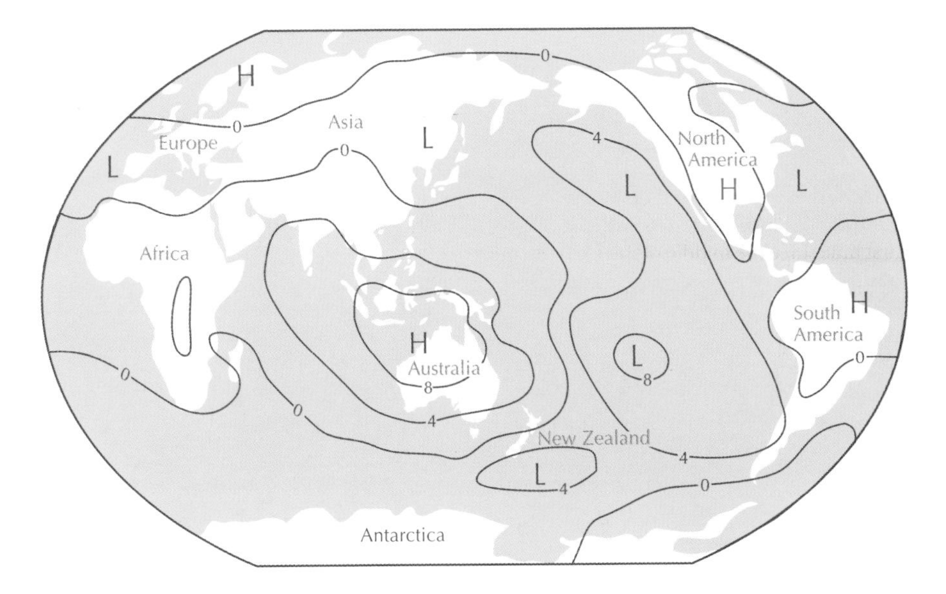

Southern New Zealand thus lies in a nodal region between climatological centres of action associated with the see-saw of the SOI. With this insight, it comes as no surprise that its weather is strongly dependent on the phase of the Southern Oscillation. When the SOI is negative, the pressure distribution indicated in Fig. 24 ensures that more south-westerly winds occur than is normal. During this El Niño condition, rainfalls tend to be above average, especially in the west and south, and it is cooler than normal. When the SOI is positive, there is a higher frequency of north-easterly winds. During these La Niña events, the overall southern New Zealand rainfall tends to be lower than average, especially in the south and west, and it is warmer than normal in much of southern New Zealand.

The quasi-biennial oscillation

A phenomenon known as the quasi-biennial oscillation (QBO) has a cycle of 25 to 28 months and is most marked over the tropics, where stratospheric winds reverse from east to west. In a complicated series of processes, it is postulated that atmospheric circulation over New Zealand displays a flip-flop from southerly to northerly tendencies, depending upon the phase of the quasi-biennial oscillation. Consequently the phase of the QBO exerts an effect on rainfall. The QBO also causes fluctuations in ozone levels in the stratosphere.

The solar cycle

Sunspots vary markedly in number over time, with a cycle of approximately eleven years. It is suggested that at times of high sunspot numbers, rainfall increases. There is some evidence to suggest that the flow of the region's largest river, the Clutha, increases about these times.

Volcanic eruptions

During a major volcanic eruption a substantial volume of dust is spewed into the stratosphere, where it resides for about four years. This diminishes the receipt of solar radiation at the surface and leads to cooling. Thus, in 1992, temperatures over southern New Zealand were substantially below average following the Mt Pinatubo eruption in the Phillipines.

The trans-polar index

The behaviour of the circumpolar westerlies may be a factor causing climate variability, albeit a poorly understood one. This great vortex of strong winds may possess a long-term wobble. From time to time it expands or shrinks, so that the westerlies either dominate or are to the south of the country. Such a wobble would exert control over the intensity of frontal activity. More intense fronts from expanded westerlies bring more rain. When the westerlies contract, fronts are weak or less frequent and it is usually dry.

Sea surface temperatures

Extensive areas of the ocean around New Zealand are sometimes a few degrees warmer or cooler than normal for time of year. These temperature anomalies represent vast changes in energy exchange between atmosphere and ocean and are known to exert profound influence on the climate of southern New Zealand. For example, during El Niño and after major volcanic eruptions, the sea surface temperatures are cooler than normal: this helps lower temperatures on the land. During La Niña, the opposite is often true. Sea surface temperature anomalies can also play a role in steering weather systems and in their intensification.

2 EL NIÑO AND THE SOUTHERN OSCILLATION

In the summers of 1982–83, southern New Zealand experienced the most pronounced El Niño for over a century. The SOI was strongly negative, and the impacts were deeply felt among farmers, fruitgrowers and holiday-makers. The south and west of the region suffered a bleak, wet and cold summer. Some dubbed it a 'Clayton's summer' – the one you are having, when not having a summer.

In the east during El Niño, drought grips the land. Everywhere the wind blows and blows, usually from the south-west. In the summer 1988–89, the weather was in complete contrast. La Niña exerted its influence for the first time since the middle of the 1970s. In this positive phase of the Southern Oscillation, the east is often wet, while the south and west of the region are dry. It was so much warmer than normal in 1988–89, that many people talked about global warming as having already come to southern New Zealand. Southland farmers did their lambing rounds in slippers, while the ski fields languished for lack of snow.

To understand what caused these marked perturbations in climate, the bigger global picture must be considered. The Pacific Ocean is so large that it creates its own circulation system that operates between South America and Indonesia. This east-west cell is known as the Walker circulation. Part of it is the south-east trade winds that blow towards the equator in the South Pacific. Over the western Pacific, where the ocean is warmer, the air gains moisture and heat, and rises. After this rising, much of this aloft air flows east, to form the completed circulatory cell. In its normal mode, the Walker circulation is characterised by low pressure over northern Australia (say, as measured by a barometer at Darwin) and high pressure over the eastern Pacific (say, as measured by a barometer at Tahiti).

Because of interactions between the ocean and the atmosphere, a disturbance occurs in the Walker circulation every two to seven years: this is known as the Southern Oscillation. During El Niño events, the ocean waters off the Peruvian coast suddenly rise by about 3–5°C. El Niño is produced by a weaker Walker circulation, which may even reverse, so that westerlies occur in the tropical Pacific, rather than the more normal easterly trades. The distribution of pressure across the Pacific reverses. It becomes low over Tahiti and high over Darwin (Fig. 23) as dry warm air subsides over the western Pacific.

During La Niña events, the Walker circulation is as for the normal case, but much more intense. Ocean waters off Peru are much colder than normal, while in the Indonesian Sea waters are warmer than normal. Pressures are lower than normal at Darwin and higher than normal at Tahiti.

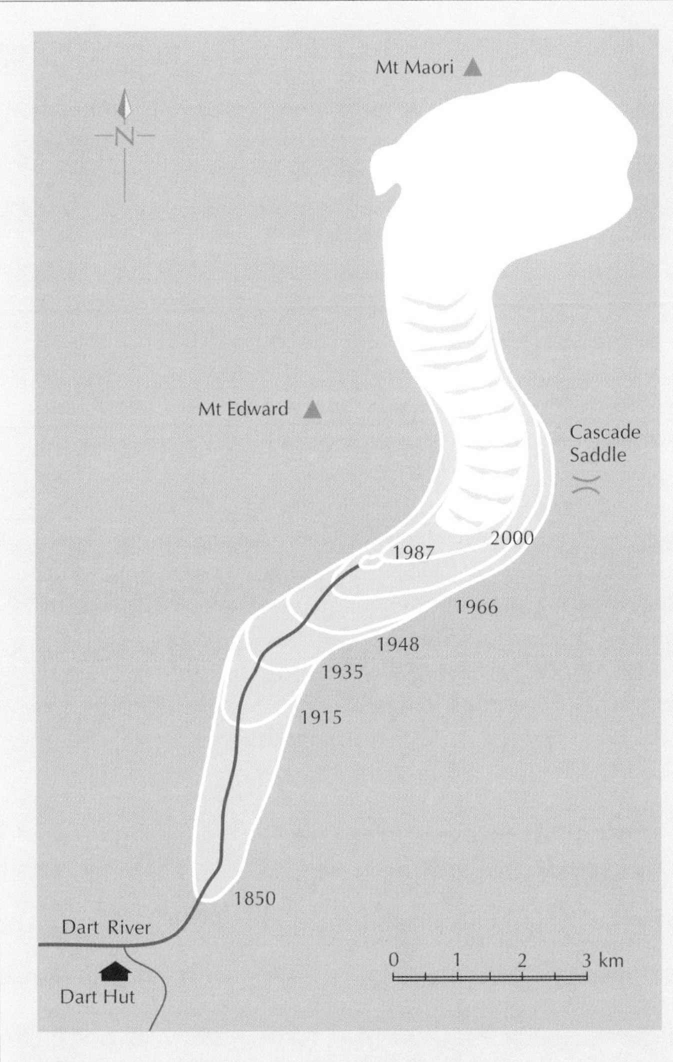

Positions of the Dart glacier terminus since 1850.
Based on a figure from Vanishing Ice *by Graham Bishop and June Forsyth (John McIndoe Ltd and DSIR)*

CLIMATE CHANGE

Southern New Zealand has undergone substantial climate change in the past. About 18,000 years ago, the region was about 6°C colder than at present. Glaciers reached beyond the present coastline in Fiordland and to the southern end of most of the inland lakes. About 12,000 years ago, rapid warming was underway. Glaciers retreated rapidly, and as warming continued, sea levels rose by up to 130 m. Subsequently, there have been several periods when the climate was both warmer and colder than that at present (see Chapter 6). The most recent cooling of the 'Little Ice Age', that began in the seventeenth century, led to a marked advance of glaciers. When the first Europeans explored the Southern Alps during the second half of the nineteenth century, they considered that the glaciers were either still advancing, or close to their maximum extent. Confirmatory evidence has come from vegetation studies on well-developed terminal moraine loops. This suggests the instrumental record, which began first in Dunedin in 1852, captured the tail end of a major cold period.

In the twentieth century, there was a warming trend of between 0.5°C to 0.8°C. This was detected in rural climate stations, far removed from the effects of urban heat islands and is consistent with known behaviour of sea surface temperatures, as measured by ships in southern waters for more than 150 years. Precipitation shows a number of quasi-periodicities, as outlined above, but no secular trend. Warming in southern New Zealand has had a devastating effect on most alpine glaciers. They have lost about one third of their volume and rapidly down-wasted. For example, a chronology of the position of the terminus of the Dart Glacier derived from photographs and trim lines shows that it has retreated by nearly 5 km and the ice surface has lowered by over 100 m since 1850 (Fig. 25). Some of the smaller glaciers in Fiordland have completely disappeared.

3 GLACIERS, SNOW AND CLIMATE CHANGE

The health of an ice mass is indicated by its mass balance. When positive, the glacier receives more snow in its neve region than it loses by melt at its terminus. If sustained over decades, the glacier grows longer and thickens. When the mass balance is negative, it will shrink. Over most of the twentieth century, glaciers of southern New Zealand had average mass balances that were negative, which accounts for their ice loss (above). However, there is evidence that mass balances are now positive in some years. No consistent measurements are available as to the volume or extent of seasonal snow over the century. However, model simulations suggest that it has been highly variable from year to year, but with no clear trend.

Glacier size and snow cover extent are controlled by climate, but the relationships are complex. While there is a significant positive correlation between retreat of the Dart Glacier and temperature, with a lag time of a few years, the link with precipitation variations is less clear. By contrast, glaciers to the west of the Main Divide seem quite sensitive to variations in precipitation. For much of the twentieth century, they both advanced and retreated, behaviour which is intimately related to the strength of the westerlies over southern New Zealand.

Different weather situations generate different amounts of energy for snow and ice melt. Radiation is important during southerly circulation patterns. Convection transports heat and condensation from the air and produces the most rapid melt. This is important with northerly circulation patterns and in combination with heavy rain events leads to large floods in Otago and Southland rivers. Solar radiation is an important source of energy for snow and ice melt during the passage of anticyclones.

Research findings link ablation of the glaciers over the last four decades with known circulation changes. There was a trend for more north-easterly and fewer south-westerly flows across the country from about 1950 to 1980, due largely to an increased tendency for anticyclones to persist east of New Zealand, rather than over Australia and the Tasman Sea. The band of subtropical high pressure also migrated south. This created weather sequences that favoured higher than usual summer melt. Since about 1980, there have been more El Niño events, expanded westerlies and fewer anticyclones. These changes have reduced the summer melting of snow and ice. As a result, some glaciers have begun to thicken or even advance again and, in some years, seasonal snow fails to completely melt before the following winter.

FUTURE CLIMATE

Global warming is underway. Latest assessments of the Intergovernmental Panel on Climate Change confirm that due to the impact of human-emitted greenhouse gases into the atmosphere, this warming will accelerate. Climate changes in southern New Zealand by the year 2100 may be considered by examining a scenario of an increase in temperature of 3°C, an increase in precipitation of 10–15 per cent and a rise in sea level of 0.5 m. This scenario is not a forecast, but a general indicator of one possible future. It is generally in line with that proposed by the Intergovernmental Panel on Climate Change, although there remains considerable uncertainty about changes in precipitation.

Effect of the climate scenario on natural systems

Most glaciers will recede at an even faster rate than at present, and many of the smaller ones will disappear. It is estimated that the ice volume of the region will shrink to less than half the present value. The snow line will rise 300–400 m, with a decrease in snow accumulation below 2300 m, but an increase higher up. The volume and area of seasonal snow will be halved.

These changes will have a big impact on hydrology of major rivers such as the Clutha, Waiau and Waitaki. Seasonal flow will change markedly, with an estimated 40 per cent more flow in winter and 13 per cent less flow in summer, although the latter could be partly offset in the Waitaki by melt water from rapidly diminishing glaciers. Annual runoff could increase by about 14 per cent. On smaller mountain streams, which are vital for local irrigation, the lower snow storage will probably lead to decreased summer flows, when some will dry up. Floods will be larger because the hydrological cycle is expected to become more intense. Thus the 'probable maximum precipitation' will increase. As the waters about southern New Zealand warm, there will be an increased likelihood of subtropical depressions and their attendant heavy rainfall.

The rise in snow line and general warming will produce a corresponding rise in ecological elevation bands. For example, in the west, alpine species will colonise emerging snow-free areas and the bush-line will be higher, while in the eastern hill country the range of tussock species will be enlarged, although they will face increasing competition from introduced grasses at elevations between 250 m and 500 m. Exotic tree species such as *Pinus radiata* will grow at higher elevation and extensive areas of the middle-elevation hill country would then be available to them, both for wild seeding and as managed plantations.

Vegetation boundaries are complex mosaics of plant communities in dynamic equilibrium, so that some may respond only slowly; their migration across the landscape may lag behind the climate changes by decades. Increased carbon dioxide concentrations in the atmosphere will enhance plant productivity and reduce water stress, an important consideration for plant communities at higher altitudes and on the semi-arid margins.

Any change in climate of the magnitude of that proposed will have a profound effect on day-to-day weather. Because the nature of the atmospheric circulation in the New Zealand area for the scenario is unknown, it is difficult to detail what these weather changes will be. However, coastal weather is likely to be more pleasant, if only because the ocean will be 3 C warmer in all seasons, so that the southerly and easterly winds which lower human comfort levels on many days will be milder. A big disadvantage to living in southern cities like Dunedin and Invercargill will then be alleviated.

The predicted sea level rise will have little effect in the west, where the coastline is generally resistant to erosion, although some protective works may be necessary at the head of Milford Sound. In the east, there are substantial lengths of 'soft' coastline and many important areas that could be flooded. Coastal hazard mapping considers 360 ha of the South Dunedin and St Kilda urban area will be at risk. Farmland on the Taieri Plain (6000 ha) and Dunedin Airport, as well as the lower Clutha (5500 ha) and about Southland estuaries, are also liable to inundation.

Warming of ocean waters will produce complex, but little understood, changes in the marine ecosystem. Warmer water species of plant and fish life may replace colder ones. The distributions of rock lobster, wet fish, shell fish and squid are all expected to alter, but the impact on the fishing industry is difficult to determine. Marine mammals and birds will also be affected.

Effects of climate scenario on resources and economy

The current economy of southern New Zealand is dominated by the primary industries of hydro-electricity production, pastoral farming (mainly for meat and wool), grain and horticultural production, tourism, forestry, mining and fishing (Table 4). All these activities are resource based and weather dependent. Export trade per head of population is well above the national average, so these activities make an important contribution to the wider New Zealand economy. Thus, any change in future climate will have important implications.

Increases in flow of major rivers will enable the output of hydro-electricity to be increased from existing generating plant. The changes in seasonal runoff (more in winter, less in summer) should also be beneficial, in that the ability to generate during the peak winter load time will be increased. Furthermore, the load itself will tend to be smaller than today (by an estimated 10–15 per cent in relative terms) because warmer temperatures

Table 4. Main production sectors of the economy of southern New Zealand and estimated changes with the climate scenario discussed in the text.

Sector	Estimated % change
Energy generation	+10
Pastoral farming	+20
Tourism	-10
Fruit and grain	+300
Forestry	+20
Mining	0
Fishing	+5

will mean less electricity demand over the whole country. Water storage in lakes may become less important, and new operating rules will need to evolve to acknowledge the changed hydrological regime and loading factors. Economically, these considerations might result in a 10 per cent increase in electricity supply from existing plant. Together with the reduced load, this will defer the need for new power stations. Most new hydro construction is slated for the Clutha or Waitaki Rivers, so benefits of the extra work to the regional economy will be delayed for a further couple of decades.

The climatic scenario will produce only small increases in pasture growth over lowland Otago and Southland, where productivity is already high. However, efficiencies will be gained because the growing season will be extended and winters will be milder. There will be less winter feed required. Larger benefits for agriculture are expected in the middle-elevation hill country, where more than 500,000 ha of tussock country will have less winter snow risk and a longer growing season. This land could then support intensive pastoral farming, and potentially lead to 700 new farms in an arc stretching from West Dome to Lawrence to the Shag Valley. Higher country, presently in snow tussock or herb fields, may also support more summer grazing, although the exposed tops of the block mountains will remain inhospitable, and strong conservation lobbies will seek to preserve their shrinking plant communities.

Fig. 24. Future climate: Dunedin as the Riviera of the Southern Ocean? *Otago Daily Times*

A major expansion of horticulture should be possible with the projected warming. At present only 13,000 ha, mainly in Central Otago with a small outlier in the Waitaki Valley about Kurow, is classified as part of the South Island 'warm zone'. This is defined as having November to April growing degree days of at least 800 for a base temperature 10°C, and is very suitable for growing high-quality stone fruit. This zone will expand to include most of the lowlands of the province and, as shown in Table 5, the climate will also become suitable for grapes (growing degree days greater than 900). At the same time, winter chill accumulation that is essential for crops such as cherries, apricots, and nectarines will remain sufficient (winter chill index of more than 30 in Table 5) to ensure successful fruiting, except near eastern and southern coasts. These profound changes could produce a trebling of present horticultural output, allow new crops and stimulate the further development of the wine industry. Further irrigation schemes would be required to serve new horticultural areas and to augment established ones, which often rely on local streams whose summer flows are predicted to decrease.

A rise in snow line and thermal belts will also benefit forestry, allowing exotic trees to be planted on present mid-altitude

Table 5. Present growing degree days and winter chill index compared with those for a future scenario, assuming a 3°C rise in temperature.

Location	Growing degree days		Winter chill index	
	Present	Future	Present	Future
Invermay, Taieri	596	1110	71	29
Dunedin	672	1198	24	10
Oamaru	720	1245	32	13
Queenstown	732	1242	145	61
Cromwell	953	1473	194	186
Ophir	703	1201	289	135
Alexandra	911	1432	203	90
Roxburgh	814	1333	108	36
Mahinerangi	371	833	206	87
Tapanui	588	1093	89	37
Milton	583	1092	105	43
Balclutha	569	1082	81	33
Invercargill	470	866	126	34
Otautau	519	898	86	22
Gore	654	1158	88	36
Te Anau	514	880	142	59
Mid Dome	579	1093	159	65

tussock land. Growth rates of existing *Pinus radiata* forests will also increase, and the present deleterious effects of frost and snow damage diminish. Consequently, a further 300,000 ha of land could be planted as production forest. Douglas fir currently grows well in forests at Naseby, Tapanui and Queenstown but, with the climate scenario, these places will have warmed close to the limits for this species, unless higher elevation sites are chosen. On the other hand, frost-sensitive species such as eucalyptus will prove more successful.

Summer tourism will be enhanced throughout southern New Zealand by the warmer temperatures, because the scenic and recreational appeal of beaches, rivers, lakes and mountains will remain. However, a shrinking snow cover will threaten the ski industry, and winter visitor numbers will probably decline in the resort towns of Queenstown and Wanaka. The frequency of poor snow years is likely to increase, especially at low ski areas such as Coronet Peak, and new fields will have to be established above 2000 m elevation. The most suitable locations are closer to the Main Divide within National Parks, so severe conflicts of interest can be expected. The scenic backdrop of snow-capped mountains that is a feature of western Otago will remain, but some views may degrade as the extent of glaciers and neve is diminished.

Overall, southern New Zealand would seem to benefit from the changes outlined in the climate scenario. Electricity, agricultural and horticultural production could all increase, and the south and east will become even more suitable for forestry. If the estimated increases in Table 4 are totalled, then the economic output would increase by 20 per cent (about $400 million per year). This should easily offset the production from land that cannot be saved from flooding. However, an estimated $10 million per year will be required for earth works and pumping, to protect the coastline about Dunedin and Invercargill, as well as strategic land on the Taieri and near Balclutha. About 100 km of roads and 15 km of railway will need to be raised above the rising sea level and may cost a similar amount.

The snow industry centred on Queenstown and Wanaka will require massive capital investment to preserve winter tourism. Suitable markets, finance, favourable cost structures and correct perception of the new opportunities will also be necessary before the increased output for agriculture and forestry arising from climate amelioration on the middle-elevation hill country can be realised. Similar factors, as well as availability of irrigation water, will govern the expansion of horticulture.

Despite all of these considerations, on balance, the region stands to gain from greenhouse gas warming, perhaps by $350 million per year. These costs and benefits are assessed in purely economic terms (and rather crudely). Changes in natural ecosystems, such as shrinking of the snow cover and loss of indigenous species, together with further exploitation of such important landscape systems as the tussock grasslands are difficult to quantify economically. They represent intrinsic natural values that are threatened by climate change.

CHAPTER 5

BIOGEOGRAPHY

MICHAEL HEADS, BRIAN PATRICK, GRAHAM WALLIS, JON WATERS

New Zealand has about forty species of woody liane, of which the genus *Clematis* is prominent. These climbing plants are rich in host-dependent insects, such as this elegant undescribed broad-nosed weevil, found associated with *Clematis marata* in the Upper Manuherikia, Central Otago. *Brian Patrick*

Fig. 1. Three species of short-horned grasshopper in the genus *Sigaus* are found in the valley floors of inland Otago and South Canterbury. While *S. minutus* is confined to the Mackenzie Basin, the newly described *S. childi* (pictured) lives in the environs of Alexandra, including the Earnscleugh Tailings. A third species is more widespread in open habitats of Central Otago and remains undescribed. A further three species of the genus are found in the surrounding hill country, up to the high alpine zone. *B.H Patrick*

Fig. 2. Underlining the long history of Central Otago landscapes is this rare spring annual of the buttercup family. Endemic to the valley floors, *Ceratocephalus pungens* appears briefly in seasonally damp areas from late August to October, before being replaced by taller herbs and grasses. *B.H. Patrick*

Fig. 3. Boulder butterflies of the genus *Antipodolycaena* present a puzzling array of taxa in southern New Zealand, mostly undescribed. Each with a distinct distribution and timing of adult emergence, but all have larvae feeding on the sprawling *Muehlenbeckia axillaris.* A tiny adult from an undescribed species is pictured from the Earnscleugh Tailings near Alexandra, where adults begin emerging from late September. *B.H. Patrick*

BIOGEOGRAPHY

MICHAEL HEADS & BRIAN PATRICK

Biogeography is the study of the distribution of plants and animals. Biogeographers seek to discover where exactly different plants and animals are found, and then search for explanations for these natural distributions. Generally these distributions are found to relate to ancient geological events or processes that happened tens to hundreds of millions of years ago.

Every region of southern New Zealand has a distinctive assemblage of native plants and animals, giving a special character to the landscape. From the open starkness of the Maniototo to the mist-shrouded valleys of western Otago, it is the flora and fauna, together with the geological features and climate, that give each area its unique appeal.

The distribution of vegetation types in the southern South Island, such as forest and grassland, partly reflects current and recent environmental conditions, especially variation in altitude, west-east gradient in rainfall and north-south differences in temperature. However, underlying these more obvious vegetation patterns the species of plants and animals show striking biogeographic patterns, which often bear no relation to present-day topography and climate. This biogeographic network involves connections with other parts of Gondwana, such as Australia and South America, and these distribution patterns can be best explained as the result of evolution of different life forms on ancient landscapes (Figs 1–3, 8, 31–33).

Means of dispersal

For many years it was thought that the distribution patterns of plants and animals were established by species evolving at a point in space and then migrating outwards from this centre of origin. Thus, the different distributions might reflect differing means of dispersal. However, it is now known that organisms with widely differing means of dispersal often have virtually identical distributions. Apparently vagile (easily dispersed) groups, like moths, have many species with the same distributions as those of relatively immobile groups such as snails and slugs, and in general potential mobility does not relate to actual distribution. For example, only two per cent of New Zealand moths are shared naturally with Australia, yet New Zealand lies down wind from Australia and we know that moths can survive being blown from there. Most species seem to 'stay put' for very long periods of time, a process consistent with the development of local endemism in species with apparently very effective means of dispersal, such as mountain daisies in the genus *Celmisia*. Plants and animals are treated together here as they form the living communities, and in practice nearly always show similar patterns of local endemism (restricted to one place) and disjunction (distribution with widely separated populations). Because both moths and seed plants have the apparent ability to disperse widely, we have used them extensively in this chapter to illustrate the fact that most flora and fauna stays put. The study of their distributions provides many clues to the geological history of the south. A distribution pattern found to be unusual or even unique in a family of 100–200 species is nearly always repeated in another unrelated group, and so in this work collaboration between zoologists and botanists is highly desirable.

Vicariance, nodes and tracks

Many distribution patterns are geographically vicariant, or complementary in space, with different species in different parts of the country. The shrubs in the genus *Pimelea* (Fig. 7) are a good example of this pattern of representation. Often vicariance is very precise. Larger groups such as genera are often vicariant by 'main-massing', that is, with the bulk of their respective species in different areas, but with overlap of some species.

Vicariance takes place at or by 'nodes'. These are points or regions where taxa or characters show any or all of the following:

- geographic boundaries in distribution;
- presence of endemic forms;
- notable absences of widespread forms;
- diverse disjunct geographic affinities.

Series of nodes are termed 'tracks'. Vicariance may be geographic, phylogenetic, ecological or a combination of these. The principal of vicariance means that a taxon, character or any biological phenomenon may be technically absent as such, but represented by a related group or phenomenon. Vicariant patterns are not due to invasion, or lack of it, by one or a series of species, but to *in situ* differentiation of descendants out of an ancestral complex that was already widespread and differentiated before the evolution of the 'modern' species.

Biodiversity and nodes

Areas are not equal in terms of the numbers of species. Some areas are species-rich, others poor. The Humboldt Mountains in north-west Otago, well known for the Routeburn Track, were recognised early on as exceptionally diverse for alpine plants. North-west Otago is the centre of biodiversity of the alpine herbs *Ourisia* (Fig. 4a), but it is also the region with most diversity in small-leaved shrubs of *Olearia* which, unlike *Ourisia*, are lowland plants. These examples show that a region can be biogeographically important for groups that are ecologically very different. North-west Otago is primarily a centre of biodiversity, and within this the species have sorted themselves ecologically into lowland and upland, and wet and dry habitats. Likewise, the Central Otago zone of biodiversity around Alexandra/Cromwell is reflected in both alpine and lowland groups (Figs. 18–22). Centres of biodiversity occur in many other parts of Otago; for example, the Kakanui Mountains is the richest region for day-flying (diurnal) moths in the genus *Notoreas*, with nine species found together (sympatric) there. The Lammermoor Range is the world centre of biodiversity for the miniature shrubs in the genus *Kelleria* (six species present) as well as the weevil genus *Irenimus* (seven species).

The Gondwanan connections of the biota

Biogeographic connections between Fiordland and south-east Australia are seen in many alpine plants, and ties between the southern South Island and Kerguelen Island (southern Indian Ocean) are shown by such diverse South Island organisms as the alpine cushion plant *Hectorella caespitosa* and the intertidal mollusc *Struthiolaria*, each having close relatives there. Among lichens, *Xanthoparmelia* species show many interesting connections between Central Otago/Mackenzie Country and parts of Australia, as does *Chondropsis viridis*, the loose yellowish-green lichen species found in Central Otago, Mackenzie Country and the Nullabor Plain, Australia. These 'tracks' probably derive from ancient biogeographic patterns that go back to the formation of the animal and plant groups in their modern form.

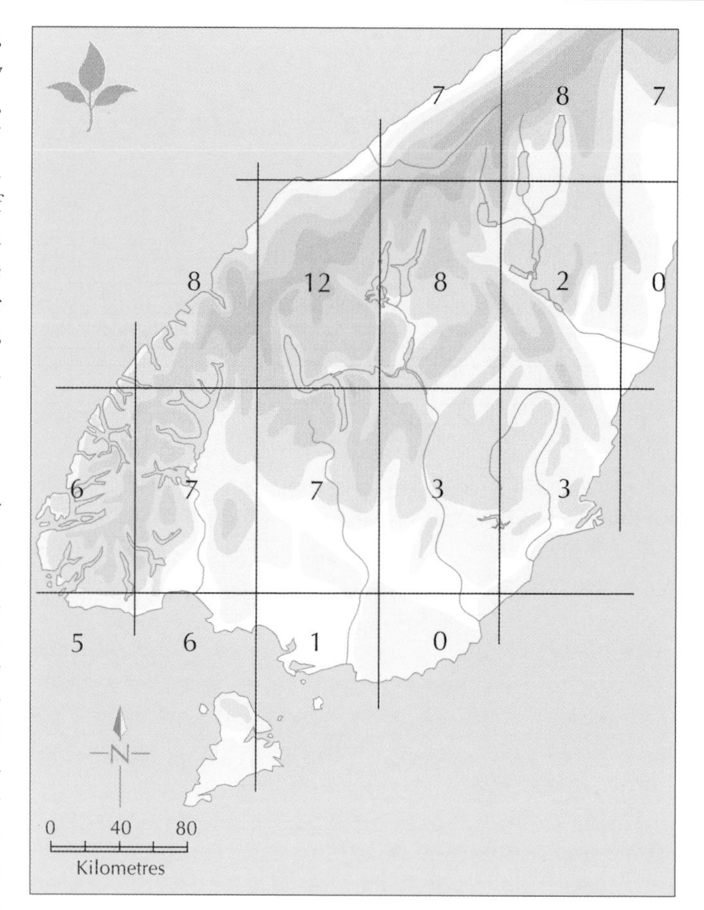

Fig. 4a.

Geological processes and biogeography

Earth and life evolve together. Historical factors such as marine transgressions often show up as traces on the flora and fauna of the landscape. The inland South Island species of the normally coastal genus *Lepidium* – Cook's scurvy grass and other cresses (Brassicaceae) – are arranged in a series of vicariant distributions around a centre near Cromwell (Fig. 22). This suggests that the species have evolved around the coastlines of steadily shrinking inland seas, rather than migrating inland from a centre of origin. Flightless chafer beetles in the genus *Prodontria* show a somewhat similar distribution pattern in southern New Zealand (Fig. 20), with sixteen species involved, and the five diurnal species in the moth genus *Paranotoreas* also show a similar pattern. In plants, Central Otago species of the shrub genus *Leonohebe* (Fig. 24) provide a very clear example of concentric distributions.

Tectonic processes in the earth's crust have also had direct effects on the earth's living 'skin', the biosphere. Over geological time, species which have evolved at sea-level may be raised to alpine heights by geological uplift. The reverse may also occur: high altitude species and communities may be lowered by down-warping. Deformation of the Otago peneplain in eastern Otago has meant a usually upland biota occurs there at sea level: for example, the diurnal moth *Eurytheca leucothrinca* at Blueskin Bay. This process accounts for what are often regarded as altitudinal 'anomalies' in many species, as well as a range of more normal altitudinal phenomena such as the very wide altitudinal range seen in ancient taxa such as the large colourful hepialid moth *Heloxycanus patricki* (Fig. 5).

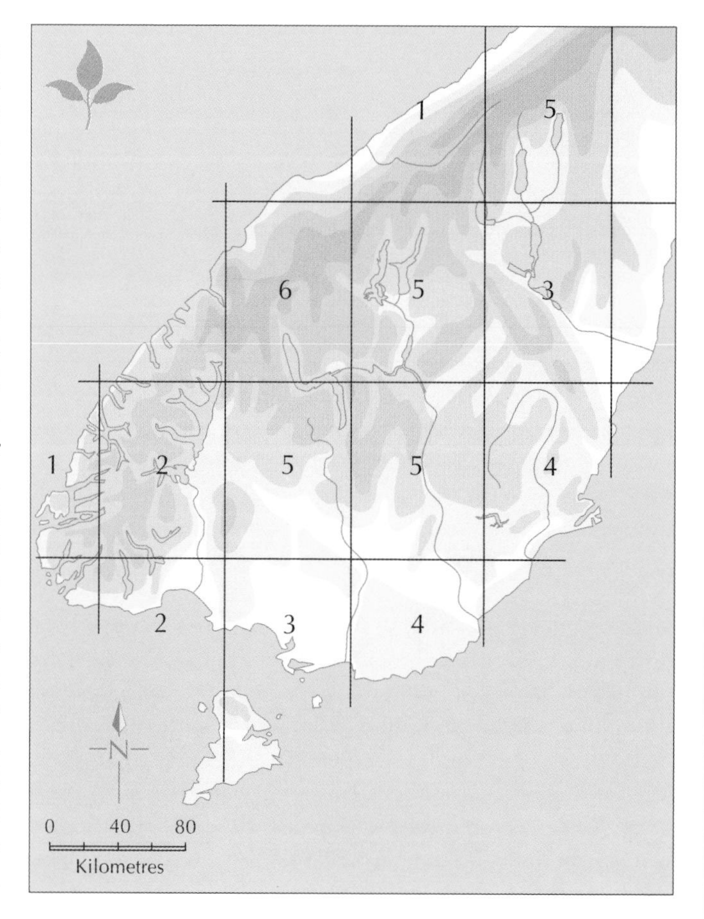

Fig. 4b.

Mapping selected species

Every region of southern New Zealand is both distinctive and connected to its neighbouring regions. By mapping the animal and plant distributions, we trace these connections and reveal much about the history of the landscape. One method biogeographers use to map distribution is to plot the number of species of a genus found per one degree of latitude and longitude. The maps in Fig. 4 have been drawn utilising this method and show the pattern of diversity for two endemic plant groups. Figs 6–7 and 9–30 go on to cover the main biogeographic patterns encountered in southern New Zealand. These are arranged in an east-to-west sequence.

Fig. 4a. Numbers of species and subspecies of the herbaceous *Ourisia* per degree square in the southern South Island. **Fig. 4b.** Numbers of species and subspecies of the small-leaved daisy shrubs *Olearia* section *Divaricaster* per degree square in the southern South Island.

COASTAL

Fig. 5. Distribution of the hepialid moth *Heloxycanus patricki* (Hepialidae). *Sphagnum* bogs from coastal areas to the alpine zone are the habitat for *H. patricki*, the sole species in the genus. It is a large moth with non-feeding adults; larvae feed on sphagnum moss from deep tunnels within it. Strangely, adults emerge only every second year, during late autumn and winter, and so escaped detection at Danseys Pass until 1978.

Fig. 6. In northern East Otago the Shag River, running along the Waihemo Fault Zone, is an important biogeographic boundary. The miniature shrub *Kelleria laxa* (Thymelaeaceae) and buttercup *Ranunculus crithmifolius* (Ranunculaceae) range, respectively, north and south of the Waihemo Fault Zone, a geological contact between the Otago Haast Schist and sedimentary rocks of the Torlesse terrane. The widespread *Kelleria dieffenbachii* shows a striking altitudinal anomaly here – throughout its wide range in New Guinea, Australia, New Zealand and the Auckland Islands it is never found below 400–500 m, other than at Shag Point, where it grows near sea level in the scientific reserve, and in the Waiau Valley, east of Fiordland.

Fig. 7. Native daphnes in the genus *Pimelea* (Thymelaeaceae) show the same boundary illustrated in Fig. 6 at the Waihemo Fault Zone, as well as illustrating endemism in western Southland.

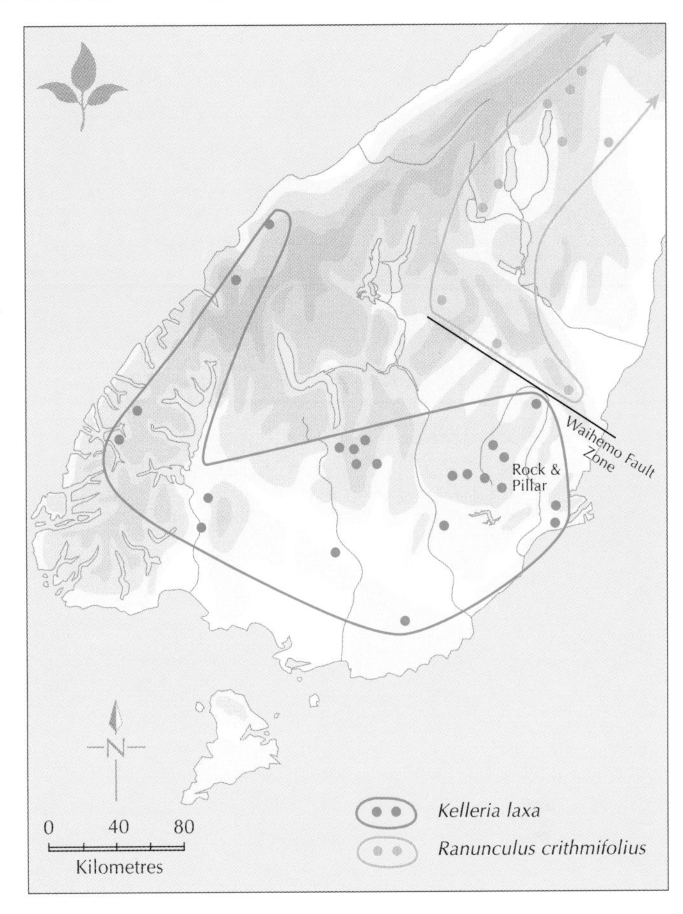

Fig. 6.

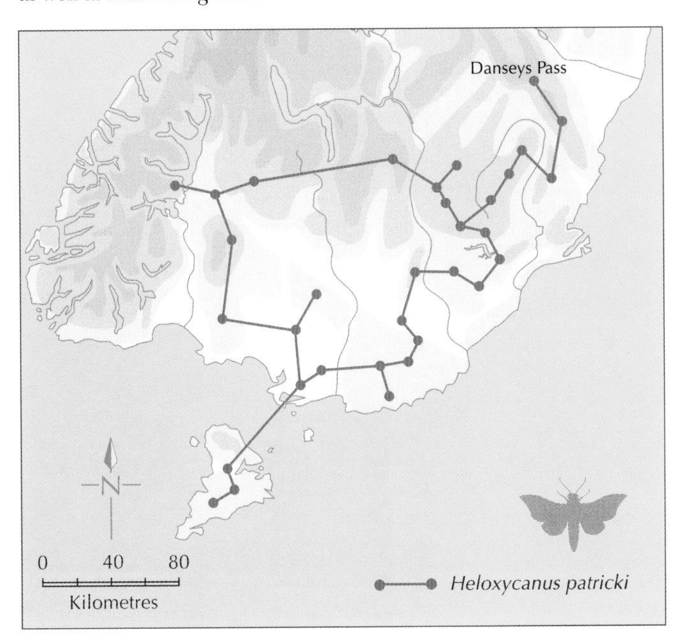

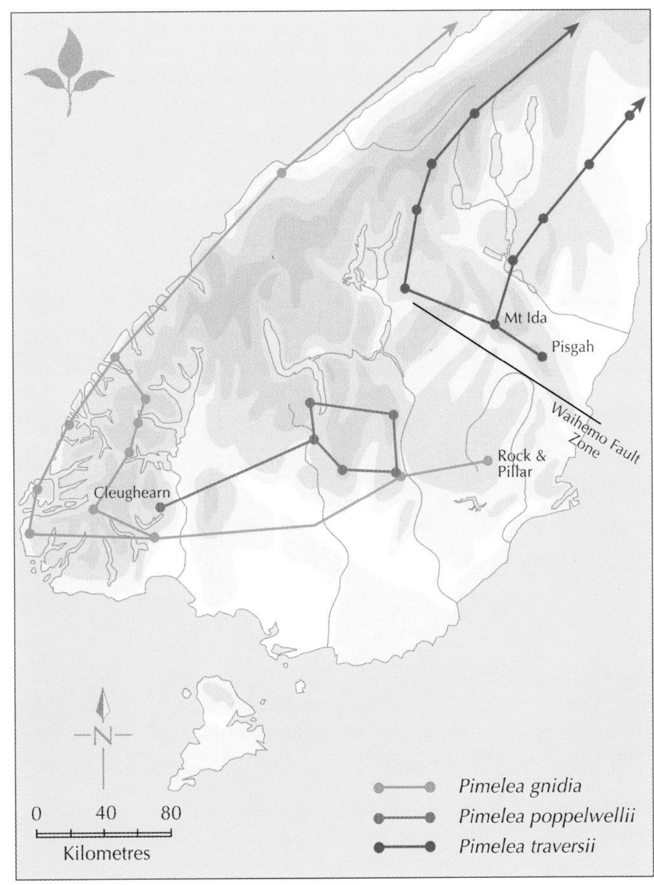

Fig. 5 (left). **Fig. 7** (above).

Fig. 8. The low sprawling shrubs or cushions of *Kelleria* are diverse in the south, with up to six of the seven species found on some mountains. Both *K. dieffenbachii* and *K. lyallii* can be found at or near sea level at strategic places: Shag Point and on the south coast near Bluff, respectively. But typically the species are alpine to high alpine, like the loose cushion of *K. croizatii* pictured here from alpine Fiordland. *B.H. Patrick*

Fig. 9. The herbaceous chenopod *Atriplex buchananii* (Chenopodiaceae) exemplifies the biogeographic connection between coastal East Otago and the heart of Central Otago. A similar pattern is seen in the grass genus *Simplicia*. Patrick (1994) lists many moth species that also display this distribution pattern.

Fig. 10. Further south, eastern connections between the Catlins and Fiordland are shown by the rare shrub *Pittosporum obcordatum*. The distribution line continues northward to Banks Peninsula, eastern North Island and Northland.

Fig. 11. Similar ties are shown by the coastal daisy *Celmisia lindsayi* and its close relative, the alpine *C. bonplandii*.

Fig. 12. Two tiny diurnal species of micropterigid moths, both in the genus *Sabatinca*, are endemic to southern New Zealand. The family Micropterigidae is considered to be a basal branch of the line leading to modern Lepidoptera, retaining some primitive features such as jaws. Larvae feed on liverworts. Because of the great age of this group (fossils are known from the Eocene, but the group is probably much older) and their limited dispersal ability, their distributions should shed light on ancient geological patterns and processes. Apart from one record of *S. quadrijuga* from Invercargill, that species and *S. caustica* have largely vicariant (non-overlapping) distributions, north and south respectively, of the Southland Syncline.

INLAND

Plants and animals on the Rock and Pillar, Lammerlaw and Lammermoor Ranges are often distinct from their nearest relatives. The alpine herbaceous daisy *Celmisia haastii* variety *tomentosa* and the miniature shrub *Kelleria villosa* variety *barbata* are both restricted to the Rock and Pillar Range and are related to forms which are widespread in the central and alpine South Island. Other examples are the flightless chafer beetle *Prodontria montii* and the large hepialid moth *Aoraia orientalis*, both found on the Rock and Pillar and Lammermoor Ranges (Fig. 19).

Fig. 13. Distributions of selected caddis species that are endemic to southern New Zealand. Larvae are aquatic and found in ponds and streams from sea level to the high alpine zone. Note the four species on the Rock and Pillar Range.

Fig. 14. Some eastern Central Otago species are closely related to others found only in the heart of Central Otago. For example, the newly described, small alpine daisy *Abrotanella patearoa* is closely related to the more widespread *A. inconspicua*.

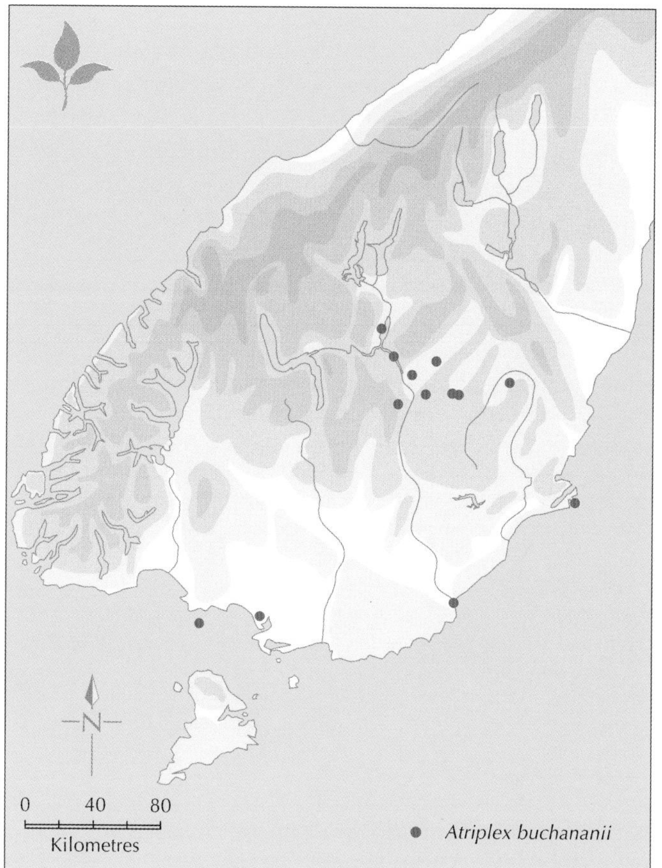

Fig. 9.

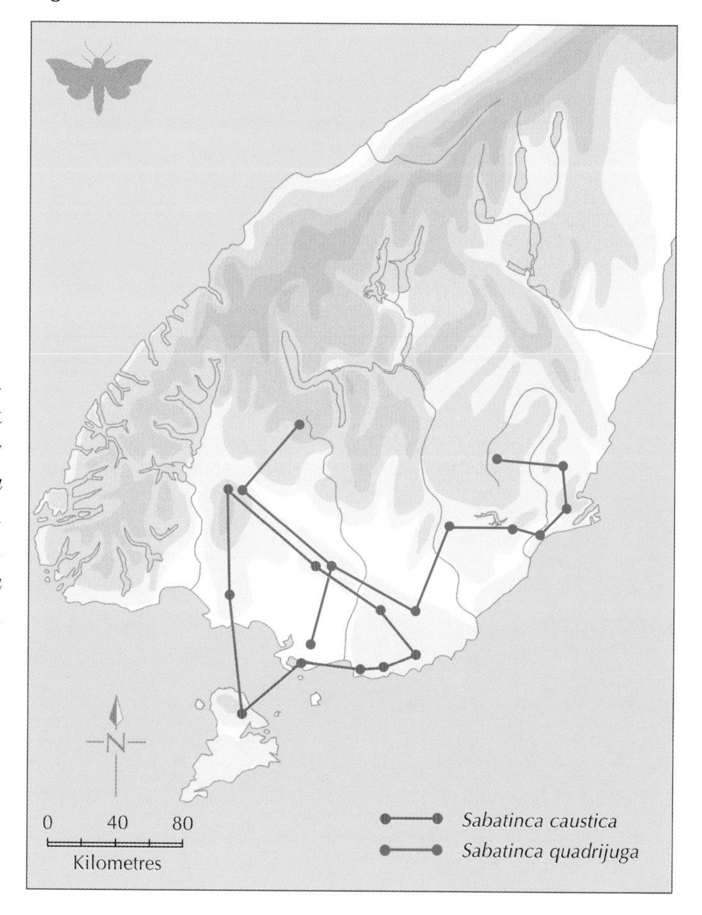

Fig. 12.

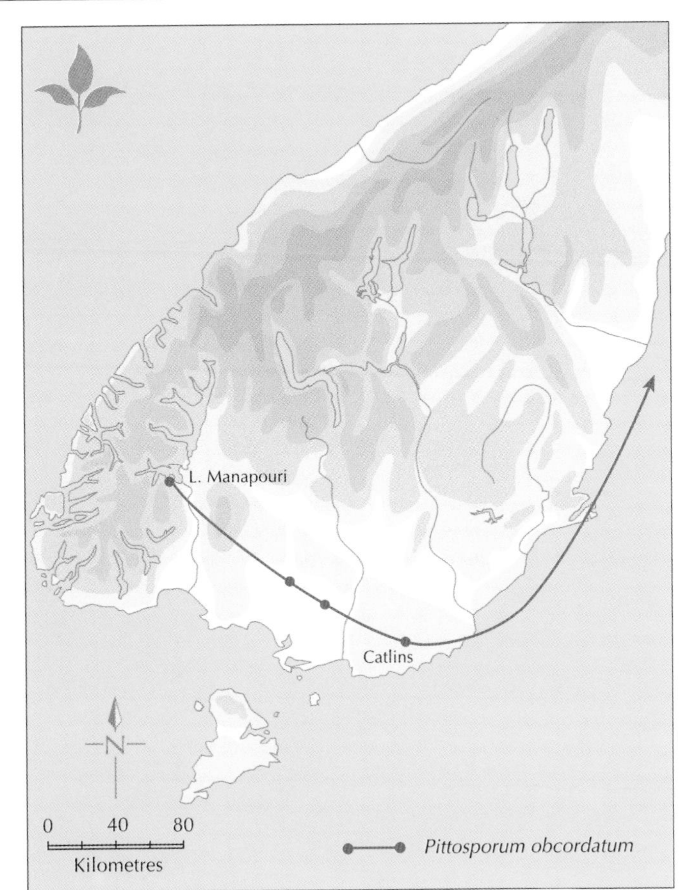

Fig. 10.

Mt Aspiring
Mt Bonpland
L. Manapouri
Nugget Point
N
0 40 80
Kilometres
Celmisia bonplandii
Celmisia lindsayi

Fig. 11.

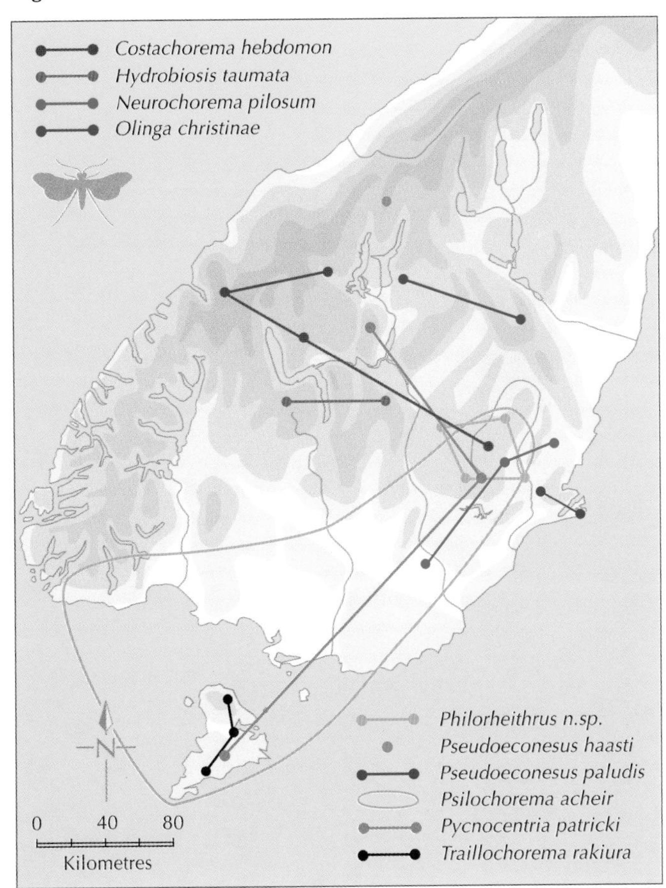

Fig. 13.

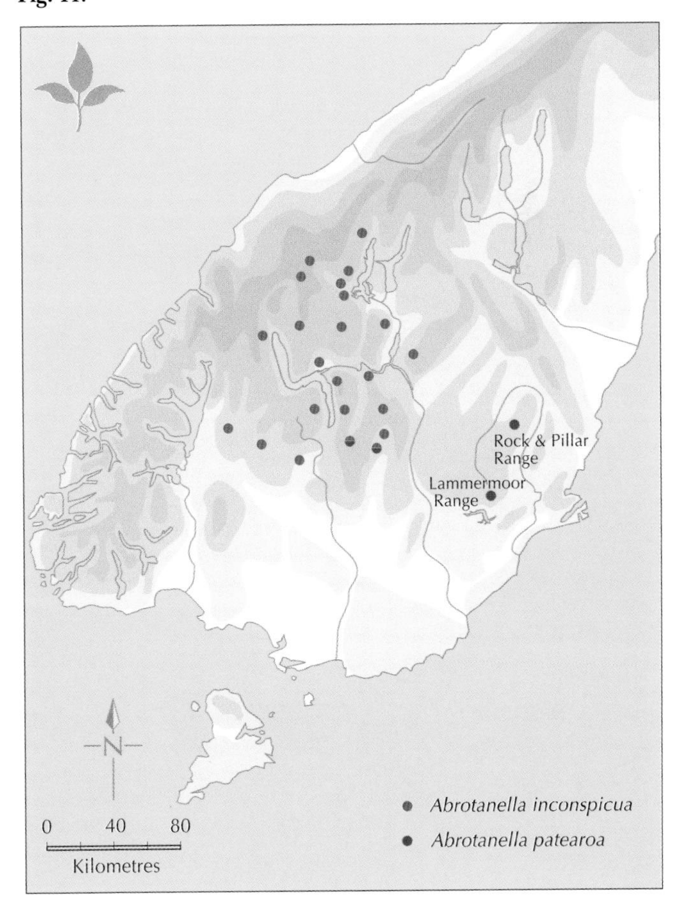

Fig. 14.

Fig. 15. In patterns like that of the alpine cushion plant *Kelleria childii,* there is no distinction between eastern Central Otago and nuclear Central Otago populations.

Fig. 16. The characteristic rock tors of Central Otago are the habitat for a group of moths in the genus *Dichromodes* (subfamily Oenochrominae) where the larvae feed on various lichens. Of the seven southern New Zealand species, six are found in Central Otago, the highest diversity of the genus nationwide. Many of the species are diurnal and are often seen flying swiftly around the warm rock faces. The three distributions mapped again indicate that speciation may have taken place around the shores of inland seas and lakes which dried up in the Miocene (15–20 million years ago). The three species are centred on an area about ten km east of Alexandra.

Fig. 17. Three alpine herbs in the genus *Hebejeebie* divide Central Otago into three sectors centred near Cromwell.

Fig. 18. The native forget-me-nots illustrate localised distributions around Cromwell, with *Myosotis cheesemanii* found only on the Pisa Range and northern Dunstan Mountains, and *M. albosericea* on the southern Dunstan Mountains above the Cromwell Gorge. These upland taxa are ecologically different from lowland species endemic to the same centre, e.g. the Cromwell chafer beetle *Prodontria lewisi* (Fig. 20), but are part of the same biogeographic centre.

Fig. 19. Twelve species of the hepialid ghost moth genus *Aoraia* are found in southern New Zealand, and ten of these are endemic to parts of this area. The main massing concentration of species is in nuclear Central Otago, with four species on the Old Man Range/ Umbrella Mountains. These species belong to an ancient family in which the adults do not feed. The larvae inhabit subterranean tunnels and many species have short-winged females incapable of flight. They are amongst New Zealand's largest moths and have a wing span of up to 7 cm. Adults emerge in autumn and are nocturnal, crepuscular or diurnal, depending on the species. In addition to the eight localised distributions shown, *A. dinodes* is found in Fiordland, Southland and Stewart Island, *A. rufivena* is in Otago and the Catlins, *A. lenis* is in the North Otago mountains and northwards and *A. aurimaculata* is in West Otago and northwards.

Fig. 20. Flightless chafer beetles in the genus *Prodontria* (Scarabaeidae) are restricted to southern New Zealand, from the Ben Ohau Range southwards to The Snares. The main concentration massing of species is in Central Otago, especially around Alexandra. Nine of the seventeen species are mapped here. In addition, *P. truncata* and *P. setosa* have localised distributions in Southern Fiordland, *P. minuta* and *P. matagouriae* in the Mackenzie Basin, *P. jenniferae* in Kawarau Gorge, *P. lewisi* at Cromwell, *P. bicolorata* at Alexandra and *P. longitarsis* on the Snares. They live in habitats ranging from sand dunes, both inland and coastal, to high-alpine cushionfields. Most have very restricted natural distributions, and many are threatened with extinction.

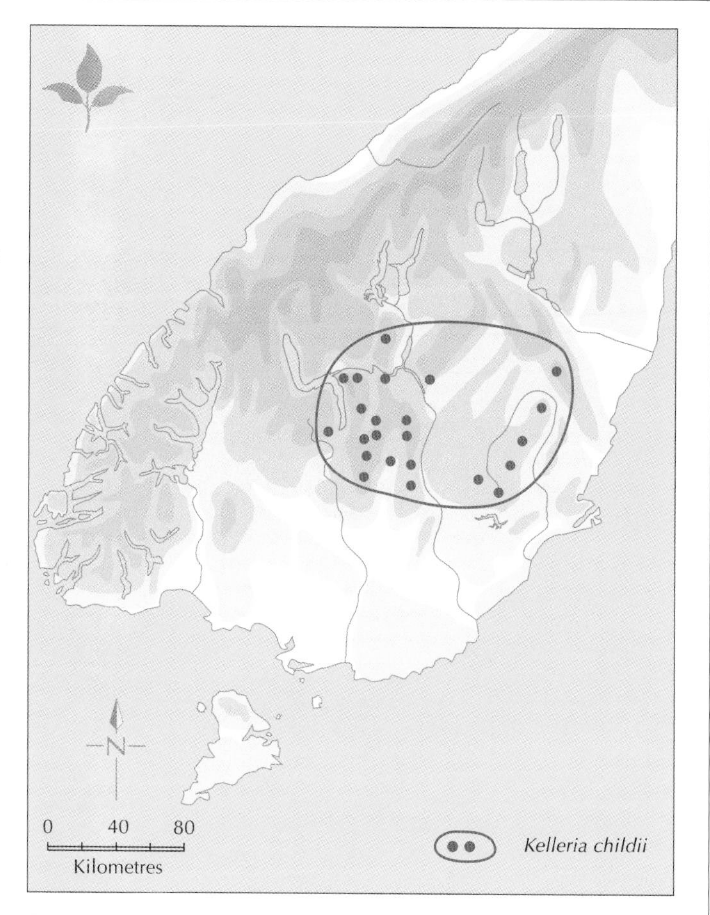

Fig. 15.

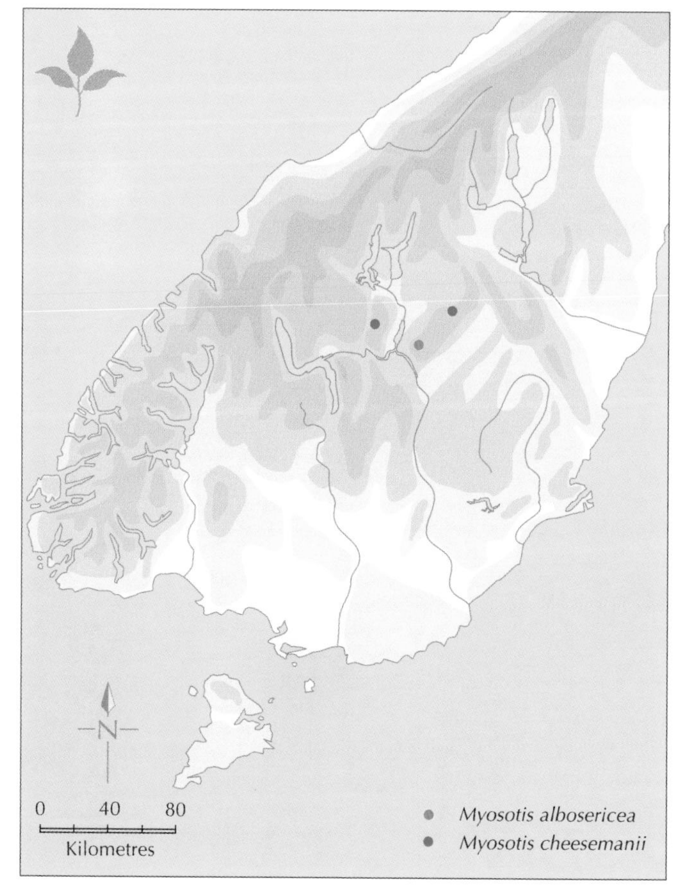

Fig. 18.

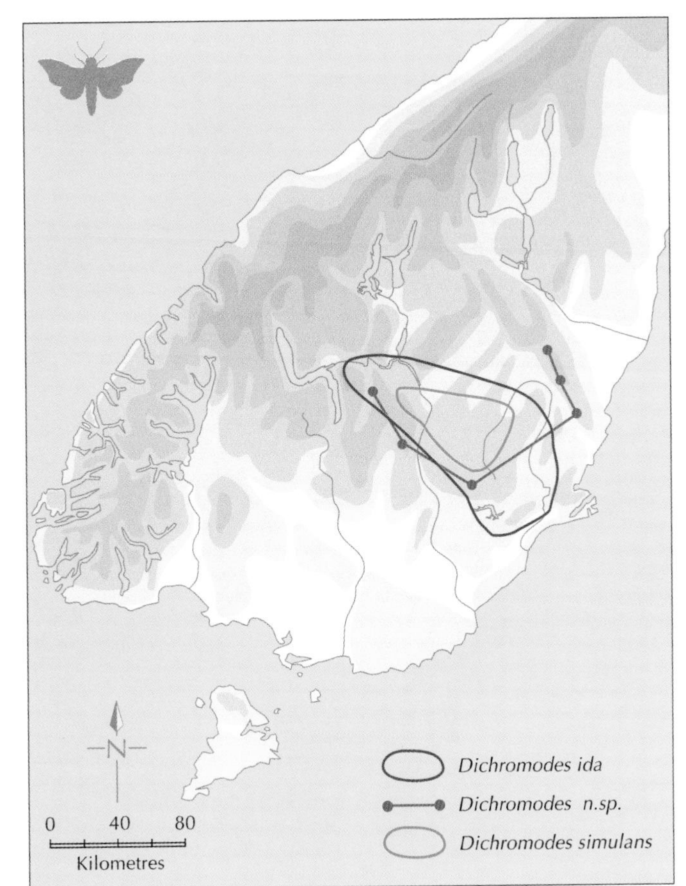

Fig. 16.

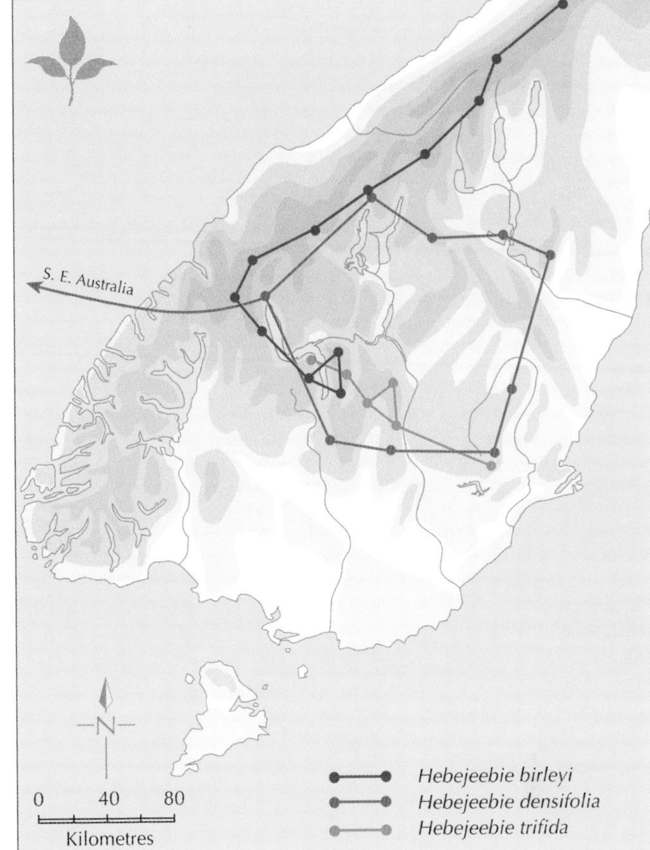

Fig. 17.

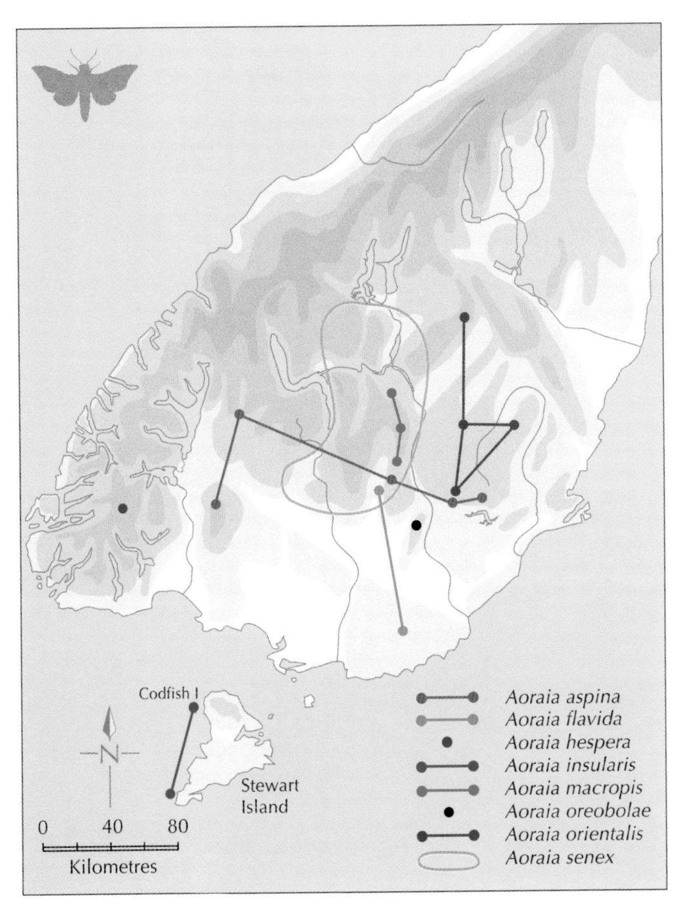

Fig. 19.

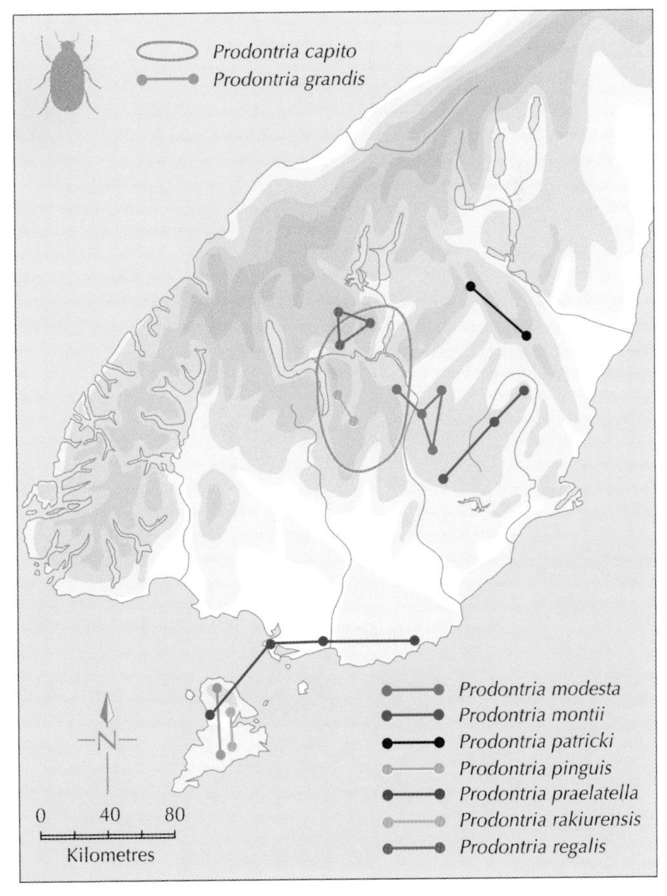

Fig. 20.

Fig. 21. Over 50 species of the moth genus *Orocrambus* (Crambidae) have been described. Of these, more than 40 are found in southern New Zealand, and seven species are endemic to parts of southern New Zealand, with main centres of endemism in the Lammermoor Range, Rock and Pillar Range and Mackenzie Country. The genus includes both diurnal and nocturnal species. Some have short-winged females. Larvae feed on the bases of grasses, cushion plants and sedges. Three species occur around Alexandra/Old Man Range, showing connections with, respectively, the Eyre Mountains, Rock and Pillar Range and North Otago.

Fig. 22. Distributions of the cresses *Lepidium* species (Brassicaceae) in the South Island. Note the 'piling up' of species round Alexandra.

Fig. 23. The localised distribution of the shrub *Hebe biggarii* highlights the biogeographically important Eyre and Garvie Mountains. A narrower distribution on the Eyre Mountains, located by the intersection of the Moonlight Tectonic Zone and the Livingstone Fault, is held by the daisies *Celmisia spedenii* and *C. philocremna*, the latter a very distinct species combining characters of different sections.

Fig. 24. Herbs in the genus *Ourisia* (Scrophulariaceae) show patterns similar to those of the shrubs *Leonohebe* and daisies *Celmisia*, with, for example, *O. caespitosa* present in eastern and central Fiordland. The Moonlight Tectonic Zone and Caples/Haast Schist boundary zone correlate with boundaries in this and related species. The central line in the distribution of *O. caespitosa* is the western limit of *O. caesipitosa* var. *gracilis*.

Fig. 25. A group of species in the shrub genus *Leonohebe* section Flagriformes illustrates the crowding of forms in southern Central Otago, as well as connections between Central Otago and Fiordland. These connections cross the 'Hokonui Assemblage' of terranes along the line of the Moonlight Tectonic Zone. *L. imbricata* is in eastern Fiordland (Mt Burns and Mt Cleughearn) and the Eyre Mountains, but in the Caples and the Haast Schist the group is much more diverse. Reading from outside to inside, *L. propinqua*, *L. poppelwellii* and *L. subulata* form a concentric series of rings with localities such as Roxburgh at the centre. This sort of dispersal again suggests form-making in a widespread ancestral complex by the shores of progressively diminishing inland bodies of water present during the early Miocene.

Fig. 26. Like these 'whipcord' members of *Leonohebe* (replaced in Fiordland by *L. hectorii* and *L. laingii*), the low shrub *Hebe buchananii* is present in Fiordland only in the east, with records at Green Lake/Hunter Mountains in the south and the Darran Mountains in the north. North of the Livingstone Fault the species is widespread.

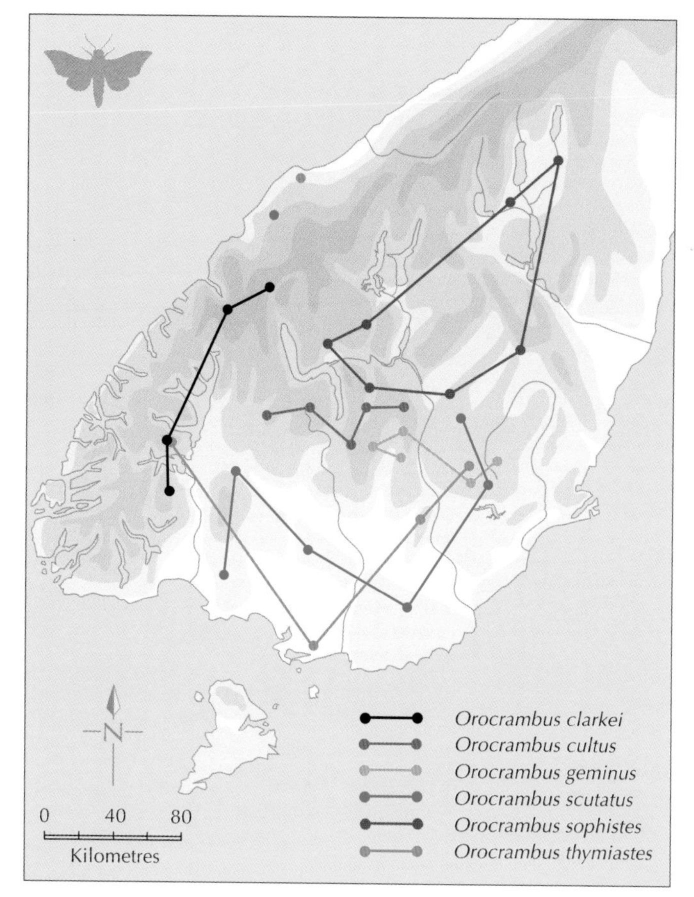

Fig. 21.

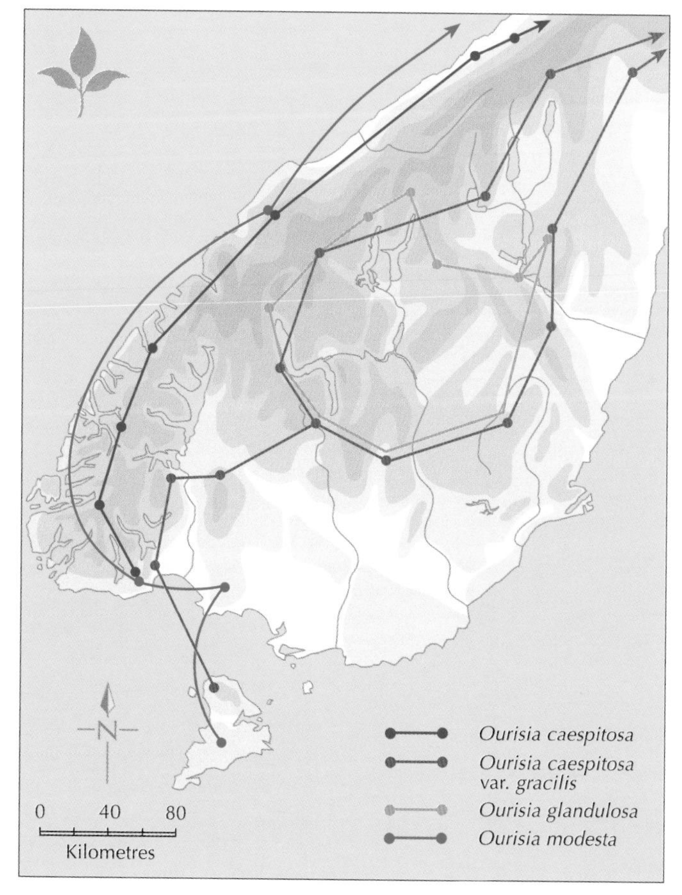

Fig. 24.

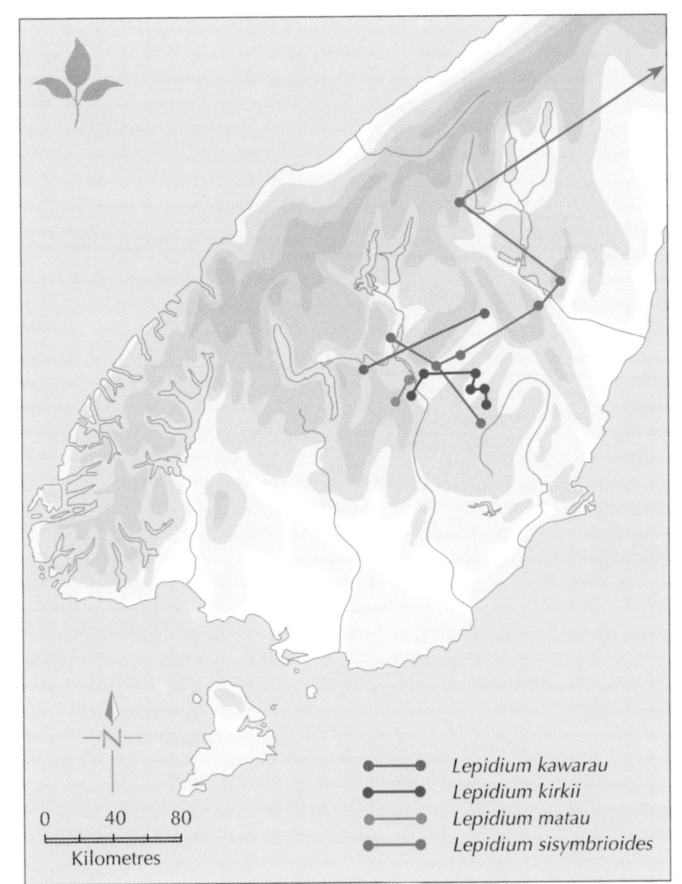

Fig. 22.

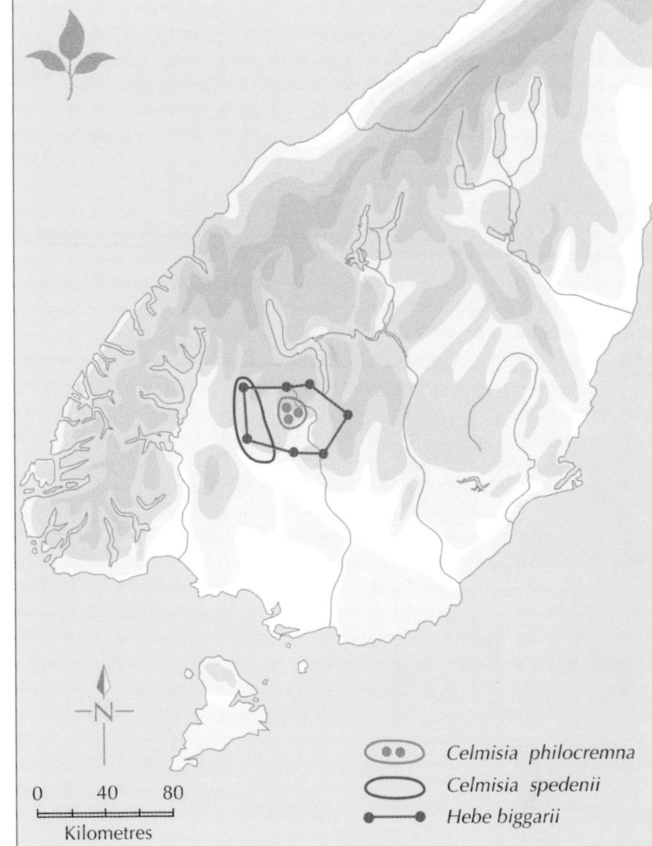

Fig. 23.

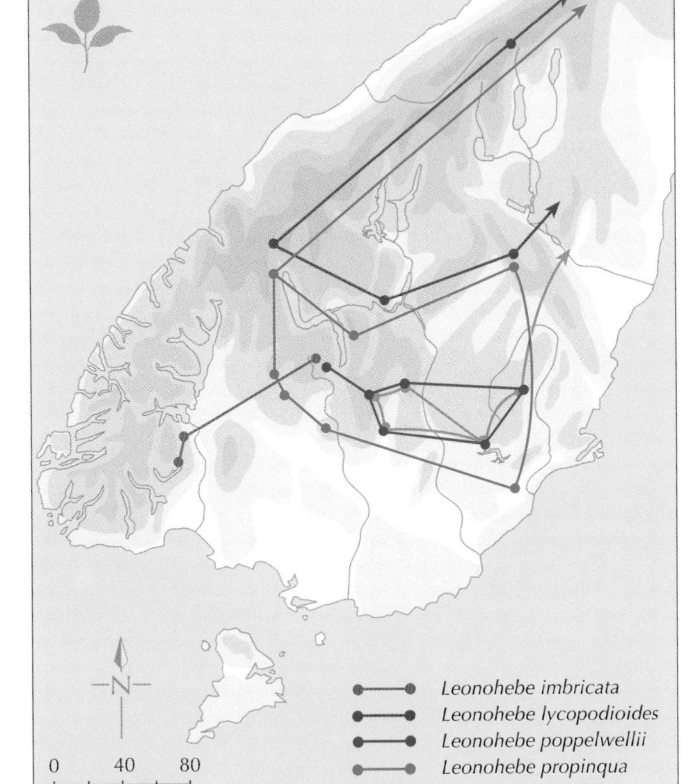

Fig. 25.

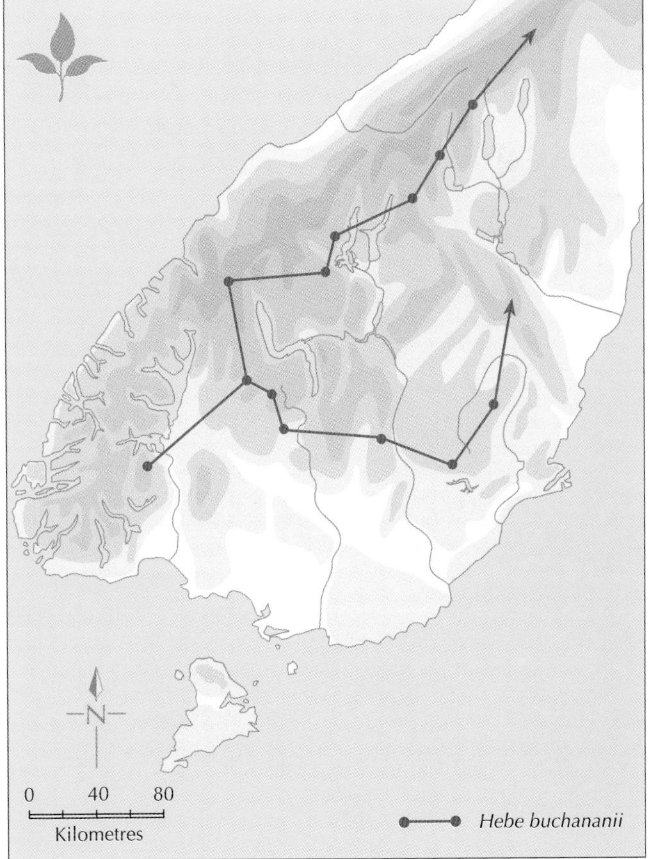

Fig. 26.

FIORDLAND

Fig. 27. Distribution of the noctuid moth *Ichneutica lindsayi* (Noctuidae). Many insect species share the central Fiordland distribution pattern of this handsome moth. The female of the species is short-winged and effectively immobile.

Fig. 28. Large diurnal geometrid moths in the genus *Aponotoreas* (Geometridae) have their centre of diversity in southern New Zealand, where eight of the nine known species are present. A selection of five species is shown here, four of which illustrate different kinds of Fiordland distribution. Most species are diurnal and many are alpine. Larvae feed on various families of monocotyledons and also the turpentine scrub *Dracophyllum* (Ericaceae s.1.), which resembles many monocots in its morphology.

Fig. 29. The herbaceous daisy *Celmisia inaccessa* occurs within the quadrangle: Thompson Sound/Caswell Sound/Doon Saddle/Mt Luxmore. This is an important pattern of endemism straddling western, central and eastern Fiordland. The Caswell Sound/Doon Valley region is particularly important biogeographically. For example, Caswell Sound is the only locality where *Leonohebe odora*, widespread in the North, South and Stewart Islands, reaches western Fiordland.

STEWART ISLAND

Fig. 30. In Stewart Island parallel arcs of endemism in the north, and concentric rings of endemism in the south are separated by the Median Tectonic Line. These patterns are illustrated here for seed-plants. Northern Stewart Island: the gentian *Gentiana gibbsii* (Gentianaceae), the speargrass *Aciphylla traillii* (Apiaceae), daisy *Celmisia aff. durietzii* (Asteraceae), the snowgrass *Chionochloa aff. flavescens* (Poaceae). Southern Stewart Island: the snowgrass *Chionochloa lanea*, the daisy *Celmisia polyvena*, the speargrass *Aciphylla cartilaginea* and speargrass *Aciphylla stannensis*.

CONCLUSIONS

This bewildering array of distributional patterns reflects the complex geological history of southern New Zealand. Mention has been made of some major geological features and their correlation with some biogeographic patterns, e.g. Waihemo Fault Zone, Moonlight Tectonic Zone, Southland Syncline and Median Tectonic Zone. This biogeographic analysis with its strong linkage to fundamental geological processes has relevance to biodiversity and its conservation.

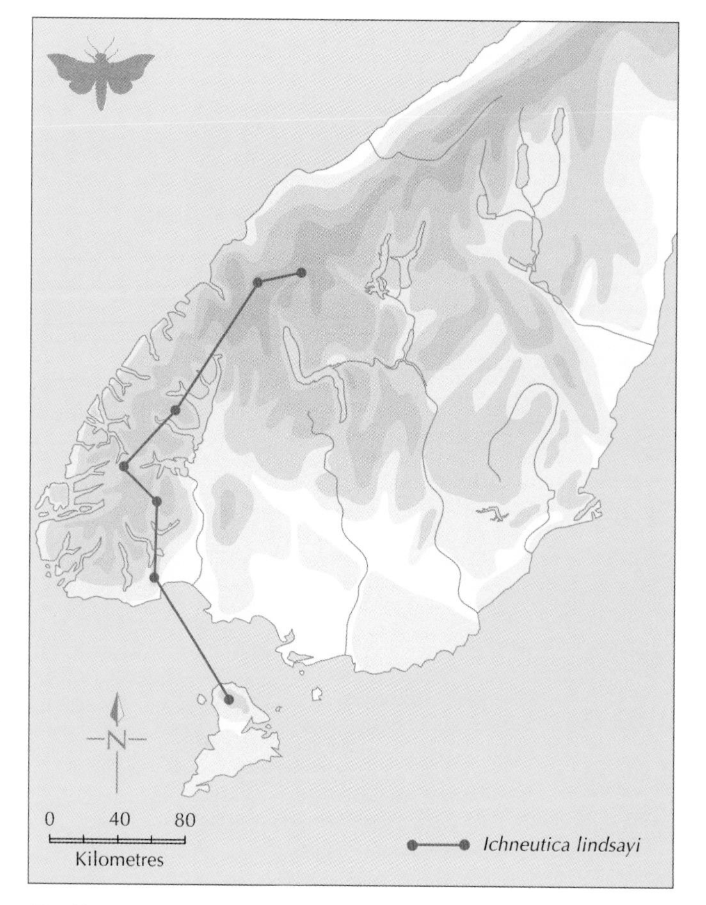

Fig. 27.

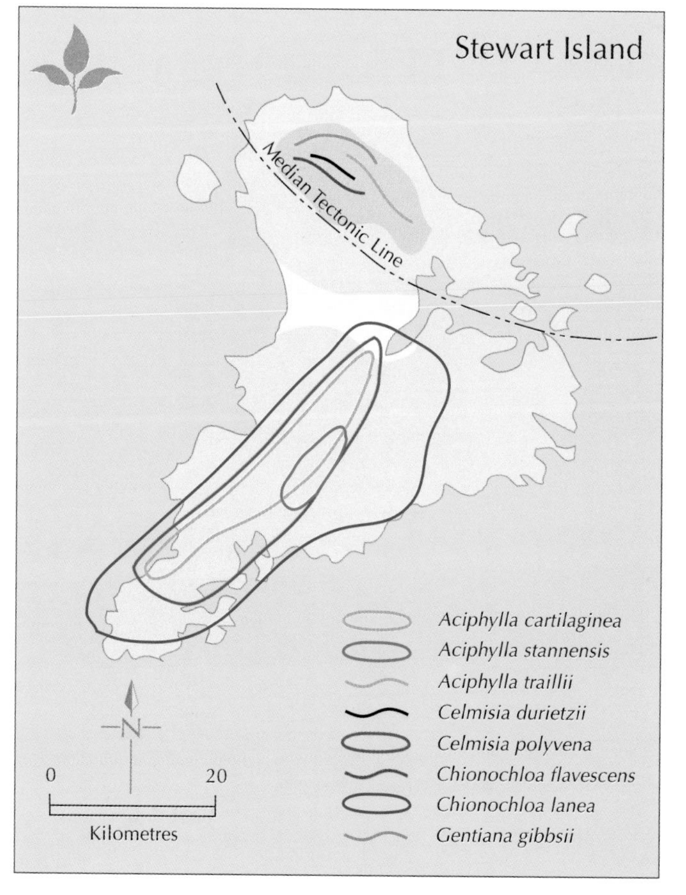

Fig. 30.

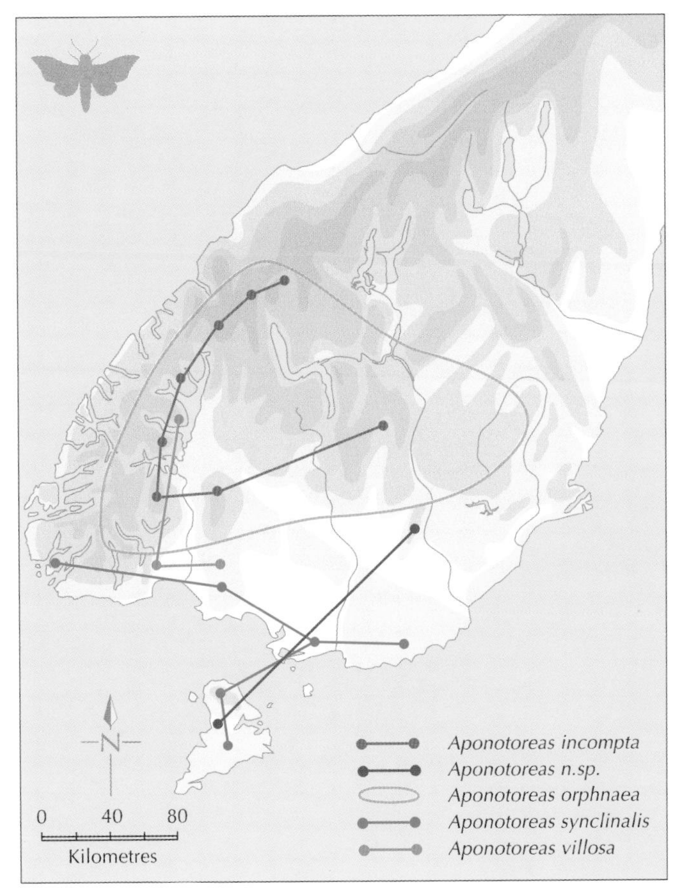

Fig. 28.

Fig. 29.

Fig. 31. Diurnal geometrid moths are a feature of southern New Zealand, from sea level to the high alpine zone. Among the most diverse is the genus *Notoreas* with over twenty-one species found in the south. Three of the species are strictly coastal, two are found in inland montane areas and the remaining fifteen are alpine. The larvae are selective feeders, eating only plants of the daphne family (Thymeleaceae) represented by the two genera *Kelleria* and *Pimelea*. Pictured is *N. mechanitis* from the western mountains. B.H. Patrick

Fig. 32. Limestone outcrops of North Otago are important refugia for a surprising array of small herbs, sprawling shrubs and grasses. Of the approximately fourteen species mostly confined to this specialised habitat, ten are undescribed at present. All are threatened with extinction including four named species – two small grasses *Simplicia laxa* and *Poa spania*, the sprawling shrub *Carmichaelia hollowayi* and slender herb *Ischnocarpus exilis*. *B.H. Patrick*

Fig. 33. Five species of endemic nettle are found in the region, each occupying a distinct community type. The subantarctic megaherb *Urtica australis* is found north to islands in Foveaux Strait, whereas the rare wetland *U. linearifolia* has been found at Seaward Moss, margin of Lakes Tuakitoto and Waihola. Forests and shrublands often provide habitat for the more widespread tree nettle *U. ferox* and the smaller *U. incisa*. Pictured is the rare *U. aspera*, a montane to alpine spreading herb, here at 700 m on the Dunstan Mountains, close to its type locality. *B.H. Patrick*

GRAHAM WALLIS & JON WATERS

THE PHYLOGEOGRAPHY OF SOUTHERN GALAXIID FISHES

Biogeography is the study of the past and present distributions of species. Its earliest conceptions date to Darwin and Wallace, who recognised distinct global biogeographic realms according to their attendant species and communities. We now know that some of the similarities across these realms date back to the time of the supercontinent 'Pangaea', over 200 million years ago (mya). Such an explanation of shared features is said to be vicariant: it explains similarity by common descent from an ancestral continent which subsequently fragmented by plate tectonics. Vicariant interpretations therefore derive from the expectation that a species' distribution remains closely associated with a specific land mass, or indeed a certain body of water. This view is characterised by the phrase 'earth and life evolving together'.

In contrast, dispersalist explanations invoke movement of organisms from one area to another to explain appearance in two locations. Both processes certainly occur. For example, the distribution of ratites (large flightless birds such as moa, kiwi, emu, ostrich, rhea, elephant bird) throughout the southern hemisphere, and their absence from the northern hemisphere, is simply explained by a Gondwanan origin. As the supercontinent Gondwana broke up, around 120 million years ago, each southern land mass evolved its own ratite species in isolation. In contrast, the shared occurrence of several flighted bird species in both New Zealand and Australia (e.g. tauhou/silvereye, poaka/pied stilt, kotuku/white heron, white-faced heron and welcome swallow) is explained by much more recent dispersal east across the Tasman, with colonisation facilitated by the development of agriculture.

Many situations, however, are less clear cut. In the sixth edition of his *Origin of Species*, Darwin drew attention to the unusual distribution of *Galaxias attenuatus* (now *Galaxias maculatus*, the inanga whitebait species, Fig. 34). Most freshwater fish species have restricted distributions: they are rarely found on two separate continents. Inanga, however, have a remarkably wide range, occurring in southern Chile and Argentina, the Falklands, Australia and New Zealand. Does the distribution of this species reflect an ancient Gondwanan distribution, or did it evolve more recently and spread by means of a marine juvenile stage? Darwin invoked a dispersalist explanation, suggesting 'dispersal from an Antarctic centre during a former warm period.' He probably preferred this explanation because he thought it unlikely that populations of a species on different continents could retain their morphological and behavioural similarity in isolation. However, it remains an open question – patterns of distribution alone might suggest likely hypotheses of origins, but distributions alone are usually open to different interpretations. We need some data that inform us more directly about the order of splitting and time since divergence.

A new approach to biogeography has emerged over the last decade or so. This approach, known as 'phylogeography', uses genetic (usually DNA sequence) data to analyse the geographic

Fig. 34. *Galaxias maculatus* or Inanga, the most common whitebait species, which occurs in southern Chile and Argentina, the Falklands, Australia and New Zealand.
Ken Miller

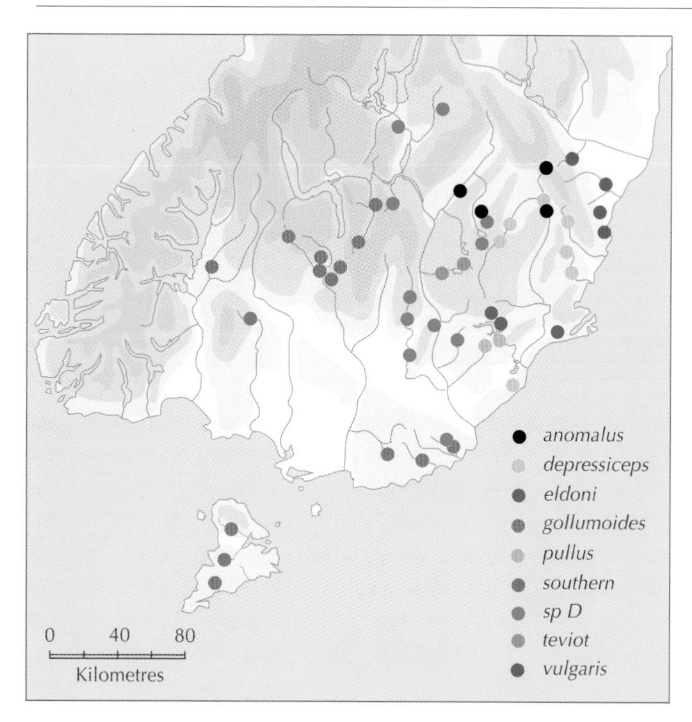

Fig. 35. Map of Otago/Southland showing currently known distributions of nine lineages of the river galaxiid (previously all *Galaxias vulgaris*). Koaro (*G. brevipinnis*) is found widely throughout the region. G. Wallis and J. Waters

Fig. 36. Three species of river galaxiid endemic to Otago, showing differences in head shape and patterning. From left to right: *G. eldoni*, *G. depressiceps*, *G. anomalus*. Ken Miller

distribution of evolutionary lineages. At the simplest level, genetic data can redraw species boundaries and reveal new forms. More interestingly, the relationships and distributions of lineages can tell us something about the way plants and animals have evolved in concert with the geologic processes that shape our land. We have been looking at galaxiid fish in this regard since 1988, using a variety of different molecular genetic techniques.

First of all, with respect to inanga, the genetic data are clear cut. Chilean, Tasmanian and New Zealand populations of inanga each form genetically quite discrete groups. At first sight, this might appear to support a Gondwanan explanation for their origin, but that is not the case. First, Tasmanian inanga are more like New Zealand inanga than Chilean inanga. This pattern is not in keeping with the order of the break-up of the continents, because New Zealand split from Australia 80 mya ago, well before the link between Australia and South America (through Antarctica) was broken (38 mya). Second, the depth of genetic difference between New Zealand and other populations indicates several million years' divergence, but nothing like as much as 80 million. Darwin could conceivably have been right about an Antarctic origin, as temperatures there were warm until the circumpolar current became established in the last 30 million years. Within New Zealand, the data are also revealing, showing no evidence for inanga population structure. Fish from the Bay of Islands are genetically no more similar to one another than they are to fish from the Cascade River on the west coast. These populations thus show no phylogeographic structure within New Zealand, suggesting that they maintain contact by dispersal.

Locally, galaxiid fish of Otago and Southland have provided a complex and fascinating case study in phylogeography. Over the last thirteen years, work at the University of Otago has focussed on the common river galaxiid (*Galaxias vulgaris*) complex, which is closely related to a whitebait species (koaro; *G. brevipinnis*). Unlike the migratory koaro, members of the non-migratory river galaxiid complex do not have a marine phase: they complete their entire life-cycle in freshwater. The river galaxiid complex is thought to be derived from a koaro-like ancestor.

Until 1996, the dominant stream-resident galaxiid fish found in rivers east of the divide was given the name *G. vulgaris*, the common river galaxiid. Genetic data show that *G. vulgaris* in Otago/Southland is better regarded as a 'species flock', composed of several different evolutionary lineages. To the north of Palmerston, and throughout the Canterbury Plains to Kaikoura, there is a single major lineage (true *G. vulgaris*). In contrast, the south houses a panoply of different forms (Fig. 35). Some of these have now been described: *G. depressiceps* (Shag, Waikouaiti and Taieri Rivers, Akatore Creek), *G. anomalus* (Taieri and Clutha Rivers), *G. gollumoides* (Catlins, Mataura, Waiau and Nevis Rivers, Stewart Island), *G. pullus* (Taieri and Clutha Rivers) and *G. eldoni* (Taieri Rivers). Additionally, there are at least three other distinct southern line-ages as yet unnamed. Galaxiids as a whole are conservative in general body plan and appearance, and this species flock is no exception. However, there are some fine details (for example, fin ray numbers, head shape, eye diameter) which can be used to distinguish species, with practice (Fig. 36).

These differences had simply not been noticed until the genetic studies prompted a closer look.

So how did these different forms arise, and why are there so many in Otago and only one in Canterbury? First, DNA data show that this group has an evolutionary history of up to 8 million years. This age is well after the Oligocene marine transgression (ending some 30 mya), and probably postdates the end of the Miocene lake system (Lake Manuherikia) that dominated Central Otago about 15 mya. There are numerous fossil galaxiids of this general type described from various locations around Central Otago dated from 9–20 mya, consistent with a precursor to the *G. vulgaris* complex. The idea is certainly attractive, as large lake systems often support huge species flocks of fishes: the African rift lakes, for instance, have several hundred species of cichlid fish. Alternatively, perhaps the rapid uplift of the Southern Alps transformed a rather topographically featureless landscape into one of distinct river systems which allowed independent evolution of new forms. This scenario requires an explanation for the lack of equivalent diversity in Canterbury. It may be that the braided river systems that dominate the plains periodically mix the fish, particularly during extreme flood conditions, preventing differences from accumulating. Alternatively, as has been widely shown in the northern hemisphere, glaciation during the Pleistocene may have reduced the amount of diversity present in this region. Unlike the Canterbury Plains, the Otago peneplain is an ancient and stable land surface, too elevated for recent marine inundation and too low for glaciation.

These ideas about origins of the river galaxiid complex are largely vicariant in nature – they explain the evolutionary splitting of the lineages by their passive association with a dynamically changing landscape. An alternative scenario is that the migratory koaro has repeatedly 'spawned off' stream-resident (non-migratory) populations, which have then diverged *in situ*. This could have happened many times in different South Island rivers. Since the complex almost certainly arose from koaro originally, the real question is: Did loss of migratory behaviour occur many times or once only? That question can be answered by using gene sequences to build an evolutionary tree of all river galaxiid lineages and koaro (Fig. 37), and seeing where and how many times on the tree koaro appears. In short, there is only one New Zealand koaro lineage, and its position on the tree suggests that there have been three losses of the migratory phase. Nevertheless, the tree indicates that many of the river galaxiid lineages probably speciated as a result of vicariance within the rivers of Otago/Southland.

The most convincing scientific evidence occurs when a specific prediction has been made at the outset, and data are collected to test the idea. Research in the field of biogeography has all too often followed the opposite pathway: data are collected first, then a story is put together to explain the data. We are working with geologists to test some specific geological hypotheses about the structural evolution of Otago/Southland river systems with the aid of fish biogeography. One compelling example involves the Nevis River, which runs north, entering the Kawarau River gorge (Fig. 38). Geological rates of uplift and positioning of fault lines suggest that the Nevis in fact used to run south, into what is now the Nokomai River, a tributary of the Mataura River. At some point since the start of southern uplift, this part of the Mataura system reversed its course to become part of the Clutha system. The Southland plain is currently dominated by two river galaxiid lineages: the round-headed *G. gollumoides*, and an un-named square-headed form (southern). (*G. gollumoides*, originally described from Chocolate Swamp on Stewart Island, was named after 'Gollum' in *Lord of the Rings*, a dark swamp-dweller with big round eyes!) Extensive sampling elsewhere in the Clutha system has revealed at least three major lineages, but neither of the two Southland forms. Our prediction was that if the Nevis River used to be part of the Mataura system running south, then it would still carry derivatives of the Southland forms rather than those of the Clutha system, to which it is now attached. DNA analysis revealed *G. gollumoides* throughout the Nevis River, in the absence of any other form. Yet fish from as nearby as Bannock Burn look like other Clutha system lineages, none of which occur in Southland. The Nevis River *G. gollumoides* has diverged from the Southland stock since the flow reversed. We can use the amount of divergence to give us a rough idea of when river capture occurred. The differences in the DNA sequences suggest that it took place about 1–2 mya, during the Pleistocene.

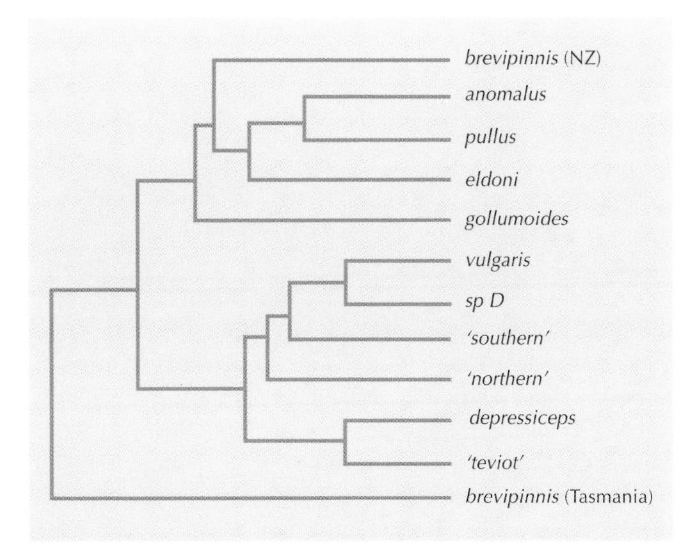

Fig. 37. Evolutionary tree of ten lineages of river galaxiid based on DNA sequence data, showing their relationships with koaro (*G. brevipinnis*) from both New Zealand and Tasmania. *G. Wallis and J. Waters*

G. gollumoides is instructive in another way that relates to an understanding of landscape change through geological time. Although we know relatively little of its basic biology, *G. gollumoides* resembles the rest of the river galaxiid complex in that it lacks a marine phase. Yet it is found on both sides of Foveaux Strait. One might propose that populations from the Southland Plain would differ from those on Stewart Island because of the seawater barrier. However, during the last glaciation, which finished only 14,000 years ago, Stewart Island was connected to the mainland, and their rivers may have merged. Genetic data suggest little if any divergence between these two regions, consistent with a recent freshwater connection.

Once again, genetic relationships are informative with respect to historical river connections, reinforcing the power of DNA analysis as a tool in biogeography.

To finish: two notes of caution. First, this type of genetic structure serves as a warning to those who move animals around New Zealand without understanding the genetic differences among locations. Whereas botanists have long appreciated that plant species exist as ecotypes or varieties specific to particular regions, similar variation in animals is not always as easy to see. In the Pool Burn region of the upper Taieri, for example, we have documented the effect of human-mediated disturbance on galaxiid fish. During the gold-mining era of the 1860s, water races were liberally constructed throughout Central Otago. A system of races connected streams of the Serpentine Diggings area (Waimonga and Totara Creeks, Taieri system) with the Pool Burn/Manuherikia area (Clutha system). This man-made connection had the effect of mixing fish (and presumably other aquatic animals) from two different river systems. Genetic analysis of the fish shows us that genes of the Pool Burn fish have crossed into the Taieri, and genes of the Taieri fish have crossed into the Pool Burn. So, in this region, the fish are hybrids between two distinct genetic types, and have lost their original species integrity. Second, and even more signifcantly, introduced brown trout have excluded native fish from large stretches of rivers through a combination of competition and predation. This displacement has occurred to the extent that, in some rivers, galaxiid populations are now found only above waterfalls where trout cannot penetrate. Indeed, searches for native fish in the Kawarau and its tributaries came up with over seventy locations where galaxiids were absent, and only six where they were present (five of these in the Nevis). Trout are abundant at many of the locations where river galaxiids are now absent; trout and galaxiids seldom co-exist.

Biogeographic research is said to be 'multidisciplinary'. It requires knowledge, understanding and methods of multiple fields of scientific endeavour, including genetics, biochemistry, zoology, botany and geology. With respect to the progress now being made in the field of New Zealand biogeography, our research exemplifies the value of this multidisciplinary approach. Similar work at Victoria University on skinks and geckos has also revealed high diversity in the Otago/Southland region. The fascinating evolutionary history of the region has provided the legacy of a unique flora and fauna for the enjoyment of biogeographers, natural historians and tourists alike.

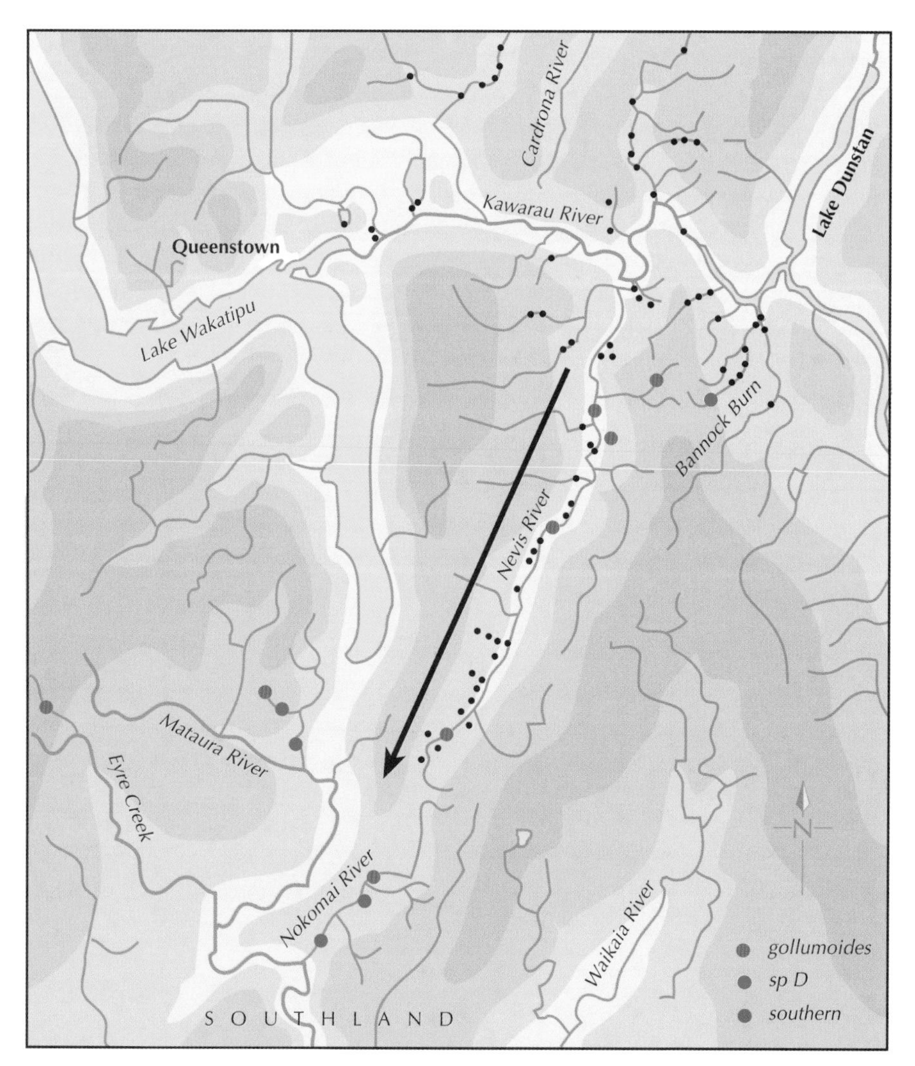

Fig. 38. Detail of the Nevis River, showing its current course into the Kawarau, part of the Clutha system. The large arrow indicates the likely direction of flow until about one million years ago. *Galaxias gollumoides* are found in the Nevis and rivers of the Southland plain. Black dots indicate sites where galaxiids are absent. G. Wallis and J. Waters

CHAPTER 6

ENVIRONMENTAL CHANGE SINCE THE LAST GLACIATION

MATT MCGLONE, PETER WARDLE, TREVOR WORTHY

Bracken fernland and invading manuka on terrain that was once forested, on the eastern side of Lake Te Anau. *Peter Wardle*

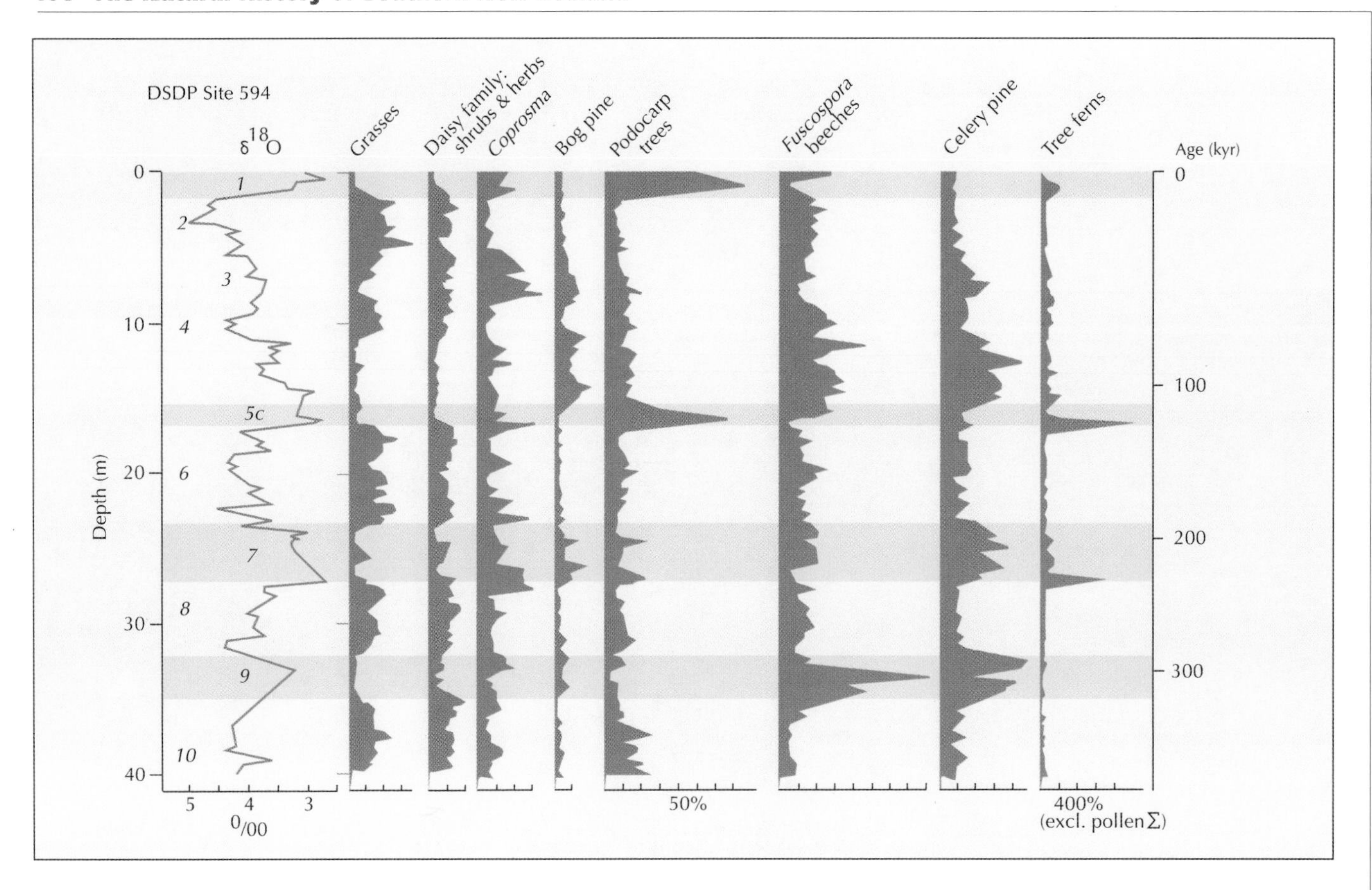

Fig. 1 (above). Deep-sea core DSDP Site 594 located in 1.2 km depth of water 300 km off the Waitaki River mouth. The oxygen isotope stratigraphy (*18O) indicates interglacials and interstadials (odd numbers) when global icesheets are at a low and sea temperatures are warm, versus glacials (even numbers) when global icesheets and glaciers are at full volume and local sea surface temperatures are cold. Glacial stages 2, 3, 6 and 8 have high proportions of silt in the sediments. *Data from Nelson et al. (1985) and Heusser & van der Geer (1994).*

Fig. 2. Southern New Zealand at the height of the Last Glacial Maximum. *Ice distribution from New Zealand Geological Survey 1973 Quaternary Geology – South Island 1:1,000,000 (1st Ed). DSIR, NZ Geological Survey Miscellaneous Series Map 6.*

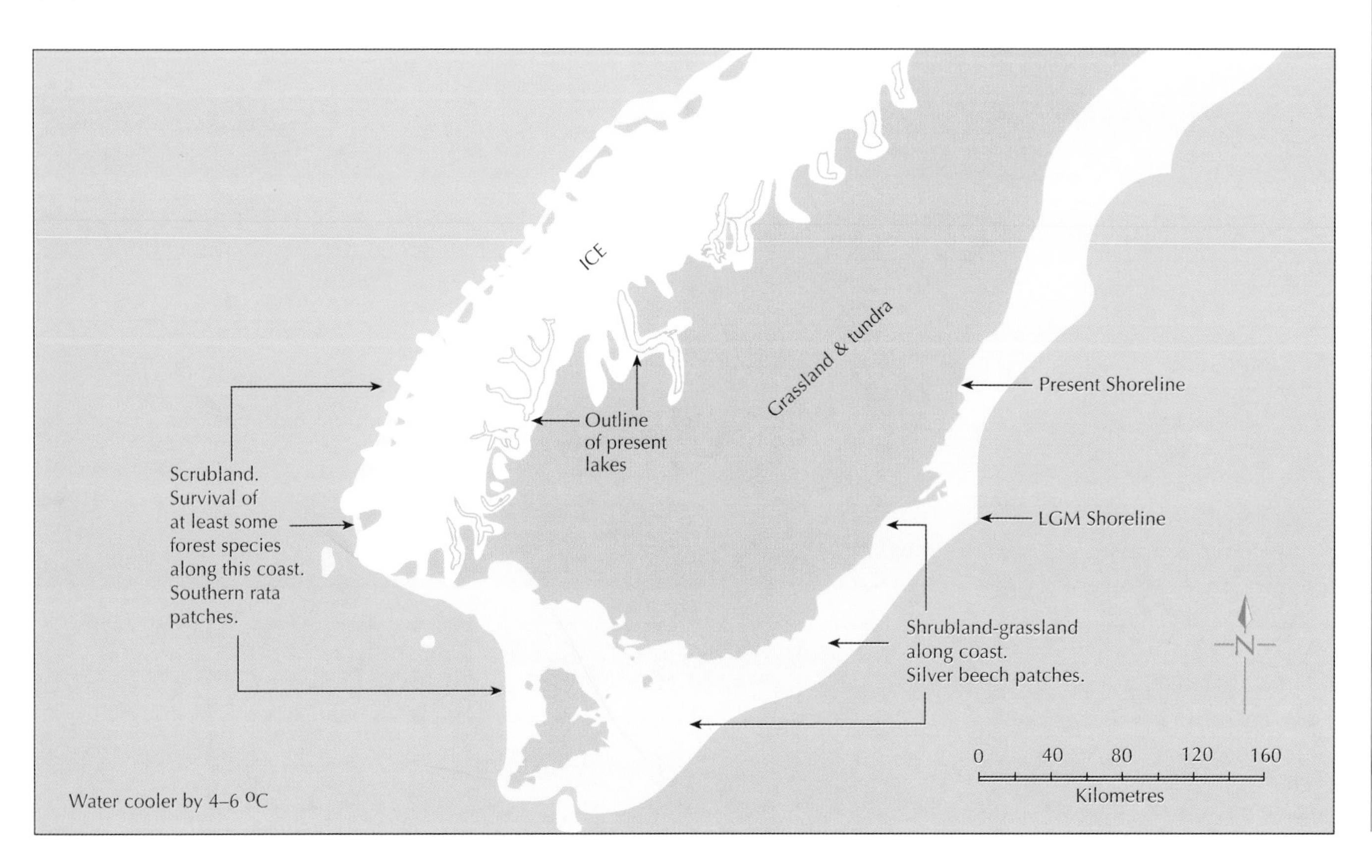

ENVIRONMENTAL CHANGE SINCE THE LAST GLACIATION

The ecology of southern New Zealand has been transformed by hundreds of years of Maori settlement and 150 years of intensive farming and pastoralism. Large areas have almost completely lost their original ecosystems and even those with a largely untouched vegetation cover have been affected by loss of birds, reptiles and insects. In our attempts to envisage what these lost ecosystems were like, we can extrapolate from the fragments of natural habitat that remain. However, these too have been altered, or have survived because of attributes which make them unrepresentative of the larger, more vulnerable whole. If we wish to gain a full appreciation of the present, we must study the past.

The fossil record of southern New Zealand (Chapter 3) extends back to when it was part of Gondwana, and documents the enormous changes in biota over millions of years. However, the history of the Tertiary flora and fauna, although fascinating in its own right, has little to do with the current biota of southern New Zealand. The distinctive southern environments did not take on their current form until about 2.5 million years ago, at the beginning of the Quaternary period, a time of global and local climatic cooling and rapid uplift of the southern mountain chain. In this chapter we will first give a brief overview of the Quaternary period to set the scene. We then discuss in more detail the environmental history of the last 20,000 years, which covers the period from the depths of the Last Glaciation through the warm interglacial of our present epoch – the Holocene. The vegetation evidence, based largely on fossil pollen, is continuous over most of this period and we will use this as a basis for discussing large-scale changes in landscape and climate. Fossil vertebrate evidence tends to be less continuous, and we will deal with only the faunas of the Last Glacial Maximum (LGM) and then those of the later part of the Holocene. Finally, we discuss the effects of human settlement on the composition and distribution of both vegetation and vertebrates.

VEGETATION AND CLIMATE HISTORY

The Quaternary

Between 5 and 2 million years ago the characteristic landscapes of southern New Zealand came into being in the course of the rapid uplift which created the axial mountain chains. A sharp depression of mean annual temperatures associated with a global cooling about 2.5 million years ago built glaciers in the South Island mountains, and led to the expansion of grassland and shrublands. The period of fluctuating climates that this cooling initiated, and which persists through to the present, is called the Quaternary. Estimates of when it began range from 2.5 to 1.6 million years ago, but the earlier beginning seems to fit the New Zealand sequences best. Quaternary climates have not been stable, undergoing marked cyclical fluctuations between warm and cold states. There have been about fifty of these warm/cold cycles in total during the Quaternary. During the cool phases (glacials) temperatures in mid latitudes fell on average by 4–6°C worldwide. Ice accumulated as huge ice sheets at high latitudes on northern continents and as valley glaciers on mountains throughout the temperate zone. The locking up of huge quantities of water in polar ice caps resulted in sea levels falling by as much 120–150 m. The warm phases (interglacials) had temperatures similar to those of the present, or a degree or so warmer, and at times sea levels were as much as one or two metres higher than now. Although warm, the interglacials have tended to be brief, lasting not much more than 10,000 years on average. Therefore, during most of Quaternary time the globe has been cooler than at present. The glacials typically terminate with a short but intense cold period, known as a full-glacial. The last one occurred between 25,000 and 15,000 years ago, and is known as the Last Glacial Maximum. After a 5000-year transition, the last Quaternary warm stage – known as the Holocene – began around 10,000 years ago.

Beyond the shallow water of the coasts, the ocean bottom slowly accumulates mud formed from the plankton living in the overlying water column and fine terrestrial material blown by wind or washed out to sea by rivers and streams. These deep-sea records are the only long continuous records of change in the New Zealand region and are therefore of great value in understanding the significance of the more fragmentary onshore deposits. Pollen and sediments from a deep-sea core 250 km east of the South Island span four complete glacial/interglacial cycles (Fig. 1). The pollen in the core shows that beech *Nothofagus* spp. forest and scrub dominated by celery pine *Phyllocladus alpinus*, bog pine *Halocarpus bidwillii* and a range of small-leaved shrubs has been the major vegetation cover of the southern South Island landscape for most of the last 450,000 years. During the brief interglacials, which averaged less than 15,000 years in length, tall podocarp forest expanded to cover the lowlands of southern South Island. During the equally brief glacial maxima, grass and scrub virtually excluded forest. Vastly extended glaciers and the bare rock slopes above them generated shattered rock which was eventually carried by rivers as gravel, sand and silt down to the coast. Water and wind eroded bare or sparsely vegetated ground in the unglaciated east. Wind-blown silt of glacial age forms the substrate for most eastern lowland soils. Vast quantities of fine silt blown or washed out to sea from the southern South Island make up 70–90 per cent of the deep-sea core sediment during glacial periods.

Major changes also occurred in the seas surrounding southern New Zealand during glacials, and these are reflected in the changing proportions of temperature- and salinity-sensitive fossil plankton in the cores, and oxygen isotopes in carbonate shells. At present, the Subtropical Convergence, where warm, salty northern water meets cool, low salinity southern water, curls around southern New Zealand and then heads eastwards. During the LGM, the Subtropical Convergence did not move, remaining pinned against the Chatham Rise. However, to the south, cool ocean water masses moved northwards by at least 5° latitude, creating a steep ocean temperature gradient around southern New Zealand. Water cooler than present by 4–6°C – possibly for brief intervals as much as 10°C cooler – surrounded the southern South Island. Icebergs from Antarctica have drifted this far north at times in the past. As the oceans to the north of New Zealand did not cool as much as those in the far south, the steepening of ocean temperature gradient intensified glacial westerly winds across the southern landmass.

In the course of the Last Glacial, ice piled up high along the Main Divide of the Southern Alps, fed by the moist westerly wind flow (Fig. 2). Along the central and southern axial mountain chain, glaciers filled and deepened the great U-shaped valleys formed by repeated past glaciations. In the west, the glaciers pushed out beyond the present coastline. In the east, the great terminal moraines at the outlets of the southern lakes mark the limit of glacial advance. Small glaciers carved cirques below the southern and eastern summits of the higher ranges in Central and Eastern Otago, Southland and Stewart Island, but otherwise these regions were not glaciated. Falling sea levels exposed the wide continental shelf that stretches eastwards and southwards from Southland, joining Stewart Island to the mainland. In the west, the Fiordland coast plunges deeply, so that the glacial coastlines were within 5 km of the present coastline. Rivers cut deep gorges that now form submarine canyons in the shelf, but at that time would have provided habitats sheltered from the winds of the open coast and cold temperatures of the interior.

The Last Glacial entered its coolest phase (LGM) around 25,000 years ago. Glaciers extended forward in a series of great pulses between 22,000 and 14,000 years ago, building giant terminal moraines. Annual temperatures, estimated from the lowering of snow lines and data from deep-sea cores, were on average 4–6°C colder than those of today. Rainfall was also much lower, perhaps by more than a third, and the drier atmosphere greatly increased the severity of winter frosts. The expanded land area must have increased the seasonal temperature range in the eastern interior, summers being warmer and winters colder than would be predicted from annual temperatures. Geomorphic evidence in the form of moraines, loess sheets, soil layers and erosion horizons also indicates a dry, cool, frosty climate.

Only a little fossil evidence dates to this period. Pollen records from the east suggest an open grass and herb-dominated landscape, with some scattered patches of low-growing shrubs, perhaps similar to a dry cold tundra (Fig. 3). Pollen of trees and conifer shrubs is rare and some of it may have been blown south from further north; no wood or charcoal dating to the LGM has yet been found. The deep-sea core pollen record for the LGM (Fig. 1) also shows that grasses, herbs and shrubs were abundant, but indicates that substantial amounts of forest (especially beech) and conifer scrub remained, possibly along the coastline. A pollen site in Preservation Inlet, Fiordland, recorded the vegetation immediately in front of the fiord glacier 18,000 years ago. Grass and scrub of *Coprosma*, *Myrsine* and five-finger *Pseudopanax* spp. were the main vegetation cover, but groves of southern rata trees *Metrosideros umbellata* also grew close by.

The distribution of plant species confined to the southern part of the South Island – the regional endemics – further helps to define the LGM climate. Since few, if any, of them could have evolved in the short (geologically speaking) span of time since the glaciers receded, they must have persisted through the period of maximum climatic rigour. Most of them are high-mountain plants, and include some gems of the regional flora – species of *Celmisia, Ranunculus, Cheesemania, Aciphylla,* among others. Those now confined to the high plateaux of Central Otago and inland Southland were doubtlessly at home in cold, dry, wind-swept glacial tundras. Others now endemic to the Fiordland tops or the highest summits of Stewart Island provide evidence of western climates that were cold, but moist. There are also endemic shrubs in the south and west, a few of which can become small trees. Examples are the neinei or 'pineapple tree' *Dracophyllum fiordense* that is endemic to the subalpine belt of western Fiordland and South Westland, and a pair of tree daisies, tete-a-weka *Olearia angustifolia* and Snares tree daisy *Brachyglottis stewartiae,* that are confined to far southern coasts. Such species are further indications that moist, mild climates persisted close to the coast in high rainfall areas.

We cannot be certain on the basis of pollen and macrofossil evidence alone whether or not tall lowland trees survived in the southern South Island during the LGM. On the assumption that temperatures decreased about 0.6°C for every 100 m in altitude, as they do today, vegetation zones should have been about 1000 m lower. At present, silver beech forms treelines as high as 1100 m in our region, and mountain beech treelines up to 1300 m. Therefore, LGM temperatures should have allowed our hardiest trees to survive in favoured localities, especially if we take the lower sea level of the time into account. The modern distribution of the beeches provides indirect evidence of such survival. At present, wide gaps without beech separate the beech forests in the south of the South Island from those in the north of the island; west of the Southern Alps, there is a stretch of 170 km that has no beech at all and, on their eastern side, a stretch of 125 km with only widely scattered stands. Fossil evidence shows that this 'beech gap' existed even before the LGM and, since beeches are not known to disperse seed over distances greater than 6 km, the modern beech forests of our region must have spread from pockets that survived the glaciation.

The main centres of forest survival, or refugia, in the east may have been in the Catlins and Otago Peninsula districts,

Fig. 3. Tussock grassland and low scrub extending to the valley floor in the head of the Earnslaw Burn at the head of Lake Wakatipu, under the influence of a cold, frosty climate: perhaps similar to those that prevailed widely during glacial times. *Peter Wardle*

where an indented glacial shoreline lay close to coastal hills. Groves of silver beech that are now scattered along the eastern foothills of Otago and Canterbury are possibly descended from these eastern refugia. In the west, coastal Fiordland and far-south Westland – despite being interrupted by a series of tidewater glaciers – still experienced the moderating influence of the ocean, and here some forest also persisted. Beyond this, how much forest survived, and where and what species, becomes speculative. Annual average temperatures experienced during the LGM suggest that few species among our taller trees could have survived. Red beech *Nothofagus fusca*, hard beech *N. truncata* and the lowland podocarps – matai *Prumnopitys taxifolia*, miro *P. ferruginea*, totara *Podocarpus totara*, kahikatea *Dacrycarpus dacrydioides* and rimu *Dacrydium cupressinum* – should not, on the basis of what we understand of their ecology, have been among them. Yet, for red beech too, our southern stands are separated from those in the north of the South Island by 250 km. If red beech survived in the south, lowland podocarp trees may have also survived in favoured locations.

We now follow the successive changes of the vegetation as it responded to altering climates and human impacts (schematically depicted in Figs 4 and 5). Representative pollen diagrams, on which these interpretations are largely based, are given in Figs 7 and 8.

The late-glacial warming (15,000 to 10,000 years ago)

The glaciers began a rapid retreat about 15,000 years ago as a result of the rapid global warming called the late-glacial. However, the climate still remained cool, probably averaging 2–3°C colder than today for the next 4000 to 5000 years. Temporary reversals of the warming trend occurred, one of which, about 11,000 years ago, pushed forward glaciers on both sides of the Southern Alps, creating the spectacular Waiho Loop moraine of Franz Josef Glacier. East of the Southern Alps, the low grassland and herbfield vegetation that dominated at the end of the glaciation was steadily replaced by shrubs and tall tussock grassland. Grasses, *Coprosma*, *Myrsine*, *Muehlenbeckia*, *Dracophyllum*, *Olearia* and other members of the daisy family (Asteraceae) dominate the late-glacial pollen rain, but occasional occurrences of more poorly represented pollen types such as *Pseudopanax*, and *Melicytus* suggest a diverse vegetation. The majority of woody species in this late-glacial vegetation in the interior were low-growing, small-leaved and shrubby, reflecting a windy, cool dry environment. Along the southern and south-eastern coasts, ground ferns increased together with the shrubs. Comparable fern-rich scrub today grows mainly on flat-floored valleys, and up on to adjacent fans and colluvial slopes. Usually, the soils are recent and well drained, but they can also be swampy, especially west of the Main Divide. Few New Zealand trees can withstand frosts of more than –10°C, especially at the vulnerable seedling stage, and severe frosts resulting from night-time temperature

Fig. 4 (right). Palaeovegetation transect for profiles in Fig. 5.

Fig. 5. Palaeovegetation cross-sections for the intervals: 18,000 years ago; 10,500 years ago; 8000 years ago; 1500 years ago; 500 years ago.

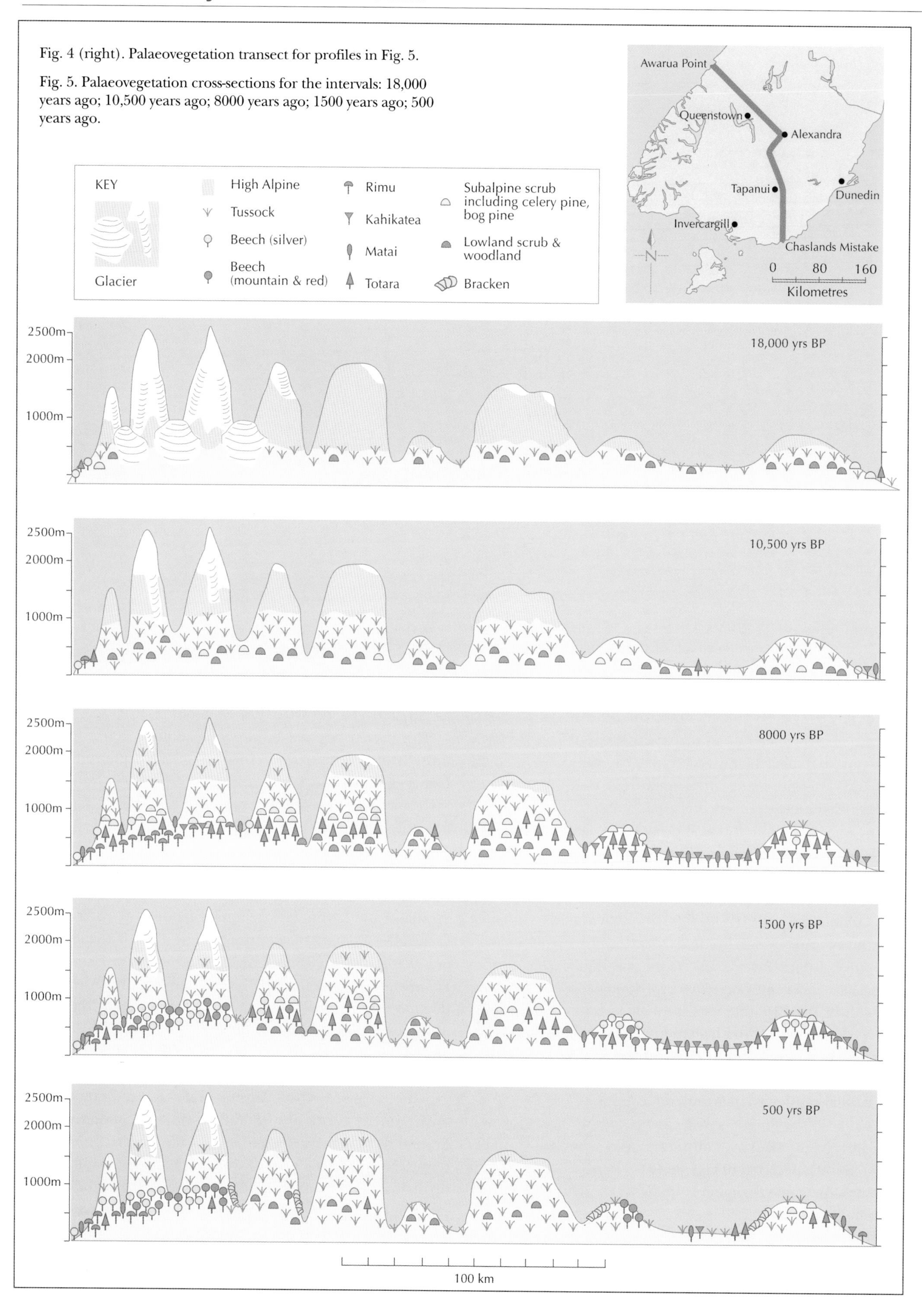

inversions keep this modern scrub free of trees. The late-glacial was therefore not only cool, but very frosty.

From around 12,000 years ago, in the later part of the late-glacial period, scrub and low forest began to replace the previous grassland and shrublands. Southern rata and tree ferns were first to spread, forming substantial forests between 14,000 and 12,000 years ago in Fiordland and Stewart Island. Between 12,000 and 9500 years ago, kamahi *Weinmannia racemosa* partially supplanted rata in these south-western forests. On the southern and eastern coasts, mountain lacebark *Hoheria glabrata* – and later lowland ribbonwood *Plagianthus regius* – became common, forming forest groves in a landscape still largely in grassland and shrubland. From 11,000 years ago, tree ferns increased in all coastal districts south of the Otago Peninsula, forming tree fern forests intermingled with scrub and ribbonwood that persisted for several hundred years. The cool, but rather moist, climates of the southern New Zealand coastal region generally favoured broadleaved trees, ground ferns and tree ferns, while the south-eastern rainshadow area of the interior remained dry, with grasses and small-leaved shrubs being dominant. The first significant stands of tall podocarp forest may have established at around 12,000 years ago on the eastern and southern fringes of the dry interior, but it is unclear whether this represents migration from the northern half of the South Island or expansion from small local pockets. In the northern South Island, first bog pine *Halocarpus bidwillii* and then celery pine *Phyllocladus alpinus* began to replace the scrub/grassland communities of the early late-glacial after 12,000 years ago. However, these hardy small trees played only a minor role in the late-glacial scrub of the south-east. This late-glacial scarcity of conifer scrub in the south-east is surprising, as celery pine and bog pine are the most cold-tolerant trees in New Zealand, withstanding winter temperatures below –20°C.

Fig. 6. Celery pine woodland near the head of the Minaret Burn, west of Lake Wanaka. This is a remnant of a vegetation type that covered much of inland Otago until it began to be overwhelmed by advancing beech forest some 2000 years ago, and was finally destroyed in fires lit by people. *Peter Wardle*

The early Holocene (10,000 to 7000 years ago)

The development of the vegetation

The late-glacial terminated with rapid warming around 10,000 years ago. Annual average temperatures became as warm as, or even somewhat warmer than, those of the present. Mountain glaciers retreated far up their valleys, and there were no major re-advances for many thousands of years.

Between 10,000 and 9500 years ago, matai, miro, totara and kahikatea increased massively over several hundred years to form the first continuous tall forests in the south and east of the South Island (Figs 7 and 8). In contrast, matai and totara did not become common in the wet, western coastal districts of Fiordland. There, the tree fern-rich kamahi/southern rata forests of the late-glacial were invaded by rimu, miro and kahikatea in the early Holocene. Stewart Island did not share in the resurgence of tall lowland podocarps. While it seems that all the tall podocarps were present on the island, with miro being relatively abundant in some areas, none were dominant in the early Holocene (Fig. 7). Instead, a low forest of kamahi, southern rata, and tree ferns prevailed in the lowlands, with scrub in the uplands. Pollen and macrofossils show that silver beech was locally dominant in parts of Fiordland from at least 8000–7500 years ago, having undoubtedly survived glaciation there.

In the semi-arid parts of Central Otago, small-leaved shrubs and grassland prevailed until 8000–7500 years ago, when abundant pollen of tall podocarps, and matai in particular, appears in nearly all sites. While small stands of matai may have occurred within Central Otago, it seems more likely that the source for this pollen, and that of most other tall podocarps in Central Otago, was the extensive matai/kahikatea-dominated forests that grew on the inland Southland plains, and extended up the fertile valley bottoms on the edge of the dry heart of Otago. At the same time that lowland podocarps spread, mountain totara *Podocarpus hallii*, celery pine and bog pine began to occupy the upper forest zone. Stands of celery pine (Fig. 6) grew throughout, accompanied by mountain totara below 900 m, snow totara *Podocarpus nivalis* above 900 m, with bog pine on terraces, gentle slopes and some rock outcrops. However, occurrence of forest charcoal is always rare in Central Otago, even that of celery pine charcoal being mostly confined to eastern mid-slopes of the ranges and sheltered slopes of narrow mountain valleys. Dense forest and scrub cover was most extensive on east-

1 EVIDENCE FOR PAST ENVIRONMENTAL CHANGE

Microfossils

Any deposit that has either been continuously wet throughout its existence (for instance, swamp and bog peats, lake sediments, wet soils, deep ocean muds), or dry (mainly cave deposits) is likely to preserve many types of small fossils. Swamp and bog deposits have been our main source of microfossils in southern New Zealand. Protozoa, diatoms, pollen, fern spores, fungal spores and hyphae, and small fragments of plant charcoal are among the common microfossils preserved in peat.

The most useful of these for charting changes in vegetation at a landscape scale are pollen and fern spores, and microscopic charcoal. The classic pollen studies of Southland and Otago, published in 1936 by Cranwell and von Post, were the first substantial work done outside Europe and were highly influential. There are now many published pollen diagrams covering the last 15,000 years from throughout New Zealand, and the broad outline of postglacial vegetation change is well understood.

Pollen and fern spore outer walls are distinctively sculptured and highly resistant to decay. While the cell contents and cellulosic inner walls vanish within a short time, the outer walls take at least several years to decay even in a biologically active soil. In wet sediments the walls are virtually immortal. A cubic centimetre of peat may contain a million or more grains of pollen. Small subsamples (usually a cubic centimetre or less) of sediments are sieved, and chemically or physically treated to remove mineral and organic material and to concentrate the pollen and spores. A random sample of the pollen grains and spores is then counted (usually more than 250) and the results variously calculated as a percentage of the total number of pollen grains in the sample, or as pollen grains per cubic centimetre of sample, or if sufficient radiocarbon dates are available, as pollen influx. All of the work discussed here is based on percentage pollen results.

Interpreting pollen diagrams is never straightforward. First, while some pollen types are produced only by one plant species (for instance miro, rimu, matai, ribbonwood, silver beech), others can only distinguish a genus (for instance, *Phyllocladus*, *Coprosma*, *Myrsine*), and yet others relate only to families (grass). Pollen and spores are produced in widely differing amounts by plants: some like matai and beech shed great clouds of pollen that are carried long distances by the wind; others like flax produce tiny amounts, and are dropped very close to the parent plants; yet others produce very large amounts of pollen that is carried hardly any distance (manuka, kamahi). In order to interpret fossil pollen results, pollen analysts have analysed large numbers of samples from surface sediments so that modern-day vegetation can be compared directly with modern pollen rain.

Macrofossils

Plants Macrofossils are undecayed leaves, stems (including tree trunks), seeds and other plant parts preserved in various deposits. While most peats and lake muds contain macrofossils, they are often poorly preserved and consist mainly of locally growing wetland plants. Well-preserved macrofossils of dryland plants that are occasionally found in soils, peats and alluvial silt are of great significance because they provide certain evidence of a plant species being present at a given location at a given time. Wood and charcoal macrofossils, like pollen, usually permit identification only of the genus rather than the species. However, logs lying on the ground are common in southern New Zealand, as are identifiable fragments of tree and shrub charcoal in the subsoil. In general, identifiable soil charcoal is preserved only if it is buried shortly after the fire; otherwise it is incorporated into the biologically active upper levels of the soil and broken down.

Subfossil subalpine woodland of bog pine and pink pine (*Halocarpus bidwillii* and *H. biformis*) exposed in an area of eroding peat on the summit ridge of Maungatua (Maukaatua) at 890 m, among mixed narrow-leaved snow tussock grassland. Bog pine still grows on the mountain but the nearest living pink pine is on Mt Cargill, much closer to the coast. *Alan Mark*

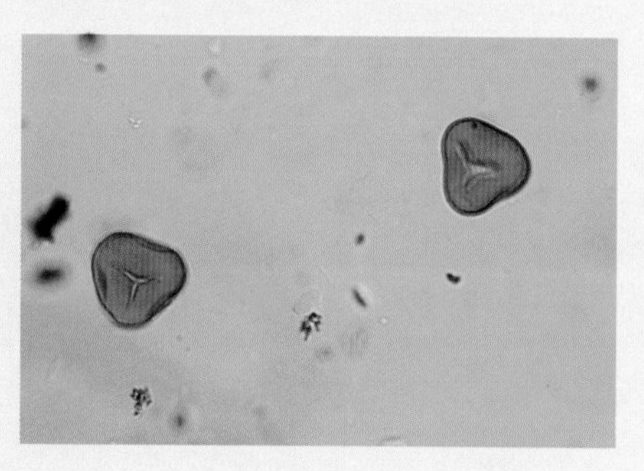

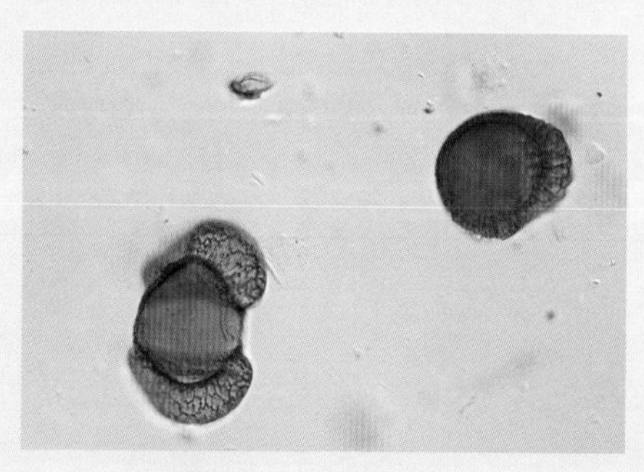

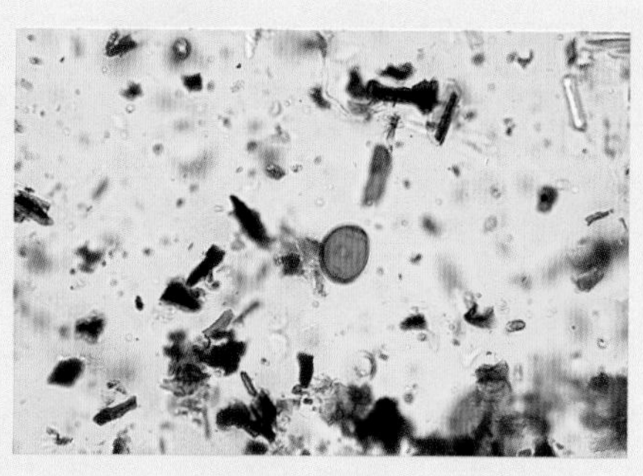

Fossil pollen grains and charcoal x 400: spores of bracken (top); matai (with prominent wings), rimu (with frilled edge) (centre); tutu pollen and charcoal fragments (bottom).

Animals While a great range of animal parts can be fossilised – including bones, feathers, skin, insect integuments and shells; only bird and reptile bones have been studied in any detail in this region. Calcareous swamps and springs, where normal peat acidity that dissolves bone is reduced, are a major source of bird bone fossils. Large birds such as moa are commonly found in these sites, presumably trapped as they ventured out on a treacherous surface to feed or drink. Some have been found with crop contents intact, a valuable source of information on their diet. Vertical cave entrances, mainly in calcareous or schistose rocks, form pitfall traps that collect and preserve bones of birds and reptiles that have fallen or fluttered in, and been unable to escape. Some caves were used by the predatory laughing owl, and accumulated regurgitated bones form a good record of the smaller prey in the area.

Because Central Otago is dry, some caves and rock shelters have contained spectacular mummified remains. Among the best finds have been the lower leg of a *Dinornis novaezealandiae*, which has all the skin and scutes complete, and the complete leg of a *Megalapteryx didinus* – that shows this species was feathered from the femur to the base of the toes. Such feathers are rare in birds, but a useful adaptation for birds that must have often foraged in snow. Several moa eggs have been found in more or less perfect condition; one even had the bones of the embryo within it. Moa are not the only animals mummified: a tuatara has also been found. Moa crop stones are widespread in the region.

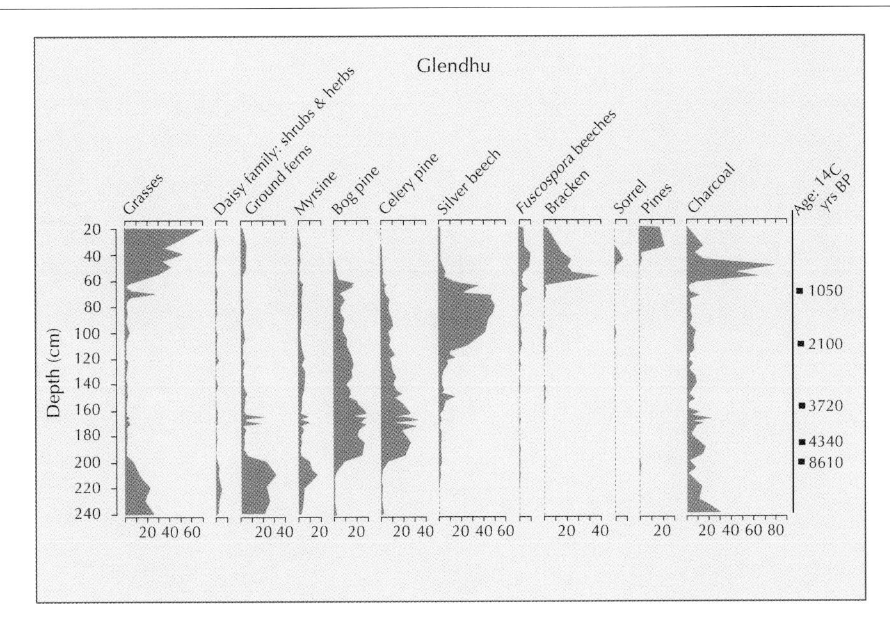

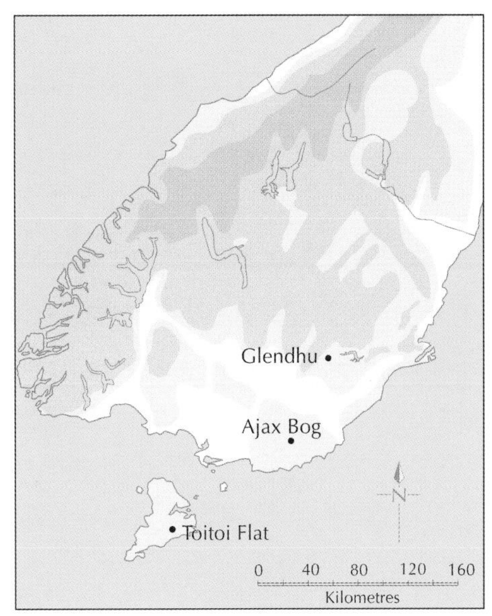

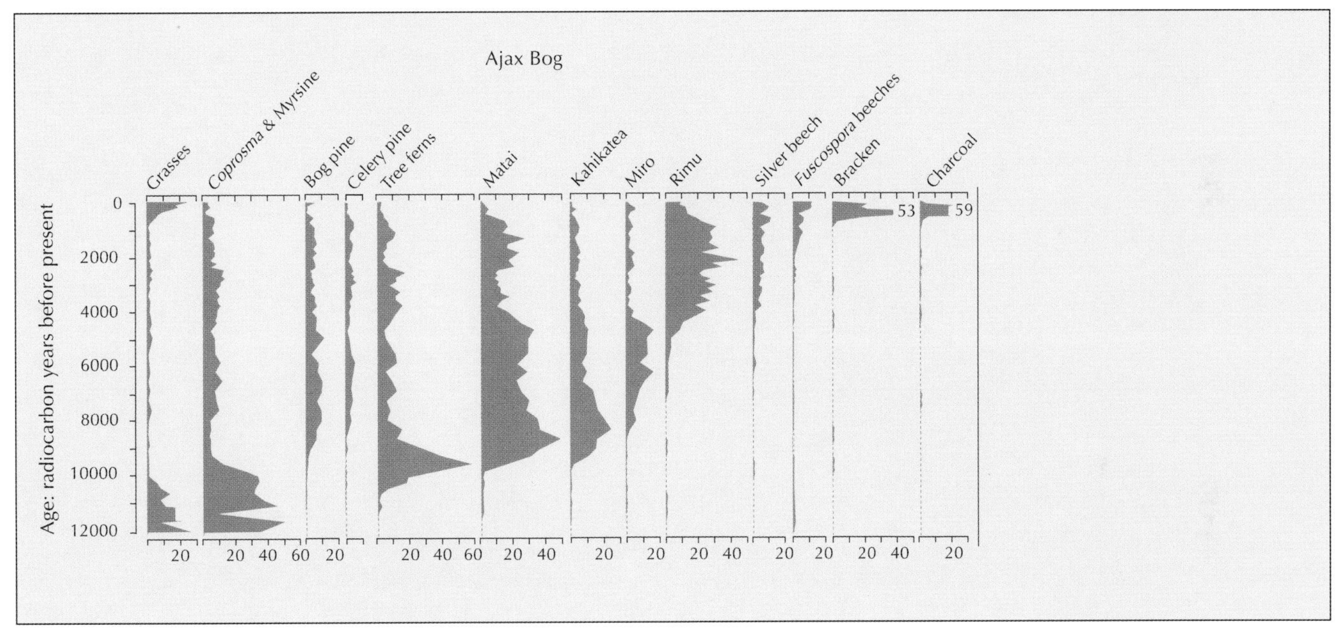

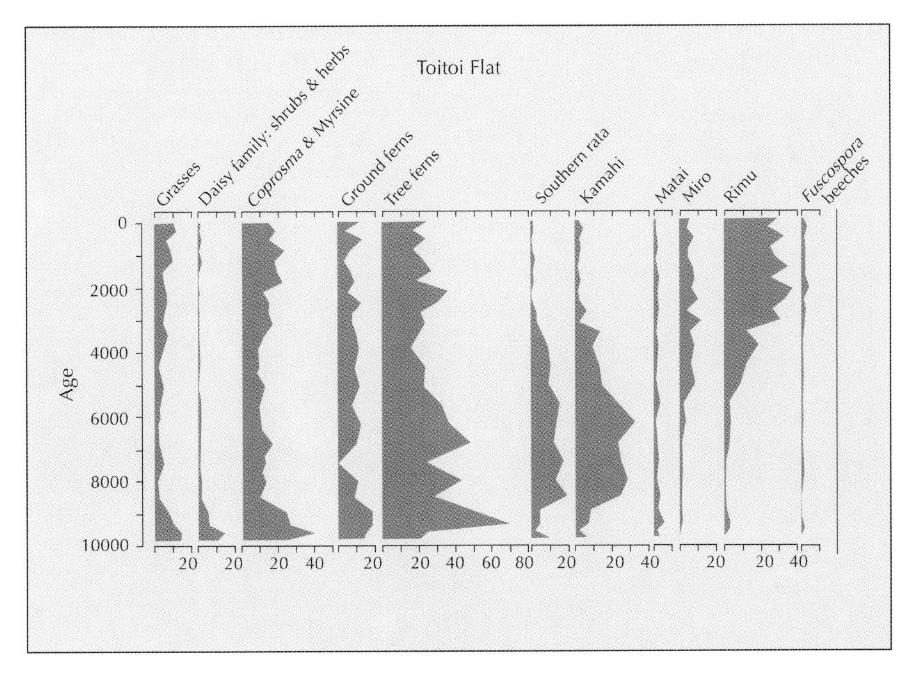

Fig. 7. Holocene pollen diagrams from coastal Southland and Otago. Toitoi Swamp (Stewart Island; McGlone & Wilson, 1996); Ajax Bog (Catlins, McGlone unpublished); and Glendhu Bog (near Lawrence, McGlone & Wilmshurst, 1999). Note the abundance of tree ferns in the highly oceanic Stewart Island setting throughout the Holocene. All three sites have a very similar timing (4000–2000 years ago) of the replacement of previous forest types by rimu at Toitoi Swamp, and by silver beech at Glendhu Bog and Ajax Bog. *Fuscospora* beech pollen appears simultaneously at around 2000 years ago in all sites, representing growth of beech forest in central and western districts.

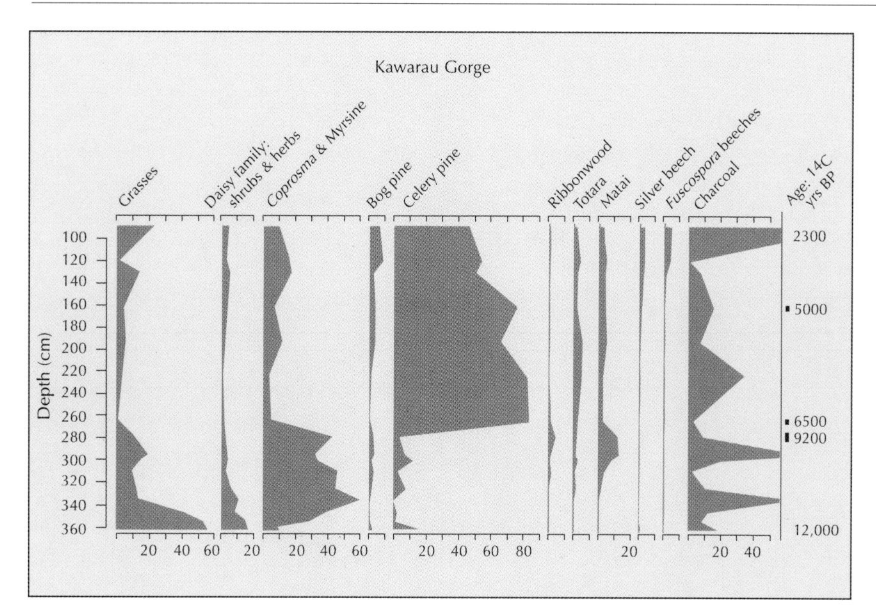

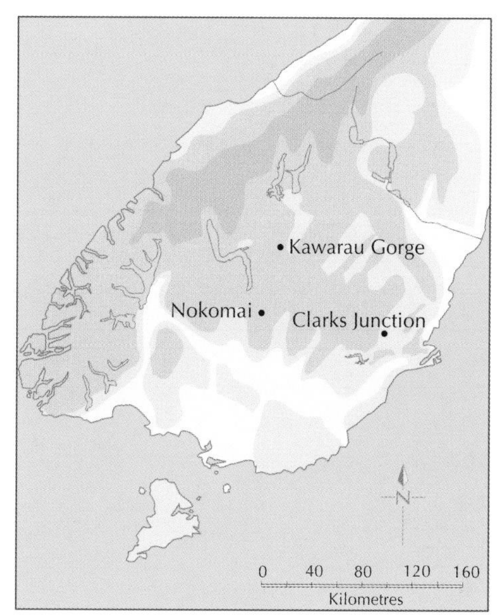

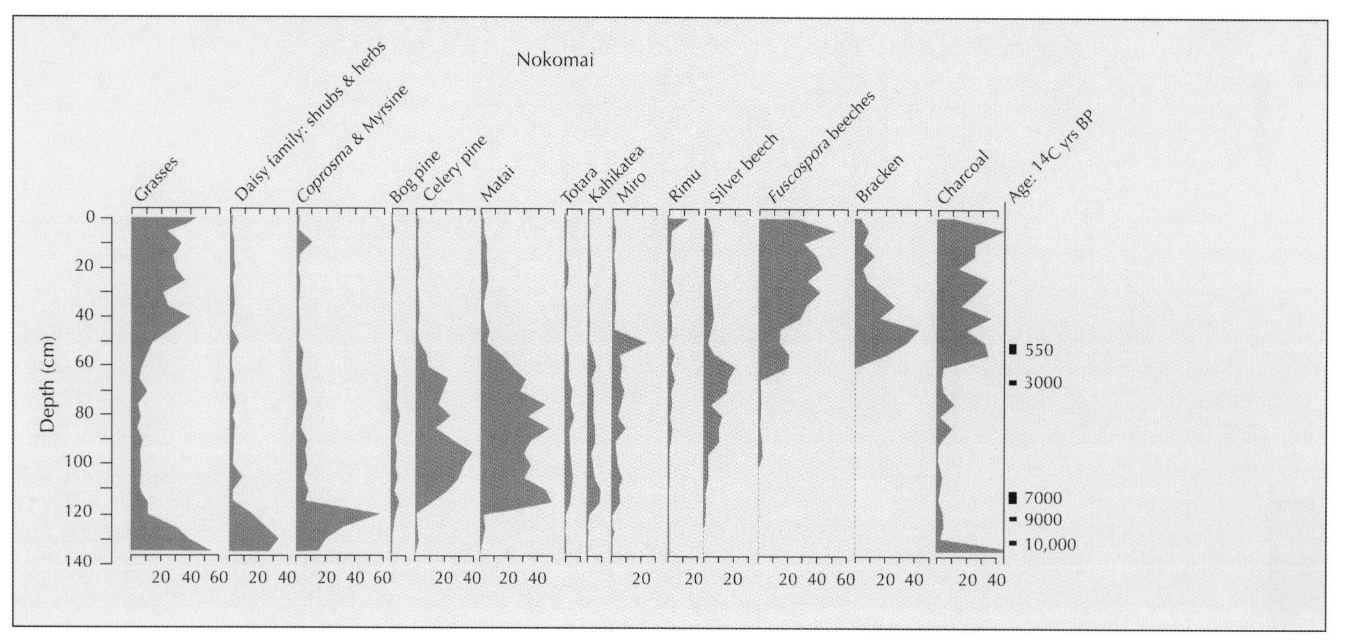

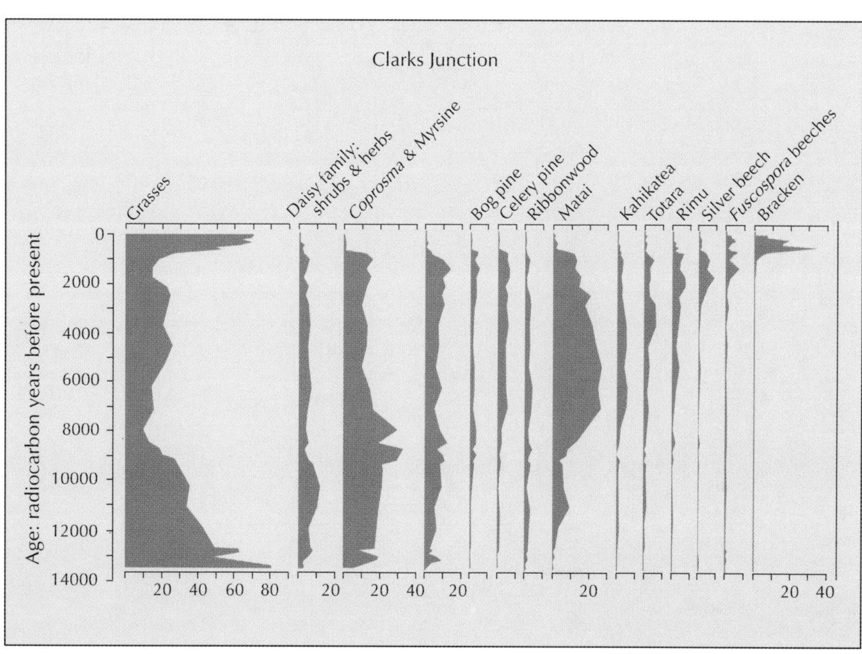

Fig. 8. Holocene pollen diagrams, inland Otago. Kawarau Gorge and Nokomai (McGlone et al. 1995); Clarks Junction (McGlone, unpublished). Note the highly discontinuous peat accumulation at these sites versus the more continuous accumulation of the coastal sites. This is probably due to the drier inland climates. Full forest cover did not develop at these sites until well after 9000 years ago, and possibly as late as 7000 years ago in the interior. Celery pine and other shrubs were important constituents of the vegetation in the interior, and fire always important. Maori fire and obliteration of dryland vegetation occurred around 700 years ago, and is clearly shown at Nokomai and Clarks Junction.

Fig. 9. Mountain totara trees growing among kanuka on a schist crag on the eastern slopes of the Pisa Range, Central Otago. *Peter Wardle*

facing mid slopes (Fig. 9), but also occurred on other aspects and descended locally to the valley floors. On dry lower slopes and valleys, and on open western slopes, rare finds of charcoal are mostly derived from shrubs such as matagouri and *Muehlenbeckia* lianes. Elsewhere, on dry valley floors and north-facing slopes, a dry scrub of small-leaved divaricating shrubs and grassland probably grew, and has left little trace of charcoal or pollen.

In the higher-rainfall inland zone closer to the Southern Alps, pollen evidence shows that bog pine and celery pine scrub spread across the landscape, accompanied by a few small stands of matai and totara forest. Charcoal of celery pine is ubiquitous in the soils but this does not necessarily indicate it was totally dominant, as this species seems to contribute to soil charcoal more readily than other trees and shrubs. We assume that it was accompanied by other woody species that have left little trace. Matai charcoal has been found at scattered localities around the shores of Lakes Hawea and Wanaka. Podocarp forests, similar to those of coastal districts, occupied the lower slopes around the southern lakes and extended up the tributary valleys towards the Main Divide. Upslope, and in the upper valley reaches, these merged, first into low-canopied forest and then into subalpine scrub, both with celery pine as a major component. In some western locations, silver beech formed part of these early forest and scrub communities.

The vegetation pattern of the southern South Island during the early Holocene suggests a lower but well-distributed rainfall and equable temperatures, with mild winters. This is part of a New Zealand-wide pattern. For instance, some trees and shrubs – for example matai, miro, kahikatea and horopito *Pseudowintera colorata* – extended their range beyond current limits into the frosty upland basins of inland Canterbury, and North Island vegetation generally reflected mild winters and moist summers.

Holocene migration or local survival?

The timing of the various phases of the reafforestation of the southern region is matched by warming sea-surface temperatures in the surrounding oceans and retreat of glaciers in the mountains. We can therefore be sure that warming climates were the primary cause of this major vegetation change. However, an intriguing question centres on whether the early Holocene scrub-forest successions that we have outlined were driven by consecutive migration of species from the north, or by expansion of species already present in the area, as climatic, soil or competitive conditions arose that suited that particular species. Some of the major tree species that probably persisted in the south throughout the LGM – kaikawaka *Libocedrus bidwillii*, silver beech, mountain beech *N. solandri* var. *cliffortioides*, southern rata and, more problematically, red beech, were not prominent in the early Holocene forests. Instead, matai, kahikatea, totara, celery pine and bog pine were the most abundant trees and tall shrubs. They are all adapted for bird dispersal, and therefore should have spread rapidly once climates ameliorated. One possible scenario is that they migrated, following warming, into the southern South Island from refugia in the north of the island. However, pollen evidence for nearly synchronous expansion of tall podocarp forest between 10,000 and 9500 years ago throughout the southern two-thirds of the South Island seems to rule out any possibility of a classic migratory wave of lowland trees moving to higher latitudes following a warming. The alternative scenario is that small pockets of many lowland species survived close to the coast and inland on relatively mild, sheltered sites, and expanded out of them as the climate became favourable. There is no macrofossil evidence for lowland forest trees surviving in the southern South Island during the LGM, but traces of forest pollen hint at small scattered patches surviving. However, the relative importance of local sources versus migration from the north in the formation of Holocene forests is not yet settled and remains a problem for further investigation.

Beech in the early Holocene

Despite the beeches most likely having survived in southern New Zealand, they are not at all prominent in early Holocene vegetation. The tree fern-rich podocarp forests of the lowlands and coastal slopes growing under ideal climates and on fertile soils would have been well able to resist invasion from scattered pockets of the relatively light-demanding beeches, and even overwhelm them. It is less easy to explain why the beeches did not spread on to the colder, subalpine slopes, or into the interior which, even during the early Holocene, must have been relatively dry and frosty. Here, time and distance may have played a part. If glacial refugia were located at low altitudes near the coast, beech may simply have been unable to spread upwards and inland across the barriers presented by forests of podocarps and associated broadleaved trees. However, we must also consider the possibility that features of the upland and inland climates may have been inimical to the beeches.

Most of the formerly glaciated valleys that extend eastwards and westwards from the Main Divide today have beech forests in their middle reaches that extend from the valley floors up to regular treelines. But in the uppermost reaches towards the valley heads, the beech forests give way abruptly to other vegetation that includes snow tussock grassland, dense subalpine scrub and even low-canopied forest. Celery pine is usually abundant in the woody communities of such valley heads, and can be dominant. Bog pine, pink pine *Halocarpus biformis*, and occasionally kaikawaka can also be present. Studies in the mountains of Canterbury have shown that the slopes enclosing these valley heads are frostier than slopes further down-valley. Minimum temperatures, both during summer and winter, fall low enough to kill foliage of mountain beech, but they are well within the tolerance of celery pine and bog pine. Summer frosts of only –4°C suffice to kill the young foliage of mountain beech, and appear to be a major factor limiting its spread up-valley, upslope and down onto the valley floors where night-time temperature inversions lead to severe frosts. Southern New Zealand may have experienced a climatic regime in the early Holocene in which damaging frosts seldom occurred near the coasts – which were buffered by a warmer ocean – but were more frequent than now at higher altitudes and in inland localities because of persistent high pressure weather systems during winter. Such juxtapositions of lowland environments that experience little or no frost and mountain environments that experience frequent frost are characteristic of highly oceanic and tropical regions.

The middle to late Holocene (7000 years ago to the present)

Between 7000 and 5000 years ago, the equable climate of the early Holocene changed. The distribution of solar radiation had been changing steadily since 11,000 years ago towards increasing summer and decreasing winter sunlight, thereby increasing the seasonal winter/summer contrast. Pronounced climatic oscillations began, marked by advances and retreats of existing glaciers. These well-dated Southern Alp advances, although puny compared with the massive glaciers of the LGM, are highly significant indicators of climate change. The first advances were about 5000 yrs ago, and resulted in moraines that, in some instances, lie a kilometre or more down-valley from the terminal moraines of the recent centuries. Similar rejuvenation of glaciers occurred in many parts of the world. The most recent significant advances took place between A.D. 1600 and 1800; in the northern hemisphere this interval has become known as the 'Little Ice Age'.

Recent research has shown that glacial advances have been more likely to occur in years characterised by strong south-westerly wind flow and a northwards movement of the subtropical high in summer. These changes in the regional atmospheric circulation often accompany El Niño events that bring generally cooler weather to the entire country, drought to eastern districts of the South Island and increased rainfall along the southern edge of the South Island.

Rimu and silver beech began to spread about 6000 years ago in eastern districts, replacing matai, miro and kahikatea on coastal plains and hills. On Stewart Island, rimu moved into broadleaved forests of kamahi and southern rata, becoming an overstorey dominant by 4000 years ago. Falling annual temperatures and increased south-westerly airflow, bringing wetter, cooler winters, especially along the coast and to the uplands of the interior, seem to have been the major stimuli for these forest changes. Soil deterioration must have also played a part as the fresh rejuvenated soils of the late-glacial became leached of nutrients and developed hard pans that restricted drainage. As studies on forests and soils in Westland show, dominance passes from matai and kahikatea to rimu and miro as soils become old and leached.

Celery pine, bog pine and their scrub associates continued to be common over large tracts of the interior, even though fire periodically swept through this highly inflammable vegetation. At one site in the Manorburn basin, fires burnt celery pine on four occasions between 8000 and 1400 years ago. Charcoal of kanuka *Kunzea ericoides* and manuka *Leptospermum scoparium* accompanies that of celery pine at many locations, showing that these shrubs and small trees were part of the association, perhaps playing a successional role after fire.

Beech spread

Spread of beech in the later Holocene was the most significant vegetation change before deforestation. The patterns of beech forest spread have long been of interest to biogeographers and ecologists and we will therefore look at them in more detail.

From around 6000–5000 years ago, silver beech began to increase throughout southern New Zealand, both in the interior and in coastal districts from the Longwood Range on the south-eastern border of Fiordland to as far north as Dunedin. In the west it may have spread through low transalpine passes from ice-age refugia on the coasts of South Westland and Fiordland. It was present near The Neck, between Lakes Wanaka and Hawea, about 4000 years ago. We do not yet know whether it spread rapidly in leaps from the west when the climate became favourable for it, or whether it had been moving steadily since the end of the LGM. However, a scenario consistent with the ecology of silver beech would be that it dispersed most rapidly after the retreat of the ice, when open sites with raw soils rich in the phosphorus-yielding mineral apatite were still extensive. By the end of the late-glacial it was established in small scattered stands throughout what was to be its ultimate range. Before these pockets had time to expand, the vigorous podocarp-dominated forests of the early Holocene hemmed them in. On the southern and eastern coastal ranges silver beech spread mainly in the uplands from small coastal refugia, replacing rata, kamahi and subalpine scrub at the forest limit.

We do not know when silver beech first appeared in Central Otago, but its charcoal on the eastern side of the Pisa Range has been dated to about 1600 years ago, and pollen evidence suggests that it became abundant sometime between 6000 and 3000 years ago. Around the perimeter of Central Otago and the upper Waitaki basin, silver beech became established within districts that otherwise were still dominated by celery pine and associated species. It still survives as widely scattered and generally small stands (Fig. 10); the charcoal record shows that some of these were once larger than they are now, but they were never extensive. On the mountain ranges of the interior of Central Otago, silver beech began to move into celery pine and bog pine scrub, forming a forest/scrub mosaic in the moist and cool subalpine zone. The forest on the drier mid to lower slopes remained in totara and celery pine, grading down to the small-leaved shrubland/grassland of the driest valley floors.

Between 4000–2000 years ago, the broad pattern of silver beech dominance that was to persist until the present was established. Despite the wide separation of tracts of silver beech forest in eastern Southland, Otago and Fiordland, the timing of population growth is more or less identical. Therefore, whatever the migration paths involved, it is highly likely that climate change was at least instrumental in creating the situation for rapid expansion.

Mountain beech, red beech and hard beech are now classified in the subgenus *Fuscospora*. Pollen of *Fuscospora* (mainly from mountain beech) had been increasing steadily in western Fiordland from around 6000 years ago. However, east of the Main Divide its pollen is barely registered until after 3000 years ago, and rapid increase of *Fuscospora* pollen began only about 2000–1500 years ago. The mountain beech forests that now form an interrupted band between the dry interior and the moist west and south are therefore surprisingly recent. So also must be the red beech forests that form an irregular star-shaped pattern centred on Lake Te Anau and the Eglinton Valley, with plants reaching the Blue Mountains, Lake Manapouri, Milford Sound, Big Bay, the Matukituki Valley and Lake Hawea.

A striking fact about the Holocene *Fuscospora* beech-spread throughout New Zealand is that it has mainly replaced upland conifer low forest and scrub growing on less fertile sites. In Fiordland, *Fuscospora* beeches spread mainly into celery pine and pink pine low forest, and in south-eastern districts into subalpine celery pine and bog pine scrub. On poor soils, these small conifers form a variable, open canopy, so that the faster growing but light-demanding beeches can enter and overtop them. In lowland Southland, *Fuscospora* beeches also replaced dominant silver beech forests in wetter districts

Fig. 10. Silver beech trees surviving at the head of a ravine in the Motatapu Valley south-west of Lake Wanaka; charcoal in the soil under the prevailing snow tussocks shows that forest was extensive here until around 700 years ago. *Peter Wardle*

The spread and increase of *Fuscospora* beeches occurred thousands of years later in the south-east than in other areas of New Zealand, including Fiordland. Why was it delayed in the south-eastern South Island? The slow spread probably resulted from a combination of factors. First, forest in general was much less common in the far south than elsewhere at the end of the glaciation, patches of beech forest in particular were rare and scattered. Beech trees have wind-dispersed seed which does not travel far, and in rugged terrain expansion must have been slow. Second, *Fuscospora* beeches are favoured by disturbance, and the persistence of settled climatic patterns and landscape stability in the south-east throughout the early and mid Holocene period made it difficult for them to invade conifer scrub and forest. And third, and possibly most importantly, the prevalence of dry, settled frosty weather in the interior of the south-eastern South Island during the early Holocene may have made it difficult for the frost-sensitive *Fuscospora* beech seedlings to invade.

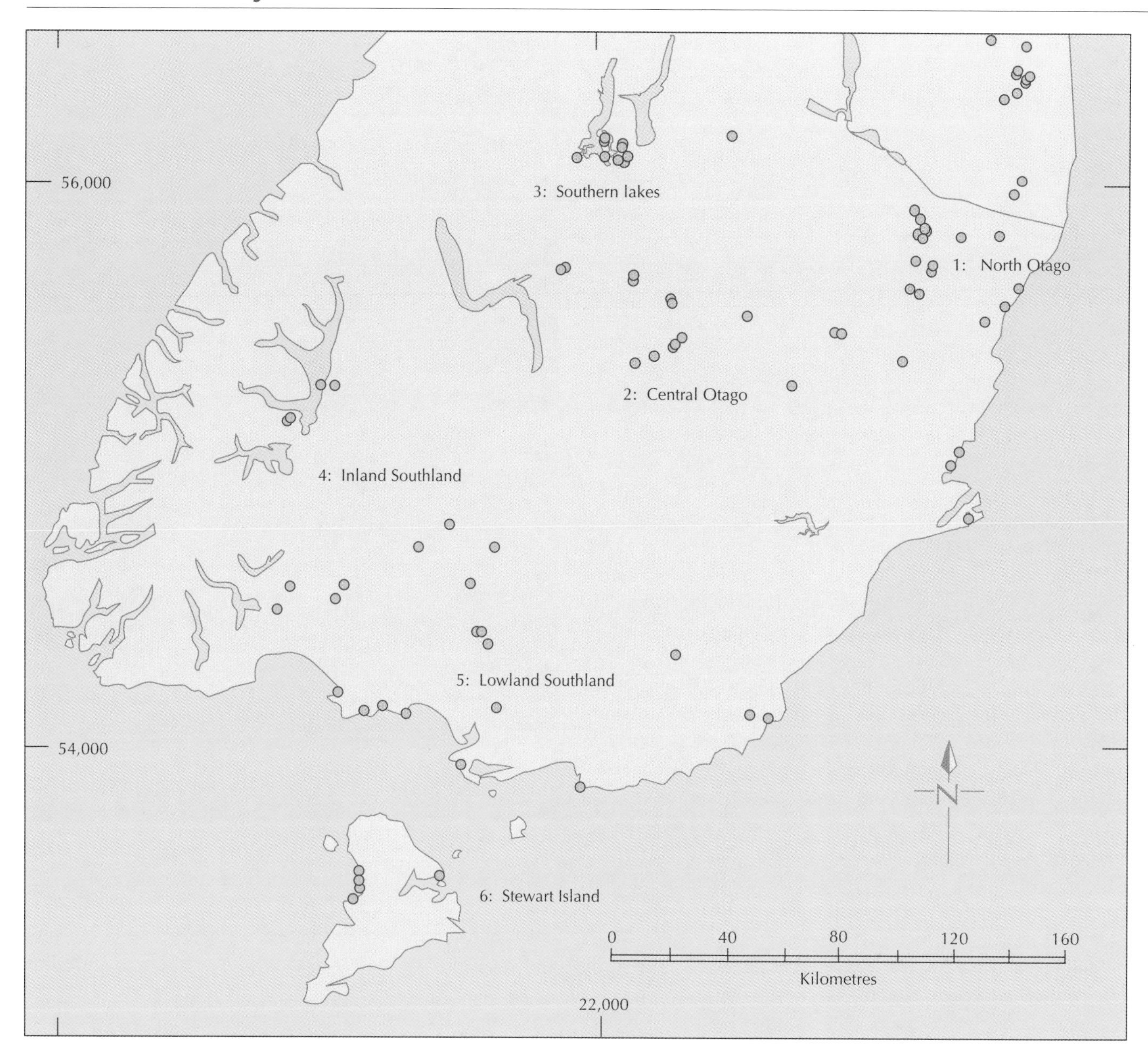

QUATERNARY VERTEBRATE BIOTA

The Last Glacial Maximum

The extremely open landscapes of the LGM supported a relatively impoverished fauna. *Pachyornis elephantopus* was the most common moa species, and there were lesser numbers of *Euryapteryx geranoides*, and rare *Dinornis struthoides* and *D. giganteus*. The dominance of *Pachyornis elephantopus* in the glacial faunas of Otago reflects the norm in other eastern glacial faunas. This most bulky and least agile of moa (1.2 m in height at its back and up to 250 kg in weight) apparently preferred the open habitats of grasslands and low shrublands under which loess accumulated during each of the coldest periods of the last glacial. It is the most common moa in loess faunas throughout the South Island. Its later fall in relative frequency is correlated with an increasingly vegetated landscape after the end of the Glacial period. From glacial faunas elsewhere, such as those from Merino Cave, Mt Cookson and Glencrieff Swamp, both in North Canterbury, we can predict that kea *Nestor notabilis*, pipit *Anthus novaeseelandiae*, quail *Coturnix novaezelandiae*, South Island goose *Cnemiornis calcitrans* and Finsch's duck *Euryanas finschi* would have also been common, and Haast's eagle *Harpagornis moorei*, New Zealand coot *Fulica prisca* and New Zealand raven *Corvus moriorum* present. In sheltered valleys where some scrub or low forest was present, rifleman *Acanthisitta chloris*, piopio *Turnagra capensis* and kokako *Callaeas cinerea* are likely to have been abundant. Species normally found in forest, such as robin *Petroica australis* and saddleback *Philesturnus carunculatus*, must have survived in isolated forest remnants in favourable inland situations or in milder locations along the lowered shoreline.

Holocene

Most finds of vertebrate fossils in southern New Zealand date to the mid to late Holocene, and it is not yet possible to reconstruct the full sequence of events from the LGM onwards. We can divide southern New Zealand into six sub-regions on the basis of their fossil vertebrate faunas (Fig. 11): North Otago – the downlands between Oamaru and Duntroon; 2. Central Otago – the Alexandra/Cromwell region; 3. Southern Lakes –

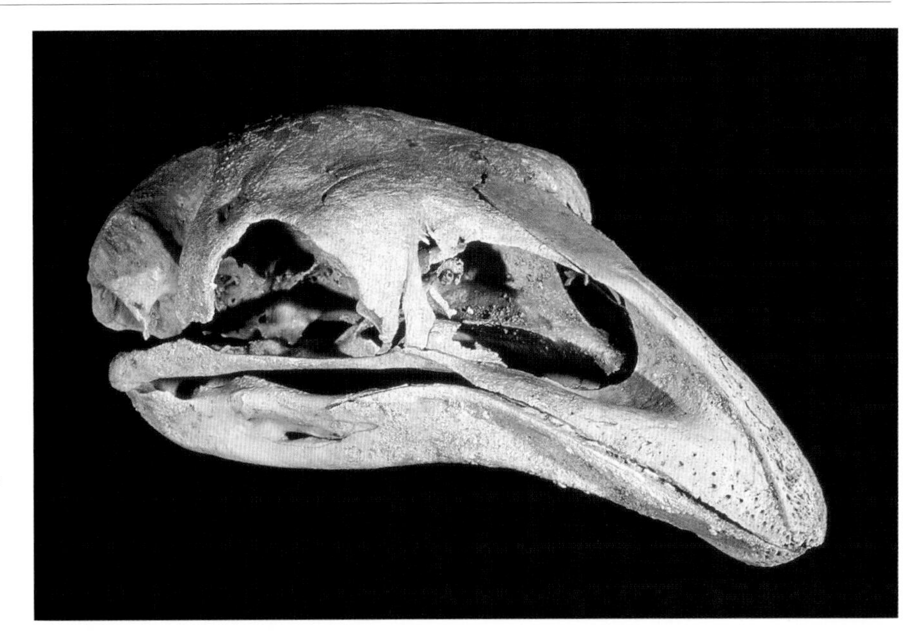

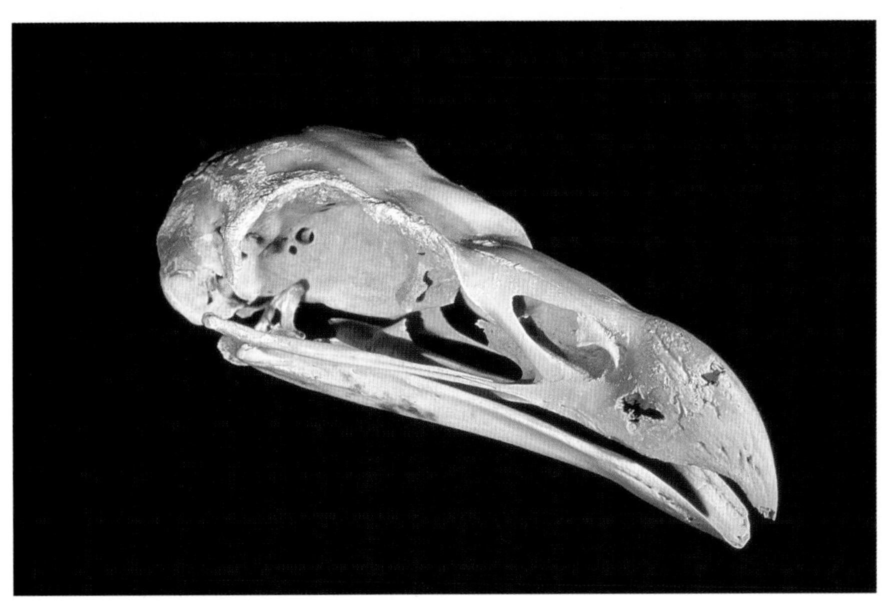

Fig. 11 (opposite). Distribution of vertebrate fossil sites in the southern South Island.

Fig. 12 (top). Skull of the slender bush moa (*Dinornis struthoides*) collected near Wanaka in 1994. *Otago Museum/Rod Morris*

Fig. 13. Skull of the small male New Zealand eagle (*Harpagornis moorei*) collected from Castle Rocks in Southland by A. Hamilton, 1892. This, New Zealand's largest predator, killed and ate moa. *Otago Museum/Rod Morris*

in the vicinity of Wanaka; 4. Inland Southland; 5. Lowland Southland; and 6. Stewart Island. Vertebrate fossil finds are concentrated in the first three sub-regions, and the last three have only a few sites between them. We will deal with each of these areas in turn. Table 1 lists all species breeding in southern New Zealand during the late Holocene.

1. North Otago

Five species of moa, more than half of the nine South Island species, lived in the wide, flat valleys and downlands of North Otago in the late Holocene. There are six major swamp deposits known from this area: Totara Swamp, Ardgowan, Five Forks, Spring Gully, Enfield and Kia Ora. Enfield is of historic importance as it was discovered in August 1891 by a J. Flett and then excavated by H.O. Forbes in the same year. Thousands of bones were obtained, and in 1896 Professor Hutton was able to analyse the species composition, using no fewer than 1031 leg bones. Unfortunately no more than 311 survive in New Zealand museums, with many lost to trading overseas. None of the other sites were excavated by specialists, the faunas being recovered after exposure by drainage operations, and very little is recorded about them. The Totara Swamp is important as it had a significant assemblage of birds other than moa associated with it – one of only three such sites in the eastern South Island. In these swamp sites, the proportions of the moa are remarkably consistent, suggesting that the sites were located in similar habitats and of a similar age. The mean frequency of moa across the six sites is: *Emeus crassus* 55 per cent; *Euryapteryx geranoides* 27 per cent; *Pachyornis elephantopus* 11 per cent; *Dinornis struthoides* 5 per cent; and *Dinornis giganteus* 4 per cent. *Emeus* is absent from the only pitfall trap* in the area (in the hills at Ngapara) but *Euryapteryx geranoides, Pachyornis elephantopus* and *Dinornis giganteus* are present. Elsewhere in the South Island the distribution of *Emeus* fossils suggests this species preferred low-altitude valley sites and

*'Pitfall trap' is here exclusively used to describe a fossil site in which animals accumulate at the base of a vertical drop into a cave or fissure; a fauna accumulated in either a swamp or a springhole is covered by the term 'swamp-mired'.

was replaced on nearby hills by *Euryapteryx*. The presence of these moa suggests a mosaic of clearings, dry forest and scrub, with little open country.

An extremely diverse fauna of a further fifty-three species of birds is recorded from the North Otago region. In the wetlands, as shown by faunas from Totara Swamp, Enfield, Prydes Gully and modern analogues, brown teal *Anas chlorotis*, grey duck *A. superciliosa*, paradise shelduck *Tadorna variegata*, stilts, banded dotterel *Charadrius bicinctus*, black-backed gulls *Larus dominicanus*, and the extinct New Zealand swan *Cygnus sumnerensis*, New Zealand pink-eared duck *Malacorhynchus scarletti* and New Zealand musk duck *Biziura delautouri* are represented by fossils; undoubtedly, dabchick *Poliocephalus rufopectus*, black shag *Phalacrocorax carbo* and white heron *Egretta alba* were also present. These swamp faunas and those from predator sites accumulated by falcon *Falco novaeseelandiae* and laughing owl *Sceloglaux albifacies* along the limestone cliffs show that along stream margins and in adjacent forests are represented: the extant brown (*Apteryx* sp. large) and little spotted kiwis *Apteryx owenii*, weka *Gallirallus australis*, parakeets *Cyanoramphus* spp., kaka *Nestor meridionalis*, kea, long-tailed cuckoo *Eudynamys taitensis* morepork, *Ninox novaeseelandiae*, falcon, rifleman, fernbird *Bowdleria punctata*, tui *Prosthemadera novaeseelandiae*, bellbird *Anthornis melanura*, robin, tomtit *Petroica macrocephala*, yellowhead *Mohoua ochrocephala*, brown creeper *Mohoua novaeseelandiae*, fantail *Rhipidura fuliginosa* and pipit; the nearly extinct takahe *Porphyrio hochstetteri*, kakapo *Strigops habroptilus* and bush wrens *Xenicus* spp.; the kokako, saddle-back and New Zealand snipe *Coenocorypha* sp. which are extinct in the South Island; and the fully extinct Haast's eagle, Eyle's harrier *Circus eylesi*, laughing owl, New Zealand owlet nightjar *Aegotheles novaezealandiae*, New Zealand raven, adzebill *Aptornis defossor*, New Zealand coot, Hodgen's rail *Gallinula hodgenorum*, New Zealand quail, Finsch's duck, South Island goose, Stephen Island wren *Traversia lyalli* and piopio.

Judging from their representation in laughing owl prey assemblages, none of the smaller species were as disproportionately abundant as, for example, parakeets were in the northern South Island. The wrens were in low numbers compared to wetter western regions, and so were weka, kakapo and kiwis. However, piopio is more common in these deposits than any others in New Zealand, and it obviously preferred these dry forest/scrub mosaics. In addition, nesting in burrows along ridge crests were at least five species of procellariiforms: Cook's petrel *Pterodroma cookii*, fluttering shearwater *Puffinus gavia*, diving petrel *Pelecanoides urinatrix*, grey-backed *Oceanites nereis* and white-faced storm petrels *Pelagodroma marina*. It is likely that mottled petrels *Pterodroma inexpectata*, fairy prions *Pachyptila turtur* and sooty shearwaters *Puffinus griseus* also nested in the area, as their remains are known from similar sites in South Canterbury.

2. Central Otago

The moa fauna of Central Otago was more diverse than that of North Otago, with seven species present. This is partly a result of a greater altitudinal range, and hence more diverse habitats. In the broad flat valleys, as typified by the historically famous site of Hamilton discovered by gold miners in the nineteenth century, and the more recently discovered Paerau Swamp deposit, and further west, sites at Moa Creek and Ida Valley, *Emeus crassus*, *Euryapteryx geranoides* and *Pachyornis elephantopus* were the most common, with a few *Dinornis struthoides* and *D. giganteus* also present. The range of all of these species extended also up the lower hill slopes, where they were joined by *Dinornis novaezealandiae* and *Megalapteryx didinus*. *Megalapteryx didinus* and *E. geranoides* were the most abundant species in hill sites.

The principal habitat of *M. didinus* throughout New Zealand was the subalpine zone dominated by shrubs and tussocks which, although extensive in Central Otago, contains no known fossil vertebrate sites. *Megalapteryx* must have been the moa analogue of the mountain goat. Its slenderness and long toes suggest an agility necessary to traverse the steep, rocky terrain that so often surrounds the pitfall traps where this species alone was common. For instance, it was common on the steep slopes of the Cromwell Gorge, as shown for example by the site termed the Station Deposit, discovered during reconstruction of the road through the gorge, and in alpine karst areas elsewhere (Fiordland, and North-west Nelson). Migration of primarily subalpine species during winter, when the mountain tops were snow covered, to the lower slopes accounts for the presence of *M. didinus* in lower altitude sites. There are much higher proportions of *P. elephantopus* in Central Otago than in North Otago, suggesting a greater area of open country without forest or tall scrub. The presence of *Dinornis novaezealandiae* in the later Holocene may indicate the establishment of tall podocarp forests, as such forests were its preferred habitat elsewhere in New Zealand.

Because other birds in Central Otago are known mainly from pitfall traps, the smaller passerine element of the fauna is poorly represented. The most famous pitfall trap is Earnscleugh Cave, which contained moa remains in which soft tissue was dried and so preserved, but most of these are now lost, including a complete head. Nevertheless, a remarkable section of a moa neck with skin and feather bases does survive. The site is also the type locality for the extinct, flightless Finsch's duck and, amazingly, dried articulated skeletal parts survive for this species as well. Some of the bones in the cave were accumulated by prey activities of the laughing owl, which also died in the site. From this, and other pitfall and a few swamp sites, we know that Haast's eagle was the most common predator, and would have fed mainly on moa. Falcon, laughing owl and New Zealand owlet nightjar were also present, and fed on smaller birds. South Island geese and Finsch's duck were very abundant terrestrial grazing birds, reflecting the open habitat. Kea, parakeets, quail, pigeon *Hemiphaga novaeseelandiae*, rifleman, robin, bellbird, saddle-back, kokako, pipit, brown and little spotted kiwis, weka, brown teal, blue duck *Hymenolaimus malacorhynchos*, New Zealand snipe, takahe, Hodgen's rail, greater short-tailed bat *Mystacina robusta* and tuatara were present. Weka and kiwi were relatively rare compared to wetter western districts. Curiously, the enigmatic adzebill was not present in Central Otago – perhaps the habitats were too dry. Overall, the assemblage suggests a dry mosaic of forest, scrub and grassland.

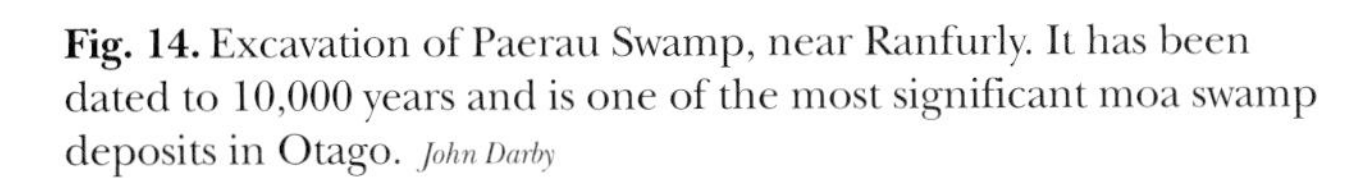

Fig. 14. Excavation of Paerau Swamp, near Ranfurly. It has been dated to 10,000 years and is one of the most significant moa swamp deposits in Otago. *John Darby*

Fig. 15 (below). Dried leg of Upland Moa *Megalapteryx didinus* found in a cave in the headwaters of the Waikaia River, Old Man Range, in 1895. It shows that this species was feathered to its toes, an adaptation to the snowy environment in which it lived. Upland rheas (*Pterocnemia pennata*) in South America also have feathered legs. *Otago Museum/Rod Morris*

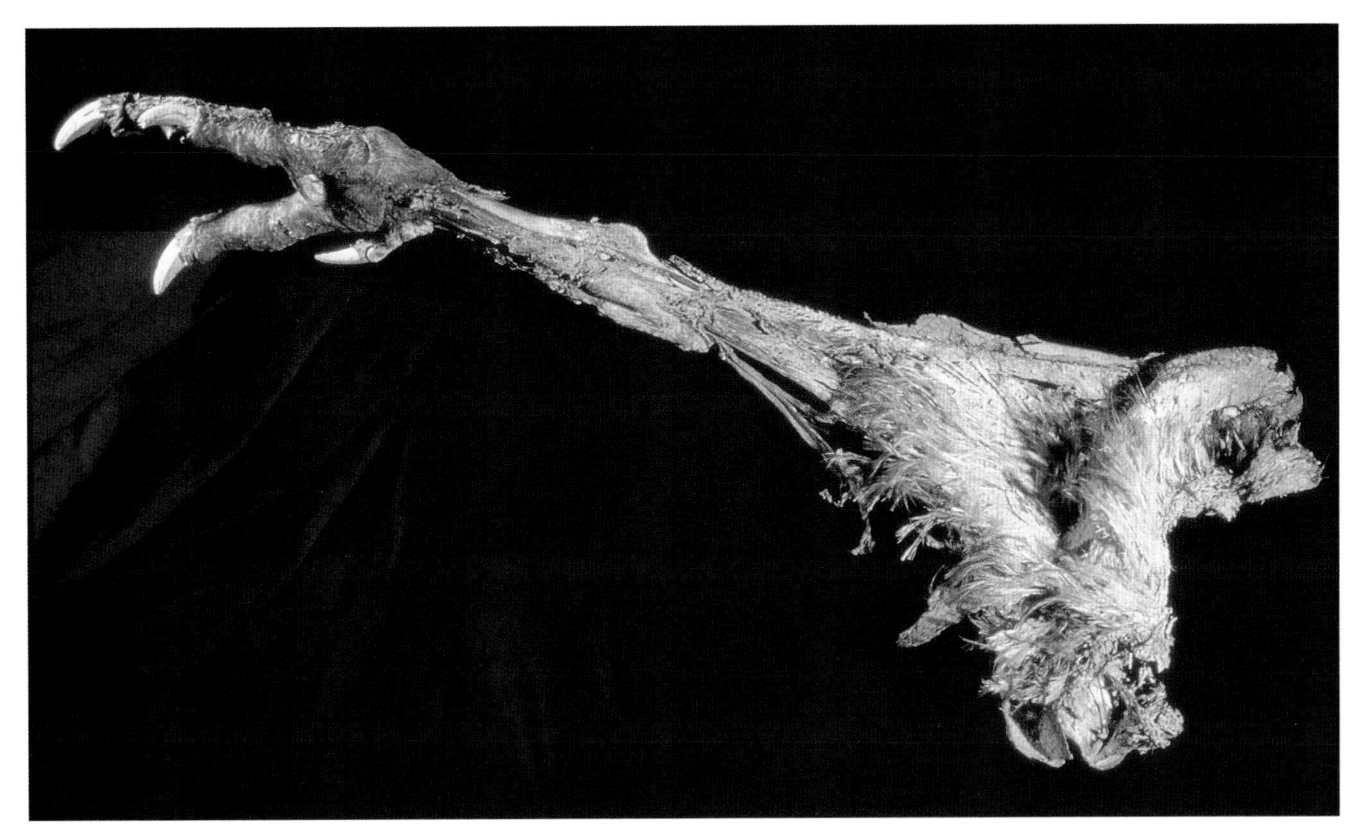

3. Southern Lakes: Wanaka

There are a number of sites near the south-west corner of the lake, two of which are swamp-miring sites of late Holocene age. *Dinornis giganteus*, *D. struthoides* and *Euryapteryx geranoides* were the main moa species living on flat land around the lake shores, but *Emeus crassus* and *Pachyornis elephantopus* were also present. On the hill slopes above the lake and river flats *Megalapteryx didinus* was the most common moa, and was associated with *D. struthoides*, *Pachyornis australis* and rare *Anomalopteryx didiformis*. The last species is typical of wet, closed canopy forest of western regions and Southland, and its presence here is explained by the beech forest that occupied the mid slopes at the time.

Associated vertebrate species are known from two pitfall traps and four assemblages accumulated by laughing owls. By day, Haast's eagle preyed on the moa, and falcons on smaller birds. At night, owlet nightjar, morepork and laughing owl were active predators. The diversity of birds in this area was much lower than in North Otago, probably reflecting greater uniformity of habitat. Kiwis, weka, Finsch's duck, extinct goose, pigeon, kakapo, kaka, saddleback, kokako, tomtit, robin, bellbird, rock wrens *Xenicus gilviventris* or bush wrens *X. longipes*, yellowhead, brown creeper, parakeets, piopio and New Zealand snipe have all been found as fossils. Lesser short-tailed bat *Mystacina tuberculata*, tuatara and various skinks and geckos were also present. Since laughing owls were opportunistic predators, limited by a maximum prey size (pigeon approximately), the rarity of wrens and robins, but abundance of yellowhead, brown creeper, saddleback and piopio in the owl deposits probably reflected the actual abundance of these birds in the surrounding forest in the late Holocene. Yellowhead is now most abundant in beech forests in valleys along the axial ranges of the South Island. Greater short-tailed bats were absent, and lesser short-tailed bats were probably rare, as they are common in owl sites elsewhere in the South Island.

4. Lowland Southland faunas

The fossil faunas of Southland are derived from dune, swamp and cave deposits. All three types are present in the lowlands around Invercargill; however, mainly moa have been collected from swamps and dunes. The dunes, probably most of the swamps and most of the cave deposits are of Holocene age so can be compared to each other. The moa composition of the three deposit types is very different, indicating substantial differences in the immediate habitats. Dunes can be assumed to be unselective as to what they preserve. Caves are also unselective with respect to moa, since they trap whatever is walking by. However, swamps trap what is attracted to them by the presence of water and succulent plants on or around them.

The dune moa fauna has a species composition very like the downlands of North Otago, with *Emeus crassus* dominating and *Euryapteryx geranoides* and *Pachyornis elephantopus* next most abundant. A seral community of grasslands, shrublands and some forest probably occupied the dunes when the fossils were being deposited.

The fauna of the cave deposits contrasts markedly with that of the dunes, with *Anomalopteryx didiformis* replacing *Emeus* as the dominant species, and all others present in minor numbers. The caves are in low hills that rise above the surrounding alluvial plains and swamplands. In the Holocene these ridges were almost certainly vegetated in tall closed-canopy podocarp forest, the preferred habitat of the little bush moa *Anomalopteryx didiformis.*

The swamp deposits that generally lie inland of the dunes have a somewhat intermediate moa species composition. *Emeus* and *Euryapteryx* dominate the fauna, suggesting that the swamplands formed a mosaic of shrub and forests, without the tall closed-canopy forests of the nearby limestone ridges.

The non moa fauna is known only from cave deposits around Winton at Browns and Forest Hill, and slightly farther inland at Castle Rocks. Brown kiwis, little spotted kiwis, and kakapo were very common in these deposits. Rails were numerous, with takahe common, and New Zealand coot, adzebill, Hodgens rail and weka all present. New Zealand quail was rare, as expected in a forested habitat. Predators included the owlet nightjar, laughing owl and morepork by night, and falcon and Haast's eagle by day. The waterfowl were dominated by Finsch's duck, but include brown teal, blue duck and South Island goose. Other birds included pigeons, parakeets, kea, kaka, robins, saddleback, piopio, kokako, tui, rifleman, stout-legged wren *Pachyplichas yaldwyni,* Stephens Island wren, bush wrens, bellbirds and snipe. The petrels nesting in this area included sooty shearwater, fluttering shearwater, fairy prion, broad-billed prion *Pachyptila vittata,* Cook's petrel, mottled petrel, and diving petrels. An interesting record from near Browns is two bones of *Dendroscansor decurvirostris,* the long-billed wren, perhaps New Zealand's rarest fossil bird, otherwise known only from parts of four individuals in North-west Nelson. It is uncertain whether the Haast's eagle remains excavated from Castle Rocks are of Holocene age, but the species was present in the Holocene dune deposits, and so certainly hunted moa in the scrublands of Southland during this period.

This diverse range of birds indicates a rich avifauna. Some elements, such as the Finsch's duck and adzebill, are indicative of the close proximity of shrublands – these species were absent from Holocene forests on the West Coast, where shrublands did not exist – and complement the rare presence of *Emeus, Euryapteryx* and *Pachyornis* in the cave deposits. The swamp and dune moa faunas indicate that it was in the dunes and on the alluvial lowlands where such shrublands and grasslands occurred. The faunas of the lowlands in Southland therefore attest to a diverse range of habitats, ranging from grasslands through to tall, closed canopy forests.

5. Inland Southland

In inland Southland faunas are known from caves on Mt Luxmore, Aurora Cave, Lake Te Anau and Takahe Valley. In the forests immediately above the lake, the moa *Anomalopteryx didiformis* is most common, but is associated with significant numbers of *Megalapteryx didinus.* The latter species dominates nearby deposits in the subalpine areas above the forest, and no doubt its range was depressed during winter so some individuals ended up in the same traps as *Anomalopteryx.* Associated species are fewer than in the lowlands, but nevertheless indicate a diverse fauna dominated by kakapo and robin. Other species present included brown kiwi, weka, mottled petrel, Finsch's duck, brown teal, blue duck, snipe, pigeon, kaka, parakeets, morepork, owlet nightjar, yellowhead, tomtit, bellbird, saddleback, and Stephens Island wren.

Archaeological sites in Takahe Valley indicate that hunters of moa penetrated such remote areas of Fiordland. On Lee Island remains indicate hunting expeditions for the feathers of birds such as kaka and kakapo, and meat from breeding colonies of seabirds such as diving petrels and mottled petrels. The crested grebe *Podiceps cristatus* was also hunted on Lake Te Anau.

6. Stewart Island

There is very little data available for Stewart Island, as its fossil faunas have not yet been studied. However, abundant moa remains of *Emeus crassus* in middens on the Neck indicate this species was common, and a couple of skeletons in natural dune deposits indicate that at least *Euryapteryx geranoides* was also present. Nothing is known about the smaller bird fauna on Stewart Island, but on nearby Native Island bones of the Auckland Island Merganser *Mergus australis* are common in eroding dunes. Associated land birds include weka, snipe, brown teal, parakeets and kaka.

2 EXTINCT BIRDS

The weird, the wonderful, and the bizarre were the first casualties when humankind encountered a fauna for the first time. New Zealand is no exception.

Moa

There was no odder bird than the moa, the only bird to entirely lack wings. The Otago moa included some of the most bizarre of all. Giant moa *Dinornis giganteus*, were the heaviest and tallest of the group at 78–140 kg, standing near 1.8 m tall at their back and with a head reach of 3 m. They browsed leaves and twigs along forest margins.

The large bush moa *Dinornis novaezealandiae*, 84–122 kg, and slender moa *Dinornis struthoides*, 59–81 kg, were successively smaller versions of the giant moa that also ate mainly twiggy vegetation. The large bush moa preferred closed canopy situations and so rarely occurred with giant moa, but the slender moa lived in a wide variety of habitat and co-occurred with the other two.

In the second family of moa (Emeidae) there was a diverse array of forms. The upland moa *Megalapteryx didinus*, a small and most slender form, only 0.5 to 0.8 m high at its back and weighing 26–51 kg, was a goat-equivalent in the mountains and high country.

At lower altitude in the closed-canopy forest of West Otago and inland Southland occurred the similar-sized (31–61 kg) little bush moa *Anomalopteryx didiformis*.

Open grassland and scrubby habitats contained the strangest forms in this family. The sharp-billed heavy-footed moa *Pachyornis elephantopus* was the most barrel-breasted bird in the world, having a weight of 66–151 kg but never standing more than a metre high. Its feet were very broad, supporting a body 0.5–0.7 m wide, yet its legs were so incredibly short that its breast feathers must have continually brushed the ground. Speed and agility was clearly not a priority for these birds.

In the same grassland/scrub habitat lived the not quite so big stout-legged moa *Euryapteryx geranoides* which, at about 1.0 m high at its back and weighing 64–96 kg, was still a massive bird. Its short, blunt bill was suited to eating only soft leafy material and berries.

And finally in the group was the eastern moa *Emeus crassus*, a smaller (36–79 kg) version of the stout-legged moa.

Note: For recent advances in the classification and biology of Moa, see Bunce, M. *et al.* and Huymen, L. *et al.* in *Nature* 425, 172–178 (2003).

Predatory Birds

Three avian predators are now extinct. The laughing owl *Sceloglaux albifacies* was, at 600 g, a much larger bird than the morepork, and the main nocturnal predator in New Zealand. It hunted opportunistically and its prey included lizards, tuatara, bats and birds ranging in size from riflemen to moreporks, pigeons, kakapo and shearwaters. Insects, large beetles and weta were taken as well.

A smaller nocturnal predator was the owlet nightjar *Aegotheles novaezealandiae*, but this longer-legged and weaker-flying version of the Australian nightjar fed primarily on insects.

Most spectacular of all was the diurnal Haast's eagle *Harpagornis moorei*, the largest eagle in the world, with the larger females attaining 13 kg in weight and a wingspan approaching 3 m. This is more than ten times the body mass of a harrier hawk and twice that of a golden eagle. They are known only from the South Island, and during the Holocene were restricted to the eastern areas of that island. As they needed to see their prey – various moa, the goose, takahe and other larger birds – they hunted the open areas of the upland zone, and the forest margins and abundant scrub and grasslands of eastern districts.

The New Zealand raven *Corvus moriorum* was found mainly around the coast, where it could scavenge carrion that was formerly abundant on beaches, and especially around seal colonies. It probably looked much like the rook *Corvus frugilegus*, the all-black bird common in parts of New Zealand now.

Waterfowl

Of the thirteen formerly-resident New Zealand waterfowl, several have become extinct. The South Island goose *Cnemiornis calcitrans* was large (14–20 kg) and flightless. It was some three times heavier than a Canada Goose *Branta canadensis*.

The New Zealand swan *Cygnus atratus* has the unique history of being the first New Zealand species to have been hunted to extinction and then to have been successfully reintroduced from populations in Australia. Until recently the members of the prehistoric population of swan known from New Zealand were named *C. sumnerensis*, but in fact the bones are not distinguishable from the living form, and so more correctly should be called Australasian swan. In prehistoric New Zealand the swan was restricted to large shallow lakes and estuaries and hence had a mainly coastal distribution.

Of less stature, Finsch's duck *Euryanas finschi* was small and flightless, or nearly so. About the size of a mallard *Anas platyrhynchos*, it was not a dabbling duck, but closely related to the Wood Duck *Chenonetta jubata* of Australia and doubtless grazed grass and short herbs in terrestrial situations. It was common in the mosaics of grassland, scrub and forest characteristic of the drier eastern areas.

Rails

The avian family most plagued by extinctions worldwide are the rails (Rallidae), and New Zealand provided no exception. Formerly rails dominated the ground-bird fauna in New Zealand, but now only two of the endemic species survive. One, the South Island takahe *Porphyrio hochstetteri*, is well known.

The South Island adzebill *Aptornis defossor* is classed in its own family (together with its North Island congener) and is now thought to be closer to trumpeters and basal cranes than rails. This large 18-kg bird was entirely flightless – its wings were so small that no external vestige would have been visible in the plumage. It had a most impressive head with a 20-cm long adze-shaped bill. The role of this bird in prehistoric New Zealand is little known but various lines of evidence – its rarity, its morphology and dietary evidence based on the isotopic composition of its bones – all point to it having been a predator. It probably filled the role that cranes do elsewhere, preying on large insects and small vertebrates.

The New Zealand coot *Fulica prisca* was quite unlike the highly aquatic coots *Fulica atra* we now see on some of our lakes. The New Zealand coot had the body form of a pukeko *Porphyrio melanotus* and, like it, could fly. It led a similar terrestrial existence and browsed vegetation along forest margins from the subalpine zone to sea level.

Hodgen's rail *Gallinula hodgenorum*, was a small waterhen that was flightless and, like the coot, an inhabitant of the eastern vegetation mosaics of grassland, shrubland and forest. It has not been recorded where there was a continuous forest canopy. As do other waterhens, it probably fed on grass, herbs and the occasional insect.

Species	Common name	Record	Status
Dinornithiformes	**moa**		
Anomalopteryx didiformis	little bush moa	**F m**	**Pe**
Megalapteryx didinus	upland moa	**F m**	**Pe**
Emeus crassus	eastern moa	**F m**	**Pe**
Pachyornis elephantopus	heavy-footed moa	**F m**	**Pe**
Pachyornis australis	crested moa	**F**	**Pe**
Euryapteryx geranoides	stout-legged moa	**F m**	**Pe**
Dinornis struthoides	slender moa	**F m**	**Pe**
D. novaezealandiae	large bush moa	**F m**	**Pe**
D. giganteus	giant moa	**F m**	**Pe**
Apterygiformes	**kiwis**		
Apteryx sp. large	'brown' kiwi	**F m**	
Apteryx owenii	little spotted kiwi	**F m**	**Re**
Podicipedidae	**grebes**		
Poliocephalus rufopectus	dabchick		**Re**
Podiceps cristatus	crested grebe	**m**	
Procellariiformes	**petrels and allies**		
Puffinus griseus	sooty shearwater	**F m**	
Puffinus gavia	fluttering shearwater	**F m**	
Puffinus bulleri	Buller's shearwater	**F**	
Pelecanoides urinatrix	common diving petrel	**F m**	
Pelecanoides georgicus	South Georgian diving petrel	**F**	
Pachyptila turtur	fairy prion	**F m**	
Pachyptila vittata	broad-billed prion	**F m**	
Pterodroma cookii	Cook's petrel	**F m**	
Pterodroma inexpectata	mottled petrel	**F m**	
Oceanites nereis	grey-backed storm petrel	**F**	
Pelagodroma marina	white-faced storm petrel	**F m**	
Sphenisciformes	**penguins**		
Eudyptula minor	blue penguin	**F m**	
Eudyptes pachyrhynchus	Fiordland crested penguin	**F m**	
Megadyptes antipodes	yellow-eyed penguin	**F m**	
Pelecaniformes	**cormorants and allies**		
Phalacrocorax melanoleucos	little shag	**F m**	
Phalacrocorax carbo	black shag	**F m**	
Phalacrocorax varius	pied shag	**F m**	
Leucocarbo carunculatus chalconotus	Stewart Island shag	**F m**	
Stictocarbo punctatus	spotted shag	**F m**	
Morus serrator	gannet	**m**	
Ciconiformes	**herons**		
Egretta alba	great white heron	**m**	
Egretta sacra	reef heron		
Botaurus poiciloptilus	Australasian bittern		
Ixobrychus novaezelandiae	NZ little bittern		**Ee**
Anseriformes	**duck-like birds**		
Hymenolaimus malacorhynchos	blue duck	**F m**	
Anseriformes	**duck-like birds** *contd*		
Anas chlorotis	brown teal	**F m**	
Anas gracilis	grey teal	**m**	
Anas superciliosa	grey duck	**F m**	
Aythya novaeseelandiae	NZ scaup	**m**	
Biziura delautouri	NZ musk duck	**F**	**Pe**
Malacorhynchus scarletti	NZ pink-eared duck	**F**	**Pe**
Tadorna variegata	paradise shelduck	**F m**	
Euryanas finschi	Finsch's duck	**F m**	**Pe**
Mergus australis	Auckland Island merganser	**F m**	**Ee**
Cnemiornis calcitrans	South Island goose	**F m**	**Pe**
Cygnus sumnerensis	NZ swan	**F m**	**Pe**
Falconiformes	**diurnal birds of prey**		
Circus eylesi	Eyles's harrier	**F m**	**Pe**
Harpagornis moorei	Haast's eagle	**F m**	**Pe**
Falco novaeseelandiae	falcon	**F m**	
Galliformes	**game birds**		
Coturnix novaezelandiae	NZ quail	**F m**	**Ee**
Gruiformes	**rails and allies**		
Porphyrio hochstetteri	S.I. takahe	**F m**	
Gallirallus australis	weka	**F m**	
Gallirallus philippensis	banded rail	**m**	
Porzana pusilla	marsh rail		
Porzana tabuensis	spotless crake		
Gallinula hodgenorum	Hodgens's rail	**F**	**Pe**
Fulica prisca	NZ coot	**F m**	**Pe**
Aptornis defossor	S.I. adzebill	**F m**	**Pe**
Charadriiformes	**waders, gulls**		
Haematopus finschi	S.I. pied oystercatcher	**m**	
Haematopus unicolor	variable oystercatcher	**m**	
Himantopus novaezelandiae	black stilt	**F**	
Charadrius obscurus	NZ dotterel		
Charadrius bicinctus	banded dotterel	**m**	
Thinornis novaeseelandiae	shore plover		**Ee**
Anarhynchus frontalis	wrybill		
Coenocorypha sp.	NZ snipe	**F m**	**Pe**
Larus dominicanus	black-backed gull	**F m**	
Larus novaehollandiae	red-billed gull	**m**	
Larus bulleri	black-billed gull	**m**	
Sterna albostriata	black-fronted tern	**m**	
Sterna caspia	Caspian tern		
Sterna striata	white-fronted tern	**m**	
Columbiformes	**pigeons**		
Hemiphaga novaeseelandiae	NZ pigeon	**F m**	
Psittaciformes	**parrots**		
Strigops habroptilus	kakapo	**F m**	**Re**
Nestor meridionalis	kaka	**F m**	
Nestor notabilis	kea	**F**	

Table 1
Indigenous breeding birds resident in southern New Zealand in the Late Holocene. In the Record column, **F** indicates that the species has been recorded as fossil in southern New Zealand, **m** indicates that it has been recorded from middens. In the 'Status' column, **Pe** indicates extinct in the Polynesian period, **Ee** indicates extinct in the European period, and **Re** indicates a species that is now regionally extinct. This list excludes seasonal migrants (charadriiforms) and occasional visitors (eg. procellariids). The appended list of new colonists gives their date of establishment and whether this was during the Polynesian period **Pp**, or European period **Ep**.

Species	Common name	Record	Status
Cyanoramphus auriceps	yellow-fronted parakeet	F m	
C. novaezelandiae	red-fronted parakeet	F m	
Cuculiformes	**cuckoos**		
Chrysococcyx lucidus	shining cuckoo		
Eudynamys taitensis	long-tailed cuckoo	F	
Strigiformes	**owls**		
Sceloglaux albifacies	laughing owl	F m	Ee
Ninox novaeseelandiae	morepork	F m	
Caprimulgiformes	**nightjars**		
Aegotheles novaezealandiae	NZ owlet nightjar	F	Pe
Coraciformes	**kingfishers**		
Halcyon sancta	kingfisher		
Passeriformes	**Passerine birds**		
Xenicus longipes	bush wren	F	Ee
Xenicus gilviventris	rock wren	F	
Acanthisitta chloris	rifleman	F	
Traversia lyalli	Stephens Is wren	F	Pe
Pachyplichas yaldwyni	S.I. stout-legged wren	F	Pe
Dendroscansor decurvirostris	long-billed wren	F	Pe
Anthus novaeseelandiae	NZ pipit	F m	
Bowdleria punctata	fernbird	F	
Mohoua ochrocephala	yellowhead	F m	
Mohoua novaeseelandiae	brown creeper	F	
Gerygone igata	grey warbler		
Rhipidura fuliginosa	fantail	F	
Petroica macrocephala	tomtit	F	
Petroica australis	robin	F m	
Anthornis melanura	bellbird	F m	
Prosthemadera novaeseelandiae	tui	F m	
Callaeas cinerea	kokako	F m	?Ee
Philesturnus carunculatus	saddleback	F m	Ee
Turnagra capensis	S.I. piopio	F m	Ee
Corvus moriorum	NZ crow	F m	Pe

Post-Polynesian impact	Self-colonisation	Status	Date
Circus approximans	swamp harrier	Pp	?
Porphyrio p. melanotus	pukeko	Pp	?
Anas rhynchotus	shoveler	Pp	?
Vanellus miles novaehollandiae	spur-winged plover	Pe	c.1932
Charadrius melanops	black-fronted dotterel	Pe	c.1954
Himantopus himantopus	pied stilt	Pp	
Ardea novaehollandiae	white faced heron	Pe	c.1865
Bubulcus ibis coromandus	cattle egret	Pe	1960s
Platala regia	royal spoonbill	Pe	1940s
Zosterops lateralis	silvereye	Pe	1856
Hirundo tahitica	welcome swallow	Pe	1958

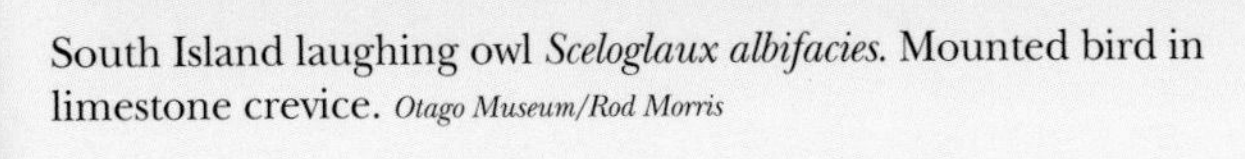

South Island laughing owl *Sceloglaux albifacies.* Mounted bird in limestone crevice. *Otago Museum/Rod Morris*

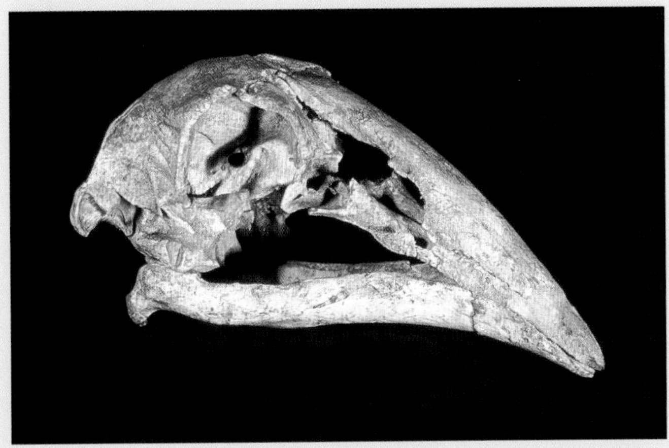

Skull of the adzebill *Aptornis defossor* from a cave near Forest Hill, Southland. This large turkey-sized bird may have been a predator on small animals, e.g. lizards. *Otago Museum/Rod Morris*

MAORI SETTLEMENT

The south just before Maori settlement

In the millennium before settlement major vegetation changes were underway, driven mainly by a long established drift towards a cooler, moister, but more seasonal and extreme climate regime. As part of this change, Southern Oscillation events – the El Niños and La Niñas – were creating year-to-year variation that swung wildly between drought and flood. In Fiordland, on Stewart Island and on still-forested ranges in the east, such as the Longwoods and the Catlins, vegetation immediately before Maori settlement was essentially identical to that of the present. Elsewhere, in very general terms, tall podocarps dominated forest on the lowland plains and hills, silver beech forest on the mid slopes of the wetter ranges, red beech forest in inland valleys and lower slopes where the rainfall was sufficient, and mountain beech forests above these. In the dry interior, celery pine, bog pine and mountain totara persisted on leeward mid slopes, and short-tussock grassland and small-leaved scrub occupied the valley bottoms. Infrequent fires burnt through wooded areas from time to time.

Rainfall and drought-proneness was highly important in controlling the species composition and openness of lowland forest. A series of sites from Dunedin northwards to Shag River shows that immediately before settlement rimu-dominated conifer-broadleaved forest grew on Otago Peninsula and a few kilometres north. Pleasant River marked a boundary between this dense southern forest and more open forest, in which matai and totara dominated, and lacebark, ribbonwood, kowhai, ngaio, ti and small-leaved, divaricating coprosmas and other shrubs of similar habit played an important role. Such forest must have extended north up the eastern lowlands through to Marlborough. This open forest/scrub was also typical of ecotonal vegetation between the dense coastal forests of Southland/Otago and the scrub of the interior. Microscopic charcoal has been recorded at these open forest sites and is found in soil profiles in the immediate district, confirming the importance of fire in maintaining an open mosaic of forest, scrub and grasses. Certainly, part of the reason for the extremely diverse birdlife of the North Otago downlands was the wide variety of vegetation afforded by the close occurrence of tall podocarp forest, open and partly deciduous low forest, scrub and small patches of grassland.

Settlement and impact

Exactly when the first Maori settled New Zealand is debated. The orthodox view has been that settlement took place at or shortly after about A.D. 1000, but some argue for very much earlier settlement at around 2000 years ago, and others for less than 800 years ago. Fossils of kiore (Polynesian rat) have been dated to 2000 years ago in both the North and South Islands. Kiore cannot have reached New Zealand unaided, so we must – if these dates are substantiated – accept early human contact. However, no archaeological sites record such an early human presence. Moa kill sites, middens, ovens and other signs of settlement appear only after 800 years ago, and the effective settlement of southern New Zealand must date from this time.

Between 800 and 600 years ago, pollen diagrams record abrupt decline of forest and scrub pollen types, and an upsurge in microscopic charcoal fragments, bracken spores and pollen of grass, spaniard and tutu (see Figs 7 and 8). Nothing similar to this widespread and near simultaneous outbreak of fires had occurred in the past and, in light of the archaeological record, it is certain that they were lit by Maori settlers.

Perhaps the most direct information on the date of these fires comes from surface totara logs. Maori-lit fires scorched or charred many totara trees but often did not consume them totally. Trunks of accompanying celery pine and most broadleaved species were either completely burnt, reduced to charcoal or subsequently decayed, although occasionally bog pine, pink pine, cedar and beech logs are encountered. Totara heartwood is highly resistant to decay, and prodigious quantities of totara logs used to litter mountain slopes from the eastern ranges of Otago, across Central Otago and west to the Southern Lakes. They are also found throughout much of the rest of the South Island east of the Main Divide. However, the stock of subfossil totara logs is now much depleted, as they make excellent fence posts and firewood. It is not a simple matter to date the time of death of these totara trees, since charring during the initial and subsequent fires, and decay of the outer sap wood, tends to result in older wood being available for dating. The youngest (and probably most accurate dates) converge on 700 years, clearly falling within the period of early Maori occupation.

Over most of the rainshadow area of the east, fires lit by the early Maori were too frequent to allow regeneration of forest. Grass, fern and scrub therefore replaced forest in virtually all but wet mountainous areas. Bracken was particularly vigorous immediately after forest fire, thriving in deep soils fertilised by forest ash. Great expanses of bracken fern developed around the southern lakes and near the coast, where climates were moist and mild, and some of these fernlands have persisted to the present (see page 105). Bracken also was initially prominent in the dry, more seasonal interior, but quickly faded. With repeated burning and deterioration of soil structure and fertility, grasses became more competitive. There is some evidence that the first grasses to spread were species of *Poa*, *Festuca*, *Rytidosperma* and *Elymus*, presumably because of their high reproduction, good dispersal, and fast growth rates, and widespread pre-fire occurrence in dry, open sites throughout. These species – in particular fescue *Festuca novae-zelandiae*, blue *Poa colensoi* and silver *Poa cita* tussocks – persisted until European settlement in semi-arid grasslands and dry shrublands. On wetter sites (over 700 mm rainfall per year), the long-lived, taller snow tussock species *Chionochloa* eventually gained dominance, mainly spreading down from the high ranges, although delayed by slow growth and sporadic seeding. Red tussock *Chionochloa rubra* ssp. *cuprea* which became especially dominant in lowland Southland and coastal Otago, was successful through having been common pre-fire on poorly drained soils. Many of the herbs associated with these grasslands were also favoured by fire, and dense speargrass or spaniard thickets formed in many places. Even in the snow tussock grasslands above the treeline, increased fire acted to reduce low-growing shrubs in favour of grasses, and species of

Celmisia were often prominent after fire. Only those woody species that could resprout or quickly regenerate on bared ground benefited from the new fire regime. Manuka, kanuka, tutu, ti, matagouri and small-leaved coprosmas were among those promoted by the increased frequency of fire.

The most recent faunal deposits show that although a reduction of bird species followed the establishment of kiore in the eastern South Island, New Zealand quail and pipit became more numerous, reflecting the expansion of their preferred open habitats.

Tall podocarp/broadleaved forest in the south-east was practically eliminated except on some hilly coastal areas such as the Catlins and around Dunedin, and as scattered patches on the Southland plains. Matai, for instance, is rare in inland districts, and in western Otago is represented only by a few scattered trees near the lakes. The stands of celery pine and mountain totara that had endured through the sporadic fires of the Holocene disappeared, apart from some scattered shrubs of celery pine protected from fire within boulder fields, and a few, mostly small stands, in the heads of valleys west of Lake Wanaka. Bog pine also was nearly eliminated, with the exception of a few stands, most notably that in the Wilderness Scientific Reserve near Mossburn. A few mountain totara survive in some localities in Central Otago, where surrounding rocky terrain has protected them from grass fires, and some of these are showing vigour, with numerous seedlings establishing, some at a distance from their parents.

Mountain beech was spreading vigorously at the time the Maori fires began. The earliest fires may have encouraged this expansion by removing conifer and broadleaved scrub. However, the dynamics of fire-induced forest regeneration are complex. Where fires burnt into wetter areas, for example around the southern lakes, beech forest has slowly recovered through spread from the margins of surviving forest. However, fleshy fruited podocarps and broadleaved trees and shrubs can disperse greater distances than the beeches, which has enabled them to leap-frog in and occupy areas previously in beech. However, beech is slowly reoccupying ground lost in the fires and their aftermath, overtopping the faster spreading but more open and lower-canopied mountain totara, broadleaved forest and scrub.

Predation by kiore on small birds, reptiles and invertebrates, by humans on larger prey such as the moa, South Island goose, and takahe, disruption of nesting sites, and wholesale reduction of the once varied forest and scrub to vast expanses of bracken and tussock that offered reduced resources, resulted in a dramatically diminished fauna. The relict distributions of large invertebrates (for example, Stephens Island weevil) and small vertebrates (for example, New Zealand snipe) were created at this time. All Holocene extinctions of vertebrates date to the last 1000 years.

Destruction of the moa and other large-sized avian herbivores within a few centuries of Maori arrival must have influenced the vegetation, but we can only guess as to what the effect may have been. It has long been noted that many podocarp forests in the south have few young trees, and there has been much discussion as to possible causes. The paucity of young trees may in many cases result from these stands having been initiated many years ago as dense stands after disturbances such as changes in river courses, landslides, extensive wind-throw or fires. Lowland podocarps live many hundreds of years, and an old stand may show little sign of the disturbance that allowed it to establish, and the dense shade and thick understorey prevent regeneration. However, grazing by deer, sheep or cattle can favour establishment of podocarp seedlings and saplings through destroying vigorous undergrowth of palatable ferns, shrubs and small broadleaved trees that would otherwise smother them. It is possible that our extinct large herbivores played a similar role of encouraging podocarp regeneration until a few hundred years ago.

Fig. 16. A log of mountain totara (*Podocarpus hallii*) lying in a damp gully amongst rushes at 880 m in the Lochar Burn catchment of the Pisa Range, Central Otago. Radiocarbon dating of the outer wood of this log gave a date of A.D. 1169 +/- 49 years. The tall unbranched form of these logs and their density in this area (about ten per hectare), much less than in an intact forest, suggests that another tree species, probably silver beech, was also present. This area has recently been protected from fire with a special covenant. *Alan Mark*

The role of birds in dispersing seeds and fruits must now be severely curtailed through extinction or reduced range and abundance. The huge flocks of parakeets recorded by the early settlers are no more; kaka, kokako, saddleback, and piopio likewise abounded before introduced predators and habitat destruction effectively eliminated them. It seems highly likely that effective loss of these dispersers cannot help but influence forest structure for many years to come.

CONCLUSIONS

At the end of the Last Glaciation, 14,000 years ago, the southern New Zealand landscape was a bare, windswept wilderness of tussock/shrubland, bare ground and ice, with small patches of forest clinging to the coast and some sheltered inland slopes. At the peak of the Holocene, forest or tall scrub covered the landscape in an unbroken sweep from coast to treeline. Pulsing of glaciers in the Southern Alps, cooler, moister conditions along the south-eastern coastline, and increased outbreaks of fire and generally more unstable conditions in the drier eastern districts are all consistent with a dramatically enhanced El Niño effect in the late Holocene. Eastern forest and scrub vegetation in the last few millennia may have been fundamentally unstable, needing only some outside influence to push it over the edge. Disruption came with Maori settlement and fire, reducing the vegetation of the dry east to mere fragments of its original Holocene extent, and once again establishing the dominance of grassland. Maori hunting and predation by rats reduced the bird fauna dramatically, and further reductions occurred when a whole suite of predatory mammals was introduced by Europeans. European grazing and farming and introduction of exotic animals and plants converted large areas to nearly completely exotic communities, and modified the remainder. Elements of all the vegetation stages that have occurred since the end of the Last Glaciation are with us still, albeit often as small patches of vegetation, or even individual trees. However, most of the bird and reptile dominants are gone. It is a daunting challenge to understand this landscape, to interpret its history and deal with its present disrupted natural systems.

The changes inflicted on our indigenous vegetation since the arrival of European colonists have been more catastrophic than anything that happened in the preceding 10,000 years. Whereas Maori fire altered the balance – usually dramatically – between various vegetation types, European settlement has led to the wholesale usurpation of certain habitats by introduced plants. Fertile lowland alluvial plains, riverbeds and lowland swamps have few traces of their original vegetation. Even when indigenous plants remain dominant – as in some lowland raupo swamps – their composition may bear little resemblance to that of the original communities. However, it is worth noting that European era modification of the natural landscape has affected the boundaries of native forest surprisingly little, except over limited areas where there has been deliberate felling and clearing. Although fire was used extensively to promote ease of travel and grazing, it nearly always burnt over areas that had been previously influenced by Maori fires. For instance, surveyors who reached the upper Clutha before 1860 – about the same time as the first sheep farmers – drew maps that show forest boundaries almost the same as those of today.

We are now well into a new era of land management, in which economic values and human usage have to be balanced against those of preservation of landscape values and the remaining indigenous biotic diversity. As well, there is now a widespread and growing desire among some landowners and custodians of public lands to restore 'natural' landscapes. But, as we have seen, over very large areas only a tiny scatter of sites preserve original vegetation. Of all the areas in New Zealand, the low rainfall areas of the south-east South Island are the most difficult in which to establish what the natural vegetation was and to learn how to manage it. Do we mean by 'natural' the situation in A.D. 1840 (the definition the Protected Natural Areas programme has adopted), or that prevailing 800 years ago, before Maori impact? In what ways was the vegetation changing before human influence became overwhelming? Should we allow these dynamics to continue, even if they mean the obliteration of much-loved and familiar landscapes? How do we factor fire into the equation, given that natural fires were widespread before human arrival? In undertaking restoration projects, should we aim to re-establish the previously dominant forests and woodlands of the lowland and montane areas, or have 700 years of vastly expanded bracken and tussock communities given them *de facto* rights? Careful, focused studies of recent history can help us explore these questions, and have an important role in guiding management.

Ultimately, however, it has to be accepted that, in the same way as the recent avian extinctions are final, many ecosystems that supported those birds have vanished just as irrevocably. The present admixture of native and introduced biota with all the problems it brings is set to continue well into the future. This, and the pervasive influence of humans, means that in southern New Zealand there is no going back to any pristine past. Nevertheless, perhaps more than anywhere else in the country there is room for choices, and knowledge of the past will help to inform these. Restoration of past ecosystems will play its part in determining the shape of future landscapes, along with productive land use, ecology, aesthetics and local memories in what seems certain to be another century of momentous change in southern New Zealand.

ACKNOWLEDGEMENTS

MSM and PW were funded by the Foundation for Research, Science and Technology (Contract CO9313); THW was funded by the Foundation for Research, Science and Technology (Contract TWO 601). We thank the Otago Museum, and John Darby, for facilitating photography of some of their specimens. We are grateful to Alan Mark and John Darby for editorial comment and assistance, and to Jill Hamel for helpful referee comments on a draft version of this chapter.

CHAPTER 7

THE HUMAN FACTOR

JILL HAMEL, RALPH ALLEN, LLOYD DAVIS, RICK MCGOVERN-WILSON, IAN SMITH, PETER PETCHEY

Nineteenth-century nurseries in Otago distributed thousands of exotic tree seedlings. *Sequoiadendron*, such as these in the Queenstown Gardens, were favourites for public parks and runholders' gardens alike. Though these trees are well over 100 years old, they are still in their 'seedling' form, and will eventually form rounded heads during their possible life span of 3600 years. *Jill Hamel*

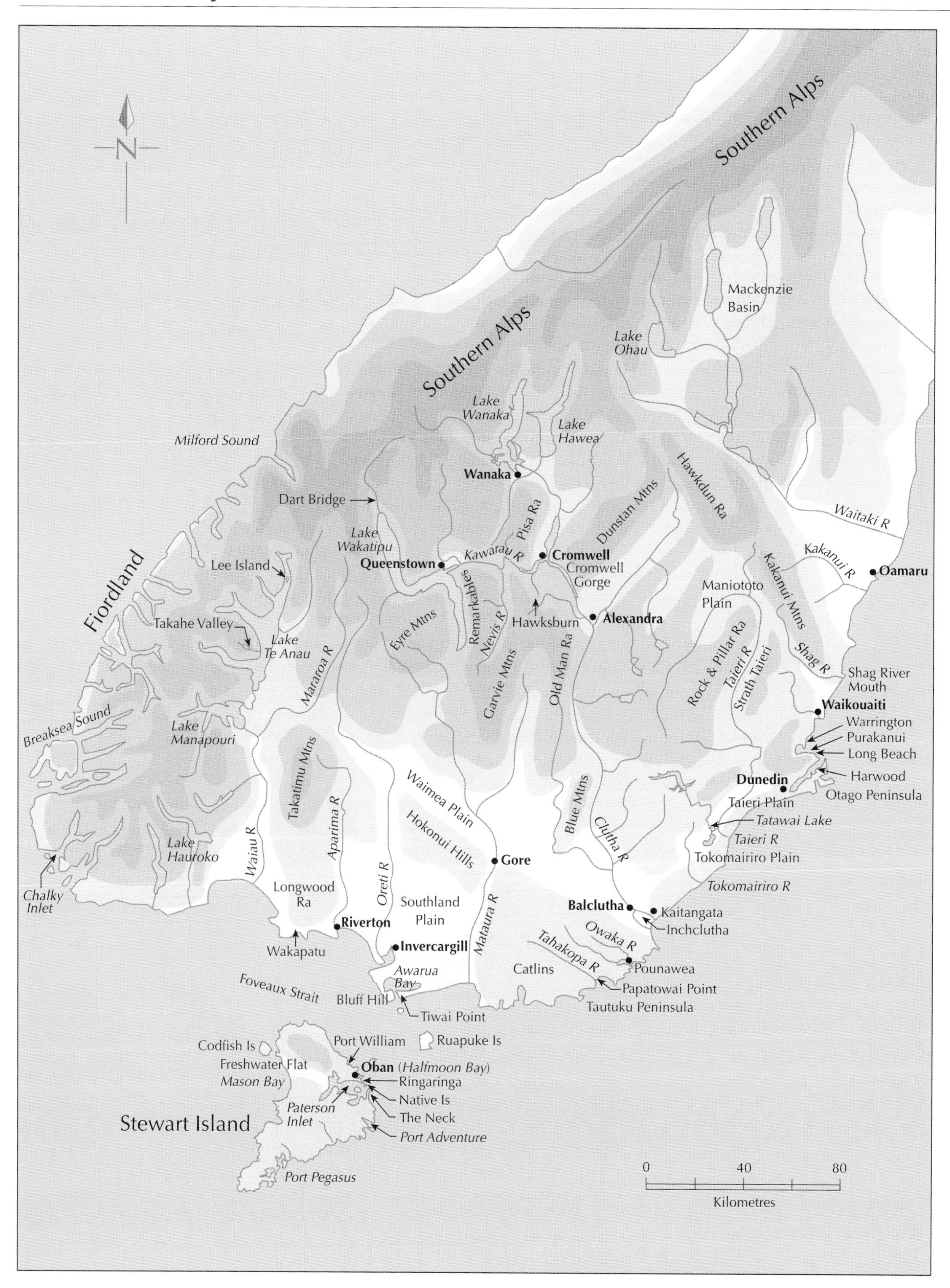

Fig. 1. Southern New Zealand, showing places and archaeological sites mentioned in this chapter.

THE HUMAN FACTOR

JILL HAMEL

THE FIRST NEW ZEALANDERS

Homo sapiens is a part of nature, and a natural history cannot avoid describing the interactions between this most tumultuous of primates and its environment. New Zealand was the largest landmass free of human interference until about 800 years ago, making this land a laboratory for investigating human impact on a pristine environment.

The first New Zealanders were a tall, hardy and robust people. They arrived in well-built, sea-going canoes, propelled by both paddle and sail, and had worked their way out into the Pacific from Asia over the previous 4000 years. They had come out of a population movement into the islands of the western Pacific as the Last Glacial Maximum relaxed its hold and sea levels rose. The big islands south from Japan to the Philippines, Indonesia and New Guinea carried populations of skilled hunters, gatherers and fishermen, who may have been forced by the constraints of island living into undertaking the tedium of cultivating crops of tubers such as taros and yams, as well as managing fruiting trees. Certainly horticultural activity in the highlands of New Guinea began soon after the end of the glaciation, although as yet it is not clear how soon it spread to other islands. It is likely that at the point when we first pick up traces of people definitely ancestral to the New Zealand Maori, they may have had a long tradition of horticulture and arboriculture, linked to a highly developed maritime economy (ocean-going boats, navigational skills and a fishing technology based on shell, bone and complex fibre processing) that was already several millennia old.

Prior to human settlement, the islands of the Pacific lacked major predators and had evolved rich populations of large flightless and near-flightless birds. From the analyses of the bones of extinct birds on Pacific islands, it is estimated that Oceania had about 9000 species of birds, more than the total surviving in the whole world at present. Most were destroyed directly or indirectly by the new colonists. Seabirds in particular had built up enormous populations at a time when they were not restricted to breeding on offshore islets or inaccessible cliffs, as at present. There were great clouds of feeding birds around each island and broad streams of migrating birds leading to and from their breeding grounds at appropriate seasons, making the islands easier to locate than today. Early Polynesian settlements are marked by middens full of bird bone.

What do we know about the baggage, physical and mental, that the ancestors of the present-day Maori brought to New Zealand? Of domestic animals, they brought dogs but apparently no pigs or hens. Out of thousands of identified bones from pre-nineteenth century middens, none have been reliably assigned to pigs or hens, but these turn up promptly in middens with the first signs of European crockery. Mysteriously, Maori also brought the Polynesian rat or kiore. Even though rats were eaten, especially from bush areas where they were largely frugivorous, the damage they caused to food crops was surely a recognised disadvantage. The fact of their arrival could be taken as an indication that the number and size of the canoes which reached New Zealand were so great that the kiore was able to stow away.

Only a few plants were brought and established. Except for kumara (sweet potato), none of the tropical plants could provide the migrants with a really productive garden crop, and even the kumara could be grown effectively only as far south as Banks Peninsula. At first this hardly mattered to the new migrants, as they harvested the abundant supplies of seals, moa and shoreline colonial nesters, such as penguins and shags.

THE EFFECTS OF COLONISATION

How would a people with a long history of gardening and fishing respond to a pristine environment?

These people had brought a wide range of skills to make effective stone and wooden tools, to work up fibres from any suitable plant, process and preserve a wide variety of foods, and construct houses, canoes, fish weirs, traps, nets and baskets. The social structure which is suited to organising gardens and defending the resultant crop is, however, different from that required to organise hunting groups and defend the more transient living sites of the hunter and gatherer. There is less need for exclusive use of land and for chiefly organisation of pa building among the latter. Two aspects of Polynesian social structure would have been useful to the hunters in southern New Zealand: the authority structures involved in building sea-going canoes and for organising fishing trips. The same sort of leadership could be used to decide the how, when and who of major moa-hunting expeditions, exploration and mining of pounamu, and trading trips north for kumara and obsidian.

Around the world, hunters and gatherers generally have a simple band-type level of social structure with no permanent villages. Southern Maori, when Europeans arrived, had a stratified tribal structure with a paramount chief, Tuhawaiki, and a very mixed settlement pattern of a few permanent villages and numerous camp sites. The villages were not, however, occupied by one homogeneous group belonging to one or two hapu (a

Fig. 2. A reconstruction of the moa-hunter site at Dart Bridge. Only two moa were consumed at this site, which contains many ovens for processing cabbage trees into kauru. It was an important camp site in the course of travelling to and from the Dart Valley nephrite source. *Painting by Chris Gaskin. Reproduction by permission of the Department of Conservation, Dunedin*

related group of extended families), as in the North Island. Instead, each camp site, hamlet or village was occupied by representatives of many hapu, maintaining the hapu rights to whatever local resource was being harvested at the time. It might be eels at Inchclutha or titi (muttonbirds/sooty shearwaters) around Foveaux Strait (Fig. 1). This system of resource allocation was known as 'mahinga kai'. We do not know what the social structure of the first-comers was, but it was sufficiently ordered for the development of some remarkably large sites based on 'big game hunting' around the shores of Otago and Foveaux Strait.

Radiocarbon chronology indicates a very rapid initial spread of the colonists, not only around the coast but inland as well, to early moa hunting sites such as Dart Bridge (Fig. 2), the Cromwell Gorge Rockfall sites, and Hawksburn (Fig. 3). Early moa hunting sites in Otago and Southland were large: at Waitaki River Mouth there were hundreds of ovens, in groups of two to five, scattered over 70 ha. At Shag River Mouth, a fourteenth-century village covered more than three hectares. Very large numbers of moa and seals were taken; somewhere between 3300 and 9240 moa were consumed over a period of about 50 years at Shag Mouth. Throughout Otago, moa were butchered and cooked in hangi or spit-roasted for immediate consumption. There is no clear evidence of preservation of moa flesh, presumably because the birds were available at all seasons. On the coast, villages were located close to seal colonies where animals were likely to be available the whole year round.

Recent work at Shag River Mouth (page 134) has clarified much of the data derived from smaller excavations at other sites around the coast. It is now thought that the initial settlement pattern was one of relatively large, transient, coastal villages, lasting about 20 to 50 years, supported by abundant large animals (moa and seals). As each seal colony declined, the village turned more to fishing for barracouta (Box 1), to the smaller species of

1 MOA HUNTERS AT THEIR FISHING CAMPS

For more than a century of academic archaeology, the people who hunted the moa were visualised as doing just that – mostly hunting moa. 'Moa Hunter' with capitals was used to describe their culture: their adzes in particular. Then sites were found with early dates and Moa Hunter tools but no moa bones. To deal with the anomaly, the culture was called Archaic, and moa hunting lost its capitals and became an activity. Once the moa became extinct, it was thought that the people turned to more intensive fishing and birding, and changed their tool types. The new culture was named Classic Maori.

This classification took another knock when the two open beach sites of Purakanui and Long Beach were excavated in the 1980s. The lowest layer at Long Beach was dated to the early fourteenth century and Purakanui to early fifteenth century, and both were obviously specialised fishing camps, mostly for barracouta. Given the maritime traditions of these first New Zealanders, it would have been surprising if they had *not* fished.

After the analysis of Shag River Mouth village (above), places like Purakanui and Long Beach can be seen in a different context. The part of the site excavated at Purakanui was occupied on three occasions over a brief span of years, and during that time at least 230,000 red cod and barracouta were caught and their remains deposited over 1000 square metres of midden. It is known from early accounts that these two species occurred in vast numbers around the Otago Peninsula, and it would have been perfectly possible for a small group to catch such large numbers in only a few years. Judging by the size of the site, though, it is likely that the 4700 kg of edible flesh was cooked, dried and taken elsewhere to eat.

At Long Beach, an area of midden only five metres square and 10–15 cm thick, dated to the fifteenth century, produced the remains of 2800 red cod and barracouta. The bones lay articulated in the sand, barely disturbed from the day 500 years earlier when the filleted bodies were thrown on the midden. The undisturbed remains here also suggest a short occupation span by a group who came, fished intensively, and cooked and dried the fish to take back to another camp. Both Purakanui and Long Beach can be envisaged as specialised fishing camps for villages nearby at Warrington and Harwood who had depleted their seal colonies and moa populations, and were nearly ready to move on.

Jill Hamel

Fig. 3. A reconstruction of the moa-hunter site at Hawksburn. The forest burning on the mid slopes of the Old Man Range never regenerated. *Painting by Chris Gaskin. Reproduction by permission of the Department of Conservation, Dunedin*

moa, and to taking more inland birds rather than coastal. Once the seal colony reached a particular level of depletion, the village was shifted a short distance to another major seal colony and to where moa could also be easily hunted. The effect on fur seal and sea lion colonies was to reduce their breeding range to the less hospitable shores of Fiordland and offshore islands. Though these initial population densities of people would have been low, the use of fire and a superior knowledge of prey life cycles and distribution, compared to other predators, made human beings a formidable enemy of naive prey. Nine moa species of a wide range of size and ecology, as well as four other large species (swan, goose, adzebill, eagle) and four species of ducks, became extinct during this period of Polynesian occupation.

Setting aside the direct effects of hunting, the remarkable terrestrial fauna of New Zealand was very vulnerable to fire and to the predatory effects of the kiore. The place of cursorial mammals here was taken by flightless birds and by invertebrates such as large weevils and grasshoppers. The latter still take the place of small herbivores in the subalpine vegetation. Even the three species of small native bats spend far more time foraging on the ground than do other bats. There were also about six species of primitive frogs, the tuatara and more than forty species of geckos and skinks. The tuatara was widespread and, like many of the skinks, seems to have lived mostly in seabird colonies, using old burrows and preying on chicks and eggs. Just as the arrival of humans and the Polynesian dog and rat was followed by extinctions among the birds (Box 2), three species of frogs and several lizards became extinct and the range of the tuatara was reduced to rat-free islands.

Moa were found in most abundance in mixed shrublands and more open forests, such as those forming bands on the mid slopes of hills around the inland Otago basins. These dry forests were easily burnt, and dates from charcoal and old logs show that most were burnt about 700 years ago, never to return. This may have been a nasty surprise to the colonists, coming from islands with tropical forests which have immense regenerative powers. Fire, to them, may have been an old and familiar tool for managing vegetation, in order to produce a rich mosaic of different seral stages, with long winding edge zones where birds were more easily hunted than in dark, dense-stemmed and high-canopied forests. At Shag River Mouth, it took less than 50 years to change the local hills from a forest of matai and totara with patches of ribbonwood to a mosaic of more open country, shrubland and forest. Natural fires had already affected this landscape over the previous 2000 years, but their frequency had not been sufficient to maintain bracken stands. Whether deliberately or not, these first people set fire to great swathes of relatively dry forests, sufficiently often to encourage bracken growth and prevent regeneration of tall forest.

There are some more subtle effects of Polynesian arrival which still require investigation. They would have had an intense interest in potential crop plants among the native flora, especially in plants with starchy stems or rhizomes and fibrous leaves. There are indications in the literature that nineteenth-century Maori

2 BIRD EXTINCTIONS

Three phases can be distinguished in bird extinctions in Otago and Southland. First, the species that were particularly vulnerable to disturbance by hunters at their breeding colonies or to rat predation went. Second, there was gradual reduction of populations by Polynesian hunting and destruction of habitat leading to extinction. Third, there was a loss and reduction of range of those species which were susceptible to hunting by European weapons, predation by black and Norwegian rats, cats and mustelids, loss of habitat by fire and from competition from introduced mammalian herbivores. The main species in each group known for Otago and Southland were:

Phase 1: A mainland race of snipe, Stephens Island wren, Finsch's duck, Auckland Island merganser, *Aptornis*, Hodgen's rail, New Zealand crow

Phase 2: All the moa (7–10 species), the eagle, New Zealand raven, New Zealand hawk, giant owlet nightjar, giant coot, flightless goose and swan.

Phase 3: New Zealand thrush, laughing owl, New Zealand quail, mainland populations of saddleback and probably the South Island kokako.

Jill Hamel

were very aware of variations in the productivity of bracken, as well as some evidence that they selected and managed some bracken patches with thicker rhizomes. The evidence for cabbage tree selection is more equivocal, but the spiritual importance of cabbage trees as markers of burial sites, especially of placentae, suggests that some sort of selection and management was being carried out. The evidence is clearest for flax, and there are collections of flax varieties being nurtured and expanded today by local iwi. One list names and describes 37 varieties. Since there is no evidence for formal protected gardens alongside southern villages, all this management was of wild populations in semi-isolated pockets close to kaika, and is likely to have started at an early stage.

Shag River Mouth: North Otago in the fourteenth century

Shag River Mouth has always been considered to be the richest of the moa hunter sites along the southern coasts, filling museum shelves with many and varied artefacts. An investigation worthy of its status was started in 1987 by local archaeologists and has resulted in a series of revelations. Excavations, totalling 114 cubic metres, were placed across the whole of the site, which lies on a sand spit between an estuary and a fruitful sea, with rocky shorelines north and south (Fig. 4).

When the site was occupied for about 20 to 50 years in the fourteenth century, the adjacent hills were covered by a forest dominated by matai and totara, with ribbonwood and kahikatea beside the river and areas of dense shrublands around the salt marsh. The rocky shoreline had one or more flourishing fur seal colonies, numerous blue penguins and colonies of Stewart Island and spotted shags. The forests carried a full suite of moa species, from the medium-sized species (*Emeus crassus* and *Euryapteryx geranoides*) to the larger ones (*Dinornis* spp.), as well as the smaller forest birds. Some open grassy areas were rich in the extinct New Zealand quail. At sea not only large shoals of barracouta could be taken, but also white-capped mollymawks.

Careful surveying of the three-hectare site suggested that the midden once occupied about 15,000 cubic metres, of which less than one per cent was excavated. The distribution of midden was extremely patchy, but there may have been a hundred times more of the materials counted from the excavations in the site as a whole. The excavations produced 57 silcrete blades, but there may have been about 6000 in the site. Other important artefacts found were five complete adzes, 59 drill points, 168 fish hooks, 59 awls and as wide a range of other prehistoric tool types as found anywhere in New Zealand. The faunal material excavated included the remains of up to 70 moa, 57 fur seals, 76 dogs, 510 small birds, 1442 fish and about 75,000 shellfish. (Over half a kilo of moa eggshell was scattered throughout the layers, and a previous excavator, Griffiths, had found a whole moa egg.) The site as a whole may have contained the remains of between 3300 and 9240 moa and similarly large numbers of the other groups. From the presence of 44 hearths, it is estimated that a population of 100–300 people consumed the 1000 or so tonnes of meat from these animals, supplemented by starch from bracken rhizomes and cabbage tree stems.

The density of bones indicates that 100 to 200 dogs were culled for food for every year of occupation. There is some evidence that the dogs were being bred for heavier shoulders to use in hunting moa and that the seals were partially used to feed them. The introduction of such large numbers of dogs into a pristine environment would have had calamitous effects on the local bird life. If some were lost on more distant hunting trips, wild packs may have been established early on. Certainly packs of wild dogs were a curse to the early runholders in the nineteenth century, and were said to have included strains of Maori dog.

For all that the site was occupied for such a short time, change in subsistence could be seen. The upper layers included fewer of the bones of the larger moa than the lower layers did. The indicators for the weight of meat from moa and seal dropped dramatically and the amount from fish rose equally dramatically in the upper layers, and there was a marked increase in shellfish. There was a switch from colonial nesting shore birds to open country birds, particularly quail. It is hardly surprising that such intensive continuous foraging depleted the local populations and encouraged the occupants of Shag Mouth to leave the site.

THE SOUTHERN COASTS AND FIORDLAND

Moa hunter sites along the coastline from the Catlins to Riverton were occupied during the fourteenth century, and could well be interpreted as transient villages, though on a smaller scale than Shag River Mouth. Pounawea and Papatowai are thought to have been two of the largest, and recent work at Papatowai has shown it was occupied for only a brief period in the fourteenth century. Other and smaller sites occur at regular intervals all along the forested Catlins Coast. Judging by the presence of numerous large matai trees within the Papatowai Reserve that are older than the site itself, the Catlins forests were relatively unaffected by fire during the Polynesian period. Clear areas along the valley floors of the Owaka and Tahakopa Rivers may have been due to a frost pocket effect, as much as the result of fire. These grassland corridors gave the coastal inhabitants easier access to the Southland plains and to sources of good flaking rock such as porcellanite, which is abundant as artefacts in the coastal sites.

West of the Catlins, moa hunting sites are fewer, possibly because the great bogs, such as those at Awarua, were less favourable habitat for moa. The largest well-documented site is that of Tiwai Point on Awarua Bay. At the end of the Awarua Peninsula, this site for making adzes from the local meta-argillites contained abundant moa remains, including articulated neck vertebrae, indicating that whole birds had been brought to the site. Of equal interest was the presence of two dense concentrations of midden, containing the remains of very young muttonbirds in one patch and older ones in the other. Judging from present-day harvests, the two sets were taken in March and April respectively. The nearest possible colony would have been one historically known to have been on Bluff Hill, but the good

Fig. 4 (opposite). Shag River Mouth from the north. The site lies along the inland side of the dunes and out on to the sand flats.
Reproduction by courtesy of Ian Smith, Anthropology Department, University of Otago

canoe landing facilities of the site indicate that the birds could also have come from an offshore island. There was no discernible stratigraphy in the site and it was probably also a short-lived large camp or small village. It indicates that muttonbird colonies were exploited from at least the fourteenth century, and mainland colonies were subsequently reduced or destroyed.

Further west along Foveaux Strait, four sites have been briefly excavated and only one, Wakapatu, showed any depth of deposit. Riverton and Wakapatu both provided fourteenth-century dates and Wakapatu had some moa bone. These sites were used for working the Riverton and Tiwai meta-argillites into adzes, and all provided clear indications of people living on local resources. Wakapatu had an unusual collection of bird remains: red-crowned parakeets and native pigeons from the local podocarp forests, as well as white-faced storm petrels, were the most abundant bird bone.

Further west again, around the shores of Fiordland, archaeological evidence is confined mostly to caves. The small bush moa was hunted from sites at Chalky Inlet and Breaksea Sound and in Takahe Valley, but there is little other evidence of early occupation. At later sites, the occupants concentrated on fishing, shell fishing and taking both coastal and forest birds. The dense wet forests are unlikely to have been much affected by fire, and the large seal colonies found by the European sealers show that there had not been the scale of exploitation of Fiordland seals that there had been on the eastern coasts.

The coastal Fiordland sites could well be similar in age to the late sixteenth-century birding site on Lee Island, Lake Te Anau, which is the most intensively examined of all the Fiordland sites. The island lies in the north end of Lake Te Anau, where it is surrounded by low mountains densely covered with podocarp, rata, kamahi and beech forests. Four long, narrow rock shelters contained unusually well-preserved wooden and fibrous materials, as well as bones and feathers from the birds processed there more than 350 years ago. Two drying racks showed that the occupants had come to preserve foods, and not just to camp overnight on their way to the west coast. The birds that had been selectively hunted were native pigeon, kaka, parakeet, kakapo, and mottled petrel. There were one or two individuals from about ten other species as well – 97 individuals in all from the areas that were excavated – but no moa bone. The presence of the mottled petrels is not so surprising, since the species still nests at Lake Hauroko. The remains of one or two subadult dogs, one or two eels and galaxiids and about 50 freshwater

mussels were also present. There were no signs of equipment for taking the eels, with which the lake is well stocked. The amount of food available suggests that this could have been an overnight camp for a small group of people or for occupation for about a week by a family of five, who took half the food away with them. Takahe Valley, in the nearby Murchison Mountains, was also a small fowling camp, and there are other sites around Te Anau which suggest that transient camps for preserving birds were fairly common in the area.

The only other properly excavated site in the western mountains is at the Dart River Bridge, where repeated short-term camp sites in the fourteenth and sixteenth centuries and perhaps in between contained the highly fragmented remains of one or two moa and numerous pits suitable for cooking cabbage trees. The main activity was cooking cabbage tree for sugar, possibly to carry the travellers through to the west coast.

It may seem strange that such a small population as the twelfth to fourteenth century Maori could have brought about the extinction of so many species of moa, ranging in size from large turkey to small deer and living in so wide a range of habitats. It is a tribute to both the hunting skills of the Maori and probably the sharp noses of their dogs. On the other side of the coin, moa have large feet and would have left even clearer trails than deer. They laid no more than one or two eggs and may have had very easily found nesting sites. Certainly moa egg shell turns up in nearly every early site. There would also have been the devastating effects of large and frequent fires destroying both birds and habitat.

After the demise of the moa, it is unlikely that the more remote mountains were used as much, other than as occasional refuges from conflict. The routes to pounamu areas and the fiords themselves seemed to have attracted regular use, probably as a means of maintaining territorial rights. The pattern of activities of Maori in the open country of Otago and Southland and along the easier coasts had been one of exploration and intense patchy exploitation, followed by specialised foraging and food preservation, but within the wetter forests of Fiordland and Stewart Island it had been one of exploration and only very light exploitation.

EFFECTS ON SMALL BIRD SPECIES

Rick McGovern-Wilson

The smaller bird species of Otago and Southland were hunted opportunistically, rather than intensively, as moa were. Bird remains in nearly all of the 49 archaeological sites from which material has been recovered for analysis indicate that a wide range of species and habitats were being exploited, with the fowler taking whatever was available. The importance of small birds in the diet at most sites appears to have been negligible.

There are a few sites in Otago and Southland, such as Lee Island and Tiwai Point, which do not fit the pattern of generalised fowling. At these sites there has been specialised hunting, whereby a small selection of species is taken in large numbers. Analysis of the body parts shows that those parts with little or no meat value, such as the heads, wings and lower legs, were discarded before the birds were brought back to the site.

The presence of small bird species in both archaeological and natural sites has been used to document the distribution of both extinct and existing species during the prehistoric period. Most species were widespread, and it is evident that the ranges of many surviving species, such as kaka, weka, takahe and kakapo, have been greatly reduced. A total of 21 species of small birds became extinct in the prehistoric period, and ten of these extinctions are documented in Otago and Southland. Rather than relying on a single cause to explain these extinctions and reductions of range, the best model would see humans as the

3 EARLY DOGS AND RATS

One of the least-understood of introduced predators is also the most familiar, the domestic dog *Canis familiaris*. It is known that the Polynesian dog, kuri, arrived here in about the thirteenth century and that they were bred in large numbers at such places as the early site at Shag River Mouth. There is a strong possibility that wild packs established soon after initial colonisation, especially in the warm dry conditions of Central Otago. Though dogs were commonly eaten when they had just reached maturity, it is highly likely that they were also deliberately bred for hunting by the first Polynesian settlers. Rock art drawings in Otago show animals with powerful forequarters. Feral packs with strong hunting instincts may have accounted for the extinction of the smaller species of moa and the last widely dispersed individuals and nests that human hunters could not economically hunt. The dogs, feral and domestic, could have done great damage to all species of ground-nesting birds. When European dogs were brought in, stray animals joined up with the wild kuri. These wild dog packs were well-recorded as a major nuisance to the runholders of Otago and Southland and were vigorously hunted to extinction in the nineteenth century. The fact that the runholders were able to kill out the packs indicates that they were living in the open and had never adapted to forest existence.

The Polynesian rat or kiore *Rattus exulans* has recently caused much dissension. Its bones are difficult to radiocarbon date, providing dates much older than other materials dated from the same archaeological layers at Shag River Mouth. Nevertheless there are claims that rats arrived centuries before the most acceptable date of human arrival about the thirteenth and fourteenth centuries. If these earlier dates are eventually accepted, they still do not demonstrate earlier settlement by Polynesians, since contact may have been transient.

Kiore is smaller than the ship and Norwegian rats which have over-run the country in the last two centuries. The only evidence available for the kiore's distribution before its big cousins arrived comes from Maori middens. Rat bones preserve well in shell middens, suggesting a coastal distribution. This could be quite true, since the nesting colonies of shoreline and oceanic birds would have provided them with large quantities of food. But inland sites such as rock shelters, where the preservation conditions are also good, also have rat bones. It is highly likely that the kiore was widespread through the lowlands of Southland and Otago by the fifteenth century. It would seem, though, that it could not withstand competition from the other two rat species, and it is now limited to islands, such as Stewart Island, and to remote shorelines.

Jill Hamel

prime movers in a complex web of inter-related multi-factor processes. These would include both habitat modification and predation, along with the impact of introduced predators, beginning with kiore and the Polynesian dog (**see Box 3**).

MAORI IMPACT ON MARINE MAMMALS

Ian Smith

Bones recovered from archaeological sites in Otago show that at least five species of marine mammals were exploited by Maori people before the arrival of Europeans. Bones and teeth of large and medium-sized whales are found only rarely, and may be refuse from artefact manufacture rather than food remains. In the few cases where these can be identified to species they derive from pilot whales (*Globiocephala* sp.) and are almost certainly from natural strandings. In contrast, smaller cetaceans such as the common dolphin *Delphinus delphis* were hunted at sea using bone harpoons, particularly around and north of the Otago Peninsula. This was, however, not a common activity and does not seem to have affected the dolphin populations.

Land-based hunting of seals was much more common and did have a measurable impact. When people first arrived in the region, the most abundant marine mammal was the New Zealand fur seal *Arctocephalus forsteri*, which maintained breeding colonies on suitable rocky shores, particularly in Fiordland, Foveaux Strait, Stewart Island, South Otago and about the Otago Peninsula. Fur seal hunting took place in the vicinity of these colonies and at haul-out stations elsewhere around the coast. This provided one of the major sources of protein in the Maori diet, in most places outweighing that provided by either moa or fish. Seal skins seldom survive in archaeological sites, making it difficult to be sure how widely they were used, but a pair of slippers and small tags of skin that appear to have been attached to cloaks have been recovered from dry cave sites in Fiordland. Fur seal teeth were made into fish hook points and ornaments, but there is only limited evidence for the use of their bones in artefact manufacture.

The importance of fur seal hunting was short lived (see Box 4). Persistent hunting in the vicinity of colonies was focused on subadult animals, presumably because they are not tied to the maintenance of territories and feeding of pups and range more widely around adjacent shores. This process restricted recruitment into the breeding age classes and reduced local fur seal populations until they were no longer viable. By the late seventeenth century, breeding colonies had disappeared from the east coast of Otago, so that when European sealing began in A.D. 1792 resident populations of fur seals were confined to Fiordland, Foveaux Strait and Stewart Island.

New Zealand sea lions *Phocarctos hookeri* were also hunted in the early years of settlement, but were never as common as the fur seal. They had regular haul-out stations about Foveaux Strait and probably also along the east coast of Otago. The presence of a few adult females in South Otago sites suggests the possibility of former breeding colonies in this area, but the evidence for this is uncertain. What is clear is that sea lions are less common in more recent sites, indicating a reduction in their distribution similar to that for the fur seal.

Elephant seals *Mirounga leonina* and leopard seals *Hydrurga leptonyx* were also hunted, but, as both these species appear to have been only sporadic visitors to the Otago coast, hunting had no measurable impact upon their distribution.

Details of change

Jill Hamel

New Zealand's first serious archaeologist, Julius von Haast in the 1870s, saw that moa bones lay below heaps of shellfish and thought that the two layers were deposited by different races living centuries apart. Teviotdale in the 1930s thought the moa

4 MEAT WEIGHTS AT SHAG MOUTH

Regular hunting at fur seal colonies rapidly reduced the size of resident populations. The speed with which this occurred can be seen at the early Maori village at the mouth of the Shag River which was occupied in the middle of the fourteenth century. When the settlement was first established (layers 8–11), fur seals hunted at the nearby Shag Point colony provided almost 40 per cent of meat consumed at the site, but by the end of the occupation (layers 1–4) this had fallen to just 12 per cent. Radiocarbon dating shows that this process took only 20 to 50 years.

Ian Smith

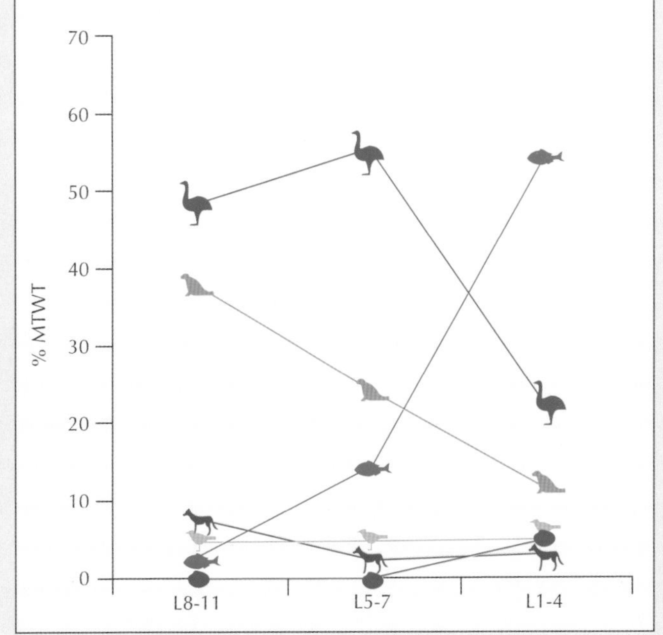

Graph of meat weights in successive layers in a deep midden in the central sand dune at Shag River Mouth. The lowest and earliest layers are to the left. *After Anderson, Allingham and Smith (1996), by courtesy of Ian Smith*

hunters came from the northern part of the South Island for seasonal hunts, but Lockerbie argued that there was gradual change from moa hunting on the coast to inland sites in the period from the tenth to fifteenth centuries, followed by a reliance on fish, shellfish and small birds after A.D.1600. The latter idea was persuasive until 1996, when the detailed evidence from Shag River Mouth was published. Careful dating of the complex layers at Shag River Mouth has shown it not only to have been occupied for only about 50 years, but also in that time the basic change in food gathering happened as well!

This idea of a few early transient villages, depending on seals and moa and moving every few decades to another site, has suddenly made sense of some past anomalies, such as the 'horizontal stratigraphy' at Pounawea, the shortened time span for Papatowai, and some curious patterns of adze types in early sites. The pattern of exploitation of this pristine environment was far from sustainable. There seems to have been neither substantial trade with other communities nor inland hunting to bring back preserved moa to the coast to eke out the depletion of food resources around coastal villages.

After about 400 years the system fell to pieces when there were no more readily accessible seal colonies and moa were close to extinction. (A similar system of over exploitation fell apart for European pastoralists when they brought the first sheep flocks into Otago – see below.) Though it seems like poor management, given the need of the new immigrants to build up their populations rapidly, it made sense to vigorously use up the patches of rich resources. Coming from islands in the tropical Pacific, these people would have had no tradition of managing large flightless birds or marine mammals.

A new system of foraging and new social arrangements had to be developed. When the Europeans arrived, they found the main coastal settlements were small villages of 20 to 30 houses, belonging to a larger tribally organised group of 3000–4000 people who claimed kinship in up to 100 hapu. A few paramount chiefs controlled the whole group, and though villages were relatively short-lived it was often the death of a chief which caused a shift, rather than the depletion of resources. These villages were the centres of seasonal foraging trips and storage of preserved foods: mostly muttonbirds along Foveaux Strait, eels and barracouta in the central districts and cabbage tree kauru (a sort of sugar) in North Otago and Canterbury (Figs 5, 6). Each village was occupied by members of many hapu, maintaining their rights to the local resource. The trading of preserved foods north for kumara and other commodities was under the control of tribal leaders. The development of this system had been relatively recent, since sites dated to the seventeenth and eighteenth centuries tend to be small camp sites inland or small coastal pa, and do not exhibit the old pattern of village development.

Even though the seventeenth and eighteenth-century populations were so dependant on barracouta, there is no evidence of depletion of stocks. What is more certain is that it has been European fishing, rather than the canoe fishing of the Maori, that has reduced the major inshore fish species favoured by Maori, such as barracouta, ling, red cod and groper. (The only range restriction shown by evidence from middens is that of snapper, which occurs in small numbers in some early sites and not in later, and is certainly rare in southern waters today.) Descriptions of the colossal barracouta shoals seen off the Otago Heads in the 1880s are a revelation, and it would be interesting to know what effect the sudden reduction of seal colonies in the fourteenth century had on the prey-predator relationships of the two species.

Barracouta are large and exciting fish to catch and were taken by the Maori with a whirling lure on the end of a short rod and line. Ling, red cod and groper are inshore fish, easily taken with a line. The other abundant fish of the southern estuaries, such as flounder and kahawai, and fish of rocky shores and the deeper waters are very poorly represented in middens. If barracouta were so easy to take from the surface with a short line, why hazard precious hooks and hand-spun line on rocky shores, except when conditions were too rough to put to sea? The lack of estuarine fish is more puzzling, but the loss of body heat involved in wading to spear flounder and the extra work needed to make nets may not have seemed worthwhile in cold southern waters. Their descendants frequently emphasise the pragmatism of these first New Zealanders and their choice of fish species may illustrate just that. Though there is plenty of evidence for freshwater fishing (whitebait, lamprey and eels) at the time of European contact, the remains of eels and some of the larger galaxiids are only occasionally found in middens. The favoured eeling lakes of the nineteenth century do not show signs of long occupation during the prehistoric period, and there have been suggestions that eeling was a relatively late development.

Another flow-on effect which is difficult to evaluate is the release of shrublands and forests from browsing pressure by moa populations. Knowledge of moa browsing is limited to deductions about heights to which they could reach, some scant evidence of plant remains in a few gizzards preserved in peat bogs, and to proposals that the divaricating habit of many shrubs and tree seedlings is a response to mitigate the effect of moa browsing. It is frustrating that the recent work on plant species favoured by takahe, kakapo, deer and possums has not thrown up a clearly delimited group of common palatable species, which we could assume to have been on the increase between the demise of the moa and the advent of introduced herbivores. Nor do there seem to be any tree species with hard fruit which are failing to regenerate because there are no longer moa gizzards to process the fruit.

Looking at the matter the other way round, we can expect that the abundant species of shrublands and forests may have evolved internal defences, such as a bitter taste, sharp spicules, and tough fibres, specifically against moa browsing. Bracken rhizomes have strong bands of fibre, copper tussock is loaded with silica spicules and cabbage trees have tough bark. Over the hundreds of millennia in which New Zealand forests and moa evolved together, the plants put considerable evolutionary energy into thwarting the browsing bird. Selection pressure for these traits was suddenly turned off about 400 years ago. We may know very little about what those traits were, but in considering the dynamics of native forests we should remember that the release happened.

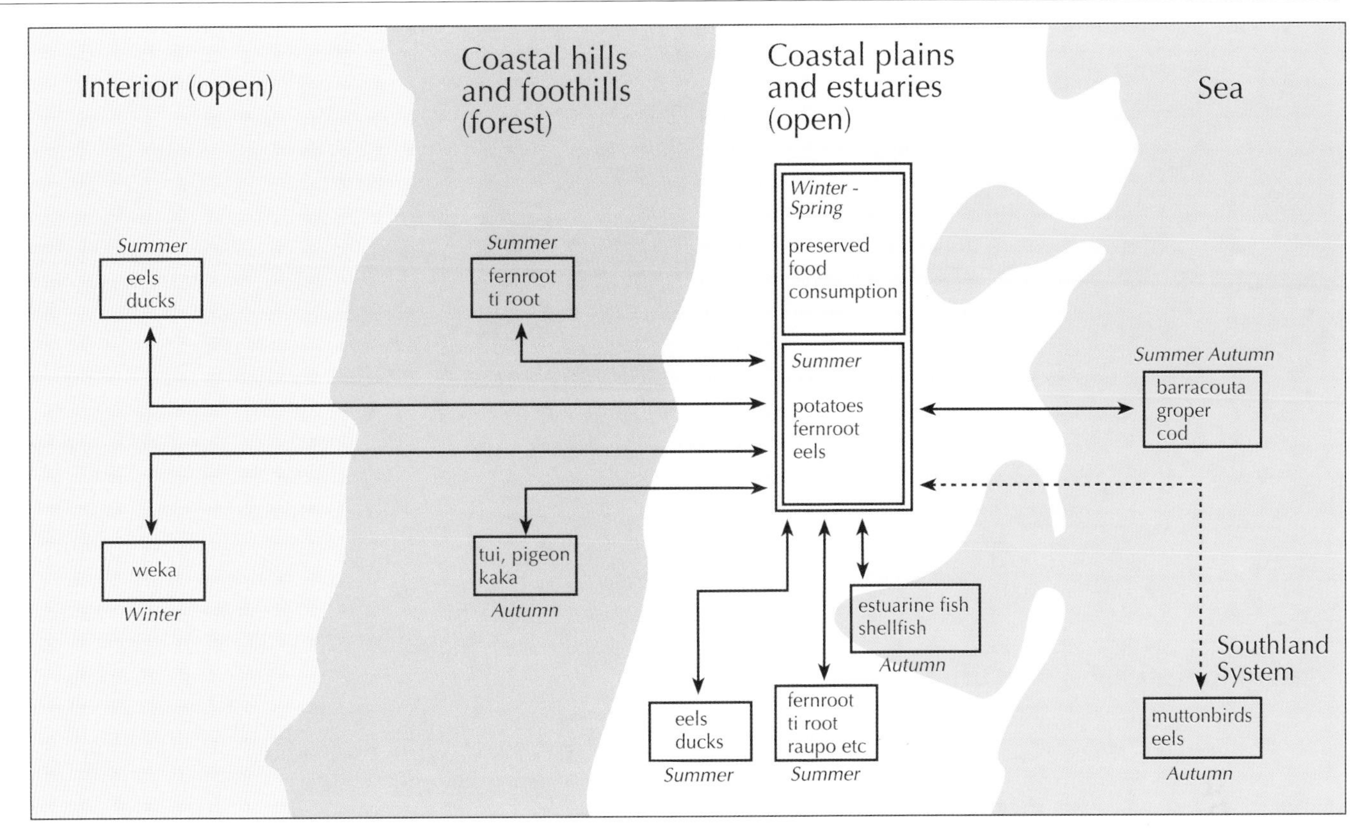

Fig. 5. Pattern of seasonal food gathering by Otago Maori in the nineteenth century. Note the greater use of fern root and the ti or cabbage tree, compared to Southland. *After Anderson (1983), by courtesy of Otago Heritage Books, Dunedin*

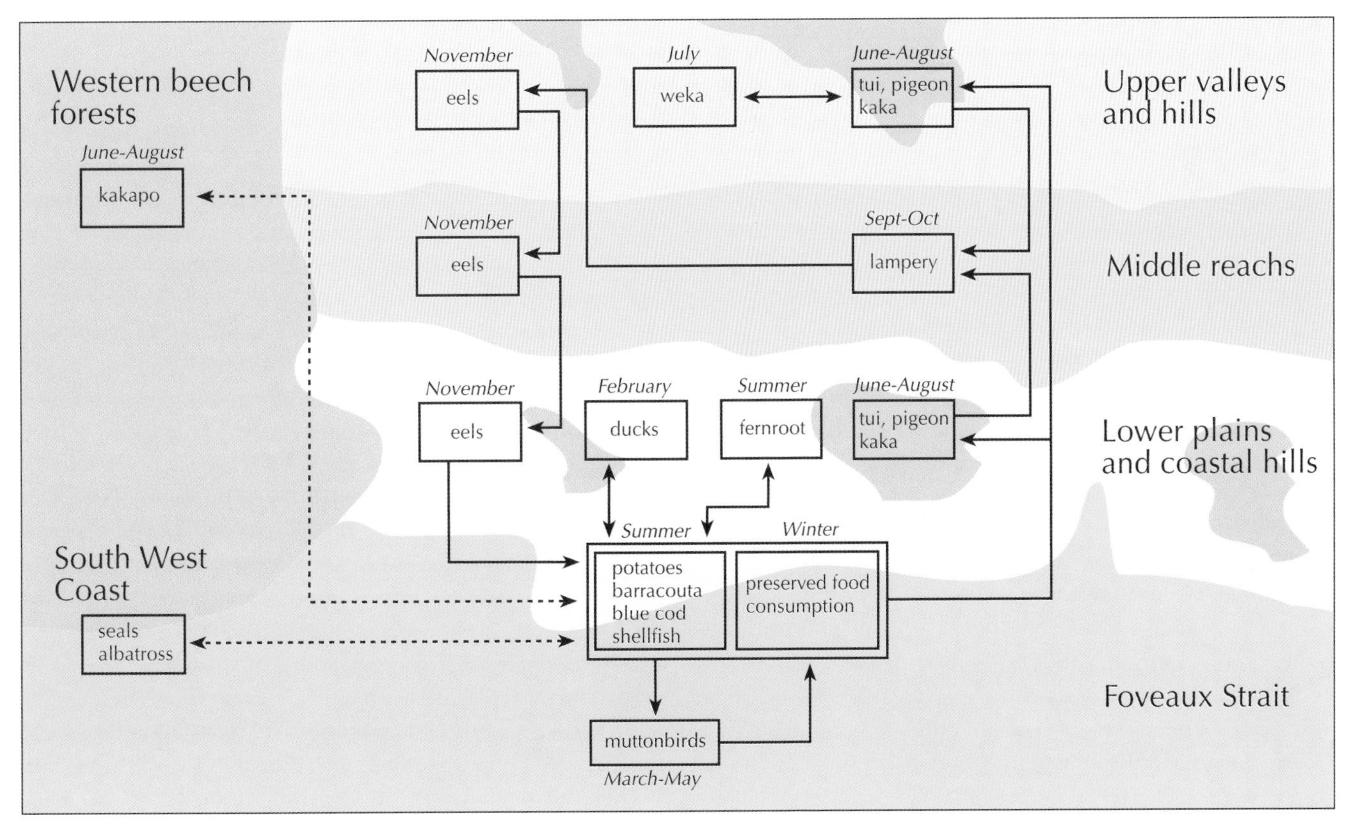

Fig. 6. Pattern of seasonal food gathering by Southland Maori in the nineteenth century. *After Anderson (1983), by courtesy of Otago Heritage Books, Dunedin*

Fig. 7. An eroding midden, of paua and other shells, marks a camp site of early Maori in the Mason Bay area. *Neville Peat*

STEWART ISLAND, THE UNKNOWN ISLE

The formal archaeology of Maori life on Stewart Island is almost non-existent. There have been no modern excavations or site surveys of the island. An excavation at the Neck, Paterson Inlet, by Les Lockerbie about 1960 has not been written up, and all we know is that midden material containing bones of a medium-sized moa (known then as *Euryapteryx gravis*) was associated with oven charcoal, adzes and flakes. (Some of this was surface-collected by David Teviotdale and Gordon Dempster.) Other sites where moa bone has been found with signs of Maori occupation are close to the Neck at Ringaringa and Native Island and also at Mason Bay. Trampers sometimes see midden layers of shell and charcoal around the beaches of Oban, Patersons Inlet and Mason Bay, along with eroding oven stones (Fig. 7).

Considering the nature of colonisation of the adjacent shores of Foveaux Strait, we can make a best guess that Maori arrived on the shores of Stewart Island about 800 years ago. Around the coastlines they found large colonies of burrow-nesting petrels, penguins, shags and breeding fur seals. The ocean-going canoes in which they had arrived, the magnificent podocarps at river mouths on Foveaux Strait for making more canoes, and their skills in using marine resources enabled these migrants from tropical Polynesia to adapt quickly to the life of sub-Antarctic hunters. Even the southern shorelines of Stewart Island were probably soon explored.

The effects of this first colonisation on the ecosystems of Stewart Island are likely to have varied from nil to locally profound. As along the east coast of Otago, subadult seals were an easy source of food. Supplementing seals at first with moa, it is likely that family groups on the eastern coasts of the island worked from one accessible colony to the next, reducing each in turn, judging by the illuminating evidence from Shag Mouth. The western coasts were more difficult to reach, and seal colonies were still relatively strong when Europeans arrived.

As at Tiwai Point, the preserving of muttonbirds was probably developed well before the last of the moa were gone. It is likely that there were once mainland colonies of petrels as dense as those of the present-day titi islands, the former being steadily reduced until uneconomic, leaving in place the more adventurous titi island harvests which continue to the present day. These harvests were of such importance that the main village of the paramount chief of the area was established on Ruapuke, from where he could command the muttonbird harvests. (It was on this small island that the first mission station in Southland was established by Rev. Johann Wohlers, because it was the centre of political power in the region.)

Life in the nineteenth century

In contrast to its prehistory, the history of Stewart Island in the nineteenth century has been extensively researched. The first European settlement was a group of sealers at Sandy Beach on Codfish Island. The settlement was short-lived from 1825 to about 1850 and had little effect on the vegetation of the island. The weka, probably introduced for food by Maori mutton-birders, had a more profound effect by killing young petrels, especially those of the rarer species.

Stewart Island itself was settled by Europeans along its eastern shores and through the Freshwater Flats to Mason Bay. The first were a group of seven unfortunate sawyers and their Maori wives, who were set down in 1826 at Port Pegasus by a rascally ex-sealer to build a 100-ton vessel for him, with only six months' supply of food. The sealer abandoned the group, who lived in some distress off the local resources, with sporadic assistance from passing whalers, for at least four years. Even in the 1930s there was little sign of their previous existence, other than an area of second-growth forest replacing the older forest that they had felled.

More permanent settlements were established at Oban and the mouth of Patersons Inlet (Fig. 8). From the 1820s, shipbuilders settled briefly in nearly every sheltered cove from Port William to Port Pegasus, through to the Norwegian shipyard in Patersons Inlet in the mid twentieth century. All made their impact on the local forests. The tin miners at Port Pegasus in the 1880s set fire to the Tin Range (Box 5), and some unknown hand has created fire scars even in the wet forests of the west coast opposite Codfish Island.

The Otago Settlements Act in 1871 provided for Special Settlements on Stewart Island, but the lack of roads and difficulties in clearing the forests meant that even the hardy Shetland fishermen who settled at Port William did not thrive. By 1885 there were only about 400 acres held under freehold, 2800 acres under lease and less than 1700 acres under cultivation or sown pasture. Sheep had been established at the Neck and Halfmoon Bay to supply the sealers in the 1820s, but the first venture into extensive pastoralism was not until 1874, when a flock was placed on the Freshwater Flats. In this area of bogs and dunes the sheep either died or became more a semi-wild resource than a managed flock. A drainage channel, 4 km long, was cut to drain the Island Hill flats into the Freshwater River. Only the moderately dry land close to Mason Bay was grazeable, and a flock was run there until about 1990. Shearing was carried out by stalking individual sheep, blade-shearing each one and carrying the wool in a sack back to the homestead. The homestead at Mason Bay had a garden which was fenced against kiwis as well as sheep, as a kiwi probing for worms was no respecter of young lettuces.

Fig. 8. Lewis Acker built his cottage (1834–36) in Halfmoon Bay from large stones brought by whaleboat across Foveaux Strait. The original roof was thick stone slabs. *Jill Hamel*

Conservation issues

The present system of conservation land began with the government botanist, T. Kirk, who suggested that timber reserves be declared for most of the island in 1884. All the Crown-owned land was declared a State Forest in 1886. A complex history of lifting and designating reserves of various status has ended in the present pattern of about 85 per cent of the island becoming a National Park (157,000 ha) with a further 10 per cent in reserves. Unfortunately, early in the twentieth century red and white-tailed or Virginian deer were liberated, and they have caused great damage to the under-storey of the forests. Wild cats have done surprisingly well and are considered to have been the major predator of the kakapo populations, even as far south as Port Adventure. Kakapo are now confined to offshore islands.

Though ecologically modified, Stewart Island landscapes show little sign of human presence. The lack of mice and mustelids, the lack of trout in the streams and minimum of modification by sheep makes it special as a place with a high potential for retention of indigenous ecosystems from dense forest to subalpine tarns.

5 STEWART ISLAND TIN MINING

Gold was discovered in Pegasus Creek in 1882. A black impurity found with it was assayed and identified as stream tin (cassiterite, tin ore). A tin rush ensued, concentrated on the streams draining from what became known as the Tin Range. To serve the new mining community, Louis Rodgers established a store, hotel and post office at Port Pegasus in1889. But while tin was present in the streams and hills, it did not exist in sufficient quantities to repay working. By 1890 the rush was over and the hotel bankrupt.

A second large-scale attempt to open the tin field was made in 1912, when a company, The Stewart Island Tin and Wolfram Lodes Ltd, was formed. At great expense the company built a wharf and buildings at Diprose Bay and constructed a tramway to the western flank of the range. But by 1917 this company had also failed, having produced only tiny quantities of tin.

Peter Petchey

The company also constructed this dam in the headwaters of McArthurs Creek, on the western flank of the Tin Range. It supplied water to sluicing operations via an iron-lined water race and pipeline. *Peter Petchey*

INTRODUCED MAMMALS

Lloyd S. Davis

Apart from seals and bats, all other terrestrially-breeding mammals in New Zealand are immigrants to this land. One of the first to arrive, humans, introduced the rest. Polynesian immigrants brought with them kuri/a dog *Canis familiaris*, and kiore/a rat *Rattus exulans*. The kuri is now gone and the kiore, itself the victim of competition from subsequent introductions of other species, is restricted mainly to offshore islands, Stewart Island and parts of Fiordland. However, the most lethal invasion of mammals occurred with the arrival of Europeans, who liberated everything from alpacas to zebras: of those species that became established, almost every one has left its destructive mark on the New Zealand environment.

Among the first mammals to be introduced were pigs *Sus scrofa*. Captain Cook and the French explorer, De Surville, released pigs at various points around New Zealand during the final part of the eighteenth century, often as gifts to local Maori. Populations of feral pigs soon became established around Maori and European settlements and their numbers and distribution grew quickly. Today they can be found in the marginal farmlands, tussock grasslands and forests of Otago, Southland and eastern Fiordland (Fig. 9). They have been eradicated from Stewart Island, but remain on Ruapuke Island in Foveaux Strait. Apart from the damage feral pigs cause to pastures and exotic plantations, they have probably had detrimental affects on our native flora such as orchids and tussock, and on ground nesting birds like kakapo and many seabirds.

In addition to pigs, from 1773 Captain Cook and, subsequently, the sealers, released goats *Capra hircus*, often to provide a source of food for castaways. Feral goat populations are now found throughout New Zealand in forested ranges and scrubby hill country, with most goats in the southern region being concentrated in Central Otago around the southern lakes district and near the coast east of Bluff. Coat colour is variable and may be either black, brown or white, or any combination of these. Both males and females have horns, but there is a marked sexual dimorphism, with males being larger, and having bigger horns and shaggier coats. Because New Zealand's native flora is not adapted to browsing, goats – along with other introduced browsing mammals – have caused immense damage to native vegetation and undermined land stability in places.

Other species of domesticated bovids, also introduced for food, became feral as well. Cattle *Bos taurus* were first brought to New Zealand in 1814. By the 1860s, feral cattle were widespread and common throughout the North and South Islands. As a result of hunting pressure, numbers of feral cattle have declined dramatically, and they are now restricted to areas of thick bush or scrub. The last population in the southern region was exterminated from the Catlins State Forest Park in 1979. Sheep *Ovis aries* were introduced to New Zealand throughout the nineteenth century, mainly from Australia. Put on to unfenced range, populations of feral sheep quickly became established and widespread. Better control and management of sheep has seen the numbers of truly feral sheep dwindle, so that now the only flocks left within the southern region are at Waipori Gorge and Hokonui Hills. Living in marginal scrub and bush habitats, these feral animals have little impact on the New Zealand biota.

Fig. 9 (left). Pig *Sus scrofa* introduced by Captain Cook. *Rod Morris*

Fig. 10 (top). Norway rat *Rattus norvegicus*. *Rod Morris*

It was the mammals that Cook, the early sealers and settlers introduced unintentionally that have had some of the most catastrophic effects on New Zealand ecosystems. For example, Norway rats *Rattus norvegicus* and cats undoubtedly got ashore when Cook moored the *Resolution* in Dusky Sound for a month in 1773. Norway rats are good swimmers and as they were the commonest shipboard rat until about 1830, it is safe to assume that there were many unescorted disembarkations of such rats upon the arrival of the early Europeans. They quickly colonised the North and South Islands, for the most part displacing kiore, and by mid-way through the nineteenth century were widespread and numerous. The Norway rat is the largest of the rats in New Zealand (e.g. Norway rats from Stewart Island average 182 g and 214 g for females and males, respectively, and large ones can approach 300 g), with a long and thick tail (12.8–18.8 cm). The coat colour is grey to greyish brown (Fig. 10). They are not good climbers and it was the ground-nesting land and sea birds that were most affected by their predation. By 1900, however, the vast numbers of Norway rats were declining rapidly: today they have disappeared from large tracts of Fiordland and, while distributed widely throughout Otago and Southland, they are really numerous only on Stewart Island. The decline in the fortunes of the Norway rat has probably occurred in response to predation from other introduced creatures (mustelids) and competition from another introduced rat.

The ship rat *R. rattus* was also introduced from ships of early European visitors to New Zealand, especially those arriving from 1850 onwards, but became widespread in the South Island only after 1890. They are intermediate in size between the Norway rat and kiore (ship rats from the Hollyford Valley in Fiordland average 131 g and 152 g, respectively, for females and males, with large individuals being up to 200 g). The tail is sparsely haired and long (16.0–21.7 cm). They occur in three colour morphs: '*rattus*' (Fig. 11) which has a black back, '*alexandrinus*' which has a grey-brown back, and '*frugivorous*' which has a grey-brown back but whitish undersides. They live in forests as well as a range of other habitats, including living alongside humans, and they are now distributed throughout the whole of New Zealand. Agile climbers, they have been particularly devastating for the small native tree-nesting birds, being responsible in many cases for a decline in numbers and in some instances, extinctions.

The other rodent to establish in New Zealand also arrived as a stowaway aboard ships. The house mouse *Mus musculus* was first recorded in New Zealand on Ruapuke Island in Foveaux Strait following a shipwreck in 1824. By 1852, mice still had not been recorded as being present in Otago, but within two years mice were seen in Dunedin and they rapidly dispersed to the Taieri Plains and across the Taieri and Clutha Rivers. Today mice are found in suitable habitats throughout the southern region with the exception of Stewart Island, where the presence of large numbers of Norway rats is thought to have prevented their establishment. They inhabit forests, pasture, croplands and subalpine tussock, as well as living commensally with humans. Their major impact has probably been on New Zealand's native invertebrate fauna, although potentially they can kill lizards and eat birds' eggs. They can also have a damaging effect indirectly, by helping to sustain other introduced predators of our native fauna.

Fig. 11 (top). The ship rat *Rattus rattus* was introduced from ships of early European visitors. They occur in three colour morphs: 'rattus' which has a black back, shown here, 'alexandrinus' which has a grey-brown back, and 'frugivorous' which has a grey-brown back but whitish undersides. *Rod Morris*

Fig. 12 (right). The European rabbit *Oryctoagus cuniculus*. *Rod Morris*

Along with the ship-borne rodents came the ships' cats *Felis cattus*. Feral cats did not become established until the mid nineteenth century, but now they are distributed widely throughout New Zealand, including Stewart Island. While they prey on introduced mammals, they have also had perhaps the most underestimated effect on our native birds; responsible for the extinctions of several species, especially on islands, they have also decimated other populations of birds, including those of kakapo on Stewart Island. Cats introduced onto Herekopare Island in Foveaux Strait in about 1925 were responsible for the disappearance of at least six species of land bird and two species of seabird. Cats also kill bats and lizards. That cats have not got the bad press they deserve, and the reason that they were liberated so freely, is probably due to a perception that they help control the numbers of other introduced mammals, principally rodents and rabbits.

Many cats were released by farmers in the mid to late nineteenth century in an attempt to control burgeoning populations of the European rabbit *Oryctolagus cuniculus*. Rabbits were often carried on ships for food, and the first known introduction was by Captain Cook in the Marlborough Sounds in 1777. Whalers and settlers released rabbits in the late 1830s at Tautuku and Riverton, but it is unknown exactly when or if populations of domesticated rabbits became established ferally. Certainly the former Acclimatisation Societies, to their subsequent shame, were involved in the importation, breeding and distribution of wild rabbits from Britain, with the first being released in Southland in 1864. The spread of rabbits was limited by the availability of suitable habitat, and their distribution followed the development of pastoral land throughout the country. The original tussock grasslands of Central Otago, unsuitable as they were for rabbits, were burnt and grazed, thereby providing a perfect environment for the rabbits, whose numbers exploded in the Mediterranean-like climate. By the 1870s, stocking rates for sheep in parts of Otago and Southland had fallen by 60 per cent as rabbits destroyed the grasses and caused soil erosion. Many run-holders simply abandoned their land. Today, rabbits are distributed at low densities throughout New Zealand, with Central Otago being one of the few areas where densities of rabbits remain high (Fig. 12). The high rainfall and dense bush of Fiordland and Stewart Island have kept those areas relatively rabbit-free. Attempts to control rabbit numbers have mostly involved introductions of predators, shooting and poisoning. In the 1950s an unsuccessful attempt was made to introduce the viral disease myxomatosis, but it did not become established, probably because of the lack of a suitable vector. While there were repeated calls from the farming industry for its reintroduction, together with a flea that could act as the vector, recently the focus for biological control shifted to another virus: calicivirus (Box 6).

6 RABBIT CONTROL

Farmers had long called for the introduction of the disease myxomatosis for rabbit control, but the call was rejected in a 1990 political decision driven largely by urban animal welfare concerns. More recently, farmers and associated agencies applied to have RHD (Rabbit Haemorrhagic Disease, also known as RCD or Rabbit Calicivirus Disease) introduced. The government had changed the law to put the decision in the hands of the professional bureaucracy advised by scientists, but the application was rejected in 1997 on the grounds that there was insufficient knowledge about how the virus would work in New Zealand.

Despite the fact that the applicant group was given a strong indication that it should try again when more was known from experience with the disease in Australia, desperate and irresponsible elements in the farming community began to spread RHD throughout the South Island high country and into the North Island during the spring of 1997. It seems clear that there was an illegal importation even prior to the rejection of the application, although spreading the disease was subsequently legitimised by the government.

Many farmers made a broth of material containing the virus and spread it on the land via baits, in much the same way as 1080 is used in carrots (the biocide method). Other farmers released small quantities of virus and allowed it to spread naturally (the bio-control method). Research commissioned by the Otago Regional Council showed that the results from the biocide method were poor, with kill rates varying between 40 per cent and less than 20 per cent. Biocontrol results were better (sometimes up to 90 per cent). Immunity to RHD has been shown in as many as 55 per cent of surviving rabbits. The use of the biocide method seems to have been a serious error, with the weakened virus on baits probably having innoculated rabbits rather than killed them.

The growing incidence of RHD immunity in the rabbit population generally is a serious concern. It is difficult to be clear about what is happening in view of the uncontrolled way the disease was released, but it is now seen to be giving uneven results irrespective of which release method was used, and a return to conventional control methods has been urged in some places. Possible factors in the apparent failing of the disease are weaknesses in the strain illegally imported, a degree of natural immunity in the rabbit population prior to release and mutations in the virus since release. Common sense suggests that a living organism like a single host virus (which RHD is) will find a way to keep its host species going.

Overall, rabbit populations are still a lot lower with RHD in circulation, and farmers have benefited from higher land productivity and lower pest control costs. One overseas scientist, however, has predicted that the farmers will be back to square one in five years!

Even if this prediction turns out to be exaggerated, the introduction of RHD in New Zealand plainly could have been managed to better effect; a valuable opportunity for more effective rabbit control and knowledge accumulation has been lost. More alarming perhaps, was the disregard by elements of the farming community of cautionary scientific advice and the flouting of laws designed to prevent the arrival in New Zealand of unauthorised organisms. The high country farming community then closed ranks to protect the original offenders.

The farming community owes much of its present productive success to scientific research and innovation, and much of its competitive advantage is based on an absence of pests and diseases that are found overseas. The attitude of defiance that many farmers still display towards the RHD issue suggests that they are out of touch with mainstream community concerns about biological control and the general expectation that the law of the land should be respected. It is important that when RHD finally loses its punch, the farming community takes a more responsible approach to the search for and application of the next rabbit control method.

Acclimatisation Societies were also responsible for introducing another lagomorph for sport: the brown hare *Lepus europaeus*. Distinguished from rabbits by their larger size, longer ears and hind feet, tawny colouration and yellow eyes, hares were first released in Otago in 1867 and Southland in 1869. They are now found throughout New Zealand except for Stewart Island and most of Fiordland. They occupy mainly grasslands or open countryside. Because they occur at low densities, the damage they cause by grazing is much less significant than that caused by rabbits, though it can be important for horticulture in some localised areas.

The devastation to grazing lands caused by wild rabbits brought demands from farmers to restore the balance by importing their natural predators. Ferrets *Mustela putorius*, were first released in 1879 in the Conway River valley. Five years later, the first importation of stoats *M. erminea* and weasels *M. nivalis* was made by a resident of Palmerston. Over the next two years, thousands of ferrets, 592 weasels and 214 stoats were released at sites including Lake Wanaka, the Makarora and Wilkin Valleys, Lake Wakatipu and parts of Southland. Unfortunately, these mustelids had little effect on the rabbit population, but by 1892 the damage weasels and stoats were causing to native birds, as they moved into the forest of Fiordland and the like, was becoming apparent. Ferrets, by far the largest of the three mustelids, with a black face mask (Fig. 13), are found throughout Otago and Southland today, but are absent from areas with high rainfall and dense bush, like Stewart Island and Fiordland. Stoats, the middle-sized member of the mustelid trio, with a straight line separating the brown fur on the back from the white underbelly and a long black-tipped tail, are found throughout New Zealand, but are absent on Stewart Island (Fig. 14). Weasels, the smallest New Zealand mustelid, with a variable line separating the brown dorsal regions from the white ventral side and a short brown tail, experienced a contraction in range following an initial irruption after release, and now weasels are absent from Fiordland but occur at low densities throughout Otago and Southland (Fig. 15). The mustelids are widely believed to have decimated populations of many native bird species and to have had a detrimental effect on others. While the extent of the carnage exacted by ferrets, weasels and stoats, in particular, has been difficult to document accurately, and granted that they may help to keep in check populations of other introduced mammals such as rats and mice, there is little doubt that the diversity and character of the New Zealand ecosystem has been forever diminished by the folly of their introductions.

But the folly did not stop there, and once again it was fuelled by the Acclimatisation Societies and the desire of settlers to have suitable animals to hunt. Deer were introduced to New Zealand beginning in the mid nineteenth century. Red deer *Cervus elaphus scoticus* were brought to New Zealand from 1851, and by 1923

Fig. 13. The ferret *Mustela putorius*. Rod Morris
Fig. 14. The stoat *M. erminea*. Rod Morris

Fig. 15. The weasel *M. nivalis*. Rod Morris
Fig. 16. Red deer *Cervus elaphus scoticus*. Rod Morris

over 1000 had been liberated. Red deer are a medium-sized, round-antlered deer with a brownish body (Fig. 16). They became well and truly established throughout the forests and high country of all the southern region, including Stewart Island. They thrive in forest and scrublands, where they have had a dramatic effect on a native New Zealand flora unadapted to browsers, eliminating or reducing plants that contribute to the forest understorey. Control measures, such as shooting deer from helicopters and, more recently, recovering deer for farming, have proved effective at reducing deer numbers and thereby limiting the amount of damage that they now cause.

The red deer were imported mainly from British stock. Eighteen individuals of a much larger subspecies, wapiti or elk *Cervus e. nelsoni* were brought from North America in 1905 and released in George Sound. The largest of the round-antlered deer, with dark brown heads, necks and legs that contrast with their light yellowish brown bodies, the descendants of that single liberation are now found in an area of northern Fiordland west of Lake Te Anau. Grazing by wapiti has a similar effect on our forests as does that by its subspecific cousins, but of course the effect is much more localised due to their limited distribution.

The former Otago Acclimatisation Society also liberated seven axis deer *Axis axis* between Oamaru and Palmerston in 1867. They became established and by 1877 numbered about 100. However, settlers had exterminated them by 1895. Not content, the Otago Acclimatisation Society released three sika deer *Cervus nippon* near Oamaru in 1885, but it seems that these were also shot out by settlers and never became established in the South Island. But liberation of other species was more successful.

Fallow deer *Dama dama*, a small species with palmate antlers, were released in the Blue Mountains near Tapanui in 1869, where they persist today. The Otago Acclimatisation Society provided stock from the Blue Mountain herd for release in the Lake Wakatipu area, and today their descendants form the second enclave of fallow deer in the southern region, concentrated in the Greenstone and Caples Valleys and around Mount Creighton. Within their localised distribution they have been responsible for major modification of native vegetation.

Nine white-tailed deer *Odocoileus virginianus* were released both on Stewart Island and in the Rees Valley at the head of Lake Wakatipu in 1905. A smallish round-antlered deer with a uniform brown coat and a long tail with a white fringe and undersides, this species quickly became established in these two distinct areas: Stewart Island, especially in the coastal forests, and at the head of Lake Wakatipu. On Stewart Island, where the deer have become locally abundant, they have browsed the palatable shrubs and ferns and interfered with the regeneration of woody trees by eating the saplings. The herd at Wakatipu is too small for such effects to be so noticeable.

Fig. 19. The brushtail possum *Trichosurus vulpecula* eating mistletoe. Rod Morris

Perhaps one of the most surprising introductions to New Zealand was that of moose *Alces alces*. After the failure of an earlier attempt, ten moose were released at Supper Cove, Dusky Sound in 1910. A small population became established in the Dusky Sound area and may still exist today, although there have not been any confirmed sightings for over 20 years. Nevertheless, a recent expedition to the area found enough sign to suggest that a few moose may still survive in the area.

Despite the negative impact of deer and their like, the prize for champion modifier of New Zealand's vegetation must surely go to one of the most ill-conceived of the mammalian introductions: the brushtail possum *Trichosurus vulpecula* (Fig. 17). A forest-dwelling marsupial, the possum was introduced from Australia to establish a fur trade. The first successful introduction was made near Riverton in 1858. Subsequently many liberations were made, mainly by the former regional acclimatisation societies. After representations made by the Southland Acclimatisation Society, the government brought the possum under the Protection of Animals Act 1880 and, despite outcries from farmers and conservationists, possums remained protected in some form or another until 1946. By then the damage was done: possums have spread through most of New Zealand and in the southern region occupy all areas of forest except parts of western Fiordland. They are selective browsers of native forest trees, causing extensive defoliation and mortality. Recent evidence shows that they also eat birds' eggs and native insects. Finally, they act as a reservoir and vector for bovine tuberculosis, thereby having the potential to both devastate our farming industry and destroy our natural heritage.

Other marsupials were also introduced to New Zealand, principally wallabies, but only one has become established in the southern region. Bennett's wallaby *Macropus rufogriseus* was liberated in the South Island in 1874 near Waimate, Canterbury. But a small population has also become established at Quartz Creek, between Lakes Hawea and Wanaka, after an escape from a zoo about 1945. There they live in silver beech forests and manuka scrub.

Introduced alpine mammals, such as thar *Hemitragus jemlahicus* and chamois *Rupicapra rupicapra* are discussed in Chapter 9 (page 235).

Of all the introduced mammals that have become widespread, perhaps the one thought to have the least impact is the European hedgehog *Erinaceus europaeus*. The first recorded liberations in the southern region occurred in Dunedin in 1885. At first hedgehogs were brought to New Zealand from Britain, to remind the settlers of their homeland, but they were actively introduced after it became apparent that they could control garden pests such as slugs and snails. Hedgehogs

Fig. 18. Transformed landscape: pastoral land replaces former wetland and forest. *John Darby*

are now distributed throughout lowland districts, including Stewart Island, but are absent from alpine areas and areas of high rainfall like Fiordland. As the hedgehog has been reported taking eggs and chicks of ground-nesting birds in parts of Europe, we cannot afford to ignore its habits in this country.

New Zealanders have a love-hate relationship with mammals. We love them because invariably they are cute and furry. Yet because they have done more damage to the country's fragile ecology than anything else in the last 65 million years, we should hate them with a passion.

Sheep, cattle and their ecosystems *Jill Hamel*

The first Europeans to follow Captain Cook to New Zealand were the sealers. As soon as they realised they would be returning to the wild shores of Foveaux Strait on their sealing trips, they encouraged local Maori to grow *Solanum* potatoes and raise pigs, sheep and cattle for them. Whalers set up shore stations in the 1830s and 1840s, the more long-lived ones being at Moeraki, Waikouaiti, Otakou, Taieri Island, Tautuku, and Riverton. These became centres of distribution for the new plants and animals.

Pigs and cattle, being more adventurous than sheep, were starting to run wild in the coastal forests of Southland by 1825. They were soon followed by stray dogs of European breeds, who joined up with the wild kuri. The howling of these wild dog packs kept the whalers awake at nights. Gardens were established at most of the stations, goats kept for milk and meat, and even rabbits introduced on to offshore islands for more delicate meat. When whaling declined, many of the whalers became small-scale farmers at places like Moeraki and Riverton, soon to be followed by large numbers of settlers and even larger numbers of predators, carnivorous and herbivorous.

Europeans introduced so many mammals that every habitat from mountain tops to the deepest forests has its own introduced herbivore. No native vegetation is exempt. The three species which are likely to have the most lasting impact are sheep, rabbits and possums – sheep because of their high numbers and the other two because of the difficulties of eradicating them in southern New Zealand. The management of sheep, cows and other domesticated stock has produced a completely new phenomenon in natural New Zealand – a widespread ecosystem of exotic pastures (Fig. 18).

The first sheep flocks in Otago were brought in by John Jones at Waikouaiti and in Southland by Captain Howells at Riverton. By 1844, Jones had 2000 merinos. The runholders made the first tracks into Central Otago to drove their sheep on to their runs, some of them renting flocks from Jones. They criss-crossed Otago by some amazing routes, one man setting off from Hindon near Dunedin to take 600 sheep almost to Riverton, arriving there a year later. The first runholders found pastures full of palatable native species, especially a blue grass (probably an *Elymus*) and anise (probably *Gingidium montanum*). The sheep creamed off the palatable plants, and the montane grasslands have never again been able to carry the stocking densities of sheep that were present in the 1870s. In 1880, the mountain tussock grasslands were carrying ten times the number of sheep carried in 1950. The men who took up these runs had access to cheap shipping, which they had just used to bring in sheep, men and gear. The wool went out the same way. Sheep were pushed into places where they had little chance of survival, such as the David Peaks at the head of the Mararoa River. The system of pushing sheep higher and higher eventually broke down; much of the high ground above 1000 m was retired during the first half of the twentieth century. A recurrent theme with many introduced mammals has been increasing numbers to a peak which was not sustainable, followed by a decline and subsequent minor or local peaks. This holds not only for rats, pigs, and red deer but also for sheep under extensive pastoralism.

In about 1840 John Jones and Captain Howells put up the first fences, to contain their stock at Waikouaiti and Riverton respectively. It was the containment of sheep and cattle behind fences, as much as ploughing and the sowing of grass seed, that created the low-biomass ecosystem of introduced pasture now covering most of lowland Otago and Southland: 3,265,000 ha in Otago and 1,567,000 ha in Southland, according to a recent census. In Otago, another 1,826,000 ha of tussock is grazed and in Southland 360,000 ha. By selective nibbling and pruning, sheep and cattle create a wholly distinctive assemblage of plants, strikingly different from the more diverse and taller vegetation on the adjacent road verges. These road edges have usually been roughly cultivated at some stage in their lives, and receive wind-blown fertiliser and seed from the adjacent pastures, but the only real difference in their treatment is that they are not grazed by stock. Through the higher rainfall areas they are dominated by cocksfoot and fescues, and in the drier parts of Central Otago by a riot of wild flowers of a wide range of families other than grasses. The same diversity can be seen developing in blocks and windbreaks shut up for forestry.

On golden-brown tussock hillsides, bands of green appear where oversowing with clover and topdressing with superphosphate create a more nutritious sward on which the sheep concentrate. Nutrients transported to sheep camp sites on tops of ridges lead to green caps on brown ridges. Mob stocking during the winter on the Southland plains creates evenly regenerated, shining green pastures with not a weed in sight. Sheep and dairy cows put different selective pressure on the species in older pasture, resulting in slightly different assemblages of grasses and herbs. But the ability of both to create ecologically depauperate landscapes should not be underestimated. Cattle do additional damage by pugging wet ground, destroying riparian vegetation and breaking down stream banks.

Along with fencing, the other most devastating mechanism for native ecosystems has been drainage. Wetlands in New Zealand generally and in Southland in particular have been reduced to a very small proportion of their former extent. Otago has lost at least two shallow lakes – Tatawai on the Taieri Plain and Kaitangata near Inchclutha. The Southland Plains had enormous swards of copper tussock swamplands when early runholders such as Roberts and the surveyor, J.T. Thompson, described them in the 1850s. Hundreds of kilometres of tile drains, made from local clays, were laid. In a good earthquake, the paddocks rattle! The loss of not only copper tussock but also raupo, large *Carex* species, smaller native wetland plants and the associated avifauna has had the effect of further reducing biodiversity and biomass.

7 THE SETTLERS' BAGGAGE

Driven by nostalgia for the landscape of their homelands as much as by practical needs, European settlers introduced hundreds of plants in the first few decades of settlement. Many more plants arrived unintentionally as contaminants in imported materials. The outstanding success of some of these introductions is perhaps demonstrated best in southern New Zealand by the dominant ecological role now played by gorse, broom and briar.

Gorse and broom, both nitrogen-fixing legumes with long-lived seeds, are the most common woody plants pioneering succession in formerly forested environments throughout the region (opposite). Waste ground, ill-managed farmland, and even predominantly native tussock grasslands are vulnerable to infestation. The settlers' hedgerows were probably the original starting points of invasions, whence seeds were carried in the coats and on the hooves of stock (and people!). Once gorse and broom establish along waterways, their seeds are rapidly transported downstream to the ideal habitat of open gravel riverbeds. From here seeds are commonly spread in excavated gravel, as can be seen in long corridors of both species along the routes of existing and abandoned railway lines and along roadsides. Thence it is easy for seeds to be dispersed in mud stuck to vehicles. Ballistic dispersal of seeds from exploding pods is effective only for a few metres and makes a relatively insignificant contribution to gorse and broom dispersal, although it results in rapid and complete occupation of vulnerable sites following initial establishment.

Unless native trees get established at the same time as gorse or broom, and that only happens if a suitable seed source is close by, the weeds can have exclusive occupation of the site for anything up to 30 years. At that stage senescence of the gorse canopy creates openings, which may be occupied by native species if their seedlings can penetrate the dense dry litter beneath, before new gorse sprouts and seedlings fill the gaps. With broom stands, the problem is more one of competition from the strong sward of introduced grasses and herbaceous weeds which can establish as the canopy thins. In either case, it is likely to be another 30 years before even a semblance of native woody vegetation is established. Unless followed up with intensive management, removal of the gorse or broom by burning or mechanical means merely perpetuates the problem, as thousands of dormant seeds in every square metre of soil germinate and start the process of occupation afresh.

Briar was often used at first as an ornamental hedge between gardens and paddocks to keep stock out of the gardens. It is most aggressive in drier inland environments, where it can occupy all but the most droughty sites. Its massive crops of fruits ensure its wide dispersal by birds, particularly the introduced blackbird, with which it evolved in its native Europe. Briar's deep-rooted habit and its fast growth rate enable it to compete successfully with native shrubs, such as matagouri, coprosmas, native brooms and olearias, although generally it does not form stands sufficiently dense to displace the natives completely. Few trees, whether exotic (some pines, Douglas fir, larch, rowan and sycamore) or native (mountain totara, perhaps mountain and silver beech) can survive in this environment. Unless there are seed sources nearby, shrubland of which briar is a conspicuous component will remain the dominant vegetation.

Ralph Allen

A COMPLEX VEGETATION

People brought new species of plants into southern New Zealand for a whole raft of reasons and the effects of those introductions were sometimes predictable and sometimes very unexpected. Sometimes the native plants went under and sometimes they fought back (see Boxes 7, 8 and 9).

Plants as icons

Both waves of migrants, Polynesian and European, concentrated at first on the cultivation of exotics. The first European gardeners were the missionaries in the North Island, and in Otago the whalers established gardens in the 1830s. The abundant wild white pea along the strand line of Lathyrus Bay behind Tautuku Peninsula is probably an escape from the Tautuku whaling station. The nurseryman, Matthew (senior), was growing and selling garden plants from his Moray Place site in Dunedin by the early 1850s. Attitudes to natives as garden plants were distinctly negative, though they were considered to be useful as shelter. For nearly a hundred years it would be considered that native trees and shrubs generally were difficult to propagate and slow to grow, and few of the smaller plants had flowers worth bothering with. Ironically, back in Europe in the nineteenth century, New Zealand plants were viewed with a different eye, and many were being introduced into collectors' gardens. Some New Zealand natives – such as the bidibid, *Acaena novae-zelandiae* – have even become weeds in Europe.

In 1874 a forester thought the native trees would not be able to resist the advances of civilisation and, like the Maori themselves, would in time almost entirely vanish! The forestry department of Lands and Survey set up nurseries in the 1890s at Tapanui, Ranfurly and Cardrona. The lists of species, from America, Asia and Europe, offered by these nurseries was mind-boggling. One of their most conspicuous introductions was the Wellingtonia *Metasequoia*, and some giant trees of this species grace gold mining towns. Douglas firs, rowans, poplars and crack willows also thrived. The historic plantings in Otago reflect a need to transplant a piece of 'home' into the new land. The planting of American trees indicates something different. Many miners came from California and, though their roots were not there, the vigour of the Douglas firs and Wellingtonias must have impressed them (see page 129).

Two or three Otago nurserymen collected natives and advertised them in the 1890s, but from photographs of town gardens it is apparent that only cabbage trees, flaxes and toetoes with their sub-tropical effects were at all popular. Duncan and Davies in the 1920s advertised the largest collection in the world of New Zealand natives, and then highlighted double-flowered pink manuka, variegated karaka, bronze cabbage tree and bronze rangiora. It was the novelty cultivars, that would blend with the European plants, that were emphasised, not the native-looking ones. David Tannock, director of Dunedin's Botanic Gardens, promoted Leonard Cockayne's book *The Cultivation of New Zealand Plants*, which was the first serious effort to promote natives for their own sake and in order to create distinctively New Zealand gardens. His advice fell on deaf ears, and it was not until Muriel Fisher and Lawrie Metcalf published their well-illustrated books that some real interest in growing natives was established. The change has been very slow, however, and some

Gorse in flower on the hills south of Dunedin is admired only by tourists. *John Darby*

8 THE INDIGENES SOMETIMES FIGHT BACK

Many a gardener and farmer, after clearing and planting, considers that they have replaced the indigenous flora with plants of their own choice. But have they? The New Zealand flora has some remarkable characteristics, but one that is not often talked about is the capacity of some species to behave like weeds – invading and growing in places where they are not wanted. Many gardeners on clay soil will recognise a little creeping Epilobium with a tiny white flower, which clings so close it has to be scraped off with the finger nails, and the least little bit left will grow. It is a native – *Epilobium nummularifolium* – from the North Island, which now grows wherever there are gardens. The native wax weeds *Hydrocotyle* have the same habit, but at least the little bitter cresses *Cardamine* are easier to pull out, though they still manage to fling their seed into your eyes.

In coastal Otago, some farms were cut out of fire-induced kanuka stands more than 130 years ago. Since then the kanuka has fought back, re-invading vigorously even into dense pasture. In some places the kanuka in the gullies has been cut for firewood three times, with fire and stock used to maintain areas of coarse pasture on the drier ridges. Much of the land was too steep to plough, but after 100 years of grazing the cover looks, at a quick glance, to be wholly introduced grasses and weeds.

In a recent analysis near Dunedin of old pasture, grazed by both sheep and cows for about 140 years but destocked two years previously, there were a surprising number of native species. *Senecio minimus* was just as abundant as its relative the introduced groundsel *S. jacobaea*, and spreading vigorously along roads and into any disturbed ground. A bidibid *Acaena novae-zelandiae* and a small carex *Carex wakatipu* were common among the cocksfoot, Yorkshire fog and brown top. Two other native carexes, three *Juncus* species and a *Luzula* occurred in several quadrats. There were occasional small plants of the ferns *Blechnum fluviatile, Hypolepis millefolium* and *Polystichum vestitum*, as well as the cosmopolitan Jersey cudweed. This group had probably resisted grazing pressure and co-existed with the pasture weeds for many decades. Most of the other native species were forest seedlings on the way back in after sheep had been withdrawn. Even they were an impressive list – wineberry, clematis, three coprosmas, tree fuchsia, broadleaf, whitey-wood, *Muehlenbeckia, Pennantia corymbosa*, lemonwood, peppertree and lawyer. In terms of plant cover, though, the introduced plants dominated, and would no longer provide a home for the native invertebrates and microflora which once lived there.

A similar story could be told for New Zealand's invertebrates. We have exported the common clothes moth, as well as a voracious planarian worm of fresh waters. Among our beetles, the grass grub *Costelytra zealandica* and porina *Wiseana* spp. have found the new pasture grasses very much to their taste and probably now occur in their native land in far higher numbers than ever before. Unlike exotic pests, grass grub and porina are still attacked by a suite of diseases which have evolved with them. The beetles seem, however, to have been largely released from predators, parasites and competitors, who perhaps could not make the transition into the new ecosystem. It may be that our native grasslands still contain some parasites which could be assisted by genetic manipulation to make the move and reduce these beetle pests.

Jill Hamel

Kanuka establishing in pasture at Pigeon Flat, north of Dunedin. *Ralph Allen*

older gardeners still mutter about 'the blasted heath effect' that they think natives give.

There are some hints that natives are becoming icons. The growth of the conservation movement has given them status, sometimes beyond their ornamental worth. Iwi are setting up projects to propagate such plants as pingao to compete with the foreign marram grass on local beaches. The magnificence of the great snow tussocks, especially when they explode into a white light of flower and ripple in the wind, are finding more and more converts. Endangered species in particular are being given special attention, and where beauty and rarity are combined, as in the Marlborough rock daisies, a distinctively New Zealand look is appearing in gardens.

Acknowledgements

Jill Hamel is grateful to Helen Leach, Ian Smith and Ralph Allen for reading a draft of this chapter and for their useful comments.

9 GOLDMINERS' WORKS AND GARDENS

Gold is very evenly dispersed through the rocks of Otago, and the effects of the nineteenth-century goldminer are likewise dispersed. These men and women came from Britain, western USA and Australia. They brought with them a different suite of plants from the pastoralists, some of which have clung around the stone ruins of their huts without spreading much further. Gooseberry, blackberry and elder are the most common, but there are also black currants, red currants, scented double daffodils, rhubarb, old fruit trees, lombardy and black poplars, and a range of garden plants such as thyme, californian poppy, periwinkle, echiums and others which have naturalised more widely. All except the Californian poppy are sentimental imports from Europe. In the dry open ground of Central Otago, dozens of species of wildflowers produce a splendour of colour in early summer not seen anywhere else in New Zealand.

Mining has had two other distinctive effects. Water races draw faint horizontal lines of gently curved tussock ridges across landscapes or were painstakingly built around steep faces. The races run from one catchment to another and allow the little native galaxiids of the upper headwaters to cross into foreign waters. Trout also use the irrigation races to travel across country. The water from these races produced the other change – the washing of enormous quantities of silt into the river systems. Our streams and rivers contain a rich invertebrate fauna, which has a natural resilience to sudden events. With the increased runoff of silt from mining and the scouring of creek beds using wing dams, there must have been much localised disturbance, followed by recolonisation from undisturbed stretches. Little is known of this pattern of events and whether the fauna has returned to anything like its previous state.

Jill Hamel

A view of Gabriels Gully during the gold rush in 1862 shows mounds and hollow tailings, and the miners living in tents up the hillsides. Traces of this type of working are now very rare, since in the following decades ground of this sort was sluiced away, using water brought by races from great distances to sluice and elevate the gravels.
Courtesy of Peter Petchey

The best line for a miners' water race often lay along steep hillsides and often needed revetting with schist slabs. This race above the Fraser Dam is unusual because, for part its length, the slabs have been stacked sideways. *Jill Hamel*

CHAPTER 8

FORESTS AND SHRUBLANDS

RALPH ALLEN, ALISON CREE, JOHN DARBY, LLOYD DAVIS, BRIAN PATRICK, HAMISH SPENCER

Kereru, the New Zealand pigeon, has adapted well to urban gardens.
John Darby

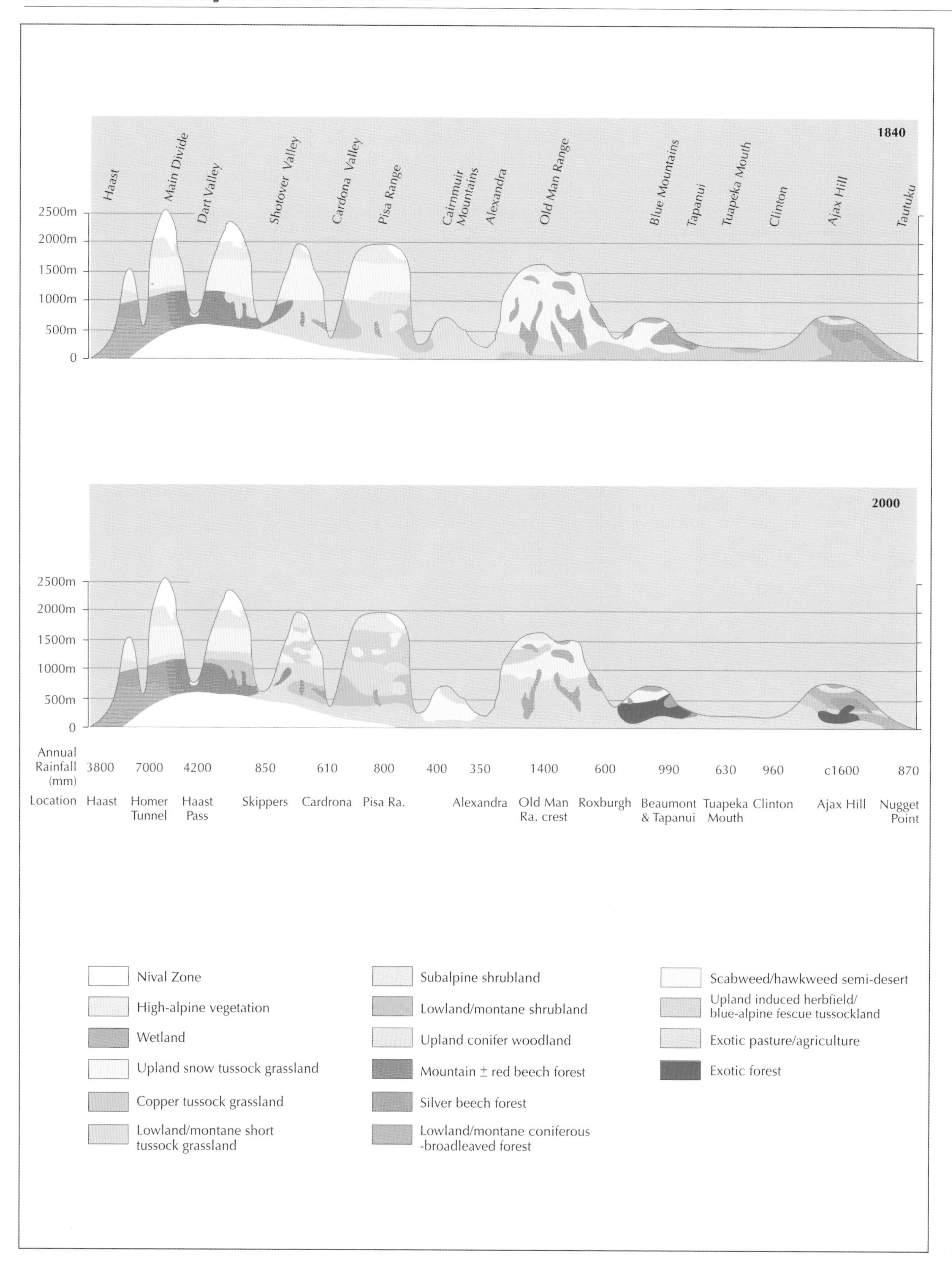

Fig. 1 (top). Profile across southern New Zealand showing the generalised vegetation pattern as it is assumed to have been at the time of European settlement, about 1840.

Fig. 2 (centre). Profile across southern New Zealand showing the generalised vegetation pattern as at present. *Both figures compiled by Ralph Allen, Alan Mark and Bill Lee*

FORESTS AND SHRUBLANDS

Almost all of southern New Zealand below about 1000 m elevation has a climate that can support forest, and there is plenty of evidence that most of the area was forested before the arrival of humans. Palaeobotanical studies of pollen and wood fragments preserved in wetland sediments, along with charcoal from topsoils, have shown us not only the distribution of these former forests, but also their composition (Chapter 6).

More than a quarter of the southern land area still supports native forest, albeit modified everywhere by introduced browsing animals, and in many places by human activities such as logging, burning and farming (discussed in several boxes below). The climatic gradient that helped shape the landforms and determined the distribution of plant communities also largely controls human influences. Thus the most extensive and least modified surviving native forests are in the wet, mountainous west, whereas with distance eastwards the drier climate and relatively gentle landforms have facilitated conversion to native grassland and, more recently, to farmland and exotic forest (Figs 1 & 2). In this chapter, we deal first with the flora of the region's forests and shrublands before proceeding to treatments of their invertebrates, reptiles, birds and mammals.

As is the case elsewhere in New Zealand, the native forests of the south can be divided into two major groups on the basis of the dominant canopy trees. Beech forests, where the canopy comprises mainly trees of silver beech *Nothofagus menziesii*, mountain beech *N. solandri* var. *cliffortioides* or red beech *N. fusca*, are largely confined to the western mountains, extending eastwards generally only as far as the Central Otago lakes (Fig. 3). Substantial outliers are located only in three areas: the Blue Mountains, the Waikaia Valley and the Catlins region of south-east Otago. However, scattered small stands occur further east towards and near Dunedin, supporting paleobotanical evidence that beech forests were once widespread through Central and eastern Otago. The beeches may be accompanied by several other broadleaved tree species, particularly kamahi *Weinmannia racemosa* and southern rata *Metrosideros umbellata*, along with podocarps, mainly mountain totara *Podocarpus hallii*, rimu *Dacrydium cupressinum* and miro *Prumnopitys ferruginea*. Kaikawaka *Libocedrus bidwillii* (a conifer of the cedar family) is a distinctive component of higher altitude forest in the Catlins, but occurs only rarely in the west.

In the absence of beech species, most of the forest remaining in the south has a canopy composed of other broadleaved trees, through which emerge the podocarps mentioned above, as well as matai *Prumnopitys taxifolia*, totara *Podocarpus totara* and kahikatea *Dacrycarpus dacrydioides*, with kaikawaka at higher altitudes around Dunedin and in the Catlins. Variously called conifer-broadleaved, conifer-hardwood, podocarp-broadleaved or podocarp-hardwood forest, this is characterised in the wetter south-west and south-east by a canopy dominated by kamahi and, in most places, southern rata (Fig. 4). These species extend north on the east coast only as far as the Taieri River, except for a stand of ancient kamahi on the eastern flank of Mt Cargill, near Dunedin.

1 THREATENED PLANTS OF SOUTHERN NEW ZEALAND

New Zealand is renowned for the high proportion of its indigenous biota that is threatened for one reason or another. Habitat destruction, the presence of aggressive plant and animal pests that have been introduced accidentally or intentionally, as well as the very limited distribution of some native species, have each been important factors in accounting for this situation.

Measures aimed at conserving and retaining these threatened biota are dealt with in Chapter 13. In each of the chapters that describe particular vegetation types, we list the vascular plants that have been placed on the New Zealand Threatened Plant Committee's list of threatened and uncommon plants of New Zealand in the following categories:

- **Presumed extinct:** no longer known to exist in the wild or in cultivation.
- **Critically endangered:** extinction considered inevitable unless there is direct conservation intervention.
- **Endangered:** in danger of extinction if the causal factors continue operating.
- **Vulnerable:** likely to move into the endangered category in the near future if the causal factors continue to operate.
- **Declining:** numerically abundant but under threat from serious adverse factors throughout their range; or widely scattered small populations undergoing decline.
- **Recovering:** populations either naturally restricted to susceptible habitats where survival is dependent on continual conservation measures, or populations once under serious threat that are recovering naturally as a result of past conservation intervention.
- **Naturally uncommon:** not under immediate or obvious threat, but sparsely distributed, transitory, or of restricted range, and have the potential to become threatened.
- **Insufficiently known:** suspected but not known to belong to one of the above categories because of a lack of information.

Ralph Allen

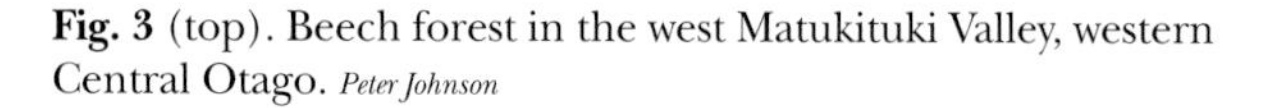

Fig. 3 (top). Beech forest in the west Matukituki Valley, western Central Otago. *Peter Johnson*

Fig. 4. Podocarp-broadleaved forest with a canopy dominated by southern rata and kamahi, Tautuku Bay, Catlins, south-east Otago. *Peter Johnson*

Fig. 5 (top). Podocarp-broadleaved forest lacking southern rata and kamahi, near Leith Saddle, Dunedin. *Alan Mark*

Fig. 6. Upland bog forest of kaikawaka and pink pine, interspersed with scrub and red tussock, Ajax Plateau, Catlins, south-east Otago. *Ralph Allen*

Fig. 7 (opposite). Relict podocarp trees, Munro Bush, eastern Southland. *Peter Johnson*

Where annual rainfall is less than about 1000 mm, on the plains and low central hills of Southland, the coastal hills of Eastern Otago, and the Kakanui Mountains of North Otago, the canopy comprises a variety of broadleaved tree species other than kamahi and southern rata (Fig. 5). On the most favourable sites podocarps may form a more or less continuous canopy, and the broadleaved trees are relegated to subcanopy status. This vegetation is perhaps more accurately described as podocarp forest, but it is now very rare (Fig. 7).

Within each of the two major forest groups are several less extensive but distinctive variants whose composition and structure are controlled by unusual combinations of substrate and climate, or by disturbance. Factors such as soil fertility, drainage and salt spray, along with fire and erosion, result in enclaves of forest dominated by species or combinations of species that are rare in or absent from the surrounding major associations. Examples include the kahikatea stands of river deltas and swamps, the mountain ribbonwood *Hoheria glabrata*dominated forest of Fiordland landslips, the kaikawaka-pink pine *Halocarpus biformis* upland bog forest of the Catlins (Fig. 6) and the kanuka *Kunzea ericoides* woodlands of Central and Eastern Otago.

Some of the small trees of our forest understorey are remarkably widely distributed, and can be found in almost any forest type. The most common are peppertree *Pseudowintera colorata*, broadleaf *Griselinia littoralis*, marble leaf *Carpodetus serratus*, lemonwood *Pittosporum eugenioides*, kohuhu *P. tenuifolium* and, where they escape the notice of deer and goats, pate *Schefflera digitata* and *Pseudopanax* species. Mapou *Myrsine australis* and its less conspicuous congener *M. divaricata*, along with stinkwood *Coprosma foetidissima* and a few other *Coprosma* species, are typical shrubs in the understorey of most forests, along with the tree ferns ponga *Cyathea smithii* and wheki-ponga *Dicksonia squarrosa*. Hen and chickens fern *Asplenium bulbiferum*, crown fern *Blechnum discolor* and kiokio *Blechnum novae-zelandiae* are characteristic ground cover more or less throughout, along with hound's tongue fern *Microsorum pustulatum* and prickly shield fern *Polystichum vestitum*, particularly in the southern and eastern forests.

Fig. 8 (top). Dense understorey of silver beech-dominated forest, Wirere Creek, southern Fiordland. *Peter Johnson*

Fig. 9. Bands of dune forest with silver beech, southern rata, kamahi and emergent podocarps, near Haast, north-western Otago. *Alan Mark*

THE BEECH FORESTS

The beech forests of the far north-western extremity of Otago and of Fiordland have been protected from exploitation by their isolation and, more recently, by National Park status. Although inhabited by deer for over half a century, they are amongst the least modified of the southern forests, and in places have yet to be invaded by possums (Fig. 8).

Silver beech is dominant almost everywhere, but red beech can oust it from the relatively rich soils of sheltered lower colluvial slopes as far south as Lake Monowai, and mountain beech copes better than silver beech under drier climate and on very shallow or infertile soils. On the rain-drenched lowland glacial outwash terraces between Haast and Jackson Bay, rimu, kahikatea and miro emerge above a canopy of silver beech, southern rata and kamahi, and vegetation similar except for its low kahikatea density occurs on Holocene dune systems in the same area (Fig. 9). Rimu and kamahi are also important at lower elevations in Fiordland's beech forests, but above about 700 m beech dominates alone.

Fiordland beech forest understorey plants are also segregated by altitude. At low elevation on the coast, characteristic frost-tender or salt- and exposure-tolerant shrubs and small trees of beech forest include mahoe *Melicytus ramiflorus*, pigeonwood *Hedycarya arborea*, *Ascarina lucida* and the tree daisies *Olearia oporina* and *Brachyglottis rotundifolia*. These rapidly drop out with distance from the coast, and at lower to mid altitudes are replaced in the subcanopy by the ubiquitous broadleaved small tree species of most southern forests. *Coprosma colensoi* is a characteristic shrub of the understorey, and the herb *Nertera villosa* joins other more common herbaceous and fern species on the forest floor. At upper altitudes, *Coprosma pseudocuneata*, *C.* sp. aff. *parviflora*, *C. astonii*, celery pine *Phyllocladus alpinus*, *Archeria traversii* and, near the treeline, *Dracophyllum uniflorum*, *D. menziesii* and leatherwood *Olearia colensoi*, become prominent understorey species. Here, prickly shield fern, filmy ferns *Hymenophyllum* species, *Astelia* species and mosses form much of the ground cover.

East of the Main Divide, in the drier mountains of western Southland and the Central Otago lakes district, beech forests tend to be simpler in structure and composition than those of the wetter west. In the headwater catchments of Lakes Hawea, Wanaka and Wakatipu, silver beech dominates towards the Divide. As rainfall declines, silver beech gives way to mountain beech, for example in the Hunter Valley (Fig. 10), around Lake Hawea, and in the mid Routeburn Valley as it turns eastwards towards the Dart Valley. Red beech is absent in the north, but co-dominant with silver and mountain beech in the lower Matukituki catchment west of Lake Wanaka. It also occurs in the Dart and Rees Valleys at the head of Lake Wakatipu, and on warm northerly-aspect slopes below mountain beech forest in the upper Routeburn Valley. Mountain totara is found throughout; kaikawaka, southern rata and kamahi west of Lake Wanaka; and the last two also in the Hunter catchment north of Lake Hawea. At upper altitudes, the understorey of mountain beech forest tends to be sparse and dominated by *Coprosma pseudocuneata*, *Archeria traversii* and celery pine. Except where the canopy is very dense, understorey density and diversity increases down slope, where several *Coprosma* species are found: *C. rhamnoides*, *C. depressa*, *C.* sp. aff. *parviflora*, *C. ciliata* and stinkwood. Prickly shield fern provides most of the ground cover. Silver and red beech forests at mid and lower altitudes have a denser understorey diversified by subcanopy small trees of broadleaf, lancewood *Pseudopanax crassifolius*, *Raukaua simplex* and sometimes other broadleaved species.

Further south, between Lakes Wanaka and Wakatipu, beech forest is confined mainly to the Shotover catchment, but extends eastwards as small stands in the Cardrona Valley, the Arrow catchment and on the Pisa Range. Mountain beech dominates, with a simple understorey of *Coprosma* sp. aff. *parviflora*, which is joined by celery pine, snow totara *Podocarpus nivalis* and *Coprosma pseudocuneata* where moisture permits. Silver beech is largely restricted to small stands in the lower Shotover catchment, although it occurs as scattered trees and small stands in the Arrow catchment and on the Pisa Range (Fig. 11). Red beech is found no further east than on the eastern slopes of the northern arm of Lake Wakatipu.

West of the Shotover River, beech forest becomes more continuous and more diverse. Red and mountain beech dominate on lower slopes, silver and mountain beech on mid slopes, and silver beech alone near the treeline (Fig. 12). Forest structure and understorey composition are similar to those west of Lake Wanaka. This pattern continues southwards between Lake Wakatipu and Lake Te Anau, except that mountain beech is found only near valley floors close to the latter. East of Lake Te Anau in the headwaters of the Eglinton, Mararoa, Oreti and

Fig. 10 (top). Sparse understorey of dry silver beech forest, Hunter Valley, western Central Otago. *Peter Johnson*

Fig. 11. Silver beech stand in Luggate Creek, Pisa Range, Central Otago. *Alan Mark*

Fig. 12. Altitudinal zonation of beech forest, Emily Pass Stream, Routeburn Valley. Mountain beech occupies the cold upper slopes and valley floor, and red beech the more mesic mid-slopes. *Alan Mark*

Mataura Rivers, mountain and silver beech are dominant, but red beech grows with them on lower slopes of relatively gentle gradient. Understorey species include *Coprosma pseudocuneata*, celery pine, three finger *Pseudopanax colensoi*, *Raukaua simplex* and kohuhu, with *Astelia nervosa* and hook sedges (*Uncinia* species) as major ground cover species.

Forest on the Takitimu Mountains is almost entirely silver beech. Mountain beech is restricted to dry and poorly drained sites, and red beech is rare. Understorey and ground cover are still similar to those found near Lake Wanaka. Silver beech alone occurs in the beech forests of western Southland hills between the south-eastern Fiordland lakes and the Aparima River. On the Longwood Range at the southern end of this region, silver beech forest alternates with podocarp-broadleaved forest, the latter occupying warmer, moister sites. At higher altitudes mountain totara and pink pine accompany silver beech, but below about 500 m broadleaved small trees, such as broadleaf, lancewood, *Raukaua simplex* and pokaka *Elaeocarpus hookerianus*, form a subcanopy over a relatively diverse under-storey including tree ferns, stinkwood, *Coprosma pseudocuneata*, *Neomyrtus pedunculata* and *Myrsine divaricata*. Ground cover is mainly crown fern and prickly shield fern, with *Hymenophyllum* species prominent on drier sites.

Fig. 13 (top). Beech forest in the Waikaia Valley, northern Southland, showing an uneven treeline caused by incursions from fires lit in the pastoral subalpine tussock grassland. *Alan Mark*

Fig. 14. Silver beech forest beside the Catlins River, south-east Otago. *Peter Johnson*

THE BEECH OUTLIERS

The westernmost of the three major outliers, in the Waikaia Valley (Fig. 13), contains forest dominated by silver beech, accompanied by mountain beech mainly along watercourses, and red beech at lower altitudes. Mountain totara is often a canopy co-dominant.The understorey is generally open, and comprises the common broadleaved tree species and shrubs, along with *Coprosma colensoi* and *C. rhamnoides*. In most places ground cover is provided by hook sedges and a few ferns. Deer and possums are present, and have considerably reduced the density of palatable understorey plant species.

Some fifty kilometres south-east of the Waikaia Valley lie the Blue Mountains, where silver beech is overwhelmingly the dominant canopy species of the second major beech forest outlier. The very open understorey is mainly peppertree and *Coprosma rhamnoides*, with broadleaved small tree species including fuchsia *Fuchsia excorticata* confined to areas little visited by deer. In the Rongahere Gorge of the Clutha River, on the eastern flank of the Blue Mountains, lie small stands of red and mountain beech, representing the easternmost distribution limit of both species. Here the three beech species are accompanied by matai, the two tree totara species and kahikatea, on sheltered fertile alluvial terraces. Most of the native forest of the Blue Mountains has been severely damaged by deer and possum browsing.

The Beresford Range of the Catlins, near the south-eastern extremity of the region, contains the third major outlier of beech forest, comprising only silver beech (Fig. 14). Although annual rainfall here is generally less than 2000 mm, it is evenly spread throughout the year, and evapotranspiration is relatively low. Thus the more or less permanently moist environment supports relatively rich and dense beech forest. Kamahi accompanies silver beech throughout this forest, and southern rata is common at higher altitudes, where, on poorly drained sites, kaikawaka, celery pine and pink pine are present as well. Rimu, miro and mountain totara are also quite common in the beech canopy. The under-storey is relatively dense, and comprises the common broad-leaved small trees and shrubs. Tree ferns are conspicuous. Crown fern, kiokio and *Nertera villosa* are the main ground cover plants.

Until the 1970s, the beech forests of the Catlins were relatively free from deer, and possum numbers were low. Consequently, palatable species such as stinkwood, tree fuchsia, broadleaf, three finger and *Raukaua simplex* are still common, and peppertree has not gained the dominance seen elsewhere.

In the Lawrence/Waipori area of Eastern Otago, a few small stands of silver beech exist in sheltered gullies surrounded by snow tussock grassland (Fig. 15). At the head of the Waipori Gorge, which joins the Otago plateau with the Taieri Plain, silver beech forest becomes continuous, and it extends as stands of varying extent around the eastern flank of the Maungatua Range to Woodside and the Traquair Burn at the north-eastern extremity of the range. Mahoe and the common broadleaved small trees form a subcanopy and sometimes reach the canopy. Rimu, miro, matai, totara and, rarely, kahikatea, are also associated with silver beech, particularly at lower altitude. The generally open understorey comprises the usual widespread

Fig. 15. A relict stand of silver beech forest in the catchment of the Nardoo Stream, near Lake Mahinerangi, East Otago. *Alan Mark*

Fig. 16. Silver beech stand (far left) emerging from kaikawaka-podocarp-broadleaved forest on Mt Cargill, Dunedin. *Alan Mark*

2 EXPLOITATION OF BEECH FORESTS

Virtually all accessible beech forest in western Southland has been logged, and much has been clearfelled for woodchip production, exotic plantation or farmland. Several hundred hectares have been committed to beech management for the production of beech sawlogs on a sustainable basis (below left). The success of this venture is yet to be established. The beech forests of the Waikaia Valley have been logged where accessible, and fire has considerably reduced their extent, particularly in the upper catchments where forest abuts tussock grassland. Several thousand hectares of beech forest on the Blue Mountains have been replaced by exotic plantation, and the remaining accessible native forests have been logged. Commercial exploitation of beech in the Catlins was constrained by access because of the steep and broken topography, so little of the present forest has been logged. Nevertheless, about 1500 ha of accessible beech forest was clearfelled and replaced by exotic plantation before 1980, and there has been continual attrition of silver beech from accessible areas of private land for woodchip production since then (below). Much of the beech forest of the Waipori Gorge and Maungatua Range has been modified by logging or hydro-electric works, and much has been cleared for farmland. *Ralph Allen*

Seed trees left after clearfelling of silver beech forest for beech production forestry, western Southland. *Alan Mark*

Silver beech forest felled and burnt for pine plantation, Catlins Valley, south-east Otago. *Ralph Allen*

shrub species of southern forests, and ground cover is provided by the common ferns, bush flax *Astelia fragrans* and hook sedges. The native forest of this area has a long history of damage by deer, feral sheep, pigs, goats and possums.

A tract of beech forest similar to that of the Waipori Gorge occupies parts of the catchments draining the Silver Peaks just north of Dunedin, and a few small stands are scattered through the podocarp-broadleaved forests of Mt Cargill, Flagstaff and Dunedin's other northern hills (Fig. 16).

3 THREATENED PLANTS OF BEECH FORESTS

- **Declining:** *Alepis flavida, Peraxilla colensoi* and *P. tetrapetala*, all mistletoe species found on beech trees.
- **Naturally uncommon:** *Drymoanthus flavus*, an epiphytic orchid.

Ralph Allen

THE PODOCARP-BROADLEAVED FORESTS

Southern rata-kamahi-podocarp forest

Although southern rata and kamahi are important in forest subcanopies in much of coastal Fiordland, they are seldom canopy dominants. However, forest dominated by southern rata and kamahi covers about sixty per cent of Stewart Island, and is distributed eastwards from the Waitutu River of southern Fiordland around the southern coast and its adjacent hills to the Catlins (Fig. 17), then as far north as the Taieri River, and extends inland to cover much of the ranges of the Catlins, as well as the Hokonui Hills. At low to mid elevations, rimu and miro are almost ubiquitous podocarps in this forest type, and mountain totara is common at higher altitudes, along with kaikawaka in parts of the Catlins (Fig. 18). A variant of southern rata-kamahi-podocarp forest occurs as small, apparently relict, stands in the upper reaches of the Waipara, Wilkin and east Matukituki Rivers in Mt Aspiring National Park.

On Stewart Island, except in dune forest close to the sea, rimu usually emerges above the canopy of kamahi, and southern rata is often also at least semi-emergent. Miro and mountain totara are common. The understorey here contains mainly the common southern small broadleaved trees and shrubs, but includes the tree daisies *Brachyglottis rotundifolia* and leatherwood close to the sea, and on sites characterised by cold air drainage, respectively. Ground cover is provided by *Blechnum* ferns, along with hound's tongue fern where deer numbers are low. Epiphytic ferns are abundant. At higher elevations and on poorly drained soils, manuka *Leptospermum scoparium* and yellow-silver pine *Lepidothamnus intermedius* are found in a low-statured variant of this forest type. On the few alluvial soils of the island, kahikatea, matai and mountain totara are charactersitically dominant over very variable understorey tiers containing peppertree and *Neomyrtus pedunculata*, as well as the species found in other rata-kamahi-podocarp forest.

Below about 400 m elevation on the Longwood Range of western Southland, kamahi and southern rata form a canopy with rimu, miro and mountain totara. Silver beech appears to be invading from adjacent beech forest. The understorey comprises the common broadleaved trees, along with wheki-ponga. Hook sedges and bryophytes provide most ground cover at lower elevation, but the contributions of the common ground ferns and *Nertera* species increase with altitude.

Pahia Hill, at the eastern end of Te Waewae Bay, supports forest in which kamahi predominates and the podocarps are represented by rimu, miro, mountain totara and matai. Southern rata is locally abundant. The diverse understorey includes dense thickets of wheki-ponga as well as the usual broadleaved trees and several *Coprosma* species. Similar forest survives in other small

Fig. 17 (top left). Red flowers indicate the canopies of southern rata trees in southern rata-kamahi-podocarp forest on the Beresford Range, Catlins, south-east Otago. *Peter Johnson*

Fig. 18 (top right). The conical crowns of kaikawaka emerge from a canopy of southern rata and kamahi near Mt Pye, Catlins, south-east Otago. An invading stand of silver beech can be seen in the left middle distance. *Ralph Allen*

Fig. 19 (right). Podocarp-dominated dune forests with a subcanopy of southern rata and kamahi, Tahakopa Bay, Catlins, south-east Otago (discussed page 162). *Alan Mark*

reserves near the southern coast as far east as Bluff Hill.

The eastern Hokonui Hills carry southern rata-kamahi forest with pokaka, rimu, matai, miro, mountain totara and kahikatea. The understorey is similar to that of the Longwood Range forests, with the addition of milktree *Streblus heterophyllus*, *Coprosma areolata* and *Melicope simplex* on warm sites.

In the Catlins, southern rata-kamahi forest with rimu, miro and mountain totara covers the mid and upper slopes of all the ranges except the Beresford Range, where beech forest is predominant, and poorly drained plateaux, where kaikawaka and pink pine are the characteristic tree species. On most lower slopes of the main valleys, pasture now occupies land formerly clothed in podocarp-dominated forest where rimu, matai, totara and kahikatea emerged from a canopy in which kamahi was the main broadleaved tree species. Southern rata and kamahi are still important in podocarp-dominated dune forests along the south-eastern coast, notably at Tautuku and Tahakopa Bays (Fig. 19). Throughout the Catlins there is an understorey of the common small broadleaved trees and shrubs and tree ferns. The common ground ferns are joined by several additional *Blechnum* species and, where browse pressure is low, by multitudes of seedlings of understorey and canopy shrub and tree species.

On easterly and southerly aspects of the drier hills north of the Catlins, the forests have a kamahi and southern rata canopy with emergent rimu, matai, totara and miro. Mahoe and mapou are important subcanopy species, and an open understorey is provided by peppertree, *Coprosma rotundifolia*, *C. linariifolia*, *C. areolata*, *C. rhamnoides* and *Neomyrtus pedunculata*, species typical of relatively low-rainfall eastern coastal sites. The common ground ferns are present in most forest remnants. Westerly and northerly aspects have a mixed broadleaved canopy, with kamahi, the common broadleaved trees and, on lower slopes and valley floors, lowland ribbonwood *Plagianthus regius*, narrow-leaved lacebark *Hoheria angustifolia* and kowhai *Sophora microphylla*. The podocarps here are rimu, matai, totara and kahikatea. Peppertree, mahoe, *Myrsine divaricata*, *Lophomyrtus obcordata*, *Melicope simplex*, milktree and, on the driest sites, *Coprosma crassifolia*, form the understorey. Tree ferns here include silver fern *Cyathea dealbata*. Crown fern dominates the ground cover on relatively moist sites, but hen and chickens fern, *Asplenium hookerianum*, *Pellaea rotundifolia* and shield fern *Polystichum richardii* are important in drier situations.

Podocarp-broadleaved forest with rimu, miro, mountain totara, kamahi and, close to the coast, southern rata, formerly clothed the coastal hills from the Clutha to just south of Taieri Mouth. Most has been replaced by pasture and scrub, and remnants are confined mainly to gullies and steep land.

Fig. 20. Kaikawaka and mountain totara emerging from a canopy of celery pine, ribbonwood and broadleaf (left middle distance), adjacent to silver beech forest, upper east Matukituki Valley, western Otago. *Alan Mark*

Kaikawaka occurs in a few places above about 300 m elevation, but is mostly moribund. Lemonwood, kohuhu, broadleaf, three finger and lancewood are common subcanopy trees, and the understorey is mainly stinkwood, *Coprosma* sp. aff. *parviflora*, *C. propinqua*, peppertree and kaikomako *Pennantia corymbosa*. Seedlings of these, with the common ferns and climbing rata *Metrosideros diffusa*, provide much of the ground cover.

In the upper reaches of the Waipara River, a tributary of the Arawata River in Mt Aspiring National Park, stands of montane forest dominated by mountain totara, southern rata and kaikawaka, with kamahi co-dominant on lower slopes, are being invaded by silver beech from the continuous beech forests below. The understorey comprises common small broadleaved trees, along with *Brachyglottis rotundifolia*, mountain holly *Olearia ilicifolia*, mountain neinei *Dracophyllum traversii*, and inaka. Prickly shield fern, *Blechnum* species, *Cyathea colensoi*, *Hypolepis millefolium* and the hook sedge *Uncinia rupestris* comprise most of the herb layer. On the opposite side of the Divide, merging with subalpine scrub in the upper reaches of the Siberia and east Matukituki Valleys west of Lake Wanaka, are similar small stands of low forest with celery pine, inaka, mountain ribbonwood and broadleaf, with emergent mountain totara and kaikawaka (Fig. 20). They form an abrupt lower boundary with silver beech forest, which appears to be invading and replacing them.

Podocarp-broadleaved forest

Away from the southern and south-eastern coasts, the Catlins, and the Hokonui Hills, southern rata and kamahi are less prominent in, or are absent from, podocarp-broadleaved forests.

Small remnant pockets of podocarp-broadleaved forest that dot the Southland plains and their adjacent low hills, for example at Edendale, Mabel Bush, Seaward Downs, Forest Hill and Titiroa, are generally characterised by kahikatea, matai, totara, rimu and pokaka. However, southern rata, kamahi and miro are present in some, usually as relatively minor constituents of the vegetation. Peppertree is ubiquitous, reflecting the long history of browsing by stock in most of these areas, but many other common broadleaved small tree and shrub species are also present, and sometimes kowhai, milktree and *Melicope simplex*. Ground cover is frequently sparse, with the usual ferns, bush flax and hook sedges.

Inland of the Clutha mouth, the relatively dry low hills of South Otago carry remnants of totara-dominated podocarp-broadleaved forest where southern rata and kamahi are usually absent. Broadleaved tree species are common only in gullies. Understorey shrubs and small trees include *Myrsine divaricata*, *Lophomyrtus obcordata*, *Coprosma areolata*, *C. colensoi*, *C. crassifolia*, *C linariifolia*, *C. lucida* and *C. rotundifolia*, and this is perhaps the only southern forest type where the main tree fern is *Dicksonia*

4 EXPLOITATION OF SOUTHERN RATA-KAMAHI-PODOCARP FORESTS

Almost all the accessible southern rata-kamahi-podocarp forest of the Longwood Range has been logged, and large areas of lowland forest have been cleared for farmland and exotic plantation.

The southern rata-kamahi-podocarp forests of the Catlins have a history of logging dating back to the 1860s. Vast quantities of timber, particularly rimu, were exported northwards to build cities such as Dunedin. The fertile valley floors towards the coast have long been cleared for farmland, and large areas of cut-over forest on poorer country were still being cleared for this purpose in the 1980s and early 1990s. Woodchip production was the primary economic motivation for the recent clearing, which seldom resulted in sustainable pasture (right). The high price and demand for rimu timber still encourage landowners to log the remaining forests in private or Maori ownership.

Ralph Allen

Southern rata-kamahi-podocarp forest felled for woodchip production then burnt for pasture development, Progress Valley, Catlins, south-east Otago. *Alan Mark*

5 EXPLOITATION OF PODOCARP-BROADLEAVED FORESTS

Podocarp-broadleaved forests once covered the plains and rolling downlands of Southland and South Otago, and extended in a broad band northwards on the coastal hills and lowlands to the Kakanui Mountains. They have been diminished over centuries by Polynesian fire and, more recently, by logging and European farming practices. Most of the formerly forested landscape is now farmland.

Ralph Allen

Fig. 21. Kahikatea, matai and totara form a more or less continuous canopy over narrow-leaved lacebark, pokaka and other broadleaved trees, Otanomomo, south-east Otago. *Peter Johnson*

fibrosa. Important ground cover species include *Asplenium hookerianum* and *Pellaea rotundifolia*, characteristic ferns of dry, open forest on the eastern hills.

The single remaining stand of podocarp-broadleaved forest on the alluvial soils of the Clutha delta, at Otanomomo, indicates that in this environment matai, kahikatea and totara formed a more or less continuous canopy, with less rimu and miro (Fig. 21). The common broadleaved trees are present, as well as narrow-leaved lacebark and pokaka. Saplings of these species also form much of the understorey, along with the usual shrubs, but tree ferns appear to be absent. Ground cover vegetation is typical of south-eastern forests.

Southern rata and kamahi are absent from the Taieri River northwards, apart from a small stand of kamahi at Graham's Bush, on the eastern slopes of Mt Cargill. Instead, on north to westerly aspects and on fertile lowland sites the characteristic broadleaved trees are narrow-leaved lacebark, lowland ribbonwood and kowhai, along with the mahoe, broadleaf, lemonwood, kohuhu, marble leaf and *Pseudopanax* species which dominate on south to easterly aspects and at higher elevations (Fig. 22).

Rimu and miro are still scattered throughout the podocarp-broadleaved forest remaining around Dunedin, with mountain totara on shady aspects and at higher elevations, and matai and totara on more equable sites. Kahikatea is rare, and confined to low elevation sites. Kaikawaka is a prominent feature of this forest on the upper slopes of the higher hills around Dunedin, including Silver Peaks, Mt Cargill, Mihiwaka, Flagstaff and Swampy Summit (Fig. 23). Celery pine and pink pine also occur in much of this forest.

North of Dunedin, in the north branch of the Waikouaiti River, the few remaining podocarp-broadleaved forest stands are characterised by kahikatea and totara, with little matai, and rare rimu and miro. Narrow-leaved lacebark, lowland ribbonwood and kowhai are the main broadleaved species except on shady and moist sites, where mahoe and broadleaf predominate. The understorey comprises shrub species typical of relatively dry eastern forests. Ground cover is sparse on sunny aspects, where only *Hydrocotyle* and prickly shield fern tolerate the high browsing pressure, but there is a higher density of common ferns and climbing rata on shady aspects, where totara and *Coprosma* seedlings are also prominent ground cover plants.

Similar forest occupies catchments of the Waianakarua River on the northern side of the Kakanui Mountains, but here the main podocarps are matai, totara and, towards gully heads, mountain totara.

6 THREATENED PLANTS OF PODOCARP-BROADLEAVED FORESTS

- **Declining:** *Tupeia antarctica* and *Ileostylus micranthus*, mistletoes found on several species of broadleaved tree.
- **Recovering:** *Pittosporum obcordatum*, a small tree of fertile alluvial sites (right).
- **Naturally uncommon:** *Anemanthele lessoniana*: a fine-leaved tussocky grass of open low forest; *Brachyglottis sciadophila*, a scrambling semi-herbaceous vine of light coastal forest; *Drymoanthus flavus*, an epiphytic orchid; *Pseudopanax ferox*, a lancewood of lowland forest in relatively dry, fertile sites.

Ralph Allen

The filiramulate (divaricating) habit of *Pittosporum obcordatum*, a rare small tree of fertile alluvial sites. *Alan Mark*

Secondary broadleaved forest

Throughout Otago and Southland, much of the land that formerly supported forest dominated by podocarps now carries either logged forest with few remaining podocarps, or secondary forest which established after clearance of the original. In both cases, the present forest is of similar composition to that which preceded it, differing mainly in its lower stature, the low density of podocarps and the reduced abundance of palatable plants.

Exotic forests

Exotic trees have been established as plantation forests in Otago and Southland since the second half of the nineteenth century, and several species have naturalised to form untended forest that has displaced native vegetation.

Major increases in the exotic forest estate took place during and after the 1920s, when readily accessible supplies of native timber were becoming scarce. The main species planted were, and still are, the north American conifers radiata pine *Pinus radiata* and Douglas fir *Pseudotsuga menziesii*, but Australian *Eucalyptus* species, particularly *E. nitens*, have become more commonly planted in the past two or three decades, along with a wider range of conifers.

Conifer plantations established in areas where the native forest was clearfelled for the purpose, for example in western Southland and the Blue Mountains, can have an understorey that contains some of the shrub, fern and herbaceous species of the former native forest. However, at best this is an impoverished reminder of the natural vegetation, and most conifer plantations contain little or no understorey. *Eucalyptus* plantations, on the other hand, have a sufficiently open canopy to allow a relatively diverse understorey of native species to develop (Fig. 24).

Most or all native plant species are completely suppressed in the conifer plantations established in tussock grasslands, for example at Naseby in Central Otago, and on the Lammerlaw Range of Eastern Otago. Remnants of the native vegetation struggle on at roadsides and around water supply dams and natural wetlands that have not been planted with exotics.

Unproductive, weed-infested and erosion-prone farmland has also been converted to exotic plantation, especially on the coastal hills of Eastern Otago. Here any understorey is usually gorse,

Opposite
Fig. 22. Kowhai flowering in broadleaved forest on Mt Watkin, Eastern Otago. The distinctive dark canopy of kanuka shows where this species is invading former pasture in the right middle distance. *Peter Johnson*

Fig. 23. Living and dead conical kaikawaka and the shaggy crowns of rimu emerge from a canopy of broadleaved tree species near Leith Saddle, Dunedin. *Peter Johnson*

Above
Fig. 24. A diverse understorey of native shrubs and small trees has established under tall *Eucalyptus regnans* at Orokonui, Eastern Otago. *Peter Johnson*

Fig. 25. Larch, Douglas fir and Corsican pine invading and displacing grassland and shrubland, Mt Aurum, Shotover River, Central Otago. *Ralph Allen*

but there may be a few native shrubs and ferns if native vegetation persists nearby.

Naturalised conifers in Otago and Southland include radiata pine, Corsican pine *Pinus nigra*, lodgepole pine *P. contorta*, larch *Larix decidua* and Douglas fir. They establish in low- statured vegetation such as tussock grassland, where they form dense stands, either singly or in combination, to the exclusion of almost all other species (Fig. 25). Naturalised conifers are particularly prominent on the high country around Lake Wakatipu, but they present a threat to upland landscapes and vegetation throughout the region, from West Dome in northern Southland to the Maungatua Range near the east coast and the Hawkdun Range of North Otago.

Naturalised broadleaved trees also have the potential to form forests in the region. At present only sycamore *Acer pseudoplatanus*, rowan *Sorbus aucuparia* and hawthorn *Crataegus monogyna* have had an impact, but ash *Fraxinus excelsior* and several *Eucalyptus* and wattle *Acacia* species can establish in the wild, and may contribute to future forests. Fortunately, none seem able to penetrate or displace native forest that retains a continuous canopy, but shrublands and grasslands are at risk of invasion.

Fig. 26. Muttonbird scrub on an exposed bouldery coastline at Big River mouth, southern Fiordland. *Peter Johnson*

Fig. 27 (opposite). Dune scrub with flax, *Ascarina lucida*, broadleaf and several other shrub and small tree species, Poison Bay, Fiordland. *Peter Johnson*

SHRUBLAND AND SCRUB

The term shrubland is used here to indicate communities in which shrubs are the tallest plants but do not form a continuous canopy. Significant areas of grasses and herbaceous species can establish between and under the shrubs. In contrast, scrub is dense-canopied, can contain a significant proportion of tall plant forms other than shrubs, for example bracken *Pteridium esculentum* and flax *Phormium* species, and generally lacks an understorey of grasses. Both vegetation types occur in Otago and Southland, either as relatively short-lived vegetation preceding the re-establishment of forest, or in habitats where there are environmental constraints on tree establishment and growth. The latter include exposed coasts, dunes, salt marsh, bogs and swamps, shallow and infertile soils, semi-arid hill country and the subalpine zone above tree limit.

Coastal

Exposed cliffs and steep bouldery shorelines carry scrub that can tolerate wind and salt spray. On the Fiordland and Stewart Island coasts, the leathery-leaved tree daisies *Brachyglottis rotundifolia* and *Olearia oporina* are prominent in this vegetation, commonly called muttonbird scrub. Flax *Phormium tenax*, *Hebe elliptica*, inaka, and, in northern Fiordland, kiekie *Freycinetia baueriana* ssp. *banksii*, may also be present, and *Olearia lyallii* is important on highly organic soils on Stewart Island. In southern Fiordland, *B. rotundifolia* dominates, with minor contributions from inaka and *Hebe elliptica* (Fig. 26). South and eastwards around the Southland coastline, *Hebe elliptica*, *Olearia arborescens*, manuka and inaka form much of the wind-shorn canopy, for example at Bluff Hill. Flax becomes increasingly dominant towards and around the Catlins coast, such as at Curio Bay and Nugget Point. Northwards on the east coast *Hebe elliptica* regains importance, and *Olearia avicenniifolia* and tree nettle *Urtica ferox* become more prominent on the coastal cliffs and headlands of Otago Peninsula. Ferns, grasses and herbaceous plants growing in association with coastal scrub include *Blechnum durum*, *B. banksii*, *Poa astonii*, *Hierochloe redolens*, *Isolepis nodosus*, *Anisotome lyallii*, *Apium prostratum* and, in the Catlins, the locally endemic daisy *Celmisia lindsayi*.

Dunes behind the sandy beaches of relatively sheltered bays also support distinctive scrub, comprising plants which can tolerate salt-laden winds, abrasion by wind-driven sand and, especially in the east, often droughty conditions. At Big Bay in northern Fiordland these include flax, *Coprosma propinqua* and tutu *Coriaria sarmentosa* on the seaward dunes, giving way inland to broadleaved small trees, with *Ascarina lucida* and broadleaf prominent. Further south at Transit Beach, dune scrub resembles that of the exposed rocky coast (Fig. 27). Variations of these two examples, sometimes with manuka, inaka and *Coprosma propinqua*, are found on dunes through to the Wairaurahiri River and thence to Bluff on the south coast. On Stewart Island, broadleaved species including mapou, broadleaf and marble leaf are prominent, often with *Brachyglottis rotundifolia*, southern rata and *Coprosma* species. Stable dune hollows of the extensive dunes at Mason Bay carry open shrubland and tussockland where the main shrubs are *Coprosma*

Fig. 28 (top). Scrub dominated by inaka and manuka, with wire rush amongst taller vegetation, Awarua Bog, Southland. *Peter Johnson*

Fig. 29. Flax and inaka dominate dune hollow wetlands and grade into silver pine and manuka woodland, then forest (left middle distance), on glacial outwash terraces near Haast. *Alan Mark*

acerosa, *Pimelea lyallii*, tauhinu *Ozothamnus leptophyllus*, inaka, mingimingi *Cyathodes juniperina* and manuka, accompanied by flax, sedges and rushes.

East of Bluff at Waituna Beach is dune scrub with flax, bracken, *Coprosma propinqua* and tauhinu, but further eastwards in the Catlins, where dunes are backed by native forest, these species are joined or replaced by stunted totara, southern rata and small broadleaved trees.

From here northwards, exotic tree lupin *Lupinus arboreus* and gorse *Ulex europaeus* dominate dunes to the exclusion of most native shrub species except the vine *Muehlenbeckia australis*. Only occasionally are flax, *Coprosma propinqua*, *Hebe elliptica*, *Olearia avicenniifolia* and poroporo *Solanum laciniatum* prominent, and they are joined by ngaio *Myoporum laetum* and the introduced native taupata *Coprosma repens* north of Taieri Mouth.

Salt marsh scrub dominated by marsh ribbonwood *Plagianthus divaricatus* and flax, usually in association with the rush-like plants *Isolepis nodosus* and *Leptocarpus similis*, is largely restricted to the southern and eastern coasts, where it occupies estuary edges and salt marshes from Invercargill to Oamaru. Toetoe *Cortaderia richardii* and *Coprosma propinqua* are commonly associated with this community.

Wetlands

Flax is a dominant plant in scrub of lowland freshwater swamps throughout Otago and Southland. Here it is joined by various combinations of manuka, *Coprosma* sp. aff. *parviflora*, *C. propinqua*, *Blechnum novae-zelandiae* and a range of sedges amongst which tussocks of *Carex secta* and *C. virgata* are prominent. Gorse and broom are serious threats to this vegetation. Examples of swamp scrub are seen in northern Fiordland at Big Bay, west of Riverton at Lake George, in the lower reaches of the Waipati and Catlins Rivers in the Catlins and at Lake Waihola on the Taieri Plain.

The peat substrates of lowland bogs support thickets of scrub with manuka, which can be accompanied by flax, inaka and red tussock (e.g. the Awarua Bog on the coastal Southland plain); inaka, yellow-silver pine, pink pine and the small shrubs *Cyathodes empetrifolia* and *Pentachondra pumila* (e.g. on Stewart Island); or bracken, tutu *Coriaria arborea*, *Coprosma* sp. aff. *parviflora*, mingimingi and bog pine *Halocarpus bidwillii*, for example in mires near Te Anau (Fig. 28). Similar vegetation is characteristic of pakihi on the glacial outwash terraces and associated dune slacks in the north-west of the region, where it grades into silver pine *Manoao colensoi* woodland and manuka shrubland, respectively, and then forest with podocarps, silver beech, rata and kamahi (Fig. 29). Wire rush *Empodisma minus*, tangle fern *Gleichenia dicarpa* and sedges, especially *Baumea tenax*, are characteristic of lowland bog scrub.

High-altitude bogs can also carry scrub. On Stewart Island, subalpine cushion bogs support manuka and various combinations of *Dracophyllum politum*, *Olearia colensoi*, inaka and mingimingi. The Ajax Bog, at about 700 m elevation in the headwaters of the Maclennan and Tahakopa Rivers in the Catlins, has scattered manuka and inaka on cushion bog, but thickets of inaka, pink pine and kaikawaka scrub on better-drained slopes, where these grade into low forest (see Fig. 6). Inaka also dominates scrub on poorly drained sites amongst cushion bog and herb moor at 700–800 m elevation on the summit ridge of Maungatua, in Eastern Otago. Here it is accompanied by tauhinu, mountain flax *Phormium cookianum*, *Coprosma pseudocuneata* and *Hebe odora*.

Subalpine scrub

Above treeline there is usually a zone of low-statured woody vegetation that can include stunted individuals of species from the adjacent forest, as well as shrubs specific to this environment. In Fiordland the latter include *Olearia colensoi*, *O. lacunosa*, *O. crosby-smithiana*, *Dracophyllum fiordense*, *D. menziesii* and *Pimelea gnidia* (Fig. 30). Some of these extend east as far as the Takitimu Mountains, but the general eastward trend is for inaka, *Dracophyllum uniflorum*, snow totara and celery pine to become dominant. On Stewart Island, equivalent communities are extensive above the altitude attained by rimu. Here, the dense vegetation, 1–8 m tall, comprises combinations of inaka, *Olearia colensoi*, southern rata, manuka, pink pine, yellow-silver pine and kamahi (Fig. 31).

East of the Main Divide, *Coprosma pseudocuneata*, *C. ciliata*, tauhinu, *Olearia nummulariifolia* and mountain flax are common in subalpine scrub, and *Hebe odora* is important in eastern Central Otago, notably on the Rock and Pillar Range (Fig. 32). Throughout this vegetation type, ground cover is provided mainly by sub-shrubs, grasses and herbs from the adjacent subalpine grasslands.

Infertile and toxic soils

Bog pine is the dominant woody plant of open shrubland growing on very infertile soils on glacial outwash gravels at the Wilderness Scientific Reserve, near Te Anau (Fig. 33). Smaller shrubs, including the native heaths *Gaultheria depressa*, three *Leucopogon* species and *Dracophyllum uniflorum*, as well as tauhinu, manuka and some *Coprosma* species, occur in the spaces between and under the canopies of bog pine plants. Ground cover is mainly a mosaic of mats of bryophytes and club moss (*Lycopodium* species). The introduced heather *Calluna vulgaris* is considered a threat to this rare and unusual shrubland.

Ultramafic soils can contain levels of iron, magnesium, nickel, chromium and cobalt that are toxic to most plants. Shrublands occupy such soils on the Olivine Range in north-west Otago, around Lake Ronald near Milford Sound (Fig. 34) and in the southern parts of the Eyre and Livingstone Mountains of north-western Southland. The Cascade Plateau, which consists of glacially deposited ultramafic rocks at the northern end of the

Fig. 30 (below left). Subalpine scrub with *Olearia colensoi* and *Dracophyllum menziesii*, The Hump, southern Fiordland. Peter Johnson

Fig. 31 (bottom left). Subalpine scrub with inaka, manuka and *Olearia colensoi*, Mt Anglem, Stewart Island. Alan Mark

Fig. 32 (below right). *Hebe odora*-dominated subalpine scrub, Rock and Pillar Range, Central Otago. Alan Mark

Fig. 33 (bottom right). Bog pine and *Dracophyllum uniflorum* are prominent in scrub on leached soils of glacial outwash gravels at the Wilderness, near Lake Te Anau. Peter Johnson

Olivine Range, carries stunted forest on its deepest soils, where mountain totara, rimu, yellow-silver pine, pink pine and silver pine are found with kamahi, rata and mountain beech. Where soil conditions prevent the establishment of forest, woodland with manuka, pink pine, kamahi and rata prevails, interspersed with open shrubland and sedgeland (Fig. 35). In the Red Hills of the Olivine Range, up to about 600 m elevation the tallest vegetation on ultramafic scree soils is low forest which contains mountain beech, yellow-silver pine, pink pine, southern rata, manuka, celery pine and kamahi. Above 600 m, manuka is the dominant species. From 700–760 m, ultramafic surfaces are very sparsely colonised mainly by grasses and herbaceous species, but as organic matter accumulates, shrubs establish, including mingimingi, *Dracophyllum uniflorum*, *Melicytus alpinus*, *Hebe odora* and *Coprosma* sp. aff. *parviflora*. In turn, shrubs are succeeded by woodland with manuka, southern rata, inaka and celery pine. Ultramafic soils on Black Ridge, at the southern end of the Eyre Mountains, support manuka woodland with mingimingi, *D. uniflorum*, *Pimelea oreophila* and *M. alpinus*.

Drylands

Shrubland dominated by small-leaved species with densely interlacing stems (filiramulate or divaricating shrubs) is characteristic of the dry Central Otago basins, where it seems to be the climax vegetation. It varies in character from savanna-like, with widely spaced shrubs of a single species on dry, gently-sloping fans and valley floors, to dense, impenetrable and diverse scrub in moist gullies (Fig. 36). Matagouri *Discaria toumatou* is the most common and widespread native shrub, but it is joined by a variety of other species, depending on habitat, that include kowhai, *Coprosma propinqua*, *Olearia lineata*, *O. odorata*, *Carmichaelia petriei*, *C. compacta* and *Melicytus alpinus*. Entanglements of lawyer *Rubus schmidelioides* are common. The exotic shrub sweet briar *Rosa rubiginosa* is a thoroughly established part of this vegetation.

Fig. 34. Shrubland with manuka, mountain beech, southern rata and yellow-silver pine, on ultramafic soils near Lake Ronald, Fiordland. *Peter Johnson*

Fig. 35 (bottom). Low heath forest with inaka, pink pine, mountain beech and southern rata, on ultramafic soils at Fiery Peak, north-west Otago. *Alan Mark*

Fig. 36. Matagouri and *Coprosma propinqua*, characteristic shrubs of semi-arid shrublands in the Kawarau Gorge, Central Otago. The sword-leaved speargrass *Aciphylla aurea*, seen flowering here, is a common constituent of vegetation in these dry, rocky environments. *Peter Johnson*

Fig. 37 (bottom). Kowhai trees flowering amongst scattered kanuka shrubs on the north-western slopes of the Dunstan Mountains, Central Otago. *Alan Mark*

Kanuka and manuka

Scrub and shrubland dominated by either or both of these species are sufficiently distinct to warrant a separate description, even though they occupy a very wide range of environments in Otago, Southland and Stewart Island.

Kanuka forms more or less climax scrub and shrublands in some of the driest environments in the upper Clutha Valley: on river terraces, the rocky northern slopes of the Dunstan Mountains and the north and east faces of the Pisa Range. In the first two of these locations, the accompanying sparse vegetation is mostly introduced grasses, scabweeds*Raoulia* species, occasional epacrid sub-shrubs, sweet briar and, rarely, kowhai (Fig. 37). On the east face of the Pisa Range, kanuka forms woodland in which mountain totara and celery pine become prominent, with some bog pine, on the middle slopes. With increasing altitude, however, kanuka is displaced in this association by manuka, which is more tolerant of the lower temperatures at these elevations.

In Eastern Otago, from the Clutha River to the Kakanui Mountains and as far inland as the southern and eastern boundaries of the dry Central Otago basins, kanuka is the dominant canopy tree in secondary vegetation that has established when seed sources for broadleaved species are scarce, or where the post-forest environment is particularly dry (Fig. 38). Manuka often accompanies it from its establishment, but is overtopped and displaced by taller kanuka after some thirty years. Typically associated with kanuka on dry, rocky sites, such as in the Taieri and Rongahere Gorges, are the filiramulate shrubs *Helichrysum aggregatum*, korokio *Corokia cotoneaster*, *Melicope simplex*, milktree, *Coprosma crassifolia* and *Myrsine divaricata*.

In the Catlins and throughout Southland, large areas of terrain were cleared of tall forest in pre-European times. Beyond the distribution limits of kanuka, this land was occupied by manuka and broadleaved scrub, which, like kanuka further north, provided a nursery for the re-establishment of forest (Fig. 39). Clearance for agriculture and exotic forestry has reduced the area of manuka-dominated vegetation to a fraction of its former extent. Manuka also dominates seral communites resulting from past burning of forest and scrub on Stewart Island.

In more mesic environments, as they grow older, kanuka and manuka stands characteristically contain an increasing diversity and density of the species of nearby podocarp-broadleaved or beech forests; indeed, they provide the nursery within which these forest types re-establish. Between ten and thirty years after kanuka and manuka establishment there is a self-thinning stage that results in a tangle of fallen dead stems under a relatively open canopy of living plants. The subsequent increase in light levels can lead to impenetrable thickets of *Coprosma*-dominated understorey, frequently interlaced with lawyer vines, but containing the saplings of trees that will form the eventual forest canopy. Few manuka plants live beyond fifty years, by which time the trees that succeed them are well established, but kanuka can persist in forest for well over a century.

7 EXPLOITATION OF KANUKA AND MANUKA

Kanuka and manuka are traditional sources of firewood in New Zealand. Near urban centres, unprotected accessible stands of these species seldom survive the chainsaw long enough to be succeeded by taller forest. The ecology of both species indicates that they could be managed sustainably for firewood cropping. However, their long rotation – a minimum of fifteen years for both to reach firewood size – and their reputation as weeds, encourage and perpetuate the view that they are a 'free' and limitless natural resource, there for the taking.

Ralph Allen

Fig. 38. Tall kanuka scrub on sites formerly in pasture, Otago Peninsula. The lone kahikatea tree in the centre of the photograph attests to the earlier presence of podocarp-broadleaved forest.
Peter Johnson

Fig. 39 (bottom). Dense manuka scrub on derelict farmland in the habitat of former podocarp-broadleaved forest with southern rata and kamahi, upper Tahakopa Valley, Catlins, south-east Otago.
Ralph Allen

Broadleaved successional scrub

If seed sources survive nearby, broadleaved forest species can appear on cleared and subsequently undisturbed land soon after forest clearance. *Coprosma* species, kamahi, wineberry *Aristotelia serrata*, lancewood, tree fuchsia, marble leaf, peppertree, lemonwood and kohuhu are common in this role. They may be accompanied by species that arrive from further afield by wind-borne seeds, such as tauhinu, manuka and kanuka, but, apart from kanuka, these seldom persist for more than a few decades. The common ferns and herbaceous plants of the forest understorey also establish relatively early in this succession. A mature forest of broadleaved trees, with their accompanying understorey species, will develop within a century or so. This pattern is commonly seen where land is abandoned after forest clearance in western Southland, in the Catlins, around Lake Wakatipu and on the hills around Dunedin.

Exotic scrub

Several exotic shrub species are naturalised in Otago and Southland, but relatively few have yet become the dominant plants of shrubland or scrub.

Broom *Cytisus scoparius* and gorse *Ulex europaeus* are the most conspicuous and widespread exotic scrubweeds. (Fig. 40) They form dense, more or less monospecific, scrub on unproductive farmland and wasteland throughout the region, and can invade tussock grasslands up to at least tree limit. Broom stands may be sufficiently open to permit the persistence of other species, but these are usually exotic grasses and weeds remaining from the former pasture. Gorse tends to form dense stands that exclude most other species. The establishment of native species is dependent on the availability of a seed source, but, even under favourable circumstances, it can be several decades before either broom or gorse is replaced by native woody vegetation. If broom or gorse or their succeeding vegetation are cleared or burnt, a fresh crop of the weed germinates en masse from a soil store of seeds, which can accumulate to thousands per square metre in a matter of a few years, and remain viable for at least fifty years.

Fig. 40. Gorse flowering amongst broom at the mouth of the Akatore Creek, Eastern Otago. *Peter Johnson*

8 THREATENED PLANTS OF SHRUBLAND AND SCRUB

- **Critically endangered:** *Carmichaelia juncea*, a native broom of lake and river margin scrub in the Lakes district.
- **Endangered:** *Pittosporum patulum*, a shrub or small tree of streamsides and margins of montane scrub and low forest, Lakes district; *Olearia hectori*, a small tree of open low forest and shrubland throughout on fertile disturbed soils such as talus slopes and alluvium.
- **Vulnerable:** *Carmichaelia curta*, a sprawling native broom of gravelly river terraces in the Waitaki Valley; *Uncinia strictissima*, a sedge of lowland scrub in Central and eastern Otago.
- **Declining:** *Carmichaelia kirkii*, a scrambling native broom of open shrubland on river terraces, streamsides and wet ground in Central Otago; *C. compacta*, a low shrub of open shrubland in the Kawarau and Cromwell Gorges; *Olearia fragrantissima*, a small tree of open low forest and shrubland in relatively warm, dry habitats; *O. fimbriata*, a small tree of open low forest and shrubland in northern Southland and eastern Otago; *Melicytus flexuosus*, a shrub of lowland alluvial valley floor scrub east of the divide and on Stewart Island; *Teucridium parviflorum*, a small shrub typically found at forest edges along riverbanks in the east.
- **Naturally uncommon:** *Coprosma intertexta*, a sprawling shrub of debris slopes in Central Otago; *Coriaria* 'Rimutaka', a sprawling shrub of dune shrubland around Mason Bay, Stewart Island; *Korthalsella salicornioides*, a small mistletoe on *Coprosma*, manuka and other shrubs.
- **Insufficiently known:** *Melicytus* 'Matiri', a shrub recorded in southern New Zealand only from the foot of a cliff in the Eyre Mountains.

Ralph Allen

Briar *Rosa rubiginosa* also forms conspicuous stands, but is more limited in distribution than broom and gorse, and co-exists with many native species from the dryland shrubland described above. It is characteristic of all but the driest parts of Central Otago, and extends westwards into much of the wetter Lakes district.

There are several other exotic woody species that can form locally important stands of scrub. For example, the driest hillsides in Central Otago, around Alexandra and Cromwell, can be completely clothed in thyme *Thymus vulgaris*. Darwin's barberry *Berberis darwinii* is prominent in parts of the Catlins, around Dunedin and in the Manapouri and Te Anau basins. The North Otago coastline and the lower Waitaki Valley have extensive stands of boxthorn *Lycium ferocissimum*. Spanish heath *Erica lusitanica* is widespread and increasing but seldom dominant in eastern tussock grasslands, and can cover roadside cuttings throughout the eastern and southern hills. Himalayan honeysuckle *Leycesteria formosa* forms dense stands in gullies and on shady faces in many places where native forest has been cleared. All of these, and several other species, have the potential to increase their range and their impact: they are indeed the weeds of the future.

INVERTEBRATE FAUNA OF SOUTHERN FORESTS AND SHRUBLANDS

Invertebrates are animals without backbones. Although this has limited the size they can attain it has not hampered them in terms of number of species, diversity of form or the large range of habitats they inhabit. This section covers the invertebrate phyla Mollusca (snails), Onychophora (Peripatus) and Arthropoda (insects, spiders, crustaceans, millipedes and centipedes). In total 26,000 species are known from New Zealand and more than 90 per cent of these are endemic; indeed, New Zealand is a major contributor to global biodiversity, with 2 per cent of the world's invertebrate species found only here. About 16,000 invertebrate species occur in the south with 5–10 per cent found exclusively in the region. The New Zealand invertebrate fauna is also characterised by some endemic families such as the batfly Mystacinobiidae, spider Huttoniidae, tiny diurnal moths Mnesarchaeidae and harvestmen Synthetonychidae, all of which are present in the south.

The flora and fauna of southern New Zealand have evolved from a biota that persisted on an ancient (tertiary) archipelago, the largest islands of which were centred in the south. Most of the present biota derives from this source and hence biodiversity increases towards the south. (The land mollusca are a notable exception, with greatest diversity in the north-west Nelson area.) More recent glaciation has resulted in a bias towards cold-tolerant taxa. For insects the most popular host plants are those that are cold-adapted, such as alpine or southern species, or genera that are found from coastal to alpine areas. As in the rest of New Zealand, the upland fauna is richer than that in lowland areas, probably because of greater human impacts in the lowlands. It is important to note that most of the conspicuous invertebrates around us are native, such as the blue-bottle fly *Calliphora quadrimaculata.* The few exotic species feature most strongly in pastoral, industrial and suburban settings.

A characteristic of the New Zealand invertebrate fauna, well illustrated in southern forests, is the large number of species that inhabit and feed on dead material including leaf litter, rotten wood and detritus. While some groups such as Peripatus (Box 9) and spiders are predators of smaller animals, most feed on dead material on the forest floor, where all depend on the high humidity and relatively undisturbed nature of the habitat.

Invertebrates are excellent 'botanists' – they have an ability to recognise particular plants or habitats that are necessary for their survival. As a result, each forest or shrubland type has its own special faunal assemblage. Although most biodiversity in each vegetation type is associated with the soil, leaf litter and rotten log layers, an abundant fauna is also found on tree trunks, understorey plants and canopy species. Surprisingly, despite woody vegetation having its own associated invertebrate fauna, some of this crosses habitat boundaries and is found above the treeline in alpine grasslands. For example, the large land snail *Powelliphanta spedeni* is distributed from eastern Fiordland eastwards across northern Southland beech forests and grasslands including those of the Eyre Mountains, Mataura Range, Umbrella Mountains and Mount Benger. The shell of *P. spedeni* is shown in Fig. 41.

9 SOUTHERN PERIPATUS

Peripatus, or 'velvet worm', is the common name for ancient invertebrates of the phylum Onychophora. They have long been described as 'missing links' because of their unique combination of annelid worm and arthropod characters. These animals superficially resemble caterpillars, with a thin, velvety cuticle, but from above there is no sign of external segmentation, and the head is not marked off from the body. All body segments are similar, each bearing a pair of limbs which end in claws.

Fewer than ten species have been described from New Zealand, but many undescribed species have recently been recognised though not yet formally named. The common, grey species in southern New Zealand is *Peripatoides novaezealandiae*, a species that gives birth to live young. But in contrast a widespread brown species with greenish spots on its back, *Ooperipatus viridimaculatus,* lays eggs. Another brownish species *Ooperipatellus nanus* lives in forest and alpine habitats on the Takitimu Mountains and is fairly common there. Forests of the Dunedin area, including the Otago Peninsula, are home to a blue-grey undescribed species that is so distinctive that it should occupy a new genus. In 1994 the Dunedin City Council acted to purchase a small area of remnant forest in Caversham Valley, a key habitat for the species. The area, an extension of the Dunedin Town Belt, is now managed to provide ideal Peripatus habitat. A species very abundant on Birch Island and in other parts of the Rongahere Gorge, Clutha River and other parts of the Blue Mountains is also a very distinct new species (see below). A third undescribed species in southern New Zealand is found in forest at Trotters Gorge. All species are predatory, subduing their prey with a sticky fluid, and grow to about 35 mm long.

Other Peripatus species are known from scattered alpine localities such as the Eyre, Hector and Kakanui Mountains making this part of New Zealand important in Peripatus biogeography and biodiversity.

Tony Harris

A new and as yet unnamed species of Peripatus. *Brian Patrick*

Another interesting feature of the fauna is that the majority of leaf-, flower- or stem-feeding species are restricted to eating just one plant species (monophagous) or a group of closely related species (oligophagous) rather than feeding on a wide variety of species (polyphagous). Specialising on one plant species or genus can become a risky business if an invertebrate's host plant is palatable to exotic mammals. The genus *Gingidia,* containing the aniseed plants, is an example of a plant genus that has declined markedly. The large larentiine moth *Gingidiobora nebulosa* (Fig. 42), which has well-camouflaged larvae that feed on *Gingidia* leaves, is now restricted to refugia such as steep rock faces.

Invertebrates play important roles in all the ecosystems to which they belong. Among these is their part in recycling nutrients by helping to break down leaf litter, rotten logs, detritus and live foliage. With the aid of bacteria and fungi, nutrients are thus released for use by plants in the future. Moreover, invertebrates are the major pollinators of most higher plants, with flies, beetles, bugs and moths conspicuous in this fundamental role. They may also play a more subtle role in maintaining the resilience of plant communities to change. By eating the seeds of invading species, invertebrates may tend to maintain the status quo; with up to ninety-nine per cent of seeds being eaten, it is extremely hard for seeds arriving in low numbers from long distance to get established. Additionally, selective feeding on plants helps to determine the composition and architecture of the plant community. Thus, some plants are heavily browsed while those that are unbrowsed are more likely to fulfil their growth potential.

The significant ecosystem roles played by invertebrates makes their conservation a priority, but they have special needs. The available habitat must be capable of sustaining each of the four stages of the insect life-cycle (and up to eight moults of other arthropods). Each stage may require quite different conditions, and these need to be available at just the right time each and every year to ensure species survival.

Invertebrates in podocarp forest

Mixed broadleaved forest with emergent podocarps contains a rich invertebrate fauna, especially in moist undisturbed leaf litter and soil. Tiny land snails, amphipods, numerous spiders and insect larvae abound in such habitats. Southern forests with relatively undisturbed rotten logs and leaf litter are home to several springtails of the ancient family Poduridae; unusual species in the genus *Ceratrimeria* lack the apparatus to leap and at 10 mm in length are particularly large, colourful and conspicuous (Fig. 43). Logs on the forest-floor are home to a large variety of other animals such as flatworms, millipedes and predatory carabid beetles (Box 11) including the large *Mecodema sculpturatum* and shiny green *Megadromus meritus,* both widespread in southern New Zealand. The forest floor also supports four large brown and white moths of the genus *Aoraia,* with wingspans over 70 mm. The subterranean larvae feed on surface leaf litter and can be infected by a fungus that kills them and forms so-called 'vegetable caterpillars'. *Aoraia dinodes* is found in Stewart Island, southern and western Southland forests

Fig. 41 (top left). The giant snail *Powelliphanta spedeni* is typically found as an empty shell, often with damage as pictured in the specimen found in the Nokomai Range (size 5 cm). Most of the other 400 or so species of land snail, which are found in the south, are tiny. *Brian Patrick*

Fig. 42. Pictured is the cryptic large adult of *Gingidiobora nebulosa,* a species rare in north-eastern Otago and elsewhere only found in Marlborough. *Brian Patrick*

Fig. 43 (top right). The springtail species in the genus *Ceratrimeria* pictured is as yet undescribed but relatively widespread in beech forests in the south. *Brian Patrick*

10 WETA: NEW ZEALAND'S FLIGHTLESS CRICKETS

Weta are special because they are so massive – easily a hundred times heavier than the average insect. In fact, the heaviest insect in the world is a weta surviving only on Little Barrier Island in northern New Zealand (the record is 71 grams). These elephants of the insect world are paralleled in southern New Zealand by kakapo, the world's largest parrot, and takahe, a large flightless rail.

Secretive, vividly patterned cave weta (Raphidophorids) have a small body but long slender legs for springing about, and long fine antennae for touching and smelling the dark damp environments in which they live. These creatures are not restricted to caves; they also inhabit forests, coastal areas and tor landscapes. Many cave weta species are still waiting to be described.

Better known are the larger tree weta *Hemideina crassidens* of the forests of northern Fiordland and the west coast, and the large weta *H. maori* in the mountains of northern Southland, Central Otago and South Canterbury. They are flightless – like all weta – and are therefore vulnerable to predation, being known only from very isolated places. The presence of *H. maori* on two islands in Lake Wanaka, free of exotic predators, hints at a much wider distribution for the species in pre-human times than the mountain-top refugia they now occupy. For insects, these species are long lived; they take up to fourteen months to mature and the adults are reproductively active for a year. Weta communication must be very complex, since males can defend territories containing many sexually active females. The males are known to fight and have fearsome heads and jaws. Both sexes make a rasping chirruping noise and try to bite if annoyed.

The Herekopare weta *Deinacrida carinata*: an adult female at night on punui (Pig Island, Foveaux Strait). *Rod Morris*

Ground weta in the genera *Hemiandrus* and *Zelandosandrus* are still to be found in many places, both urban and rural. Perhaps it is because they are smaller and inhabit soil burrows that they can persist where many potential enemies exist above ground. The chunky *H. focalis* is widespread in alpine areas of the south, while an undescribed species lives in cushionfield habitat near Alexandra, Cromwell and in the upper Waitaki Valley.

Tree weta species and the ground weta species are in the family Stenopelmatidae. Two more stenopelmatids are special in southern New Zealand, the Herekopare weta *Deinacrida carinata* and the scree weta *D. connectens*. Sturdily built weta, with a covering of tough pitted plates and spiny hind legs, they cannot jump and are called giant weta. Herekopare weta are only known from two islands in Foveaux Strait. Scree weta are widespread in the high mountain screes of the South Island but more local in the south, with populations on mountains in North and Western Otago and on the Takitimu Mountains. A newly described species *Deinacrida pluvialis* lives in the low-alpine zone of the Matukituki Valley where it was discovered in the 1990s. It must be hardy, surviving a contrast of freezing and intense solar heating conditions and living on plants that are just as hardy. These giant invertebrates persist in their mountain strongholds whereas any lowland populations have long since disappeared.

Eric Edwards

11 CARABID BEETLE DIVERSITY

Carabid beetles are the primary carnivores of the insect world. They have large jaws with which they attack their prey and defend themselves if necessary. The larvae of most species are also predators and feed on soil-dwelling invertebrates. Some carabid beetles feed on plant material and seeds, but most species have a predatory life style. Carabid beetles vary in size from the large *Mecodema* and *Megadromus* species up to 40 mm long to the tiny, soil-dwelling, often blind and unpigmented Anillini which are mostly under 2 mm long. They occupy just about all terrestrial habitats over the entire altitude range and some could be considered almost semi-aquatic, living under rocks at the edges of fast-flowing streams.

In southern New Zealand over a hundred carabid species are found, all with discrete distributions. Some of these have become less abundant as a result of habitat loss and agricultural development of the lower altitude tussock grasslands and forest margins that they inhabit. *Mecodema chiltoni* is 36–38 mm long, and was thought to be endangered, but a large population has recently been discovered at the southern end of the Eyre Mountains. Another magnificent species, *M. laeviceps*, described from the Ida Valley has been recently rediscovered in Waikaia Forest and on the Kakanui Mountains. The large *M. rex*, up to 35 mm long, is found on the Longwood Range in Southland, and the similarly large *M. laterale* is quite common in Mount Aspiring National Park. A spectacular species in the genus *Megadromus* is *M. bullatus*, up to 30 mm long with metallic green reflectance. This species was described from near Queenstown, but is widespread in Otago and Southland, although nowhere common. *Holcaspis* is a genus of over thirty species of smaller, 10–15 mm long, black carabids which have diversified throughout New Zealand but particularly in Otago and Southland, with one species confined to Stewart Island, *H. stewartensis*.

Because they are generally quite fast-moving, surface-dwelling predators, carabids are easily caught in pitfall traps, which simply consist of a container sunk into the ground, and the collector can quite quickly gain an indication of the diversity of species present.

Barbara Barratt

Mecodemia chiltoni. *Brian Patrick*

Migas goyeni, a tree trapdoor spider

This species makes a tunnel with a strong and resistant silk into which is interwoven small pieces of bark so that, when finished, it merges with the tree on which it is built.

Clubiona, a hopping spider

This little *Clubiona* spider, about 7–10 mm in length, is easily recognised by its amber-coloured cephalothorax and speckled abdomen. Other species of clubionids have brown or greenish abdomens. All feed on small insects and are characterised by fast running movements interspersed with sudden hops. Found under stones, in clumps of grass, or inside rolled-up flax leaves, they seem equally at home in dry or swampy areas. All of them spin silken retreats in which to moult but it is here, too, that males often mate with newly moulted females. Their neat and rounded eggsacs, thinly coated with silk, are laid within these retreats. These native spiders are common throughout New Zealand.

The green crab spider, _Diaea ambara_

These small spiders belong to a group whose members are generally green with amber-coloured legs, although some have abdomens which are yellowish or mottled with darker markings. As might be expected, they live amongst foliage where they are well camouflaged. The sideways orientation of their legs links them to the crab spiders and, like crabs, they can sidle quickly in any direction if disturbed. As sit-and-wait predators, they grab small insects and other tiny creatures, particularly those which suck the sap or feed on the pollen of plants. If you hold a white tray or cardboard lid beneath a shrub and shake a branch, several of these tiny spiders may then be found moving about in your tray and some of the creatures they feed on will probably be amongst the debris also. After mating, females look for curled-up leaves or partly rolled-up flax or *Astelia* swards. Here they enclose themselves within thinly woven silk cocoons and lay their eggs. The mother guards them until they hatch some three weeks later.

The lichen spider, _Cryptaranea subcompta_

This orbweb spider is found only in native forest, usually near the coast or where the forest is thinner so that sunlight filters through and light breezes prevail. The web is built more or less at right angles to a lichen-encrusted tree, the large orb-shaped snare standing out from the trunk and held in place by a nearby branch or sapling. At night the spider sits at the hub of its web waiting for flying insects, such as crane flies or moths, to become trapped in its snare. The spider then rushes to the struggling insect, and wraps and bites its prey, before dangling the bundle from its spinnerets and carrying it back to the hub to eat. During the day, this spider, its body patterned, coloured and tufted like lichen, rests amongst the lichen it resembles, invisible to all eyes, or so one would imagine. But there is one sharp-eyed insect, *Pison spinolae*, the mud-dauber wasp, which is able to distinguish this spider from its surroundings. This daytime parasitic wasp catches spiders to stock its nests and nourish its young and there amongst the larder its success in locating *Cryptaranea subcompta* is recorded.

The chevron spider, _Pianoa isolata_

Pianoa isolata is one of the rarest spiders in New Zealand, being found only in the silver beech of Waikaia Forest in Southland. It also has the distinction of belonging to an ancient lineage which can be traced back to Gondwana. Several features make this spider special. Like the very earliest spiders, it breathes with four booklungs instead of the two booklungs and tracheae that are common to most of the recent spiders. Its fangs move diagonally, whereas in trapdoor spiders the fangs move up and down and in recent spiders they move from side to side. Despite these primitive features, it is a hunting spider which does not make a snare and so moves about in search of food. Pale fawn in colour, *Pianoa* is easily recognised by the chevron markings on its abdomen. Amongst decaying logs, leaf litter and moss, these spiders make their home. Here they catch their prey, preferring flies and moths and even other spiders from amongst the large variety of invertebrates that live there. Within cavernous spaces beneath tree stumps and ageing logs, the female lays her eggsacs, suspended from above with long thin stalks.

Jumping spiders

Of all the spiders, the Salticidae or jumping spiders have the best eyesight. They move around in daylight looking for prey and when they see a likely insect, they stalk it in much the same way as a cat stalks a mouse. Like most spiders, salticids have eight eyes, six of which are quite large compared to other spiders. Four of these face forwards and the other four face to the side and the back. When the rear eyes detect movement from the back or side, the spider swivels round to face the source of the movement. The two middle front eyes examine the object that moved and, if prey, the spider creeps towards it, this stalking behaviour being directed by the front eyes. If the insect runs away, the spider chases it, the speed and direction of the insect's escape route being monitored by the forward facing side eyes. When about five centimetres away, the spider crouches and jumps, immediately embedding its fangs. Jumping spiders are found everywhere, from the seashore, to pastures and forests, in leaf litter, and high mountain tops such as the snowy peaks of the Remarkables in Central Otago. The jumping spider illustrated here belongs to the genus *Trite* but it has no species name as it has not yet been officially described.

Ray and Lynn Forster

Migas goyen

Clubiona sp.

Cryptarenea subcompta

Diaea ambara

Pianoa isolata

Trite sp.

Fig. 44. The large green shield bug *Oncacontias vittatus*. Brian Patrick

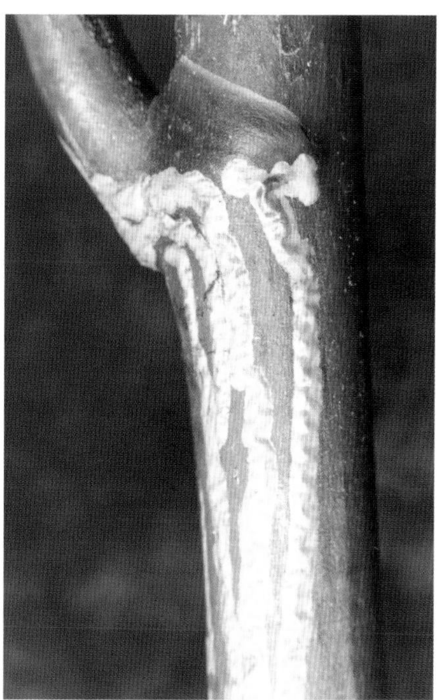

Fig. 45. The bark mine of *Acrocercops panacitorsens* on *Pseudopanax colensoi*. This species is one of five of this elegant moth genus which mine various *Pseudopanax* species. Brian Patrick

Fig. 46. Pictured are the adults of *Tatosoma agrionata* whose larvae defoliate yellow mistletoe. Brian Patrick

and is well known as a host of the *Cordiceps* fungus. The similarly large *A. rufivena* and *A. aurimaculata* are found in Otago forests and grasslands, emerging in autumn.

Native cockroaches abound in these forests, on the ground, on tree trunks and amongst foliage. Cockroaches undergo incomplete metamorphosis (the larval stage closely resembles the adult). The flattened adults are conspicuous, running swiftly when disturbed. The female often has a purse of eggs being gradually extruded from her abdomen.

An important habitat for invertebrates is provided by the surfaces of tree trunks, which are particularly complex in podocarp forest because of the great age of many of the trees. Loose bark inevitably shelters a variety of spiders, together with their silk snares and egg sacs. These include tree trapdoor species in the genus *Migas* and jumping spiders, Salticidae (Box 12), which prey on adult moths and bugs. Algae on the bark are eaten by a variety of cased larvae of moths in the genera *Reductoderces* and *Grypotheca*; the fragile adult males may be seen flying on calm, frosty, spring mornings, seeking out a wingless female clinging to her pupal case, in which she later lays eggs. Arboreal land snails and tineid moth larvae are also frequent inhabitants of podocarp bark, mostly feeding on detritus.

A wide variety of species are associated with the foliage of forest plants. For example, forest-floor herbs in the genera *Acaena, Coprosma* and *Urtica* support an assemblage of leaf-tying moth larvae, while the multitude of ferns provide food for elegant geometrid moth larvae of the genera *Ischalis, Sarisa* and *Sestra.*

True bugs are numerous in all vegetation types in the south. As both adults and larvae, most pierce plant cuticle and suck sap. The large green shield bug *Oncacontias vittatus* (Fig. 44) is especially common on rimu, other podocarp foliage and lianes such as *Muehlenbeckia australis.*

Larval stages of many insects live almost their entire lives within leaf or soft tissue on branches or trunks. There they are protected from the adversity of climate and are relatively secure from predators. The leaf-mining strategy is widespread among larvae of wasps, flies, beetles and moths. In southern New Zealand it is best exemplified by the tiny, elegant moths of the genus *Acrocercops*. Different species mine the leaves of *Pittosporum, Parsonsia* and *Coprosma*, while another five species mine *Pseudopanax* species in particular ways (Fig. 45). The adults lay eggs on the stem or leaf surface, with the emerging larvae tunnelling into the plant tissue, forming tortuous mines or scribbles as they feed, move and grow. Most pupate within the mine and emerge from the mine as an adult to mate and disperse. Some species spend their entire larval life within tiny leaves as small as a kowhai leaflet.

Extremely well camouflaged amongst the foliage of podocarps are stick insects, with the larger females of species such as *Argosarchus horridus* reaching 15 cm in length. Similarly cryptic looper caterpillars of the genus *Pseudocoremia*, such as *P. suavis* and *P. leucelaea*, feed on podocarp foliage, along with leaf-rollers of the genus *Pyrgotis* and pseudo-leafminers of the genus *Chrysorthenches.*

Mistletoe species such as *Ileostylus micranthus* and *Peraxilla colensoi* are more common in the south than elsewhere in New Zealand. They support polyphagous (broad diet) weevils such as *Catoptes censorius*, as well as a small group of moth species (Fig. 46) that are oligophagous (narrow diet) and cause defoliation on this small group of related mistletoes. The small moths *Zelleria sphenota* and *Z. maculata* have larvae that first mine, then bore into the buds, flowers and seeds of both mistletoe species. Their presence is signalled by thin scribbles on the leaves.

No New Zealand native plant is richer in specialist native insects than the large liane *Muehlenbeckia australis*. Stick insects, bugs, beetles, flies, over thirty native moths and over fifteen copper butterflies depend on it for larval food. Their larvae mine the leaves, feed within the stems or feed on the flowers, leaves and leaf litter. In return for shelter and food, the adult insects often pollinate the flowers of the host. Despite its poor

13 CHOOSY FEEDERS ON NATIVE DAPHNES

In coastal to alpine settings, native daphnes of the genus *Pimelea* are host to the larvae of a number of colourful moths in the genera *Notoreas, Merophyas, Ericodesma* and *Meterana*. Pictured is the striking adult of *Meterana meyricci* which feeds on *P. oreophila* in the alpine zone of the Rock and Pillar Range. Other hosts within the genus include *P. poppelwellii, P. prostrata, P. lyallii* and *P. traversii*. Brightly coloured adults and larvae are characteristic of species that feed on this genus, advertising the defence they aquire by consuming the toxins contained in the *Pimelea* leaves. This host plant depends on natural disturbance to provide a substrate for its seedlings.

Brian Patrick

Meterana meyricci. *Brian Patrick*

Fig. 47. A larva of the larentiine moth *Hydriomena purpurifera* feeding on *Epilobium* on a wet bank. *Both photos Brian Patrick*

Fig. 48. The tiger beetle *Neocicindella parryi* from Eastern Otago.

popular image, *M. australis* is more important to local insect biodiversity than all of the trees it drapes over and may even have a role to play in protecting forest remnants, by providing shelter for both mature trees and seedlings.

Wet banks provide a special habitat that supports a wide range of invertebrates, including the glow-worm, the larvae of a native fly. The light emitted by the maggot attracts small flying insects that are then trapped by sticky beads hanging close by and later eaten. Glow-worms are common in the south on damp banks or in caves. Herbs of damp banks and streamsides support a wide variety of larentiine moth larvae, feeding on *Gunnera, Hydrocotyle* and *Ranunculus* species. Fig. 47 shows the distinctive larva of *Hydriomena purpurifera*, which feeds on *Epilobium penduncularе* in such wet habitats. The activities of humans often result in the drying out of habitats, so wet banks are often difficult to find in a natural state.

Clay banks on forest margins are the distinctive habitat for tiger beetles (Fig. 48). Adults are ferocious predators running fast with intermittent flights to catch smaller insects. The similarly predatory larvae live in holes on the bank, catching prey that settles nearby on hot days. The large-winged antlion *Weelius acutus*, one of seven lacewing species found in the south, is an uncommon inhabitant of dry banks on podocarp forest margins in Eastern Otago and the Catlins. It is also present in western Otago in the Skippers area.

Invertebrates in beech forest

Despite the apparent lack of floral diversity in beech forests, which are generally dominated by a single *Nothofagus* species, there is a distinctive and relatively rich invertebrate fauna. Red, mountain and silver beech forests each have a specialist invertebrate fauna consisting of mites, scale insects (Fig. 49), leaf-mining or shoot-galling weevils, gall-midges and foliage-feeding moth larvae. Silver beech has at least eight specialist moth species that mine, bind or feed on the leaves. For example, well camouflaged amongst silver beech foliage can be found the mature larvae of noctuid moth species of the genus *Meterana* that are specialist on this species (Fig. 50). But it is mountain beech foliage that can sustain large-scale damage from the attention of the larvae of the moth *Proteodes carnifex*. Complete defoliation occurs periodically on a large scale, such as in the forests near Te Anau, but is eventually brought under control by natural predators.

Recently, the first South Island record of the endemic New Zealand family of flies associated with short-tailed bats was made in the Eglinton Valley. Batflies were found on the bats and feeding amongst the guano in their roosts. These wingless, spider-like flies cling to the bats and are transported from roost to roost.

The large, slow-moving bear weevil *Rhyncodes ursus* is a conspicuous beetle of southern beech forest. These are regularly seen on tree trunks, while beetles of the genus *Philpottia* feed in the canopy of the beech trees. The legless fat grubs of bear weevils are sometimes parasitised by the giant ichneumon wasp *Certonotus fractinervis*, as they feed deep within silver beech wood.

Southern forests of all types support New Zealand's largest beetle, the huhu *Prionoplus reticularis*. The noisy adults fly at night while the helpless fat larvae feed in rotten logs on the forest floor. The larvae of stagbeetles also feed on decomposing wood. The magnificent stagbeetle *Geodorcus helmsi* is widespread in western forests, including those of Fiordland and Stewart Island, but less common close to Invercargill, while the related *G. philpotti* is endemic to the Waitutu Forest, west of The Hump. Their presence in these places is possibly threatened by mice, rats and stoats that eat forest insects. The smaller stagbeetle *Ceratognathus parrianus* is widespread in the south (Fig. 51).

A feature of southern forests from mid summer till autumn is the loud singing of several species of cicada (Fig. 52). Largest of the four forest species and least often seen is *Amphipsalta zelandica*, as it sings from the forest canopy. In keeping with other species, males do most of the singing as they compete for a mate. Females use a sharp ovipositor to deposit their eggs in rows in outer branches of trees and shrubs. After the eggs hatch, the larvae drop to the ground to begin a year's feeding underground, where they tap the juice of roots and often cause branches to die.

The noctuid moth genus *Tmetolophota* has proliferated on monocotyledons, with different species feeding as larvae on flax, *Microlaena*, *Libertia*, *Carex* and *Astelia* species in the forest setting. Larvae of the moth *T. purdii* (Fig. 53) notch the leaves of various *Astelia* species by night, remaining secreted deep within the basal leaves by day.

Fig. 49. A male native scale insect of the family Pseudococcidae. While the males fly, the mealy-like females crawl over tree trunks and foliage, feeding on leaf juices. *Brian Patrick*

Fig. 50 (left). Camouflaged amongst silver beech foliage in Rongahere Gorge on the Clutha River are the mature larvae of one of three noctuid species of the genus *Meterana*. *Brian Patrick*

Fig. 51. The stagbeetle *Ceratognathus parrianus*. *Barbara Barratt*

Fig. 52. *Kikihia rosea*, a common Dunedin forest cicada. *Brian Patrick*

Fig. 53. Larvae of the noctuid moth *Tmetolophota purdii*. The adult has a wingspan of 50 mm. *Brian Patrick*

Invertebrates in shrubland

Reflecting the diverse array of shrubland communities in the south is a huge variety of attendant invertebrates. Subalpine shrublands at treeline are particularly interesting, with an array of colourful diurnal bugs, beetles, flies, wasps and moths. Botanically, these shrublands are dominated by various *Dracophyllum, Leonohebe, Coprosma, Olearia* and *Hebe* species, together with *Ozothamnus vauvilliersii* and mountain toatoa *Phyllocladus alpinus*. All of these have distinctive associated faunas. For example, the colourful mirid bug *Romna bicolor* is common on *Hebe* foliage in Fiordland at about 900–1000 m (Fig. 54). The larvae of the various insect species feed in a variety of characteristic ways on a certain part of the host plant at a particular time of year. Larvae of plume moths in the genera *Stenoptilia* and *Platyptilia* (Fig. 55) bore into *Leonohebe odora* buds in many low-alpine sites in southern New Zealand, including Stewart Island; the adults often fly at dusk over the shrubs, mating and laying eggs. Typical of the many diurnal geometrid moths that fly swiftly over alpine shrubland is *Dasyuris transaurea* (Fig. 56). Its brown larvae feed on *Anisotome* foliage on many mountains of southern New Zealand but not Stewart Island. An additional 55 other diurnal geometrids are found under similar circumstances in southern New Zealand.

Drier shrublands in valley floors, gorges and coastal areas consist of a range of smaller-leaved shrubs, some of which are deciduous. Most important in terms of invertebrate hosts are *Olearia, Coprosma, Corokia, Carmichaelia, Pimelea, Melicytus* and the liane *Muehlenbeckia*. Some of the richest of these shrublands, in terms of invertebrates, occur at inland sites in Central Otago. Here, because of the severe winter, larval activity is restricted to the warmer months. The dry leaf litter layers of shrubland contain a modest fauna, of which the diurnal moth *Hierodoris frigida* is typical, but foliage-feeding larvae and adults are numerous amongst the flies, bugs, beetles and moths. New Zealand's only native praying mantis *Orthodera novaezelandiae* has its only southern population in Central Otago, where it hunts for its prey on various shrubs including the diminutive exotic thyme. Remnant shrubland in gullies, riparian zones and roadsides can support a surprising array of native invertebrates, out of proportion to its stature. One fast-moving bug that is widespread in these shrublands is *Romna scotti*, often joined in this habitat by several species in the genus *Chinamiris*. Both larvae and adults pierce the plant leaves and stems to suck juices. Among the most important shrubs for insects in Central Otago is the deciduous *Olearia odorata*. The plant supports many species of beetle, bugs, flies and stick insects, but it is the moths that are

Fig. 54. The mirid bug *Romna bicolor* is common on *Hebe* foliage in Fiordland (Fig. 8.inv.16). *Brian Patrick*

Fig. 55 (bottom). Larvae of plume moths in the genus *Platyptilia*. *Brian Patrick*

Fig. 56. A diurnal adult geometrid *Dasyuris transaurea*. *Brian Patrick*

Fig. 57 (bottom). A full-grown larva of the large noctuid moth *Meterana grandiosa*, from the Kawarau Gorge, feeding on *Olearia odorata*. *Brian Patrick*

best known. Among them is the large noctuid *Meterana grandiosa* (Fig. 57) and the rare geometrid *Pseudocoremia cineracia*, two of over forty species whose larvae feed on the shrub. *Pseudocoremia cineracia* is known only from Moke Lake and the Kawarau Gorge in the south. Elsewhere it is known from the MacKenzie Country.

Coastal shrubland, consisting of *Hebe elliptica*, supports insect activity all year round. Many insect species are confined to this zone, with several defoliating moth larvae being prominent. Some coastal shrubland communities, such as that on Tiwai Peninsula, are of a subalpine composition and contain subalpine insects. Ngaio, a small tree, grows naturally on the coast as far south as Brighton on the Otago coast and supports a small but distinctive insect fauna. This includes a sap-sucking mirid bug, a bark-boring tortricid moth larva, flower-feeding caterpillars in the genus *Chloroclystis* and a leaf-mining fly on its foliage.

Diverse coastal forest at the mouth of Akatore Creek, South Otago, is home to many interesting plant-insect interactions. A small bright orange day-flying moth *Cephalissa siria* flies around the forest and shrubland edge in October. Its slender green larvae feed on the rare climbing fuchsia *Fuchsia perscandens*, which is present as a small population. Other plants rich in specialist insects are totara, *Helichrysum lanceolatum*, *Carmichaelia petriei*, *Muehlenbeckia australis*, *Rubus* spp. and *Olearia fragrantissima*.

14 NATIVE MOTHS AS BIOCONTROL AGENTS

Several introduced weeds in the south are kept in check to some degree by native insects. Among the native biocontrol agents is the diurnal magpie moth *Nyctemera annulata* whose hairy orange and black larvae ravage ragwort leaves and flowers. The species' natural food plants are a range of related native *Senecio* species that grow from coastal to montane sites. The brightly coloured cinnabar moth from Europe, with its striped larvae, has been introduced to assist the magpie moth, but has only persisted in the south in localised parts of western Southland.

Attacking the buds and shoots of blackberry are the fat larvae of a native moth *Heterocrossa adreptella*. Its natural food plant is lawyer (*Rubus* species). Although it can devastate blackberry over wide areas, its effect is fortunately temporary.

Two moths, which have been introduced accidentally, fortuitously attack problem weeds. Larvae of *Agonopterix alstroemeriana* roll the leaves of hemlock, and move into and feed on the flowers, and the shining white *Leucoptera spartifoliella* has stem-mining larvae that feed on exotic broom. Since the 1980s, both species have become well established and widespread in the south.

Brian Patrick

REPTILES OF SOUTHERN FORESTS AND SHRUBLANDS *Alison Cree*

Many species of lizards are found in southern New Zealand. Before discussing them in detail, it is important to note that there have been considerable changes to the taxonomy of New Zealand lizards over recent years. In particular, the use of modern genetic techniques, in addition to traditional procedures of morphological comparison, have helped to clarify relationships and identify subtly different species. As a consequence, the scientific names used here may differ from those in older texts.

Currently, three genera of lizards are represented amongst the forests and shrublands of southern New Zealand. These are the gecko genera *Hoplodactylus* and *Naultinus* and the skink genus *Oligosoma*. All these genera are endemic to New Zealand, although their respective families (Gekkonidae and Scincidae) are well-represented in other parts of the world. The two types of lizard can easily be distinguished from each other. Geckos have large eyes that cannot blink, and dull and velvety skin without overlapping scales. Skinks, in contrast, have relatively small eyes with movable eyelids, and their skin is shiny with overlapping scales. Skinks also tend to be more slender and fast-moving than geckos. Male geckos have a large swelling at the base of the tail on the undersurface; this hemipenial sac encloses the two intromittent organs (hemi-penes) used in copulation. Male skinks also have hemipenes, but they are not enclosed in such an obvious swelling.

Naultinus, the genus of 'green geckos', is represented in New Zealand by seven species. Until recently, the five South Island species were placed in a separate genus, *Heteropholis*, but they are now grouped within *Naultinus*. Unidentified specimens of *Naultinus* have been reported from several islands in the Foveaux Strait/Stewart Island region and from Fiordland, but little is known about these populations.

The jewelled gecko (*N. gemmeus*) is the best-known species of green gecko from southern New Zealand. It is found in Otago and Southland, and also extends into Canterbury. Like other species of *Naultinus*, *N. gemmeus* is a diurnal (day-active) and arboreal (tree- and shrub-dwelling) species. Its back is coloured a striking, emerald green, and a bold white or yellow stripe runs down each side of the back from the eye to the end of the tail (Fig. 58). Often this stripe is broken into a series of diamonds. The slender, clawed toes are well suited to gripping twigs, with the long and finely tapering tail assisting as a 'fifth leg'. As with all New Zealand geckos, it is a live-bearing (viviparous) species, giving birth to two young in late autumn (May).

Jewelled geckos have been recorded around Lake Mahinerangi, a few other inland parts of Otago, and in the Catlins.

Fig. 58. The jewelled gecko *Naultinus gemmeus* demonstrates its agility among branches of *Coprosma propinqua*. Alison Cree

However, the species is best known from the Otago Peninsula, where it is conveniently placed for study by students of the University of Otago. These studies indicate that it is most frequently seen on *Coprosma propinqua* (mingimingi), a native shrub with divaricating branches and small green leaves. On the peninsula, the species has also been observed on manuka and kanuka, and rarely on broom and mistletoe. One specimen was seen many metres above ground in a macrocarpa tree on the edge of a reserve of *C. propinqua*. Elsewhere, jewelled geckos have also been reported from beech forest.

The dense canopy of divaricating branches of *C. propinqua* provides shelter, basking sites and food (both insects and fruits) for jewelled geckos. The translucent, blue-flecked fruits are eaten by the geckos, which probably play a role in seed dispersal. Indeed, small-leaved *Coprosma* species from low altitudes, such as *C. propinqua*, have evolved several features that seem to favour seed dispersal by lizards rather than birds. These include non-red fruits (New Zealand birds are thought to favour reddish fruits), and placement of the fruits within a dense tangle of twigs where they would be easily accessible only to small animals.

15 STEWART ISLAND: A CENTRE FOR LIZARD DIVERSITY?

As ectotherms or 'cold-blooded' animals, lizards depend ultimately on the sun's warmth to raise their body temperature. Activity is strongly temperature-dependent. At high temperatures, metabolic processes such as digestion, growth and reproduction tend to occur rapidly, whereas at low temperatures, torpor results. As a consequence of this pervasive influence of temperature on activity (and perhaps also because of their own preference for warmth!), reptile biologists have tended to focus on hot environments, such as deserts, for studying lizards.

Now the tide is turning. Searches by more adventurous biologists have revealed the presence of many lizards in cool temperate climates, including the Stewart Island region. At about 47° S, Stewart Island must be one of the world's more challenging environments for lizards. Frosts are rare (thanks to the buffering effect of the nearby ocean), but sunshine hours are low, rainfall is frequent and winds can be strong. Remarkably, at least six species of lizards are known from the Stewart Island/Foveaux Strait region.

Undoubtedly the most eye-catching is the harlequin gecko *Hoplodactylus rakiurae* (the specific name refers to a Maori name for Stewart Island, Rakiura). Shades of bright green, brick red, white, brown, grey and/or yellow form a herringbone pattern on the back, with no two individuals seeming identical. Although the pattern looks striking against a plain background, harlequin geckos are in fact extremely cryptic amongst the shrublands, herbfields and granite rock outcrops of their habitat.

As if living in a cold environment was not enough, harlequin geckos are also active during the coldest part of the diel cycle – at night. Nocturnal foraging has been observed at air temperatures of 8–13°C, although individuals also bask by day. Sun-basking is, however, much more obvious among the skinks with which harlequin geckos share their habitat. Two such skinks, *Oligosoma notosaurus* (southern skink) and *O. stenotis* (small-eared skink), have been recognised only since the early 1990s, their distinction from other small, brown forms of 'common skink' being based on close morphological examination supported by modern genetic techniques.

As further attention is given to the Stewart Island fauna, it is possible that more species will be discovered. In addition, our meagre understanding of existing species – including their habitats and how they survive in their cold environment – will improve. One suggestion that has yet to be confirmed is that harlequin geckos may primarily be a forest-dwelling species. This idea stems from the observation that most of Stewart Island was once forested and that the extensive shrublands currently occupied by harlequin geckos are fire-induced.

Regardless of such details, it is clear that the Stewart Island region has much to offer reptile biologists. In the words of conservation geneticists Geoff Patterson and Charles Daugherty, the Stewart Island region must now be considered as 'a significant centre of reptilian diversity for New Zealand'.

Alison Cree

Top left: Harlequin gecko – close-up. Bottom left: Southern skink – close-up. Below: Habitat of harlequin geckos. *Alison Cree*

On fine, sunny days (especially mornings), jewelled geckos emerge to bask on the canopy. Some individuals regularly use the same basking sites. On occasions, individuals have been observed to move tens of metres between successive captures (including movement across open grass between shrubs). Ability to home over distances of at least 100 metres has been noted.

The shrubland habitat in which *Naultinus gemmeus* is often seen has been greatly fragmented by pastoral and residential develop-ment. Although the species has been reported in recent years from twenty-six sites on the Otago Peninsula, most sites are small and separated from each other by over one kilometre of pastureland. One small (0.85 ha) site that supported about 70–140 geckos was purchased in 1993 with funding from the Royal Forest and Bird Protection Society, the Save the Otago Peninsula Trust and the Department of Conservation; it is known as the Every Scientific Reserve in recognition of the interest and assistance of its former owners.

The genus of so-called 'brown geckos', *Hoplodactylus*, is represented in the forests and shrublands of southern New Zealand by at least four species. One, which is not brown, is the harlequin gecko from Stewart Island (Box 15). Others, including *H. granulatus*, *H. nebulosus* and members of the *H. maculatus* species complex, are more fittingly referred to as 'brown'.

The forest gecko *Hoplodactylus granulatus*, New Zealand's second largest gecko with a length of up to 190 mm, is widespread with known locations in southern New Zealand including the Catlins, Fiordland and the southern coastline of Southland. True to its common name, it inhabits forest (including beech) as well as shrublands (including manuka). It has been reported feeding from rata flowers, and could thus play a role in pollination. Its dorsal surface has large, irregular blotches of brown and grey, with additional markings of black, white and sometimes red or yellow (Fig. 59). By day, it clings to branches, or hides under loose bark or in tree crevices. Its colouration gives it effective camouflage. Although basking has been reported, it is primarily nocturnal.

A similar, grey-brown species, the cloudy gecko, occurs on islands in the vicinity of Stewart Island. Although some older texts treat this as *Hoplodactylus granulatus*, recent genetic studies support its separation as *H. nebulosus*. Little is known about this species, except that its snout is more blunt and rounded than in *H. granulatus* and there may be more green on the dorsal surface. It has been found on shrubs and under stones.

Several other brown-grey geckos in southern New Zealand belong to the *Hoplodactylus maculatus* species complex. Until the early 1990s, *H. maculatus* was considered to be one species, widespread throughout New Zealand and occurring in forest, shrubland, grassland and boulder beaches. However, morphological and genetic differences identified by Rod Hitchmough (Victoria University of Wellington) and colleagues indicate that there are several species within this group. As taxonomic descriptions and distributions have yet to be finalised, the biology of the forest-shrubland forms is not considered further here (the biology of some better-known grassland forms is discussed in Chapter 9, however). Geckos that are probably not members of the *H. maculatus* species complex and that also do not fit any other known species description continue to be found from parts of southern New Zealand, and it is likely that new species will be described in coming years.

New Zealand skinks in the genus *Oligosoma* were classified until 1995 with related Australian species in the genus *Leiolopisma*. However, genetic studies indicate that the relationship between the New Zealand and Australian forms is not close, and thus the New Zealand species have been placed in a separate genus.

Ten species of *Oligosoma* have been described from southern New Zealand, of which several are sometimes found in or among shrubs. Two of these, *O. notosaurus* and *O. stenotis*, are discussed in Box 15. Among the others, the green skink *O. chloronoton* is sometimes seen in shrubby coastal vegetation as well as in boulder fields and tussock grassland. It is known from Otago, Southland and Stewart Island. The cryptic skink *O. inconspicuum* has also been found in or among shrubs in several locations in Otago and Southland. As both of the latter species are best known from tussock grassland, they will be considered in more detail in Chapter 9.

Fig. 59. The forest gecko *Hoplodactylus granulatus*. Tony Whitaker

16 REPTILES AND AMPHIBIANS LOST FROM SOUTHERN LANDS?

If this book had been written just prior to the arrival of humans in New Zealand, it undoubtedly would have been very different. Since humans arrived, landscapes have been greatly modified and many alien species introduced. As a consequence, some endemic species have disappeared from the southern region of New Zealand, and some have become totally extinct.

Tuatara *(Sphenodon* spp.) are an example of the former. The two surviving species superficially resemble lizards, but are in fact members of a different though related order – the sphenodontians (wedge-toothed reptiles). Sphenodontian fossils have been found in many parts of the globe and appear to have been relatively common reptiles about 190 million years ago. However, for many millions of years they have survived only in New Zealand. Within New Zealand, tuatara were once widespread over both the North and South Islands (their subfossil remains have been found in both Otago and Southland, although not Stewart Island). However, they appear to have declined dramatically following the arrival of the first humans, probably as a consequence of predation and competition from introduced mammals such as the Polynesian rat (kiore) as well as habitat modification and human consumption. By European times, tuatara were virtually extinct on the North and South Islands. Today, they are found only on about thirty offshore islands. By analogy with present-day habitats, tuatara in southern New Zealand may have been most numerous in coastal forest in close association with nesting seabirds. However, they also occurred well inland, including near Alexandra and Wanaka.

Although we are fortunate that tuatara still survive elsewhere in New Zealand, total extinction seems to have been the fate of two rather special frogs. They belonged to the family Leiopelmatidae, which is considered to be amongst the most anatomically primitive of frog families. Like the tuatara lineage, the lineage of leiopelmatid frogs appears to have been isolated in New Zealand since the break-up of Gondwana. Two extinct species have been described by palaeoecologist Trevor Worthy from subfossil material found near Te Anau. One, *Leiopelma auroraensis*, is known with certainty only from this site, whereas the other, *L. markhami*, occurred in other parts of New Zealand. Both were robust species about 60 mm long. Analysis of the relative lengths of their leg bones suggests that they were frogs that walked rather than hopped. They appear to have been more closely related to *L. hochstetteri*, a surviving leiopelmatid species from North Island forests, than to the other three surviving species. Both *L. auroraensis* and *L. markhami* appear to have become extinct since human arrival (within the last 1000 years). Given that they were probably inhabitants of damp forest, and that this habitat is still present in the Te Anau region, their disappearance is attributed to the impacts of introduced species such as kiore. Perhaps – just perhaps – a population of leiopelmatid frogs could still survive undetected in some remote part of Fiordland.

Less likely, but even more dramatic, would be the discovery of a surviving population of the gigantic gecko known as *Hoplodactylus delcourti*. In 1979 a stuffed specimen was found in a French museum. With a snout-vent length of 370 mm, it is 54 per cent larger than the next largest species of gecko. Although no collection site was recorded for the only known specimen, morphological comparisons have led taxonomists to place it in the New Zealand genus *Hoplodactylus*. Possibly it represents the kawakaweau, a creature of Maori lore that was reported to early Europeans as being about two feet long, as thick as a man's wrist, and living beneath the bark of trees. Most accounts of kawakaweau come from the North Island. Although two bones found in Central Otago were once thought to have come from a lizard of this size, one, a reptilian jaw bone, has since been lost, and the identity of the other remains contentious. Recent examination of extensive subfossil material from the South Island by Trevor Worthy and others has yielded no definitive evidence that *H. delcourti* did occur here.

Alison Cree

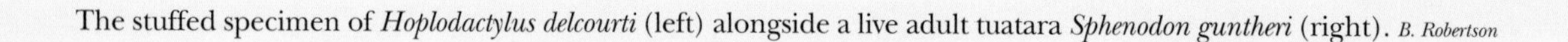

The stuffed specimen of *Hoplodactylus delcourti* (left) alongside a live adult tuatara *Sphenodon guntheri* (right). B. Robertson

BIRDS OF FOREST AND SHRUBLANDS

Hamish Spencer and John Darby

Southern New Zealand's native forests are home to a number of New Zealand endemic birds, the best known of which is the South Island brown kiwi *Apteryx australis australis*. This nocturnal worm-eater is widespread in Fiordland, although its onomatopoeic shrill whistle is heard far more often than the bird is seen. The larger Stewart Island kiwi *A. a. lawryi* (Fig. 60), however, is often about in daylight and some populations forage for invertebrates on sandy beaches. A second flightless endemic, the weka *Gallirallus australis* is found in the same areas but is active in the daytime as well as at dusk. Some wekas in Fiordland, known as black wekas, are much darker than the usual form, which also occurs in Westland and Nelson. Wekas are inquisitive, bold birds, flightless but with strong beaks and legs. Just why their range has shrunk considerably since the arrival of Europeans is unclear; until recently large numbers occurred around human habitation in such places as Gisborne, and they are still common south of Westport.

Parrots were once well represented in the southern forests, but the kakapo *Strigops habroptilus* (Fig. 61) is extinct on the mainland and survives only in island refuges. Box 17 describes the species now extinct or endangered in southern New Zealand. Kaka *Nestor meridionalis* populations are now restricted to Stewart Island, Fiordland, west of the Southern Alps and south-western Southland, and they appear to be declining. These birds nest in tree holes and feed on fruits and insects, which they excavate from decaying wood with powerful beaks. Kea *N. notabilis* forage in the higher mountain forests of Fiordland, especially in the

Fig. 60. Stewart Island kiwi. *E. Brinkmann*

winter. Persecuted for many years as sheep killers, they were afforded partial protection in 1970 and have been fully protected since 1986. The red-crowned parakeet (kakariki) *Cyanoramphus novaezelandiae novaezelandiae* is very much reduced in numbers (if not extinct) in its former strongholds in the South Island, although it survives in good numbers on Stewart Island and a number of subantarctic islands. The yellow-crowned parakeet (kakariki) *C. auriceps* is still found in good numbers in areas of undisturbed beech forest, especially in Fiordland's Eglinton Valley and Waikaia, as well as on Stewart Island (Fig. 62). Both kakariki species are omnivorous, but the red-crowned feeds mostly on seeds, fruits, flowers, shoots and leaves and the yellow-crowned takes mostly invertebrates.

Fig. 61. Male kakapo *Strigops habroptilus* on forest floor at night. *Rod Morris*

Populations of kakariki and of mohua or yellowhead *Mohoua ochrocephala* can fluctuate markedly in response to beech mast years, in which huge amounts of seed are produced. Insect, rat and mice numbers, along with those of kakariki and mohua, boom and subsequently populations of predators such as the stoat *Mustela erminea* also increase. In the year following the mast year, incubating females are often predated by hungry stoats and population numbers collapse. These birds nest in tree holes (Fig. 63) from which escape is impossible; their populations now show male-biased sex ratios.

Mohua and the closely related brown creeper *M. novaeseelandiae* feed in noisy family groups, vigorously investigating any crevices in the tree tops and, in the case of brown creeper, lower branches for insects. Brown creepers are more widespread, however, occurring even in second-growth scrub, for example, on Flagstaff near Dunedin. Flocks of brown creeper in the bush are often accompanied by one or two fantails *Rhipidura fuliginosa*. Up to twenty per cent of populations of this species are the melanic form, known as the black fantail (although it is really dark brown below). Fantails catch their food on the wing and are thus accomplished aerial acrobats (Fig. 64). Together with the grey warbler *Gerygone igata*, they are also found outside forests in suburban gardens with mature trees. Grey warblers, more often heard than seen, usually forage in pairs, picking insects off leaves and twigs. The pear-shaped nest of this species is suspended high up a tree and anchored from below to prevent it swinging. The entrance is on the side near the top.

Southern New Zealand has good populations of the South Island robin *Petroica australis australis*, both in native forest, especially Fiordland, and in exotic forest, such as Douglas fir near Dunedin (Fig. 65). A very tame species, the birds feed on insects and worms uncovered in the leaf litter on the forest floor. The Stewart Island form *P. a. rakiura* is found away from settled areas. The smaller South Island tomtit *P. macrocephala macrocephala* is sometimes called the yellow-breasted tomtit because of the male's colouration, although its breast ranges from an intense orange through pale yellow to almost white. It, too, is a bird of the lower forest, feeding on insects spotted on tree trunks and branches while perching quietly. These two *Petroica species* are thought to be descended from separate invasions of Australian stock. The smallest bush bird (and the smallest of all birds in New Zealand) is the rifleman *Acanthisitta chloris* (Fig. 66), one of two extant members of the endemic family Acanthisittidae, now considered to be only distantly related to other passerine families. The male of this species, unlike most passerines, makes a larger contribution to parental care of the young (especially daytime incubation and feeding) than does the female. They feed in small family groups, which keep in contact with soft high-pitched calls, gleaning insects off leaves, bark and twigs at all levels in the forest. The now completely extinct bush wren *Xenicus longipes longipes* probably died out in Otago and Southland some time in the 1960s. Stead's bush wren *X. l. variabilis* survived on Stewart Island up to 1951 and Big Cape Island until the year following the accidental introduction of ship rats in 1964. Six birds were transferred to Kaimohu Island where they survived until 1972.

Fig. 62 (top left). Kakariki. *J. Newman*

Fig. 63. Mohua male at nest hole in red beech. *M.F. Soper*

Fig. 64 (left). Pied male South Island fantail with outspread tail. *Rod Morris*

17 RECENT EXTINCTIONS AND ENDANGERED BIRD SPECIES

New Zealand has lost much of its avifauna since the arrival of humans. Many species became extinct after the arrival of the Maori, either as a direct result of hunting (e.g. several moa species) or indirectly, as a result of habitat destruction (e.g. the giant eagle *Harpagornis moorei*) or predation by introduced predators such as the kiore *Rattus exulans* (e.g. several species of rail), or some combination of these factors.

A second wave of extinctions occurred as a result of European settlement. Habitat destruction, the introduction of mammalian predators and avian diseases via introduction of exotic species have all played a role. Here we list and briefly discuss the species that are known to have occurred in southern New Zealand since 1800, but are now either extinct or severely endangered here.

Little spotted kiwi *Apteryx oweni*: Probably extinct in Fiordland, surviving in good numbers only on Kapiti Island off the Wellington coast. Recently introduced onto several small islands off the North Island. Mammalian predation is the most likely cause of decline.

Southern crested grebe *Podiceps cristatus australis*: Almost extinct in Fiordland, with about twenty birds known in Otago and Fiordland and about forty in Westland. Larger, but still very reduced populations survive in Canterbury (and significantly greater numbers occur in Australia).

New Zealand dabchick *Poliocephalus rufopectus*: Extinct in the South Island, but not uncommon in the North Island, especially near Rotorua.

Yellow-eyed penguin *Megadyptes antipodes*: Recent catastrophic decline appears to have been arrested by habitat restoration and predator trapping.

Mottled petrel *Pterodroma inexpectata*: Once a widespread mountain breeder, but probably extinct in Otago. Survives on islands off Fiordland and on islands around Stewart Island, including Codfish and Big South Cape. There is a breeding colony on an island in Lake Hauroko. Many other seabird species would also have once bred on the mainland of the South Island. Mammalian predation is the most likely cause of decline.

New Zealand little bittern *Ixobrychus novaezelandiae*: A poorly known species, now extinct. One of the approximately twelve known specimens came from Lake Wakatipu, but most were from Westland.

Brown teal *Anas chlorotis*: Fewer than fifty birds survive in Fiordland. Disappeared from Stewart Island about 1972. Fiordland birds are possibly subspecifically distinct from those of the North Island.

New Zealand quail *Coturnix novaehollandiae*: Extinct since about 1870, probably because of mammalian predation or disease. A grassland species, once especially abundant east of the Southern Alps.

South Island takahe *Porphyrio hochstetteri*: Fewer than 200 birds survive in the Murchison Mountains of Fiordland. An extensive egg-incubation programme has been successful at hatching eggs, but birds introduced to offshore islands have usually failed to raise young. Fossil and subfossil remains suggest the alpine habitat is a relict one, although reasons for the shrinking of the range are unclear.

Southern New Zealand dotterel *Charadrius obscurus obscurus*: Fewer than 200 of this recently differentiated subspecies survive cat predation where they breed on remote mountain tops in Stewart Island. It is thought to have once bred in similar places on the South Island.

Shore plover *Thinornis novaeseelandiae*: Poorly known from the eastern South Island last century. Now confined to Rangitira (South East Island) in the Chathams.

Stewart Island snipe *Coenocorypha aucklandica iredalei*: Known last century from several islands off Stewart Island, but extinct on the last of these (Big South Cape) by the 1960s. Extirpation caused on some islands by introduction of weka, on Big South Cape by ship rats *Rattus rattus*.

Black stilt *Himantopus novaezealandiae*: Now found in very small numbers on the Ahuriri River in Otago. Mammalian predation and, more recently, competition with the pied stilt *H. himantopus leucocephalus* for mates are the main threats. An intensive breeding programme in Twizel has had some success, but chick survival in the wild is low.

Kakapo *Strigops habroptilus*: Extinct in southern New Zealand, except for Codfish Island. All surviving birds are descended from those caught on Stewart Island in the early 1980s. Survived in Fiordland until about this time. Attempts last century at conservation on Resolution Island by Richard Henry failed because of the ability of mustelid predators to swim to the island. One of New Zealand's most endangered species. A novel supplementary feeding programme for adult birds, aimed at breeding a higher proportion of female to male chicks, produced startling results in 2002, and may well be the factor that ultimately ensures the survival of this very rare species.

Orange-fronted parakeet *Cyanoramphus malherbi*: Until recently considered a form of the yellow-crowned kakariki *C. auriceps*, this species now occurs only in small parts of Canterbury. It was never common in European times, but records exist for Otago and Fiordland.

Laughing owl *Sceloglaux albifacies*: Extinct throughout New Zealand since the 1920s, probably due to mustelid predation. Once a common bird in open country in Otago.

Bush wren *Xenicus longipes*: Disappeared from the South Island gradually. Probably extinct since the late 1960s. The Stewart Island subspecies, Stead's bush wren *X. l. variabilis* disappeared after ship rats were introduced to its last stronghold on Big South Cape Island.

South Island kokako *Callaeas cinerea cinerea*: Almost certainly extinct, in spite of persistent unconfirmed reports from Stewart Island and parts of Fiordland. Last confirmed sightings on the mainland from the early 1960s.

South Island saddleback *Philesturnus carunculatus carunculatus*: Reduced by the 1950s to the South Cape Islands off Stewart Island. Rescued from certain rat-induced extinction in 1964, they are now present (in southern New Zealand) on some small islands off Stewart Island and Breaksea Island in Fiordland (from which mammalian predators have recently been removed).

South Island piopio *Turnagra capensis*: Once an abundant and tame bird, now extinct (since some time in the first half of the twentieth century), probably because of mustelid predation. Now thought to be related to Australian bower birds.

Hamish Spencer and John Darby

Adult male takahe in snow. *Rod Morris*

18 RECENT BIRD ARRIVALS

The exchange of bird traffic between New Zealand and Australia tends to be a one-way process, with twenty-five families of Australian birds represented in New Zealand. Many of New Zealand's well-known species are either shared with Australia (e.g. the fantail) or are derived from Australian ancestors (e.g. the New Zealand robin). Many other species have arrived more recently, however.

The **white-faced heron** *Ardea novaehollandiae* is now the most common heron in New Zealand, despite the fact that it was not recorded as breeding until 1941. Found in almost every wetland habitat to the bushline, it is most frequently seen on rocky and sandy shores around inlets and estuaries.

Also from Australia are a number of birds that are regular but non-breeding visitors. These include the **cattle egret** *Bubulcus ibis coromandus.* Much smaller than the kotuku or white heron, it is most often associated with cattle and sheep, as it follows them and feeds on insects disturbed by their feet. Sometimes found with cattle egrets is the **glossy ibis** *Pelgadis falcinellus,* although this species also feeds in wetter areas such as coastal lagoons. Yet another white heron, the **little egret** *Egretta garzetta* is also a regular Australian visitor. The **white-winged black tern** *Chlidonias leucopterus* is recorded almost annually in coastal lagoons and sometimes inland (e.g. on the Waitaki River).

The **royal spoonbill** *Platalea regia* is a much more recent introduction to Otago. First recorded as breeding at Okarito on the West Coast, a small breeding group was discovered on Maukiekie Island off Moeraki in the early 1980s. Shortly after they were found to be breeding on Green Island off Otago Peninsula, where there is now a resident population of 20 or more nests. They have been breeding on Omaui Island in Southland since 1992.

The **Australasian coot** *Fulica atra australis,* confirmed as breeding in New Zealand at Lake Hayes in Otago in 1958, has since spread throughout much of New Zealand, although the population is still small. They are quite distinctive, with their white frontal plate, but sadly are frequently shot during the duck hunting season.

The **black-fronted dotterel** *Charadrius melanops* is a self-introduced species from Australia and has been recorded as breeding on the Manuherikia in Otago and by 1980 in Southland on the Aparima, Oreti, and Mataura Rivers, and in the Te Anau area on the Mararoa and Lower Whitestone Rivers.

The first **spur-winged plovers** *Vanellus miles novaehollandiae* to arrive in New Zealand appeared to have made landfall near Invercargill in 1932. By 1950 there were over 100 within a 16-km range of Invercargill and by 1965 they were well established in Otago and are now distributed throughout the North Island. Their raucous cry and yellow wattle from eye to base of bill provide a distinctive key to their identification.

The swift darting flight of the **welcome swallow** *Hirundo tahitica neoxana* still arouses interest in many Otago and Southland members of the public, attesting to its relatively recent arrival in these parts. Although a straggler was first seen on the Auckland Islands in 1943 it was not until September 1953 that it was seen on Stewart Island. Sightings in Southland in 1963 were apparently of a juvenile bird, and it was not until the early 1970s that the species was considered to breed in Otago and Southland.

One of the most attractive birds of garden and food station, the **silvereye** *Zosterops lateralis* became the subject of an extensive pioneering banding programme by Professor B.J. Marples of the University of Otago in the late 1930s, followed up by an equally interesting survey and banding programme by Dr J. Kikkawa, who estimated that the winter population of silvereyes was boosted by up to 30 per cent by those generous souls who maintained bird tables. Known as tauhou (stranger) by the Maori, silvereyes were not known to breed in New Zealand until 1862. Since then they have become widespread throughout the country.

Two species of small grebes from Australia were first recorded breeding in the southern South Island. The **Australian little grebe** *Tachybaptus novaehollandiae* first appeared on Lake Wanaka in 1968, and the **hoary-headed grebe** *Poliocephalus poliocephalus* in Southland in 1976. The first species still occasionally breeds in Otago, and has become well established in parts of the North Island, but the second remains uncommon. The hoary-headed grebe is closely related to the native New Zealand dabchick *P. rufopectus,* which is now extinct in the South Island.

Although about 100 **black swans** *Cygnus atratus* were introduced from Australia between 1864 and 1868, the rate of increase in population size suggests that natural immigration was also occurring. This species is now abundant in the coastal lakes and lagoons of Otago and Southland. The **chestnut-breasted shelduck** *Tadorna tadornoides* arrived in New Zealand from Australia some time in the 1970s, and there are now regular widespread reports of this species from many parts of the region (e.g. Hawksbury Lagoon, Waimatuku). The **white-eyed duck** *Aythya australis,* closely related to the endemic New Zealand scaup *A. novaeseelandiae,* shows the opposite pattern: there are numerous records from the nineteenth century, but just one from the region (Southland) in the past hundred years (1991). Similarly, the **red-necked avocet** *Recurvirostra novaehollandiae* was widely reported in the 1800s (e.g. Invercargill), but there are no recent sightings from Otago or Southland.

Individual arrivals

Many Australian species have been recorded in the region on the basis of a few individuals, but have subsequently failed to become established. Such species are the Australian pelican *Pelecanus conspicillatus* in Southland in 1977, Australian white ibis *Threskiornis molucca* at the Southland lagoons in 1957, nankeen night heron *Nycticorax caledonicus* at Owaka in 1980 and the upper Taieri in 1988, plumed whistling-duck *Dendrocygna eytoni* at Kaitangata in 1871, Cape Barren goose *Cereopsis novaehollandiae* in Fiordland in 1947, 1967 and 1990 and the Ahuriri River in 1966, Australian wood duck *Chenonetta jubata* at Wanaka and Wairaki in Southland in 1910 and 1944, dusky moorhen *Gallinula tenebrosa* on Lake Hayes in 1968, black-tailed native-hen *G. ventralis* at Colac Bay in 1923, oriental pratincole *Glareola maldivarum* on Stewart Island in 1963 and Ruapuke Island in 1988, gull-billed tern *Gelochelidon nilotica* along the northern coast of Foveaux Strait in several years since 1955, oriental cuckoo *Cuculus saturatus* at Te Anau in 1902 and Kapuka in 1983, pallid cuckoo *C. pallidus* at Craig Flat on the Clutha River in 1941 and Omarama in 1990, fan-tailed cuckoo *Cacomantis flabelliformis* at Wanaka in 1990, channel-billed cuckoo *Scythrops novaehollandiae* at Invercargill in 1924, spine-tailed swift *Chaetura caudacuta* at Milford and Horseshoe Bay, Stewart Island in 1942, the Hokonui Hills in 1968, Hampden in 1977, the Catlins in 1978–1979 (when at least 60 birds were present) and Stewart Island in 1979, fork-tailed swift *Apus pacificus* near Invercargill in 1960, dollarbird *Eurystomus orientalis* in Fortrose Southland in 1967, Australian tree martin *Hirundo nigricans* at Oamaru in 1893 (when a pair apparently bred), Taieri Plain in 1981–1984 and the Eglinton Valley in 1983, fairy martin *H. ariel* also on the Taieri Plain in 1981–1984, black-faced cuckoo-shrike *Coracina novaehollandiae* at Invercargill in 1870 and 1976 and Tarras in 1990, white-winged triller *Lalage tricolor* on Otago Peninsula in 1969, and white-browed woodswallow *Artamus superciliosus* and masked woodswallow *A. personatus* at Naseby in the early 1970s.

John Darby and Hamish Spencer

Two species of cuckoo are nest parasites of several of these small bush birds. Interestingly, both cuckoos are migrants, the commoner shining cuckoo *Chalcites lucidus* wintering in the Solomon Islands and the larger falcon-like long-tailed cuckoo *Eudynamis taitensis* over a wide area of the South Pacific. They arrive in southern New Zealand and Stewart Island in September-October and depart again in February. Shining cuckoos apparently prefer to lay their eggs in grey warbler nests, whereas long-tailed cuckoos choose the nests of brown creeper and mohua. Both species have also been known to lay in the nests of tomtit, silvereye *Zosterops lateralis*, fantail and some introduced passerines. The shining cuckoo is found in bush as well as in gardens and scrub, whereas the long-tailed cuckoo is mostly restricted to the bush, where it is sometimes mobbed by tui and bellbirds.

The kereru or New Zealand pigeon *Hemiphaga novaeseelandiae novaeseelandiae* is another species without close relatives outside New Zealand. Although usually thought of as a bush bird, this species has adapted to European settlement, and is commonly seen in city suburbs with mature trees (page 153). It feeds on a wide variety of fruits, flowers and young leaves. Some native plant species, including miro, tawa, taraire, karaka and puriri, are dependent on kereru for their seed distribution.

The bellbird *Anthornis melanura melanura* is a second bush bird common in town areas with sufficient tree cover (Fig. 67). The tui *Prosthemadera novaeseelandiae* is also widespread, but less common in Dunedin and some parts of Fiordland dominated by *Nothofagus*. Both of these honeyeaters forage for nectar, fruit and insects, and aggressively chase away competitors, including silvereye. This last species is one of the few self-introduced Australian bush birds and remains the only successful self-introduced bush bird: the first Otago flocks date from 1860, although the first record is from Milford Sound in 1832. Less common in large stands of bush, it can be abundant in almost any area with trees. In winter, large flocks form along the Otago coast and they will readily come to feed at bowls of sugar-water. Other recent arrivals are discussed in Box 18.

Few artificially introduced birds are common in native forest, with those present often restricted to the fringes, or disturbed areas. Such species are the dunnock *Prunella modularis* and blackbird *Turdus merula*, which keep close to the ground, foraging for insects and other invertebrates, and the redpoll *Carduelis flammea*, which flocks around the tree tops. The most successful European species is undoubtedly the chaffinch *Fringilla coelebs*, which can be found in the deepest bush and almost every other habitat from farmland to suburban gardens and beaches.

Birds of prey are not often seen in the New Zealand bush. The endemic New Zealand falcon *Falco novaeseelandiae* is present in small numbers in Fiordland, but is commoner outside the bush in the tussock lands of Central Otago. The Australasian harrier *Circus approximans* is sometimes seen soaring above the canopy, but seldom enters the forest. The most common avian predator in southern New Zealand and Stewart Island, although rather rare in the Dunedin area, is the morepork *Ninox novaeseelandiae*. Often mobbed by flocks of tui, bellbirds and smaller bush birds during the day, it feeds from dusk on insects and spiders and occasionally small birds, rats, mice and lizards.

Fig. 65 (above). Male South Island robin on forest floor. *Rod Morris*

Fig. 66 (below). The rifleman is New Zealand's smallest bird. *M.F. Soper*

Fig. 67. Endemic songster, adult male bellbird on flax flowers. *Rod Morris*

MAMMALS OF FORESTS AND SHRUBLANDS

Seals (Chapter 12) and bats are the only terrestrially breeding mammals that are native to New Zealand. Even then, we have a relatively depauperate collection of both. Of the three species of bat (Order Chiroptera), one is already extinct (greater short-tailed bat, *Mystaina robusta*), one is found now only in remnant pockets of its original distribution (lesser short-tailed bat, *M. tuberculata*), and the only one with anything like a New Zealand-wide distribution (New Zealand long-tailed bat, *Chalinolobus tuberculatus*) is rarely seen by humans.

Populations of the lesser short-tailed bat exist on Codfish Island and Solander Island and a small enclave has been recently discovered in Fiordland. The long-tailed bat is more widely distributed but mainly in Fiordland and in a relatively small area of forest east of Bluff. It is a small bat, the length of its head and body being no more than 5 cm and its wingspan less than 28 cm. A tail that is almost as long as its body distinguishes it from the short-tailed bats, as do its relatively small ears. Its coat colour is dark brown to black. The bats roost in trees or in the crevices of caves, with accumulations of guano often being the only obvious field sign of their presence. They can be seen flying from dusk to dawn when they capture insects such as moths and mosquitoes on the wing.

All other mammals in New Zealand are foreigners. They are discussed in Chapter 7.

19 STUDIES OF NATIVE BATS IN FIORDLAND

Research in the Eglinton Valley, Fiordland, is helping to unravel the many myths that surround New Zealand's only native land-breeding mammal. Until recently, only the long-tailed bat *Chalinolobus tuberculatus* was known to exist in the Eglinton Valley. However, early in 1997 researchers based at Knobs Flat discovered short-tailed bats *Mystacina* sp. living nearby, representing the first records of these bats in Fiordland since 1871. This makes the Eglinton Valley only the second region in the South Island where short-tailed bats are known to exist, the other being Kahurangi National Park near Nelson. Compared to lesser short-tailed bats *M. tuberculata* on Codfish and Little Barrier Islands, the Fiordland bats are heavier, have larger wings, smaller ears, and are sexually dimorphic. In February, roosting groups numbered from 107 to 279 individuals and the bats ranged over 130 square km of the valley.

The Eglinton Valley represents one of the few sites in New Zealand dedicated to long-term studies of the ecology of long-tailed bats. Researchers have been investigating factors which potentially limit long-tailed bat populations and have gathered detailed information on the population structure, breeding biology, movement, social structure, habitat use and roosting ecology. Long-tailed bats forage along forest edges, in canopy gaps, and in summer, over manuka shrubland. Social groups range over large geographic areas when foraging (over 100 square km), often flying 10–15 km to their favoured feeding areas. Small groups of bats (usually fewer than 70 individuals) select small cavities in old red beech and standing dead trees for roosting and raising their young. Female long-tailed bats begin breeding at 2–3 years of age and give birth to a single young each year. Long-tailed bats are unusual in that they move to a new roost tree virtually every day.

The Eglinton Valley has also been an important study site for scientists investigating the echolocation abilities of New Zealand bats. Much of the work carried out has helped scientists to develop and refine acoustic survey techniques. Research carried out includes studies of how well sounds in the frequency range used by bats travel through different habitats, how survey results might be affected by the type of equipment used to 'listen' for echolocation calls (known as 'bat detectors'), and how the activity patterns of bats may be better quantified. Recordings made of the echolocation calls of long-tailed bats have also been used to investigate why echolocation calls change as a bat approaches and subsequently attacks a flying insect, thus allowing scientists further insight into how bats use sound to capture prey.

Below: How do the echolocation calls of a bat change as it approaches a potential prey item? The calls recorded here were from a long-tailed bat as it approached and attacked an insect. After detection of the insect (A) the calls become shorter but more broad-band, allowing the bat to gain as much information about the target as possible. If the bat decides the target is worth attacking (B) it begins to produce terminal phase calls. Terminal phase calls tend to be more narrow-band and shorter in duration than either search or approach phase. This change in call structure provides the bat with important information about the relative velocity of the target. Note how call repetition rate also increases as the bat moves towards the target thus providing as much information about target movement as possible. Call repetition rate in the buzz is so high that individual calls cannot be distinguished.

Stuart Parsons, Colin O'Donnell and Jane Sedgely

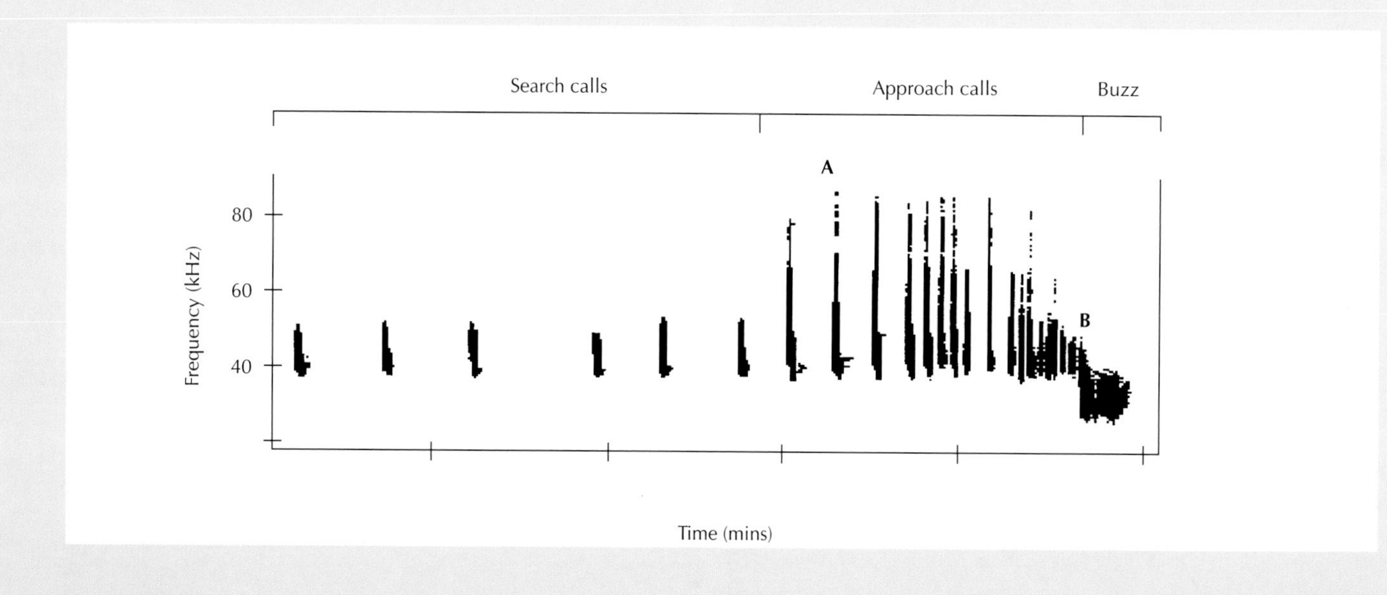

CHAPTER 9

TUSSOCK GRASSLANDS AND ASSOCIATED MOUNTAIN LANDS

ALAN MARK, WILLIAM LEE, BRIAN PATRICK, ALISON CREE, JOHN DARBY AND HAMISH SPENCER

The spectacular tussock grasslands of Danseys Pass. *Neville Peat*

1 CLIMOSEQUENCE OF SOILS, BASED ON THE OLD MAN RANGE IN CENTRAL OTAGO

Brown soils (yellow-brown earths)
(brown subsoil colours)

- Precipitation usually more than 800 mm except on stony valley floors.
- Naturally moderate to high organic matter contents.
- Strongly leached. Low contents of basic cations. Commonly acidic.
- Brown colour due to iron oxide dispersed throughout the soil mass.
- Rooting depths generally not physically restricted and usually free draining.
- Low natural fertility with deficiencies of nitrogen, sulphur, phosphorus and molybdenum.
- Production limited by low temperatures.

Allophanic brown soils
Acidic, very strongly leached.
The non-crystalline mineral allophane imparts a porous, low density, crumb structure.

Acid brown soils
Acid, pH<4.8 in subsoil

Orthic brown soils
Ordinary brown soils

Pallic soils (yellow-grey earths)
(pale soil colours)

- Rainfall between 500 mm and about 800 mm/year.
- Moderately leached, Droughty summers, moist winters.
- Naturally low to moderate organic matter level.
- Production is limited by water deficits.
- Low cation exchange capacity.
- Moderate fertility. Sulphur and phosphorus and molybdenum commonly deficient.
- Rooting depth and drainage limited by high density subsoil or by rock, except on many steeplands where rooting is not restricted.

Semi-arid soils (brown-grey earths)
(semi-arid climate)

- Rainfall 350 to 500 mm/year, weakly leached.
- Naturally low organic matter content with low nitrogen and sulphur reserves.
- Low cation exchange capacity.
- High availability of nutrients to plants, but nutrients in reserve are limited due to low organic matter.
- Rooting depth is commonly limited by moderate to high density subsoils.
- A layer of calcium carbonate commonly occurs in deeper subsoils.
- Production is severely limited by water deficits.

MOUNTAIN TOR LANDSCAPE

RIPPLE LANDSCAPE

FOOTHILL TOR LANDSCAPE

1500

1000

500

1. Alpine herbfields and cushion plants; Obelisk soils.
2. High altitude snow tussock; Carrick hill soils.
3. Mid-altitude snow tussock; Tawhiti soils.
4. Low-altitude fescue tussock with scattered matagouri; Blackstone soils.
5. Semi-desert dominated by scabweed and depleted fescue tussock; Conroy hill soils.

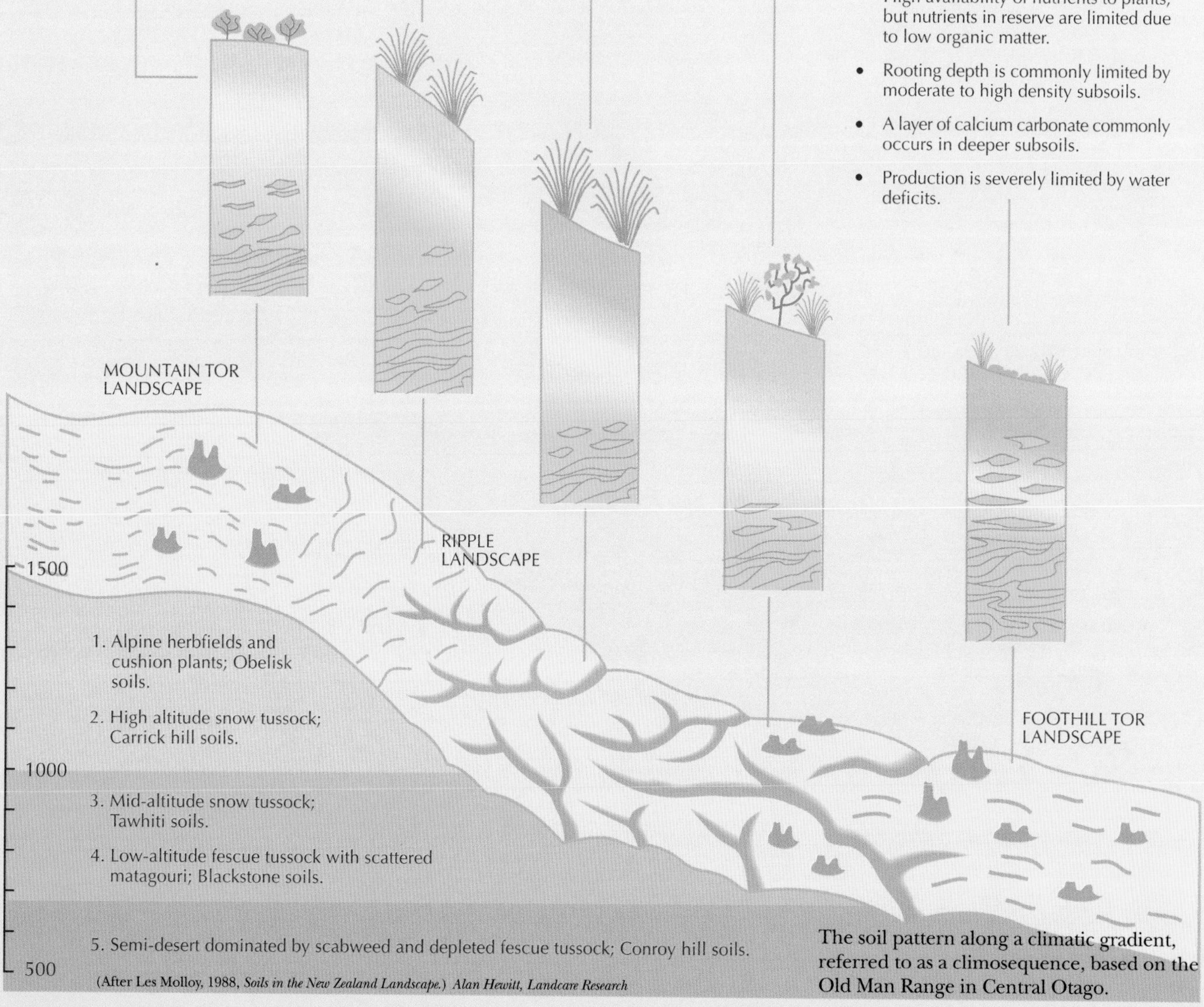

(After Les Molloy, 1988, *Soils in the New Zealand Landscape.*) *Alan Hewitt, Landcare Research*

The soil pattern along a climatic gradient, referred to as a climosequence, based on the Old Man Range in Central Otago.

TUSSOCK GRASSLANDS AND ASSOCIATED MOUNTAIN LANDS

Indigenous grasslands, dominated by species of *Chionochloa, Poa* and *Festuca,* exhibit their maximum extent and diversity in southern New Zealand. They generally occur wherever environmental conditions are beyond the tolerance of woody species, usually associated with relatively low temperatures, greater moisture extremes and strong maritime influence. Thus natural grasslands occur in the mountains above tree limit and on seasonally frosted valley floors. Above the limits of the alpine tussock grasslands, the short plant cover of the high-alpine zone varies largely in relation to climate, parent rock material and topography. Grasslands are present in the semi-arid zone and may also occupy poorly drained soils within forest.

Climates suitable for forest do not preclude the occurrence of grasslands, however. Moisture and temperature are the most critical aspects, with the mean air temperature during the warmest month needing to reach at least 10°C to support a forest cover. Tussock grassland communities cover the slopes and basins of the rainshadow mountains east of the Main Divide, in areas climatically suitable for forests and where there is clear evidence (see Chapter 6) for their displacement of woody vegetation. These forests were destroyed through the influence of periodic burning, mostly during the last millennium when fires associated with, and probably resulting from, human occupation were both more widespread and frequent than natural fires during the pre-human period. Evidence from pollen records and charcoals from several sites in Central Otago indicates such grassland establishment dating back some 2500 years. Tracing the local and more general extent of woody vegetation at the time of human settlement relies on a range of evidence. Such information includes surviving relict stands of mostly beech forest, remains of forest in the form of buried charcoals, surface wood of mainly durable Hall's or mountain totara *Podocarpus hallii* (now much less evident because of its use for fuel and fence posts) and forest 'dimples' (surface mounds and adjacent depressions left by upturned trees). Above the limit (about 1050 m) of the now-destroyed forest, the lower reaches of the alpine zone were naturally dominated by tall tussock grassland that persists, at least locally, in the rainshadow regions and more generally in the higher rainfall regions along and west of the Main Divide where forest has generally persisted.

Determining the pre-human boundary between woody vegetation and the low-alpine tussocklands throughout the rainshadow region of the Southern Alps is now difficult. With much of the woody cover removed, the transition from the semi-arid lowland grasslands to the alpine communities is blurred through upward migration by the former and downward migration by the latter. The soil pattern in this region, however, continues to reflect the climatic pattern (Box 1) and may also assist with the interpretation of vegetation changes associated with human settlement (Box 2).

The picture is no less modified in some of the higher rainfall regions, especially at low altitudes. Extensive areas of forest were destroyed by fire, probably lit by the early Maori settlers, on the plains and surrounding hills of Southland. Tussock grassland, mostly dominated by copper tussock *Chionochloa rubra* ssp. *cuprea,* eventually replaced most of the destroyed forest, spreading from its natural habitats on permanently moist, often peaty depressions below treeline. Such areas persist, typically as natural bogs scattered across the Southland Plains and also in the Murchison Mountains of Fiordland, notably Takahe Valley (Fig. 1).

It is clear that throughout the eastern grasslands there was substantial modification to the natural plant cover following Polynesian settlement, especially through fire. Even so, the plant cover at this time was entirely indigenous. European settlement in the 1840s proved extremely disruptive, due to land development combined with the introduction of a wide range of plants and animals, several of which have become major pests. For these reasons we base our account on the vegetation pattern that prevailed in the 1840s (Fig. 1 in Chapter 8), prior to the major influences that have moulded the varied landscapes we see today (Fig. 2 in Chapter 8). Later in the chapter we describe the fauna associated with grasslands and mountain lands.

Fig. 1. Copper tussock grassland on the floor of Takahe Valley at about 900 m, Murchison Mountains, Fiordland National Park. The grassland occupies the permanently moist depression which experiences severe frosts in winter. Note the sharp lower treeline and the upper treeline at about 1100 m. *William Lee*

2 CHANGES IN THE VEGETATION PATTERN OF THE RAINSHADOW REGION ASSOCIATED WITH HUMAN SETTLEMENT

The pattern of indigenous vegetation in the rainshadow region to the east of the Southern Alps has changed over the last thousand years in response to two separate phases of human settlement. The overall patterns of change can be determined with a reasonable degree of certainty and are of some interest in interpreting the vegetation patterns we see around us today. The time scale begins at approximately A.D. 1000, with the probable pattern immediately prior to human occupation when the vegetation was closely related to the major physical factors, notably temperature and rainfall, as well as with the associated soil types. These factors can still be assessed today and have changed little over the last millennium, despite theories of climate change invoked by some workers. The present environmental patterns are shown on the right side of the diagram below.

The major influence of fire during the period of Polynesian occupation from about the twelfth century, has been broadly documented from a range of evidence, including buried charcoals, surface logs, forest dimples, soil profile patterns and pollen preserved in peat bogs. We now know that forest was once widespread over all but the driest parts of the intermontane valleys, i.e. where yearly rainfall is less than about 400 mm and soils have the features of brown-grey earths, and also above the treeline which occurs where the mean air temperature for the warmest months is close to 10°C. Treeline elevation in the region decreases from about 1200 m to 800 m with increasing latitude and proximity to the coast, or decreased continentality.

Descriptions of the changing pattern of vegetation during the early decades of European settlement are limited and only a generalised pattern of the present vegetation and causal factors can be presented in a diagram of the scale shown here. Also included is the accepted baseline of 1840 (immediately pre-European) for the assessment of nature conservation values aimed at a more representative system of protected natural areas.

Alan Mark

Changes in the vegetation pattern of Otago, eastwards from the Lakes District, associated with human settlement. Major climatic and soil factors are shown (mean annual precipitation, mean air temperature for warmest month, and soil classification): brown-grey earths (BGE), yellow-grey earths (YGE), yellow-brown earths (YBE), upland yellow-brown earths (U YBE) and alpine yellow-brown earths (A YBE)) as they relate to the pre-European vegetation pattern of 1840. The generally accepted baseline for the Protected Natural Areas Programme (PNAP) is also shown.

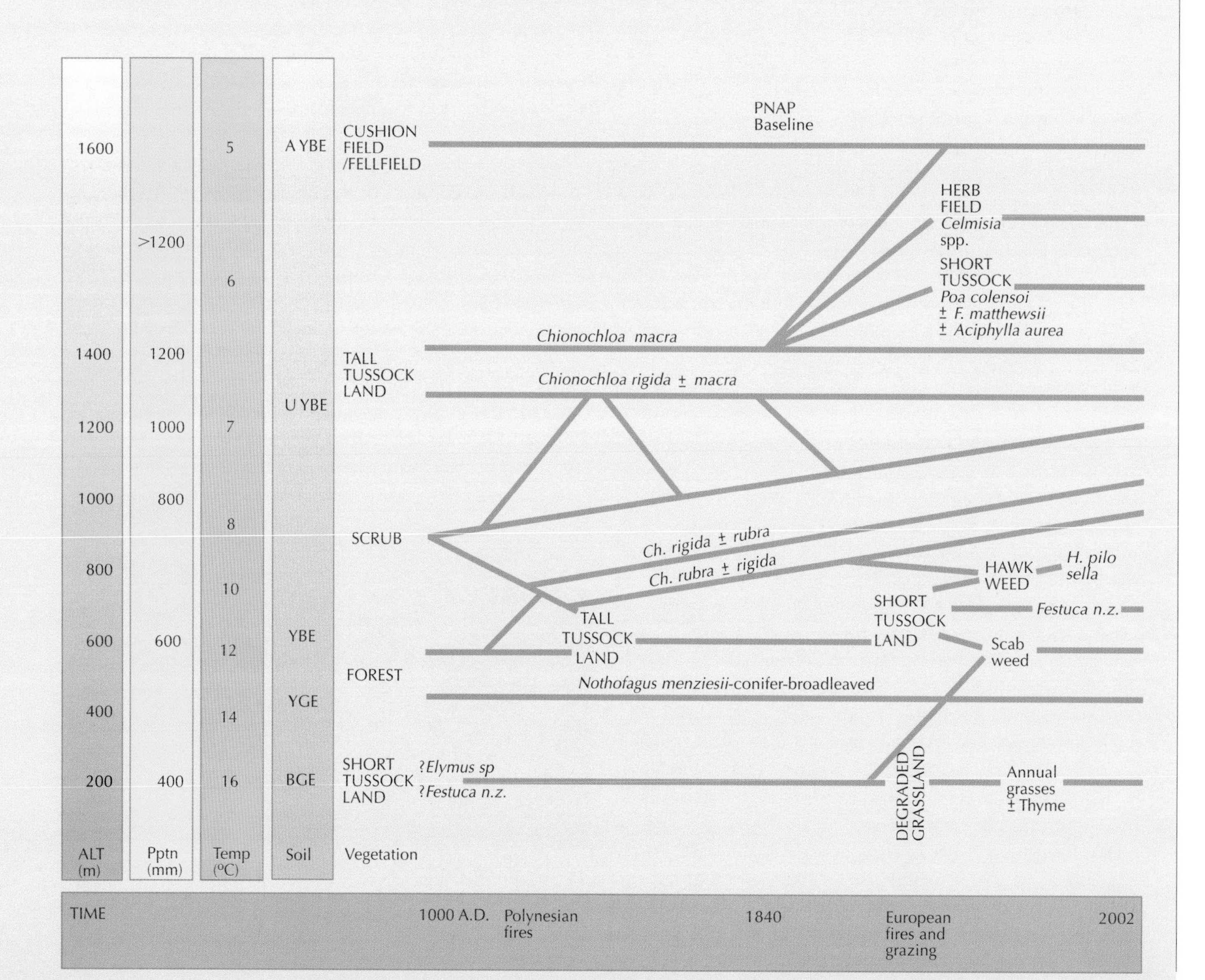

GRASSLAND AND MOUNTAIN LAND VEGETATION PATTERNS

The extent, patterns and composition of the indigenous grasslands and associated mountain lands are described in two main sections to reflect substantial differences in climatic conditions and broad vegetation patterns, albeit largely induced by early human influences (Fig. 1 in Chapter 8). The drier rainshadow region to the east of the Main Divide is discussed first, followed by the generally wet western regions close to the Main Divide, including Stewart Island, before we turn to a treatment of the nival zone (permanent ice and snow cover).

The rainshadow region

The rainshadow region of southern New Zealand (Fig. 2) has a good representation of the three main types of indigenous tussock grassland that were present at the time of European settlement. These occur from the lowland to low-alpine zones. The high-alpine zone is also represented here with some distinctive plant communities (see Box 3 for details of the overall pattern).

Lowland-montane short tussock grasslands

The dry valley floors of the region's interior basins and the adjacent lower mountain slopes were dominated by short tussock grasslands, probably to a modest elevation of perhaps 500 m. Lowland hard or fescue tussock *Festuca novae-zelandiae* and silver tussock *Poa cita* were probably among the main dominants, as they are today, grading into tall tussock grassland on the mid and upper slopes. However, early accounts in the late 1800s suggest that the highly palatable blue wheatgrass *Elymus solandri* was also prominent in this grassland type. Major changes probably occurred in these grasslands at a very early stage in response to the initial pastoral practices of repeated burning and the grazing of domestic stock (sheep and cattle), as well as the first wave of devastation by rabbits. These practices favoured the less palatable plants, such as the hard and silver tussocks, while the more preferred plants such as blue wheatgrass would have declined. Combined with this, the grazing of palatable and nutritious new growth of the narrow-leaved snow tussock *Chionochloa rigida* after fire would have caused loss of vigour and widespread tussock death. Such effects over the last century have led to an advance upslope by the short tussock species which also tolerate fire but, in contrast to snow tussock, remain quite unpalatable to stock.

Blue wheatgrass is currently increasing in importance but only locally, in areas protected from stock grazing, such as on the lower slopes of the Eyre Mountains and at Flat Top Hill Conservation Area (Fig. 3), on the Clutha Valley floor near Alexandra. The replacement of snow tussock by fescue tussock continues (Fig. 4), to an elevation of 1000 m or more on the drier sunnier sites, as on the north-west slopes of the Dunstan Mountains. In some places, snow tussock persists close to the valley floors but only very locally, as on the cool southerly aspect slopes of the Rock and Pillar Range, at less than 600 m elevation.

The valley floors of the drier interior basins were not always covered entirely in grassland. Indeed, the earliest accounts describe shrublands or woodlands where few persist today.

Fig. 2. View west up the Matukituki Valley to Mt Aspiring and the Main Divide showing the reduction in forest cover associated with the increased desiccation of the rainshadow region. Narrow-leaved snow tussock *Chionochloa rigida* grassland dominates in the foreground on the slopes of Mt Alta, at about 1200 m elevation. *Alan Mark*

Fig. 3. Flat Top Hill Conservation Area just south of Alexandra, as viewed from State Highway 6, showing a range of vegetation types, with both indigenous (e.g. scabweed) and exotic (purple thyme, bright green stone crop and sweet brier) plant communities on different sites in a typical lowland (about 500 m) Central Otago landscape. Photographed soon after its acquisition for conservation in December, 1992. *Alan Mark*

Fig. 4. Mixed fescue-snow tussock grassland at about 800 m on the lower slopes of the Old Man Range, Central Otago. The smaller and paler hard tussock, being much less palatable to stock than the snow tussock, even after fire, is gradually displacing it over time as a result of pastoral farming. *Alan Mark*

3 PRESENT PATTERN OF TUSSOCK GRASSLANDS AND ASSOCIATED MOUNTAIN LANDS IN THE RAINSHADOW REGION OF OTAGO

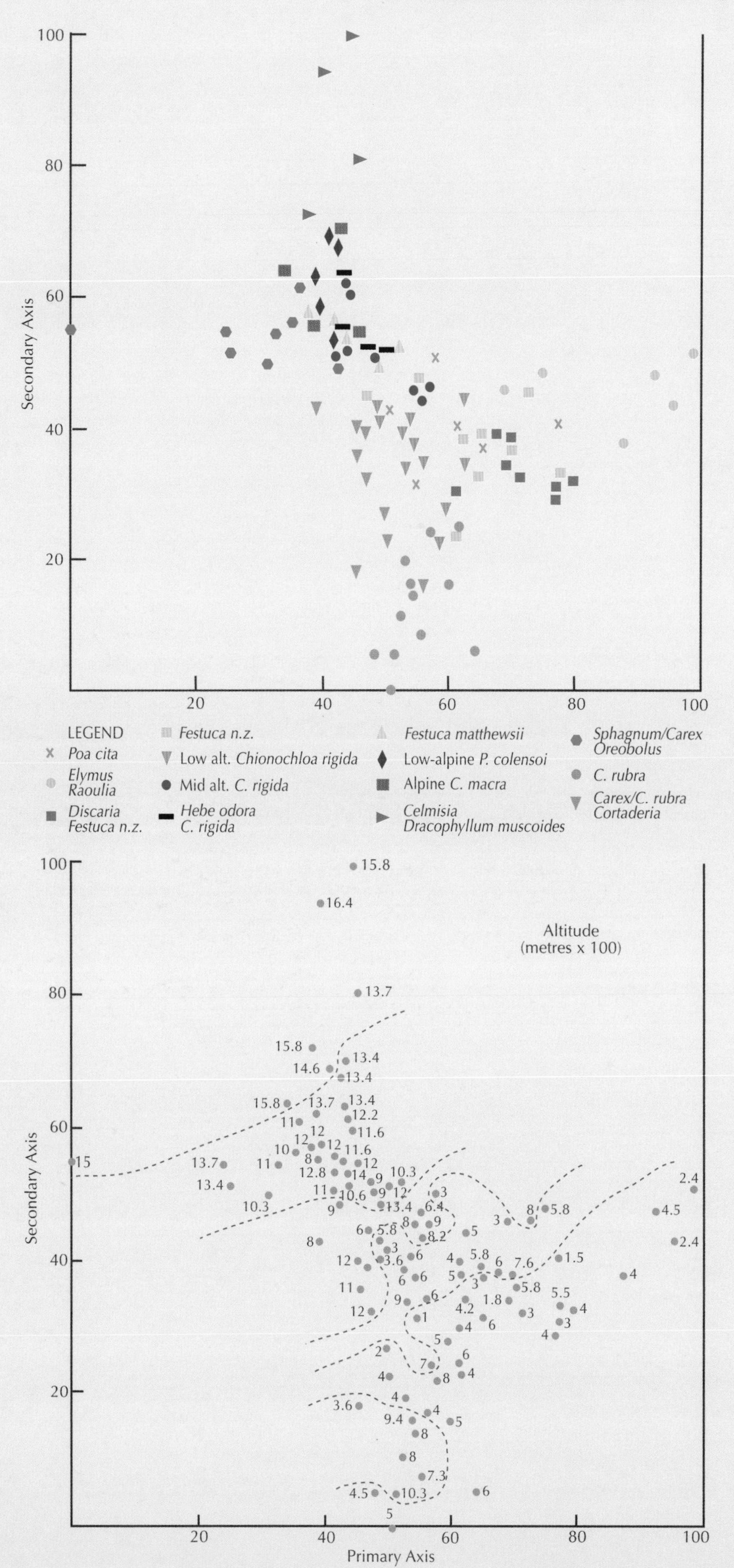

A broad-brush study of tussock grasslands and associated mountain land communities in the rainshadow region of Otago by Gavin Daly, and one of the first in New Zealand to use the ordination method of analysing vegetation information, was based on sampling a representative series of 99 stands. Habitats ranged from arid to semi-aquatic and also had a very wide temperature range (reflected in elevation: from 90 m to 1650 m). Species and their cover were both recorded and soil moisture and other environmental factors assessed. The 99 stands were then ordered in a two-dimensional pattern of 'vegetation space', with the primary (horizontal) axis expressing the greatest amount of variation in the data and the secondary (vertical) axis reflecting the greatest amount of the remaining variation.

The 14 vegetation types recognised on the basis of the dominant species showed a fairly consistent pattern, with red tussock grassland located at the base of the secondary axis, low- and mid-altitude narrow-leaved snow tussock grassland nearer the centre and high-alpine *Celmisia* herbfield-cushionfield communities near the top. By contrast, *Sphagnum-Carex-Oreobolus* wetlands were near one extreme of the primary axis and a blue wheatgrass-scabweed community was near the other (top). When scores for altitude (bottom) and soil moisture (opposite) were plotted on the same ordination, their correlation with each of the two vegetation axes was confirmed. Altitude, which showed a somewhat skewed pattern, was assumed to be acting through its influence on both temperature (affecting mostly the primary axis) and precipitation (particularly as it affected soil moisture that largely determined the secondary axis).

Alan Mark

Top: A two-dimensional pattern of 'vegetation space' for the tussock grassslands and associated mountain lands of the rainshadow region of Otago. The symbols identify the positions of the 14 vegetation types recognised among the 99 stands that were sampled. They are named on the basis of their dominant species.

Bottom: Altitude for each of the 99 stands of vegetation, plotted on the same two-dimensional pattern shown above. Since altitude influences both temperature and rainfall, it shows a relationship to both the primary and secondary axes.

Opposite: Moisture scores shown for each of the 99 stands of vegetation, plotted on the same two-dimensional pattern as shown above. Moisture was scored subjectively on a range of 0 (arid, excessively drained) to 12 (semi-aquatic, poorly drained). It can be seen to have a strong relationship with the primary axis.

Species of the native broom *Carmichaelia*, *Coprosma* and *Hebe*, as well as *Corokia cotoneaster*, *Olearia virgata*, manuka *Leptospermum scoparium* and matagouri *Discaria toumatou* were recorded (or predicted to have grown here) alongside streams. These woody species were in demand as fuel, especially during the gold rush in the nineteenth century, as the name of Firewood Gully close to Cromwell testifies. Some of these stands remain, but the present shrubs are hardly of a size to convert to charcoal for commercial use, as occurred in the 1860s. Relict stands of tall open matagouri woodland, such as probably occurred along these valley floors, persist today in the lower Dart Valley near Paradise, giving us a picture of the likely stature of the original communities.

Pastoral practices of the early European settlers caused an extension of the range of short tussock grassland upslope, at the expense of snow tussock grassland, but the natural dry core of the short tussock grassland was in turn being displaced. A sparse cover of native scabweed, cushion species of the *Raoulia* daisy (mostly *R. australis*) that naturally inhabited gravel riverbeds spread, together with a range of hardy exotic species. Aided by the first major eruption of rabbits in the region, the driest short tussock habitats became a so-called semi-desert (Fig. 5).

Along the wetter margins of the interior tussock grasslands, fire was used by early pastoralists to displace shrubland that had been invading at the time of European settlement. This was to promote additional grazing for the rapidly increasing number of sheep and cattle. The use of fire in pastoral farming was condemned for Otago as early as 1868 by John Buchanan, when he wrote:

Fig. 5. View upslope across short fescue tussock grassland from about 600 m on the northern end of the Old Man Range, showing greyish scabweed vegetation established on the driest northern aspect slopes and scattered bushes of matagouri *Discaria toumatou* on the moister slopes and depressions (above), and a close-up view to show details of the scabweed community in the vicinity (below).
Alan Mark

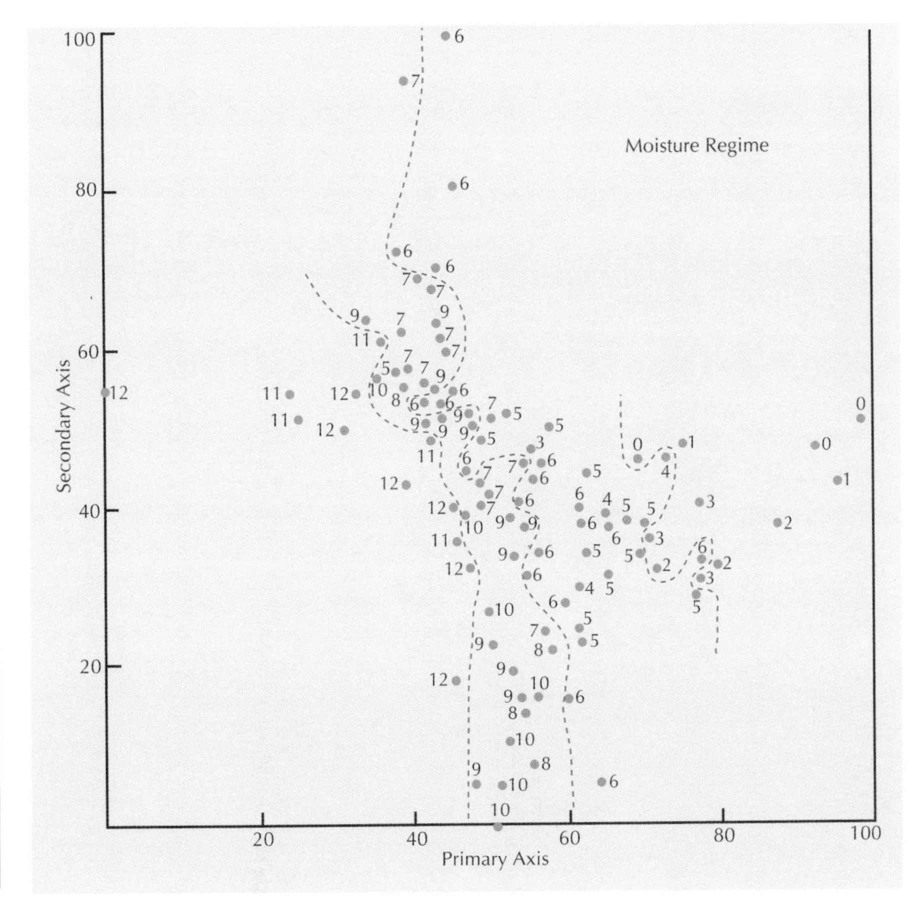

4 SALTPANS OF CENTRAL OTAGO

An adult of the day-flying moth *Loxostege* n. sp. This species has wriggly larvae that feed on *Atriplex buchananii* growing on the saltiest of sites. *Brian Patrick*

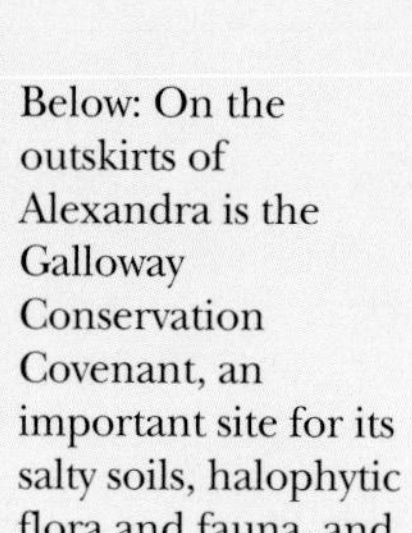

Below: On the outskirts of Alexandra is the Galloway Conservation Covenant, an important site for its salty soils, halophytic flora and fauna, and its rare herb *Lepidium matau*. *Brian Patrick*

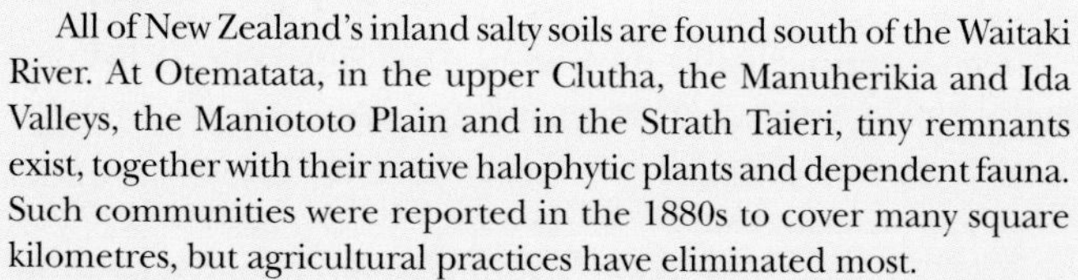

All of New Zealand's inland salty soils are found south of the Waitaki River. At Otematata, in the upper Clutha, the Manuherikia and Ida Valleys, the Maniototo Plain and in the Strath Taieri, tiny remnants exist, together with their native halophytic plants and dependent fauna. Such communities were reported in the 1880s to cover many square kilometres, but agricultural practices have eliminated most.

A distinctive native turf flora is associated with the saline soils, forming an obvious zonation around the barest, most salty sites. Mats of the glaucous grey *Atriplex buchananii* are the most salt tolerant among the natives, followed by the sparse, thin-leaved turfs of the saltgrass *Puccinellia raroflorens* and the thread-like cress *Lepidium kirkii*. Exotic halophytes, such as *Plantago coronopus* and *Puccinellia fasiculata*, are common and can dominate large areas, whereas the natives *Selliera microphylla*, *Samolus repens*, *Plantago spathulata*, *Lepidulum sisymbrioides*, *L. matau*, *Sarcocornia quinqueflora* and *Isolepis basilaris* are localised, some restricted to only one inland saltpan. A few saltpans have luxuriant herbfields containing the native celery *Apium* n.sp., *Chenopodium ambiguum* or tiny *Myosurus minimus*. Some of these species are rare and nationally threatened, such as the three *Lepidium* species, *I. basilaris* and the unusual mat-forming *Polygonum plebeium*, found only in one area near Otematata on a saltpan margin.

The antiquity of these inland salty areas is underlined by the number of native plant species that are confined to them, including *Lepidium kirkii* and perhaps *Puccinellia raroflorens*. A similarly distinctive insect fauna, particularly bees, bugs, beetles and moths, is known from saltpans. Because of the extreme harshness and openness of the sites, the fauna is diurnal. It includes native bees and the agile tiger beetle *Neocicindela dunedinensis*, both of which form holes in the saline soil in which larvae live and feed. The ubiquitous bug *Nysius huttoni* and several species of tiny black beetles are common over the barest salty soils in spring. But it is the diurnal moths *Paranotoreas fulva* and an undescribed species of *Loxostege* (top left), both with larvae feeding on *Atriplex buchananii*, that are the most conspicuous insects of the saltpans.

An exposure of salty soils at Galloway supports populations of *Lepidium kirkii* and *L. matau*, in addition to saltgrasses and *Atriplex buchananii*. This site (left) is protected by a conservation covenant. Sutton Salt Lake is set amid tors in the Strath Taieri with the typical mid-summer water level leaving exposed extensive cracked mud-flats sprinkled with the sprawling *Chenopodium ambiguum*. This is New Zealand's only inland salt lake, and is within a 140 ha scenic reserve (below); its water is about half as salty as seawater.

Brian Patrick

New Zealand's only inland salt lake is found near Sutton in the Strath Taieri region of eastern Central Otago. The photograph shows the shrunken lake in mid-summer with an array of halophytes occupying the cracked mud shore. *Brian Patrick*

> Nothing can show greater ignorance of grass conservation, than the repeated burning of the pasture in arid districts, which is so frequently practised. ... Much of the grass lands of Otago has thus been deteriorated since its occupation, by fire, and it is no wonder that many of the runs require eight acres to feed one sheep, according to an official estimate. It is a fallacy to suppose that grass country requires repeated burning to clear the surface of the excess of plants, as the old and withered grass forms shelter for the young shoots protecting them from parching winds, sun and frosts.

Despite repeated messages of such concern, the controversial practice continues, albeit more restricted, as does also the debate on its effects.

Saltpans, which are a special feature of the short tussock grassland in the semi-arid zone in lowland Central Otago, associated with saline soils, are discussed in Box 4.

Upland tall snow tussock grasslands

At the time of European settlement, the uplands in the rainshadow region of the southern South Island were dominated almost everywhere by tall snow tussock grassland, up to about 2000 m elevation. Only the most exposed sites, on the plateau summits of the Central Otago mountains, would have been too extreme for this grassland. It was also absent or sparse where snow persists well into the growing season, such as in the high basins and in snowbanks on the upper leeward slopes of the Central Otago mountains, as well as on the higher rocky rubble slopes and the most exposed areas on the plateaus of the northern greywacke ranges where fellfield and screes, typical of Canterbury, extend into North Otago.

Sufficient remnants of the upland grasslands remain to reconstruct both their extent and pattern at the time of European settlement. Narrow-leaved snow tussock grassland up to 1.2 m tall, and only locally with tall shrubs and large herbs (Fig. 6), would have dominated the grassland up to 1200–1400 m elevation, depending largely on aspect. Locally this species also descended to sea level, as at Shag Point near Palmerston, where it was continuous with inland higher-altitude snow tussock grasslands of the Horse and Kakanui Ranges. Narrow-leaved snow tussock gets shorter with increasing elevation and grades into slim snow tussock *Chionochloa macra* grassland, a community that still extends locally to 2000 m as on the Remarkables, Eyre Mountains and Treble Cone (Fig. 7). Similar in appearance, these two snow tussock species can be distinguished on whether the sheaths of the outer leaves that have died remain intact (as in *C. macra*) or whether they fracture into chaffy sections (as in *C. rigida*).

During the early decades of pastoral farming, extensive modification also occurred at these higher altitudes. In most areas the more vulnerable slim snow tussock grassland was largely displaced by a much shorter cover of less palatable herbs, including grasses and cushion species, to create what has been referred to as 'indigenous-induced' communities. These patterns were well established by the time the first comprehensive descriptions of Central Otago vegetation were compiled by Leonard Cockayne in the early decades of the twentieth century. These induced communities are described later in this section.

Fig. 6. View upslope over narrow-leaved snow tussock grassland from about 1200 m on Mt Cardrona, Central Otago, showing complete dominance of the tussocks with abundant leaf litter and a sparse intertussock plant cover. Photographed during a mast flowering in February 1979. *Alan Mark*

Fig. 7. Treble Cone summit (2088 m), showing the upper limit of slim snow tussock *Chionochloa macra* at about 2000 m elevation on the relatively snow-free sites, with the shorter snow patch tussock *C. oreophila*, here at its eastern limit, occupying depressions (foreground) where snow persists. *Alan Mark*

Fig. 8. Narrow-leaved snow tussock *Chionochloa rigida* grassland, April 1997, in the Rock and Pillar Scenic Reserve at about 1100 m on the eastern slope of the range, where shrubs of *Hebe odora* have increased in cover since the area was reserved in 1989. *Alan Mark*

Relatively unmodified narrow-leaved snow tussock grassland forms a dense cover with scattered shrubs of cottonwood *Ozothamnus* (*Cassinia*) *leptophyllus* and locally of boxwood hebe *Hebe odora* (Fig. 8) or of inaka *Dracophyllum longifolium* on poorly drained peaty hillocks. Mountain flax *Phormium cookianum* and the silvery astelia *A. nervosa* are often prominent in moist lower elevation sites, while golden speargrass *Aciphylla aurea* is more widespread. There is a large array of smaller herbs and shrubs, notably blue tussock *Poa colensoi*, *Raoulia subsericea*, heaths including snowberries (*Gaultheria macrostigma* and *G. depressa*), as well as the dwarf heath *Pentachondra pumila*, together with a range of small herbs including *Anisotome aromatica*, the ground lily *Herpolirion novae-zelandiae* and *Gonocarpus micranthus*. The species mix varies with altitude and other factors, but there is no striking change where slim snow tussock replaces the narrow-leaved snow tussock at around 1200–1400 m, being generally higher on warmer sunny slopes and usually associated with some hybrid snow tussocks.

Fig. 9. Abrupt upper limit of narrow-leaved snow tussock *Chionochloa rigida* grassland at about 1300 m on the eastern slope of the Old Man Range, Central Otago. The more palatable slim snow tussock *C. macra* has been displaced above by shorter blue tussock *Poa colensoi* and *Celmisia viscosa* that are more tolerant of pastoral farming practices. *Alan Mark*

Fig. 10. View south over an area of relatively undisturbed slim snow tussock *Chionochloa macra* grassland at about 1300 m on the southern Old Man Range showing a dense cover of tussock and prominent flower heads of the speargrass *Aciphylla scott-thomsonii* (the green heads are of female plants, the yellow ones are of male plants). The Waikaia Bush Road is visible in the distance. *Alan Mark*

The replacement of narrow-leaved snow tussock by slim snow tussock with increasing elevation in the southern South Island has generally been attributed to a greater cold tolerance of the latter. However, soil factors may also be important here as in mid Canterbury, for example, slim snow tussock occupies somewhat younger soils of a higher nutrient status at lower elevation. In many areas today the extent of the slim snow tussock is considerably reduced, reflecting its much greater vulnerability to the pastoral practices of burning and grazing. Where the species has been displaced, the upper limit of narrow-leaved snow tussock is usually quite abrupt, giving way to induced blue tussock grassland or *Celmisia* herbfield. Such a pattern can be seen on much of the Carrick, northern Garvie, Old Man (Fig. 9) and Rock and Pillar Ranges. Where slim snow tussockland does persist, as on the southern Old Man Range and northern Dunstan Mountains, the ground layer is usually dominated by blue tussock, loose mats of *Raoulia subsericea*, trailing subshrubs of the heaths *Gaultheria depressa* and *Pentachondra pumila*, as well as gentians *Gentiana bellidifolia*, rosettes of *Anisotome flexuosa* and *Brachyglottis bellidioides*. There is usually a range of alpine mountain daisies, most notably *Celmisia viscosa*, *C. brevifolia* and *C. laricifolia*, as well as some areas with co-dominant speargrass that is conspicuous when in flower (Fig. 10).

Tall copper tussock grassland

The southern South Island red tussock *Chionochloa rubra* has recently been described as subspecies *cuprea*, which refers to its copper hue when sunlit, and is the basis for its new common name, copper tussock. It is widespread throughout the southern region, but is generally confined to areas below the climatic treeline that are seasonally or permanently wet (Fig. 1). Once widespread on the Southland Plains, development for farmland has meant that copper tussock is today of limited extent. Reasonable cover remains only in north-western Southland, on the Eastern Otago uplands of the South Rough Ridge-Lake Onslow section of the Manorburn Ecological District, and the western section of Waipori District, covering a range of landforms from about 750 m to 1000 m altitude.

Today, the condition of copper tussockland is quite variable, with those remnants in the best condition considered to be nationally important, as in parts of the upper Manorburn catchment, where the Minister of Conservation in 1992 attempted to prevent its deterioration through planned burning and associated grazing that had been approved by the Otago Regional Council under the Resource Management Act. Here dense copper tussock, up to 1.2 m tall, occupied gentle summits, depressions and valley floors where drainage is generally poor (Fig. 11). Intertussock cover is mostly sparse, but consists of several species, including mosses *Sphagnum* and *Polytrichum*, grasses (blue tussock, *Deyeuxia avenoides*) and *Pimelea prostrata*, *Raoulia subsericea*, as well as sedges, rushes and some low heaths (species of *Leucopogon*, *Pentachondra* and *Gaultheria*), plus scattered emergent shrubs of *Coprosma ciliata* and *Hebe rakaiensis*. Sweet vernal *Anthoxanthum odoratum*, browntop *Agrostis capillaris* and catsear *Hypochoeris radicata* are the only common exotic

plants, though hawkweeds (*Hieracium* spp.) have become established and are obviously increasing. Here narrow-leaved snow tussock often co-dominates with copper tussock on the better drained spurs where their hybrids may also occur (plants with transverse fractures on their leaf sheaths – a *C. rigida* feature, but blades that crack longitudinally when unrolled – a *C. rubra* feature). In the Manorburn District, narrow-leaved snow tussock generally replaces copper tussock as altitude increases above about 800 m, and as soils become thinner due in part to the lower content of wind-blown loess. The transition between the two grassland types appears to be close to the original treeline and it has been suggested that areas of copper tussockland were forested prior to human settlement, while narrow-leaved snow tussock generally dominates beyond the limits of the previous forest (Fig. 12).

At lower altitudes in the Manorburn District, copper tussock grassland generally becomes quite open and the short fescue tussock increases in prominence, together with a wide range of exotic plants of which mouse-ear hawkweed *Hieracium pilosella* is usually the most conspicuous. Indeed, in many parts of the district, hawkweed species (mouse-ear, king devil *H. praealtum*) and tussock hawkweed *H. lepidulum* are highly invasive.

Small pockets of copper tussock grassland, sometimes with some narrow-leaved snow tussock or hybrids, also persist in wet depressions throughout the Otago uplands, often as the only remaining natural features of otherwise developed farmland. The persistence of copper tussock reflects its low palatability, plus the relatively poor soil conditions in these habitats which makes them the least attractive for agricultural development. The wet peaty plateau of Swampy Summit (739 m) overlooking Dunedin also supports a copper tussock grassland and hybrids with narrow-leaved snow tussock.

There are more extensive stands of copper tussockland on peat in northern Southland. Two occur alongside the main highway near Pukerau, where wirerush *Empodisma minus*, umbrella fern *Gleichenia dicarpa* and *Sphagnum* moss are abundant. Podocarp wood in the underlying peat here testifies to a previous forest cover. Both stands are reserved, one being recently gifted by Tasman Agriculture Ltd.

Fig. 11. Copper tussock (*Chionochloa rubra* ssp. *cuprea*) grassland dominating the gently rolling topography at about 850 m in the Manorburn District, with wetlands localised in the depressions. Narrow-leaved snow tussock *C. rigida* grassland occupies the distant summits above about 900 m. *Alan Mark*

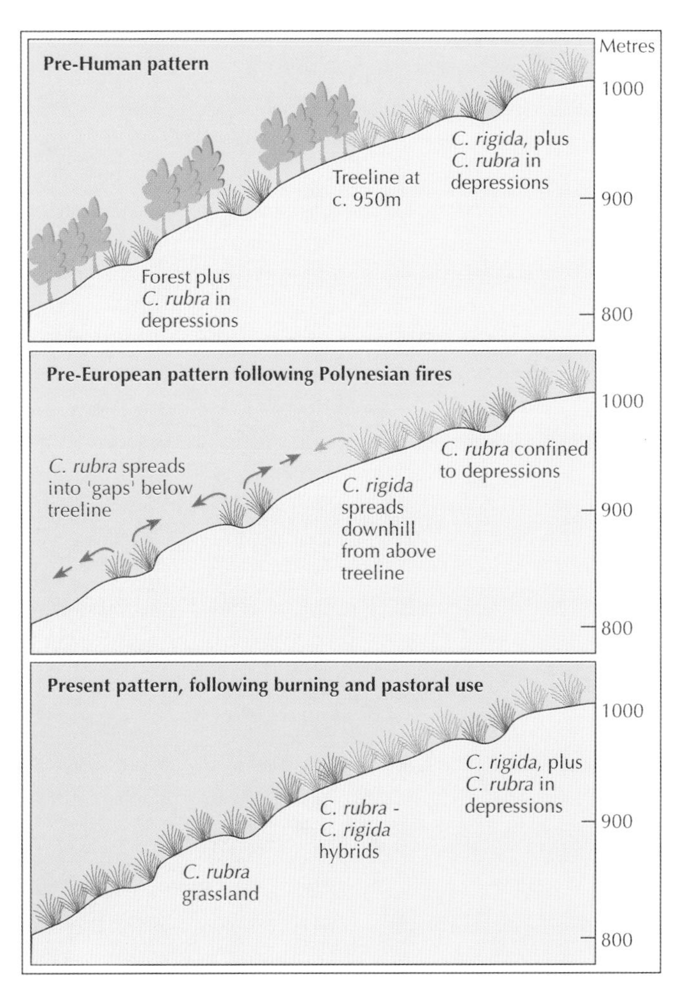

Fig. 12. Probable vegetation history of the Manorburn District in relation to the present vegetation pattern of narrow-leaved snow tussock *Chionochloa rigida* and copper tussock *C. rubra* ssp. *cuprea* grassland, as suggested by Dr Peter McIntosh, based on soil and other features of the area.

Fig. 13. Copper tussock grassland at the Burwood Conservation Area, northern Southland, showing the irregular border with mixed beech forest that is indicative of the grassland having originated through fire, probably in pre-European times. *Alan Mark*

The most extensive remaining stands of copper tussock grassland occur in north-western Southland, on outwash gravels such as at Burwood (Fig. 13), Mavora Lakes-upper Mararoa Valley and Dunrobin south of Mossburn, as well as in other valleys draining the slopes of the Takitimu Mountains. These areas have experienced varying amounts of pastoral use, but some stands still appear to be little modified and several are reserved.

Copper tussock is generally sparse in Fiordland National Park due to the persistence of extensive forest. However, copper tussock occupies permanently wet and seasonally cold valley floors on the eastern edge of the Park, as in the Murchison Mountains (Point Burn and Takahe Valleys, Fig. 1) and on the Hunter Mountains (Borland and Grebe Valleys), surrounded by beech forest. These stands undoubtedly reflect the natural pattern and role of the copper tussock grassland in southern New Zealand.

Very localised areas of copper tussock grassland are common on poorly drained coastal to lower subalpine flats and stable dunes. These communities are often found with flax (*Phormium* spp.), manuka and wire rush, such as at Tiwai Point, and on Stewart Island at Mason Bay, where small stands of silver and fescue tussock also occur in the lowlands on well-drained sandy soils and stabilised dunes. The fescue tussock grassland here has a generally similar composition to those of the South Island, according to Hugh Wilson.

Indigenous-induced communities

Extensive areas of tussock grassland have been cultivated or converted to exotic forestry, in some localities up to 700 m elevation. However, as previously mentioned, pastoral practices of periodic burning combined with grazing have resulted in a wide range of plant cover. These changes in plant cover appear to have a significant effect on water yield, particularly from the uplands (see Box 5, page 204). At lower elevations, aerial oversowing and topdressing since World War II have contributed to the modification. Only the least modified and more common of these systems are discussed here. Inevitably, the less palatable species that are also tolerant of fire have succeeded in displacing the original indigenous tussock cover. Among the indigenous species, golden speargrass, blue tussock, alpine fescue tussock *Festuca matthewsii* and several mountain daisies, most notably *Celmisia semicordata*, *C. lyallii*, *C. viscosa* and *C. densiflora*, as well as Maori onion *Bulbinella angustifolia*, have been the most successful.

Golden speargrass was probably a minor component of narrow-leaved snow tussock grassland over most of its range prior to pastoralism. However, the species has increased in many areas, becoming dominant in several places where snow tussock has been severely depleted. Such effects can be seen on parts of the Carrick and Pisa Ranges and the floor of the upper Nevis Valley in Central Otago, usually in association with blue tussock and alpine fescue tussock (Fig. 14).

Although blue tussock is probably the most widespread species of snow tussock grasslands, its dominance in many upland areas today seems to reflect the loss of snow tussock cover through pastoral farming. On many of the rainshadow ranges, induced blue tussock grassland now forms the upper zone of grassland, often occurring as a belt about 100 m elevation wide. Extensive examples can be seen on the Pisa, Dunstan, Carrick, Old Man, Rock and Pillar (Fig. 15) and Hector Mountains, and more locally on the Lammerlaw, Lammermoor, Kakanui, Hawkdun, St Bathans, Cardrona, Remarkables, Eyre, Takitimu, Richardson, Thomson and Livingstone Mountains, reflecting the widespread influence of pastoral farming activity. Blue tussock is often associated with, and sometimes mistaken for, alpine fescue tussock. Both have similar form and colouration. However, alpine fescue may be distinguished from blue tussock by being somewhat taller (25 v. 15 cm) and by a flower head which carries small awns or bristles on the florets, while those on the blue tussock are awnless. Alpine fescue tussock dominates in some areas, such as on the Old Man and Carrick Ranges and in the upper Nevis Valley. Here, as in many of these induced short grassland communities, the dwarfed creeping daisy *Raoulia subsericea* is the commonest associate.

Induced herbfields of mountain daisies are prominent today on many of the rainshadow ranges. On the Rock and Pillar, Old Man and south Dunstan Mountains, *Celmisia viscosa* is the most

Fig. 14. A cover of golden speargrass *Aciphylla aurea*, together with short tufts of blue tussock *Poa colensoi* and alpine fescue tussock *Festuca matthewsii* induced from snow tussock grassland depleted through the pastoral practices of burning and grazing, eastern slopes of the Pisa Range at about 1300 m. *Alan Mark*

Fig. 15 (right). View north-east from McPhee's Rock at about 1310 m along the crest of the Rock and Pillar Range, showing extensive areas of blue tussock *Poa colensoi* grassland induced from slim snow tussock grassland through pastoral farming practices of burning and grazing. Relic snow tussocks persist, together with areas of dark green shrubland on southern-aspect slopes. *Alan Mark*

Fig. 16 (below, right). The upper eastern slopes of the Rock and Pillar Range at about 1200 m, showing a blue tussock-herbfield of *Poa colensoi* and *Celmisia viscosa* induced from alpine slim snow tussock *Chionochloa macra* grassland through pastoral farming practices. *Alan Mark*

conspicuous species (Fig. 16), whereas on the somewhat drier northern Dunstan, Carrick, Garvie, Hector and Harris Mountains, *C. lyallii* is more prominent. Where the climate is wetter, as on parts of the Garvie, Old Man, Lammerlaw, Umbrella and Eyre Mountains, the larger *C. semicordata* has become abundant (Fig. 17). There is usually a variety of other native alpine species present in these induced herbfields, as well as thinly scattered individuals or small stands of relic snow tussock, indicating their earlier dominance in such areas.

The semi-arid lowlands of Central Otago today, have a generally sparse cover of scabweed – isolated or merging, low dense cushions of *Raoulia*, chiefly the common scabweed *R. australis* – among a sparse cover of mostly exotic annual grasses, flat weeds and bare eroding soil or stone pavement. This depleted landscape has replaced short tussock grassland, itself reduced by the heavy grazing of rabbits and stock from the early days of pastoral farming. The scabweed cover is best developed on the lower hills and sunny slopes up to about 1000 m in the Clutha catchment from Fruitlands through to Bannockburn, Oturehua and Tarras. Nonetheless, the community is now less prominent, since a wide range of exotic plants and some recovering short tussocks have displaced the scabweed in many areas (Fig. 18).

Fig. 17 (bottom left). Eastern slope of the Garvie Mountains at about 1220 m near Titan Rock, showing extensive areas of herbfield dominated by *Celmisia semicordata* that has been induced from snow tussock grassland through pastoral farming practices. Blue tussock *Poa colensoi* is common among the daisies and a few relic snow tussocks persist. *Alan Mark*

Fig. 18. Area of scabweed *Raoulia australis* being invaded by fescue tussock *F. novae-zelandiae*, together with a range of exotic weeds and annual grasses at about 600 m on North Rough Ridge, Central Otago. The Ida Valley and Raggedy Range are beyond. *Alan Mark*

5 WATER YIELD IN RELATION TO LAND USE IN THE UPLAND SNOW TUSSOCK GRASSLANDS OF EASTERN AND CENTRAL OTAGO

Upland areas generally receive much more precipitation than the lowlands and are therefore the source of most of the water we use for a range of purposes – domestic supply, irrigation and hydro-electric generation, as well as recreation and conservation. Several recent studies in the upland snow tussock grasslands of Eastern and Central Otago indicate that the amount of precipitation, as well as the type and condition of vegetation, affect both the quantity and quality, as well as the pattern, of water yield. A paired catchment study was established in 1979 at 460–650 m in Glendhu Forest in the upper Waipori catchment on the Lammerlaw Range to assess the effects of converting snow tussock grassland to an exotic pine forest (Figs 1 and 2).

For the three-year period prior to planting, both catchments recorded unusually high and sustained dry weather flows. Of the mean annual precipitation of 1305 mm, 835 mm or 64 per cent became runoff while only 210 mm or 16 per cent, about half that predicted from climatological values, was transpired. The balance, 260 mm (20 per cent) was lost by interception. Over the first six years following the planting of *Pinus radiata* at 1230 stems per hectare in the larger (310 ha) catchment, there was little change in the annual water balance and also a negligible difference between this and the control (unplanted) catchment (218 ha), but subsequently the annual runoff from the planted catchment has continued to decline. By 1993, just nine years after planting, water yield from the planted catchment was 31per cent less than from the tussock grassland catchment. The difference was expected to increase as the forest matures, due to increased transpiration and interception loss, although it had not exceeded 35 per cent by 2001 (Fig. 3). Summer low flows have decreased from the planted catchment, as well as the magnitude and intensity of floods, but water quality is essentially unchanged.

Other studies, using specially designed tanks or lysimeters of various sizes and sophistication (weighing and non-weighing types), have measured the water balance of upland snow tussock grasslands in relation to the effects of burning and grazing, as well as its replacement by shorter blue tussock grassland, *Celmisia viscosa* herbfield or exotic pasture. All these studies have consistently revealed the relatively high water yield associated with least modified snow tussock grassland, the actual amounts varying largely in relation to location and reflecting the local climate.

Water balance values were derived from a large weighing lysimeter containing snow-tussock grassland with nine mature tussocks. This was placed at 570 m on the broad ridge between the paired catchments at Glendhu (Fig. 4) and the water balance was measured over a one-year period. Results confirmed the low evapotranspiration of the snow tussock grassland and recorded a generally similar water balance to that of the whole catchment for the same period.

A series of 84 smaller non-weighing lysimeters were designed to represent a unit area of snow tussock grassland, containing a single mature snow tussock (Fig. 5) or a range of other vegetation types that may replace snow tussock grassland in different areas (Fig. 6). These lysimeters were used to obtain water balance information from seven sites at 490–1340 m on the Rock and Pillar and Lammerlaw Ranges. Following a six-year pilot study at 1000 m on the Rock and Pillar Range, lysimeters were operated over a two-year period at the seven sites and confirmed the relatively high water yield associated with the least modified snow tussock grassland. This vegetation produced the greatest water yield at all six sites above 700 m. Values ranged from 80 per cent of the measured 1372 mm of annual precipitation on the southern Lammerlaw Range to only 12 per cent of 510 mm at the lowest site (490 m) at the northern end of this range. At the other sites, all above 700 m, the annual water yield from snow tussock grassland exceeded 25 per cent, and was generally in the range of 60–80 per cent, of the measured precipitation at the particular site. Water yields in the snow-free six months from the lower sites, calculated as a percentage of the measured rainfall, were somewhat less than for the whole year but they increased at the higher and/or foggier sites, reaching 86 per cent at 870 m on the upper plateau of the southern Lammerlaw Range. Water yields from all other types of cover, including bare soil, at the six sites above 700 m were consistently less than that from snow tussock grassland at the same sites. Indeed, water yield from bare soil, a blue tussock turf and *Celmisia viscosa* herbfield were generally similar within sites, but ranged between sites from 16 to 62 per cent of measured annual precipitation. Water yield from a pasture sward at the lowest site (490 m) was negligible, the low value being related to the lower precipitation and also to the relatively high rates of transpiration from this type of plant cover, compared with any of the types of native vegetation used in this study.

The increased water yield associated with snow tussock grassland appears to be due only partly to the low water consumption that results from the special features of its leaves. It also seems to reflect interception gains from wind-driven fog (and rain) made by the long fine foliage of a snow tussock (Fig. 7). This was subsequently confirmed using a different approach. The composition of stable heavy isotopes of hydrogen and oxygen from samples of fog and rain at three sites on the Otago uplands was consistent with those from other parts of the world. Fog had a significantly higher proportion of these isotopes (being an early stage condensate of saturated air) than did rain (a later stage condensate) from the same site. The isotopic composition of groundwater sampled at the same sites was intermediate, indicating it contained a subequal mixture of fog and rain. The yields of water are greatest from the most fog-prone sites, where individual tussocks may filter up to half a litre of water an hour from a dense wind-driven fog. The contribution of fog to water yield by certain types of vegetation in particular environments, although accepted as significant in several other parts of the world, in a range of vegetation types, remains contentious on the southern uplands and is worth further study.

AlanMark

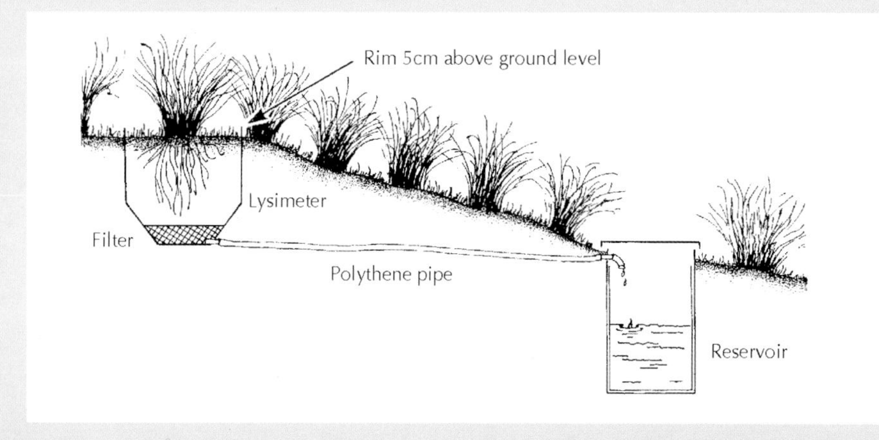

Fig. 5. Cross-section of a non-weighing lysimeter showing a snow tussock grassland treatment and slope orientation allowing gravity drainage to a reservoir.

Fig. 1. View up the Glendhu catchment planted in *Pinus radiata*, showing the V-notch weir installed to measure water yield at the bottom of the catchment. *Alan Mark*

Fig. 2 (top right). View up the control catchment at Glendhu, showing the V-notch weir and the snow tussock cover beyond. *Alan Mark*

Fig. 3 (right). Comparative water yields from the paired catchments at Glendhu from 1980 to 1998. Differences in water yield are expressed as mm of equivalent rainfall for the particular year. The zero line indicates no difference in yield. *Redrawn from Barry Fahey's unpublished 1998 report.*

Fig. 4 (below). The large weighing lysimeter installed at Glendhu on the broad ridge at 570 m between the control and planted catchments (seen behind, three years after planting). *Alan Mark*

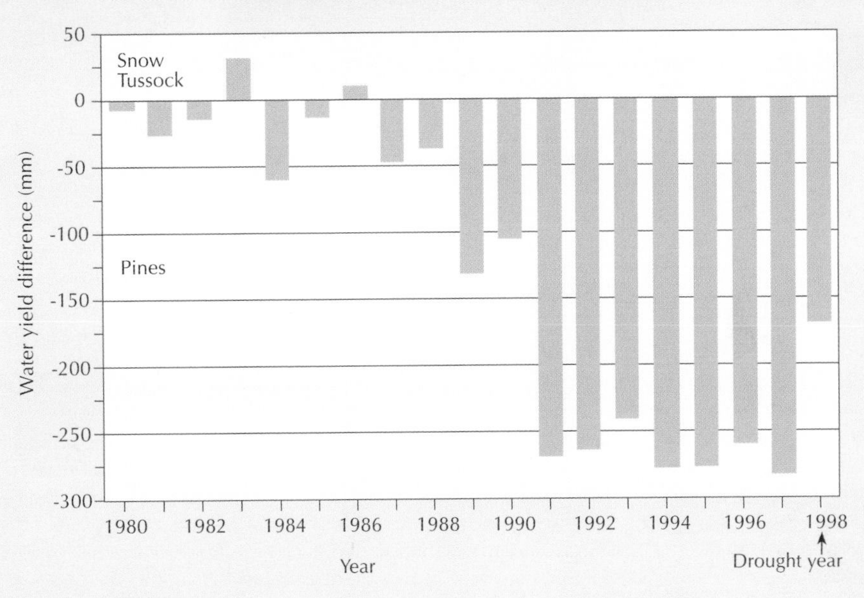

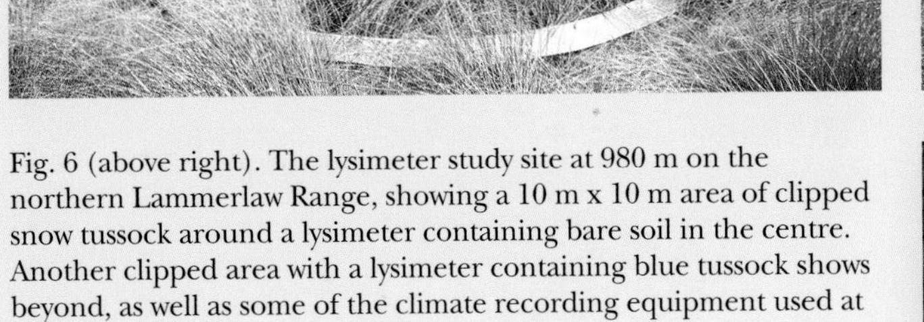

Fig. 6 (above right). The lysimeter study site at 980 m on the northern Lammerlaw Range, showing a 10 m x 10 m area of clipped snow tussock around a lysimeter containing bare soil in the centre. Another clipped area with a lysimeter containing blue tussock shows beyond, as well as some of the climate recording equipment used at this site. *Alan Mark*

Fig. 7 (right). Water droplets intercepted from fog on the fine foliage of narrow-leaved snow tussock. *Alan Mark*

6 THE TORS OF CENTRAL OTAGO

One of the spectacular summit tors on the Old Man Range. Note the cavernously weathered upper section reminiscent of the lowland forms. *Barry Fahey*

A tor outcrop in a lowland setting at Butchers Dam near Alexandra. *Barry Fahey*

The Central Otago landscape is one of the most spectacular in New Zealand. Away from the coast, the rolling downlands are replaced by stark fault-block ranges separated by extensive plateau areas and depressions. Common to all these terrain types are block-like outcrops of bedrock called tors (above), a Celtic word meaning an up-standing rock tower. To understand the formation of these unique landforms we need to know something about the geologic history of Central Otago.

About 200 million years ago rocks across much of the region were buried to depths of over 10 km in the earth's crust, and altered by heat and pressure to form the distinctively layered schist bedrock that underlies much of Central Otago. Gradual uplift exposed the schist to a prolonged period of erosion under subtropical conditions. Remnants of the soils that developed during this period are still preserved in some parts of Central Otago today. The developing surface of low relief became deeply weathered as erosion rates declined, but structural and mineral variations in the schist were instrumental in preserving some areas beneath the surface, especially where rock joints were widely spaced or the mineral layers in the schist were more resistant. These unweathered remnants were gradually exposed as the more weathered schist around them was removed, forming a tor landscape like that to be seen between Roxburgh and Alexandra. We can envisage their origin taking place in two stages: deep weathering beneath the surface but with the preservation of unweathered blocks, followed by the removal of the weathered material to eventually expose the tors.

Those tors found on the summits and upper flanks of the Central Otago mountains such as the Rock and Pillar and the Old Man Ranges (above left) differ markedly from their lowland counterparts (above right). Although less numerous, they are taller and much sharper in outline. Originally they were thought to be lowland forms that had survived uplift during the formation of the fault block ranges. However, their larger size and sharper outline also suggest a much more recent origin through frost processes in an alpine environment. Local frost conditions would have been even more severe when glaciers occupied Lakes Wanaka, Hawea, and Wakatipu some 12,000 years ago. It is unlikely, however, that frost processes alone could have caused the weathering and removal of up to 15 m of schist that would have been required to expose these tors on the summits during and after the last glacial episode (bottom). Indeed, it could be just as strongly argued that current frost action on the summits is in the process of destroying the tors rather than forming them.

There is some evidence to suggest that tors could have formed under more temperate conditions than encountered on the summits today, perhaps during warmer interglacial phases. Some, for example, have a characteristically weathered appearance, especially on their upper surfaces, and others have large pits etched into their surfaces and into those of adjacent detached blocks that we know take many thousands of years to form. It is also possible that the contrasting processes of chemical and physical weathering have both been able to produce these landforms, and some tors may even have a composite origin. However, it is less likely that they could have survived intact through a few million years of uplift to become enlarged and sharpened by frost action, and thereby converted to upland forms on the summits of the emerging fault blocks. Thus, while there is general agreement on the formation and antiquity of the lowland tors, those found on the summits remain something of an enigma.

Barry Fahey

Stages in the origin of Central Otago's tor landscape: a long period of weathering leaving more resistant segments preserved beneath the surface (a), followed by removal of the weathered products to expose the tors (b). *Barry Fahey*

Vegetation patterns of the high-alpine zone

High-alpine communities are those which occur above the natural limits of the tall snow tussock species characteristic of the low-alpine zone. This transition is associated with a mean air temperature for the warmest month of about 5°C.

Four types of high-alpine communities can be recognised in the rainshadow region of southern New Zealand. The distinction is based on both physiography and flora. Snowbank and fellfield communities occur wherever mountains reach the necessary elevation. They are present in both the rainshadow and western high-rainfall regions, though differing in many of their component species between these two regions. The last two, cushionfield and talus or scree, however, are confined to the rainshadow region. Cushionfield is associated with the high plateau summits and tor landscapes of the schistose Central Otago mountains (see Box 6). Scree is a feature of the steeper slopes on less highly metamorphosed rock types. These are predominantly greywacke in North Otago and low-grade semi-schist of the Otago Lakes District, as well as andesites and associated rocks on the Takitimu Mountains in western Southland.

Snowbank vegetation

For some distance below the permanent or summer snowline, snow will accumulate and persist well into the normal growing season, in depressions and other sheltered sites. Such accumulation occurs particularly on leeward slopes, as is common in cirque basins that are sheltered from the prevailing westerly winds. The pattern of plants within a snowbank depends largely on the patterns of snowlie and snowmelt, which are generally consistent from year to year, and the tolerance of species to a shortened growing season. However, there are some differences between the snowbank vegetation throughout southern South Island in relation to differences in plant distribution. Snowbanks in Central Otago are usually dominated by mountain daisies. *Celmisia haastii* indicates where snow lies for about 200 days or till mid-December (Fig. 19). The bright green *C. prorepens*, endemic to Central Otago, characterises the mid snowbank, with about 163 days of snow cover, the snow thawing by late November, while *C. viscosa* dominates the margins where there is little delay in the thaw. Here, snow usually lies for up to 130 days, thawing generally by early November. *Celmisia hectorii* replaces *C. prorepens* further west and snowpatch tussock *Chionochloa oreophila* is the most conspicuous snowbank plant in the wetter western regions, though it extends eastward as far as Treble Cone (Fig. 7) and The Remarkables.

Several alpine plant genera have eastern and western snowbank species in southern South Island, the most notable being *Ranunculus* (*R. pachyrrhizus* / *R. sericophyllus*), *Psychrophila* (=*Caltha*) (*P. obtusa* / *P. novae-zelandiae*) and *Kelleria* (*K. childii* / *K. croizatii*). Widespread snowbank plants include blue tussock, *Coprosma perpusilla*, *Gaultheria nubigena* (= *Pernettya alpina*), *Carex pyrenaica* var. *cephalotes*, *Raoulia subulata* and *Plantago lanigera*.

Fellfield

These are sparsely vegetated communities where the plants are apparently wedged among relatively stable rocks or boulders with very little soil. Like snowbanks, there are some notable differences between the relatively dry fellfields of the rainshadow mountains and those of the wet western regions.

The most distinctive dry fellfields are those with one or more species of 'vegetable sheep'. These extend southwards from Canterbury on to the higher greywacke mountains of North Otago, notably the Hawkdun Range. Here, the best known and most widespread species of vegetable sheep, *Raoulia eximia*, is locally common on the rocky upper slopes near the centre of the Range around Mt Ida (Fig. 20). A distinctive cushion speargrass *Aciphylla dobsonii*, along with *A. gracilis*, also reaches its southern limit here. Several lichens, notably species of *Umbillicaria* and *Neuropogon*, are also conspicuous on the boulders here and elsewhere in the high-alpine zone. For more details on lichens in the alpine zone, see Box 7 (next page).

Fig. 19. A late snowbank on the upper eastern (leeward) slope of the Old Man Range at about 1550 m, showing dominance of *Celmisia haastii* in the snow-melt and flush areas, surrounded by slim snow tussock *Chionochloa macra* grassland. Late November 1992. *Peter Johnson*

7 LICHENS OF THE TUSSOCK GRASSLANDS AND ASSOCIATED MOUNTAIN LANDS OF SOUTHERN NEW ZEALAND

Lichens, although commonly referred to as plants, are in fact fungi with a special physiological lifestyle – symbiosis with an alga or cyanobacterium. Although they appear to be independent organisms, they are correctly small ecosystems consisting of a dominant fungal partner and one or more photosynthetic partners (a green alga or a cyanobacterium) living together in a complex, controlled, mutually beneficial symbiosis and representing one of the most successful examples of a symbiotic association in the living world. The relative fragility of the physiological balance between the symbiotic partners (the dominant fungal partner is a consumer of carbon, while the photosynthetic symbionts are carbon producers) is easily disrupted by a variety of environmental effects (e.g. atmospheric pollution, heavy metals, acid rain), making lichens sensitive biological indicators and barometers of both short- and long-term environmental change.

There are thought to be between 15,000 and 20,000 different lichens in the world. New Zealand has about 1350 species in 308 genera (of which 29 are genera of lichenicolous fungi), a figure which probably represents less than 80 per cent of the total to be found here. Even with many more new species to be discovered, New Zealand has for its size one of the richest lichen mycobiotas in the world.

A rich diversity of lichens and lichen communities is found in the tussock grasslands and associated mountain lands of southern New Zealand. From treeline to the highest exposed rocks of the nival zone, a variety of habitats support some 100 lichen genera. Of these genera, 20 per cent have the potential for nitrogen-fixation (through their symbiotic cyanobacteria), and include mainly foliose (rarely fruticose) lichens of high biomass, notably *Coccarpia, Peltigera, Placopsis, Pseudocyphellaria, Psoroma, Solorina, Stereocaulon,* and *Sticta.* Lichens of major habitats are briefly discussed below.

Tussock grasslands: Here lichen diversity is generally low but several macrolichens appear well suited to this biome and may produce a considerable biomass on soil in low-light, high humidity sites, such as at the bases of tussocks or small shrubs. Major lichens present are *Alectoria nigricans*, species of *Cladia, Cladina, Cladonia, *Collema, Hypogymnia lugubris, Lecanora epibryon* ssp. *broccha, *Leptogium, Micarea, Ochrolechia xanthostoma, *Peltigera, *Pseudocyphellaria, *Psoroma, *Stereocaulon, *Sticta* and *Thamnolia vermicularis.* Most (*) are nitrogen-fixers and potentially capable of adding substantial amounts of organic nitrogen to the grassland nitrogen budget.

Shrubs in grassland: Shrubs such as *Hebe* and *Dracophyllum* have lichen communities adapted to high light and high UV radiation, and generally produce high levels of photoprotective cortical compounds such as atranorin or usnic acid. Prominent lichens on shrubs include: *Haematomma alpina, Hypogymnia lugubris, Lecidella elaeochroma, Lemanelia inactiva, Menegazzia globulifera, M. inflata, M. testacea, Parmelia cunninghamii, P. sulcata, Physccia stellaris, Pseudocyphellaria coronata, P. crocata, P. degelii, P. glabra, P. pickeringii, Pyrrhospora laeta, Teloschistes fasciculatus, T. velifer, Usnea contexta* and *U. pusilla.*

Rock outcrops: Rock outcrops are usually almost completely covered by a complex mosaic of crustose and/or closely attached foliose lichens. Crustose lichens are from four main families, viz. Lecanoraceae (*Carbonea, Lecanora, Lecidella, Protoparmelia, Ramboldia, Pyrrhospora*); Lecideaceae (*Lecidea*), Pertusariaceae (*Ochrolechia, Pertusaria*); Porpidiaceae (*Immersaria, Labyrintha, Poeltiaria, Porpidia*). Foliose lichens are mainly from the family Parmeliaceae (*Flavoparmelia, Neofuscelia, Parmelia, Xanthoparmelia*) and species of *Umbilicaria*, black or grey lichens attached by a central umbilicus. Fruticose lichens are mainly *Stereocaulon caespitosum* and *Usnea torulosa.*

Bogs: Subalpine and low-alpine peat bogs are scattered among mainly coastal ranges and hills in Fiordland, Eastern Otago (Swampy Summit near Dunedin, Maungatua, Blue Mountains, Catlins) and Southland (Awarua Plain, Longwood Range) to Stewart Island. They have characteristic lichen communities, dominated by the genera *Cladia, Cladina, Cladonia, Siphula* (the rooting structures of which consolidate peat), with the rare subantarctic lichen *Knightiella splachnirima* (above) occasionally present.

High-alpine habitats: The habitats of fellfield and cushionfield vegetation, including soil stripes, hummocks and solifluction terraces and lobes, which are especially well developed on the summits of the Central Otago mountains, support the highest diversity of alpine lichens in New Zealand and are richer in both genera and species than more recently glaciated surfaces close to the Main Divide. A number of lichens found here are bipolar, being well-known components of alpine vegetation in alpine and arctic habitats of the Northern Hemisphere, while also occurring as widely separated disjuncts on the summits of South Island mountains and sometimes, too, in the mountains of Tierra del Fuego. The bipolar element in the New Zealand lichen mycobiota is most strongly marked in the high-alpine areas of Central Otago.

Soil: A variety of lichens grow on exposed soil in high-alpine habitats, providing there is sufficient moisture, and include: *Alectoria nigricans, Arthrorhaphis alpina, A. citrinella, Brigantiaea fuscolutea, Catapyrenium cinereum, Cetraria aculeata, C. islandica* ssp. *antarctica, Cetrariella delisei, Cladia aggregata, Cladina confusa, C. mitis, Cladonia bimberiensis, C. capitellata, C. sulcata, Collema durietzii, Degelia neozelandica, Diploschistes muscorum* ssp. *bartletti, Lecanora neglecta, Leptogium laceroides, Micarea isabellina, Omphalina ericetorum, Peltigera didactyla, P. dolichorhiza, P. subhorizontalis, Pseudocyphellaria crocata, P. degelii, P. maculata, P. pickeringii, Placopsis trachyderma, lpsora decipiens* (only on Fairfax Spur, Dunstan Mountains, where it is well developed on calcareous schist), *Psoroma buchananii, P. fruticulosum, P. hypnorum, P. rubromarginatum, Siphula complanata, S. decumbens, S. dissoluta, S. foliacea, S. fragilis, Siphulastrum mamillatum, S. triste, Solorina crocea, Stereocaulon* sp., *Sticta martinii, Thamnolia vermicularis* and *Trapelia coarctata.*

Cushionfield: Prostrate shrubs and dead tussock bases are habitats for a number of lichens including: *Alectoria nigricans, Brigantiaea fuscolutea, Candelaria concolor, Candelariella vitellina, Haematomma alpinum, Hypogymnia lugubris, Lecanora epibryon* ssp. *broccha, Lecidella elaeochroma, Leptogium limbatum, Menegazzia inflata, M. lucens, M. testacea, M. ultralucens, Micarea austroternaria, Ochrolechia xanthostoma, Pertusaria dactylina, P. gymnospora, Pseudocyphellaria degelii, P. crocata, P. glabra, P. maculata, Psoroma hirsutulum, Pyrrhospora laeta, Sticta martinii, Teloschistes fasciculatus* and *Usnea contexta.*

Rock: Rock outcrops, from the spectacular schist tors of the Central Otago mountain tops to smaller outcrops, boulders or shattered rock plates in fellfield carry an often complex mosaic of crustose, foliose and fruticose lichens. Lichen communities are often densely developed on tors (above the level of winter snow cover) which provide a variety of ecologies, from exposed tops and sides to sheltered, dry overhangs and crevices, to joints and channels having intermittent water flow. These allow a complex set of communities to develop. Taxa well developed on tors include *Alectoria nigricans, Bryoria austromontana, Caloplaca* spp., *Candelariella* spp., *Cladia aggregata, Cladina* spp., *Cladonia* spp., *Coccocarpia palmicola, Diploschistes* spp., *Hypogymnia* spp., *Immersaria athroocarpa, Lecanora* spp., *Lecidea* spp., *Lecidella* spp., *Lepraria* spp., *Leproplaca lutea, Menegazzia* spp., *Neofuscelia* spp., *Neuropogon* spp. (below), *Pannaria hookeri, Parmelia signifera, P. sulcata, Pertusaria* spp., *Physcia caesia, P. callosa, Placopsis* spp., *Poeltiaria* spp., *Porpidia crustulata, Protoparmelia badia, Pseudephebe pubescens, Pseudocyphellaria* spp., *Psoroma* spp., *Pyrrhospora* sp., *Ramalina fimbriata, Ramboldia petraeoides, Rhizocarpon* spp., *Rimularia* spp., *Schaereria fuscocinerea, Steinera sorediata, Teloschistes fasciculatus, Tephromela atra, Toninia bullata, Trapelia* spp., *Umbilicaria* spp., *Usnea torulosa, Xanthoparmelia* spp., and *Xanthoria elegans.*

Alpine streams: These provide habitats for a number of submerged pyrenocarpous lichens such as *Staurothele fissa, Verrucaria austroschisticola, V. margacea* and *V. rheitrophila.*

Adaptations of alpine lichens: Lichens in alpine environments have evolved several strategies for dealing with growth in often inhospitable environments. Three of these strategies have considerable ecological and biological importance:

Photoprotection: Since high-altitude lichens are adapted to grow in environments exposed to high as well as low temperatures, to high light levels and to often damaging levels of UV radiation, over their long evolutionary history they have developed sophisticated chemical mechanisms using secondary metabolites to protect damage both to genes and to the photosynthetic apparatus. Protective cortical compounds, such as atranorin and usnic acid, act as light filters and as UV-B protection. These lichen 'sunscreens' absorb damaging short wavelength radiation and re-emit this as fluorescence at longer wavelengths which can then be transferred, without harm, to the photosynthetic pigments.

Soil consolidation: A number of soil lichens produce root-like structures known as rhizomorphs. These are linear aggregations of fungal hyphae that may extensively penetrate soil or rock (or wood) substrata and are involved both in substratum colonisation and anchorage of lichens. As lichen colonies develop, the developing net of soil rhizomorphs help stabilise soil, usually by binding small pebbles and soil together, and preventing erosion by wind and water. Species of *Siphula* produce the most obvious and complex rhizomorphs, but other genera such as *Placopsis, Psoroma, Siphulastrum* and *Solorina* are also useful soil consolidators.

Nitrogen fixation: Many alpine lichens are diazotrophic, that is they are able to fix nitrogen from the air into organic nitrogen by means of their cyanobacterial photobionts, which produce the enzyme complex nitrogenase in specialised cells known as heterocysts. Diazotrophic lichens thus act as important biological fertilisers in low-nitrogen habitats. In alpine grasslands they include: *Coccocarpia, Collema, Degelia, Labyrintha, Lecidoma, Leptogium, Pannaria, Parmeliella, Peltigera, Placopsism, Placynthiella, Pseudocyphellaria, Psoroma, Siphulastrum, Solorina, Steinera, Stereocaulon* and *Sticta.*

David Galloway

Opposite top: *Knightiella splachnirima*, in peat on Swampy Summit, Dunedin. *Peter Johnson*

Above: *Neuropogon* on alpine schist outcrop, Treble Cone, overlooking Lake Wanaka. *Peter Johnson*

Fig. 22. Two endemic fellfield species from the Eyre Mountains: *Celmisia thomsonii* (above) which has about five per cent of the population with mauve flowers, as here, and *C. philocremna* (below). *Alan Mark*

Fig. 20 (top left). View across scree and fellfield at 1560 m on the steep slopes below Mt Ida on the Hawkdun Range, North Otago. The pale cushions of vegetable sheep (*Raoulia eximia*), up to 1 m across, are prominent, together with slim snow tussock (*Chionochloa macra*) on relatively stable fellfield, while loose stony debris characterises the sparsely vegetated areas of scree. *Philip Grove*

Fig. 21 (bottom left). High-alpine fellfield at 2050 m on the crest of the St Bathans Range, North Otago. Dark, lichen-covered boulders, some up-ended as 'gravestones', surround circular frost boils of finer material where current frost activity is indicated by their lack of a lichen cover. Brown cushions of the speargrass *Aciphylla dobsonii* are a distinctive feature of this fellfield. *Alan Mark*

These dry fellfields also have extensive areas of pavement and boulderfield composed of large rocks, some arranged as upright 'gravestones' and others into stone nets up to 3 m across, with currently active frost boils of finer material in their centres (Fig. 21) that are indicative of the prevailing periglacial climate. Fellfield is not always clearly distinguishable from plant communities found on the rock bluffs that are widespread through the high mountains of southern New Zealand.

There are also extensive fellfields developed mostly on low-grade schist, as on The Remarkables, Hector, Garvie, Pisa, Richardson, Eyre, Thomson and Takitimu Mountains, in the rainshadow region. Fellfields on these mountains are also distinctive, but those of the Eyre Mountains particularly so, with both high diversity and endemism. Four locally endemic fellfield species – *Celmisia philocremna*, *C. thomsonii* (with up to five per cent of its population having mauve petals – Fig. 22), the cushion speargrass *Aciphylla spedenii* and *Ourisia spathulata* occur here, the first three on the threatened plant list. (For details of the threatened plants from the tussock grasslands and associated mountain lands of southern New Zealand, see Box 8). Other distinctive plants are *Cheesemania wallii*, the generally rare *Epilobium purpuratum*, the spectacular and highly palatable *Ranunculus buchananii* (see Fig. 41) and the strictly high-alpine species *Parahebe birleyi* (see Fig. 40).

Fig. 23. View south across extensive dwarfed cushionfield on the exposed plateau summit of the Pisa Range at about 1830 m, with scattered rock tors and, on the upper leeward slope, steep cirque basins (left). *Alan Mark*

Fig. 24. Upper eastern slope of the Old Man Range at about 1600 m, showing a group of large rock tors in an area of cushionfield vegetation with numerous small solifluction terraces that have *Celmisia viscosa* herbfield present in their sheltered lee. Occasional relic plants of slim snow tussock *Chionochloa macra* still persist. *Alan Mark*

Cushionfield

This is a distinctive vegetation type of the southern South Island high-alpine rainshadow region, characterised by an extremely dwarfed tundra-like plant cover. It dominates the extensive windswept plateau summits of the schistose mountain ranges of Central Otago above about 1400 m (Fig. 23). This landscape is dotted with often massive rock tors (Fig. 24) that add to its special features (see Box 6 on tors). An extremely severe periglacial climate and associated patterned ground are other distinctive features of cushionfield. Windblown silt and fine sand (loess) has accumulated to produce variable soil depth related to exposure. The usually very strong westerly winds, cold summers and frequent freeze-thaw cycles, especially during the snow-free periods (Fig. 25), keep plants close to the ground.

Cushion, mat or low creeping plants predominate here with some eighteen different genera, each with the smallest of their species represented. Even though quite unrelated, many of the plants are remarkably similar and, with their equally small flowers, this cushionfield has been referred to as a 'botanist's nightmare'. Given the frequent high wind speeds, slight differences in exposure are sufficient to produce changes in both cover and composition of the cushionfield. The most common and widespread species is the brown moss-like *Dracophyllum muscoides* that tolerates all but the most exposed sites. The associated cushion species generally include several alpine daisies (species of *Raoulia, Abrotanella, Celmisia,* and *Leptinella (=Cotula)),* cushion species of *Anisotome, Chionohebe, Phyllachne* and *Colobanthus,* as well as dwarfed species of sedge (*Carex*) and grass (*Agrostis* and *Poa),* including the ubiquitous but here extremely dwarfed blue tussock plus the cosmopolitan *Trisetum spicatum.* The biogeographically isolated high-alpine cushion plant, *Hectorella caespitosa,* is also common here. There are also many lichens among the cushions, notably *Cetraria islandica, Alectoria nigricans* and *Thamnolia vermicularis* and several mosses, all inconspicuous. On the most exposed sites, few plants can cope and there may be only loose mats of *Raoulia youngii,* an as yet undescribed white *Raoulia* cushion and perhaps *Colobanthus buchananii.*

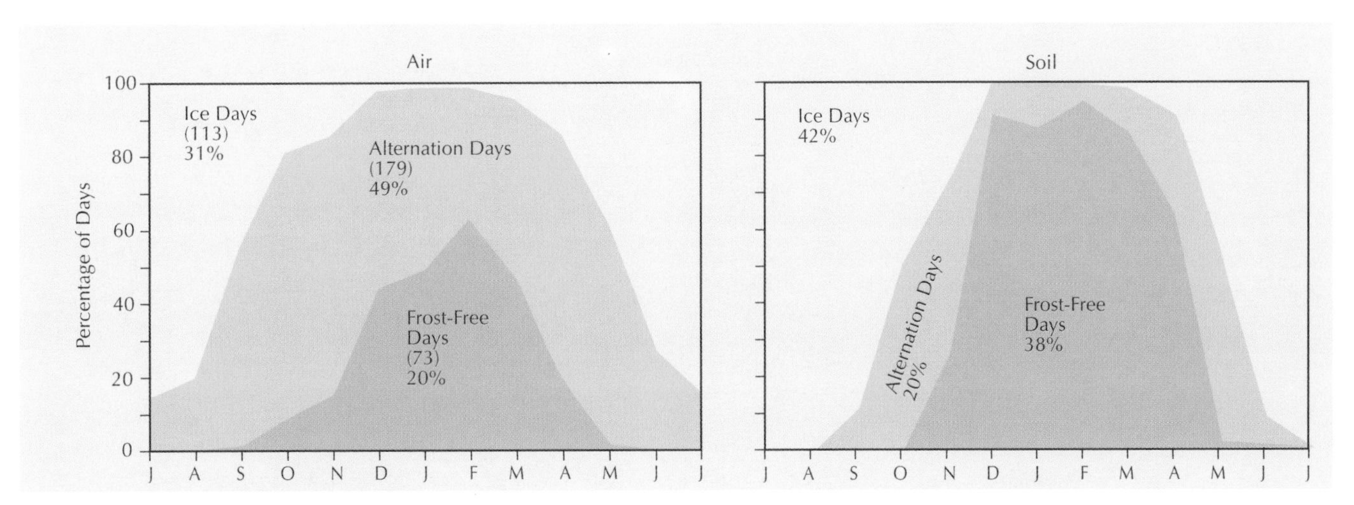

Fig. 25. Seasonal temperature patterns from the high-alpine zone of the Old Man Range recorded over a 5-year period. Monthly and annual values for percentage of ice days, frost-free days and freeze-thaw (alternation) days are shown for both air (1.2 m) and soil (-10 cm).

8 THREATENED PLANTS FROM THE TUSSOCK GRASSLANDS AND ASSOCIATED MOUNTAIN LANDS OF SOUTHERN NEW ZEALAND

New Zealand is renowned not only for the high proportion of its land area that has been allocated for nature conservation and related uses but unfortunately also for the high proportion of its indigenous biota that is threatened for one reason or another. Very limited natural distribution of certain species, habitat destruction, or the presence of aggressive plant and/or animal pests that have been introduced accidentally or intentionally, have each been important factors in this.

Measures aimed at conserving threatened biota are dealt with in the final chapter. Here we merely list all of those higher (i.e. vascular) plants for the tussock grasslands and associated mountain lands of southern New Zealand, as for the forest and shrublands, wetlands and coastal communities elsewhere, that have been placed on the threatened plant list, published in 1995, in one or more of seven categories as listed for forests and shrublands (Chapter 8).

For each species the distributions within and outside the region are given, plus a brief note on its habitat. Numbers in brackets indicate the number of plant taxa within this category, nationally.

Extinct (9)

Stellaria elatinoides. North and Central Otago; Hawkes Bay, Ashburton: bare places in grassland.

Critical (20)

Carmichaelia kirkii s.l. Otago; Canterbury, Nelson, inland Canterbury: lowland to montane open ground and grassland-shrubland.

Lepidium sisymbrioides ssp. *matau*. Central Otago: salt pans.

Endangered (37)

Carex inopinata. Central Otago: dry montane short tussock grassland-shrubland.

Ischnocarpus novae-zelandiae. Inland Otago-Southland; Nelson, Canterbury: dry rocky sites in lowland and montane short tussock grassland.

Lepidium kirkii. Central Otago: salt pans.

Olearia hectorii. Inland Otago-Southland; Nelson, Marlborough, Canterbury: montane open grassy shrubland.

Pseudognaphalium 'compactum'. Inland Otago-north central Southland: tarn margins in montane short tussockland.

Simplicia laxa. Inland Otago: montane to subalpine tussocklands, on or near rock outcrops.

Vulnerable (62)

Carmichaelia curta. North Otago: open gravelly river terraces, Waitaki Valley.

Chenopodium detestans. Inland Otago; Marlborough, Canterbury: lowland to montane short tussockland.

Deschampsia spitosa. Widespread: in locally wet grasslands and lake margins.

Hebe cupressoides. Otago Lakes District; Marlborough, Canterbury: local on river terraces in tussock-shrublands.

Iphigenia novae-zelandiae. Inland Otago-northern Southland; Canterbury: lowland to montane damp red tussocklands and wetlands.

Lepidium sisymbrioides ssp. *kawarau*. Central Otago: salt pans.

Luzula celata. Coastal and Central Otago: open sandy grassland-scabweed.

Olearia fragrantissima. Inland and eastern Otago-Southland; mid-south Canterbury: grass-shrubland of forest margins.

Triglochin palustre. Inland Otago; inland Canterbury: damp hollows in montane short tussockland.

Rare (79)

Carex uncifolia. Central Otago-inland Southland; Volcanic Plateau, inland Marlborough-Canterbury: valley tussocklands

Chionochloa spiralis. Fiordland: rock bluffs

Chionohebe glabra. Central Otago-inland northern Southland: snowbanks.

Hebe 'Takahe'. Fiordland: upland limestone bluffs near treeline

Lepidium sisymbrioides ssp. *sisymbrioides*. Inland Otago-Southland; inland Canterbury: lowland to montane short tussockland

Leptinella albida. Central Otago: cushionfield

Luzula crenulata. Central Otago: cushionfield

Myosotis albosericea. Dunstan Mountains, Central Otago: blue tussock grassland-cushionfield

Myosotis 'glauca'. Inland Otago-northern Southland: rocky places in montane short tussock grassland

Myosotis oreophila. Dunstan Mountains, Central Otago: cushionfield

Myosurus minimus ssp. *novae-zelandiae*. Inland Otago-Southland; eastern South Island: seasonally damp hollows in lowland to montane short tussockland

Olearia fimbriata. Coastal-inland Otago-Southland: lowland to montane open grassy shrubland

Plantago obconica. Central Otago; mid Canterbury: moist turf in alpine snow tussockland

Poa senex. Central Otago-north central Southland: cushion bog and snowbanks

Puccinellia raroflorens. Central Otago: damp areas on salt flats

Ranunculus haastii subsp. *piliferus*. Central northern Southland: screes

Senecio dunedinensis. Coastal and inland Otago-Southland; Canterbury: lowland to montane rocky tussockland

Montigena (Swainsona) novae-zelandiae. North Otago; Marlborough, Canterbury: scree and rocky tussockland.

Insufficiently known

Chionohebe myosotoides. Central Otago: cushionfield

Deschampsia pusilla. Otago Lakes District: damp alpine tussocklands and snowbanks

Myosotis cheesemanii. Central Otago: cushionfield

Myosotis glabrescens. Otago Lakes District: cushionfield

Rytidosperma tenue. Eastern-Central Otago: snow tussockland

Uncinia purpurata. Eastern-inland Otago-Southland; inland Canterbury: snow tussockland

Local

Aciphylla montana var. *gracilis*. North Otago; South Canterbury: rocky snow tussockland.

Aciphylla stannensis. Stewart Island: snow tussock-shrubland.

Brachyscome humilis. Central-eastern Otago: snowbanks

Carex edgariae. Otago Lakes District-northern Southland: damp areas in montane short tussockland.

Carmichaelia compacta. Central Otago: open rocky short tussockland-shrubland.

Celmisia haastii var. *tomentosa*. Rock and Pillar Range, Central Otago: snowbanks.

Celmisia hookerii. Coastal-inland Otago-northern Southland: subalpine-rocky snow tussockland.

Celmisia philocremna. Eyre Mountains: fellfield.

Celmisia spedenii. Eyre Mountains: ultramafic outcrops.

Celmisia thomsonii. Eyre Mountains: fellfield.

Ceratocephalus pungens. Central Otago: seasonally wet hollows in semi-arid short tussockland.

Cheesemania wallii. Otago Lakes District-north central Southland: fellfield.

Coprosma intertexta. Central Otago; inland south Canterbury: rocky open tussockland-shrubland.

Crassula multicaulis. Central Otago-northern Southland; Inland South Island: damp areas in short tussockland.

Epilobium purpuratum. Otago-north central Southland rainshadow high mountains: fellfield and snowy rock ledges.

Gentiana gibbsii. Stewart Island: wet snow tussock-moorland.

Gentiana lilliputiana. Dunstan Mountains, Central Otago; inland south Canterbury: damp flushes in snow tussockland.

Geum pusillum. Central Otago-north central Southland: snowbanks.

Hebe murrellii. Fiordland: fellfield.

Ourisia spathulata. Northern central Southland: rocky snow tussockland and fellfield.

Pleurosorus rutifolius. Central Otago; widespread but uncommon: sunny rock faces in semi-arid lowlands.

Poa pygmaea. Central Otago: cushionfield.

Ranunculus brevis. Inland Otago; South Island rainshadow region: damp areas and river terraces in tussockland.

Ranunculus maculatus. Central Otago-north central Southland: wet flushes in herbfield.

Ranunculus scrithalis. Eyre Mountains: screes.

Ranunculus stylosus. Stewart Island: windswept rocky subalpine snow tussockland-herbfield.

Stipa petriei. Central Otago; inland south Canterbury: rocky bluffs in dry lowland short tussockland.

Alan Mark and William Lee

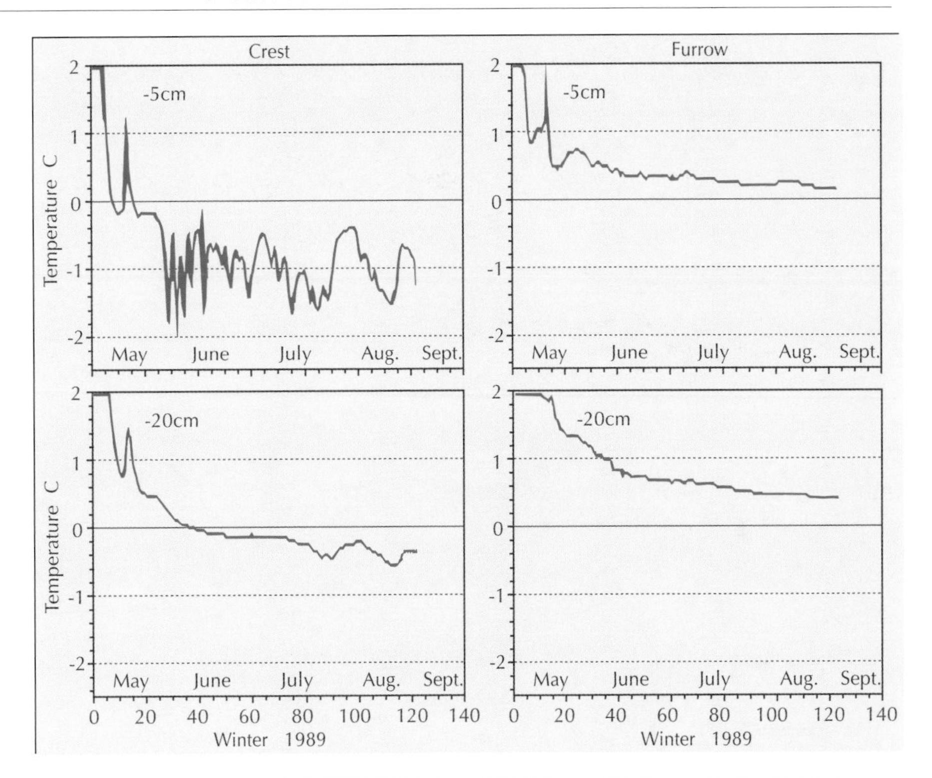

Fig. 26. Daily ranges of soil temperatures at two depths (5 and 20 cm) beneath the crest (left column) and furrow (right column) of a soil stripe, similar to that shown in Fig. 27, from the Old Man Range, Central Otago, during the 1989 winter. Values above 2°C recorded during the first few days are not shown. Numbers along the base are days from 1 May.

Fig. 27 (centre right). Earth or soil stripes about 1.2 m across and up to 30 cm high, developed in high-alpine cushionfield vegetation on a gentle south-aspect slope at about 1850 m on the Pisa Range, Central Otago. *Alan Mark*

Patterning of the ground surface into hummocks, stripes and solifluction terraces adds to the intrigue of these cushionfields. Soil or earth hummocks up to 2 m across and 30 cm deep on near-flat ground, that grade into stripes of similar dimensions as the slope increases beyond about 5°, can be found on most of the Central Otago mountain summits (Fig. 27). Solifluction terraces up to 1.2 m high and many metres long occur singly or in suites of broad steps down leeward slopes of 2–10°, where snow usually accumulates to provide copious meltwater in spring (see Fig. 24). Smaller crescentic solifluction lobes may occur on steeper slopes within many of the larger snowbanks. All of these features are vegetated with patterns that reflect the importance of microtopography in controlling several environmental factors, particularly exposure and associated snow lie. The hummocks and stripes had been considered to be relics of an earlier period, when the climate was more severe than today. However, some of the terraces and lobes give clear signs of activity, in the form of collapsed over-steepened fronts and buried soils that have been over-ridden by down-slope movement of soil. Recent studies have confirmed down-slope movement of the larger terraces, amounting to about 3.5 cm a decade, while the hummocks and stripes have been found to experience differential freezing in winter, apparently sufficient to maintain them. Soil on their crests freezes to at least 20 cm over winter while in the depressions it remains unfrozen at this depth, freezing only intermittently at 5 cm depth (Fig. 26).

Cushionfield has obviously become more extensive as a result of pastoral farming, having replaced slim snow tussockland on many of the Central Otago mountains (Fig. 28). Distinction between the natural and induced cushionfields is no longer possible except where remnant slim snow tussocks persist. The most exposed and higher altitude areas without snow tussocks are probably natural.

Fig. 28. View south along the crest of the Garvie Mountains at about 1740 m, showing an extensive area of slim snow tussock *Chionochloa macra* grassland which locally has been transformed into a cushionfield through removal of the dominant tussock, probably several decades ago, through pastoral practices of burning and grazing. Rather subdued soil stripes can be seen where the cushionfield and grassland merge. *Alan Mark*

Fig. 29. The scree buttercup *Ranunculus haastii* (top), one of the most conspicuous scree plants when flowering, has been excavated to show the large perennial underground stem with its numerous roots, plus the pale-coloured current season's stem which has penetrated the mobile layer of surface stones. The plant at bottom is subspecies *piliferus* from the Eyre Mountains at about 1370 m. *Alan Mark*

Scree

Talus slopes or screes are characterised by a mantle of loose, sometimes quite mobile, stones 10–15 cm thick. These occur over a moist sandy substrate on uniformly steep slopes of about 32° angle, the so-called 'angle of repose', and have a very sparse cover of highly cryptic, specially adapted plants. They are a distinctive feature of the greywacke mountain ranges of the northern and central South Island from central Marlborough to North Otago. The Kakanui, Hawkdun and St Bathans Ranges on the northern perimeter of the region have some typical screes. Several of the more widespread scree plants can be found here including the colourful scree buttercup *Ranunculus haastii* (Fig. 29, top) together with *Epilobium pycnostachyum*, *Lobelia roughii*, *Stellaria roughii*, *Poa buchananii*, *Hebe epacridea*, and the penwiper plant *Notothlaspi rosulatum*, as well as the generally rare legume *Montigena* (*Swainsona*) *novae-zelandiae*.

There is also extensive development of coarse debris slopes on the high mountains of the Otago Lakes District, associated with low-grade schist. These have some features, including their plants, of scree. The Eyre Mountains have a few widespread scree species: *Epilobium pycnostachyum* and *Stellaria roughii*, as well as some local endemics confined to such habitats. *Ranunculus scrithalus* occurs locally on fine scree, usually with surface clay, and *R. haastii* ssp. *piliferus* occurs on coarse debris slopes in the vicinity (see Fig. 29). The latter also occupies a similar habitat on the upper western slopes of East Dome.

THE WET WESTERN MOUNTAINS AND STEWART ISLAND

Low-alpine snow tussock – herbfield

Where forest persists virtually intact in the wet western mountains along and close to the Main Divide, tussock grasslands remain restricted to sites that are too cold and/or wet for woody vegetation. These areas prevail mostly in the low-alpine zone, apart from three local situations below treeline. Permanently wet, seasonally cold depressions in valley floors may be occupied by copper tussockland or mixed shrub-tussockland, as described above for areas in eastern Fiordland (Fig. 1). The second situation can be found on generally well-drained broad valley floors close to the Main Divide which may also be in grassland, presumably because of periodically very severe winter frosts that the forest trees cannot tolerate. While most of these valleys are also periodically subjected to severe floods, the grasslands generally extend above the normal flood level to the change of slope on the valley sides. Where the forest margins are dominated by the most cold-tolerant of the local tree species, notably celery pine *Phyllocladus alpinus*, mountain beech *Nothofagus solandri* var. *cliffortioides* and silver beech *N. menziesii*, the beech trees along the actual margin may show obvious frost damage following the most severe winters. Grasslands on the more accessible of these open valley floors have been modified by domestic stock and, at earlier times, most have been grazed by large herds of red deer. The plant cover varies not only with grazing but also with altitude, as well as elevation above river level which affects frequency of flooding. Relatively flood-free areas below treeline are often dominated by several tussock species. These include silver tussock, alpine fescue tussock or, on higher ground in the more remote valleys, by narrow-leaved and/or curled snow tussock *Chionochloa crassiuscula*, sometimes with copper tussock and usually several smaller tussock grasses such as blue tussock. Associated species are sedges (*Carex* spp., *Uncinia* spp.) and rushes (*Luzula* spp., *Schoenus pauciflorus*), as well as several herbs: species of orchid *(Microtis)*, ferns *(Ophioglossum)*, several daisies (species of *Helichrysum*, *Leptinella* and *Celmisia)* plus *Plantago*, *Ranunculus*, and some exotics such as sheeps sorrel *Rumex acetosella*, browntop, Yorkshire fog *Holcus lanatus*, mouse-ear chickweed *Cerastium fontanum*, white clover *Trifolium repens* and ragwort *Senecio jacobaea*. Wetlands may occupy local depressions on these valley floors. The third situation of coastal moorlands is described later.

The pattern of low-alpine vegetation above the treeline has been described in general terms for Mt Aspiring and parts of Fiordland National Parks, as well as Stewart Island. The general vegetation pattern of the low-alpine zone changes gradually as rainfall increases westward towards the Main Divide. The dominant snow tussocks change in composition and overall plant diversity increases, notably and conspicuously among the larger herbs and co-dominant shrubs: hence the name snow tussock-herbfield. This transition can be seen on the Eyre Mountains in north-central Southland, 60–70 km east of the Main Divide. Here six species of *Chionochloa* and twenty of *Celmisia*, embracing ones of both eastern and western distribution, as well as several endemic species, occur in a range of snow tussock grassland and associated mountain land communities.

Along the southern part of the Main Divide, the pattern of low-alpine snow tussock-herbfield communities is reasonably consistent among the four widespread species of alpine snow tussock (*Chionochloa rigida, C. pallens, C. crassiuscula, C. oreophila*), but with some increase in diversity in Fiordland, where four endemic snow tussocks are present (*C. teretifolia, C. acicularis, C. ovata, C. spiralis*). These four are assumed to reflect origin and survival somewhere in the Fiordland region through the Pleistocene glaciation.

The low-alpine zone begins at a generally abrupt treeline of silver and/or mountain beech at 900 m–1200 m, decreasing mainly with increasing latitude and proximity to the coast. A tall mixed snow tussock-shrubland with the western subspecies of narrow-leaved snow tussock (*C. rigida* ssp. *amara* based on Henry Connor's recent revision of *Chionochloa*; previously referred to in this region as *C. flavescens*) and several tall shrubs, usually dominate within 200 m of treeline (Fig. 30). The shrubs include *Coprosma* cf. *pseudocuneata, Dracophyllum uniflorum*, celery pine, snow totara *Podocarpus nivalis* and *Hebe* spp. Several large herbs are also usually present: *Aciphylla horrida, Astelia* spp., *Phormium cookianum, Ourisia macrophylla* and *Celmisia* spp. With increasing altitude, the tall shrub component decreases or becomes restricted to relatively snow-free slopes and ridge crests. Narrow-leaved snow tussock is replaced by the somewhat shorter curled and midribbed snow tussocks (*C. crassiuscula* ssp. *torta* and *C. pallens*) with several large herbs. The altitude of this transition varies in relation to aspect and steepness of slope, which largely reflects length of snowlie. These two snow tussock species usually occupy a well-defined band virtually to the upper limits of the low-alpine zone, but their proportions may vary depending generally on topography and the past stability of a particular site (Fig. 31).

Site stability affects a range of physical and chemical soil properties. Young or immature soils usually occupy stony or rocky, relatively unstable slopes. They are coarse textured with only weakly differentiated horizons, and free-draining, with low organic content yet relatively high nutrient status. These features contrast with more mature soils on stable sites, that generally have well-developed horizons, are fine textured and strongly leached, with generally poor drainage, high organic content and low levels of nutrients. Midribbed snow tussock predominates on the less stable sites and curled snow tussock on more stable areas, with intermediate sites varying in their proportion of these tussocks. Associated plants also vary among the range of sites. Some are more often associated with midribbed snow tussock: co-dominant shrubs such as *Hebe hectorii*, large herbs such as *Ranunculus lyallii, Anisotome haastii, Ourisia macrocarpa* and *Aciphylla crenulata* as well as many smaller herbs, e.g. Anaphaloides (*Helichrysum*) *bellidioides, Geranium microphyllum, Oreomyrrhis colensoi, Viola cunninghamii.*

Fig. 30. Treeline of silver beech at about 950 m near Key Summit, Fiordland National Park, showing tall mixed snow tussock-shrubland that usually dominates up to 200 m above treeline. The shrubs here are mostly of turpentine scrub *Dracophyllum uniflorum*, while the western subspecies of narrow-leaved snow tussock and speargrass are also conspicuous. Lake Gunn in the Eglinton Valley can be seen beyond. *Alan Mark*

Fig. 31 (left). Low-alpine snow tussock grassland at about 1300 m in the Stillwater Valley, Fiordland National Park, showing community patterns related particularly to topography. The tall tussocks, mostly midribbed snow tussock *Chionochloa pallens* with some narrow-leaved snow tussock *C. rigida* ssp. *amara*, dominate the more exposed rocky sites with some co-dominant shrubs, mostly *Dracophyllum uniflorum* and large herbs, while curled snow tussock *C. crassiuscula* ssp. *torta* dominates most of the depressions. *Alan Mark*

Fig. 32. Low-alpine snow tussock grassland in the upper Chester Burn, Murchison Mountains, Fiordland National Park. Differences in topography here are associated with variation in past stability of the landscape, which is also reflected in differences in the snow tussock communities. Midribbed snow tussock *Chionochloa pallens* dominates with shrubs, mostly *Hebe* spp. on the debris fan (right) while needle-leaved and curled snow tussocks (*C. acicularis* and *C. crassiuscula* ssp. *torta*) occupy the permanently wet, stable moraine mounds (left) and mixed narrow-leaved (*C. rigida* ssp. *amara*)-curled snow tussock grassland dominates the sites of intermediate stability and drainage, between. (See Fig. 33 for diagramatic representation.) *Alan Mark*

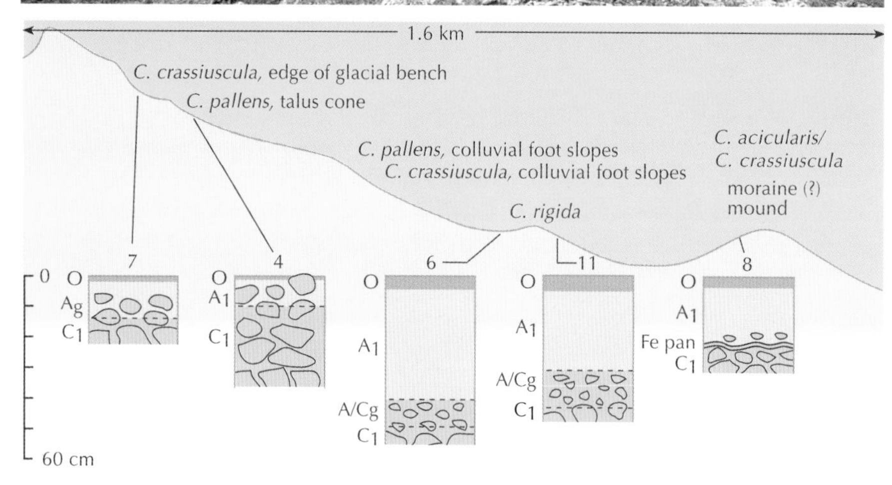

Fig. 33. Diagram showing the relationship between landscape and associated soil features, plus the changing pattern of dominance among the local snow tussock species, based on a study by Peter Williams. (See Fig. 32 for photographic representation.)

Other species characterise communities dominated by curled snow tussock on more stable sites. Again there is a mix of co-dominant shrubs such as *Dracophyllum uniflorum*, large herbs like *Astelia nivicola* and many smaller plants such as *Oreobolus pectinatus*, *Phyllachne colensoi*, *Forstera sedifolia*, *Myrsine nummularia*, *Lycopodium fastigiatum*, *Celmisia durietzii*, *C. glandulosa* and *Psychrophila* (*Caltha*) *novae-zelandiae*. Other more widespread species may be associated with both types of grassland: *Poa colensoi*, *Celmisia petriei*, *Coprosma cheesemanii*, *C. perpusilla*, *Anisotome flexuosa*, *Oreobolus impar*. Two of the three Fiordland endemic snow tussocks, *C. teretifolia* at moderate altitudes (eastern Fiordland) and needle-leaved snow tussock *C. acicularis* at relatively low altitudes (western/northern Fiordland), are both associated with mature peaty soils on stable poorly drained sites such as glacial benches or moraine mounds (Figs 32 and 33).

Needle-leaved snow tussock descends to sea level locally near West Cape, where it dominates on wet peaty coastal moorlands among manuka shrubland and woodland on a rolling peneplain of apparently non-glaciated marine terraces (Fig. 34). Umbrella fern *Gleichenia dicarpa*, *Lycopodium ramulosum*, wire rush, several orchids (*Aporostylis bifolia*, *Prasophyllum colensoi*, *Thelymitra venosa*, *Pterostylis graminea*) and small heaths (*Pentachondra pumila*, *Cyathodes empetrifolia*), as well as several other bog species, are associates in this moorland type grassland community.

The fourth widespread species of alpine snow tussock, snowpatch tussock *Chionochloa oreophila*, is largely restricted to snowbanks and other high-alpine communities and will be discussed with them later.

Fig. 34. Needle-leaved snow tussock *Chionochloa acicularis* dominating wet coastal moorland on rolling non-glaciated terrain near sea level at West Cape, Fiordland National Park. Manuka *Leptospermum scoparium* is the commonest shrub but pink pine *Halocarpus biformis*, yellow-silver pine *Lepidothamnus intermedius* and inaka *Dracophyllum longifolium* are also present, while tall woodland and forest occupy the better drained slopes and valleys. *Alan Mark*

Fig. 35. Snow tussock-cushion moorland at about 950 m near the summit of Mt Anglem, Stewart Island. The pungent local subspecies of curled snow tussock (*Chionochloa crassiuscula* ssp. *crassiuscula*) dominates the area but a plant of the western subspecies of narrow-leaved snow tussock (*C. rigida* ssp. *amara*) is on the left and cushions of the pale green endemic *Raoulia goyenii* and brown *Dracophyllum politum* are also obvious. *Alan Mark*

Fig. 36. Cushion herbmoor at about 870 m on the crest of Maungatua, in Eastern Otago, surrounded by mixed narrow-leaved snow tussock-shrubland dominated by *Chionochloa rigida* and *Dracophyllum longifolium*. The cushion plants are mainly *Donatia novae-zelandiae* and comb sedge *Oreobolus pectinatus*, with a matrix of white shoe-string lichen *Thamnolia vermicularis*. *Alan Mark*

Stewart Island

There is sparse development of low-alpine snow tussock grassland on Stewart Island, and also on the ultramafic outcrops of both the Red Hills region in north-west Otago and the Livingstone Mountains in northern Southland. On Stewart Island, the highest elevation (980 m) is considered barely sufficient to reach the alpine zone but abundant moisture and very poor drainage on gentle slopes in the subalpine zone above 400–500 m generally favours an herbaceous cover. This vegetation ranges from tussock grassland (or moorland) to cushion bog and includes three snow tussock species. The local subspecies *crassiuscula* of curled snow tussock (previously known as *C. pungens* for its pungent leaf tips) is confined to the Island, as is *C. lanea*, together with the western subspecies (*amara*) of narrow-leaved snow tussock (referred to as *C.* 'Mt Anglem *flavescens*' by Hugh Wilson). *C. lanea* tall tussockland and tall tussock-shrublands, usually with some curled snow tussock, are most extensive in the Table Hill/Mount Allen area, on wet peaty slopes at 550–650 m. Tall narrow-leaved snow tussockland and tall snow tussock-shrubland, often with some curled snow tussock, are largely restricted to the summit ridges of Mt Anglem, also on wet peaty soils. In both areas, a range of cushion species occur with the snow tussocks, as well as forming dense turves of cushion herbmoor (Fig. 35). Herbmoor, usually with a sparse cover of curled snow tussock overlying peat, may be extensive on the many gentle crests and is essentially continuous above 850 m on Mt Anglem. Cushions of *Donatia novae-zelandiae* and comb sedge *Oreobolus pectinatus* are the most common, but there are many associates, including several of limited distribution: *Astelia subulata*, *Abrotanella muscosa*, *Raoulia goyenii*, *Celmisia polyvena*, *C. clavata*, as well as many bryophytes.

Cushion herbmoor is also a feature of similar sites in the low-alpine zone among snow tussock grassland and tussock-shrubland in the rainshadow region on Maungatua (870 m, Fig. 36), the Blue Mountains (998 m) and on Mt Pye/Ajax Hill (720 m) in the Catlins Region.

Ultramafic vegetation

Soils derived from ultramafic rocks (igneous rocks composed of ferro-magnesium minerals) have a distinctive chemical composition with high levels of magnesium and nickel, and low concentrations of calcium, phosphorus and potassium. In southern New Zealand, they occur discontinuously from Red Mountain in north-west Otago, south to the Livingstone Mountains in northern Southland. Fiordland National Park contains ultramafic rocks around Anita Bay and Lake Ronald, south of Milford Sound, and a small outcrop at the summit of Mount Luxmore, in eastern Fiordland.

Because of the extreme soil conditions, vegetation development on ultramafic rocks is noticeably different from that on adjoining rock types. In the montane zone, tall beech forest on schist rock abruptly gives way to a sparse cover of mixed heath forest on ultramafic rock, usually of small conifers: yellow-silver pine *Lepidothamnus intermedius*, pink pine *Halocarpus biformis* and celery pine, and a few tolerant broadleaved trees (mountain beech, southern rata). Vegetation in the low-alpine zone is even more sparse (Fig. 37), usually comprising low-growing small-leaved shrubs: *Melicytus alpinus*, pygymy pine *Lepidothamnus laxifolius*, *Dracophyllum uniflorum*, *Myrsine nummularia*, amongst open grassland areas with *Chionochloa rigida* ssp. *amara*, the short blue tussock and bristle tussock *Rytidosperma setifolium* (Fig. 38). The sedge *Schoenus pauciflorus* is widespread, especially in seeps, together with a few small herbs such as *Wahlenbergia albomarginata* and *Celmisia gracilenta*. The only vascular plant species endemic to the southern ultramafics, *Celmisia spedenii*, is common on outcrops in the Livingstone and Eyre Mountains. The nationally rare *Carex uncifolia* is also abundant in this habitat.

Fig. 37 (left). View west down Simonin Stream, which forms an abrupt contact between the red-coloured, essentially bare ultramafic rocks on the left and the schist basement rock which supports a normal pattern of vegetation to the right. *Alan Mark*

Fig. 38 (right). Sparse mixed shrub-tussock grassland in the foreground on ultramafic soil at Simonin Pass (about 1000 m), Red Mountain, north-west Otago. Silver beech forest occupies schist terrane in the distance. *William Lee*

Fig. 39 (below). High-alpine fellfield at about 1620 m near Tutoko Saddle, Fiordland National Park, showing a sparse cover of several species – *Aciphylla congesta* (in flower), *Celmisia verbascifolia, Dolicoglottis scorzoneroides* and *Chionochloa oreophila* growing among the stable rock. *Alan Mark*

High-alpine vegetation patterns

High-alpine fellfield and snowbank communities are common on the wet western mountains, where the patterns and processes are generally similar to those of the rainshadow mountains. Diversity tends to be higher and the communities more colourful because of generally larger plant and flower sizes in the high rainfall region (Fig. 39).

The wet western fellfields share many of the species listed for the Eyre Mountains, but there are usually several additional species of *Aciphylla* and *Celmisia* among many other genera.

Snowbank communities, as described for the Humboldt Mountains, also share many species with the Eyres, but snowpatch tussock usually provides most of the plant cover. One of the most distinctive and colourful snowbank species of the wet western snowbanks, *Ranunculus sericophyllus,* is notably absent from the Eyre Mountains.

The nival zone

In southern New Zealand, the nival zone, characterised as elsewhere by permanent ice and snow, begins at about 2000 m or where the mean air temperature for the warmest month is close to freezing. A limited range of plants may form a very sparse cover here, confined to the few sites that are either too steep or exposed to be permanently covered by snow.

Only a few peaks reach the nival zone in the rainshadow region of southern South Island – Double Cone (2324 m), Ben Nevis (2234 m) and James Peak (2019 m) on the Remarkables and adjacent northern Hector Mountains, and Jane Peak (2035 m) in the Eyre Mountains, but in the wet western region from northern Fiordland northwards along the Main Divide, extensive areas reach the nival zone.

There are only minor regional differences in the flowering plants that may be present in the nival zone. One, *Parahebe birleyi* (Fig. 40), is essentially confined to this zone, while a few others may extend up a short distance from the high-alpine zone below, most notably from snowbank sites: blue tussock, *Poa novae-zelandiae, Carex pyrenaica* var. *cephalotes, Hectorella caespitosa, Raoulia subulata, Celmisia hectorii* and *Gentiana divisa,* together with several lichens and mosses. Some species are restricted to the higher rainfall western ranges: *Ranunculus buchananii* (Fig. 41), the cushion plant *Chionohebe ciliolata, Ourisia sessilifolia* ssp. *splendida,* and, perhaps the most photographed of all our alpine species, *Ranunculus sericophyllus* (Fig. 42).

Fig. 40 (above). *Parahebe birleyi*, which has the highest overall elevation among the alpine flowering plants. It is essentially confined to the nival zone, with a range between about 2000 m and 2930 m. *Alan Mark*

Fig. 41 (top right). *Ranunculus buchananii*, a spectacular and extremely palatable herb of high-alpine fellfield in the wetter western mountains of southern South Island. *Alan Mark*

Fig. 42 (right). *Ranunculus sericophyllus*, a common and much-photographed small herb of high-alpine fellfield and snowbanks in the wet western mountains of the South Island. *Alan Mark*

EFFECTS OF EXOTIC PLANTS AND ANIMALS ON GRASSLANDS AND ASSOCIATED MOUNTAIN LANDS

Invasive weeds threatening tussock grasslands

Tussock grasslands are threatened by several groups of introduced plants that have greater invasiveness and environmental tolerance than native shrub and tree species. The establishment of most of these species is associated with human disturbance, but some can also invade relatively intact grassland.

Lowland and montane short tussock grasslands are inherently vulnerable, due to the short stature and open character of the vegetation. Hawkweeds, especially mouse ear *Hieracium pilosella* and king devil *H. praealtum* have spread during the last few decades into fescue tussock grassland in semi-arid areas (Fig. 43, page 220) and now dominate many sites to the exclusion of native species. The roles of grazing, fertiliser additions and climate in this process have been debated for some time, but there is little doubt that hawkweeds can displace native short tussock grassland. The conservation management of short tussock grasslands may be complex, perhaps requiring light seasonal grazing to limit successional processes. Around Lake Wanaka and in the Matukituki catchment to the west, including Mt Aspiring National Park, tussock hawkweed *H. lepidulum* is equally aggressive (Fig. 44, page 220).

Some woody species also threaten both the short and tall tussock grasslands in the rainshadow areas. These include several species representing life forms scarce in the New Zealand flora, including deciduous sweet brier *Rosa rubiginosa* and the nitrogen-fixing shrubs broom *Cytisus scoparius* and gorse *Ulex europaeus*. The latter group are particularly resistant, due to their fire and grazing tolerance and large persistent seed banks in the soil. The improved soil nitrogen levels associated with the establishment of broom and gorse generally favour exotic species with faster growth rates than native plants. On wet or impoverished soils supporting red tussock, introduced heaths – notably spanish heath *Erica lusitanica* and heather (e.g. *Calluna vulgaris*) – and birches (e.g. *Betula pendula*) may be locally invasive. However, perhaps the greatest threat to eastern grasslands are introduced conifers from the northern hemisphere. In southern areas lodgepole pine *Pinus contorta*, Corsican pine *P. nigra*, Douglas fir *Pseudotsuga menziesii* and larch *Larix decidua* are actively spreading into tall tussock grasslands and are capable of displacing extensive areas of these communities in the montane to low-alpine zone.

Control of invasive woody weeds is dependent on removing outliers and pioneer plants. Although sheep and cattle grazing may limit the spread of exotic weeds, this process is difficult to manage for conservation values, as the grazing intensities required to reduce the densities of woody weeds are usually too high to maintain the major tall tussock species and certainly the more palatable members of the indigenous grassland communities.

Vegetation of the high-alpine zone and mountain grasslands of the wet western regions are protected by their remoteness and/or the limited range of exotic plant species capable of growing in these environments. However, some of the introduced conifers and heaths could establish in these areas. The long-term conservation of our distinctive and diverse southern tussock grasslands and associated mountain lands will require continued vigilance to ensure that introduced woody weeds are eliminated, at least in protected natural areas.

Threats of exotic animals to native flora and fauna

Effects of pastoral farming practices, particularly grazing by stock associated with periodic burning, together with aerial oversowing and topdressing on the lower slopes of pastoral runs in the rainshadow region, clearly have had a variable but in many areas major impact on the indigenous plant communities. These issues continue to be debated, particularly in the context of the overall desire for, and national promotion of, sustainable manage-ment of natural resources in terms of the 1991 Resource Management Act. The highly modified areas, however, are less relevant to the natural history of the region than the effects of several introduced wild mammals, notably red deer and chamois, on the two large national parks, Fiordland and Mt Aspiring (Box 9), and the many other smaller areas that are formally protected for their nature conservation and related values. Despite the degrading influences of most of these wild animals on the natural ecosystems, most have their supporters among recreationalists, and debate continues as to the significance and permanence of the impacts, as well as on desirable levels of control.

Long-term monitoring is essential to provide information on vegetation and animal trends over time, as well as to indicate similar trends on areas of previous pastoral land that has only recently been acquired for conservation management through tenure review of pastoral leasehold land or other means (see Chapter 13). A limited amount of monitoring was initiated in the early 1970s but much more has been set up more recently, the results of which will be useful for indicating any future management needs.

Rabbits are in pest numbers only in the drier parts of the rainshadow region, but mostly in areas that have been modified to the extent that they are now of limited value for nature conservation. An exception is Flat Top Hill Conservation Area near Alexandra, where the effects on nature conservation values of effective rabbit control that has been attained with the illegal release of Rabbit Calicivirus (Rabbit Haemorraghic Disease) in 1997 remain uncertain (see Chapter 13 for more details).

Hares are much more widespread but generally at relatively low densities; hence they are considered to have lesser impacts on conservation values. Goats, by contrast, are more local but are of continuing importance in some areas, for example the upper Shotover catchment, despite some major control efforts. Possums, although widespread, are in relatively low numbers in the tussock grasslands and associated mountain lands, and have probably had only minor impacts on the vegetation here.

The arrival of humans and the consequent flood of introduced mammals has also had its impacts on the native fauna. One effect has been the elimination of a suite of mainly large-bodied invertebrates from the lowlands, many of which have survived in refugia in alpine areas (where although predators

Fig. 43. Mouse-ear and king devil hawkweeds (*Hieracium pilosella* and *H. praealtum*) displacing short tussock grassland (left), including the dominant fescue tussock (*Festuca novae-zelandiae*), at about 400 m on the western slope of the Rock and Pillar Range near Patearoa. The larger narrow-leaved snow tussocks appear to be more resistant to the invasion. The close-up (right) shows a weakened fescue tussock bordered by the pale mouse-ear hawkweed with the single flower heads (foreground) and the bright green king devil with multiple flower heads (behind the tussock). *Alan Mark*

Fig. 44. Tussock hawkweed (*Hieracium lepidulum*) dominating extensive areas in mixed snow tussock-shrubland above a naturally depressed treeline in the upper Rob Roy catchment at about 800 m, Mt Aspiring National Park, November, 1994. *Alan Mark*

9 MONITORING THE EFFECTS OF WILD UNGULATES ON THE TUSSOCK GRASSLANDS AND ASSOCIATED MOUNTAIN LANDS OF MT ASPIRING AND FIORDLAND NATIONAL PARKS

Monitoring of the alpine tussocklands and associated mountain land vegetation in both parks was initiated about the time commercial hunting of deer (including wapiti hybrids in central Fiordland) from helicopters began in the late 1960s. Substantial reductions in the previously very large populations, particularly in remote areas, where recreational hunting was ineffective, were rapidly achieved. Chamois, which are generally thinly spread through the alpine zone of the wet western mountains, were similarly hunted, while thar are essentially absent, since the official management plan for this species restricts them to areas immediately north of the region covered in this book.

The Fiordland monitoring study has been concentrated in the Murchison Mountains (Takahe Area) and adjacent Stuart Mountains (Wapiti Area), where the alpine grasslands are inhabited by the endangered takahe. Vegetation recovery in the Stuart Mountains has been described from records of 174 permanent plots set up in 1969–70, where plant cover and stature, as well as deer faecal pellets, were recorded. Fifty-seven of the plots which covered the range of grassland types were re-measured in 1984.

In Mt Aspiring National Park, some 88 permanent photographic points with vegetation descriptions were established in 1970–73 to represent the range of vegetation and its condition throughout the Park at this time. Of these sites, 53 were spread among six types of low-alpine snow tussock grassland, with 12 in high-alpine communities and a further seven in valley grasslands. Repeat monitoring at 5–9 year intervals over a 29-year period, and coinciding in the early stages with the period of assessment in Fiordland, showed very similar results in the two areas. Greatest recovery occurred in the low-alpine grasslands, most notably in the midribbed snow tussockland, with significant increases in height and cover of the dominant tussocks (left). Several of the larger more palatable herbs have also shown obvious recovery, in particular *Ranunculus lyallii* (right), *Anisotome haastii*, *Ourisia macrophylla*, *Dolichoglottis scorzoneroides* and *Aciphylla scott-thomsonii*.

By contrast, non-palatable or tolerant species that had increased with high grazing pressure (*Celmisia armstrongii*, *C. walkeri* and blue tussock) became less obvious as the snow tussocks resumed dominance (left). There was less spectacular recovery of the narrow-leaved and curled snow tussocklands which had also sustained somewhat less initial modification by deer because of the lower palatability of their dominant tussocks. By contrast, some other low-alpine species, both herbs (*Astelia nivicola*) and shrubs (*Dracophyllum uniflorum*), showed negligible changes over the period, as did the high-alpine fellfield and snowbank communities, including the snowpatch grass *Chionochloa oreophila*, which is common in both communities.

In the valley grasslands of Mt Aspiring National Park there was obvious recovery in size of most species of tussock (narrow-leaved snow tussock, blue tussock, silver tussock, *Uncinia affinis*) plus some exotics (sweet vernal, ragwort) and shrubs (*Coprosma propinqua*, *Hebe subalpina*), with some recruitment of the very palatable leguminous shrub *Carmichaelia grandiflora*. Similar patterns have been observed in the Murchison Mountains of Fiordland. Counts of deer pellets in both parks have confirmed the drastic reduction in numbers of deer over this period, particularly in non-forest areas, where they are extremely vulnerable to aerial hunting.

Left: Low-alpine midribbed snow tussock grassland (*Chionochloa pallens*) at about 1250 m in the upper Beans Burn Valley, Mt Aspiring National Park, photographed in 1968 (top left) when the dominance of the large *Celmisia armstrongii* and the trailing *C. walkeri* was in response to heavy grazing by red deer. Drastic reduction in deer numbers soon after, through commercial hunting using helicopters, allowed recovery of the palatable midribbed snow tussock which, by 1977, was dominant (centre), as it would have been prior to the introduction of deer. There has been little change since this time, up to the last survey in 1999. *Alan Mark*

Fig. 2 (top right). Impressive recovery of the great mountain buttercup *Ranunculus lyallii* in low-alpine midribbed snow tussock grassland that had been heavily modified by red deer, at about 1200 m in Snowy Creek, Dart Valley, Mt Aspiring National Park. January 1980. Only a few *Ranunculus* seedlings were present here when the vegetation of the park was surveyed in 1969. *Alan Mark*

Future trends will obviously depend on the extent to which deer can be maintained at the extremely low levels achieved during the 1980–90s. This is likely to depend chiefly on the market demand for venison in competition with the supply of farm-raised animals which now has grown very substantially as those from the wild have dwindled.

Alan Mark

are sometimes present they have less effect on the populations). The many alpine areas of different rock types and origins in southern New Zealand have become centres of biodiversity since the demise of low-altitude ecosystems. The region's reptiles and birds have also suffered as a result of the introduction of exotic animals. We now turn to a treatment of the fauna of grasslands and mountain lands.

THE FAUNA OF GRASSLANDS AND MOUNTAIN LANDS

Brian Patrick

Invertebrates

The grasslands and mountain lands of southern New Zealand have an exceptionally rich invertebrate fauna; species richness and the rate of endemism is rivalled within New Zealand only by north-west Nelson. Over the past forty million years, the south has had more dry land for longer and for more continuous periods than elsewhere, and the land surfaces, especially in Central Otago, are as old as anywhere in New Zealand. The richness of the fauna is perhaps explained in part by the fact that substantial parts of Otago were unaffected by intense glaciation.

Fig. 45. The flightless female of the crambid moth *Orocrambus sophistes*. *Brian Patrick*

Lowland grasslands

Valley-floor grasslands are important in southern New Zealand, especially in Central Otago, where because of local conditions they have persisted below the treeline for a considerable time. They are often associated with saline soils, which support a small, but distinctive invertebrate fauna (see Box 4).

Dry native grasslands that once surrounded the saline soils in the valley floors of Central Otago are now seriously degraded, with the loss of taller or more palatable grasses in the genera *Elymus, Poa* and *Festuca.* Diminutive native grasses (e.g. *Rytidosperma maculatus, Poa maniototo, Agrostis muscosa*), cushion plants (*Raoulia* species) and herbs are still numerous in places and support a diverse array of insects. Prominent invertebrates are black ants *Monomorium antarcticum,* tiny bethylid wasps, the diurnal bug *Nysius huttoni,* diurnal grey wolf spiders (Lycosidae) and an occasional trapdoor spider of the genus *Misgolas.* Conspicuous shield bugs of these dry grasslands are the yellowish-brown *Rhopalimorpha lineolaris* and *Dictyotus caenosus,* both widely distributed. Crickets, *Pteronemobius nigrovus* and *P. bigelowi,* and grasshoppers, particularly *Phaulacridium otagoensis*, abound. An abundance of small flies and diurnal moths, such as the elegant *Eurythecta zelaea,* are also a feature of these open cushionfields, where they are often seen on the flowers of *Raoulia australis,* the commonest cushion species. A native katydid *Conocephalus bilineatus* sings from grasslands and is especially common in autumn.

Grasslands in inland Otago are drier and of shorter stature than those in peripheral areas of the south. They contain a distinctive fauna of invertebrates that is under immediate threat of extinction due to ongoing land-use change. Insect species with immobile females that are specialised in such habitats are particularly vulnerable. The flightless female of the crambid moth *Orocrambus sophistes* (Fig. 45) exemplifies well the plight of our short tussocklands and invertebrate fauna; the species still has viable populations in the Nevis Valley and parts of the MacKenzie Country but has not been found again in the Ida Valley, where it was first discovered and described. Flightless

10 CHAFER BEETLES

Attitudes to chafer beetles (Scarabaeidae) range from a strong desire to destroy them, for example grass grub (*Costelytra zealandica*), an important pasture pest, to a resolute commitment to preserve them, as in the case of *Prodontia lewisi,* the famous Cromwell chafer, which is restricted to less than 100 ha in Central Otago.

The flightless chafer beetle *Prodontria bicolorata.* *Brian Patrick*

While adult chafer beetles tend to feed on a wide range of plant species, larvae or grubs live in soil and feed on the roots of plants. Some complete their development in a year, while others take longer to pass through the three larval stages, before they pupate near the soil surface and emerge as adult beetles.

There are quite a large number of species of chafer beetles in southern New Zealand, belonging to four main genera, *Costelytra, Pyronota, Odontria* and *Prodontria.* The first three in this list can fly and are widespread througout the region. *Pyronota* species, the metallic green manuka beetles, are unusual in that the adults fly during the day, whereas most other genera are nocturnal. *Odontria* are larger, often flying in large numbers at night and feeding on the leaves of trees and shrubs. *Odontria striata,* the striped chafer, is common in gardens, where both adults and larvae can cause considerable damage. *Prodontria* is perhaps the most interesting because it is unable to fly; isolation as a result of geological events has promoted speciation. There are seventeen species of *Prodontria,* all of which occur in the southern half of the South Island, many being confined to relatively small geographical locations. They inhabit a wide range of environments, from coastal sand dunes of Southland and Stewart Island (e.g. *P. praelatella*), dry intermontane grasslands of Central Otago (e.g. *P. modesta*), to alpine environments of the Central Otago Block mountains (e.g. *P. capito, P. patricki* and *P. pinguis*). River terraces of the Clutha River, west and north of Alexandra, are home to an endangered species of flightless chafer *Prodontria bicolorata* (left). Its habitat is threatened by changing land use and degradation, but a zone of overlap with the black *P. modesta* has been made a reserve near Conroys Dam.

Barbara Barratt

chafer beetles are also highly vulnerable (Box 10). Much of our short tussockland could be described at best as semi-natural in terms of flora, but many native invertebrates have been more resilient to the changes wrought on this important ecosystem. Studies around Alexandra have revealed a diverse native insect fauna, including over 230 moth species, surviving in unploughed grasslands. While many continue to feed on their original host, perhaps protected in refugia, the majority have enlarged their host range to include exotic plants, some of which are closely related to the original host. In some cases, native species have even achieved 'pest status' on introduced pasture plants (Box 11).

Gelechiid moths are diverse and common in the drylands of Central Otago, with several species, including *Kiwaia lithodes* (Fig. 46), attached to *Raoulia* cushions in which the larvae feed. This species displays a common pattern of southern invertebrates, being found in coastal, riverbed and montane grasslands, but not the intervening areas, matching exactly the distribution of the larval host genus *Raoulia*. The small black cicada *Maoricicada campbelli* is another example.

Fig. 46. The gelechiid moth *Kiwaia lithodes*. *Brian Patrick*

Tor insects

Of all the features of Central Otago, it is the erect schist tors that best epitomise the landscape. With their diverse and colourful covering of lower plants, tors provide a refuge for vascular plants, many of which have become rare in the surrounding grasslands. A rich and highly distinctive insect fauna is found on tors, feeding on lichens, mosses, vascular plants and detritus on ledges and in cracks.

While spiders and beetles are common on tors, it is the moths that are the most specialised to this habitat. At least three species of casemoth (Psychidae), including *Scoriodyta suttonensis*, feed on algae, more than six oecophorids feed on detritus or lichen (including the genera *Phaeosaces, Izatha* and *Tingena*) while several scopariines, including *Eudonia manganeutis*, feed on mosses. Among the geometrids, *Helastia* spp.and *Dichromodes* spp. contain cryptic larvae that feed on mosses and lichens, respectively. For *Dichromodes*, Central Otago is the centre of diversity in New Zealand. where up to four of the six species may be found breeding on a single tor in some localities. The adults are normally diurnal and cryptically coloured to match the larval foodplant. This is true of adult *D. gypsotis*, whose larvae, similarly coloured, lie motionless by day amongst its foodplant lichen (Fig 47 a,b). The species is widespread below 900 metres in open areas of Otago, being found on volcanic rocks to sea level at Dunedin. At least three species, *D. ida, D. simulans* and an undescribed species, are endemic to East and Central Otago.

At night, with a torch, many beetles can be seen feeding on and around the lichens and mosses on rock tors. These include species of shining, globular byrrhids and tenebrionids such as species of *Artystona*. Rock bluffs, such as in the Kawarau and Taieri Gorges and east of Galloway in Central Otago, are home to a large and elusive black clapping cicada *Amphipsalta strepitans*.

11 NATIVE INSECT PESTS

Europeans brought many plants to New Zealand, including forage plants such as species of clover and grasses. These plants were introduced into native grassland, and areas of bush and shrubland which had been cleared by logging and burning. Many of the native insects that inhabited these areas had to retreat to what natural vegetation remained, but some actually responded to the new pasture plant species in a very positive way, finding the new species palatable and nutritious, rather like the grazing animals for which the plants were intended. Plant diversity decreased as white clover and grass species were established over large tracts of land, and so the balance that existed before in the diverse natural vegetation was lost. The result of this imbalance was that the invertebrates that benefited from the new pasture species increased numerically to become pests in the absence of the predators, parasites and competitors that previously kept them in check.

The main examples of native species that have responded in this way are:

Grass grub *Costelytra zealandica:* This is the common name given to the larval stage of the chafer beetle that feeds on the roots of pasture plants, causing large bare patches in pasture into which weeds can become established.

Porina: This is the common name for a group of hepialid moths in the genus *Wiseana*. Six southern species have discrete distributions and their adults emerge at particular times of year and frequent particular habitats. *W. mimica* is the first to emerge in spring and is found in wetlands from sea level to alpine areas, whereas the orange *W. jocosa* emerges later in spring from forest glades and pasture. Again it is the larval stage that causes pasture damage, by 'grazing' on the vegetation at night. The larger larvae build tunnels in the soil, where they remain during the day. New pastures seem to be particularly susceptible to porina, possibly because cultivation tends to remove natural disease organisms from the soil.

Striped chafer *Odontria striata:* Tends to prefer more dense ground cover than grass grub, and so prefers the less intensively grazed improved tussock grassland to the more closely grazed lower altitude pastures. Larvae are very mobile and often move over the surface. This means that they have a less clumped distribution, and damage is more evenly spread. Populations of up to 700 per square metre have been recorded.

Manuka beetles, particularly the metallic green *Pyronota festiva*, are similar to grass grub, but tend to be found in new pasture developed from bush. Like grass grub, the larvae feed on the roots of pasture species, but unlike grass grub, adults are diurnally active, feeding on shrubs, but not only manuka.

Barbara Barratt

Fig. 47. Adult (left) and larva (right) of the moth *Dichromodes gypsotis*. *Brian Patrick*

This noisy species has a clapping sound interspersed within its call. It is not common in Otago, being at the southern limit for the species in New Zealand.

Alpine grassland invertebrates

In the alpine zone it is the invertebrates that are the most important grazers of vegetation. Among them are grasshoppers and certain moth larvae that may eat their body weight in vegetation in a day. The palatability of each plant species determines what will consume it and differential feeding by invertebrates on alpine foliage ultimately determines the structure of the vegetation. In turn, predatory invertebrates may keep the herbivores in check.

A suite of adaptations to alpine living include increased hairiness and dark colouring (melanism), decrease in size, and reduced wings. While these adaptations are not confined to alpine species, occurring in lowland species under certain circumstances, it is the widespread adoption of these characters that distinguishes the alpine invertebrate fauna.

Most alpine invertebrates are exclusively alpine, but some species live and breed both below and above the treeline. Examples in the south include the endemic giant snail *Powelliphanta spedeni,* the large carabid beetle *Mecodema chiltoni* and giant hepialid moth *Aoraia rufivena.*

Extensive low-alpine grassland, dominated by snowgrass (*Chionochloa*) with an accompanying understorey of herbs, smaller grasses and mosses, contains a large insect fauna with many orders represented. Prominent are two species of tussock butterfly *Agyrophenga antipodum* and the faster flying *A. janitae,* the big weevils *Anagotus lewisi* (Box 12), whose larvae feed on tussock tiller bases, the brown cicada *Kikihia angusta* and the small distinctive flat bug *Kiwimiris niger.* All have larvae that feed on the snowgrass in some way. Because snowgrass can live for several hundred years, it is analogous to forest and no less diverse. Similarly its leaf litter is an important resource for invertebrates such as millipedes, and fly, beetle, stonefly and moth larvae. But it is the inter-tussock herbs that support the most invertebrate species, with *Anisotome, Celmisia* and *Aciphylla* important.

Open areas within grassland or boundary habitats with wetlands or streams can have rich herbfields with a commensurate insect fauna of diurnal beetle, fly, bug and moth species. Worthy of note is the lygaeid bug *Rhypodes anceps,* the shield bug *Cermatulus nasalis,* the large broad black chafer beetle *Scythrodes squalidus* and the chunky ground weta *Hemiandrus focalis.* The diurnal moth genus *Gelophaula* is particularly noteworthy; in the south the genus has at least eight species with larvae that bore into the rosettes of *Celmisia* species. Cushion plants, especially when they are flowering, are often dotted with large numbers of small broad-nosed weevils, 3–5 mm long. The weevils feed on pollen, which provides the necessary nutrients

Fig. 48. The noctuid moth *Ichneutica ceraunias.* *Brian Patrick*

Fig. 49. The larentiine moth *'Dasyuris' callicrena.* *Brian Patrick*

12 SPEARGRASS WEEVILS

There are more species of weevils in the world than of any other group of living organisms. In southern New Zealand we have a large number of species, including a number of spectacularly large ones. Most trampers will have come across large black-and-white striped weevils (in the genus *Lyperobius*), often feeding in the middle of speargrass plants; they are restricted to plants in the family Apiaceae, such as *Aciphylla* and *Anisotome* spp. The flightless speargrass weevils, which may be up to 30 mm long, often have their rostrum (the long snout characteristic of weevils) covered in a gluey gum exuded by the damaged plant. Their defence against predators (and curious fingers) is to drop into the centre of the plant. The legless white grubs feed in the main roots of the host plants, or on the thick, fleshy leaf bases, and grow up to 32 mm long in larger species. When large numbers of weevil larvae have been feeding, the plant may be damaged quite severely. After completing development, the larvae pupate and emerge as adults in a chamber in the soil, staying there for up to eight months before emerging.

Among the sixteen species of *Lyperobius*, eleven species are found here, of which eight are confined to native grassland and alpine areas of southern New Zealand, including *L. barbarae*, *L. townsendi* and *L. patricki*. In the same tribe as *Lyperobius*, the Molytini, *Hadramphus stilbocarpae* of south-western Fiordland feeds nocturnally, unlike *Lyperobius* which feed during the day. Another genus of large weevil, *Anagotus*, occurs in southern New Zealand in tussock grassland, alpine herbfield, Fiordland beech forest margins and on off-shore islands. Many have very distinctive protuberances and tubercles over the dorsal surfaces (bottom). All these large-bodied weevil species are threatened by the introduction of rodents and loss of habitat from grazing by introduced herbivores.

Barbara Barratt

Top: The mountains of North Otago harbour two species of speargrass weevil of the genus *Lyperobius*. One of these, *L. barbarae* (length 22 mm), from the summit of Mt Kyeburn, is pictured wandering over a miniature forest of *Dracophyllum muscoides* by day. Brian Patrick

Bottom: The large weevil *Anagotus lewisi*. Note the distinctive protuberances. Brian Patrick

to mature their eggs. In wetter areas, and alongside streams, many species of chysomelid beetles can be seen on cushion plants. The metallic bluish-black *Allocharis* species, some with a band of orange along the sides of the wing cases, are particularly distinctive. New Zealand's rarest butterfly, *Erebiola butleri*, has its southern limit on the Humboldt Mountains in alpine grasslands of snowgrass, the larval foodplant. The dark-brown adults fly in late summer and have a sporadic distribution in Western Otago. A rare flightless shield bug *Hypsithocus hudsonae* is restricted to central and western Otago mountains.

Large, distinctively patterned noctuid moths of the genus *Ichneutica* are common and diverse in the snowgrass zone of southern New Zealand. Of the more than eight species present, it is *I. ceraunias* (Fig. 48) that is most widespread, even being found at sea level at Seaward Moss near Invercargill. The larvae feed on snowgrass and are important defoliators, while the adults fly by night in early summer. In some populations the females are brachypterous (short-winged), severely curtailing the dispersal ability of the population. Blue-grey species in the genus *Aletia* are also diverse in the alpine zone but, in contrast to *Ichneutica*, their larvae are herb-feeders.

Larentiine moths, mostly diurnal as adults, are typical of the low-alpine zone. Most have herb-feeding larvae that remain concealed. Fig. 49 shows *'Dasyuris' callicrena*, an early-emerging species occurring over much of southern New Zealand. Its cryptic larvae feed on hebes, which are common as an inter-tussock shrub species.

New Zealand's most spectacular fly, the batwinged fly *Exsul singularis* (Fig. 50), was described from Milford Sound in 1900. but is not common in Fiordland. To the north on the Mark Range it has been found in large numbers, frequenting low-alpine meadows, where it sunbathes on rocks in streams from where it launches attacks on the adults of emerging aquatic insects.

From coastal to high-alpine grasslands, woolly-bear caterpillars belonging to the three species of tiger moth in the genus *Metacrias* are often conspicuous. The densely hairy golden-brown larvae are great travellers. They feed on a variety of grasses and herbs but prefer *Bulbinella* and *Dolichoglottis* foliage when these are available. When ready to pupate, the larva seeks out a dry, sheltered place and builds a nest of silk and hairs. On hatching, the colourful diurnal male seeks out a female by tracing the pheromone she releases. Fig. 51 shows a male *M. huttoni* mating

Fig. 50. The batwinged fly *Exsul singularis*. Brian Patrick

with a buff-coloured short-winged female amid a cloud of fluff, on the Old Man Range. While all three species are found together in the Wakatipu area, *M. erichrysa* is distributed mainly in the western mountains, *M. huttoni* in the eastern mountains and *M. strategica* in the eastern lowlands and low-alpine zone.

The alpine zone on Stewart Island has a depauperate insect fauna, with only about 125 species recorded, but it has some very interesting species, including many local endemics. Some endemics are found only on Mt Anglem, in the north, or only on Mt Rakeahua/Table Hill and mountains to the south. For example, the diurnal flightless chafer *Prodontria rakiurensis* (Fig. 52) is confined to the south. On the other hand, the larger *P. grandis* is found in both areas. The colourful *Paprides dugdali* (Fig. 53) is the only alpine grasshopper on Stewart Island; this species and an undescribed ground weta in the genus *Hemiandrus* are the only alpine orthoptera there. Only 32 beetle species, of which eleven are weevils, are recorded from alpine Stewart Island, while the carabid beetle *Holcaspis stewartensis* is endemic to the island. Moths are relatively more diverse, with eleven of the 56 species found nowhere else. Amongst them is an undescribed *Notoreas* (Fig. 54), one of four diurnal species of the genus found in the alpine zone of the island, all with larvae feeding on the plant genus *Kelleria.*

High alpine mountain lands

Beyond the limits of the continuous snowgrass communities are extensive cushionfield, fellfield, scree, snowbanks and diminutive shrublands only centimetres high. These are the home of an extraordinary range of invertebrates that jump, crawl and fly low to the ground to avoid being blown away by the, at times, severe winds. 'Patterned ground', best developed on the Central Otago high mountains, provides micro-environments that invertebrates exploit. Hollows can provide some shelter from the wind or, under the winter snow, can be several degrees warmer than adjacent humps.

High altitude invertebrates have adapted to the long period of snow lie (June to September) by a variety of strategies. All these involve staying put; there is no migration to lower altitude, such as that practised by some alpine birds. Many synchronise their life cycles in such a way that dormant eggs lie hidden beneath the snow and hatch as the snow retreats in early summer. Another common strategy, especially for species with incomplete metamorphosis (adults resembling larvae), is for a protracted life cycle that may involve taking several seasons to reach maturity. Under this regime, the life cycle is less synchronised and all stages occur concurrently. Alpine grasshoppers (Box 13) are an example of a group with protracted life cycles; large adults

Fig. 51. A male *Metacrias huttoni* mating with a buff-coloured short-winged female. *Brian Patrick*

Fig. 52. The diurnal flightless chafer beetle *Prodontria rakiurensis*. *Brian Patrick*

Fig. 53. The alpine grasshopper *Paprides dugdali* from Stewart Island. *Brian Patrick*

Fig. 54. An undescribed moth in the genus *Notoreas*. *Brian Patrick*

emerge from beneath the melting snow, together with hundreds of young nymphs that have hatched from eggs laid underground in autumn. Studies in Australia and elsewhere have shown that many invertebrates may be active during winter, feeding and breeding under the snow. This subnivean fauna includes beetles, spiders, wasps, springtails, cockroaches, moths, flies and bugs. To date, no studies have been done to discover whether such a fauna exists in New Zealand. Mechanisms of cold hardiness displayed by a range of invertebrates are discussed in Box 14.

Screes are best developed on the Takitimu and Livingstone Mountains in western Southland and again in the mountains of North Otago, such as the Ida Range. Among the invertebrates characteristic of this habitat are the giant weta *Deinacrida connectens*, the black butterfly *Percnodaimon merula* and the grasshopper *Brachaspis nivalis*.

Fellfield areas also have substantial bare rock and share some of these insect species, but a generally flatter topography often means there is more vegetation in the form of cushionfield and patches of low grasses. Jumping and wolf spiders are a feature of this zone, together with a number of diurnal moths, such as *Aponotoreas anthracias* and *Tawhitia glaucophanes*. In Central Otago, where cushionfield is more continuous, flightless chafer beetles such as *Prodontria capito* and weevils such as *Zenographus metallescens* and *Lyperobius hudsoni* are active by day on the cushions. Another weevil, *Anagotus latirostrus*, can be found under rocks on many Central Otago mountains.

Snowbanks also have rich cushionfields, accompanied by dense herbfields. They yield a range of insect species that can overlap with the cushionfield and fellfield areas. Among the moths, an undescribed black and white gelechiid in the genus *Hierodoris* is widespread. The highest species diversity of larentiine moths in the genus *Notoreas* is found in this habitat on the Rock and Pillar and Lammermoor Ranges, also the site of the highest diversity for one of the two larval host genera, *Kelleria*. These exquisite moths resemble butterflies in their habits and bright colours. Each part of southern New Zealand has its own mix of species, with local endemics mixed with more widespread species. For example, Stewart Island has five *Notoreas* species and southern Fiordland seven species including an undescribed species (Fig. 55); its green and pink larvae feed on *Kelleria croizatii*. Adults can be particularly common in the high-alpine zone, delighting to settle momentarily on a patch of bare ground close to the host plant.

Several species of cockroach live in this harsh environment, too. The orange-spotted *Celatoblatta quinquemaculata* is common in the high-alpine zone of Central Otago while an undescribed

13 GRASSHOPPERS

The Acrididae or short-horned grasshoppers are common in southern New Zealand. They are found where the native vegetation is low and open. Above the treeline on summer days they can be seen leaping high above the tussock. However, where the native plants give way to farmland, the numbers of grasshoppers diminish. Up to three of the ten species recorded in southern New Zealand may coexist, but generally they are altitudinally separated.

Any one species usually comes in a variety of colours that match its background, from deep green to dark red. Those found in tussock have long thin stripes, whereas those among the tumbling lichen *Chondropsis viridis* in Central Otago have appropriate markings. They cannot reliably be identified by colour, but the shape and structure of the pronotum is useful, as are the male and female reproductive structures.

The larger grasshoppers are principally from the genus *Sigaus*, which is widespread through the central and southern regions of the South Island. *S. australis* is associated with tussock up to 2000 m in Otago and Fiordland. *S. campestris* is widespread in low alpine areas throughout southern New Zealand, except for Fiordland and Stewart Island. The robust and hairy *S. obelisci* occupies the high places, usually above 1500 m in Central Otago and western Southland. The recently named *S. childi* is the rarest grasshopper in southern New Zealand (right). It is adapted to the semi-arid climate of Central Otago and is only found in the presence of the cushion plant *Raoulia australis* on both sides of the Clutha River in the Alexandra area. The sleek green *Paprides dugdali* is common in the south-east of the island and is the only grasshopper found on alpine Stewart Island. It is not found in Fiordland, whereas *Alpinacris tumidicauda* is common in the eastern part of Fiordland and the Central Otago mountains.

Representatives of the genus *Brachaspis* are found in North Otago. The large and threatened *B. robustus* was formerly in the lower Waitaki region but is now locally extinct, while *B. nivalis* is found on screes on the Ida and Hawkdun Ranges and Kakanui Mountains.

Grasshoppers from the genus *Phaulacridium* are small and sprightly. *P. marginale* is found in open grasslands throughout New Zealand, while another smaller species *P. otagoensis* is found only in Central Otago, frequenting the most open vegetation. Both are very active: from January to March mating couples are common. *Colleen Jamieson*

Below: *Sigaus childi*, the rarest grasshopper in southern New Zealand. Recent research indicates that the species, which is centred on Alexandra, has its largest population within the Earnscleugh Tailings. *Brian Patrick*

14 COLD HARDINESS IN ALPINE INSECTS

The Rock and Pillars in winter seem, at first glance, devoid of animal life. But if a schist rock slab is turned over, hundreds of cockroaches will rapidly scurry away from sight, or a large male weta may be found with his harem of several females. Amongst the hazards with which these animals are faced is exposure to temperature extremes. Temperatures recorded at the surface of rocks may be as high as 40°C or as low as –10°C, with a temperature range of as much as 30°C in a single day. Perhaps the rocks under which they live provide some protection, acting as storage heaters that release the heat gained during the day. However, this is not enough to prevent the animals from freezing during the winter. Cockroaches and weta can be frozen solid with the temperature under their rock below –3°C (below). Both are freezing tolerant and can survive ice formation within their bodies.

There are two ways by which insects can survive low temperatures: they are either freezing tolerant or they avoid ice formation altogether, maintaining their body water as a liquid at temperatures at which it would normally freeze (freeze avoiding). The weta *Hemideina maori* is the largest freezing-tolerant insect in the world. Other freezing-tolerant invertebrates include alpine grasshoppers, millipedes, springtails, cicadas, beetles, dipterans, spiders, slugs, moths and mites. We know little about how these animals survive low temperatures.

In the weta, the ice is confined to the body cavity. Unlike nematodes and some dipteran larvae that can survive intracellular freezing, weta cannot survive the freezing of their cells. The weta produces an ice nucleating protein that ensures its blood (haemolymph) freezes first. The freezing process is slow and it takes several hours for the insect to freeze completely, with 82 per cent of its total body water converted into ice. During this time, freeze concentration results in the water being drawn out of the cells, preventing the cells from freezing. Particularly intriguing is the presence of a bright blue protein in the weta's haemolymph. The identity and function of this protein is unknown, but perhaps it plays a role in freezing tolerance.

The weta can survive temperatures down to –10°C at any time of the year. The cockroach, however, can increase its cold tolerance during the winter, with lower lethal temperatures of –5.4°C in summer and –8.6°C in winter. We do not yet know how they do this, but it could involve the production of proteins or other substances that manipulate the way in which ice forms in the body.

David Wharton and Brent Sinclair

Below: The large weta *Hemideina maori* (right) and *Celatoblatta quinquemaculata* (an alpine cockroach) as found frozen solid under a rock during winter (left). *Brent Sinclair*

Fig. 55. An undescribed species of moth in the genus *Notoreas* from southern Fiordland. *Brian Patrick*

Fig. 56 (a, above). Thirty mm long larva of *Dasyuris hectori* from the Remarkables; (b, right) an adult from the Eyre Mountains. *Brian Patrick*

fully black species in the same genus is found on the Blue and Umbrella Mountains.

The most majestic of high-alpine moths in southern New Zealand is the large *Dasyuris hectori*, which flies like a red admiral butterfly, agile and elusive. It frequents boulder fields where the larval host grows – *Anisotome* species. While the adult has an underside pattern like a zebra, the larvae are shades of brown. All other true *Dasyuris* species feed on plants of the carrot family too, with *D. octans* endemic to Fiordland. The most widespread is *D. anceps*, which is also found on Stewart Island, while the smaller *D. micropolis* is confined to western mountains. Fig. 56a shows the 30 mm long larva of *D. hectori* from The Remarkables and Fig. 56b an adult from the Eyre Mountains.

Only in New Zealand mountains do cicadas extend to the high-alpine zone. Seven diurnal black cicada species sing from rock outcrops in the alpine zone of southern New Zealand. Each displays its own distinctive call and the ability to throw its voice, so evading predators. Larvae feed for many years underground on roots prior to adult emergence in mid-summer. Fig. 57 shows the small *Maoricicada nigra nigra* from 1350 metres on Gertrude Saddle, Fiordland. North Otago mountains echo to the sound of *M. clamitans* and *M. phaeoptera*, while subspecies of *M. otagoensis* and *M. nigra* live in the mountains of Central Otago to the Takitimu Mountains.

Epitomising the vulnerability of the New Zealand fauna to predation by introduced mammalian predators are the flightless females of alpine hepialid moths. These are the largest of New Zealand's moths with some female *Aoraia* being the size of a small mouse. Of the twelve species found in southern New Zealand, at least nine breed in alpine grassland, snowbanks or wetlands. All have vulnerable flightless females that emerge from the larval and pupal subterranean tunnels in late summer or autumn. Fortunately, introduced predators cannot exert sufficient pressure in high-altitude areas to impact significantly on the populations of these special alpine insects. Fig. 58 shows male and female *Aoraia senex*, a species of western and Central Otago high-alpine areas.

A feature of the high-alpine insect fauna in the south is the number of species restricted to one, or adjacent groups of mountains. These local endemics occupy, in effect, alpine islands, separated from each other by an inhospitable 'sea' of lowlands. An example is the mirid bug *Chinamiris zygnotus*, which is restricted to the Rock and Pillar Range, where it is common on shrubs including *Leonohebe odora*.

Fig. 57. The small cicada *Maoricicada nigra nigra* from the Gertrude Saddle in Fiordland. *Brian Patrick*

Fig. 58 (above). Male (top) and female (bottom) of the alpine hepialid moth *Aoraia senex*. *Brian Patrick*

Fig. 59. Schist outcrop/grassland habitat of lizards. *Alison Cree*

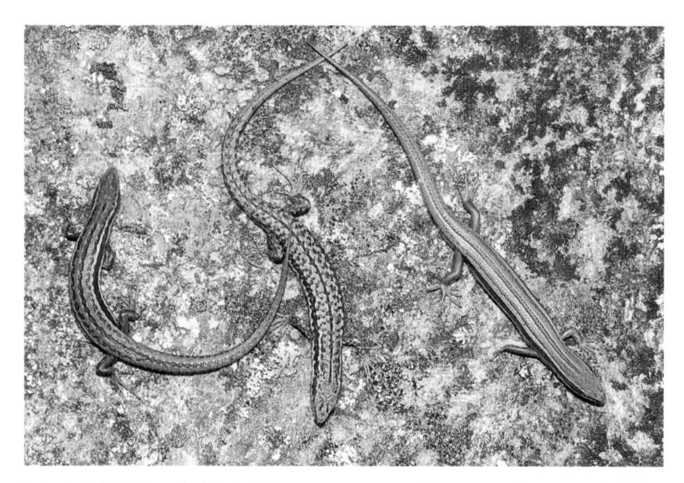

Fig. 60. McCann's skink *Oligosoma maccanni* (two specimens on left) and the common skink *O. nigriplantare polychroma* (specimen on right). *Alison Cree*

Fig. 61. The green skink *Oligosoma chloronoton*. *A.H. Whitaker*

REPTILES OF GRASSLANDS AND MOUNTAIN LANDS

The drive inland from Dunedin to Central Otago reveals a vast area of grasslands punctuated by bizarrely shaped rock outcrops (tors). This habitat harbours a diverse fauna of lizards.

The schist tors are extensively cracked and fissured, and loose slabs often lie scattered about (Fig. 59). Thus, there is an abundance of crevices for lizards to seek shelter in. On sunny days, northward-facing rock surfaces provide sites for lizards to sunbask. Invertebrate food is found on and around the outcrops and, where shrubs are present, their berries are also consumed. Although the schist changes to greywacke rock in northern Otago, this area also supports a variety of lizard species where rock screes are present.

Given the availability of suitable habitats, it is hardly surprising that the tussock grasslands and mountain lands of southern New Zealand support the most diverse assemblage of lizards anywhere on mainland New Zealand. At least eight (19 per cent) of the 42 described species of New Zealand lizards are found in this region.

Probably the most frequently seen lizards are the small, brown skinks that until recently were grouped together as one species, the 'common skink' (see Chapter 8 for distinguishing features of skinks and geckos). Prior to 1990, these were considered as a single species widespread throughout much of New Zealand (including a subspecies on the Chatham Islands). However, if there is one important message to come out of recent studies on New Zealand lizards, it is that the term 'common' is largely a product of ignorance.

In the case of the 'common skink', closer examination by Geoff Patterson and Charles Daugherty has resulted in its separation into four species represented in southern New Zealand, and a fifth species of similar general appearance has since been described. All belong to the genus *Oligosoma* (known as *Leiolopisma* until 1995; see Chapter 8). Two, the southern skink

(*O. notosaurus*) and the small-eared skink (*O. stenotis*), occur only in the Stewart Island region and were mentioned in Chapter 8. Three others are present in the tussock grasslands of Otago and Southland. One, the cryptic skink *O. inconspicuum*, is known only from Otago and Southland. McCann's skink, *O. maccanni*, occurs in Canterbury, Otago and Southland (Fig. 60). The most widespread form, *O. nigriplantare*, occurs throughout the South Island and the lower North Island as the subspecies *O. n. polychroma* (another subspecies occurs on the Chatham Islands). This species still goes by the name of the common skink, but its division into further subspecies or species is considered possible.

The biology of the common skink, McCann's skink and the cryptic skink has been studied by Geoff Patterson in an area of the Rock and Pillar Range and Lammermoor Range of Otago at about 1000 m above sea level. The dominant plant in this area is the narrow-leaved snow tussock *Chionochloa rigida*. There are many similarities in the biology of the three skinks. All are of similar size (up to 70–77 mm snout-vent length) and all give birth to about three or four young in late summer. They are all day-active, sunbasking lizards that feed on invertebrates and fruits. However, some partitioning of the microhabitat is evident. The common skink occurs most frequently in the grassland between outcrops, where its striped appearance offers it camouflage amongst tussocks. The cryptic skink is most frequently seen amongst herbs and shrubs, or using rocks that rest on soil. McCann's skink is most frequently seen amongst herbs and shrubs, and to a lesser extent on rock outcrops (rocks lying on rocks). Both the latter two species have (to humans!) a surprising ability to shelter within the spiny-leaved spaniards (speargrasses, *Aciphylla* spp.).

The remaining skinks of southern grasslands are larger and more colourful, but less likely to be seen. The sight of any of these handsome creatures basking in the sun or moving undisturbed through its wild habitat is one to be treasured. The green skink, *Oligosoma chloronoton*, reaches a snout-vent length of 108 mm and has a deep green back, flecked with lighter green and black, and a copper-brown stripe along the side of the body (Fig. 61). It is found in Otago, Southland and Stewart Island, sometimes in coastal vegetation. In Otago grasslands, it has been found in both schist and greywacke areas. It seems to favour moist, well-vegetated sites (e.g. amongst *Chionochloa* tussocks, matagouri, divaricating *Coprosma* spp., *Melicytus alpinus*) where stones or boulders are also present.

In scree slopes of northern Otago, including those in the Ida and St Bathans Ranges, lives another large and handsome skink, the scree skink *O. waimatense* (Fig. 62). Its back is yellow-brown or grey streaked with black, and it reaches about 107 mm snout-vent length. Sometimes considered a 'form' of the Otago skink *O. otagense* (Box 15), it was confirmed as a separate species in 1997. Scree skinks live in the interstices between greywacke fragments. The screes themselves are mobile and often have little or no vegetation. However, shrubs (e.g. *Coprosma* spp., *Muehlenbeckia axillaris*) are present among the surrounding tussock grassland (*Chionochloa rigida*, *Poa laevis* and *Poa colensoi*), and the shrubs are visited for their berries. As with other fruit-eating lizards, scree skinks contribute to seed dispersal. The range of scree skinks also extends into Canterbury and Marlborough.

Another abundant lizard of the tussock grasslands of southern New Zealand, the common gecko *Hoplodactylus maculatus* (Fig. 63), seems likely to be put through the same process of subdivision into separate species that the small *Oligosoma* skinks have recently experienced. Research by Rod Hitchmough (Victoria University of Wellington) and colleagues reveals that this brown, olive-green or grey gecko, currently widespread throughout much of New Zealand, is represented in Otago and Southland by several forms that may represent separate species. In 1994 these were provisionally named as the Eastern Otago species, the Western Otago species, the Danseys Pass species, the Southern Alps species and the 'Southern mini' species, although scientific names, species descriptions and ranges are yet to be published.

Like the skinks of southern tussock grasslands, common geckos can be found at fairly high altitudes (e.g. to about 1700m above sea level). However, unlike the skinks, the geckos are primarily nocturnal. The Eastern Otago geckos, for instance, which are the only ones whose biology has been closely examined, shelter in crevices and under loose slabs on their schist rock outcrops by day, and emerge onto the rock surfaces and surrounding grasslands by night. Thus, these southern lizards have an impressive ability to be active at cold temperatures (Box 16).

Fig. 62. The scree skink *Oligosoma waimatense*. A.H. Whitaker

Fig. 63. The common gecko *Hoplodactylus maculatus* from Macraes Flat. Alison Cree

15 GRASSLAND LIZARDS IN DECLINE?

Left: The grand skink *Oligosoma grande* tends to have a more upright posture than the Otago skink. *Marcus Simons*

Top: The Otago skink *O. otagense*. *Alison Cree*

Bottom: Streamside outcrops where both species are present. *Alison Cree*

Of the seven New Zealand skinks found in southern grasslands, four are endemic to this part of New Zealand. The two rarest, the grand skink *Oligosoma grande* and the Otago skink *O. otagense* (above left and right) are found only in Otago – and then in just two areas. These are the Lindis Pass/Lake Hawea area, and the Macraes Flat/Middlemarch area.

Widespread surveys suggest that suitable habitat exists for these skinks throughout much of inland Otago, yet their known range occupies only about 8 per cent of this area. Local extinctions have occurred at several sites during the last few decades. Both species are now very low in numbers (about 1500–5000 individuals each) and the subjects of a recovery plan implemented by the Department of Conservation.

Both species are sunbasking lizards that live on schist rock outcrops among subalpine tussock grasslands (shown). Otago skinks are larger than grand skinks (snout-vent lengths up to 130 mm and 109 mm, respectively) and are probably the largest species surviving today on the New Zealand mainland. Otago skinks tend to occur on more massive bluffs than grand skinks, but both can be found together. Some individuals are known to have moved hundreds of metres through the grassland between outcrops. The two species share their rock outcrop/grassland habitat with several other lizards, including common skinks, McCann's skinks, cryptic skinks, green skinks and common geckos.

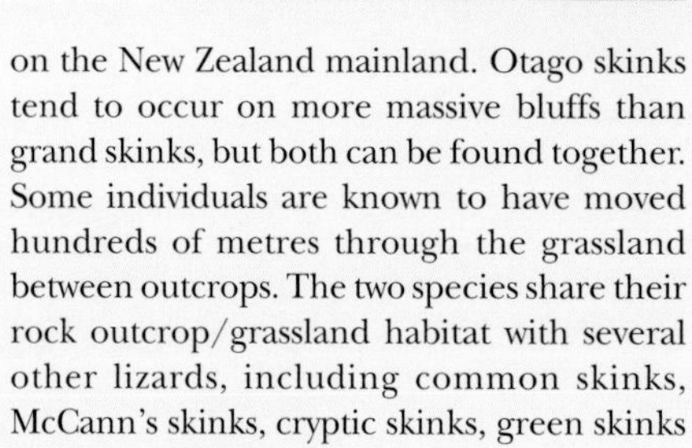

Why are the grand skinks and Otago skinks so rare? Although a definitive answer is elusive, research by Tony Whitaker and others implicates the modification of the tussock grassland habitat that has occurred with agricultural development. In the century or more since European settlement, tussocks have declined in stature with repeated burning, and in some places they have disappeared as the land is ploughed and replanted with introduced grass and clover. These practices allow more sheep and cattle to be farmed, but also change the food supplies and reduce the shelter for skinks. Rabbit control may result in unintentional poisoning of skinks, and introduced mammals and birds (including cats, ferrets, stoats and magpies) prey on lizards of various species. The impact of cats, in particular, may be severe on a local scale: one gutted feral cat had 49 lizards in its stomach.

The future for grand skinks and Otago skinks is uncertain, and their decline is a warning to be observant for declines in other species of lizards. Two areas containing grand and Otago skinks have recently been purchased as reserves by the Department of Conservation. With research to determine suitable management, and the cooperation of landowners where populations survive on private land, perhaps recovery is possible.

Alison Cree

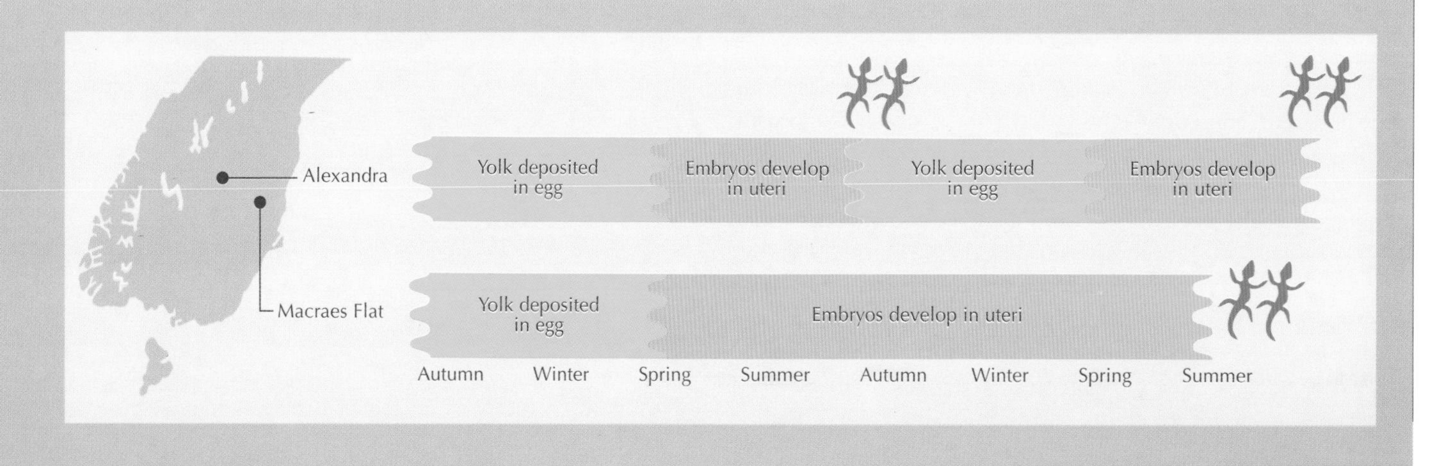

BIRDS AND MAMMALS OF GRASSLANDS AND MOUNTAIN LANDS

John Darby and Hamish Spencer

The birds of tussock grasslands and mountain lands are an admixture of native and introduced, as is the situation within most habitats in New Zealand. The Australasian harrier *Circus approximans* is one of the most ubiquitous of birds of the countryside, as it quarters grassland, hill and road side with equal ease. Prior to the deforestation of much of inland Otago and Southland, the harrier was probably nowhere near as common as it is today. Moreover, early importations such as the rabbit and possum have provided a ready supply of foods, often as road kills, that were not available in pre-European times. During the breeding season harriers defend large territories that may range up to eighty-one square kilometres, although they concentrate their hunting in an area of about nine square kilometres.

The harrier is easily distinguished from its endemic raptor relative, the New Zealand falcon *Falco novaeseelandiae* by its larger size, lighter plumage and especially its white rump. As with most falcons, the New Zealand species is variable, although the females are always larger than males. The eastern form is mostly found in open country east of Lake Te Anau, throughout Otago and most of eastern Southland, whereas the southern form also occurs in bush in Fiordland, Stewart Island and in the Auckland Islands. They are fierce hunters of introduced and native birds. Nesting (Fig. 64) begins in mid September and is extended to early December. Three eggs are usually laid. Fully protected only since 1970, it is believed that the species is declining, from a combination of habitat loss, indiscriminate shooting and the extensive use of biocides.

The California quail *Lophortyx californicus* and chukor *Alectoris chukar* are both introduced species, brought to this country as sporting birds. The quail became established between 1862 and 1880, after its first liberation in Papakura in the North Island. With a great deal of help from Acclimatisation Societies it has become well established in the South Island, particularly in open farmland and scrub country wherever cover is available and rainfall is no more than 1500 mm a year. Two subspecies of chukor were introduced into New Zealand, the Indian form *A. c. chukar* in 1926 and six years later the Persian form *A. c. koroviakovi*. The two have since merged through interbreeding. The population peaked in the early 1940s, but since then has declined. The reduction in population numbers has been attributed to the widespread use of poisoned baits for rabbits.

Fig. 64. New Zealand falcon near its nest. *Rod Morris*

The South Island takahe *Porphyrio hochstetteri* (page 187) is legendary for its discovery, 'extinction', and rediscovery in 1948. Until its rediscovery, the specimen at the Otago Museum was the only one on display anywhere in the world. Initial estimates of the population suggested that about 200 birds survived in tussock lands on the tops of Fiordland's Murchison Mountains and nearby ranges. By 1980, however, barely 120 were alive and recent estimates suggest that it continues to decline. Although competition for food from the introduced red deer has been thought to be the main cause of its decline, predation by stoats is probably an additional factor.

Three species of wader breed commonly on our southern grass- and tussock-covered lands. The South Island pied

16 LIVE-BEARING LIZARDS IN THE COLD

Imagine being pregnant for 14 months – with twins. On top of that, imagine doing it through winter, without heating, when air temperatures are close to freezing.

Sound challenging? Yet female geckos from Macraes Flat in coastal Otago do just that. Like all New Zealand geckos, the eastern Otago form of the common gecko (*Hoplodactylus maculatus* species complex) gives birth to live young, not eggs. However, unlike common geckos from other places in New Zealand (which give birth annually), females at Macraes Flat produce babies only every two years (biennial reproduction).

To produce baby geckos, the mothers first have to manufacture the yolk that will later provide food for the embryos. This process begins in late summer and is completed by the following spring. The eggs, now distended with yolk, then pass from the ovaries into the oviducts where they are fertilised with sperm from an earlier mating. Each embryo then begins its prolonged development in its separate uterus. Just over a year later, the babies (one or two) are born.

Why is pregnancy so long for the Macraes Flat geckos? Why do geckos from the same species in Alexandra have a much shorter pregnancy (four months)? Research by Alison Cree and Jennifer Rock (University of Otago) suggests that the most likely answer lies in the temperatures experienced by the pregnant females. Macraes Flat is cooler than Alexandra during spring, summer and autumn, when females are pregnant. Under laboratory conditions, cooler body temperatures result in slower embryonic development.

Cool temperatures and slow reproduction are likely to have consequences for other aspects of a gecko's life history. The baby geckos at Macraes Flat are extremely slow-growing and require several years to reach maturity. Once mature, they probably live for several decades. Slow reproduction limits the population's replacement rate and is thus a feature to be considered in the geckos' conservation.

Alison Cree

The map opposite shows the location of Macraes Flat and Alexandra and the horizontal bar-graph compares the duration of the yolking phase and pregnancy in the two populations. *Data from Alison Cree*

oystercatcher *Haematopus ostralegus finschi* (Fig. 65), colloquially known as the SIPO, breeds at elevations up to 1800 m as well as on riverbeds and lake shores. It is often seen feeding on farmland during the breeding season of early spring to summer, before it moves to the coast around all parts of New Zealand for the winter. Birds keep in contact with shrill whistles, often heard as they fly overhead. The spur-winged plover *Vanellus miles novaehollandiae* prefers farmland for breeding, where its piercing call is a familiar sound (Fig. 66). In the autumn it may be found in large flocks, although there is no large-scale migration as seen in the SIPO. (The self-introduction of this species from Australia is described in Chapter 8.) The smaller banded dotterel *Charadrius bicinctus* breeds from sea level up to about 1800 m, although the higher elevations are abandoned in winter when most southern birds fly to south-eastern Australia. Small flocks may sometimes be encountered on the tops of Otago mountains, such as the Rock and Pillar Range, as well as on shingle riverbeds, such as those in the Eglinton Valley and many areas of Southland. This endemic is a handsome species in its breeding dress, with a clean white chest crossed by a wide chestnut breast band and a narrower black stripe across the underside of the neck. The Southern black-backed gull *Larus dominicanus* is another widespread species, breeding from sea level to about 1500 m.

New Zealand possesses the world's only alpine parrot, the kea *Nestor notabilis* (Fig. 67). This highly intelligent bird is known for its clowning antics, as well as its destructive curiosity. Small flocks of adolescent males often congregate in places where human activity provides a source of food or entertainment: the Homer Tunnel on the Milford road is a reliable place to see them. Actively persecuted by high-country farmers for its rare killing of sheep, numbers have declined last century to approximately 5000 birds. It breeds in rock crevices in the alpine zone, just above the bush line, below which it often descends to feed.

A second New Zealand specialty, the rock wren *Xenicus gilviventris* (Fig. 68) also lives above the bush line, in alpine herbfields such as at the entrance to the Homer Tunnel. This species and the bush-dwelling rifleman are the only living representatives of the endemic family, Acanthisittidae. Almost without a noticeable tail, this tiny bird has a high-pitched piping call, and can be remarkably tame, staring fixedly at an intruder and bobbing up and down. Females are a duller brown than the olive-green males. Pairs remain on their territories all year round.

Fig. 65. The South Island oystercatcher *Haematopus ostralegus finschi* breeds inland at elevations up to 1800 m. *J. Darby*

Fig. 66 (left). Spur-winged plover. *Rod Morris*

Fig. 67 (above). Sub-adult male kea on a wrecked car in the mountains. *Rod Morris*

Fig. 68. Rock wren on an alpine herbfield. *Rod Morris*

The New Zealand pipit *Anthus novaeseelandiae* also breeds in high-country areas, up to about 1500 m, although apparently

not where mean annual rainfall is below 760 mm. This species has close relatives in many parts of the world, including Australia and South Africa, but New Zealand populations are not migratory. Superficially like the introduced skylark *Alauda arvensis*, it forages in rougher open country, constantly wagging its tail. The pipit also lacks the skylark's erectile head crest.

Two introduced passerines are also commonly found on high-country tussock lands and even alpine herbfields. The yellow-hammer *Emberiza citrinella* and the hedge sparrow or dunnock *Prunella modularis* are equally at home from about 1600 m down to sea level (see Chapter 8 for a fuller discussion of these).

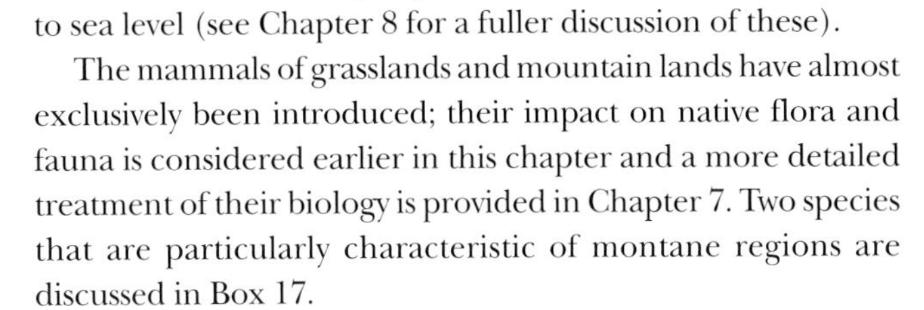

The mammals of grasslands and mountain lands have almost exclusively been introduced; their impact on native flora and fauna is considered earlier in this chapter and a more detailed treatment of their biology is provided in Chapter 7. Two species that are particularly characteristic of montane regions are discussed in Box 17.

17 ALPINE MAMMALIAN INVADERS

Many of the mammals introduced into New Zealand's lowland areas (see Chapters 7 and 8), such as stoats and red deer, eventually made it into montane regions, at least at the extremes of their ranges. But in the climate of largely unfettered introductions that existed at the turn of the century, fed by the desires of hunters and Acclimatisation Societies, the New Zealand government was not about to let niches in the high country go unfilled for long. It released two species of high country ungulates near Mount Cook. Himalayan thar *Hemitragus jemlahicus* were first liberated in 1904 and chamois *Rupicapra rupicapra* in 1907. Both went through an irruptive phase of colonisation and population growth, the former now extending in the alps as far south as Lake Wanaka and the latter as far as Milford Sound. Thar have a long coat and short stout horns that curve backwards (below). There is pronounced sexual dimorphism, with males being larger and having a shaggy mane. Chamois are smaller, with a shorter coat and horns that stand erect but are hooked at their distal end. Chamois and thar are grazers and together they have contributed to erosion and modification of our native alpine vegetation.

Lloyd S. Davis

One of the region's two species of high country ungulates, Himalayan thar. *Rod Morris*

CHAPTER 10

INLAND WATERS AND WETLANDS

ALEX HURYN, CAROLYN BURNS, RICHARD ALLIBONE,
BRIAN PATRICK, DONALD SCOTT

Wanaka, one of southern New Zealand's glacier-sculpted, deep, nutrient-poor lakes.
Laurel Teirney

Fig. 1. A selection of stream types.
Far left: A spring stream forms the headwaters of Stony Creek in the Lammermoor Range, Eastern Otago, altitude 1100 m.
Left: The steep channel of a tributary of the Cleddau River, a forest stream in Fiordland, is composed of large and stable boulders.
Lower left: The braided channel of the Oreti River in Southland is unstable, being formed from shifting cobbles and gravel.
Below: At Serpentine Flats in Eastern Otago, the Taieri, a grassland river, gently meanders through a broad floodplain. Numerous ponds and 'oxbows' can also be seen on the floodplain. *Alex Huryn*

Fig. 2. A range of Otago's freshwater lakes and ponds. Left: Ross Creek Reservoir, Dunedin. Below: a farm pond on Saddle Hill, near Dunedin. Opposite left: Gem Lake, Umbrella Range 1400 m. Opposite right top: Lake Hayes, Central Otago. Opposite right bottom: bog ponds, Umbrella Range. *Carolyn Burns, R. Nichol and Ken Miller*

INLAND WATERS AND WETLANDS

At first glance, southern New Zealand hosts a bewildering diversity of rivers, streams, lakes and ponds (Figs 1 & 2). This is largely because of its dynamic geological history, that has resulted in a rugged and mountainous terrain, and its location within the Roaring Forties latitudes of the remote south Pacific. Together these factors produce a complex pattern of rainfall, ranging from ten metres per year in parts of Fiordland to ten centimetres in areas of Central Otago. This phenomenal variation in rainfall across an equally variable topography is largely responsible for the diversity of inland freshwater habitats that characterise the landscape. Types of freshwater habitats are often consistent within regions of uniform topography and rainfall, however, so their diversity becomes less bewildering, and more manageable when viewed from a geographical perspective.

Sandy-bottom streams with gentle gradients, for example, occur on the mountain tablelands and Southland Plains (Fig. 1). Stony streams, with their familiar riffles, pools and bedrock runs, and shingle streams, with braided and ever-changing channels, are common in the high country. Steep and bouldery mountain cascades are a hallmark of the Southern Alps. Less familiar stream types include an abundance of sub-alpine seeps, springs and meltwater streams, streams that regularly 'dry up', and the meandering river-floodplain systems of inland valleys and coastal plains (Fig. 1). The difference in altitude and size among stream types is as daunting as their diversity, ranging from more than 1600 m elevation to sea level, and from seeps that may produce a cup of water each second to the Clutha River, New Zealand's largest river in terms of discharge. Finally, the overall picture is complicated by changes in flow caused by day-to-day, seasonal and annual patterns of rainfall, and the slow pace of climatic and geological change. This sketch of stream diversity in southern New Zealand identifies major physical characteristics that define broad categories of stream types. Such physical characteristics, in turn, underlie similar contrasts in the natural history of the different stream types.

The lakes and ponds of southern New Zealand offer a diversity of form that is as spectacular as that observed for its rivers and streams, ranging in size from Lake Te Anau (347.5 km^2), New Zealand's second largest lake, to small tarns and ephemeral ponds that are just a few metres in diameter. The glaciers which sculpted much of southern New Zealand in the Pleistocene have left a legacy of steep-sided nutrient-poor (oligotrophic) lakes along the eastern slopes of the Southern Alps, among them the six deepest lakes in New Zealand, namely Hauroko (maximum depth 462 m), Manapouri (444 m), Te Anau (417 m), Hawea (384 m), Wakatipu (380 m) and Wanaka (311 m). Two smaller,

shallower lakes near Lake Wakatipu (Hayes – 33 m, Johnson – 28 m, Fig. 2), legacies of the same glacier that formed Lake Wakatipu, have drainage basins subject to more human development and high levels of plant nutrients in the water.

Above the treeline in Fiordland, and further east on the flattened ridges of mountain ranges in Otago and Southland (e.g. Umbrella Range, Rock and Pillar Range, Blue Mountains) there are small cirque lakes, tarns and acidic bog ponds (Fig. 2). At lower altitudes, further east, natural lakes become more sparse. Most of those present today are reservoirs formed by damming rivers to provide drinking water for humans or stock, irrigation, or sluicing for gold (e.g. Onslow, Poolburn, Blue Lake, Manorburn, Butcher's Dam, Ross Creek Reservoir, Fig. 2). Others were created primarily for the purposes of generating electricity, including the Waitaki lakes (Benmore, Aviemore, Waitaki), the Clutha lakes (Roxburgh, Dunstan), Mahinerangi and Great Moss Swamp. Lake Wilkie, nestled in native bush on old sand dunes on the Catlins coast, Southland, is a rare example of an acidic, brown water lake in eastern New Zealand. Acidic, brown water wetlands occur also in Stewart Island, but their ecology has not been studied. Throughout Otago and Southland, farm ponds and small dams, especially those with well-developed stands of wetland plants, provide habitats for a variety of aquatic animals.

Along the meandering paths of the upper Taieri and Mataura rivers, there are shallow, oxbow lakes (also known as 'cut-off meanders', or billabongs, Fig. 1) that are subject to flooding and infilling. Further east, two shallow natural lakes on the lower Taieri plain, Waipori and Waihola, are remnants of a formerly extensive wetland that occupied much of the floodplain before it was drained for agriculture. Lake Tuakitoto, near Balclutha, is another shallow lake remnant of a once-extensive wetland near the mouth of the Clutha River. At Waikouaiti, on the Otago Peninsula, and elsewhere along the eastern and southern coast of the South Island are slightly brackish, shallow, freshwater lagoons and marshes, often near the mouths of rivers or behind coastal dunes (e.g. Hawksbury, Tomahawk Lagoons, Fig. 2).

To simplify the treatment of the diversity of freshwater habitats in southern New Zealand, streams and rivers, lakes and ponds, and wetlands are treated separately in this chapter. One must keep in mind, however, that lakes and rivers are linked in many ways beyond the simple passage of water – consider the effect of an impounded reservoir on the temperature patterns and nutrient status of the river downstream, and the role of a meandering stream channel in forming oxbow lakes. We cannot too strongly emphasise that the dichotomy implied by the structure of this chapter is a product of convenience rather than a reflection of real-world processes.

STREAMS AND RIVERS

Biodiversity

To make biological sense of the physical diversity of streams throughout southern New Zealand, we first turn to a diverse and well-studied group of organisms, the invertebrates. This is a particularly good group for beginning a study of stream ecology, because invertebrates play a central role in stream food webs by linking the primary energy base (algae, fungi, bacteria and decomposing plant tissue) to the top predators (the fish and birds).

Stream invertebrates of the South Island

Geographical patterns of invertebrate distribution among streams of the South Island and Stewart Island have been studied by Jon Harding and Michael Winterbourn (University of Canterbury). The South Island was divided into geographically uniform regions and invertebrates were sampled from ten streams in each region. The results of the study showed clear geographical patterns. Invertebrate distribution was most strongly influenced by vegetation type, biogeographical events in the remote past and changes in land use over the past century or so. More invertebrate species are found in forested regions where tree canopies completely cover streams ('closed canopy'), compared to pastoral, tussock grassland and alpine regions where canopies do not cover streams ('open canopy'). Forest streams in Westland and Nelson are richest in terms of numbers of species (75–96+ on average). Streams of the south-eastern forests (Catlins and Stewart Island) have a lower richness (54+ species). A number of invertebrate species are shared among these regions, however. For example, the snail-cased caddisfly, *Rakiura vernale*, is known only from Nelson, Westland and Stewart Island. The lowest number of invertebrate species are found in the open-canopy streams of the Southern Alps, Southland Plains, Central Otago (upper Clutha region) and the Canterbury Plains (39–49+ species). The open-canopy streams of the high country, which refers to the eastern foothills of the Southern Alps and numerous minor eastern mountain ranges, are intermediate in number of invertebrate species (71+ species). It is important to realise that the effort in collecting invertebrates in this study was carefully controlled to ensure comparable results among regions. Intensive collections from single streams over long periods may result in lists of more than 150 species. In addition, current information on the distributions of stream invertebrates

Common name	Scientific name	Number of species	% NZ fauna
Caddisflies	Trichoptera	108	48
Stoneflies	Plecoptera	64	58
Mayflies	Ephemeroptera	20	60
Dobsonflies	Megaloptera	1	100
Scorpionflies	Mecoptera	1	100

Table 1: Species richness of selected aquatic insect orders in southern New Zealand, ranked in order of relative richness. The true flies (Diptera) are undoubtedly the most numerous order in terms of abundance and number of species. However, the paucity of knowledge about this group precludes its fair treatment here.

is incomplete because of limited access to streams throughout much of the South Island. Many new species and biogeographical surprises are expected in the near future, especially from Fiordland, Westland and Nelson.

Stream invertebrates of southern New Zealand

South of the Waitaki River, streams of the high country yield the greatest richness of invertebrates and should be considered important centres of biodiversity. In these mainly stony streams, mayflies (e.g. *Deleatidium, Coloburiscus*), stoneflies (e.g. *Zelandoperla, Austroperla*), dobsonflies (*Archichauliodes*), caddisflies (e.g. *Aoteapsyche, Olinga,* Hydrobiosidae), midges and sandflies (e.g. *Maoridiamesa, Austrosimulium*) are predominant. On average, the number of species of caddisflies and true flies recorded from high country streams is higher than streams elsewhere in the South Island. Streams of the south-eastern forests are also important centres for diversity. Streams in this region have unusually high abundances of amphipod and isopod crustaceans. Several of these, *Paraleptamphopus subterraneus* for example, are found only in springs and subterranean waters elsewhere in New Zealand.

The pastoral regions of the Southland Plains and Central Otago are notable in the low richness of the stream invertebrates compared to other regions, and in the representation of different groups. Compared to the Southern Alps, high country and south-eastern forest, mayflies and stoneflies are poorly represented, whereas annelid worms (e.g. *Stylodrilus*), snails (e.g. *Potamopyrgus antipodarum*), beetles (e.g. Elmidae) and midges (e.g. *Eukiefferiella*) are well represented. Although invertebrate richness is low, streams of the Southland Plains often have high abundances of individuals, ranging to 24,000 animals per square metre, or more than ten times the abundances reported for the Southern Alps (excluding protozoans, rotifers and nematodes).

The conversion of native grassland and forest to intensively managed pasture is nowhere more apparent than the Southland Plains. Such land conversion is accompanied by changes in physical, chemical and biological characteristics of streams. A low richness of invertebrate species is usually reported for pasture streams where high sediment concentrations, increased temperature and nutrient levels, reduced oxygen concentrations, and potential contamination of streams by pesticides and herbicides can be expected. In areas where pasture management is not intensive (e.g. less than about 30 per cent of the entire catchment area), an increase in invertebrate abundance and biomass may occur but species composition may remain unchanged. Under conditions of intensive management (e.g. more than 30 per cent of catchment), as seen in the Southland Plains and many areas of Central Otago, the richness and abundance of mayflies, stoneflies and sensitive caddisflies is usually reduced.

Stream invertebrates endemic to the region

All the major aquatic insect orders and families that characterise New Zealand freshwater are found in this region (Table 1). Twenty of 33 presently known species of New Zealand mayflies (Ephemeroptera) occur south of the Waitaki River; however, none are endemic to the region (Box 1). In contrast, 33 stoneflies (Plecoptera) are endemic to southern New Zealand (Box 2). Ten species are found only in Fiordland and Western Otago, a key area for biodiversity of stoneflies in the country.

1 MAYFLIES IN SOUTHERN NEW ZEALAND

Lift almost any stone in a permanently flowing stream and you will find the larvae of several insect species. Probably the most numerous will be larvae of mayflies. Known as nymphs, and beloved by fly-fishing anglers, mayflies can be recognised by their three tail filaments and the rows of small paddle-like or feather-like gills attached to each side of the abdomen.

Almost the whole of the lifespan of the mayfly, from the hatching egg through growth and development (from about 2 months to 2 years, depending on species) is spent under water. Usually in the late spring or early summer, the larva, now about 1–2 cm long (depending on species), climbs from the water by way of a rock or vegetation trailing in the water, sheds its larval skin and emerges as the winged adult.

The adult mayfly has two large and two small wings. The larger forewings are held upright over its back. Unlike other insects, these cannot be folded down further over the abdomen. Mayflies are unique in that the newly emerged adult, the sub-imago, is only a weak flier and within a few minutes or hours again sheds it skin to become the fully mature adult, known as an imago. Usually within a day or so the males swarm aloft with a characteristic up and down dancing motion, often near the stream of their origin. Females fly into the swarm, mate, and lay their eggs by dipping the tips of their abdomens through the surface of the water. Within a day or so, both sexes die and the life cycle is completed.

Southern New Zealand has a relatively rich mayfly fauna. At least 20 of the 33 presently named New Zealand species are found in its waterways. Others remain to be described. The smallest permanent trickles and the largest rivers, from just above the limits of brackish water in river mouths to altitudes of nearly 2000 m, provide suitable habitats as long as dissolved oxygen levels are high and the water is unpolluted. Needless to say, good mayfly water is good trout water, although New Zealand mayfly species tend not to be found in lakes, except in shallow areas that are well oxygenated by wind action.

The most commonly reported species in the south are three species of the genus *Deleatidium. Deleatidium lillii* and *D. vernale* are commonly reported from streams at lower altitudes. *Deleatidium myzobranchia* is commonly reported from streams at altitudes above 600 –700 m. Other widespread species include *Coloburiscus humeralis, Deleatidium autumnale, Austroclima jollyae* and *Nesameletus ornatus.*

Terry Hitchings

A sub-imago of *Delatidium vernale* that emerged from a stream at 500 m on Crawford Hills, Ida Valley, Central Otago. The wings of the imago will be clear and glassy rather than opaque. *Brian Patrick*

2 STONEFLY LIFESTYLES

In southern New Zealand, stonefly nymphs usually live in cool, well-oxygenated streams – *Megaleptoperla*, *Zelandobius* and *Stenoperla*, for example. Others, such as the southern *Halticoperla tara*, live in water films and shallow fast-flowing water along the sides of waterfalls or torrents. Species of *Notonemoura* prefer water films away from such fast water. Species of *Cristaperla* live in pockets of sediment in small forest streams. Here their long body hairs trap detritus and provide camouflage. Larvae of *Spaniocercoides cowleyi* live in the hyporheic zone (in the gravels beneath the beds of stony streams).

Larvae of most stoneflies feed on plant material either swept into streams or growing on the stream bottom. Some shred large fragments; others browse on fine particles. Species with terrestrial larvae, such as *Vesicaperla*, feed on fungi obtained by scraping the surface of plant stems or even the bones of deer. Other stoneflies are omnivorous. Larvae of *Notonemoura* are largely herbivorous, but may occasionally capture and eat other macro-invertebrates. Larvae of *Stenoperla* – the green stonefly – are largely carnivorous but may also feed on filamentous algae. *Austroperla cyrene* – the black stonefly – may eat wood, strip soft decaying tissue off leaves, graze on fine organic particles and algae or scavenge carcasses of other invertebrates.

During winter and spring, larvae of *Acroperla* and *Nesoperla* take advantage of cold and humid conditions to leave the stream. Depending on the species, larvae may travel hundreds of metres from water although remaining within the stream's flood plain. During these excursions larvae feed on detritus under stones.

New Zealand's largest stonefly, *Holcoperla magna*, has terrestrial larvae and wingless adults, characteristics common to many alpine stoneflies. *Holocoperla magna* lives at high altitude (1600 – 2000 m) in the mountains of Western Otago, such as in humid cavities in talus on Mt Headlong, Mount Aspiring National Park. *Holcoperla angularis* is widespread in alpine Fiordland.

Perhaps New Zealand's oddest stonefly is *Rakiuraperla nudipes* from Stewart Island. This stonefly lives beneath stones on the bare granite knob of Bald Cone. It has unusually short antennae and tail filaments (cerci), a squat body, and adults lack pads (empodia) between the claws. Apparently these unusual features have developed because of the harsh climate. The only other stonefly with similar characteristics is from southern Chile.

Ian McLellan

Small shrub- and tussock-lined streams in the alpine zone of the Mark Range in the extreme north-west of the region are home to this distinctive and recently named species of stonefly *Acroperla christinae*. The green-bodied adults are probably flightless (wingspan 14–19 mm). *Brian Patrick*

The terrestrial larva of *Vesicaperla celmisia* feeds on streamside plants at altitudes of 1650 m on the Old Man Range in Central Otago. *Brian Patrick*

Male of *Holcoperla angularis* (length 35 mm). Gertrude Saddle, Fiordland, 1420 m elevation. Adults are active by day on granite talus slopes near running water. *Brian Patrick*

The stonefly fauna in Fiordland is especially notable because it contains an unusually high proportion of species that are wingless as adults (Box 3). Eastern Otago, with 12 species, is the centre of biodiversity for *Zelandobius*. Different species are found in high-alpine seepages and torrents as well as slower moving streams and rivers. *Zelandobius auratus*, the short-winged *Z. mariae* and *Z. alatus* are known only from Central Otago. Twenty species of caddisfly (Trichoptera) are found only within the study area. For example, the free-living predators *Edpercivalia harrisoni* and *Tiphobiosis salmoni* are found only in streams of Fiordland, *Traillochorema rakiura* is restricted to Stewart Island, while *Neurochorema pilosum* (Fig. 3) and the horny-cased caddisfly *Olinga fumosa* are found only in Eastern Otago. The true flies (Diptera) also are well represented among New Zealand stream fauna, and there are a number of species endemic to southern New Zealand (Box 4). This group, however, is poorly studied compared to other major aquatic insect orders.

Fig. 3. The caddisfly *Neurochorema pilosum* (length 7 mm), an Eastern Otago endemic, is known from swift stony rivers from the Shag Valley south to the Pomahaka River. The aquatic larva, a free-living predator, transforms to the terrestrial adult stage in November. *Brian Patrick*

3 FLIGHTLESSNESS AND STREAM INSECTS IN SOUTHERN NEW ZEALAND

Reduction in wing size leading to flightlessness is widespread in the adults of southern New Zealand stoneflies and caddisflies, particularly but not exclusively species that live at high altitudes.

Adults of the stoneflies *Apteryoperla, Holcoperla, Rakiuraperla, Rungaperla* and *Vesicaperla* are completely wingless. Many other genera of stoneflies, however, show different degrees of wing reduction or 'brachyptery' among different species. For example, *Zelandobius* is represented by five brachypterous species and one species, *Z. brevicauda* from Fiordland, that is totally wingless or 'apterous' in the adult form. Similarly, *Vesicaperla* has two apterous species. Other species may show varying degrees of brachyptery that appear to be correlated with elevation. For example, different populations of the widespread stonefly, *Zelandobius foxi*, show greater degrees of brachyptery at higher altitudes. Adults from forested streams in Eastern Otago have wingspans of up to 26 mm. Adults from alpine streams of Stewart Island, western and Central Otago, however, may have wingspans of only 10 mm. Populations of *Austroperla cyrene* and *Zelandoperla pennulata* also show increasing brachyptery of adults with increasing altitude.

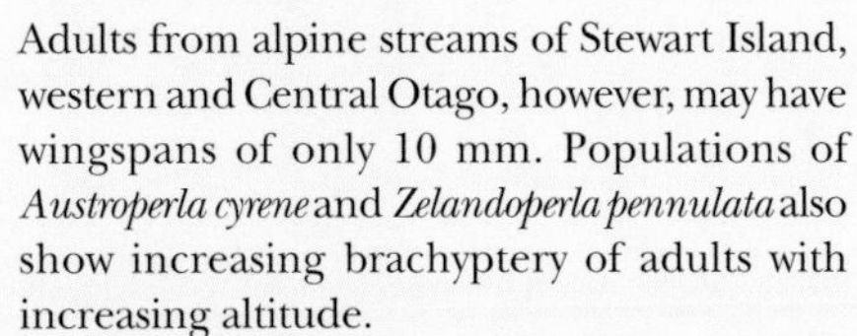

Several species of caddisflies also have flightless adults. While most of these live at high altitude (e.g. *Costachorema hebdomon* and *Hydrobiosis torrentis*), others such as *Pseudoeconesus paludis* and *Oeconesus angustus* inhabit seepages and small streams at low altitudes. All are characterised by very little loss in wing size. Some species, including the tiny black *Tiphobiosis childi* with free-living predacious larvae, have shorter wings in the high-alpine seepages than populations in the alpine zone. In contrast, an undescribed species of *Philorheithrus* in Eastern Otago is short-winged in both lowland and alpine habitats. The stout-legged adults can be found running about stream margins by night.

Brian Patrick

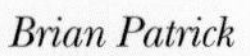

Adults from some populations of the stonefly *Zelandobius alatus* may have tiny wings, as evidenced by this specimen from the Old Woman Range in Central Otago at 1600 m elevation. The tiny adults (length 6–8 mm) are diurnal, and can be found running about the margins of snow-melt streams. *Brian Patrick*

The stonefly *Zelandoperla pennulata*, confined to a region extending from Central Otago through Eastern Fiordland, is characterised by brachypterous populations at high altitudes. *Brian Patrick*

The caddisfly *Hydrobiosis torrentis* is found along high altitude streams in Central Otago. Although the wings are not noticably reduced in size compared to flighted caddisflies, *H. torrentis* is flightless. *Brian Patrick*

An undescribed, brachypterous species of the caddisfly *Philorheithrus* from Eastern Otago. The nocturnal, strong-legged adults (length 8–12 mm) can be found running about the margins of small streams at low to high altitude streamsides by night, searching for a partner. *Brian Patrick*

4 THE TRUE FLIES – A NEGLECTED GROUP

Among the many families of true flies (Diptera) with aquatic larvae, the best known are the annoying sand flies or black flies (Simuliidae) (larva shown in Fig. 8). Of 10 species known in New Zealand, southern New Zealand is host to seven species. Among these species, *Austrosimulium stewartense* and *A. dumbletoni* are endemic to the region. *Austrosimulium dumbletoni* is New Zealand's largest sand fly and is known only from Jackson's Bay, where it bites Fiordland Crested Penguins. Species most likely to bite humans are *A. australense* and *A. ungulatum.*

The non-biting midges (Chironomidae) (Fig. 7) are undoubtedly the most numerous and diverse group of insects inhabiting New Zealand streams. Compared to other major insect orders, however, their taxonomy and biology is poorly known. Other groups of non-biting flies that are widespread in southern New Zealand are the dixid, net-winged and droop-winged midges. Larvae of the dixid midges (Dixidae) have a pair of flat paddle-shaped lobes on the tip of their abdomen. These larvae inhabit surface films along the edges of streams, swamps and lakes. They move within the surface films with a rapid 'sidewinder' motion. Nine species have been described from New Zealand and five are known from southern New Zealand. Larvae of the net-winged midges (Blephariceridae) (Fig. 8) are common in torrential streams where larvae use a central row of suckers on their undersides to maintain their position in the fast flow. Adults can be seen flying in the spray of cascades or walking along the wet surfaces of stones at the water's edge. Of 11 species known from New Zealand, six species occur in southern New Zealand, and four species are endemic to the region. The droop-winged midges (Thaumaleidae) are known from Eastern Otago and Southland, and further study will probably show that they occur throughout much of the region. Larvae live in shaded waterfilms less than a millimetre deep, on the edges of seepages over steep rock-walls and in the splash zones of streams. Larvae have a unique mode of sideways movement. Before moving, the body forms a 'U', with one end being anchored while the other moves in the desired direction.

Ian McLellan

Native stream fishes

Twenty-eight species of native fishes are known from streams and rivers of southern New Zealand. This census includes the extinct New Zealand grayling *Prototroctes oxyrhynchus*. Compared to elsewhere in the country, the southern region has a unique assemblage of non-migratory *Galaxias* (galaxiids) species (Box 5). Other notable regional distinctions are the absence of mudfish *Neochanna* spp., and the remarkably patchy distribution of the bullies *Gobiomorphus* spp., torrentfish *Cheimarrichthys fosteri* and shortfin eel *Anguilla australis*. The mudfishes and torrentfish are absent from Stewart Island.

The longfin eel *Anguilla dieffenbachii* is the most widespread native fish in southern New Zealand, being found in all major river systems. Barriers to upstream migration (such as hydro-electric dams) have reduced its occurrence in inland areas. Its close relative, the shortfin eel *A. australis* is more restricted and is generally found in lowland lakes and wetlands. Similar in form, but unrelated to the eels is the lamprey. Juvenile and adult lamprey *Geotria australis* are found in most lowland rivers, although they are secretive and rarely seen.

Ten species of non-migratory galaxiids (Fig. 4) are known from southern New Zealand; two are widespread while two are represented by minor populations at the extreme southern limits of their ranges. The remaining six species are known exclusively from a few river systems in Otago (Box 5). Populations of non-migratory galaxiids are most often found in streams where populations of trout have not been established – apparently a consequence of predation by trout on the smaller galaxiids.

All five of New Zealand's migratory galaxiids (the 'whitebait' species) are found in southern New Zealand. The rare short-jawed kokopu *Galaxias postvectis* is known from a few streams in Fiordland. The banded kokopu and koaro (*G. fasciatus* and *G. brevipinnis* respectively) are abundant in Fiordland, Stewart Island and the south-eastern forests, and occur occasionally elsewhere. The giant kokopu *G. argenteus* is locally abundant on Stewart Island, and small numbers occur in lowland areas of Otago and Southland. Inanga *G. maculatus*, the most common whitebait species, is widespread in lowland rivers and streams throughout southern New Zealand. Landlocked populations of koaro are common in large lakes and their tributaries throughout the region. In these populations, juveniles reside in the lake before migrating upstream to the adult habitat. Populations of giant kokopu also occur as landlocked populations in Lake Monowai and some of the smaller lakes of Southland, an unusual situation for this fish.

Of the southern New Zealand bullies, the common bully *Gobiomorphus cotidianus* has the widest distribution and is characterised by migratory and non-migratory populations. However, some of the lake populations have not been naturally established but are rather the result of introductions. The redfin bully *G. huttoni* and the upland bully *G. breviceps* (Figs 5 and 6) are locally common in coastal and inland streams, respectively. The giant bully *G. gobiodes* and bluegill bully *G. hubbsi* are relatively uncommon, but can be found in tidal areas and braided rivers, respectively.

Both of New Zealand's smelts are found in the region. Stokells

Fig. 4. Non-migratory galaxiids from the Taieri River drainage in Eastern Otago: *Galaxias pullus* (top); *Galaxias depressiceps* (centre left); *Galaxias eldoni* (centre); *Galaxius anomalus* (bottom).
Richard Allibone

Fig. 5. Redfinned bully *Gobiomorphus huttoni.*
Courtesy of the Zoology Department, University of Otago

Fig. 6. Upland bully *Gobiomorphus breviceps.*
Courtesy of the Zoology Department, University of Otago

5 NON-MIGRATORY GALAXIIDS IN SOUTHERN NEW ZEALAND – A RICHNESS UNEQUALLED

Eight described and a further two undescribed species of non-migratory *Galaxias* occur south of the Waitaki River – a richness unequalled world-wide. Two species (*Galaxias vulgaris* and *G. paucispondylus*) are also found in areas north of the Waitaki River. The remaining species are endemic to southern New Zealand and are sometimes referred to as the Otago galaxiids.

The distributions of all the Otago galaxiids are poorly understood and this is complicated by the patchy occurrence of these fish and the fact that they are often restricted to small, remote streams. *Galaxias gollumoides* and *Galaxias* sp1 are widespread, occurring in streams in Otago, Southland and Stewart Island. Other species, however, are restricted in distribution. *Galaxias cobitinis* is the rarest galaxiid in New Zealand and is known only from a 14 km section of the Kakanui and Kauru Rivers in North Otago. *Galaxias eldoni* (Fig. 4) is found in tributaries of the Taieri, Waipori and Tokomairiro rivers of Eastern Otago. *Galaxias pullus* (Fig. 4) occurs in the upper Waipori River, small tributaries of the adjacent Teviot, Beaumont, Waitahuna and Tuapeka Rivers and a tributary of the upper Taieri River. *G. depressiceps* is restricted to the Taieri River and areas to the east. *G. anomalus* is another species restricted to the Taieri River and just a few tributaries of the Manuherikia River. A further undescribed species is known from the Teviot River area. *Galaxias gollumoides* and all the undescribed species have only been recognised since 1997 and very little is known about their biology.

Fossil evidence from the Taieri River region indicates that galaxiids, probably the ancestors of present-day, non-migratory species, occurred in this region as early as the Miocene (25 to 13 million years ago) and Pliocene (13 to 2 million years ago). *Galaxias vulgaris*, the Otago galaxiids (Fig. 4) and a migratory species, the koaro (*G. brevipinnis*), are believed to have evolved from a single and migratory ancestor. *Galaxias paucispondylus* and *G. prognathus* form a separate branch of the evolutionary tree, having apparently evolved from a different ancestor. Populations that switched from migratory to non-migratory behaviour have become genetically isolated, thus facilitating evolution of new species. Further isolation of different populations of non-migratory species apparently underlies the unusually rich present-day fauna.

Knowledge of the life histories of the Otago galaxiids varies, ranging from considerable understanding for some of the named species to almost nothing at all for the galaxiids from the Teviot River. We do know that all non-migratory galaxiids spawn in late winter or spring, and that spawning habitats differ among species. For instance, *G. anomalus* deposits eggs among loose cobbles and gravels. Spawning females locate 'spawning sites' established in patches of porous gravel where sticky eggs are deposited, adhering to rocks and one another. Numerous males congregate at each spawning site and wait for spawning females; males may remain at spawning sites for over a month. Because of this behaviour, spawning sites, which may have an area of about 0.1 square metres, become communal nests. *Galaxias vulgaris*, on the other hand, creates saucer-like depressions within gravels of riffles and individual males mate with females as they deposit eggs within the depressions. *Galaxias depressiceps* nests under large slabs of schist or boulders in riffle areas, where females cement sheets of golden-coloured eggs to the underside of the rock. Several females may lay eggs in different clusters on the same rock.

Relatively little is known about the spawning habits of *G. eldoni* or *G. pullus*. The only nest of *G. eldoni* found so far consisted of eggs cemented to substrata around and beneath a large rock in a riffle; the eggs did not form distinct clusters and were not cemented exclusively to the underside of the rock, unlike *G. depressiceps*. *G. pullus* lays eggs among submerged roots exposed beneath under-cut banks. Regardless of species, females of the non-migratory *Galaxias* lay 150–1800 eggs depending on their body size (maximum length of *G. depressiceps* is about 130 mm, other species attain 160 mm). Eggs hatch about one month after spawning. Larval fish, described as 'a 1-cm thread of cotton with a pair of eyes', can be seen in backwaters and pools of streams by early summer.

Richard Allibone

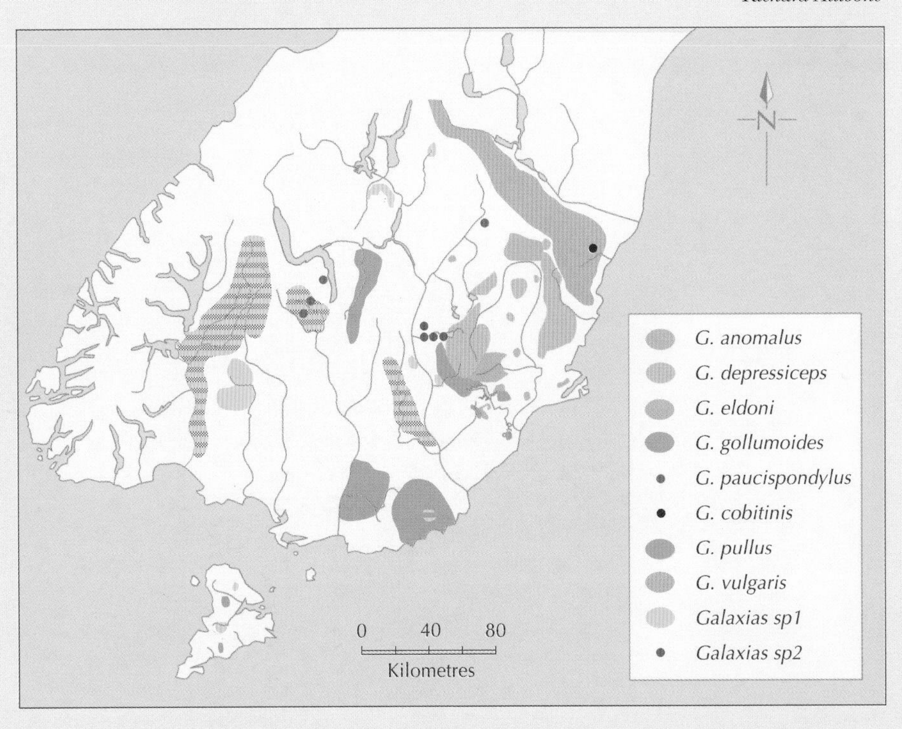

Distribution of non-migratory galaxiids in southern New Zealand.

smelt *Stokellia anisodon* is restricted to the Waitaki River and areas to the north, whereas the common smelt *Retropinna retropinna* is widespread throughout Otago, Southland and Stewart Island. Both species enter fresh water to breed, and their occurrence in rivers and streams is not continuous year round.

The most unusual freshwater fish is the black flounder *Rhombosolea retiaria*. Unlike most flounders, the black flounder penetrates far inland in low gradient rivers, with upstream limits usually being set by high gradient sections that it is unable to ascend. It is likely to be widespread, but is rarely encountered.

Finally, there is the torrentfish whose closest relative is the blue cod. This migratory fish is uncommon in most of the region but can be abundant in rivers of North Otago, where it lives in the gravel bed of fast-flowing riffle areas.

Exotic stream fishes

When the Europeans colonised New Zealand, they found a land that was both similar to, and different from the land from which they had come. Much of the country was covered with unfamiliar vegetation, but the essential structures of mountain ranges, river valleys, lakes and broad plains were similar to those of Europe. The streams and rivers had abundant fish that were easy to catch for food. However, the colonists soon decided to import fish to provide more familiar food and sport. Introductions of exotic fishes to streams and rivers of Otago and Southland began in 1864, and by 1868 self-sustaining populations of brown trout *Salmo trutta* and perch *Perca fluviatilis* had been established. At present, eight species of exotic fishes are found in the streams and rivers of southern New Zealand (Box 6).

Some species were particularly good colonisers – brown trout, rainbow trout *Oncorhynchus mykiss*, chinook or quinnat salmon *O. tschawytscha*, perch – while others were less so. The reasons underlying successful colonisation are several. First, with the exception of the giant kokopu and the relatively slow-swimming eels, there were few large predatory fishes in New Zealand. Therefore the introduced fishes assumed the role of top predator, with little competition from other species. Second, because of New Zealand's position in the Pacific Ocean, the period of optimal temperature conditions for the growth of salmon and trout (8–20°C) was considerably longer than in the Northern Hemisphere.

Brown trout are common in estuaries, rivers and streams throughout the southern South Island. Their presence in streams appears to be limited mainly by the presence of impassable waterfalls more than 3m high. In the South Island, rainbow trout are strongly associated with lakes. Chinook salmon are common throughout the Waitaki and Clutha River systems, but sporadic in other coastal systems. Brook char *Salvelinus fontinalis* occur in isolated headwaters of the Mataura, Taieri, Clutha, and Waitaki River systems, where they replace brown trout. Perch can be abundant in low gradient and weedy rivers. Since they are salt sensitive their abundance declines toward estuaries. Other exotic fishes found in southern New Zealand are listed in Box 6.

6 CHRONOLOGY AND SYNOPSIS OF INTRODUCTIONS OF EXOTIC FRESHWATER FISHES TO SOUTHERN NEW ZEALAND

Species	Date[1]	Distribution
Brown trout (*Salmo trutta*)	1868	widespread in rivers, lakes and coastal waters
Perch (*Perca fluviatilis*)	1868	widespread in lakes and rivers east of Main Divide
Tench (*Tinca tinca*)	1869	confined to a small area south of Oamaru
Brook char (*Salvelinus fontinalis*)	1885	widespread in headwaters, notably upper Manuherikia
Rainbow trout (*Oncorhyncus mykiss*)	1895	widespread in Waitaki and Clutha catchments
Chinook or quinnat salmon (*O. tshawytscha*)	1901–1907	Waitaki and Clutha catchments, other coastal waters
Sockeye salmon (*O. nerka*)	1902	Waitaki catchment
Altlantic salmon (*Salmo salar*)	1908–1911	periphery of the Waiau catchment

[1]*Date refers to date of first successful introduction.*

Donald Scott

Above (left to right): adult brown trout; adult brook char; smolt stage of chinook salmon. Young salmon transform from the parr stage to the silvery smolt stage shortly before migrating from nursery streams to the ocean. Right: adult perch. *Donald Scott*

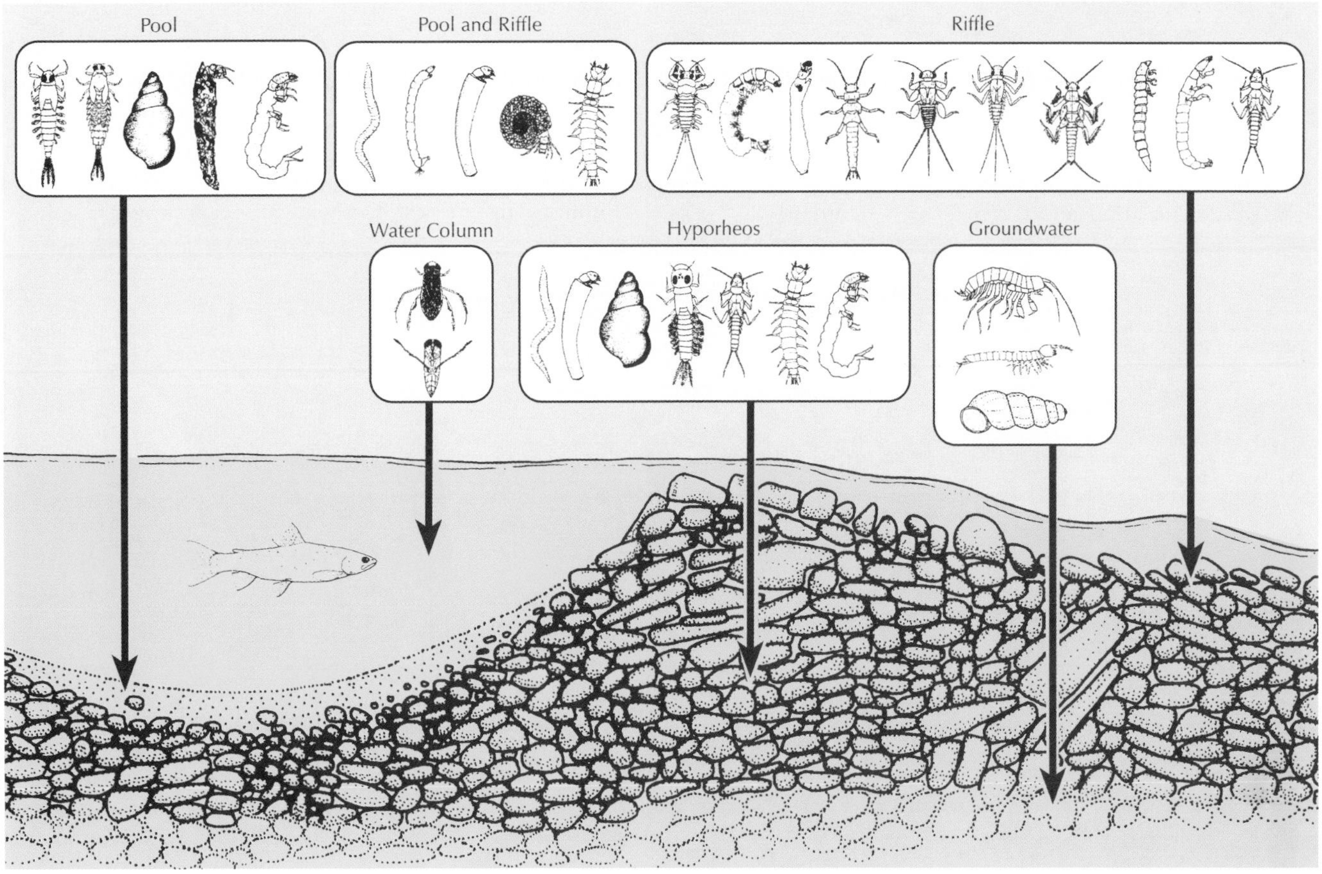

Fig. 7. Diagram showing habitat preferences for invertebrates commonly found in southern New Zealand streams. (br = browser, shr = shredder, pr = predator, ff = filter-feeder.)
Pool (left to right): *Nesamaletus* (br), *Oniscigaster* (br), *Potamopyrgus* (br), *Hudsonema* (br, pr), *Polyplectropus* (pr).
Pool and riffle: Oligochaeta (br), *Chironomidae* (br, pr), *Olinga* (br, shr), *Helicopsyche* (br), *Archichauliodes* (pr).
Riffle: *Coloburiscus* (ff), *Aoteapsyche* (ff), *Austrosimulium* (ff), *Austroperla* (shr), *Austroclima* (br), *Deleatidium* (br), *Zelandoperla* (br), Elmidae (br), *Hydrobiosis* (pr), *Stenoperla* (pr).
Water column (top to bottom): *Sigara* (br), *Anisops* (pr).
Hyporheos: Oligochaeta (br), *Olinga* (br, shr), *Potamopyrgus* (br), *Amelotopsis* (pr), *Stenoperla* (pr), *Archichauliodes* (pr), *Polyplectropus* (pr).
Groundwater: amphipod *Phreatogammarus* (br), syncarid shrimp (br), snail *Saganoa* (br).
A brown trout is shown in the pool. *Figures of invertebrates redrawn from various sources, primarily Pendergrast and Cowley (1966), Sinton (1985) and Winterbourn, Gregson and Dolphin (2000). Figure of brown trout redrawn from McDowall (1990).*

Food and feeding

Stream animals in New Zealand tend to be generalists in food preference. Most invertebrate species browse algae and other microflora from the stream bed. Others feed on decaying organic material that is either on the stream bed or suspended in the current. Many invertebrate species are predators, as are all stream fishes with the exception of the grayling and lamprey larvae. The grayling apparently was a browser, but is now extinct. Lamprey larvae feed on suspended organic matter.

Browsers and biofilms

Browsers include members from most of the major invertebrate groups common to New Zealand streams. The mayfly *Deleatidium* (Box 1 and Fig. 7) is probably the most abundant and widespread browser. Other browsers are shown in Figs 7 and 8. Elsewhere in the world, fishes can be important browsers of biofilms in streams and, as such, can have large effects on other members of their communities. As the only large vertebrate browser known from New Zealand streams, it is probable that the grayling played a similar role. The effect of its recent extinction on local stream ecosystems, however, will remain unknown.

To most people, the food of browsers – biofilms – is probably less familiar than the invertebrates themselves. Biofilms are thin layers of algae, bacteria and fungi that cover the bottoms of all streams and rivers. The proportion of the different components varies depending on stream type. Light conditions of open-canopy streams are often sufficient for maximum photosynthesis, so biofilms in these streams usually have high proportions of algae, usually diatoms and blue-green algae. In some streams, large mats of filamentous algae may also be apparent. However, this type of alga is not a biofilm component and is not readily consumed by most stream invertebrates.

The types of algae found in a given stream are controlled by nutrient concentrations (nitrogen, phosphorus and probably micronutrients, such as molybdenum), patterns of flow and temperature, light conditions and browsing. These factors are strongly influenced by geology, vegetation and land use. Algae can be lost by the scouring effects of stones tumbling along the stream bed. If scouring floods are frequent (e.g. return interval less than three months), then other factors, such as browsing, may be of little importance. In the high country of Eastern Otago,

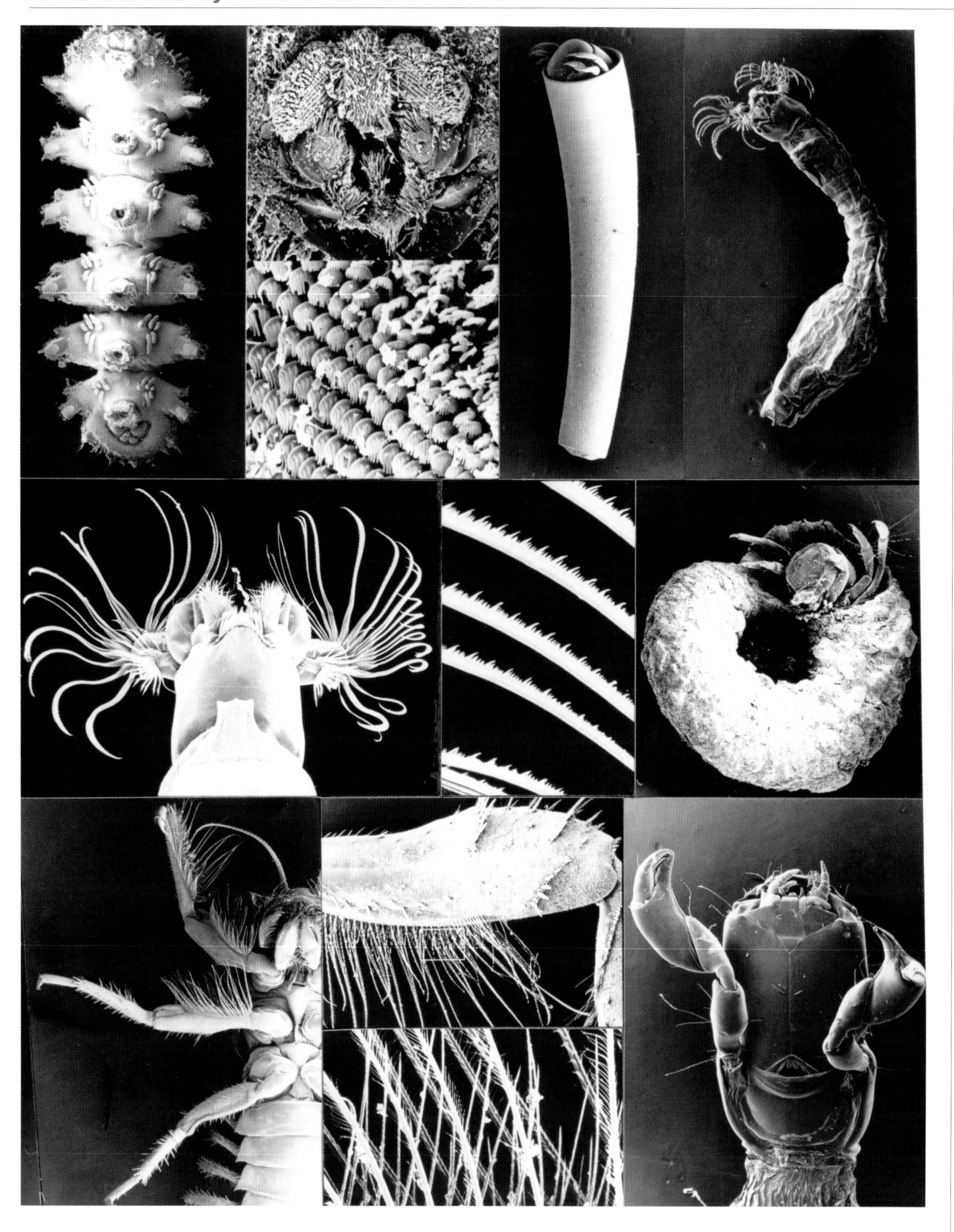

the productivity of algae in clear and shallow streams seems to be influenced by concentrations of nitrogen, apparently because a large amount of phosphorus is indirectly contributed to streams by widespread topdressing with superphosphate fertiliser. The productivity of algae in larger rivers seems to be limited mainly by light because of high amounts of suspended sediments and, consequently, low clarity.

In heavily shaded forest streams, algae form a minor component of biofilms because low levels of light may hamper photosynthesis. Here bacteria, fungi and protozoans are the usual biofilm organisms because they are able to live on dissolved organic matter from the water: a process that does not require light energy. 'Dissolved organic matter' refers to organic molecules that occur in all natural waters. Dissolved organic matter comes from many sources, both within the stream and from the surrounding catchment, and reaches greatest concentrations in low gradient streams with forested floodplains. These streams are often called 'brown water streams' because high concentrations of dissolved organic matter are visible as a dark-brown stain – much the same as a strong cup of tea. Brown water streams are common in areas of the south-eastern coast (e.g. Catlins River), Fiordland (Hollyford Valley), Westland and throughout Stewart Island. Surface foams, commonly observed in high country streams, are evidence of moderate levels of dissolved organic matter.

Biofilms are an exceedingly important component of stream food webs everywhere. Biofilms occur on all surfaces and in all streams and are absolutely essential to the conversion of light energy and the energy locked up in dissolved organic matter into a form that can be readily consumed by invertebrates.

Fig. 8. Scanning electron micrographs of selected insect larvae from Sutton Stream, a tributary of the Taieri River in E astern Otago.
Opposite top
Left: Underside of larva of the net-winged midge *Neocurupira* showing single row of 6 suction cups used to anchor larva to stream bottom (7 mm total length).
Left centre (upper): Mouthparts of *Neocurupira*, showing rows of combs used to scrape algae from stream bottom. The combs are enlarged in the inset immediately below.
Right centre: Side view of larva of the horny-cased caddisfly *Olinga* (5 mm total length). The smooth case is made of silk produced by the larva.
Right: Larva of the sandfly *Austrosimulium* (3 mm total length).
Opposite middle
Left: Head of larva of the sandfly *Austrosimulium*. Filter-feeding fans are extended from anterior margins of head. The corkscrew-shaped filament extending from the mouth is a piece of freshly extruded silk used as a safety line as the larva moves about in rapid currents.
Centre: Detail of the sandfly's filtering fans showing comb-like structure used to capture fine suspended particles.
Right: Underside of the larva of the snail-cased caddisfly *Helicopsyche* (3 mm total width). The case is made of sand particles glued together with silk.
Opposite bottom
Left: Underside of larva of the filter-feeding mayfly *Coloburiscus*, showing rows of filter-feeding hairs on front (tibia and femur) and middle (femur only) legs (8 mm total length).
Centre (upper): Detail of filtering hairs on foreleg. A few hairs are enlarged in the inset below to show feather-like filtering meshwork.
Right: Head, prothorax and forelegs of *Costachorema* – a predacious free-living caddisfly larva (12 mm total length). The forelegs are formed into pincers for grasping prey. *S. Johnstone and Alex Huryn*

Detritus and detritivores

Detritus refers to dead plant and animal tissues together with active populations of decomposers – bacteria and fungi. Compared to freshly dead plant tissue, plant detritus tends to be more palatable and nutritious. Two major types of detritus are recognised in New Zealand streams: small particles with dimensions on the order of a sugar grain, and large particles such as whole leaves and wood. Invertebrate faeces, particles abraded from biofilms, and wind-blown particles from terrestrial sources comprise the pool of small particles. Many types of invertebrates feed on fine particles that accumulate on the stream bed – these are called 'deposit feeders' (Fig. 7). However, fine particles are also continuously suspended in the current.

Some invertebrates collect suspended particles with filtering devices. The larva of the filter-feeding caddisfly *Aoteapsyche* builds a capture net from silk produced from glands below its mouth (Fig. 9). Larvae of the filter-feeding mayfly *Coloburiscus* and the

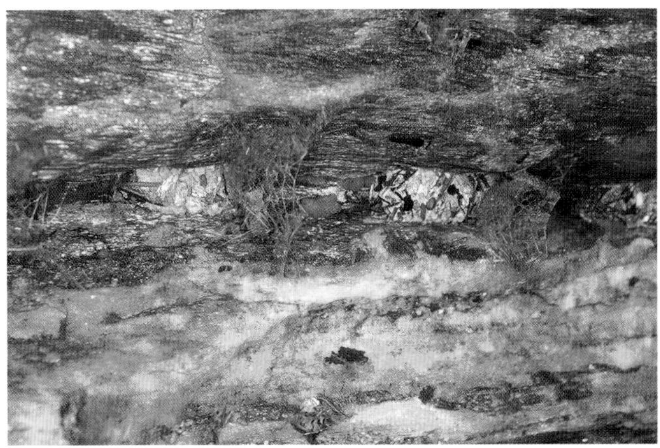

Fig. 9. Silken nets used to capture food by larvae of the filter-feeding caddisfly *Aoteapsyche* (Taieri River, Lammermoor Range, Eastern Otago). **Top:** The remarkable mesh-work of the capture net of this specimen, removed from the stream, is supported by a circular hoop of moss fragments (diameter of net is 8 mm). The tube-like structure formed from gravel opens beneath the base of the net. This is the larva's home or 'retreat'. The larva checks its net for food only periodically. Most of the time it remains sheltered within the retreat. **Bottom:** Two nets and retreats photographed in place; the water was flowing from right to left at about 50 cm/s. The retreats were sheltered in a crevice in schist bedrock. The nets were constructed near the top of the crevice, where they filtered food from the current. Two larvae of the net-winged midge *Neocurupira* can be seen above the right-hand retreat. *Alex Huryn*

sandfly *Austrosimulium* capture drifting particles with special structures located on their legs and heads, respectively (Fig. 8). The freshwater mussel *Hyridella* feeds on fine particles that are filtered from water that is pumped across its gill surfaces. Among the freshwater fishes of New Zealand, only larval lampreys (ammocetes) filter feed. Ammocetes bury themselves tail-first into the muddy bottom of pools until only their heads are exposed. From this position they filter feed in a fashion similar to that described for the freshwater mussel. Compared to browsers, there are relatively few species of filter-feeders. Filter-feeders may be very abundant, however. The population of adult biting sandflies that emerge from rivers in Fiordland provides a dramatic example.

Although usually not as prevalent as other feeding groups, some invertebrates obtain their food by feeding on decaying tussock grass or tree leaves. Invertebrates in this feeding group are called 'shredders' because their feeding breaks down large detritus particles to small particles, as faeces. The faeces, as fine particles, are consumed by filter-feeders and deposit-feeders. Like dissolved organic matter, an important source of detritus is the terrestrial vegetation surrounding the stream. The dependence of some stream invertebrates on terrestrial detritus is an example of the pervasive link between stream and terrestrial ecosystems.

In southern New Zealand, shredders are best represented by selected case-bearing caddisflies and stoneflies (Figs 7 and 8). In very small streams of the high country, tussock grasses may form dense canopies over the streams. Invertebrates that are able to shred decaying grass leaves, such as larvae of the stonefly *Austroperla* and amphipods, are often abundant in these streams. A similar situation occurs in forested streams that receive large amounts of detritus as tree leaves. Stoneflies that have abandoned the normally aquatic lifestyle as larvae have retained the usual larval diet – detritus. However, a newly discovered species of *Vesicaperla* from the high-alpine zone of the Old Man Range in Central Otago feeds on streamside plants such as *Carex* and *Celmisia haastii* (Box 2). Although case-bearing caddisflies and *Austroperla* are probably the most widespread shredders, the freshwater crayfish, *Paranephrops zealandicus*, is probably the most active shredder in streams where it occurs.

Predators and prey

Most predacious stream invertebrates capture other invertebrates by actively foraging about the stream bottom. This type of feeding is most commonly found among stoneflies, dobsonflies, and free-living caddisflies (Figures 7 and 8). The caddisfly *Polyplectropus* (Fig. 7) spins an irregular web from which it ambushes entangled prey. In this way it is very similar to a spider. Other unusual stream-dwelling predators are larvae of the semi-aquatic lacewing *Kempynus*, the mayfly *Amelotopsis* (Fig. 7), and the aquatic scorpion fly *Nannochorista* (Fig. 10).

With the exception of the lamprey and the extinct grayling, all stream fishes in New Zealand are predators of invertebrates. The larger species – eels, kokopu, koaro, common and giant bullies – also prey on other fish, amphibians and even small waterfowl. Most galaxiids feed by capturing invertebrates that are on the stream bottom or drifting in the current. Terrestrial invertebrates that fall onto the water surface from overhanging vegetation are another important food resource.

'Anti-predator' behaviour

Trout tend to be more active during the day, while galaxiids tend to feed both day and night. This is because trout rely mainly on vision to detect prey, whereas galaxiids detect prey using vibrations caused by their movements. As a consequence of this fundamental difference, the introduction of brown trout to southern New Zealand has caused changes in the behaviour of their invertebrate prey.

Larvae of the mayflies *Deleatidium* and *Nesamaletus* (Fig. 7) can detect chemicals released into the water, probably from the trout's mucous coating. When trout chemicals are absent, larvae will forage on the upper surfaces of cobbles and will disperse by drifting downstream during both day and night. When trout chemicals are present, larvae tend to remain beneath cobbles and will drift only during the night. Studies in Eastern Otago and Stewart Island have shown that this behaviour has apparently evolved since the introduction of the brown trout to Otago in 1868 (Box 6).

Omnivores and feeding generalists

Many stream invertebrates feed on animals as well as biofilms and detritus ('omnivores'). The filter-feeding caddisfly *Aoteapsyche*, for example, feeds on both fine detritus and small invertebrates drifting in the current. The predacious stonefly *Stenoperla* (Fig. 7) usually feeds on invertebrates, but may also consume algae. The cased-caddisfly *Hudsonema* (Fig. 7) feeds on detritus and small invertebrates as it forages about the stream bottom. *Hudsonema* and the familiar horny cased-caddisfly *Olinga* (Figures 7 and 8), which usually feeds by browsing or shredding, will also scavenge carcasses of trout and drowned skinks. Crayfish are perhaps the ultimate feeding generalist. They are not only exceedingly omnivorous, but can also be legitimately placed in just about all recognised feeding groups. Crayfish may function as deposit and filter feeders, shredders and predators.

Fig. 10. *Nannochorista philpotti* (wingspan 16 mm) is the only species of aquatic scorpionfly known from New Zealand. Although the scorpionflies (Mecoptera) are worldwide in distribution, aquatic species are known only from the Southern Hemisphere. *N. philpotti* was first described from the Longwood Range. Like many stoneflies and caddisflies in New Zealand, adults from alpine populations tend to have reduced wings. *Brian Patrick*

Life cycles

Invertebrate life cycles

Although most people are familiar with the longevity of an adult mayfly – about one or two days – few are familiar with the details of the larval (= 'nymph') stages. Knowledge of the duration and growth pattern of the larval stage is needed to understand differences among stream communities. For all stream insects except the beetles, adults emerge from the stream to mate. Eggs are usually deposited either directly on the stream bottom or onto the surface of the water. Females of many caddisflies, *Aoteapsyche* for example, return to the stream and cement layers of eggs beneath stones. Females of other caddisflies, true flies and many mayflies and stoneflies deposit egg masses on the water's surface. The egg masses quickly break apart and the individual eggs settle into the stream bottom. Most other major groups of invertebrates – worms, crustaceans, molluscs for example – remain in the stream throughout the entire life history. Once eggs hatch, the larvae must undergo a period of growth and development, the duration and pattern of which is determined by species, food availability and temperature.

Details of life cycles are known for invertebrates inhabiting only a few streams in southern New Zealand. The following examples are from Sutton Stream, a tributary of the Taieri River in Eastern Otago. Although details will vary in other regions of New Zealand, especially in the warmer northern regions, the life cycles observed in Sutton Stream represent all the common types. The most common is typified by the mayfly *Austroclima* (Fig. 11). Adults emerge in summer and deposit eggs. The eggs hatch directly, and larval growth occurs during the autumn and spring. Growth is slow during winter and adults emerge the following summer. On average, one generation is produced each year. An unusual variant of this life cycle is observed for the filter-feeding caddisfly *Hydrobiosella* (Fig. 11). Although one generation is produced every year as in *Austroclima*, the eggs undergo a long resting stage, during which larvae are absent from the stream. Larvae first appear in early summer and grow rapidly to appear as adult caddisflies in autumn.

The mayfly *Deleatidium* has a complex life history that is based on different types of life cycles, depending on season (Box 1 and Fig. 11). During summer, larval growth is rapid and the period required to develop from egg to adult is only three months. During winter, however, growth is slow and larvae that hatch during late autumn may require nine months before they are large enough to transform to the adult stage. Because of a relatively flexible life cycle, *Deleatidium* may produce two generations each year. Many of the midges and sandflies have life histories similar to *Deleatidium*.

The final type of life cycle is shown by the filter-feeding mayfly *Coloburiscus* (Fig. 11). *Coloburiscus* grows very slowly and two years are required for larvae to reach the size needed to transform to the adult stage. Therefore, on average, one generation of *Coloburiscus* is produced every two years. During the same period, *Deleatidium* may produce four generations. Although unusual for mayflies, long life cycles are expected for large predacious invertebrates such as the dobsonfly *Archichauliodes*, which may require four years to complete growth and development. Freshwater crayfish in forested streams near Dunedin may live for more than ten years, but reach reproductive age in about five years. The freshwater mussel *Hyridella* may live for twenty years or more. Compared to other stream invertebrates, the life cycle of *Hyridella* is remarkable because larval stages are parasitic on fishes.

Differences among invertebrate life cycles will determine which species occur in different stream types. Streams that are often disturbed by floods that tumble stones along the stream bed and destroy invertebrates are usually inhabited by 'weedy' species with flexible life cycles, such as *Deleatidium*. The short generation time and long period of adult emergence will enable rapid recovery of populations reduced by disturbance. Species with long life cycles, such as *Coloburiscus*, will not be able to recover as efficiently. Fast growth rate and the ability to produce several generations each year also results in high annual surpluses of *Deleatidium* that quickly replace larvae consumed by predators. The effects of predation on populations of invertebrates with long life cycles should be more pronounced. Finally, life cycles that include a resting egg stage, such as *Hydrobiosella*, provide a mechanism for escaping the effects of disturbance and predation during winter.

Fish life cycles

Although some freshwater fishes in southern New Zealand have life cycles that are similar to those of longer-lived invertebrates, the majority of species have life cycles that are quite different, being structured by migrations to and from the sea. This type of life cycle is known as 'diadromous'. The presence of diadromous fish in a given stream depends on the absence of physical barriers (waterfalls, dams) to movement between stream and ocean.

The life cycles of freshwater fish species in southern New Zealand can be roughly broken into four groups. The first group includes the non-diadromous fishes, or those that complete their entire life cycles in fresh water. Eight species of non-migratory ('non-diadromous') galaxiids, and the trout, brook char, upland bully and perch comprise this group. Non-diadromous galaxiids and trout spawn once each year; the upland bully spawns several times. Regardless of species, non-diadromous fish in southern New Zealand usually become sexually mature within one or two years after hatching and may live for as long as four (upland bully, *Galaxias anomalus*) to ten-plus years (other non-diadromous *Galaxias*, trout).

Spawning by brown trout occurs in the autumn; rainbow trout nest later in the year. Breeding females search for areas of swift current and suitable gravel size. The female, accompanied by one or more males, excavates a depression in the gravel by a sweeping action of her tail. After reaching a depth of about 15 to 30 cm (depending on fish size), she lays her eggs. The eggs are heavier than water so they fall to the bottom of the depression, where they are immediately fertilised by milt released from the attending male(s). After the eggs are fertilised, the female moves upstream of the depression and repeats the sweeping process to excavate additional gravel that buries the

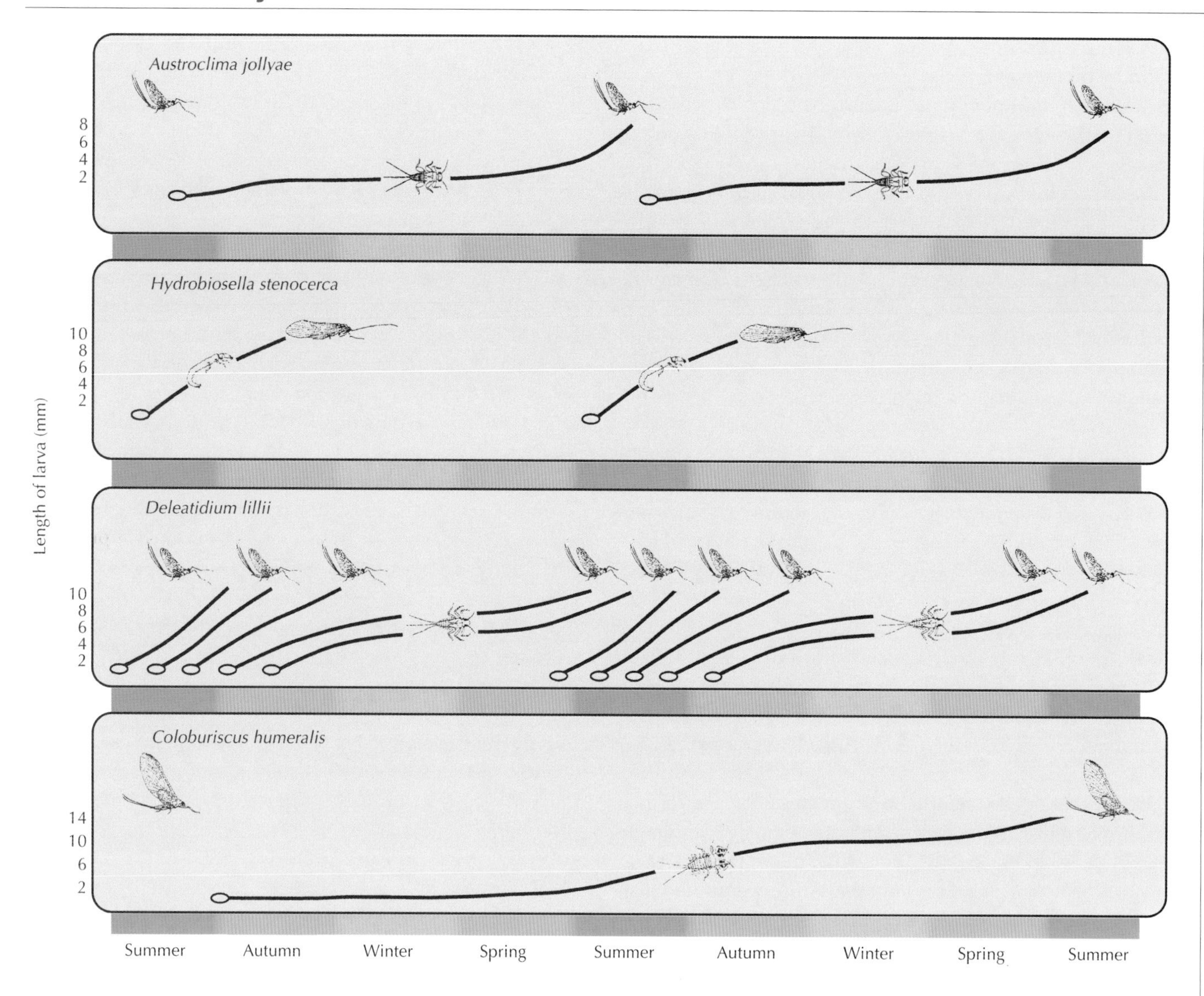

Fig. 11. Stylised diagram showing common types of life cycles for stream insects in southern New Zealand. Presence of eggs is indicated by small ovals. Lines extending from the egg indicate changes in larval length, from hatching to cessation of growth. The slope of the line indicates rates of growth – steep slopes during the summer, for example, indicate rapid growth; shallow slopes during winter indicate slow growth. Lines end when adults emerge from the stream. *Figures of invertebrates redrawn from various sources, primarily Winterbourn, Gregson and Dolphin 2000.*

eggs to complete the nest or 'redd'. The redd protects eggs from predators and most floods until hatching, a period of five weeks to four months, depending on temperature. After hatching, the young fish remain in the redd for a short period, before seeking suitable habitats above the stream bed. Non-diadromous trout and brook char may remain near the same location for their entire life cycle, or they may move considerable distances within the stream system during spawning migrations. Perch deposit sticky strands of eggs on underwater vegetation; the eggs hatch in about one week. Larval perch are planktonic for a short period before becoming free swimming. In comparison to exotic species, relatively little is known about the nesting behaviour of non-migratory galaxiids and bullies. Recent studies, however, indicate a richness of spawning behaviours among species (Box 5).

The second group includes four of the whitebait species (giant, banded and shortjawed kokopu, koaro), four bullies (giant, common, redfinned and bluegilled) and the torrentfish. These fish usually require access to both fresh and salt water to complete their life cycles. Spawning occurs in freshwater. Shortly after hatching, the larval fish (3–8 mm long) are swept to sea, where they feed for several months before running up streams and rivers as whitebait. After returning to fresh water, whitebait develop into adults and may spawn each year for as many as five years for bullies, and possibly as long as ten years for kokopu and koaro. Some populations of kokopu and bullies are land-locked and complete their entire life cycle in fresh water. Sea-run brown trout have a similar life cycle. Spawning takes place in fresh water, as described for non-diadromous trout. After one or two years, however, the young trout migrate to sea, where they feed and grow. After maturing, the adult trout run up coastal streams and rivers to spawn. After spawning, the adult trout may remain in fresh water or return to sea.

The third group includes inanga, lamprey and diadromous salmon. These species spawn in fresh water or estuaries, but complete growth and development in the sea. Unlike the whitebait species and bullies, however, mature fish usually die soon after spawning. Inanga spawn in estuaries during autumn. Upon hatching, larvae migrate to sea, where they feed for about six months and then return to coastal streams and rivers as whitebait. Inanga comprise the major proportion of the whitebait catch. Once in freshwater, the surviving fish continue to feed for another six months before returning to the estuaries to spawn. Lamprey spend much of their adult life feeding in the sea and return to coastal streams and rivers to spawn. After hatching, lamprey larvae feed in the nursery stream before transforming to adults and migrating to sea (see 'Detritus and detritivores' above). During autumn, diadromous salmon run up coastal streams and rivers to spawn, as described previously for non-diadromous trout. Migration back to the sea by young chinook salmon may occur immediately after hatching.

The fourth group includes the longfinned and shortfinned eels. Among all diadromous fishes in New Zealand, the eels alone spawn in the sea. Adult eels migrate from coastal streams and rivers into the Pacific Ocean where they rendezvous to spawn and then die in deep water somewhere between Tonga and Fiji. After hatching, larval eels (leptocephali) drift in ocean currents until they reach coastal waters, a journey of probably a year or more. Once in coastal waters, leptocephali transform into 'glass eels', small transparent eels about 10 cm long, and begin to migrate into rivers and streams, where they become pigmented 'elvers'. The elvers feed and grow in coastal streams and rivers before undertaking extensive migrations upstream. Migrating longfinned eels penetrate to the headwaters of many rivers throughout southern New Zealand; shortfinned eels tend to remain in wetland swamps and lakes. Following upstream migration, eels may feed and grow for as long as 100 years before returning to the Pacific.

Physical factors controlling stream communities

Habitat

A thoughtful survey of the fauna of any stream will quickly show that their distribution is not uniform, being dependent on 'habitat'. Most streams are composed of two familiar habitats – pools and riffles. Pools and riffles can then be subdivided into minor habitats, such as the tops and bottoms of stones, or tangles of aquatic mosses. Other important habitats, probably less familiar, are seepages, waterfalls and the sediments beneath the stream bed.

Pools and riffles

Streams and rivers are often a series of deep and tranquil pools separated by riffles, or shallow regions of rapid and turbulent flow. This type of channel results from movements of sediment during floods, as it becomes arranged in an alternating pattern of scoured regions and filled regions. After flood waters subside, the scoured regions form pools that continuously spill over the filled regions, to form riffles (Fig. 7). During normal flow conditions, pools are continuously filled with fine sediments, while riffles are eroded. This type of channel form has important consequences for the types of invertebrates that live in the different habitats. Deposit feeders and browsers are most abundant in pools because of the abundant supply of fine organic particles deposited on the bottom (Fig. 7). In some streams, backswimmers and water boatmen may live in the water column of pools. Filter-feeders, because they are dependent on organic particles and small invertebrates suspended by rapid and turbulent flow, occur only in riffles. Shredders are also often found in riffles, but beneath stones or undercut banks where tussock grass or tree leaves become lodged. Predators tend to be evenly distributed among riffles and pools, although the species vary (Fig. 7). Finally, there are numerous 'habitat generalists', usually deposit-feeders, browsers or predators, that are equally abundant in either habitat (Fig. 7). The larger stones of riffles are often covered by aquatic mosses, an important habitat for stream invertebrates as well as an indicator of a stable stream bed. The stonefly *Zelandoperla* (Fig. 7) is especially common in mosses, for example. Most invertebrates do not feed directly on mosses, but consume algae that grows as epiphytes on the plant's surface, or upon fine organic particles trapped by the dense thicket of 'stems' and 'leaves'.

Like the invertebrates, different fish species also tend to be habitat specialists. Giant, banded and shortjawed kokopu feed in pools and backwaters. The non-migratory galaxiids (Box 5) feed in pools as well as riffles, but may feed only in riffles in the presence of larger fish species such as trout. Common, upland and giant bullies forage for invertebrates on the bottom of pools; bluegilled and redfinned bullies forage in riffles. Eels are bottom feeders. Longfinned eels tend to be general in habitat preference, whereas shortfinned eels commonly feed in pools.

Altitude

Although aquatic insects abound in streams and rivers throughout southern New Zealand, it is the smaller tributaries that generally contain most species that are special to the region. While some species inhabit streams throughout a broad range of altitudes (e.g. the stonefly *Zelandobius foxi*, Box 3), many are limited to specific altitudes. For example, the day-active adults of the yellow caddisfly *Pycnocentria patricki* are found only in the alpine zone of Central Otago and Stewart Island, whereas the golden stonefly *Zelandobius auratus* is known only from low altitude wetlands near Alexandra in Central Otago. High-altitude seepages that form the headwaters of mountain streams are an especially important habitat. The larvae of the case-bearing caddisflies *Oeconesus similis* and a new species of *Philorheithrus* live in seepages below the summits of mountains of eastern Central Otago. At the highest altitudes, a habitat composed of rock slabs and snow-melt and devoid of vegetation harbours some of the largest stonefly species in New Zealand (Box 2).

Waterfalls

At altitudes of 500–1000 m in Western Otago, waterfalls provide habitats for a yellow and brown jumping stonefly *Halticoperla tara*, a rare dark-brown caddisfly *Pseudoeconesus haasti* and the day-active adults of the filter-feeding caddisfly *Aoteapsyche philpotti*. Elsewhere in southern New Zealand, adults of the free-living predacious caddisfly *Tiphobiosis cataractae* are conspicuous around waterfalls. Predatory larvae of the rare aquatic lacewing *Kempynus incisus* inhabit spray zones near waterfalls and stream margins where they capture fly larvae.

Larvae of the net-winged midge *Neocurupira* can often be found in the torrential flows associated with waterfalls. They maintain their position in the current by a series of true suction cups on their undersides (Fig. 8).

Hyporheic zone

The hyporheic zone, literally meaning 'zone beneath the stream', refers to areas below the stream bed that are permeated by water downwelling from the stream channel. If stream water downwells, it also must eventually upwell. Therefore, stream beds not only have water flowing over them, but in and out as well. All stream beds that are composed of deep layers of cobbles and gravels with sufficient space between particles for unrestricted flow of oxygenated surface water should harbour a hyporheic fauna.

Measurements of the hyporheic zone in high country streams of Eastern Otago indicate depths less than half a metre. However, in braided shingle-streams, the Oreti River for example (Fig. 3), the hyporheic zone is expected to extend to several meters. In the high country, the invertebrate fauna of the hyporheic zone is a subset of that commonly found on the stream bed. The most frequently encountered species are the cased-caddisfly *Olinga*, the snail *Potamopyrgus*, oligochaete worms, chironomid larvae and the predacious mayfly *Amelotopsis* and dobsonfly *Archichauliodes* (Fig. 7). In other areas of New Zealand, larvae of the mayfly *Deleatidium* have been collected from interstitial water from beneath fields 100 metres from the nearest stream channel, which indicates extensive lateral penetration of oxygenated surface water. The importance of the hyporheic fauna to the dynamics of invertebrate communities near the stream bed is unclear at present, but it is thought that the hyporheos may provide a refuge against the effects of bed disturbance during storms, and against predation by fishes. Hyporheic invertebrates probably feed on fine detritus that sifts down from the stream bed, and on biofilms based on dissolved organic matter from groundwater that enters the hyporheic zone from below.

Groundwater

Confined to layers of permeable gravel or rock (aquifers), groundwater enters stream channels through the hyporheic zone. If a stream receives most of its discharge as groundwater, it is called a spring. However, most streams receive the bulk of their flow from water flowing through shallow soils rather than directly as groundwater. The unique invertebrate fauna of New Zealand groundwaters – flatworms, annelid worms, crustaceans, blind beetles, and snails – was discovered over a century ago and has been well documented in Canterbury and Nelson.

Recent efforts to sample hyporheic fauna in Eastern Otago have revealed a widespread groundwater fauna. Blind amphipods, *Paraleptamphopus subterraneus* and *Phreatogammarus fragilis* (Fig. 7), minute syncarid shrimps, mites and the groundwater snail, *Saganoa* (Fig. 7) have all been recovered from sediments about half a metre below stream beds of the Taieri River drainage. *P. subterraneus* and the syncarid shrimps are widely distributed in southern New Zealand. *Saganoa* has previously been recorded only from Nelson. *P. fragilis* also occurs in Canterbury and Nelson. At present, other than a rather spotty knowledge of their distribution, little is known about the ecology of the groundwater fauna. Apparently they feed on biofilms that are dependent on dissolved organic material from the groundwater. The ground-water fauna commonly enters the lower hyporheic zone and may be found in surface waters in regions of strong upwelling. The distribution of the groundwater fauna is of great biogeographical interest, since it apparently originated in the aquifers of Gondwanaland.

Disturbance, food webs and productivity

Because most stream fauna in southern New Zealand are rather general in food preference, their feeding relationships are relatively simple. Food webs in grassland streams are used as an example because they have received the most attention in the region (Fig. 12). The productivity of the primary consumers (e.g. browsers, deposit-feeders, filter-feeders, Fig. 12.1) is dependent on biofilms (Fig. 12.2). Predacious invertebrates (Fig. 12.3) consume other invertebrates. Fishes, as top predators, feed on stream invertebrates (Fig. 12.4), and invertebrates that fall into the stream from surrounding vegetation (Fig. 12.5), and other fishes. Consequently, the overall productivity of grassland streams is controlled by the biofilm – although terrestrial sources may be important in some cases. The biofilm, as the basal 'trophic' level, supports two other trophic levels: primary consumers and predators.

Productivity of biofilms in southern New Zealand is controlled by two important categories of factors: 'top-down' and 'bottom-up'. Top-down factors operate within the food web, and in the case of grassland streams, refer to the removal of biofilm by browsers. Bottom-up factors are external to the food web and include limiting concentrations of nutrients, or light (e.g. shading by valley walls and water turbidity, Fig. 12/6), and floods that are sufficient to disturb the stream bed and remove biofilm mass.

Top-down control and non-migratory *Galaxias*

Compared to brown trout, non-migratory *Galaxias* are relatively inefficient predators and grow slowly. Therefore populations of these fishes have relatively low energy demands and, consequently, are a minor source of mortality for populations of invertebrate prey. Predacious invertebrates similarly seem to provide only a minor source of mortality for their prey. Experiments by Alex Flecker, Colin Townsend and Angus McIntosh in the Shag River in Eastern Otago, showed that, in the presence of *Galaxias*, populations of primary consumers quickly reach abundances sufficient to deplete biofilms to levels that reduce overall stream productivity (Fig. 12). Therefore, production of biofilms in *Galaxias* streams may be controlled by top-down processes, and little algal biomass should be apparent on the stream bottom.

Bottom-up control and the brown trout

Compared to the non-migratory *Galaxias*, brown trout are exceedingly efficient predators, with rapid growth rates and high energy demands. Consequently, they are a major source of mortality for stream invertebrate populations. Thus, brown trout

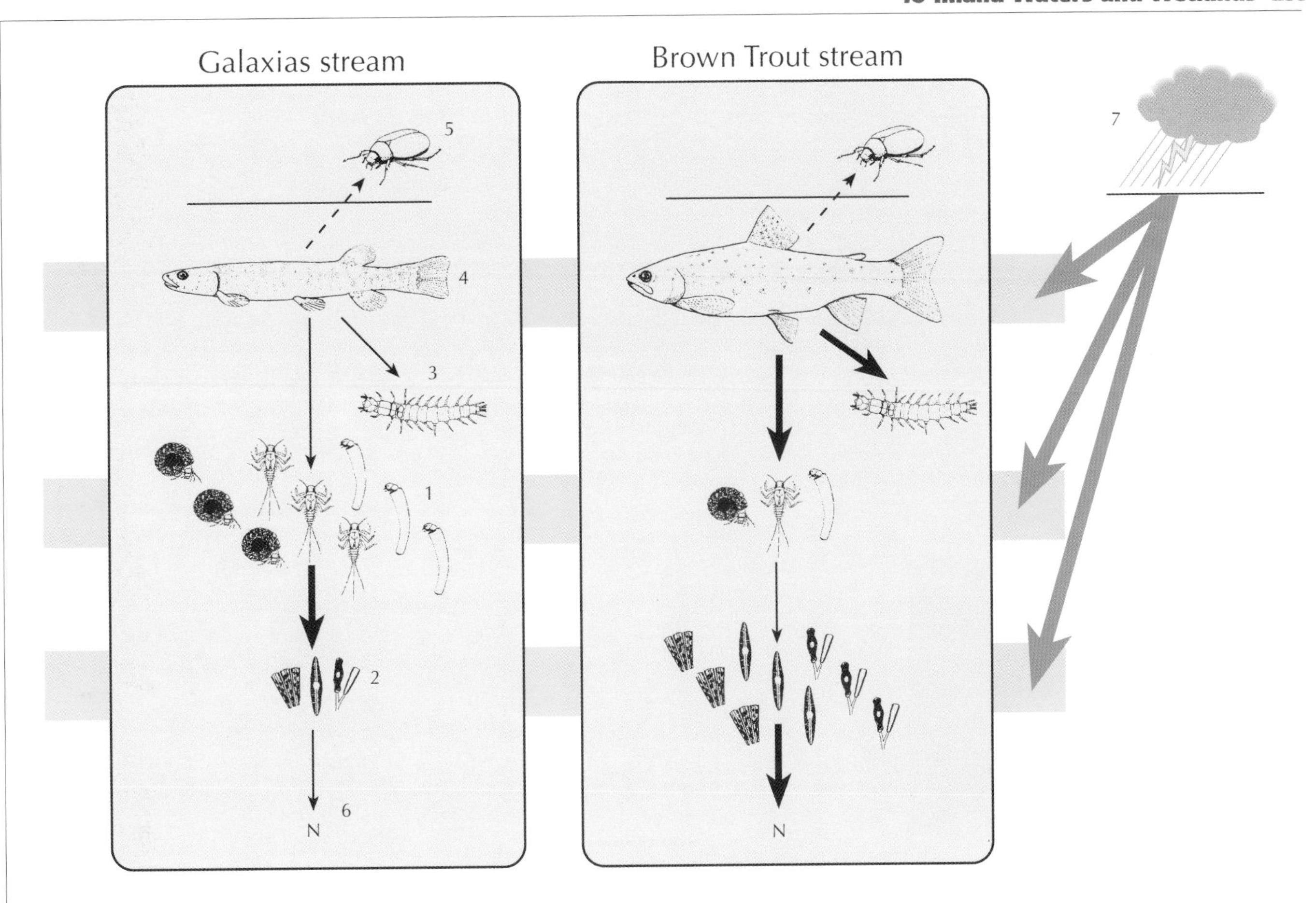

Fig. 12. Conceptual model showing how the introduction of brown trout to southern New Zealand may have affected the balance of major components of food webs in streams. In grassland streams, the productivity of browsing and filter-feeding invertebrates (1) depends on biofilms (2). Predacious invertebrates (3) consume other invertebrates. Fishes (4), as top predators, feed on stream invertebrates, invertebrates that fall into the stream from surrounding vegetation (5), and other fishes.
The numbers of individuals pictured at each trophic level indicate the relative productivity expected for that level in the different streams. Trophic levels are indicated by the gray horizontal bars. The width of the arrow connecting trophic levels indicates the strength of the effect consumers at a given trophic level exert on adjacent levels. Arrows point from consumer to food source, or nutrient source (6) in the case of biofilm organisms. In *Galaxias* streams, for example, production of browsing invertebrates is only weakly controlled by predation from fish. Browsing invertebrates, however, greatly reduce biofilm organisms which, as a consequence, have a minor effect on nutrients.
The opposite situation is seen in brown trout streams. In the absence of heavy browsing, or 'top-down control', productivity of biofilms will be controlled by 'bottom-up' factors, such as availability of nutrients (6). Floods (7) of an intensity sufficient to move a high proportion of stones composing stream beds will cause unpredictable losses of biofilm and fauna at all levels, as indicated by open arrows extending from the storm icon to all trophic levels. *Figures of invertebrates redrawn from various sources, primarily Winterbourn and Gregson (1991). Figures of fish redrawn from McDowall (1990).*

streams may have populations of invertebrates that are reduced to low levels because of trout predation (Fig. 12). In addition to their direct effect on the abundance of their invertebrate prey, trout may also have important effects on their behaviour. In the presence of non-migratory *Galaxias*, invertebrates browse continuously, spending little time in concealment. In the presence of trout, however, common invertebrates such as *Deleatidium* and *Nesamaletus* remain concealed beneath rocks during daylight, venturing out to browse or disperse only at night when trout are less effective predators. The reduction of both the abundance of invertebrates, due to predation, and the time they devote to browsing may result in the release of biofilm production from top-down control in brown trout streams. In such a case, abundant algal biomass may become apparent on the stream bottom and its production will be controlled by 'bottom-up' factors external to the food web, such as nutrient concentrations, light levels, or disturbance.

The overriding effect of disturbance

It should be clear that the different effects of trout and non-migratory *Galaxias* on stream productivity will require sufficient time for biological processes to stabilise. Since the beds of many streams throughout the South Island are often disturbed by unpredictable floods, intervals sufficient for stabilisation may occur only rarely. Floods that are of an intensity sufficient to move a high proportion of the stones composing the stream bed will cause unpredictable losses of biofilm and fauna at all trophic levels (Fig. 12.7). Consequently, variation in the frequency and intensity of flooding is probably the primary factor responsible for differences in community structure, food webs and productivity among streams throughout southern New Zealand streams.

7 MICROBIAL FOOD WEBS IN LAKE MAHINERANGI AND LAKE WAKATIPU

Some of the most important steps in production, transfer and breakdown of organic matter in the open waters of lakes involve organisms too small to be seen without the aid of powerful microscopes. They include bacteria and minute algae, 0.2–2 µm in diameter – or, 'picoalgae' (below) – and various, somewhat larger, flagellated, ciliated and amoeboid protozoa that are smaller than 200 µm (bottom). Together, these microorganisms make up the microbial food webs of lakes (below right). They are so abundant that their total biomass generally exceeds that of the larger crustacean zooplankton and rotifers.

The dissolved organic matter and nutrients in the open waters are incorporated into bacteria which are consumed by protozoa which, in turn, release nutrients and organic matter and are eaten by zooplankton (below). These zooplankton may then fall prey to fish such as larval perch, trout, koaro and bullies. The photosynthetic picoalgae are particularly abundant in the deep, clear lakes of southern New Zealand (Wakatipu, Manapouri, Te Anau); in Lake Manapouri, for example, one teaspoonful of water may contain more than 1 million picoalgae. These algae are too small to be eaten directly by copepods and so, like bacteria, they are generally consumed first by protozoa. Studies of the microbial food webs in Lakes Wakatipu and Mahinerangi show differences between the lakes in the relative abundances of components of their microbial food webs; picoalgae and flagellated protozoa are very important in the surface waters of Lake Wakatipu in summer, whereas bacteria and ciliated protozoa are more abundant in Lake Mahinerangi.

Carolyn Burns

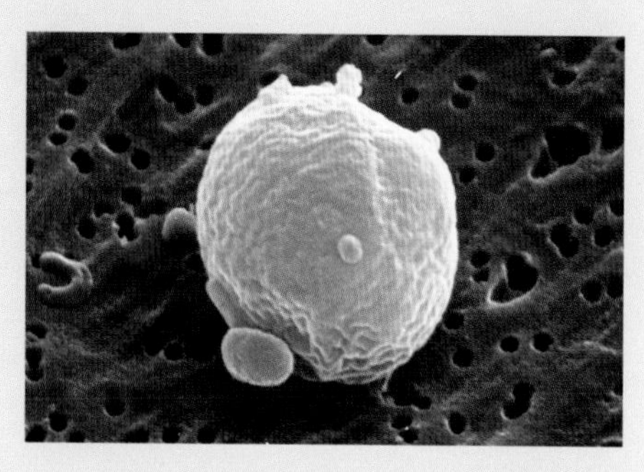

A typical picoalgal cell and smaller bacteria from the surface waters of Lake Te Anau.
Scanning electron micrograph by Carolyn Burns

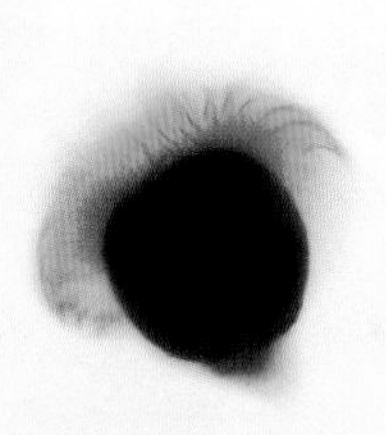

A typical ciliate (oligotrich type with anterior 'crown' of cilia) stained with an iodine stain.
Carolyn Burns

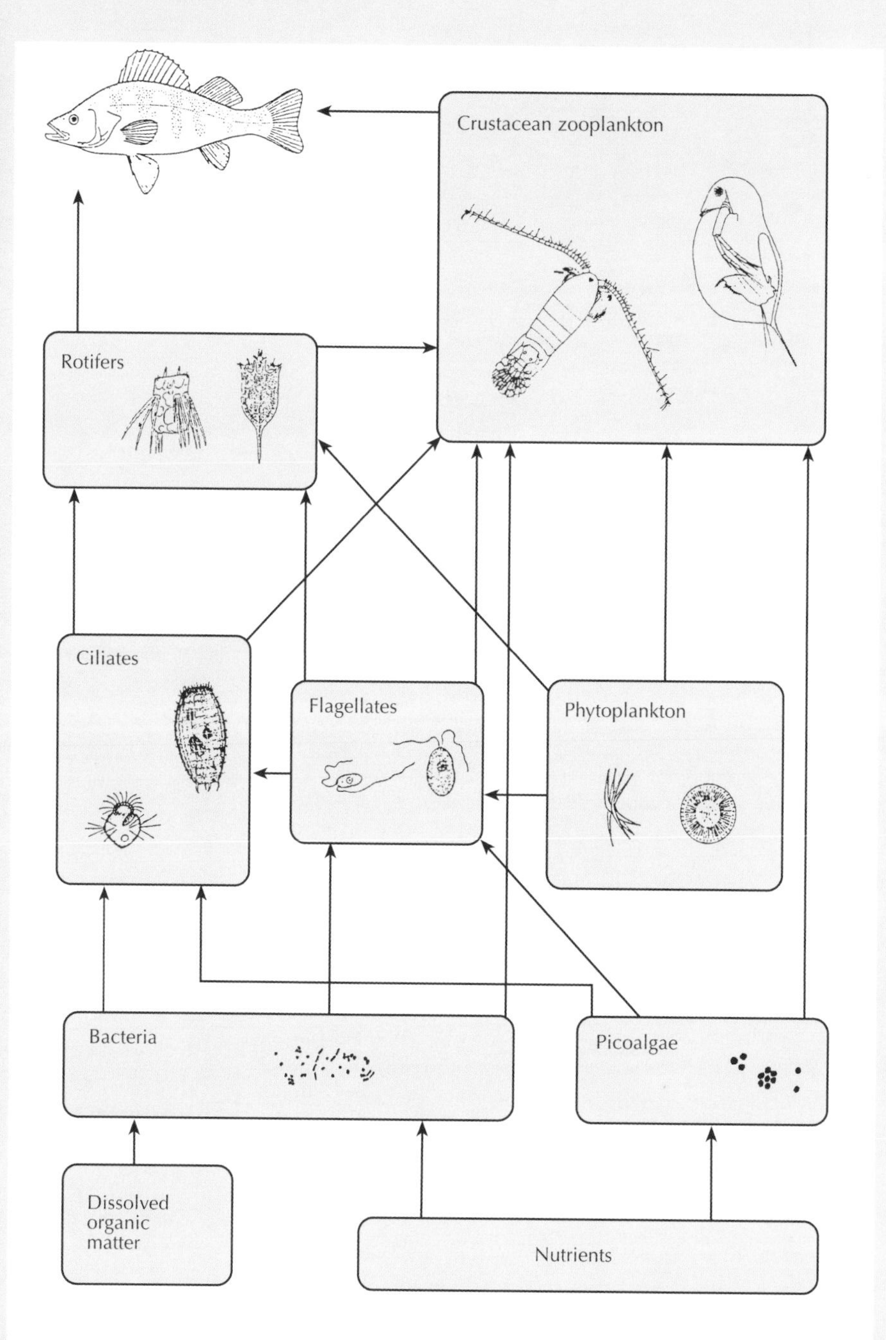

Right: The microbial food web of a lake.

LAKES AND PONDS

Biodiversity

The wide range of types of wetlands in southern New Zealand offers a rich diversity of habitats for animals and plants (Fig. 2). In the open waters of lakes and ponds are planktonic algae (phytoplankton), animals (zooplankton) and other microorganisms that make up microbial food webs (Box 7). These planktonic organisms, many of them microscopic, serve important functions in producing organic matter and recycling organic material and nutrients. Clinging to the surfaces of rocks, the submerged stems and leaves of water plants and other objects, are communities of attached algae, fungi, protozoans and invertebrates. A similar range of organisms inhabits the shingle, sand, detritus and mud on the bottom of the water body.

Algae

The species of phytoplankton in lakes in Otago and Southland are typically found elsewhere in New Zealand, and many are worldwide. New Zealand is notable, however, for a high abundance of green algae in the group called desmids, particularly species of *Staurastrum* with its several arms, and needle-like *Closterium* which dominates the plankton of Lake Hayes for much of the year, imparting a sheen to the water. Diatoms, for example species of *Cyclotella*, are also common in southern lakes such as Lakes Hayes, Wakatipu and Mahinerangi, and the filamentous cyanobacterium *Anabaena* (formerly called a 'bluegreen alga') is present in small amounts in many lakes and ponds. In Lakes Hayes, Johnson, upper Tomahawk Lagoon and Butcher's Dam, the amount of *Anabaena* often increases in summer to such an extent that unpleasant 'blooms' and scums are formed in calm weather, when millions of filaments rise to the surface. The attached algae in lakes and ponds are largely diatoms and filamentous algae, which occur also in streams.

Invertebrates

Rotifers are microscopic animals that derive their name from a wheel-like structure at the front end that rotates as they move about and feed (Box 7). The diversity of New Zealand's rotifer fauna has yet to be documented fully, but more than 95 per cent of a current total of 428 taxa are found throughout the world. More than fifty species of rotifers have been identified from lakes and ponds in and around Dunedin, including acidic bogs on Swampy Hill and the Maungatua Range, but only one species in Otago, *Notholca pacifica*, recorded from a pond on Allans Beach (Otago Peninsula), is endemic.

More is known about the diversity and ecology of freshwater crustaceans. The diversity of crustaceans in the open waters (zooplankton) appears to be low compared to that in lakes and ponds in Australia and the Northern Hemisphere, with most of the species distributed widely throughout New Zealand.

Geographic distribution of *Boeckella*

There have been few systematic studies of invertebrate distribution in lakes, ponds and wetlands of southern New Zealand, and so information on geographic distribution is fragmentary. At this stage, there are no obvious patterns in geographic distribution of planktonic freshwater organisms except for calanoid copepods. These small crustaceans are abundant in the open water of many lakes and ponds. Three of the nine species in New Zealand occur in the southern region (Fig. 13). The endemic species, *Boeckella dilatata*, is present in lakes and ponds of the glaciated regions of the southern and central South Island and in associated drainage systems, such as

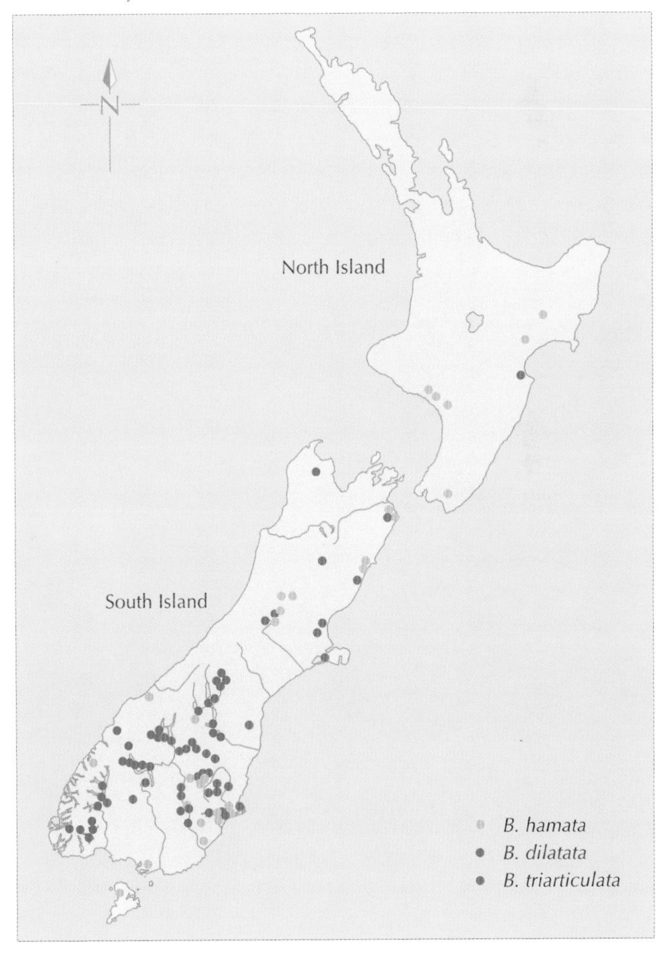

Fig. 13. The calanoid copepod, *Boeckella*. Three species of this Southern Hemisphere genus occur in lakes and ponds in southern New Zealand, *B. triarticulata* (left), *B. dilatata* (middle) and *B. hamata* (right). These females are carrying eggs; the length of the largest animal is approximately 1.8 mm and that of the smallest 1.2 mm. Their distributions (below) appear to reflect patterns of glaciation and the ability of the species to colonise and reproduce in new habitats. *Carolyn Burns*

the Waitaki lakes. In deep lakes at low elevations, such as Hayes, Wakatipu, Johnson and the Waitaki lakes, this species is virtually colourless (Fig. 14). On the other hand, populations at higher altitudes, such as those in Gem Lake on the Umbrella Range (1400 m) and Lake Alta in the Remarkable Mountains, are bright red due to the presence of carotenoid pigments that provide protection against the damaging effects of intense sunlight.

Coastal lakes and reservoirs, such as Tomahawk Lagoon, Lake Mahinerangi and Ross Creek Reservoir (Fig. 2), are inhabited by another endemic species, *Boeckella hamata*. This species is widespread throughout the South Island in comparable lowland habitats. Small ponds, particularly temporary ponds and farm dams such as those on the slopes of Saddle Hill (Fig. 2) and among the schist tors near Middlemarch, are likely to contain a large-bodied species, *Boeckella triarticulata*. Populations of this species may vary from pale pink through scarlet to dark purple over the course of a year, possibly in response to diet and turbidity of the water, which affects their exposure to sunlight. By producing resting eggs, this species can survive periods when ponds dry out; in fact, *B. triarticulata* was first described from specimens hatched and cultured from samples of dried mud. This species, which occurs also in Australia and Mongolia, is thought to have evolved in Australia and made its way to New Zealand, possibly as resting eggs caught on the feet or feathers of migrating water birds. With one exception so far (a pond in Napier), *B. triarticulata* is confined to the South Island. The co-occurrence of more than one species of *Boeckella* in a lake or pond is rare in the country and has not been recorded in southern New Zealand.

Fig. 14 (above). *Boeckella dilatata*. This endemic freshwater copepod occurs in glaciated lakes and ponds in southern New Zealand. Generally, copepods lack colour or are bluish. Individuals at high altitudes, such as the adult male from a bog pond in the Umbrella Range (at 1400 m) are vivid red and often larger (1.4 mm) than those from populations at lower altitudes such as the adult male from Lake Hayes (329 m). *Carolyn Burns*

Fig. 15 (right). A 'crested' female *Daphnia carinata* from a pond on the Taieri Plain. The crest is a transparent leaf-like expansion of the head that develops in response to substances released into the water by the backswimmer, *Anisops wakefieldi*, although other factors may also be involved in stimulating the response. Backswimmers prey on *Daphnia* and the presence of the crest and elongated tail spine may make it more difficult for them to do so. Length, excluding spine, approximately 4 mm. *Carolyn Burns*

Invertebrate assemblages

Coastal habitats

Coastal ponds and lagoons vary in salinity from fresh to brackish, depending on rainfall, the influence of salt spray, and occasional influxes of seawater during storms. The organisms that live in them comprise an interesting mix of freshwater species that can tolerate some salinity and estuarine species that can tolerate fresh water. For example, the tiny ostracods ('seed shrimps') *Cypridopsis jolleae* and *Diacypris thomsoni* can tolerate some salinity and live in slightly brackish ponds associated with saltmarshes at Allans Beach, Otago Peninsula. In upper Tomahawk Lagoon the choride levels vary twenty-fold , reflecting its closeness to the sea. The open waters of the lagoon contain rotifers (*Brachionus, Asplanchna, Keratella* and *Filinia*) and crustaceans (*Daphnia, Bosmina, Boeckella*) that are common also in inland lakes, while tiny shrimp-like amphipods more typical of coastal habitats (*Paracalliope, Paracorophium*) live on the bottom of the lagoon, from which they make frequent sorties into the overlying water. A dark, slater-like endemic isopod, *Austridotea*, closely related to marine species, lives in crevices and under rocks around the shore of the lagoon, which is the habitat also for snails *Potamopyrgus* and, occasionally, small leeches. The soft, nutrient-rich sediments harbour red larvae of the common non-biting midge *Chironomus zealandicus*, oligochaete worms, snails, pea-mussels (Sphaeriidae), and cases of larval caddisflies (*Oecetis, Paroxyethira*) that have matured and departed. In the soft sediments of coastal lakes and rivers with access to the sea, such as Lakes Waipori and Waihola, polychaete worms *Scolecolepides freemani* may also occur.

Inland habitats

New Zealand's only inland saline waters occur in Otago. Sutton Salt Lake, near Middlemarch, occupies a shallow, bedrock depression in which salts have concentrated by evaporation. The depth and salinity of the lake vary enormously with rainfall, and it dries out completely at times. Few organisms tolerate the harsh

conditions, but crustaceans typical of coastal ponds (*Diacypris*, *Microcyclops*), and an oligochaete worm, have been found in it.

Open water assemblages

In freshwater lakes and ponds there are, typically, three to five species of planktonic crustaceans, including a species of *Boeckella*, and several species of rotifers.

A common inhabitant of the open waters of ponds is the native water flea, *Daphnia carinata* (Fig. 15), which occurs also in Australia and India. Individuals vary greatly in size, and may be more than 5 mm in length in ponds that lack fish, such as one on Saddle Hill (Fig. 2); these 'giants' are two or three times larger than those in lakes such as Lakes Hayes, Mahinerangi and Wakatipu. *Daphnia*, particularly specimens that are larger than 1 mm, are readily seen and caught by several species of fish. Another, smaller species of British origin, *Daphnia obtusa*, is confined to a few ponds near Dunedin, at Otakou and Whare Flat, where populations appear to have been established for more than forty years. Two smaller crustaceans, *Bosmina meridionalis* and *Ceriodaphnia dubia*, are widespread in lakes and ponds throughout Otago and Southland, including those in Fiordland. *Chydorus sphaericus* is common also in summer in the open waters of productive lakes and ponds.

Daphnia and *Ceriodaphnia* are normally almost transparent, but in oxygen-poor habitats, such as shallow ponds with a lot of decaying vegetation, they are often pink due to the presence of haemoglobin in their blood. Those that live in ponds at high altitudes (e.g. at 1400 m in the Umbrella Range and in the Kepler Mountains) are dark brown, or black, due to melanin deposits in their exoskeleton that protect them from the damaging effects of high radiation.

Less obvious in the open waters are cyclopoid copepods, possibly because they tend to live near the sediments and are often abundant only for periods of a few weeks. *Eucyclops serrulatus* and *Macrocyclops albidus* occur in Lakes Hayes and Johnson, and *Macrocyclops* and *Acanthocyclops serrulatus* have been found in Lake Mahinerangi. Their small size makes them difficult to see in a sample of lake water, but their characteristic swimming with rapid, jerky movements distinguishes them from the slow gliding movements of calanoid copepods.

In productive lakes and ponds, such as Lake Hayes and Ross Creek Reservoir, brownish-black water mites, *Piona exigua*, can be seen scurrying through the water in summer. Very occasionally, conditions in Central Otago are conducive to the mass development of medusae of a cosmopolitan species of freshwater jellyfish, *Craspedacusta sowerbyi*. When this happens, as in Lake Hayes in February 1989, the upper waters for several days are the scene of a mesmerising ballet of small, pulsating 'bells' as medusae rise and sink. The 1–2 cm diameter medusae are carnivorous, feeding largely on small cladocerans *Ceriodaphnia, Bosmina*.

Lake margins

The margins of lakes and ponds with well-developed beds of rooted and floating water plants provide a variety of habitats for aquatic organisms. There appear to be fewer species in New Zealand wetlands than in comparable habitats in the Northern Hemisphere, and the species are widespread, so that assemblages of invertebrates in lakes and ponds in the southern region are similar to those elsewhere in the country. Water beetles (*Rhantus*), backswimmers (*Anisops*) and water boatmen (*Sigara*) dive in and around the plants, while many small crustaceans can be found among the stems and leaves of submersed and emergent water plants, on and under rocks, and in the sediments. They include several species of tiny chydorid cladocerans, ostracods and cyclopoid copepods, and a large, *Daphnia*-like cladoceran, *Simocephalus*, which rests frequently against vegetation. Here also there are brightly coloured water mites, including, occasionally, the large red mite *Eylais*. Molluscs, worms and the larvae of midges, damselflies (*Austrolestes colensonis*, *Xanthocnemis*) and dragonflies (*Aeshna*, *Procordulia*) are common, and a small 'hairy' crustacean, enchantingly named *Ilyocryptus sordidus* (from Greek, *ilus* = slime, dirt; *kryptos* = hidden, and, Latin, *sordidus* = squalid), can sometimes be found covered in the sediments in which it grovels.

Lake bottoms

The native filter-feeding mussel *Hyridella menziesi* occurs in Lakes Waipori and Tuakitoto and in lakes throughout New Zealand. The mussels grow slowly and large specimens (longer than 10 cm) may be more than fifty years old. In shallow Lake Tuakitoto, mussels are so abundant that they effectively remove all of the algae from the overlying water and undoubtedly contribute to retaining the water clarity of this nutrient-rich lake.

Freshwater crayfish *Paranephrops zealandicus* may erode the soft earthen banks of dams and reservoirs (e.g. Lake Mahinerangi), but appear to be more common in streams; their distribution in lakes of southern New Zealand is not known.

At greater depths in lakes there is less vegetation, and light and oxygen may be limited (Fig. 16). Fewer organisms live in these conditions. In Lake Wanaka at 3–5 m, for example, the dominant animals are pea-mussels *Sphaerium novaezelandiae*, red midge larvae *Chironomus zealandicus* and oligochaete worms, with a few caddisfly larvae and some species of snails. Sponges as large as 30 cm in diameter have been reported at depths of more than 10 m in Lake Wakatipu. Several species of midge larvae occur in the bottom sediments of Lake Hayes, including *Chironomus zealandicus* which is abundant to a depth of 25 m, except in summer when it is restricted to depths of less than 16 m because the deeper water lacks oxygen (Fig. 16). In the shallow parts of the lake, *C. zealandicus* has two to three generations from late spring to autumn, followed by an over-wintering generation. Larval midges feed on organic matter that settles to the bottom of the lake and their pupae, which rise to the surface for the adult to emerge, are a favourite food of brown trout in the lake. The *Anabaena* blooms that often occur in the lake in summer increase the amount of organic matter available to sustain larval midges and may partly account for increased swarms of adult midges around the lake after the decay of algal blooms.

Life in the dark, cold, silent depths of New Zealand's deepest lakes – the profundal fauna – has not been studied, but is likely to be sparse and to consist largely of oligochaete worms, although molluscs and sponges may also be present.

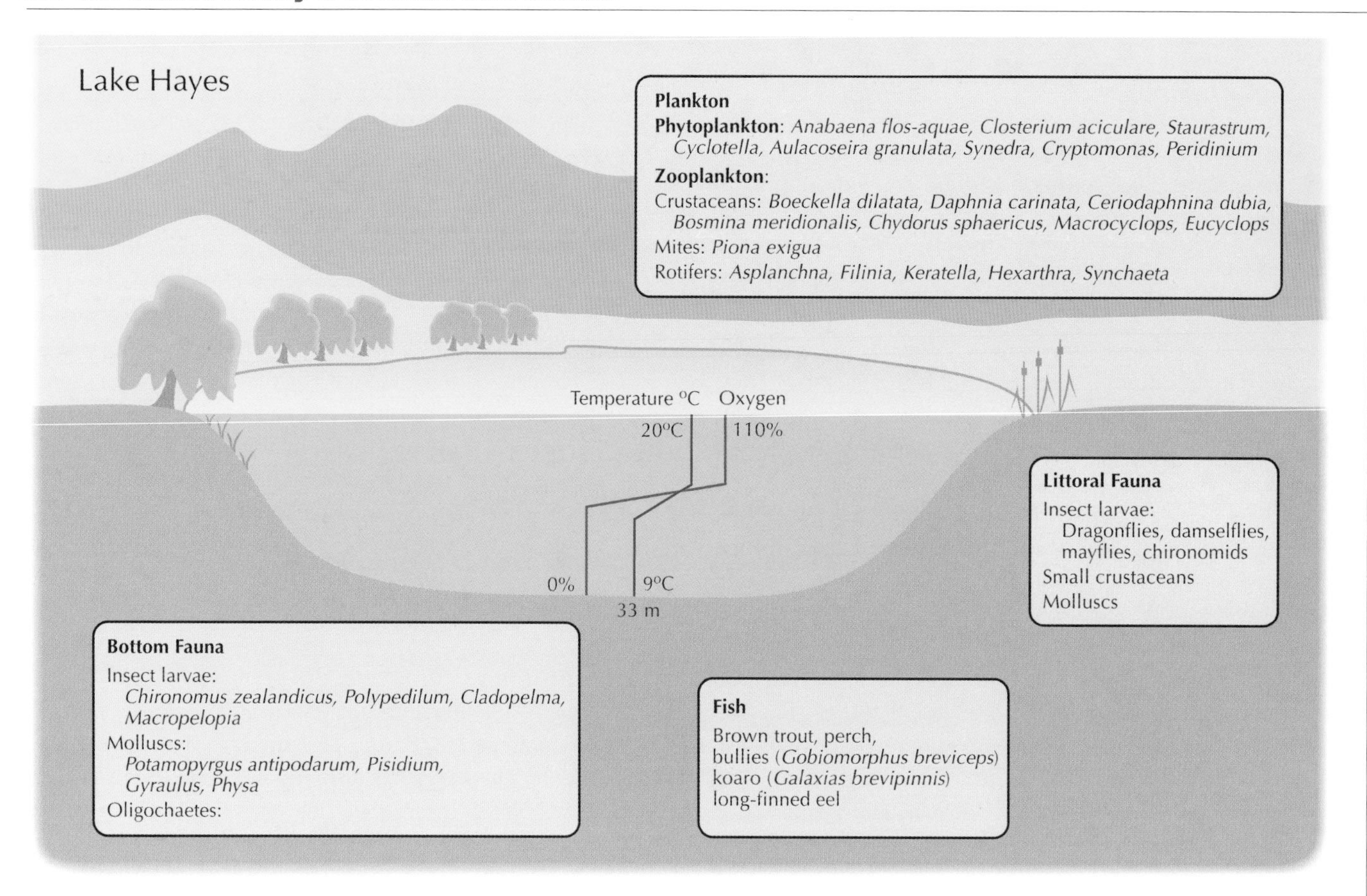

Fig. 16. Diagrammatic cross-section of Lake Hayes, showing a depth profile of water temperature and dissolved oxygen in summer, and the distribution of dominant fauna and flora in winter. In winter, animals occur throughout the lake to its maximum depth of 33 m. In summer and autumn (late January to end of May) animals are confined to the top 16 m of the lake owing to an absence of oxygen in the water below this depth.

Food and feeding

Suspension feeders

The phytoplankton, protozoa, bacteria and detritus suspended in the open waters of lakes and ponds are a major food resource for rotifers and planktonic crustaceans (Box 7). To collect these food particles, *Daphnia* and *Ceriodaphnia* generate currents towards their bodies by rapidly beating their paddle-shaped thoracic legs. Food is concentrated with the aid of sieve-like 'hairs' and bristles, pushed between stout mandibles and swallowed. So effective is this method of feeding that four large *Daphnia carinata* (4 mm) could remove all the algae from one litre of water in a day. For this reason, *Daphnia* are very effective 'clarifiers' of water in lakes and ponds. Juvenile *Daphnia* and smaller species of cladocerans are less effective, although when present at high densities they can also increase water clarity.

Boeckella are very selective feeders and obtain food in several ways. Fine particles suspended in the water are collected by the use of currents generated by their mouthparts, but larger algae and particles of detritus are seized and tasted before they are eaten, or rejected. Hungry *Boeckella* can even 'clip' filamentous cyanobacteria (*Anabaena, Nostoc*) into manageable portions, which they clear from the water at rates comparable to those of *Daphnia*, provided the filaments are not too dense. However, when *Anabaena* is so abundant that it forms blooms, *Boeckella* tend to switch to eating more nutritious algae. Protozoan prey, primarily ciliates which they capture very effectively, supplement their herbivorous diet (Box 7).

Predators

New Zealand lacks many of the predatory zooplankton that are present in lakes and ponds elsewhere. There are no predatory cladocerans (the group to which *Daphnia* belongs), few predatory copepods and no phantom midges (Chaoboridae), the larvae of which are voracious predators of zooplankton in lakes throughout the world, including Australia. The main invertebrate predators in the open waters of lakes in southern New Zealand are a few rotifers (e.g. *Asplanchna, Synchaeta*) which consume other rotifers and small crustaceans, the water mite *Piona* and the cyclopoid copepods (*Macrocyclops, Mesocyclops*). Predatory rotifers are rarely abundant and are present for short periods in the plankton, so their impact on populations of small crustaceans is likely to be small. *Piona* consumes cladocerans, particularly juvenile *Daphnia, Ceriodaphnia* and *Chydorus*, but is rarely sufficiently abundant to decrease crustacean populations significantly. *Macrocyclops* tends to live at the bottom of lakes and is rarely encountered in the open waters. For this reason, it is unlikely to have much impact on zooplankton of the open waters, although this aspect has not been studied.

Fish are probably the major predators of zooplankton, but surprisingly little is known about their impact on plankton in South Island lakes. Some lakes in southern New Zealand are

8 SOCKEYE AND CHINOOK SALMON

A shipment of sockeye salmon *Oncorhynchus nerka* eggs was made from Canada in 1901. They came from migratory parents in Shushwap Lake, British Columbia, and were released in the Waitaki system with a large proportion going to Lake Ohau. Like the Atlantic salmon (Box 9), they developed as a lake-based population with the main centre being Lake Ohau. There was no evidence of a sea migration and this species also appears to be on the decline, probably as a consequence of hydro-electric developments. Sockeye make complex migrations in the North Pacific, possibly using gyres (rotating currents), and the explanation for the failure of the Atlantic salmon as a sea-going population (see Box 9) can be applied to this species also. In contrast, chinook or quinnat salmon *Oncorhynchus tshawytscha*, which came from California in 1901–1907, did not make long sea migrations in the Pacific, and do not appear to travel far from the coast in New Zealand. They are never far from the scent of their native rivers and are unlikely to get lost. What is distinctive about chinook in the Clutha system is that, as well as establishing sea-migrating populations, they have developed lake-limited populations in Wakatipu, Wanaka and Hawea. The young fish (smolts) descend from the tributaries, but behave as if the large lakes were the sea, some remaining until maturity. A further migration may also take place downstream out of the lakes when the fish are 30–40 cm long, as if the migration pattern was still not complete. See also Box 6.

Donald Scott

9 THE ATLANTIC SALMON (*SALMO SALAR*)

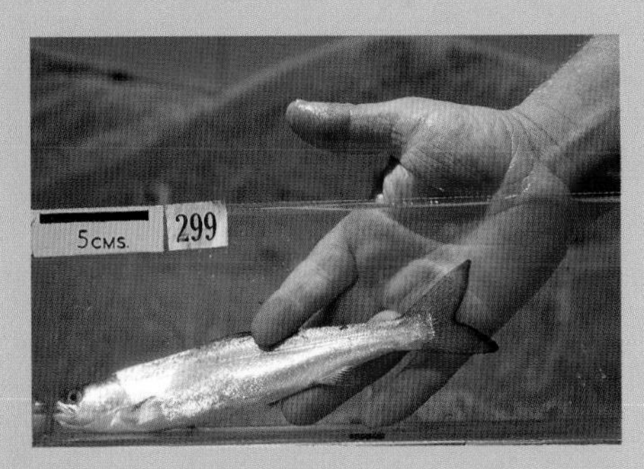

Smolt of the Atlantic Salmon. *Donald Scott*

This species has been the subject of more attempts at introduction than any other exotic fish, over 5 million eggs having been brought into Otago and Southland from 1868 to 1910. There was no evidence of success until a concerted effort was made by L.F. Ayson, who concentrated efforts on the Waiau system: 1.9 million eggs were brought from Canada, UK and Germany from 1908–1911 and distributed in the lakes and their tributaries. They appeared to flourish, but there was no evidence of a sea migration and their development as a lake-based population continued through the 1920s and 1930s. By the 1960s they had decreased significantly, possibly as a result of competition with rainbow trout. They are now on the verge of extinction, clinging to the periphery of the catchment. The freshwater environment was suitable for growth and development, and the clue to the failure appears to lie in the length and complexity of the sea migration. This migration of thousands of kilometres evolved in the Northern Hemisphere, and the navigation signals used by the young fish at sea (magnetic fields and suncompass) are reversed in the Southern Hemisphere. The young fish (smolts) reach the sea, but their inherited navigation signals are irrelevant, they cannot reach appropriate feeding areas and they cannot make a return migration.

Donald Scott

unusual in having populations of introduced salmonids that are not present elsewhere (Boxes 8 and 9). Among the introduced fishes, perch, rainbow trout and sockeye salmon are planktivorous (eat plankton) at some stages in their life cycles. Native fish that are known to eat zooplankton are smelt *Retropinna*, larval koaro *Galaxias brevipinnis* and larval bullies *Gobiomorphus* spp.

Life cycles and population dynamics

Rotifers

Rotifers reach maturity in just a few days and reproduce parthenogenetically (without males) so that high population densities can be attained within a few weeks. Declines in population density tend also to be rapid, so that there is a succession of different species throughout the year. In Lakes Hayes and Johnson, for example, the large, predatory rotifer *Asplanchna* is generally absent in winter and reappears in spring with pronounced short-lived peaks of abundance until late summer. In a pond on Saddle Hill, *Synchaeta* was abundant in spring and early summer, followed by *Conochilus* in mid and late summer, and both were absent in winter. In upper Tomahawk Lagoon, *Pompholyx campanulata* and two species of *Keratella* were present throughout a year and were also the most numerous rotifers, whereas *Asplanchna brightwelli*, *Filinia terminalis*, *Epiphanes*, three species of *Brachionus* and two species of *Synchaeta* were more seasonal in appearance. Seven other species appeared sporadically in small numbers. The results of these studies suggest that similar patterns of species diversity and succession are likely to occur in productive, non-acidic ponds and lakes elsewhere in the region.

Cladocera

In lakes and ponds that do not freeze in winter or dry up in summer, crustacean zooplankton breed throughout the year. Populations of the dominant cladocerans, *Daphnia*, *Ceriodaphnia* and *Bosmina*, contain only females which reproduce parthenogenetically for most of the year. When food is abundant and the water is warm (20°C) the *Daphnia* mature within a week of hatching and continue to reproduce every few days for several weeks; the process is slower in winter, when temperatures are cooler and their life span may last several months. Males appear occasionally in the population and there is a phase of sexual activity that culminates in the production of two resting eggs that are protected in a purse-like fold of thickened, pigmented exoskeleton. These eggs sink to the bottom of the lake or pond where they can survive for many years even if the habitat freezes or dries completely. When conditions are suitable they hatch into young parthenogenetic *Daphnia*. Potential stimuli for the production of resting eggs by Northern Hemisphere species of *Daphnia* include crowding, declining daylength in autumn, insufficient food and the presence of predatory fish, but the extent to which the same stimuli may apply to *D. carinata* populations in New Zealand awaits study. Studies of populations of *D. carinata* over more than one year in Lakes Johnson, Hayes, and Mahinerangi, Tomahawk Lagoon, and in three fishless ponds on the Taieri Plain, show that enormous variations in population size can occur from one month to the next, and between years, in response to shifts in the amount and quality of food available, and possibly also to the impacts of predators (fish, backswimmers). When well-fed, large *D. carinata* may carry more than 100 eggs in a clutch, but when food is sparse, they grow slowly and produce only one or two eggs per clutch. In Lakes Hayes and Mahinerangi, *Daphnia* have peaks of abundance in late spring and late autumn, but are scarce in summer and winter, possibly because they fall prey to plankton-eating fish at these times.

Copepods

Progress through six juvenile stages is followed by five 'adolescent' stages before copepods mature into adult males and females. More is known about the life cycles and population dynamics of calanoid copepods in Otago's fresh waters than about those of other zooplankton. In southern lakes that do not freeze in winter, *Boeckella* breed throughout the year and produce successive clutches of eggs over several weeks or months. When good quality food is abundant and the water is warm (about 16°C) they can reach maturity in just over three weeks; in winter when the water is cold (6°C) and phytoplankton is less abundant, they may take more than four months to do so. Consequently, the *B. hamata* population in Lake Mahinerangi and the *B. dilatata* population in Lake Hayes have just over three generations in a year. Relative to their copepod counterparts in Northern Hemisphere temperate lakes, these populations of *Boeckella* have low rates of biomass turnover and low productivity, possibly because, in New Zealand, they reproduce slowly all year round and may also suffer less predation from fish. The populations are controlled at times, however, by an unusual fungal parasite, *Aphanomyces ovidestruens*, that attaches to the abdomen of adult female *Boeckella*. Thread-like hyphae from the fungus penetrate and kill all the eggs of each successive clutch of an infected female, thereby decreasing the birth rate of the copepod population considerably.

Subalpine populations of *B. dilatata* in bog ponds on the Umbrella Range are very different from those in Lakes Hayes and Wakatipu. The copepods are large and bright red (Fig. 14); adult females may be twice the size of their transparent, lowland counterparts. Scarlet eggs are produced in a few, large clutches, some of which do not hatch immediately but are resting eggs that drop to the bottom of the pond, thereby allowing the populations to persist through the harsh winter. When the ponds thaw in spring, these eggs hatch into juveniles that develop into adults by early summer. Subsequently, one or two more generations are compressed into the remaining ice-free season. Despite their striking differences in appearance and reproductive behaviour, the populations of *B. dilatata* on the Umbrella Range are sufficiently similar genetically to those in lowland lakes to be considered the same species, but in the early stages of speciation.

Physical and chemical factors controlling lake processes

The nutrients that are available to sustain organisms in lakes and other wetlands determine the productivity of the ecosystem. Productivity is determined by the size and depth of the lake, climate, soil, the size of the drainage basin and the activities that occur there (Box 10). The temperature of the water affects the solubility of plant nutrients, the rates at which bacteria and fungi break down dead organic matter and the rates of metabolism of aquatic organisms; when the water is warm, these rates increase. There are two types of lakes in southern New Zealand: those in which the water mixes continually from the surface to the bottom, and those which stratify thermally. Shallow lakes, such as Tomahawk Lagoon and Lake Waipori, both of which are less than 1 m deep, are continually mixed, but so too are some deeper lakes that are very exposed to wind, like Mahinerangi (maximum depth 31.2 m). Phytoplankton, which cannot swim, rely on this mixing to keep them in the light near the surface, and to supply them with nutrient-rich water from deeper in the lake.

10 EUTROPHICATION OF OTAGO LAKES

The plant nutrients, phosphorus and nitrogen, enter lakes and other wetlands in streams and diffuse run-off from the drainage basin. Once in the lake, these nutrients stimulate and sustain the growth of phytoplankton, sometimes to the extent that water clarity diminishes and algal 'blooms' may occur. The use of land for agriculture increases the chance of nutrient enrichment (eutrophication) of wetlands in the drainage basin. Lake Mahinerangi, a 18.6 km² reservoir lake, was formed by the damming of the Waipori River in 1923 to provide electricity for Dunedin. Prior to 1964, more than half of the drainage basin was covered in tussock grassland, with pine plantations on the remaining areas; the levels of available nitrogen and phosphorus in the water were low and algal abundance and productivity were low also. In the mid 1970s, a large proportion of the non-forested catchment was progressively developed for agriculture, with the application of phosphate fertiliser, oversowing, stocking with cattle and sheep and planting of root crops for stock fodder. As these developments occurred, the levels of phosphorus in the lake water rose and the biomass and productivity of the phytoplankton increased; two potentially bloom-forming algae, *Anabaena flosaquae* and *Aulacoseira* (formerly *Melosira*) *granulata*, were recorded in the lake for the first time, and the levels of dissolved oxygen became more variable.

The drainage basins of Lake Hayes (Fig. 2) and Lake Johnson, Central Otago, are largely developed for agriculture. Nutrient-rich run-offs into these lakes have contributed to their eutrophication to the point where algal blooms have been occurring in summer for more than thirty years. Studies of these Otago lakes, and others elsewhere in New Zealand, have shown that lakes with more than 30 per cent of their drainage basins developed for agriculture are potentially sensitive to eutrophication from run-off.

Carolyn Burns

11 'SWITCHING' IN SHALLOW LAKES

As lakes become enriched with nutrients from human activities in their drainage basins the water becomes increasingly turbid owing to an increase in the growth of phytoplankton. One effect of this change is to decrease the light reaching the plants growing on the lake bottom so that, in shallow lakes, these plants may suddenly collapse. This, in turn, sets off a chain of reactions. Nutrients are released from the decaying plants, which further stimulate the growth of phytoplankton. The sediments, previously stabilised by the plants, become more susceptible to disturbance by waves, further reducing light penetration and providing yet more nutrients to the phytoplankton. Thus, the lake may become fixed in a new phytoplankton-dominated state.

While plant beds are present they suppress the growth of phytoplankton by a variety of mechanisms which are the reverse of those above, and by others that are only now becoming understood. For example, in many Northern Hemisphere temperate lakes the plants provide refuges for large zooplankton from predation by fish, and the zooplankton populations remain high enough to suppress the phytoplankton by grazing. Thus, there are two, alternative semi-stable states: a turbid, phytoplankton dominated state, and a clear-water state in which rooted plants dominate.

Otago provides two striking examples of lakes which alternate between these two states. In Tomahawk Lagoon No. 2, dense algal blooms in summer may lead to intense shading and collapse of the rooted plants, but if the bloom is less intense, the plants increase through the summer and persist with high biomasses for 1–5 years. During such periods phytoplankton are reduced to 5–10 per cent of the previous levels. In Hawksbury Lagoon, the changes are even more dramatic, with the biomass of water plants varying more than 100-fold from year to year, and with long periods in which insufficient light reaches the bottom to sustain them, although the lake is only 40 cm deep. These changes have major consequences for the populations of waterfowl that feed on large plants (Box 12). In both lakes, changes in depth of only a few centimetres may have major effects on sediment resuspension by waves; one key to effective scientific management of their water quality may lie in manipulating the lake levels.

Stuart Mitchell

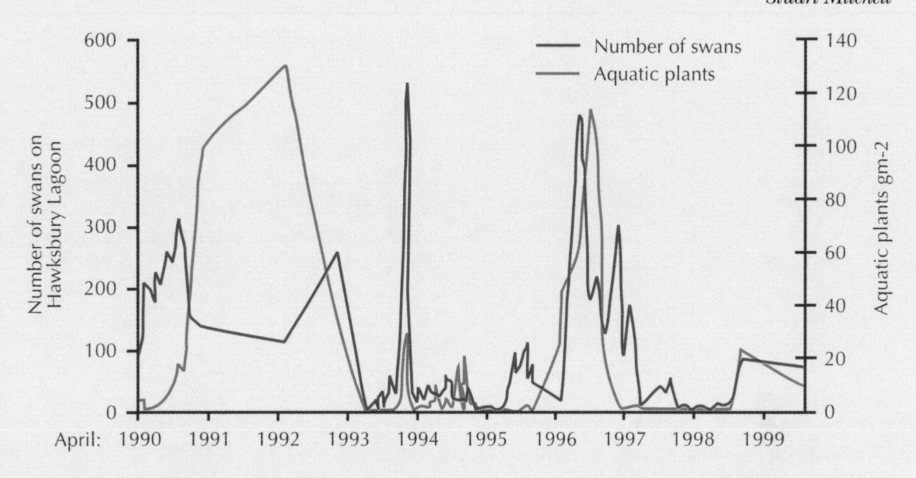

Changes in the swan population and aquatic plants at Hawksbury Lagoon, 1990–1999. When aquatic plants were abundant (1991–92 and 1996–97), the swans that feed on them were also abundant; when the lagoon was dominated by phytoplankton (1993–95 and 1998–99), swans were rare.

Shallow lakes

In shallow lakes, wind-induced mixing stirs sediments on the bottom into the overlying water, thereby releasing nutrients to stimulate the growth of algae and bacteria. At the same time, however, the increased turbidity of the water reduces the light available for plant growth. In some years, shallow lagoons such as Hawksbury and Tomahawk are choked with water weeds, and the overlying water is clear; in others, the water is like pea soup and water weeds are virtually absent. The reasons for these sudden switches reveal complex interactions that involve wind and black swans *Cygnus atratus* (Box 11). Other birds characteristic of freshwater habitats are described in Box 12.

Deep lakes

The deep, glaciated lakes in southern New Zealand do not freeze in winter and, in Lakes Johnson and Hayes, cold water mixes to the bottom of the lake at temperatures of 6–7°C. During spring and summer, the surface water warms up and a steep thermal gradient (thermocline) develops in the water column that effectively separates a layer of warm, surface water from cold water at the bottom. In midsummer, the thermocline is located at a depth of approximately 12 m in Lake Hayes (Fig. 16), at 6 m in nearby Lake Johnson, which is smaller and more sheltered, and at approximately 60 m in the middle arm of Lake Wakatipu, although it has been recorded at depths exceeding 100 m near each end of this lake. Thermoclines in the large, glaciated lakes of southern New Zealand are among the deepest in the world and are rivalled only by those at similar latitudes in southern South America.

Acknowledgements

We thank Mike Scarsbrook, Richard Montgomerie, Roger Young, Nathan Whitmore, and Eric Edwards for unpublished observations; Jon Harding for access to unpublished manuscripts detailing species richness of aquatic macroinvertebrates in New Zealand; and Sue Johnstone for SEMs of aquatic insect larvae.

12 BIRDS OF FRESHWATER HABITATS IN SOUTHERN NEW ZEALAND

Several bird species breed on the braided riverbeds of Otago and Southland. All three of New Zealand's gulls do so, usually in large colonies, established afresh each spring, although often in the same place for several years on end. The endemic black-billed gull *Larus bulleri* is the gull most often seen inland (below), especially on the large lakes, such as Te Anau, Wakatipu and Hawea, but also foraging on recently ploughed farmland. Like many riverbed breeders, this species moves to the coast in winter. The larger southern black-backed gull *L. dominicanus* is also found on and off the coast all year around, and has greatly increased in numbers over the past hundred years, presumably because of the increase in human rubbish, through which it scavenges. This species is what is known as a four-year gull: adult plumage is not attained until a bird is in its fourth winter. Southern black-backed gulls are also known to breed in small colonies on rock tors that form small islands in the upper Taieri River. The red-billed gull *L. novaehollandiae scopulinus*, superficially similar to the black-billed gull, and the white-fronted tern *Sterna striata* also breed here, but are more commonly found on the coast and at sea (see Chapter 11). The endemic black-fronted tern *Chlidonias albostriatus* is also a colony-breeder on riverbeds, sometimes in mixed colonies with the black-billed gull. The black-fronted tern has a bewildering variety of plumages: young birds are mottled, breeding adults have a jet black cap and an orange-red bill, whereas the winter adults lose the cap, retaining just ill-defined dark marks on the head.

Several waders also breed on these riverbeds, but in isolated pairs rather than in colonies. The commonest is the banded dotterel *Charadrius bicinctus*, which also nests on the tops of Otago ranges such as the Rock and Pillar. The wrybill *Anarhynchus frontalis* has its largest populations in Canterbury but also breeds on the upper Waitaki and Ahuriri Rivers and sporadically further south. Both species move to form large flocks on the coast in winter, some banded dotterel as far as south-eastern Australia and the wrybill mostly to the Firth of Thames and the Manukau Harbour in the North Island. The wrybill is unique among waders, in having the end of its bill curved rightwards, a shape that aids foraging for invertebrates under small stones. A recent arrival from Australia, the black-fronted dotterel *C. melanops* numbers about 100 in Otago and Southland. For a brightly marked bird it can blend in remarkably well with the shingle when frightened. The South Island pied oystercatcher *Haematopus ostralegus finschi* nests on braided riverbeds, but also on tussock and pastureland (see Chapter 11).

The blue duck (*Hymenolaimus malacorhynchos*) is unique in New Zealand in preferring fast-flowing streams. Now confined in our region to rivers and streams in large areas of unmodified bush, especially Fiordland, this species is strongly territorial, even driving away its own young at the end of the breeding season. It feeds on aquatic insects, using its bill with flexible lobes on each side to forage around rocks in rapids.

The introduced mallard *Anas platyrhynchos* can be found on almost any body of open water. They have almost completely replaced the native grey duck *A. superciliosa* in the region. Slightly smaller and with the distinctive eponymous bill, the New Zealand shoveler *A. rhynchotis variegata*, subspecifically distinct from the Australian form, is widespread. Unlike the mallard and grey duck, which eat a wide variety of aquatic plants and animals, the shoveler feeds almost exclusively on invertebrates. Both the endemic paradise shelduck *Tadorna variegata* and the grey teal *A. gracilis*, which is also found in Australia, have substantially increased their populations this century. In the summer, paradise shelduck moult in large flocks on open farmland. On deeper lakes, especially Wanaka, Wakatipu, Te Anau and Manapouri, but occasionally on small ponds, the endemic New Zealand scaup *Aythya novaeseelandiae*, our only diving duck, is found. Many of these species are thought to be the definitive hosts for the parasites that cause 'duck itch' (schistosome dermatitis) in humans, who are accidental hosts (Box 13). The Canada goose *Branta canadensis*, introduced from North America, is found on farmland and tussock country, usually near lakes and tarns. Abundant east of the Southern Alps, it has become an important game bird, with some 20 per cent of the population estimated to be shot each year.

The black swan *Cygnus atratus* is an important algal consumer in many of the shallow lagoons and lakes along the coast (see Box 11). Black swans were introduced from Australia last century, but there is some evidence that natural immigration occurred at about the same time. Black cormorants *Phalacrocorax carbo novaehollandiae* can sometimes be seen fishing in these lakes, but the populations of this species are still recovering from the (legal) shooting and nest destruction that continued into the 1970s. They breed in colonies, usually in large trees near or over water, but sometimes on cliffs as in the Taieri Gorge. The smaller little pied cormorant *P. melanoleucos* also occurs in these habitats, sometimes fishing in small flocks. Along the edges of bodies of water of almost any size, the pied stilt *Himantopus himantopus leucocephalus* breeds and feeds. This species has undergone a significant population increase since last century. Pied and black stilts will frequently interbreed, producing hybrids. Many inland breeders migrate to North Island

Right: The wry-billed plover breeds on braided rivers in Otago and Canterbury. *M. Soper*

Below: The breeding range of the black stilt and its numbers (about 100 individuals) has declined dramatically. They breed only in the upper Waitaki River system. *NZ Wildlife Service*

harbours in the winter. The New Zealand kingfisher *Halcyon sancta vagans* is not as common in the south as it is in warmer regions, and in winter migrates to the coast, where it feeds on marine mudflats. The welcome swallow *Hirundo tahitica neoxana*, self-introduced from Australia in the 1950s, is often seen hawking for insects over water and farmland. Pairs build their solitary nest of mud and grass high up on a vertical surface, often an artificial one such as a bridge or the eaves of a building.

The white-faced heron *Ardea novaehollandiae*, another recent Australian immigrant, has become a very common species on farmland and wetland areas, as well as on the coast, feeding on a wide variety of small animals. A tree-nesting species, it was first confirmed breeding in New Zealand in Otago in 1940. Less common is the cosmopolitan kotuku or white heron *Egretta alba*, seen in Otago and Southland near fresh water (even suburban goldfish ponds) mostly in winter, after it has finished breeding. The Australasian bittern *Botaurus stellaris poiciloptilus*, which stalks the dense rush and sedge beds fringing some of the larger lagoons, such as Lake Waihola and those to the north of Foveaux Strait, is much more difficult to spot. The male's territorial booming call is probably better known to many people than the bird itself. Almost certainly in the same places, but not well documented because of their secretiveness, are two starling-sized rail species, the spotless crake *Porzana tabuensis plumbea* and the marsh crake *P. pusilla affinis*. The fernbird *Bowdleria punctata* is found in scrub and reed beds fringing both salt and fresh water. This species can, however, also be found away from water, for example in scrub at the top of Mt Cargill, near Dunedin, or on Bluff hill in Southland and it is widespread in dense scrubby coprosma throughout Southland.

In wet areas of farmland, the pukeko *Porphyrio porphyrio* breeds in small groups of unrelated birds. By contrast, in northern New Zealand, this species breeds in groups of related birds. Variation in the quality of habitat and climate are thought to cause this difference. Yet another recent Australian immigrant, the spur-winged plover *Vanellus miles novaehollandiae*, was first noticed breeding in Southland in the 1930s. It too prefers damp farmland, around the edges of small lakes and ponds. After breeding, large flocks of several hundred birds sometimes form on suitably wet farmland.

Hamish Spencer and John Darby

Compared with its close cousin the black stilt, the pied stilt is relatively common. *M. Soper*

13 DUCK ITCH IN LAKE WANAKA

Lake Wanaka has some popular bathing beaches. However, sometimes during the summer, children and other bathers are affected by an irritating rash and can be so upset that they will cut short their summer holidays. The rash is caused by the larvae of a parasitic flatworm called a schistosome. Schistosomes cause important parasitic diseases in humans in the tropics. The adult schistosomes found in Lake Wanaka, however, normally develop in birds. Recent research at the Department of Zoology, University of Otago, has confirmed that a duck, the New Zealand scaup *Aythya novaeseelandiae* is the definitive host of the parasite (hence the name 'duck itch'). The schistosome appears to be an undescribed species. The parasite has a second or intermediate host, a freshwater snail that lives in the lake *Lymnaea tomentosa*. The larval parasite multiplies within the snail and then emerges as a free-swimming cercaria; it is this cercaria that infects the scaup by burrowing in through its skin. If humans come into contact with the cercariae they become accidental hosts of the parasite and the cercariae die within the tissue beneath the skin after penetration, producing a tissue reaction that causes the rash and itching. This problem occurs in many parts of the world, where it may be called 'swimmers' itch', 'sawah itch' in Japan, 'clam diggers itch' on Long Island (USA), 'sedge pool itch' in Michigan or, of course, 'duck itch' in New Zealand. It is more correctly known as cercarial dermatitis.

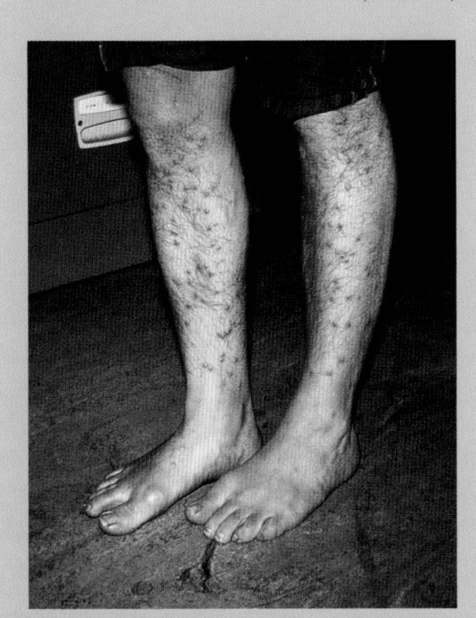

An example of a duck itch infection. *Warren Featherston*

An attempt to develop a control against duck itch in Lake Wanaka was conducted in 1989 by Norm Davis, in a trial funded by the Queenstown District Lakes Council. This attempted to remove the snails, and hence interrupt the life cycle of the parasite, by broadcasting a molluscicide (a chemical that kills snails). Snails in the treated area were killed, but within a few months the area was recolonised by snails from nearby untreated regions. Controlling duck itch by eliminating snails is difficult and expensive in a large lake.

Lymnaea tomentosa, the snail intermediate host of the parasite responsible for duck itch. *Norm Davis*

During his studies, Davis turned his attention to a more biological solution to the problem. The snail intermediate host of the schistosome is also infected by echinostomes, which are another group of parasitic flatworms. The echinostome larvae within the snail are aggressive feeders and will eliminate any schistosome larvae present by eating them. Thus, it may be possible to manipulate the situation to produce a biological control for duck itch. Canada geese (*Branta canadensis*) are the main definitive host of echinostomes. These are regularly culled and could provide a source of eggs, which could be accumulated and broadcast at a critical time of year to suppress the schistosomes and thus possibly control the duck itch problem.

David Wharton and Norm Davis

CHAPTER 11

THE COAST

MIKE BARKER, TONY BRETT, JOHN DARBY, JENIFER DUGAN, DAVID HUBBARD, PETER JOHNSON, PHILIP MLADENOV, BRIAN PATRICK, BARRIE PEAKE, KEITH PROBERT, ABIGAIL SMITH, HAMISH SPENCER, SUSAN WALKER

Otago Harbour encompasses a range of sheltered marine habitats and represents a highly significant feature of the region's coastal ecology. And, as a major site of Maori and European settlement and a key port, it is also of vital importance to the region for cultural and commercial reasons. *P.K. Probert*

Fig. 1. Indented coastline of the Catlins coast, South Otago, with headlands and bays, indicating alternating layers of more or less resistant rocks. *R. E. Fordyce*

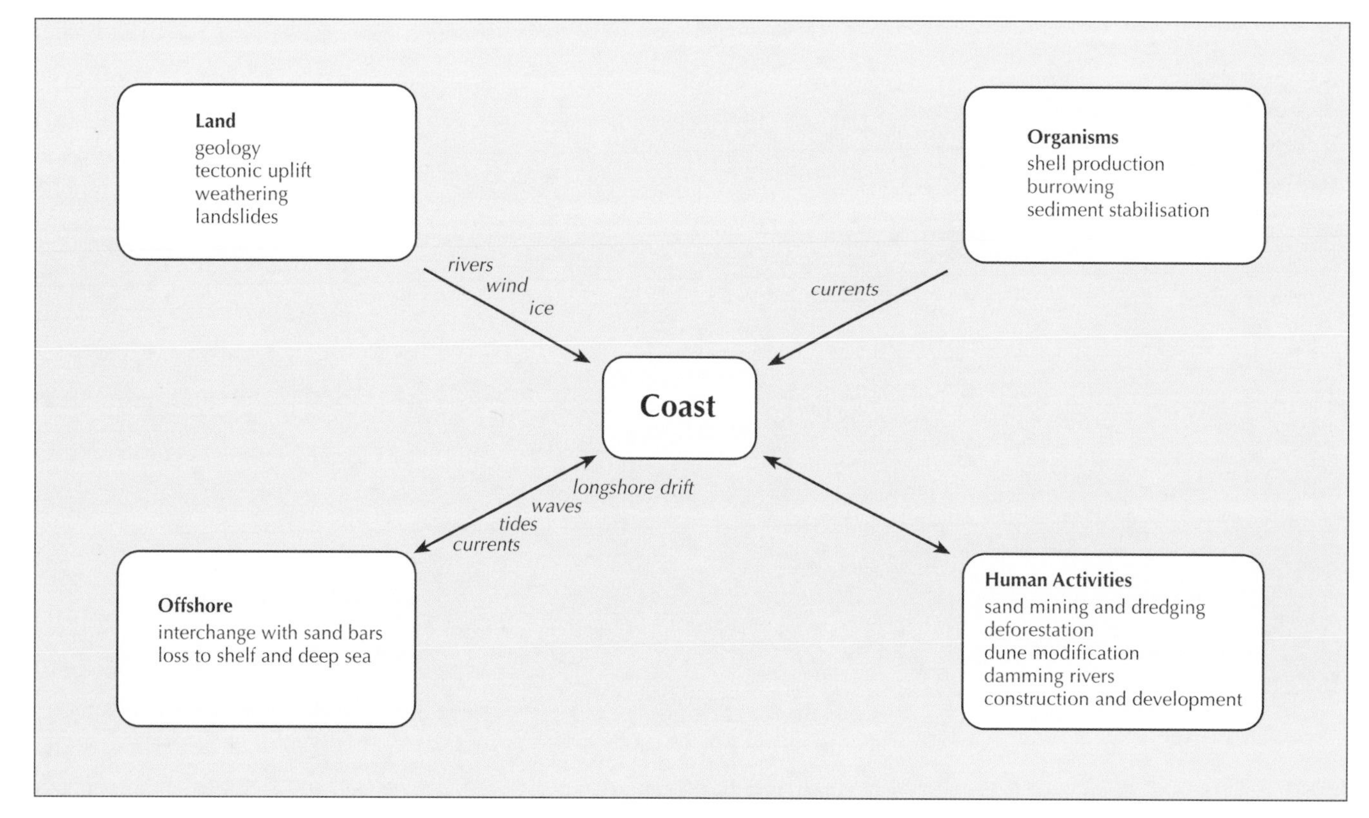

Fig. 2. A coastal sediment budget. Arrows indicate the direction in which sediment moves between systems, with transport processes in italics.

THE COAST

THE SOUTHERN COASTLINE

Abigail Smith

From its rugged cliffs and deep fiords to its active sandy beaches and rippled tidal flats, the southern New Zealand coastline is dynamic and diverse. The complex boundary where the land meets the sea is always changing and has always done so. What we see today on a map as the outline of the land is only a single picture, frozen for a moment in time, of a highly dynamic system. The modern coastline is the sum of all the processes of the past, acting on past shorelines. And, in turn, modern processes are working to make the present shoreline into some future coast.

Shaping the coastline

Three major factors determine coastal shape and type. Underlying basement rock is the foundation. Its strength, orientation, fracture pattern and composition all influence coastal shape. Sediment supply to the area is also a determining factor. If supply outstrips removal, then a soft shore, such as a beach, may grow. If sediment is removed more rapidly than it is deposited, beaches will begin to erode, or indeed there may be no beach at all, just rocky cliffs. Finally, the movement of waves, currents and tides affect the distribution of sediments and the weathering of rocks. Taken together, these three produce the diversity of coastal landforms we see today.

Rocks determine in large part the character of a landscape, even if they are hidden below vegetation, soil and water. Uplift and earth movements, too, shift and fold rocks, and thus the shape of the land. At the coastline, rocks are exposed to the forces of the sea. The composition and structure of the rocks determine how they will respond. Some rocks are strong and difficult to erode. Others are soft and easily destroyed by coastal processes. In some areas, highly resistant rocks alternate with weaker counterparts, sometimes cut by faults, such as where the Southland Syncline reaches the Catlins coast. The result is a strongly serrated coastline (Fig. 1). Otago Peninsula and, to a lesser extent, Cape Wanbrow, are examples of isolated highly resistant rocks that form headlands of strong volcanic basalt protruding from nearby receding shores of less resistant mudstones, schists and limestones. In contrast, Paterson Inlet on Stewart Island has formed in an area of rocks weakened by faulting.

Supply and removal: the sediment budget

In many areas, coastal sediments blanket the underlying rock foundation, and reflect the balance between erosion and deposition within the system. Where supply is plentiful, for example just north of the Clutha River, depositional features such as beaches, barrier islands, spits and deltas form. Where marine erosion exceeds supply, then the coast exhibits retreating cliffs, bare shore platforms and sea stacks, such as those to the south of the Waitaki River.

New Zealand is a country with high rainfall and a marked relief that reflects active tectonic uplift. The result is erosion. The land provides abundant sediment to the marine environment. In some areas, sediment becomes trapped before it reaches the coast, as in fiords and major lakes. But much of the sand drifting along the southern New Zealand coast is weathered from the Southern Alps and delivered to the sea by rivers. Major contributors include the Clutha, Oreti and Aparima Rivers which together deliver about 2.8 million tonnes of sediment to the ocean each year. Cliff erosion and land slips occasionally provide additional material. Other sediments, such as wind-blown dust, volcanic ash, shell material from organisms and chemical precipitates, are far less common.

Sediment supply is also affected by human activities. Construction of ports, harbours, ramps, breakwaters and bridges can increase or decrease sediment supply to particular environments. Deforestation, by removing a protective cover of vegetation, increases the sediment delivered to the sea, as may damage to dune vegetation. Dredging removes sediment from a given system (such as Otago Harbour), but delivers huge volumes of sediment to some other environment (such as the dump grounds just north of Otago Harbour mouth). Damming of rivers, such as the Roxburgh Dam on the Clutha River, reduces sediment delivery from land to sea. Coastal mining of sand, as has occurred at Tomahawk Beach, Dunedin, removes sediment directly. Major factors affecting coastal sediment budgets are summarised in Fig. 2.

The movement of water

Once sediment reaches the coast, it becomes susceptible to transport and re-distribution by waves, currents and tides. Coastal sediment is in constant motion, forming and reforming ripples, dunes, bars and flats. Change may be gradual, as the slow creep of sediment onshore during summer, or it may be dramatically sudden, such as extensive erosion caused by storm waves and surges.

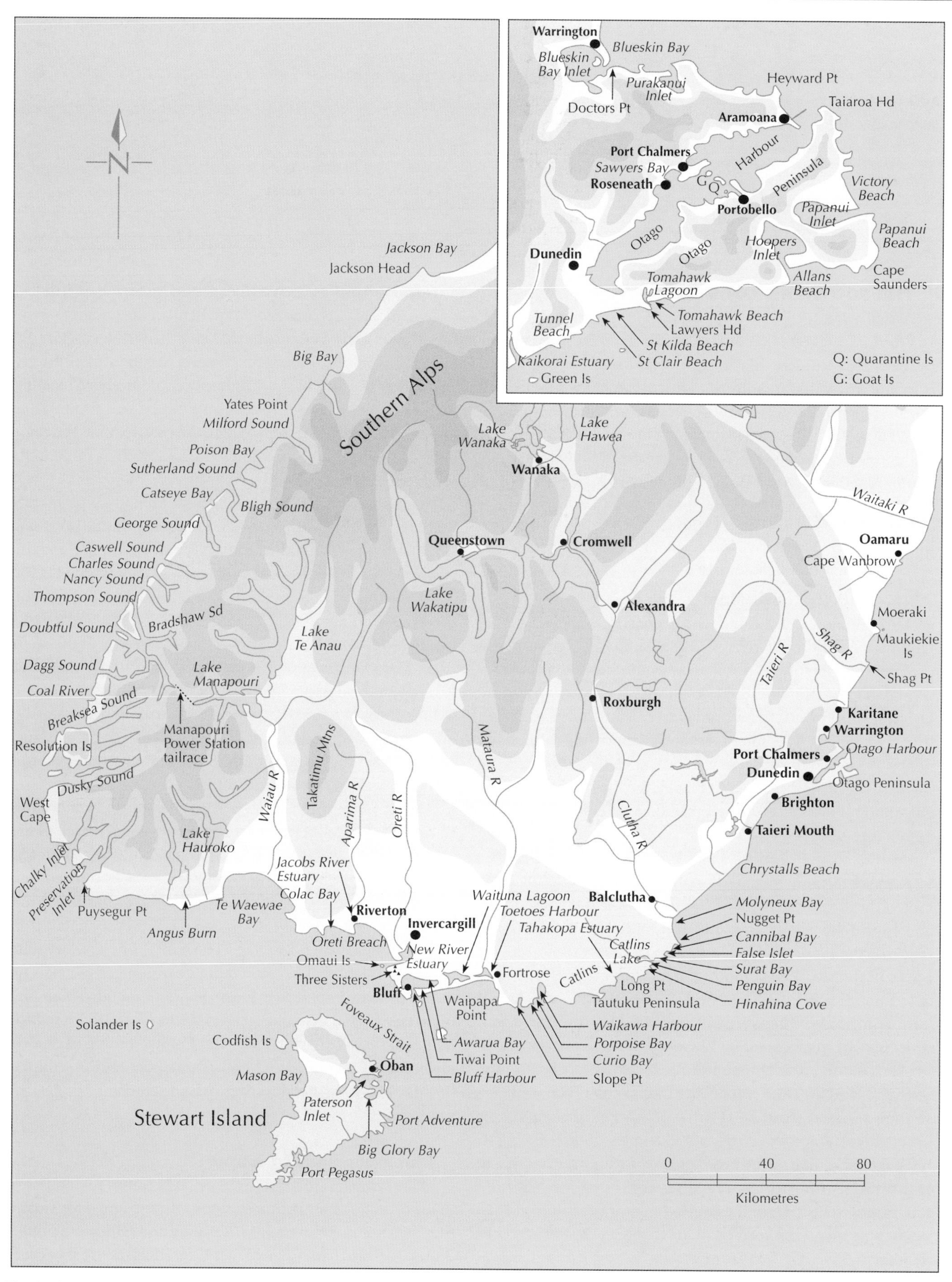

Fig. 3. Places referred to in the text.

Fig. 4 (right). The coast of southern New Zealand showing water movements and coastal types.

In southern New Zealand, most deep-water waves arrive from a southwest to westerly quarter. Thus coasts facing west and south are exposed to high-energy wave attack, whereas north and northeast-facing coasts are more protected. Fair-weather waves tend to be constructive, bringing sediment closer to shore, building up beaches and flats. It is the less common storm waves that do erosive damage to the coast, causing cliff retreat and beach scour.

As waves approach land, they 'feel bottom' and slow down. One result of this slowing is wave refraction: the straight wave fronts bend to match the contours of the seabed. Such refraction is only partial, however, and therefore waves transfer water energy along the coast, forming a longshore current. Around southern New Zealand, a powerful longshore flow transports sediment dominantly northward and eastward (Fig. 4).

On southern New Zealand shores there is about a two-metre difference between high and low water on spring tides, decreasing to about a metre at neap tides. In some areas, notably Foveaux Strait, the incoming and outgoing tides produce very strong currents, which influence sediment supply. Tidal channels are particularly important in tidal flats, estuaries and tidal inlets.

The interaction of these hydraulic processes determines coastal landforms. Wave-dominated coasts with abundant sediment supply are characterised by beaches, whereas sediment-starved coasts are cliffed. Tide-dominated coasts include tidal flats, estuaries and tidal sand ridges. Coastlines where both processes are important form tidal deltas and tidal inlets.

The role of climate

Climate has a profound and direct effect on the coast. About 18,000 years ago, New Zealand had a very different coastline from that of today. Sea level was some 130 m lower than it is now, and the South Island was thus much wider (although, in fact, much of the southern South Island was under ice). For about 12,000 years, sea level rose in a more or less continuous way to its present position about 6500 years ago. That drowned shoreline has had about 6500 years to adjust, and it is continuing to do so. The impact of human activities on climate (notably, global warming and sea level rise) has been recorded mainly in the last hundred years.

Climate has indirect effects, too. Chemical weathering, rainfall and erosion, river discharge, wind-blown dust, wave strength and direction, and storm surges are all influenced by climate. Biological influences on the coast, such as shell production, burrowing and encrustation, are strongly dependent on climate, especially temperature. Climate also determines the distribution of plants which are so important in stabilising tidal flats and sand dunes.

A diverse coastline

Southern New Zealand is remarkable for the diversity of its shoreline. In a short distance, we can find most major coastal types and landforms. In Fiordland hard gneisses and granites are highly resistant. This recently glaciated coast consists of steep cliffs and deep fiords. Almost beachless and without shore platforms, the coast is subject to crashing waves from the west. Sediments from the land are trapped in the fiords, sometimes in large deltas. The nearby Alpine Fault adds uplift to this volatile environment. Southeastern Southland, too, is made up of resistant rocks (sandstone and mudstone) that provide a straight, steep shore with little cliff erosion. Softer cliffs near the Taieri River, and near Oamaru, however, are retreating, some in spectacular fashion (0.5–2.5 m per year). They have narrow cliff-foot beaches or small shore platforms which protect them somewhat from fair-weather waves, but storm waves undercut the cliffs, causing collapse.

In contrast, the Invercargill coast and Stewart Island are characterised by rocky headlands alternating with small sandy bays, especially where rocks are fractured perpendicular to the shore, as in the Catlins. Along the Otago coast, shore platforms are common, and there are some small estuaries. The sedimentary regime is almost entirely erosional, with a few local spots of deposition at Blueskin Bay, Taieri Mouth and Clutha Mouth. Beaches in Southland and Otago are mainly of sand, although there are cobble beaches at Waitaki Mouth.

There is not an environment, on land or under sea, that is as changeable and dynamic as the coastline where they meet. It is a place of contrasts. We can walk along crumbling cliffs, then suddenly find ourselves at a strong headland. We can be on quiet tidal flats, when just around the corner high waves pound on a breakwater. Ever changing both in space and in time, the coastline today is fleeting evidence of a series of processes that are always building and moving and destroying the edge of the land, the edge of the sea.

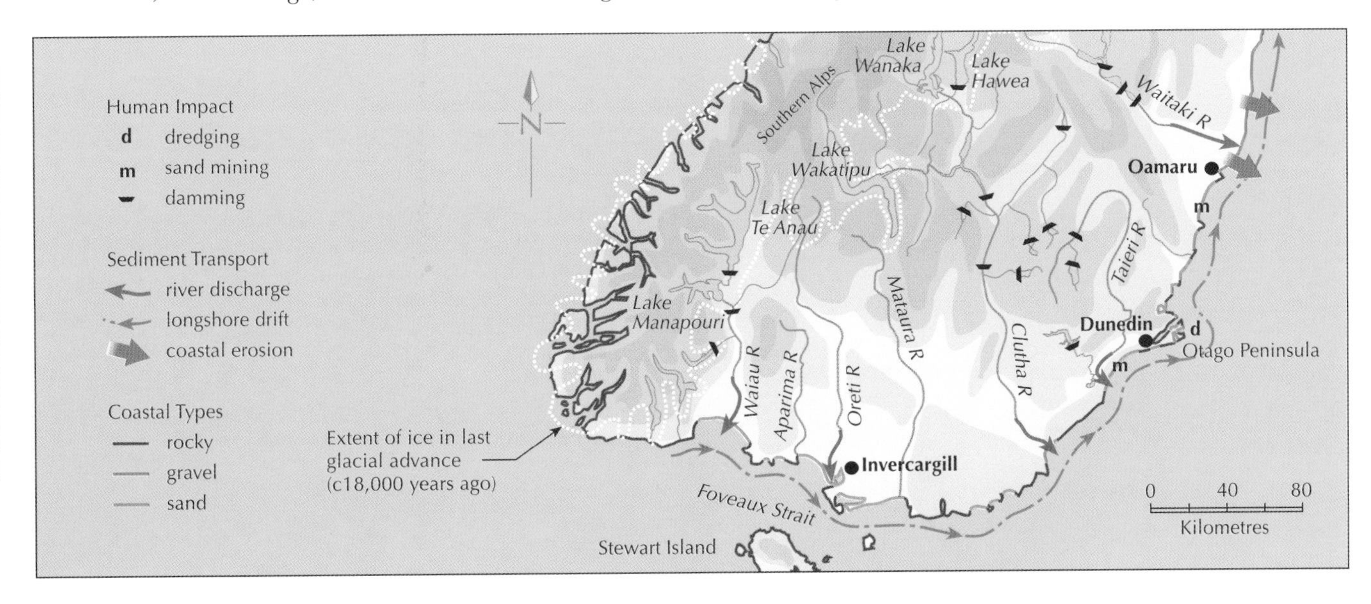

ROCKY SHORES AND KELP BEDS

Mike Barker

Rocky shores are intertidal habitats in which the substratum is mostly made up of hard bedrock, although there may also be quantities of loose material of varying sizes, from small pebbles and boulders to large, almost immobile, slabs. Below the intertidal zone rocky substrata often merge into soft sediments of sand or mud. However, in situations where wave action or strong currents prevent the deposition of sediment, the bedrock may continue into deeper water, and in such situations kelp forests are often found.

Rocky shore zonation

Rocky shores of southern New Zealand range from those of sheltered embayments, such as in Otago Harbour, to shores that face directly into the surf and swells of the Southern Ocean. When the tide retreats, we see that the plants and animals of the rocky shore are not randomly distributed, but occur in a characteristic and quite predictable pattern of intertidal zones. The width and height of these zones above low water mark is amplified by the tidal range and the degree of wave exposure the shore experiences. Wave exposure may even vary considerably between sites separated by just a few kilometres, and this has a profound influence on intertidal communities. However, the major types of plants and animals found, and their vertical position on the shore, are essentially the same for temperate shores throughout the world, and this has led to the development of a universally recognised scheme for zonation (Box 1).

For a coastline as extensive as Otago and Southland there is considerable variation in shore types, and the habitats and species present. Inevitably, therefore, the following description can cover only the main species and habitats.

Sheltered shores

Rocky shores found within harbours or protected by headlands have zonation patterns more influenced by tidal rise and fall than by wave action. This results in intertidal zones that are restricted in vertical extent (Fig. 5).

Low on the shore the sublittoral fringe algae comprise a range of species. Common are the brown algae *Cystophora scalaris, C. retroflexa* and *Hormosira banksii* or Venus' necklace, along with the green *Codium fragile* and the sea lettuce *Ulva reticulata.* Becoming more common in some localities, such as Otago Harbour, is the introduced Asian kelp *Undaria pinnatifida* (Box 20). On headlands and other sites where tidal currents are strong, the bladder kelp *Macrocystis pyrifera* can generally be seen, the long stipes and lamina often festooned over the lower shore when the tide is out (Fig. 6).

This low-shore region is inhabited by a rich assemblage of animal species, of which only the most common can be mentioned here. Grazing in and around the kelp and small boulders occur gastropod molluscs: the catseye *Turbo smaragdus,* the black ducksbill limpet *Scutus antipodes* and, under stones or in crevices where sand and silt have accumulated, the deposit-feeding turret snail *Zeacumantus lutulentus.* At this low shore level, most common at sites with significant current flow, is found the large red or maroon, solitary, stalked ascidian or sea tulip *Pyura pachydermatina* (Fig. 7), often attached in large numbers to the stipes of *Macrocystis.* The undersides of rocks and small boulders are encrusted with a highly diverse fauna including the red keeled serpulid tubeworm *Galeolaria hystrix,* the green chiton *Chiton glaucus,* compound ascidians and sponges and the green eight-legged half crab *Petrolisthes elongatus.* Also found under rocks, or less commonly in the open, is the small dividing starfish *Allostichaster insignis* and the maroon brittlestar *Ophiomyxa brevirima,* often with cream stripes down its five arms. The cushion star *Patiriella regularis* may also be encountered. In rock pools and also sheltering under stones at low tide may be found several

1 MAJOR ZONES OF TEMPERATE ROCKY SHORES

Tidal level	Zone	Indicator organisms
	Maritime zone	Terrestrial vegetation, orange and green lichens.
Approximate extreme high water of spring tides	Littoral fringe	Covered with water only during high spring tides, although spray and swash keep the area damp, especially on exposed shores. Few marine animals are able to exist this far up the shore because of the general lack of moisture, although small grazing molluscs, notably the periwinkles, are adapted to survive the physiological extremes.
	Midlittoral zone	Covered and uncovered at every tide. Upper limit of barnacles. A wide range of gastropod molluscs grazing on microalgae together with a range of filter feeders such as oysters and tubeworms, especially towards the lower midlittoral.
Approximate extreme low water of spring tides	Sublittoral fringe	Maximum upward extent of brown algae.
	Sublittoral zone	Biota never uncovered by the tide.

Mike Barker

Fig. 5. A rocky headland in Otago Harbour. The pattern of zonation on such sheltered shores is more restricted in vertical extent than that on more wave-exposed shores. *P.K. Probert*

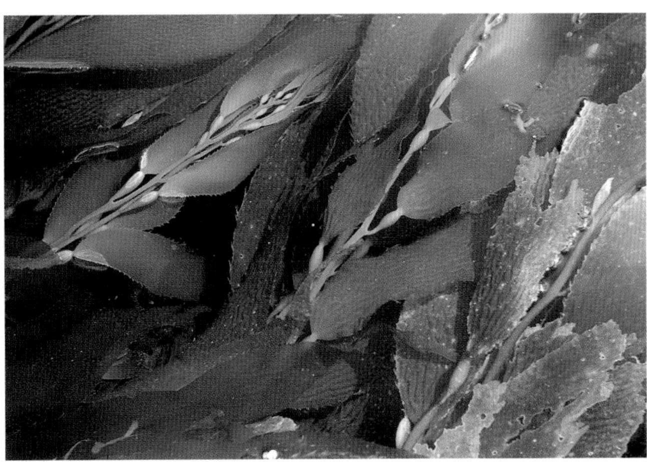

Fig. 6. Bladder kelp *Macrocystis pyrifera*, a fast-growing seaweed characteristic of the more sheltered southern coasts. *E.J. Batham*

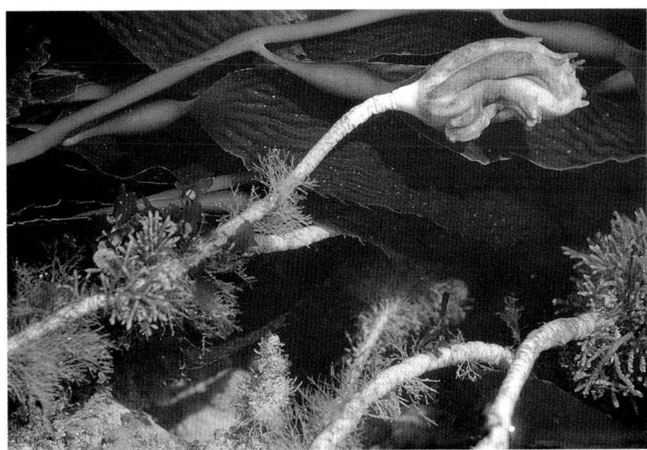

Fig. 7. Sea tulip *Pyura pachydermatina* among bladder kelp. Despite its appearance and common name, the sea tulip is a highly developed invertebrate with a free-swimming larva that resembles a tadpole. The stalks of sea tulips provide an attachment surface for other organisms, including algae, hydroids and bryozoans (as here). *E.J. Batham*

2 MARINE CADDIS FLY

Very few groups of insects have been able to exploit marine habitats, but one family of caddis fly (Chathamiidae) has done just this. The family is centred on New Zealand and one species, *Philanisus plebeius*, occurs around the South Island and Stewart Island coasts. The adults look like sandy-brown moths, about one cm long, with the wings held back against the body and antennae pointing forward. They rest up during the day under rocks or in shore vegetation, becoming active at night.

Like all caddis flies, the adults feed very little; all the feeding and growing are done by the larvae which live on rocky shores, such as at sites in Otago Harbour where they are abundant. The larva builds a portable tubular case of seaweed and sand grains, and feeds on algae. It pupates inside the case, first attaching it to seaweed. When ready to emerge the adult cuts its way out of the case.

There are more than 200 known caddis fly species in New Zealand and the larvae all have life histories similar to *P. plebeius*, but in freshwater rather than the sea.

J. B. Ward

Top: Adult of the marine caddis *Philanisus plebeius*. *B. Patrick*

Bottom: The larva of the marine caddis in its tubular case. *B. Patrick*

small fish species including the very common black rockfish *Acanthoclinus littoreus*, the little rockfish *A. rua* and the olive rockfish *A. fuscus*, as well as the blenny-like triplefins – the common triplefin *Forsterygion lapillum* and variable triplefin *F. varium*. On exposed rock, especially in rock pools or wet areas, pink encrusting or 'coralline' algae occur in both 'paint' and 'turfing' growth forms.

Moving up the shore, the lower midlittoral is marked by the appearance of barnacles. In the most sheltered sites the harbour barnacle *Elminius modestus* occurs, but on headlands and rock faces with greater degrees of wave exposure may be found the small rounded barnacle *Chamaesipho columna* and the larger *Epopella plicata*, although generally only a few per square metre (Fig. 14). Other attached filter feeders at these lower shore levels are the blue mussel *Mytilus galloprovincialis* (Fig. 8), the ribbed mussel *Aulacomya atra maoriana*, the oyster *Tiostrea chilensis* (Fig. 9), and the polychaete *Spirobranchus cariniferus* with its crown of blue tentacles. This tubeworm is less common on more southerly shores, but often forms a distinctive band on shores north of Shag Point. Also low in the midlittoral occur the limpets *Notoacmaea parviconoidea* and *Siphonaria australis* and the snakeskin chiton *Sypharochiton pelliserpentis* (Fig. 10). Grazing throughout the midlittoral zone and up into the lower littoral fringe are various gastropod molluscs: the topshell *Melagraphia aethiops*, the limpets *Cellana ornata* and *C. radians* and the chiton *S. pelliserpentis*, most often found in cracks and crevices.

The top of the midlittoral zone coincides with the top of the barnacle zone and, on southern shores, with the occurrence of the mossy red alga *Stictosiphonia arbuscula* (Fig. 5), often luxuriant after winter and spring growth but sun-scorched and patchy by late summer. Towards the upper limit of the midlittoral zone periwinkles first appear and extend upshore to where the first land plants mark the maximum vertical extent of the littoral fringe. Two periwinkle species are common, the larger brown *Nodilittorina cincta* and the smaller *N. antipodum*, distinguished by the blue band on a light background around the central whorl of the shell (Fig. 11). Lichens may also occur in encrusting patches in this zone, the dark tufted *Lichina confinis* and the

Fig. 8. The blue mussel *Mytilus galloprovincialis*, an attached filter feeder characteristic of the lower midshore zone. *P.K. Probert*

Fig. 9. The oyster *Tiostrea chilensis* occurs in different forms. It can live intertidally attached to rock, as shown here, but also in deeper water lying free on the bottom, as in Foveaux Strait where it is the basis of the Bluff or dredge oyster fishery. *P.K. Probert*

Fig. 10. Chitons or coat-of-mail shells are a distinctive group of molluscs with a shell of eight overlapping plates. They are well represented in New Zealand. The snakeskin chiton *Sypharochiton pelliserpentis* is a common rocky shore species with an obvious snakeskin girdle. *M.F. Barker*

Fig. 11. These small snails, periwinkles, are among the few animals that inhabit the upper zone of rocky shores, typically huddled in cracks and crevices. Two species are common, the larger brown *Nodilittorina cincta* and the smaller *N. antipodum*, distinguished by the blue band on a light background around the central whorl of the shell. *P.K. Probert*

smoother *Verrucaria* sp. Few other invertebrates occur in the littoral fringe, although the limpet *Notoacmaea pileopsis*, easily distinguished by its clean rounded profile and mottled green patterned shell, is quite common, generally in cracks and crevices and especially underhangs.

Wave-exposed shores

Some species, tolerant of wide variations in exposure, can be found on a broad range of different shore types, although their vertical extent, abundance and size may vary. More commonly, however, there are marked changes in the types of organisms present under different exposure regimes. These differences are caused by a complex suite of physiological and behavioural factors, including biological interrelationships of competition and predation, the abilities of species to tolerate wave action, and variations in recruitment, growth and mortality.

Examples of wave-exposed shores are found throughout the Otago and Southland coastline. Indeed, in terms of extent, much more of this rocky coast is exposed than sheltered. Very exposed shores that occur close to Dunedin include those on the southern side of Otago Peninsula, such as at Cape Saunders and Allans Beach (Fig. 12). Further south, Nugget Point, the Tautuku Peninsula and all major headlands facing directly into the southern swell would also be classified as exposed.

The sublittoral fringe of these shores is dominated by two species of bull kelp. *Durvillaea antarctica* has thick blades arising from a thick stipe. The blades have internal air-filled honeycomb-like cavities, allowing them to float. *D. willana* is generally found 30–50 cm lower on the shore and has thinner more solid blades that do not float, and which may also arise from side branches on the main stipe (Fig. 13), an arrangement never seen in *D. antarctica*. *D. willana* may extend below low tide to depths of 3–4 m. Although *Durvillaea* dominates the lower levels of the shore, other algae also occur in this zone, including the small red seaweeds *Pachymenia lusoria* and *Apophlaea lyallii*, these latter species also extending higher on the shore. Among the *Durvillaea* holdfasts, and often extending a metre or so below, mussels form a distinctive and usually tightly packed band. Three species are

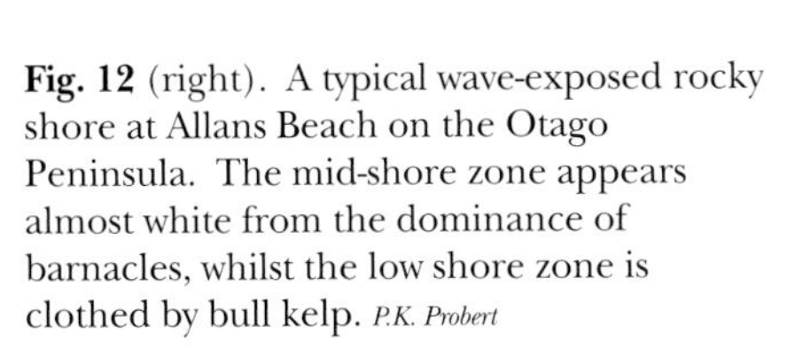

Fig. 12 (right). A typical wave-exposed rocky shore at Allans Beach on the Otago Peninsula. The mid-shore zone appears almost white from the dominance of barnacles, whilst the low shore zone is clothed by bull kelp. *P.K. Probert*

Fig. 13 (below). Two species of bull kelp characterise wave-exposed rocky coasts of southern New Zealand, as at Curio Bay in the South Catlins, shown here. In the foreground is *Durvillaea willana*, with side branches arising off the main stipe, and in the background is *D. antarctica*. *E.J. Batham*

Fig. 14. The midlittoral zone on wave-exposed shores is dominated by barnacles, filter-feeding crustaceans with a series of calcareous plates enclosing the body. Particularly abundant are the small *Chamaesipho columna* and larger *Epopella plicata*. *E.J. Batham*

common: the green-lipped mussel *Perna canaliculus* and the blue mussel *Mytilus galloprovincialis*, with occasional specimens of the ribbed mussel *Aulacomya atra maoriana*. A rich fauna of sessile species occupies any free substratum – encrusting and turfing coralline algae, sponges, hydroids (especially attached to mussel shells), bryozoans including erect branched colonies of *Elzerina binderi*, anemones, such as the large *Isocradactis magna*, and a large acorn barnacle *Balanus campbelli*. In deep channels and pools which are partially protected from the direct forces of the swell, the sea tulip *Pyura pachydermatina* may be seen attached directly to the rock surface. Chitons, particularly the large *Acanthochitona violacea* and *Cryptoconchus porosus*, are common below and among the *Durvillaea* holdfasts. Few species of gastropod molluscs can withstand the battering of the waves, although on southern Otago shores small individuals of the brown periwinkle *Nodilittorina cincta* occur well below the littoral fringe in pits and crevices down into the upper *Durvillaea* holdfasts. Why this littorinid should occur so low on southern shores compared to more northern localities is puzzling.

Above the sublittoral fringe the lower midlittoral is occupied by a suite of filter feeders. Green-lipped and blue mussels reach their upper limits here and are replaced towards the middle of this zone by the smaller mussel *Xenostrobus pulex*, characterised by a shiny black nacreous layer on its shell. In damp places, such as the edges of rock pools, the bright red anemone *Actinia tenebrosa* may be found, and in small pools the green anemone *Actinia olivacea* and the brown *Anthopleura aureoradiata* are quite common. In spring and early summer the delicate fronds and rosettes of the red seaweed *Porphyra columbina* may form a slippery coating on the rock surface.

The midlittoral, however, is dominated by barnacles, the small *Chamaesipho columna* and larger *Epopella plicata* (Fig. 14), often so closely packed that bare rock is scarcely visible. Predatory whelks, *Lepsithais lacunosus* and *Lepsiella albomarginata*, are also common here, feeding on the larger barnacles. Spaces between the barnacles are generally occupied by the periwinkle *Nodilittorina cincta*. Where there is bare rock, a range of grazing limpets and chitons occurs. Limpets include the common *Cellana radians* and, less abundantly, *C. strigilis* (distinguishable by marked radiating ridges on its shell), as well as several smaller species: *Notoacmaea parviconoidea*, *Patelloida corticata* and the pulmonate *Siphonaria australis*. The chitons *Sypharochiton pelliserpentis* and *Plaxiphora caelata* are also common from the lower to upper reaches of the midlittoral. In cracks and crevices, especially in shaded positions, underhangs and occasional caves, the stalked barnacle *Calantica spinosa* may be found. Towards the mid to upper midlittoral occur occasional specimens of a large pulmonate limpet, *Benhamina obliquata*, its presence indicated in late spring and summer by coils of jelly-like egg masses attached to the rock. As on sheltered shores, a band of the red alga *Stictosiphonia arbuscula* occurs towards the upper midlittoral. The littoral fringe hosts almost the same group of species as seen under sheltered conditions, but instead of a band of 1–2 m, the littoral fringe may extend very high on these exposed shores. *Nodilittorina cincta* is generally more common than *N. antipodum* on these shores.

Shores of intermediate exposure

Shores of intermediate exposure occur, for example, at Warrington and Karitane north of the Otago Peninsula. Such shores are not subjected to continuous wave action, although isolated storm events may occasionally generate surf conditions. The dominant alga tends to be the bladder kelp *Macrocystis pyrifera*. Other macroalgae commonly found include *Splachnidium rugosum*, *Codium adhaerens*, *Ulva reticulata*, *Cystophora scalaris*, *C. retroflexa*, *Xiphophora chondrophylla* and the long narrow blades of *Lessonia variegata*. Depending on the degree of exposure, other algae (including *Durvillaea*), and the invertebrates present, may be more representative of sheltered or exposed shores. Often on shores of moderate exposure, especially when the shore is broad, the intertidal may be covered in boulders providing shelter for many species. Low on the shore in the sublittoral fringe the yellow-foot paua *Haliotis australis* can often be found on the lower sides and undersides of boulders, generally covered with seawater. Among the boulders higher on the shore, small topshells reach their greatest diversity, with a number of closely related and morphologically similar species: *Melagraphia aethiops*, *Diloma arida*, *D. zelandica*, *D. bicanaliculata* and, at the top of the shore where kelp has been stranded, *D. nigerrima*. On some shores, as at Brighton, just south of Dunedin, the broad nature of the shore and its varied topography produce considerable variability of habitat exposure in a small area. On this shore, *Hormosira banksii*, *Xiphophora chondrophylla*, *Macrocystis pyrifera*, *Lessonia variegata* and *Durvillaea antarctica*, indicators of conditions from extreme shelter to extreme exposure, may be found within a few metres.

Rocky habitats beyond the shore

Below the lowest point uncovered by the tide, the attached invertebrate fauna closely resembles that seen on low spring tides. On *Durvillaea*-dominated exposed headlands and rocky promontories, shore platforms frequently grade into coarse highly mobile sand a few metres below the surface. When swell and waves are high, the water, loaded with this sediment, has a considerable scouring effect on the rock. As the water depth increases, sediment movement is reduced and soft sediments of one type or another become the dominant habitat to the edge of the continental shelf. However, in more sheltered situations, especially on *Macrocystis*-dominated shores, the rocky substratum extends offshore and extensive kelp beds are found, sometimes to a water depth of 30 m (Fig. 15). In shallow water (1–3 m), *Cystophora scalaris*, *C. retroflexa* and *Carpophyllum maschalocarpum* are the most common algae in sheltered sites, while in more exposed situations *Lessonia variegata* can be very common. At greater depths southern kelp forests are dominated by *Macrocystis*, with isolated specimens of *Ecklonia radiata* (the dominant subtidal kelp in northern New Zealand) and a number of small filamentous red and green algae.

Kelp beds are extremely important and productive habitats. Although some work has been done on *Ecklonia* kelp beds in northern New Zealand, little is known of the ecology of southern kelp forests. In almost all kelp beds, algal grazers are abundant and may play a key role in removing plants. Gastropod molluscs have some importance here, although species such as the black-foot paua or abalone *Haliotis iris* (Fig. 16) Cook's turban shell *Cookia sulcata* and smaller catseye *Turbo smaragdus* do not graze living kelp but feed on broken blades and stipes drifting over the bottom. The main grazers in kelp forests in all parts of the world, including New Zealand, are sea urchins, sometimes completely stripping a reef of algae then persisting in the area, often for several years and creating encrusting coralline-dominated areas known as 'barren grounds'. In most parts of New Zealand the large green sea urchin or kina, *Evechinus chloroticus*, has this grazing role (Fig. 17). However, on the Otago coast from Dunedin north to at least Shag Point, *Evechinus* is surprisingly rare. Large old individuals occur in small numbers, often on the sides of large boulders, but not the large numbers that carpet the bottom in other parts of the country. South of Nugget Point, *Evechinus* becomes more common and, around Stewart Island and in the southern fiords and coastal area, is extremely common. Other grazers found within kelp forests include two fish, the butterfish or greenbone *Odax pullus* (Fig. 18) and the marblefish *Aplodactylus arctidens*.

Many non-herbivorous fish species frequent southern kelp forests and only the commoner ones can be mentioned here. Often encountered in shallow water, sometimes with the tail wrapped around the stipe of a kelp, but almost always found within or close to seaweed is the seahorse *Hippocampus abdominalis*. Often behaving in a similar way is the southern pigfish *Congiopodus leucopaecilus*. Near the bottom, often in holes or crevices are scarlet wrasse *Pseudolabrus miles* and the red banded perch *Ellerkeldia huntii*. In more open areas the banded wrasse *Notolabrus fucicola* and the ubiquitous spotty *Notolabrus celidotus* are frequently encountered browsers. Another benthic fish predator, generally found on or close to the bottom, is the schooling blue moki *Latridopsis ciliaris*, which feeds on a wide range of invertebrates such as polychaetes, crabs and sea urchins. Sometimes swimming with moki is its close relative the common trumpeter *Latris lineata*. Another curious kelp forest fish is the leather jacket *Parika scaber*, and just beyond the kelp forest fringe is found the ever inquisitive bottom-dwelling blue cod *Parapercis colias*. The sea perch, or Jock Stewart *Helicolenus percoides*, is seen occasionally, but becomes more common in deeper water

Fig. 15. Luxuriant beds of bladder kelp *Macrocystis pyrifera* provide an important habitat for a wide range of invertebrates and fish. *C. Hepburn*

Fig. 16. Blackfoot paua or abalone *Haliotis iris* is an important grazer of drift kelp, but has become less abundant as a result of harvesting. *R.J. Street*

Fig. 17. Intensive grazing by kina *Evechinus chloroticus* can strip an area of its algae, resulting in 'barren grounds'. In turn, the sea urchins are preyed upon by seastars such as *Coscinasterias muricata*, shown here. *M.F. Barker*

Fig. 18. A number of fish species frequent kelp forests, such as the greenbone *Odax pullus*. *M. Francis*

beyond the depth of kelp. Feeding off the bottom, in mid water, often in small schools above the kelp, is the butterfly perch *Caesioperca lepidoptera*.

Rocky reefs within kelp forests are often highly dissected, sometimes with scattered boulders or deep fissures. Sessile species, such as sponges, bryozoans, compound and solitary ascidians, and coralline algae occupy much of the available free space, and diversity is high. Within this rich encrusting community are mobile crustaceans, echinoderms, polychaetes and gastropod molluscs, many of them predatory, feeding on each other and the sessile fauna.

Kelp forests are extremely important coastal ecosystems, providing a wide variety of ecological niches for a rich array of species. Whilst the high diversity is in part due to the variable bottom topography, the kelp plants themselves, standing up in the water column often to the surface, provide a separate group of niches available to fish and a range of invertebrates either attached to, or closely associated with the kelp. In addition, the plants themselves are highly productive, converting dissolved nutrients and sunlight into organic matter at a rate as great as, or greater than, the most productive ecosystems on land.

Fig. 19. Extensive gravel beaches occur along the North Otago coast, but the constant movement and grinding action of the particles excludes most shore organisms. P.K. Probert

BEACHES

Jenifer Dugan and David Hubbard

Beaches make up about a third of this country's coastline. Around southern New Zealand, beaches and surf zones cover an impressive range of types and harbour a rich and unique fauna. They range from tidally dominated protected beaches in harbours and inlets to wave-dominated surf beaches on outer coasts. The ecology and morphology of sandy beaches result from dynamic interactions between waves, tides, wind and sediments.

In southern New Zealand, the sediments of surf beaches vary dramatically in size and sorting, leading to a diversity of beach types. Boulder beaches are the coarsest (with particles more than 250 mm in diameter). Few animals attach to and live on the surface of these boulders since the shifting and rolling of large rocks during high surf conditions can crush fauna. Boulder beaches occur in a number of locations, but in general cover a minor part of the southern coast.

Extensive stretches of gravel and cobble beaches (grain diameter of 2–250 mm) occur on the eastern coast of the South Island (Fig. 19). The coarseness and looseness of gravel prevent the development of a rich faunal community. Gravel and cobble sediments are moved so often that organisms cannot attach to the rocks as on boulder beaches, and the coarse sediments also exclude burrowing animals. Distinctive storm berms often form high on these beaches. The sound of surf crashing into the berm of these cobble beaches and draining back into the sea is unforgettably loud, reflecting the energy dissipated on these beaches.

More typical sandy beaches form on open coasts where sediment grain size is finer (0.06 to 2 mm in diameter). These beaches are often associated with coastal dunes, essentially wind-controlled terrestrial habitats. Physical processes affecting sandy beaches and dunes are closely linked. Sand is moved between subtidal bars, the intertidal beach and dunes, as surf and wind conditions change, shifting the entire shore profile. Wave energy moves sand along the beach and between the beach and the surf zone. When conditions change, the beach shape responds rapidly and often dramatically. During storms or periods of large waves, sand from the intertidal beach is saturated with water, suspended, and washed out into the surf zone, flattening the beach. When wave energy decreases, sand is moved from subtidal bars to the beach face, gradually building a berm and steepening the beach. Together, beaches and dunes disperse the energy of waves and storm surges, buffering and protecting the shore during extreme events such as onshore gales.

Open coast sandy beaches are often classified in terms of their sand grain size and wave regime. So-called 'reflective beaches' have coarse sand, steep slopes, narrow surf zones and small waves (less than 0.5 m). At the other end of the spectrum are 'dissipative beaches' with fine sand, flat slopes, wide surf zones and large waves (more than 2 m) (Fig. 20). Between these extremes lie the 'intermediate beaches', the most variable and abundant beach type along the southern New Zealand coast (Fig. 21).

On beaches protected from the open sea, tides and freshwater

Fig. 20. Oreti Beach, Southland, has the very fine sand, flat slope, wide surf zone and large waves typical of a 'dissipative' type beach.
J. Dugan

Fig. 21. St Clair and St Kilda beaches, which form Dunedin's main beach, are an example of an 'intermediate' beach, the most abundant type of open coast beach along the southern New Zealand coast.
P.K. Probert

runoff are the most significant types of water movement. Here, grain size can also vary dramatically, from gravel and shell hash to fine muds and silts (grain diameter less than 0.06 mm). These beaches are found in the river mouths, harbours, inlets and estuaries of southern New Zealand.

Beach communities

Unlike rocky intertidal or estuarine shores, the rapidly shifting sediments of open coast sandy beaches do not support attached plants. Biological communities on surf beaches are dominated by mobile animals that move up and down the beach with tides and seasons.

Animals living on surf-swept beaches are generally rapid burrowers that construct temporary burrows. On more protected tidally dominated beaches, animals such as crabs, ghost shrimp and mantis shrimp may construct more elaborate semi-permanent burrows. These burrows may also house numerous commensal species that feed or find shelter with the host animal.

Three major types of invertebrate animals dominate the macrofauna of sandy beaches: the molluscs, crustaceans and polychaete worms. The relative abundance of these groups varies with beach type: on protected beaches, worms tend to predominate, whereas crustaceans are dominant on exposed beaches.

Sandy beach animals can be extremely abundant, with thousands of individuals per metre of shore. The diversity, abundance and biomass of beach animals are related to the type of beach. Dissipative beaches are home to the greatest number of species and the highest numbers and biomass of animals. At the opposite end of the physical spectrum, few species and low numbers of animals inhabit reflective beaches.

The sandy beach animals occur in a wide range of sizes. The largest beach animals in southern New Zealand are clams: the toheroa *Paphies ventricosa* and tuatua *P. subtriangulata*. These bivalve molluscs burrow in the sand using a muscular foot that extends from the shell. The smallest animals on sandy beaches are very tiny organisms (less than 1 mm), such as nematode worms and copepod crustaceans that live between sand grains or attach to the surface of sand grains.

The beaches of southern New Zealand harbour a remarkably rich mollusc fauna, with up to eight endemic species of bivalves inhabiting the intertidal to surf zone (Fig. 22). The polychaete fauna also contains endemic species, such as the bloodworm *Euzonus otagoensis*, as does the crustacean fauna, notably the upper beach sandhopper *Talorchestia quoyana* (Fig. 23). The upper beach is also home to some fascinating insects, including the flightless beetle *Chaerodes trachyscelides* (Fig. 23). Populations of this beetle take on the colour of the local beach sand. The sandy shores of southern New Zealand are not well studied and it is likely that numerous invertebrate species remain to be discovered.

Few vertebrate animals are permanent residents of sandy beaches. A few types of fishes live in the surf zone most of their lives. A number of vertebrate species use sandy beaches as resting grounds. Seabirds such as gulls commonly roost on sandy beaches, whilst yellow-eyed penguins traverse sandy beaches to onshore roosts and nests. Hooker's sea lions haul out on sandy beaches.

Fig. 22. The shells of many species of clams and other molluscs are stranded on wave-exposed beaches in southern New Zealand. The clams shown here all inhabit the intertidal or subtidal zones of exposed sandy beaches. Clockwise from top: toheroa *Paphies ventricosa*, frilled venus shell *Bassina yatei*, ringed dosinia *Dosinia anus*, tuatua *P. subtriangulata*, trough shell *Mactra discors*, tuatua *P. donacina*, and triangular trough shell *Spisula aequilatera*.
J. Nordstrand

Beach food web

Very little primary production occurs on surf-swept sandy beaches. Beach food webs depend mainly on production from outside sources, notably ocean plankton composed of microscopic algae and animals, and stranded kelp and animals (Figs 24, 25 and 26).

On sheltered soft shores, freshwater runoff and tides deliver oceanic plankton and terrestrial and marine detritus. In addition, a significant benthic microbiota can flourish in the form of diatoms and mats of cyanobacteria (blue-green bacteria that can photosynthesise).

A variety of feeding types are found among the sandy beach macrofauna. Primary consumers include suspension-feeders such as the clams, as well as deposit-feeders such as various polychaete worms and the ghost shrimp, and herbivores, such as sandhoppers, beetles and fly larvae, feeding on algal wrack. Scavenging species including isopods, amphipods, crabs and polychaetes are often encountered. Predators occur in a greater variety of forms on sandy beaches. Birds are the largest, most visible predators on sandy beaches, many foraging during lower tide in the swash zone. Southern black-backed gulls, red-billed gulls, South Island pied oystercatchers, variable oystercatchers and a variety of waders are regularly seen on beaches of southern New Zealand. Notable, but less commonly observed, are Stewart Island brown kiwi, which actively feed on sandhoppers around mounds of kelp wrack and in buried kelp on open coast beaches at night (page 185). Numerous species of fishes move onto the beach from deeper waters to feed intertidally during high tides. Paddle crabs, *Ovalipes catharus*, feed on intertidal bivalves and polychaetes during high tides (Fig. 27).

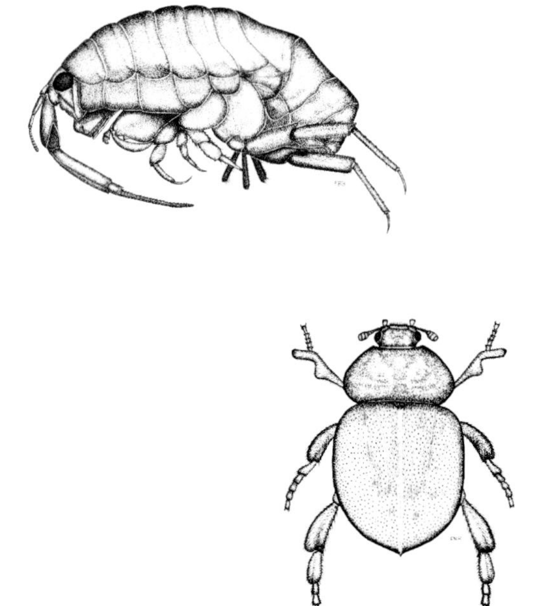

Fig. 23. The bloodworm *Euzonus otagoensis* (opposite), the sandhopper *Talorchestia quoyana* (top) and the beetle *Chaerodes trachyscelides*, are all usually abundant on open ocean beaches, especially those with deposits of drift algae. *Illustrations by P.B. Batson*

Fig. 24. Beach food webs often depend heavily on production from outside sources. Here at Karitane, large quantities of kelp (mainly bladder kelp, *Macrocystis pyrifera*) have been washed up. *P.K. Probert*

Fig. 25 below. Juvenile of the squat lobster *Munida gregaria* ('red krill'). *P.B. Batson*

Fig. 26 right. In summer, red krill often strand in huge numbers on the shore, as here at Doctors Point, north of Dunedin, and provide an important contribution to intertidal food webs. *W. Harrex*

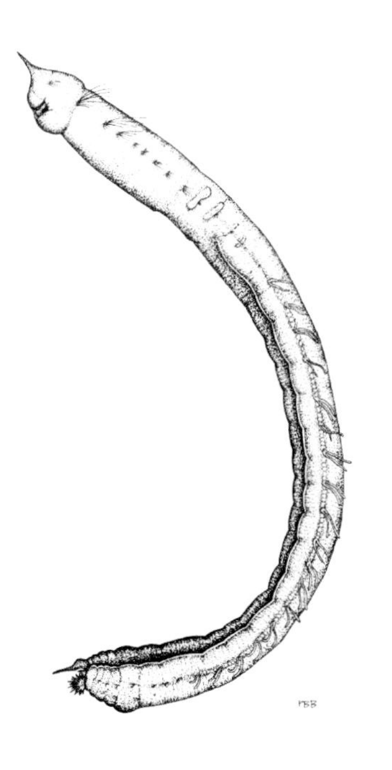

Fig. 27. The paddle crab *Ovalipes catharus* is an important predator of other invertebrates on open sandy beaches. It buries itself into the sand, but also uses its paddle-like pair of hind legs for swimming. *B.S. Batson*

Zonation of beach biota

Beaches provide a wide variety of habitats, including the dry upper beach, the swash zone or mid beach and the low intertidal beach. Because of the ever-changing shape of the beach, the biological zonation of a sandy beach is very dynamic. The mobile fauna adjust rapidly to changes in beach profile and biological zones expand and contract quickly.

Three intertidal zones are generally recognised on surf beaches. On the upper beach, where the sand is usually dry or only slightly damp, we find the upper extent of storm debris and the greatest accumulations of stranded kelp. Most of the upper beach is above the mean high tide line. Semi-terrestrial animals such as sandhoppers, kelp flies, beetles and spiders are common here (Box 3). In the mid-beach and upper swash zone where the sand is damp to saturated, a variety of deposit-feeders and scavengers are found. In the lower swash zone and shallow subtidal, the sand is almost always saturated with water. Here, the beach harbours many clams including tuatua and toheroa. The subtidal end of this zone is always submerged and contains the low intertidal clam species, predatory fishes and paddle crabs.

3 INVERTEBRATES OF STRANDED SEAWEED

Stranded seaweeds on sandy beaches are a particularly rich habitat for invertebrates, with sandhoppers most conspicuous. These laterally flattened crustaceans burrow in the sand below dead wood or algae, feeding on the dead and dying material.

Meanwhile, enormous numbers of kelp flies (*Chaetocoelopa littoralis* and *Maorimyia bipunctata*) breed as larvae within the decaying plant material, ensuring that it does not persist for long.

Scavenging rove beetles are the most diverse group of coastal beetles, with up to five species found on a single pile of dead seaweed. These distinctive beetles are characterised by short elytra exposing their abdomen and their habit of running fast. One of the larger species (up to 2 cm long) is the devil's coach-horse *Creophilus oculatus*, a predator that readily devours kelp fly larvae.

Brian Patrick

Human use

Beaches are the primary areas for coastal recreation. Visitors flock to beaches for walking, bathing and fishing. Numerous edible clams occur in the intertidal to shallow subtidal zone of beaches inlets and harbours. On ocean beaches, three species of clams, including the highly prized and protected toheroa *Paphies ventricosa* and two species of tuatua *P. donacina* and *P. subtriangulata* have been the subject of commercial and recreational harvests (Fig. 53). On more protected beaches the pipi *Paphies australe* and the cockle *Austrovenus stutchburyi* are popular with fishers (Fig. 38).

Fig. 28. An exposed coast of eroding Caversham Sandstone near Tunnel Beach, just south of Dunedin, capped with a thin layer of soil and a compact cover of salt-tolerant turf herbs.
P.N. Johnson

Fig. 29. Vegetation patterns, caused by degrees of exposure to coastal wind and salt, along the Fiordland coast at Resolution Island. Coastal grasslands and sedgelands appear yellowish on the most exposed faces and headlands. Bright green scrub of *Hebe elliptica* and *Olearia oporina* indicates slightly more sheltered sites. Wind-shorn forest of mountain beech is confined to the upper slopes and hill crests.
P.N. Johnson

COASTAL VEGETATION AND INVERTEBRATES

Peter Johnson, Susan Walker and Brian Patrick

Coastal environment

On the sunniest of summer days we can appreciate the warmth that coastal habitats provide. Even with just a gentle onshore breeze, the surf creates a hazy atmosphere, evidence that the ocean's saltiness is being carried landward to help determine the special nature of coastal habitats and vegetation. On those other days when storm waves are crashing against headlands, or when stinging sand is being whipped through sand dunes by the wind, most of us would choose to be somewhere else. But with no such choice, coastal plants have developed ways to cope with being forever salted by the sea or bashed about by the wind (Fig. 28). Temperature, salt, and wind help determine the particular patterns of plant distribution (Fig. 29) and, in turn, their associated fauna. Southern New Zealand boasts some distinctive and varied examples of coastal vegetation.

Although proximity of the ocean makes for a mildness of climate, it nevertheless gets cooler as you go south and many of southern New Zealand's coastal plants respond to this gradient. Warmth-loving plants become less common and more strictly confined to the coast, until they reach their southern limit of distribution. On the Otago coast, sea rush *Juncus kraussii* subsp. *australiensis* reaches its southern limit at Purakanui Inlet, the sedge *Bolboschoenus caldwellii* at Tomahawk Lagoon, and the coastal tree, ngaio *Myoporum laetum* probably a little south of Dunedin. On the western side, we see a sequence of southern limits being reached as we move down the Fiordland coast, including puka *Griselinia lucida* as far south as about Milford Sound, the giant umbrella sedge *Cyperus ustulatus* to Poison Bay, the climbing rata *Metrosideros perforata* to Catseye Bay, kiekie *Freycinetia baueriana* subsp. *banksii* to Dusky Sound, scarlet rata vine *Metrosideros fulgens* to West Cape, and hutu *Ascarina lucida* to Preservation Inlet. Conversely, there are strictly southern plants that peter out as they approach their northern limits of distribution, such as the large-leaved southern nettle *Urtica australis* found as far north as Resolution Island, and the Stewart Island coastal forget-me-not *Myosotis rakiura* recorded as far north as Dunedin.

Salts from seawater can affect coastal plants in many ways. Saltmarsh plants are regularly dunked in the tide and many have succulent, water-storing stems and leaves. Some, especially glasswort *Sarcocornia quinqueflora*, can even tolerate hypersaline conditions and grow in parts of an estuary where evaporation further concentrates the salt.

One of the main ways in which salt affects coastal plants is through scorching and killing the foliage. During windy weather, droplets of sea spray become small salt crystals that can be transported far inland and driven into the surfaces of leaves, especially if the wind is strong and there is no rain. Coastal plants often have stout leaves with a thick, glossy, protective cuticle. And if they do succumb to defoliation, many coastal shrubs readily resprout. It is this combination of wind and salt that produces the often contorted forms of coastal trees and shrubs, and which accounts for the typical zonation patterns, with the height of the plants gradually increasing inland, each zone of vegetation offering some protection to those beyond.

On the region's most southern coasts, saltiness encourages

the formation of peaty soils. Cool temperatures, cloudy skies, and frequent rainfall assist this process, so that litter and other organic matter do not completely decompose but accumulate as peat, just as they would in a bog or swamp. On Stewart Island and its smaller islands, and on the low tablelands of southwest Fiordland, much of the coastal vegetation sits on black peat.

Salts can also add valuable nutrients to coastal soils, as can stranded seaweed and other flotsam, while sea birds and seals, often ashore in large numbers, can make some coastal soils more fertile than the average vegetable garden.

Landforms

The great diversity of landforms and substrates that make up the coastline of southern New Zealand provide for a wide range of coastal vegetation types. The coast from Oamaru to Nugget Point is a series of cliffs, headlands and sandy beaches, carved mainly from volcanic rocks. Sand spits mark the entrances of estuaries, lagoons and river mouths, and dunes and sand plains may extend some distance inland. Farther south, the many cliffs and headlands of the Catlins coast are derived from mudstone and other sedimentary rocks of the Southland Syncline, and here the climate is sufficiently moist for the sandy beaches and estuaries to be fringed with native forest, where this has not been replaced by pasture. Westward from Waipapa to Tiwai Point, a narrow strip of gravel beach backs onto the extensive peatland of cushion bog, fern, sedges and low shrubs which has developed since the last glaciation upon the edge of the Southland Plains. Under the moist, windy and cloudy conditions of the southern coast, both the peatland and the coastal communities contain plants and insects found elsewhere only at higher altitudes. Bluff Hill shares a granite base rock with Stewart Island and its surrounding isles, and has similar coastal moor and scrub vegetation. Much of Stewart Island's shoreline comprises low headlands and sheltered inlets and bays, but wide sandy beaches such as at Mason Bay have well-developed dune systems.

West from Invercargill, a series of sweeping curved beaches – the sandy Oreti Beach and Colac Bay, and the gravelly Te Waewae Bay – are separated by low headlands and backed by pasture. Along the south Fiordland coast, gently sloping platforms of mudstones and other sedimentary rocks capped with outwash gravels meet the coast in low cliffs, and coastal turf communities run into the muttonbird scrub and forest behind (Fig. 30 and Box 4). North of windy Puysegur Point, low islands and tablelands rise from the coast as far as Resolution Island, and then hard igneous rocks form a steep outer coast, dotted with remote small beaches, often driftwood-strewn, and punctuated by fiords. Along the precipitous fiord walls, land vegetation can descend virtually to high-tide level. Because of the very high rainfall, a layer of almost fresh water can overlie ocean water in the fiords, so that the bay-head marshes have fewer salt-tolerant plants than would usually be found in saltmarshes. North of Milford Sound and up to Jacksons Bay, scattered sandy beaches occur along a coast that is mainly a gravel and boulder fringe against the base of forested hills.

Fig. 30. Much of the south coast of Fiordland (this is near the Angus Burn) has a low mudstone cliff capped with outwash gravels. Shrubs of shore koromiko *Hebe elliptica* perch near the clifftop, where coastal turf vegetation runs back into wind-shorn muttonbird scrub *Brachyglottis rotundifolia* then storm-battered forest of kamahi and southern rata. *P.N. Johnson*

4 SOUTHERN COASTAL TREE DAISIES

Coastal scrub and low forest of southern New Zealand coasts is often dominated by several species of tree daisy of the genera *Olearia* and *Brachyglottis* (the latter formerly included in *Senecio*). They are much-branched small trees, with compact crowns, having tough leaves that are glossy green above and woolly white below, resistant to wind, and able to re-sprout or regenerate readily from seed after disturbance.

Peter Johnson

Top: *Olearia lyallii* covers many of the titi or muttonbird islands south-west of Stewart Island. The peaty soil is densely burrowed by seabirds, especially muttonbirds (sooty shearwaters), undermining the short-lived olearia trees which topple in the wind, regularly creating new light-gaps which are recolonised by olearia seedlings and by coastal grasses, in this case *Poa tennantiana*. *P.N. Johnson*

Bottom: The yellow-flowered *Brachyglottis stewartiae* is one of the most attractive coastal tree daisies, but is restricted to just four southern islands, including Solander Island where it is seen (centre) among small trees of *B. rotundifolia*. This latter species – muttonbird scrub – is readily seen on Stewart Island shores, its oval leaves once serving as southern postcards as the white undersides provided a writing surface. *P.N. Johnson*

Beach vegetation

Where sand is cast ashore onto beaches and blown inland to form dunes, successional sequences of vegetation extend back through different habitats from the strandline, across foredunes, through dry or moist hollows and slacks, over crests and across sand flats. On most Otago beaches, the vigorous exotic marram grass *Ammophila arenaria* now dominates on foredunes, where the golden native pingao *Desmoschoenus spiralis* was once the principal sand binder. Behind the marram grass, a wide zone dominated by tree lupin *Lupinus arboreus* may contain scattered natives such as flax and poroporo *Solanum laciniatum*, together with exotic grasses and herbs, sometimes patches of boxthorn *Lycium ferocissimum* and often gorse *Ulex europaeus*. Taller shrubs of elder *Sambucus nigra* replace lupin on older dunes, and are often smothered by the native creeper *Muehlenbeckia australis*. Given freedom from disturbance, the sequence proceeds to low forest variously dominated by ngaio, mahoe *Melicytus ramiflorus*, kanuka *Kunzea ericoides* or totara *Podocarpus totara*. Nowadays, pasture or pine plantations tend to have replaced native forest on the inland parts of sand country.

Sandy flats of low relief may contain moist hollows or seasonally wet slacks, lined with turf herbs and fringed with rush-land plants such as *Juncus* spp., oioi *Apodasmia similis* and knobby clubrush *Isolepis nodosus* (Box 5). Dry sand plains may be dotted with rosettes or cushions of sand geranium *Geranium sessiliflorum* var. *arenarium*, dwarf forget-me-not *Myosotis pygmaea* var. *drucei*, and the southern coastal pimelea *Pimelea lyallii*.

Beaches of gravel, pebbles, cobbles, stones or boulders occur where alluvial plains of glacial outwash gravels meet the sea, near river mouths, or are dotted along shores of solid rock. An extensive parallel series of old gravel beach ridges occurs on the Tiwai Peninsula. The younger surfaces are partly covered in mat-forming small-leaved shrubs, such as sand coprosma *Coprosma acerosa*, creeping pohuehue *Muehlenbeckia axillaris* and shore gentian *Gentiana saxosa*. Farther inland there are sedge/grasslands of silver tussock *Poa cita*, knobby clubrush, and tufts of *Libertia peregrinans*, then taller grassland of copper tussock *Chionochloa rubra cuprea* with flax *Phormium tenax* and shrubland of *Coprosma* species.

In south-east Otago, Fiordland and on Stewart Island there are examples of beaches still largely unmodified by exotic species, and where forest occupies the older sand surfaces (Box 6). Foredunes typically have dense pingao, grading back to dune crests with a mixture of pingao, sand fescue *Austrofestuca littoralis*, toetoe *Cortaderia richardii*, and sand coprosma. Zones of flax, bracken fern *Pteridium esculentum*, coprosma or olearia scrub, and of low forest occupy older sand surfaces with progressively better developed soils. The oldest dunes, still recognisable on the ground by their hummocky topography, hold tall forest, usually of southern rata *Metrosideros umbellata*, kamahi *Weinmannia racemosa* and rimu *Dacrydium cupressinum*, with dense tree ferns *Cyathea smithii* and *Dicksonia squarrosa*, lianes (especially supple-jack, *Ripogonum scandens*) and the crown fern *Blechnum discolor*.

5 A MARRAM DUNE, WITH MOIST HOLLOWS

Most of the coastal sand dunes of Otago and Southland have been substantially modified by the invasion of marram grass. But some pockets of mainly native communities can still be found. *Peter Johnson*

Left: The coastal sand dunes at Surat Bay, near the Catlins River mouth, have been substantially modified by marram grass. However, moist hollows within the dunes have patches of the orange-brown oioi *Apodasmia similis*, a rush-like plant otherwise found fringing tidal estuaries and some inland lake shores, silver tussock *Poa cita* and flax *Phormium tenax*. *P.N. Johnson*

Right: The intervening turf vegetation is home to numerous plant species, including the recently described *Mazus arenarius*, a rare plant known from only a few coastal sites in Otago and Southland, including Stewart Island. *P.N. Johnson*

6 SAND BINDERS OLD AND NEW

Partly because sand is inherently mobile, sand dunes are readily disturbed by natural processes, and also vulnerable to human influences and invasion by weeds. Most New Zealand dune systems are now highly modified, but some of the most intact remaining examples are in the south, especially in Fiordland and Stewart Island. In these places, native plants still predominate, and there are intact sequences of vegetation from foredune sand-binding sedges and grasses, back to moist dune hollows, various scrub types and then to forest on the oldest, inland dunes.

Peter Johnson

Left: Coal River in Fiordland has a beach of black, mainly hornblende sand. It is one of the few remaining dune systems which retains a predominant cover of the orange, native, sand-binding sedge pingao *Desmoschoenus spiralis*. On progressively older dunes, pingao gives way to sedge/grassland of *Carex flagellifera* and toetoe *Cortaderia richardii*, *Olearia aviceniifolia* scrub, then forest of southern rata, kamahi, silver beech and rimu.

Right: Even remote localities are susceptible to invasion by the naturalised marram grass *Ammophila arenaria*, dispersing on ocean currents as seed or rhizome fragments from other New Zealand coasts. Here at Coal River, marram grass is evident as the pale green patches, initially colonising the head of the beach, then spreading inland where, if not controlled, it would replace most of the pingao. Compared with dunes of relatively gentle profile formed under pingao, those created by marram grass, which is more effective at trapping moving sand, are steeper, and therefore more prone to subsequent erosion by wind and wave action. *Both photos P.N. Johnson*

7 THREATENED AND LOCAL PLANTS OF SOUTHERN COASTS

- **Endangered:** *Lepidium oleraceum* (Brassicaceae), Cook's scurvy grass, fleshy cress of fertile coastal slopes; *Puccinellia raroflorens* (Poaceae), small salt-marsh grass of a Stewart Island tidal bank, otherwise on inland saline soils.
- **Vulnerable:** *Lepidium tenuicaule* (Brassicaceae), small rosette cress of coastal turf; *Mazus arenarius* (Scrophulariaceae), creeping herb of sandy coastal turfs; *Ranunculus recens* (Ranunculaceae), compact rosette buttercup of dune plans and headland turf.
- **Declining:** *Atriplex billardierei* (Chenopodiaceae), mealy-leaved herb of sandy beach heads; *Austrofestuca littoralis* (Poaceae), tussock grass of sand dunes; *Euphorbia glauca* (Euphorbiaceae), native sand spurge of dunes and coastal banks; *Libertia peregrinans* (Iridaceae), herb with stiff leaf tufts forming patches in coastal swards; *Sonchus kirkii* (Asteraceae), puha, fleshy rosette herb of coastal cliffs and rubble slopes.
- **Recovering, Conservation Dependent:** *Desmoschoenus spiralis* (Cyperaceae), pingao, coarse-leaved golden sedge of sand dunes; *Gunnera hamiltonii* (Gunneraceae), ground-hugging sand herb of Southland/Stewart Island; *Stilbocarpa lyallii* (Araliaceae), punui, megaherb of southern islands.
- **Naturally Uncommon, Sparse:** *Drosera pygmaea* (Droseraceae), tiny sundew of damp coastal peat; *Lepilaena bilocularis* (Zannichelliaceae), delicate submerged herb of coastal lakes and lagoons.
- **Naturally Uncommon, Range Restricted:** *Anistome acutifolia* (Apiaceae), robust coastal herb, restricted to Snares Islands; *Helichrysum intermedium* 'var. *tumidum*' (Asteraceae), whipcord daisy shrub of Otago Peninsula cliffs; *Stilbocarpa robusta* (Araliaceae), megaherb of the Snares and Solander Islands.
- **Insufficiently Known:** *Myosotis pygmaea* var. *pygmaea* (Boraginaceae), tiny forget-me-not of coastal headland turf.
- **Taxonomically Indeterminate, Sparse:** *Coriaria* 'Rimutaka' (Coriariaceae), undescribed tutu shrub, recorded from Stewart Island; *Oreomyrrhis* 'minutiflora' (Apiaceae), tiny rosette herb of coastal turf.

Peter Johnson

The megaherb punui *Stilbocarpa lyallii* was once common in coastal vegetation on southern islands. Here it is growing with the fern *Blechnum durum* as an understorey in tree daisy scrub on Solander Island. *P.N. Johnson*

A number of plant species of dune systems and other coastal habitats of southern New Zealand have particular importance to conservation because they are threatened through loss of habitat or have very localised distributions (Box 7).

Dune invertebrates

Invertebrates of dunes also have to cope with the harsh conditions. Species requiring high humidity live concealed, mostly underground or under driftwood or in leaf litter. The open habitat of dunes is, however, ideal for diurnally active adults of species such as wolf spiders, ants, weevils, flies and moths, but many other species are active only by night, gaining additional protection from predators.

Important habitats for invertebrates within the coastal sand system are the fore-dunes and back-dunes, driftwood, stranded dead seaweeds and turf vegetation in dune-slacks or at the mouths of streams. Although much modified in terms of topography and plant cover, the sand dunes of the south have a very distinctive invertebrate fauna compared to other areas of New Zealand and it is a predominantly native fauna. Fortunately, invertebrate species have survived on heavily modified dunes because of host switching, as for example with the larvae of the noctuid moths *Tmetolophota phaula* and *Agrotis innominata*, which now feed on the introduced marram grass, or because of persistence of their special needs, such as driftwood for the larvae of the sand scarab *Pericoptus truncatus*. But more importantly, a number of key dune sites have survived almost intact with a diverse array of invertebrates, such as at Mason Bay on Stewart Island, Chrystalls Beach, Cannibal Bay, Tautuku Beach, Fortrose Spit, Tiwai Peninsula, Three Sisters and numerous dunes around the coast of Fiordland. These dune systems have a distinctive fauna that consists of, for example, a flightless chafer beetle *Prodontria praelatella* (Box 8), many longhorn beetles, several darkling beetles including the larger *Mimopeus elongatus* and the weevil-like *Cecyropa modesta*. Among the 120 moth species known from southern coasts are the endemics *Kiwaia glaucoterma* with flightless jumping females, an undescribed *Pimelea*-feeding noctuid of the genus *Meterana*, and four geometrid moths

Fig. 31. The moth *Orocrambus xanthogrammus* (perched on *Raoulia*), a conspicuous species of southern coasts. *B. Patrick*

8 CONSERVATION OF COASTAL INSECTS

Conservation of native insects often translates into conservation of native plants. Native daphnes, small shrubs of the genus *Pimelea*, have declined markedly nationwide, particularly coastal species. Of the five *Pimelea* species known from the southern coast, *P. arenaria* is locally extinct together with its dependent fauna. *P. lyallii* is endemic to the south and supports a similarly rich endemic fauna. Sand dunes on both sides of Foveaux Strait are home to a flightless chafer beetle *Prodontria praelatella*. The 14-mm long adults emerge in spring and feed by night on *Pimelea lyallii*. Its larvae feed on roots of various shrubs and grasses. The beetle and its habitat are of conservation importance as habitat modification has fragmented the species' former distribution.

Three undescribed species of geometrid moth of the genus *Notoreas* are found on the coast of southern New Zealand. Typically for this mainly upland genus, the adults are extremely active by day in small localised areas where the larval foodplant *Pimelea* is found. The largest of the three (wingspan 24 mm) is found in only a very small area of steep bluffs south of the Shag River Mouth, the purplish-pink larvae feeding on the developing buds of *Pimelea* cf. *urvilleana*. Dunes of *P. lyallii* and open areas of *P. prostrata* on both sides of Foveaux Strait support the other two species. These smaller species are known from just four sites and, like the other coastal *Notoreas* species, are vulnerable to extinction.

Brian Patrick

The flightless chafer beetle *Prodontria praelatella* pictured at Fortrose Spit. B. Patrick

An undescribed species of moth of the genus *Notoreas* found on steep cliffs south of the Shag River mouth. B. Patrick

9 THE RED KATIPO SPIDER

The only poisonous spider native to New Zealand is the red katipo *Latrodectus katipo*. Confined to coastal regions, this spider has been found as far south as Doctors Point, just north of Dunedin. Its webs are commonly found at the base of marram grass clumps, but it also seeks warmth and dryness under driftwood, corrugated iron, cardboard boxes and even inside discarded cans. Built from both dry and sticky silk, the tent-like webs become coated with sand, so are seldom seen from above. The spider lives within this sheltered little sand cave, but its catching threads stretch out beyond. Insects, such as sand scarabs, brush against these silk trap lines, alerting the spider, which rushes to the spot. After touching the beetle to determine its edibility, the spider squirts sticky silk all over it, then bites its prey a number of times, wraps it with more silk and attaches it to its spinnerets. Having secured its meal, the spider hauls the dangling bundle back to its lair. Few people, however, have been bitten because of the spider's restricted habitat and retiring nature.

R.R. and L.M. Forster

The red katipo spider. R.R. Forster

(*Asaphodes stephanitis, A. frivola* and two undescribed species of *Notoreas*). Important host plants include sand daphne *Pimelea lyallii*, pingao, *Raoulia* cushions and sand coprosma, with the moths *Scoparia tetracycla, Orocrambus xanthogrammus* (Fig. 31) and the tiger moth *Metacrias strategica* also conspicuous and typical. A fierce predator of coastal invertebrates, the giant shore earwig *Anisolabis littorea* is a widespread dune or rocky shore species fairly common in North Otago, but with a patchy distribution south of Dunedin. New Zealand's only native poisonous spider, the red katipo *Latrodectus katipo*, has a stronghold in marram-dominated dunes near Karitane (Box 9).

Estuaries

In a typical southern estuary, the subtidal and lowermost intertidal zone has a sward of eelgrass *Zostera capricorni*, its grassy strap-like leaves lying relaxed on the wet substratum when exposed at low tide. Upslope, this grades to saltmarsh, with the very salt-tolerant glasswort in its many hues of red, purple or green, seablite *Suaeda novae-zelandiae* and sea primrose *Samolus repens*. An upper marsh, where tidal inundation is of lesser duration, has a turf where remuremu *Selliera radicans* is abundant, often with one or other species of *Leptinella*. The resemblance of this vegetation to that of a bowling green is more than superficial because many greens are made of *L. dioica* from coastal marsh, as well as *L. maniototo*, from inland moist turfy sites.

Tall sedges, rushes or restiads, often in pure stands, grow where their bases are submerged by the highest tides (Box 10). In Otago and Southland it is oioi which usually plays this role. As the oioi grades into flax, swamp or scrub communities it is joined by shrubs of saltmarsh ribbonwood *Plagianthus divaricatus*, the wiry stems of which are usually clothed with lichens, especially yellow *Teloschistes* and greyish *Ramalina* species.

Open, drier zones of salt meadow characteristically support a wide variety of diurnal fly, wasp, bug and moth species. Larvae of most of these rely on native plants for food but, in turn, are consumed by birds and a wide range of spider species. Insects inhabiting salt meadows may have to be adapted to tolerate inundation by sea water, but most breed in places that are reached only by spring tides. For example, glasswort is chewed by the larva of the moth *Ectopatria aspera* and *Selliera radicans* leaves are mined by the moth *Eutorna symmorpha*. Taller sedges and shrubs that form the backdrop to the salt meadow habitat are richer in dependent insect species and include a tiny undescribed moth of the genus *Megacraspedus* that feeds within seedheads of oioi, and larvae of *Pseudocoremia lactiflua* feeding on foliage of saltmarsh ribbonwood. Characteristic of saltmarsh and coastal turf areas and endemic to southern New Zealand are the diurnal moths *Merophyas paraloxa* and *Protithona potamias*. Both are bright orange and sunbathe on prostrate herbs such as *Selliera radicans*.

Steep coastal places

Coastal cliffs provide a multitude of plant habitats; their many ledges, crevices, platforms, wind-eroded concavities, seepages and talus slopes all vary in exposure to salt spray, sun and wind. On the east coast, the native ice plant *Disphyma australe* and, near habitations, the naturalised ice plants (*Carpobrotus* species) root on ledges and hang in festoons. Salt-drenched platforms and crevices are home to small tufted grasses of the genus *Puccinellia* (both native and introduced), as well as to herbs such as *Colobanthus muelleri*, native celery *Apium prostratum* and sea spurrey *Spergularia marginata*. Species of *Isolepis* and *Crassula*, a small succulent, grow in moist seeps at the foot of cliffs or on rock platforms, and the scrambling plants of steep places include the chenopod *Einadia allanii*, climbing aniseed *Scandia geniculata* and native spinach *Tetragonia trigyna*.

Grass and sedge communities occupy coastal habitats which

10 ESTUARY VEGETATION

Around the edges of estuaries, zones of vegetation occur at different levels of tidal influence. *Peter Johnson*

Below: This view of the New River Estuary, at Bushy Point, Invercargill, shows a sequence from rushland (mainly oioi, *Apodasmia similis*, brownish, at left), to scrub of manuka *Leptospermum scoparium* of gradually increasing canopy height, to low forest with cabbage trees *Cordyline australis*, then taller forest with kahikatea *Dacrycarpus dacrydioides*, rimu *Dacrydium cupressinum* and pokaka *Elaeocarpus hookerianus*. Foreground patches of oioi, each one gradually expanding at the margins, illustrate a phase of vigorous growth, as silty sediment is trapped by the rushland. *P.N. Johnson*

Below: In close-up view we see vegetation patches on a smaller scale: oioi rushland at left and beyond; saltmarsh turf of remuremu *Selliera radicans* at front left and sea primrose *Samolus repens* front right; and, in the centre, the introduced cord grass *Spartina anglica*, which has become a problem weed in several southern estuaries. *P.N. Johnson*

Fig. 32. Most species of *Celmisia* are mountain daisies, but on the south-east Otago coastal cliffs, *Celmisia lindsayi* is a distinctive and common local endemic. Here it grows at False Islet with the coastal tussock *Poa astonii*. On the cliffs beyond are green patches of native ice plant *Disphyma australe* and a white coating of mainly *Pertusaria* lichens on the uppermost rocks. *P.N. Johnson*

Fig. 33. A sedge endemic to Fiordland, *Carex pleiostachys* grows close to the coast where it tolerates being lain on by fur seals. This is at Yates Point, just north of Milford Sound. *P.N. Johnson*

are too blasted by wind and salt or too thin-soiled to support woody species. *Poa foliosa*, a tall, lush grass of the Auckland and Antipodes Islands, reaches its northern limit on the small islands about Stewart Island, as dense swards especially on steep peaty talus slopes. The fine-leaved tussock grass *Poa astonii* perches on cliffs and rocky sites (Fig. 32), and is also a saltmarsh plant in Fiordland. In Otago the larger silver tussock covers high headlands, together with several sedges (*Carex* species). Other species of *Carex* occupy specific coastal habitats in the south: *C. trifida* on moist coastal banks, *C. pleiostachys* on coastal platforms in Fiordland (Fig. 33), and *C. fretalis* on the Foveaux Strait and Stewart Island coasts.

Faces, platforms, and stacks of coastal rock can be such harsh environments that neither seaweeds nor typical land plants can grow on them. Instead, lichens tend to cover much of the exposed rocky zone where land meets sea. Lichens are slow-growing and can cope with being repeatedly dried and wetted. Those with a crustose form – virtually glued to the rock – are well able to resist being abraded or washed away. Like many coastal organisms, lichen communities array themselves in zones away from the sea, and particular lichens are characteristic of shady nooks, seepages, headlands enriched by bird droppings and so on (Box 11).

Exposed headlands in the south are home to many invertebrates living among the tight low turf of plants such as plantains. On warm calm days many diurnal species are active, sunbathing on bare ground among the turf. The small bug *Nysius huttoni*, the geometrid moth *Arctesthes catapyrrha*, the tiger beetle *Neocicindela tuberculata* and several small weevil species are characteristic and widespread in the south. A different fauna lives in the long grass adjacent to these often localised turfs.

Coastal rock faces are important habitats for invertebrates, especially beetles and moths. The invertebrates feed on the diverse lichen and moss flora as well as on higher plants. Some moth species are confined to lichen-covered rocks, including the green and black larvae of *Dichromodes gypsotis* on Otago Peninsula, where it is joined by several species whose larvae feed on detritus in cracks and ledges such as the southern endemic *Tinea furcillata*. But such species are better represented from Nugget Point south.

Most numerous on coastal rocks are the algal-feeding larvae of four species of bag moth (Psychidae), two of which (*Scoriodyta patricki* and *Reductoderces* new species) are endemic to southern New Zealand. Characteristically, each species has a distinctly shaped larval case that is affixed to the substratum once larval feeding has ceased. Males are small fragile moths that usually fly on calm frosty mornings in early spring while the wingless females cling to the case and, after mating, lay their eggs back in the safety of the larval case. One species (*Rhathamictis* new species) has a short-winged female that leaves the case to lay her eggs elsewhere on the rock face.

Rock faces in the Catlins have a rich vascular plant flora that supports a moth fauna closely related to alpine areas. The daisy *Celmisia lindsayi* (Fig. 32) supports two leaf-mining moths (*Stigmella oriastra* and *Apatetris* new species) and a seed-feeding species *Chloroclystis nereis* in addition to a southern New Zealand endemic leaf-roller *Planotortrix puffini*. Underlining the upland affinity of the flora and fauna is the presence of the *Astelia*-feeding zig-zag moth *Charixena iridoxa*, usually a species of montane and alpine zones. The larvae feed over several years in the base of the plant, causing a distinctive zig-zag pattern on the elongating leaves. The strikingly patterned and gleaming adults are day-flying but rarely seen.

The exposed western shore of Wairakei Island, where rocky shore grading into scattered grassland has an almost complete cover of mainly 'snowpake lichen', *Pertusaria graphica*. *P.N. Johnson*

11 COASTAL LICHENS

These two pictures near the outer coast of Breaksea Sound, Fiordland, show some of the lichen cover that is typical of many southern coasts.

In close-up, some of the common upper zone lichens are rosettes of *Xanthoria ligulata* (bright yellow), *Rinodina thiomela* (pale yellow) and *Tephromela atra* (white), upon a matrix of smaller, darker crustose lichens. Alongside are the coastal spleenwort fern *Asplenium obtusatum* and the coastal tussock grass *Poa astonii*. *P.N. Johnson*

Peter Johnson

INLETS AND ESTUARIES

Keith Probert

Southern New Zealand is well endowed with large sheltered inlets. Most of these are fiords: deep, rock-walled estuaries with their own special characteristics and biota, as described later in this chapter. Less well represented are the more typical shallow-water inlets and estuaries. Largest and best known are Paterson Inlet (100 km^2), Otago Harbour (46 km^2), New River estuary (43 km^2) at Invercargill, Bluff Harbour (34 km^2) and Awarua Bay (21 km^2). Others of significance include Blueskin Bay Inlet, the Catlins estuaries including the Catlins Lake, Waikawa Harbour, Toetoes Harbour at Fortrose fed by the Mataura and Titiroa Rivers, and the Jacobs River estuary at Riverton. The Waituna Lagoon (between Toetoes Harbour and Awarua Bay) is impounded by a gravel bar, opening to the sea only periodically.

Fig. 34 (above). Sheltered conditions in estuaries and inlets enable plants to gain a foothold and snails to roam on the sediment surface, as here in the Tahakopa Estuary, the Catlins. *P.K. Probert*

Fig. 35 (bottom left). The intertidal flats of estuaries and inlets are often highly productive. Much of the animal life is beneath the surface, but evident from the numerous burrows and other signs of sediment reworking. Conspicuous here are the burrows and casts of the lugworm *Abarenicola affinis*. *P.K. Probert*

Fig. 36 (below). The mud snail *Amphibola crenata* is often abundant on the upper shore of estuaries and inlets. It is a pulmonate snail, so has a lung-like cavity in the mantle instead of gills as in typical marine snails. *P.K. Probert (and Fig. 37)*

Fig. 37 (opposite left). The stalk-eyed mud crab *Macrophthalmus hirtipes*, often the most conspicuous crustacean of sheltered flats.

Fig. 38 (opposite right). The cockle *Austrovenus stutchburyi*, often the dominant shellfish of sheltered sandflats. *P.K. Probert*

Sheltered conditions favour the deposition of fine-grained sediments and organic detritus. Inlets and estuaries are thus predominantly soft-bottom habitats, often with extensive flats exposed at low tide. For instance, 70 per cent of Waikawa Harbour is mudflat at low tide. Sheltered shores are gently shelving and waterlogged, and with dark anoxic sediment typically just below the surface. With less wave action and a more stable substratum, animals such as crabs and snails are able to roam the sediment surface and eelgrass and saltmarsh plants can gain a foothold (Fig. 34). Nutrients from the adjacent sea and catchment support abundant primary producers, including microscopic benthic algae living on the sediment surface, seaweeds, and the flowering plants just mentioned. Most of this plant material is not consumed directly by herbivores, but becomes organic detritus to fuel highly productive animal communities. Estuarine sediments typically support dense populations of macroscopic invertebrates, notably polychaete worms, molluscs and crustaceans, many of them active burrowers that vigorously rework the sediment (Fig. 35). In addition, the sediment supports vast populations of minute invertebrates, principally nematode worms and benthic copepods. Given this abundance of potential food it is not surprising that sediment flats represent important feeding grounds for fish and wading birds.

Life in an inlet is not, however, without its stresses. In particular, where there is significant freshwater input, organisms have to cope with a wide range of salinity. This is a particular problem of living in an estuary, typically the lower reaches of a river mouth, where freshwater mixes with sea water. Relatively few species can tolerate the degree of environmental variability in estuaries, so that although these productive habitats support high densities of animals and plants they are places of low species diversity.

Usually typical of the upper shore is the tunnelling mud crab *Helice crassa*, its burrow entrance marked by a small pile of excavated spoil. Also common at this level is the mud snail *Amphibola crenata*, an air-breather feeding on surface organic matter and leaving a tell-tale faecal string in its wake (Fig. 36).

The mid to lower shore often supports dense populations of small tube-building amphipods *Paracorophium excavatum*. But often the most conspicuous crustacean here is the stalk-eyed mud crab *Macrophthalmus hirtipes*, larger than the tunnelling mud crab – up to 3 cm across its more oblong carapace – and with longer eye stalks (Fig. 37). It excavates shallow temporary burrows but, like *Helice*, feeds mainly on organic detritus. There are two other sizeable crustaceans that can be abundant on sheltered flats, but which build deeper burrows and will not be seen on the surface when the tide is out. The white and pink ghost shrimp *Callianassa filholi* produces volcano-like mounds from sediment expelled from its complex burrow system and sifts out organic particulate matter for food. The mantis shrimp *Heterosquilla tricarinata* builds a deep vertical burrow and is an active predator, seizing prey with barbed legs, like its insect namesake. Often an abundant burrower of the mid to lower shore is the lugworm *Abarenicola affinis*, its presence indicated by casts of expelled sediment (Fig. 35).

Dominant bivalves of these sheltered sandflats are usually the cockle *Austrovenus stutchburyi*, a filter feeder (Fig. 38), and the wedge shell *Macomona liliana*, a surface deposit feeder (Fig. 39). Inlets of southern New Zealand support some very large cockle beds, with population densities of 500–1000 per square metre being common. Such large populations of filter feeders may have an important influence on the suspended particulate load of the overlying water. Cockles are an important prey of oystercatchers. They also provide an attachment surface for the mudflat anemone *Anthopleura aureoradiata*, and the small limpet *Notoacmea helmsi*. The wedge shell is a more deeply burrowing bivalve with long siphons that extend up to the sediment surface. It uses its inhalent siphon to suck organic debris off the sediment surface, which produces a characteristic mark resembling a bird's footprint. Important gastropods on these shores include the whelk *Cominella glandiformis* (Fig. 39), an active scavenger, and two herbivores – the small mud snail *Zeacumantus lutulentus* and the topshell *Diloma subrostrata*.

The mid- to low-shore zone of sheltered sandflats is often colonised by eelgrass *Zostera capricorni*, a flowering plant (it even has underwater pollination) that can form extensive dark green meadows (Fig. 40). Eelgrass plays an important role in the ecology of inlets, as a habitat for many associated species, a source of organic detritus and as a sediment stabiliser. At the top of these sheltered soft shores, the vegetation may give way to a well developed saltmarsh (as described earlier in this chapter).

Many species of fish make use of New Zealand's southern inlets and estuaries. In the case of Otago Harbour, some 80–90 species have been recorded, though a number of these are

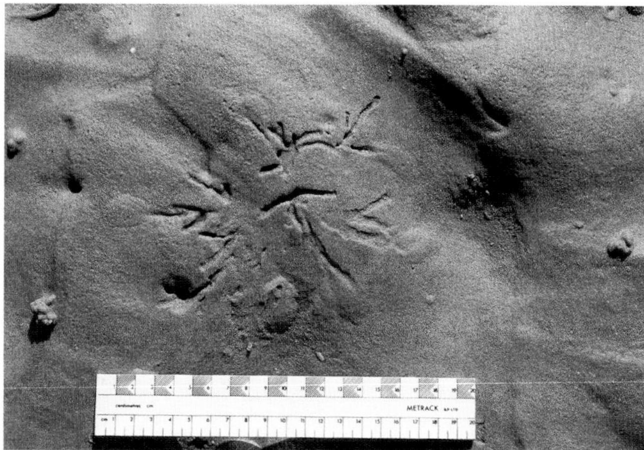

Fig. 39. The wedge shell *Macomona liliana* is a surface deposit feeder. At the bottom of the left-hand picture is the scavenging whelk *Cominella glandiformis*. Using its long inhalent siphon to suck organic detritus off the sediment surface, the wedge shell produces characteristic bird-print feeding traces. P.K. Probert

infrequent visitors. Many fish show marked seasonality in abundance, in particular being more common in summer (e.g. warehou, moki, greenbone, blue cod). As sheltered, productive habitats, inlets and estuaries are ideal as fish nursery areas. Southern inlets are, for instance, vitally important for juvenile flatfish (*Peltorhamphus latus*, *Rhombosolea plebeia* and *R. tapirina*). Inanga *Galaxias maculatus* and other galaxiid species spawn in estuaries, the juveniles migrating upstream a few months later as whitebait (see Chapter 10).

Rich feeding grounds also attract large numbers of waders and other water birds. Common species include gulls (red-billed, black-billed and black-backed), oystercatchers (variable and South Island pied), pied stilt, bar-tailed godwit, banded dotterel, little shag, white-faced heron, mallard and black swan. The Southland estuaries (Jacobs River, New River, Awarua Bay, Waituna Lagoon, and Toetoes estuary) are especially significant in providing a network of feeding grounds for waders and waterfowl, including species that migrate from the Northern Hemisphere.

Paterson Inlet

Southern New Zealand's largest shallow inlet, Paterson Inlet on the east coast of Stewart Island, is a drowned river system, mostly with water depths of 15–25 m. It has a diversity of habitats and a number of biological features of interest (Fig. 41).

Major benthic habitats range from clean sandy and gravelly bottoms at the mouth of the inlet, to muddy sediments in the middle and inner reaches, and extensive mudflats at the western end. Near the mouth of the inlet, where gravelly sediments occur in areas of moderate to strong current, are mounds formed by the bryozoan *Cinctipora elegans*, similar to those on the Otago shelf (page 323). Reef-like structures are formed too by the serpulid tubeworm *Galeolaria hystrix* whose dense aggregations of calcareous tubes also provide a habitat for many associated species.

Paterson Inlet represents one of the world's richest habitats for living brachiopods, or lamp shells (Box 12). *Neothyris lenticularis*, *Terebratella sanguinea* and *Calloria inconspicua* are characteristic of sandy areas at the entrance and of muddier areas over much of the central part of the inlet. Brachiopods are in general sensitive to increases in sedimentation, and it may be that low sedimentation is a prime reason why brachiopods thrive in the inlet. The native podocarp forest surrounding it is still intact and this, together with the granite catchment, appears to result in an unusually low sedimentation rate to the inlet floor. A fourth species, the black ribbed brachiopod *Notosaria nigricans* is associated more with rubbly areas and rock walls. Rock wall communities cover 50–60 per cent of the shoreline, and also provide habitat for reef fishes such as blue cod, wrasses and moki, as well as kina and paua. Indeed, the inlet is well known for its fish and shellfish (Box 19).

Paterson Inlet is also noted for the high diversity of its algal flora. Some 270 species of seaweeds have been recorded from the inlet, with the red algae being particularly rich. Extensive meadows of red algae (notably species of *Lenormandia* and *Rhodymenia*) grow in the central part of the inlet. Interestingly, there may be a relationship between the red algal communities and the brachiopod assemblages. By stabilising the finer fractions of the sediment, the algal meadows may help to control

Fig. 40. The eelgrass *Zostera capricorni* is a relatively small species of seagrass, with leaves up to 15 cm in length, but it nevertheless plays an important role in the ecology of many southern inlets. P.K. Probert

Fig. 41. Paterson Inlet (east of Golden Bay), Stewart Island. This is southern New Zealand's largest non-fiordic inlet, and includes a wide diversity of marine habitats. *P.K. Probert*

sedimentation and enable brachiopods to live on mud in areas of low current velocity.

Port Adventure (8.5 km^2) and Port Pegasus (27 km^2) on the south coast of Stewart Island are less well known, but share some similarities with Paterson Inlet.

Otago Harbour

Otago Harbour is a 23-kilometre long inlet effectively divided into inner and outer basins by peninsulas at Port Chalmers and Portobello, and the islands in between – Quarantine and Goat Islands (page 267). With no sizeable rivers flowing into it, the harbour lacks any significant estuarine habitat. Apart from the dredged shipping channel along its western shore and tidal scour holes between the islands, the harbour is mostly very shallow, with water depths of only a few metres.

Otago Harbour supports a wide range of habitats, including kelp *Macrocystis pyrifera* beds and rocky subtidal areas with diverse encrusting communities, eelgrass beds, algal meadows, cobble shores and muddy embayments.

Sediment bottoms predominate and, in broad terms, there is a gradation from muddy organic sediments in the upper harbour where current speeds are low, to sands and shell-sands containing little organic matter in the outer harbour where higher tidal velocities prevail. Shell gravels occur in tidally scoured channels, and often support large numbers of the turret shell *Maoricolpus roseus*. Species composition of the benthic communities is influenced particularly by sediment type, amount of algal cover, shell material and water depth.

Major features of Otago Harbour are the extensive intertidal sandflats, amounting to nearly 30 per cent of the harbour area, mainly in the middle and outer reaches. The benthos of these areas is usually dominated by the cockle *Austrovenus stutchburyi*, and associated species described earlier.

Rocky shore communities in Otago Harbour are typical of sheltered southern locations. Often similar in composition to these are the plants and animals that colonise wharf piles and other harbour structures. This community is composed pre-dominantly of algae, ascidians, barnacles, bryozoans, hydroids and serpulid tubeworms. Many of these fouling species are common elsewhere in New Zealand and some are cosmopolitan.

12 BRACHIOPODS (LAMP SHELLS)

Brachiopods have a shell consisting of two valves like a clam, but their resemblance to bivalve molluscs is purely superficial as their internal anatomy is quite different. Brachiopods were extremely important in ancient seas, reaching their greatest diversity some 400–500 million years ago, and many thousands of fossil species have been described. There are now only a few hundred living species, but about 30 of these occur in New Zealand waters and several frequent shallow-water habitats such as Paterson Inlet.

Keith Probert

Brachiopod *Neothyris lenticularis*. *P.B. Batson*

FIORDS

Philip Mladenov

The mountainous southwestern coast of the South Island, known regionally as Fiordland, is deeply indented by a series of thirteen large sea inlets, or fiords (Figs 3 and 42). They average about 24 kilometres in length and encompass a surface area of some 756 km^2. These fiords constitute a unique marine ecosystem, the result of an unusual combination of climate, topography and terrestrial vegetation, as well as physical and biological oceanographic processes. The fiords were once glaciated valleys and, like fiords elsewhere in the world, are narrow and steep-walled, and have deep inner basins (with water depths of more than 400 m in some) separated from the adjacent sea by shallow sills ranging from 30 m to 145 m in water depth. The New Zealand fiords open onto a narrow continental shelf and are thus in close proximity to the oceanic influences of the Tasman Sea.

Climate

The mountains surrounding the fiords intercept moisture-laden westerlies, resulting in prodigious rainfall that drains rapidly through the forest-clad catchments into the fiords. Rainfall varies from about three to more than 7 m per year. Despite the heavy rainfall, the input of terrestrial sediment to the fiords is small, except after floods, because of the dense undergrowth in the forest. The concentration of suspended particulates in fiord waters is, therefore, generally very low, equivalent to that found in the open ocean rather than in an enclosed body of water. Storms have an important influence on the physical environment of the fiords. Freshwater input to Doubtful Sound averages, for instance, about 100 cubic metres per second (cumecs), but can reach well over 5000 cumecs during a severe storm.

Physical oceanography

The fiords represent specialised estuarine environments. The large input of freshwater creates a thin layer of buoyant, dilute seawater, the Low Salinity Surface Layer, or LSL, that lies on top of more saline seawater beneath (Box 13). The LSL flows seawards and mixes with the higher salinity water beneath. This drives a slower return flow of salty water that moves from the open sea up to the head of the fiords. The result is a two-layered, or estuarine, circulation pattern. A still layer of oceanic water fills the basins of the fiords. This deep water is generally well-oxygenated, although in some isolated basins it may become anoxic, as sometimes happens in Deepwater Basin in Milford Sound. The deep water in the fiords is replaced periodically, probably in winter, when cold, dense shelf water flows over the sills at the entrance to the fiords.

Productivity

Little is known about the factors that drive marine production in the fiords. Nutrient levels seem to be rather low and this, in combination with low light levels and a stable water column, may mean that primary production is comparatively low. Microbes may contribute in a substantial way to the productivity of the surface waters and may form an important component in the food web. The zooplankton community in the fiords is composed mainly of copepods and the larval forms of bottom-living invertebrates such as sea urchins. Zooplankton can become concentrated at the heads of the fiords as a result of their behaviour and the subsurface countercurrent.

Fig. 42. Doubtful Sound, Fiordland. In the foreground is Deep Cove which receives the freshwater discharge from the Manapouri power station. *Neville Peat*

Intertidal habitats

The tidal range throughout the fiords is similar to that on the open coast – about two metres on spring tides and one metre on neap tides. The intertidal habitats are made up almost entirely of steep shores of solid rock overhung by terrestrial vegetation. Observations of the intertidal zone of Doubtful Sound made by Elizabeth Batham in the 1960s are generally applicable to other New Zealand fiords. The intertidal habitat of the inner reaches of the fiords is exposed to low surface salinities and typically possesses a continuous plant covering from high to low water with pronounced zonation (Fig. 43). On the upper shore, the black lichen *Verrucaria maura* forms a continuous horizontal band that extends upwards to the lowermost branches of the adjacent forest. Below the black lichen zone is a band of a tufted red alga, *Stictosiphonia vaga*. This alga often occurs in association with the barnacle *Elminius modestus* which forms a distinct zone throughout most of the fiords. A layer of the crisp, moss-like, deep-green alga *Wittrockiella lyalli* is dominant at around mid-tide level and provides refuge for a few abundant species of invertebrates, including the small, black, freshwater snail *Potamopyrgus antipodarum* and amphipods. Attached to the *Wittrockiella* can be found a number of other algae, notably species of bright-green *Ulva* and *Enteromorpha*. *Hormosira banksii*, the Venus' necklace, is found in the lower intertidal, and the blue mussel *Mytilus galloprovincialis* is patchily abundant at the low tide level.

The intertidal habitats of the outer parts of the fiords are less influenced by freshwater and are more representative of other southern New Zealand intertidal communities, as described earlier in this chapter. The black lichen zone is more irregular and the upper barnacle zone contains scattered *Chamaesipho columna*, although the estuarine barnacle *Elminius modestus* still dominates. In this zone occur a variety of marine invertebrates typical of rocky shores elsewhere in New Zealand but absent from inner fiord shores, especially grazing molluscs including high shore periwinkles (*Nodilittorina cincta* and *N. antipodum*), the snakeskin chiton *Sypharochiton pelliserpentis*, topshells (*Diloma* spp. and *Melagraphia aethiops*) and limpets (*Cellana ornata, C. strigilis* and *Siphonaria zelandica*). Other low-tide brown algae, in addition to the widespread *Hormosira banksii*, occur here including *Cystophora scalaris, Xiphophora chondrophylla, Durvillaea antarctica* and *Macrocystis pyrifera.*

Subtidal habitats

The steep rock walls of the fiords in the shallow subtidal to a depth of about 4–5 m are strongly influenced by the presence of the LSL. This is a relatively bare, low diversity zone possessing only species that can tolerate repeated exposure to water of reduced salinity, such as blue mussels, barnacles (*Chamaesipho columna* and *Elminius modestus*) and various sponges, bryozoans, tubeworms and algae. This zone is separated from the zone beneath by a distinct horizontal boundary.

Below the shallow subtidal zone, and to a depth of 30 m or more, is a region consisting of large patches of remarkably high

13 THE LOW SALINITY SURFACE LAYER – A UNIQUE ENVIRONMENT

The Low Salinity Surface Layer (LSL) is variable in thickness. During dry periods it becomes very thin, while it may be 10 or more metres deep just after floods. The high runoff from the catchments washes large quantities of 'yellow substances' (fluvic and humic acids) from the surrounding forests into the fiords, which make the LSL quite brown.

The brown-stained LSL has a profound effect on biological processes in the fiords. It stabilises the water column and forms an absorbing layer that reduces light levels in the seawater below. This combination of still dark water provides an opportunity for certain forms like black coral and snake stars, normally occurring only in deeper water on the continental shelf, to thrive in the fiords. It also results in a less prolific growth of kelp in the fiords compared to that on the adjacent open coast. Mobile creatures, such as sea urchins, sea stars and fish, that cannot tolerate prolonged exposure to freshwater, are excluded from the freshwater layer. This layer also repels the delicate larval forms of marine animals, thus limiting their ability to recruit into the shallow regions of the fiords. Periodic thickenings of the LSL following heavy rainfall may result in damage or death to attached marine invertebrates such as black coral.

Philip Mladenov

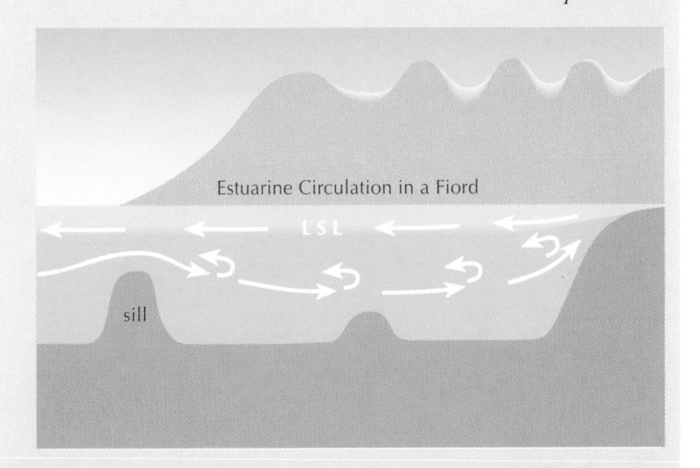

Fig. 43. Rocky shore community in Doubtful Sound in the 1960s, before the Manapouri power station came into operation. At the top of the shore is a zone of the black lichen *Verrucaria maura* and, in crevices, light brown patches of the alga *Stictosiphonia vaga*, beneath which are zones of the estuarine barnacle *Elminius modestus* (white), green algae, and the Venus' necklace weed *Hormosira banksii.* The latter is a characteristic low-shore species of sheltered locations, but now appears to be absent from the Sound, most probably on account of the increased freshwater input from the power station tailrace. *E.J. Batham*

14 BLACK CORAL – SLOW GROWING, LONG-LIVED – AND PROTECTED!

The large colourful snake star *Astrobrachion constrictum* is often found intertwined around the branches of black coral colonies and seems to live nowhere else. *B.G. Stewart*

Black coral colonies are the visually dominant species on the subtidal walls of the fiords between water depths of 10 and 35 metres.

Philip Mladenov

The black coral *Antipathes fiordensis* (right) has been found only in the fiords. It has been protected under New Zealand Fisheries Regulations since 1981. Growth rates are very slow, about 2–4 cm per year. The largest colonies can be very old, perhaps more than 300 years, and they are over 45 years old before they reach sexual maturity. Colonies spawn over a few weeks in late summer. *B.G. Stewart*

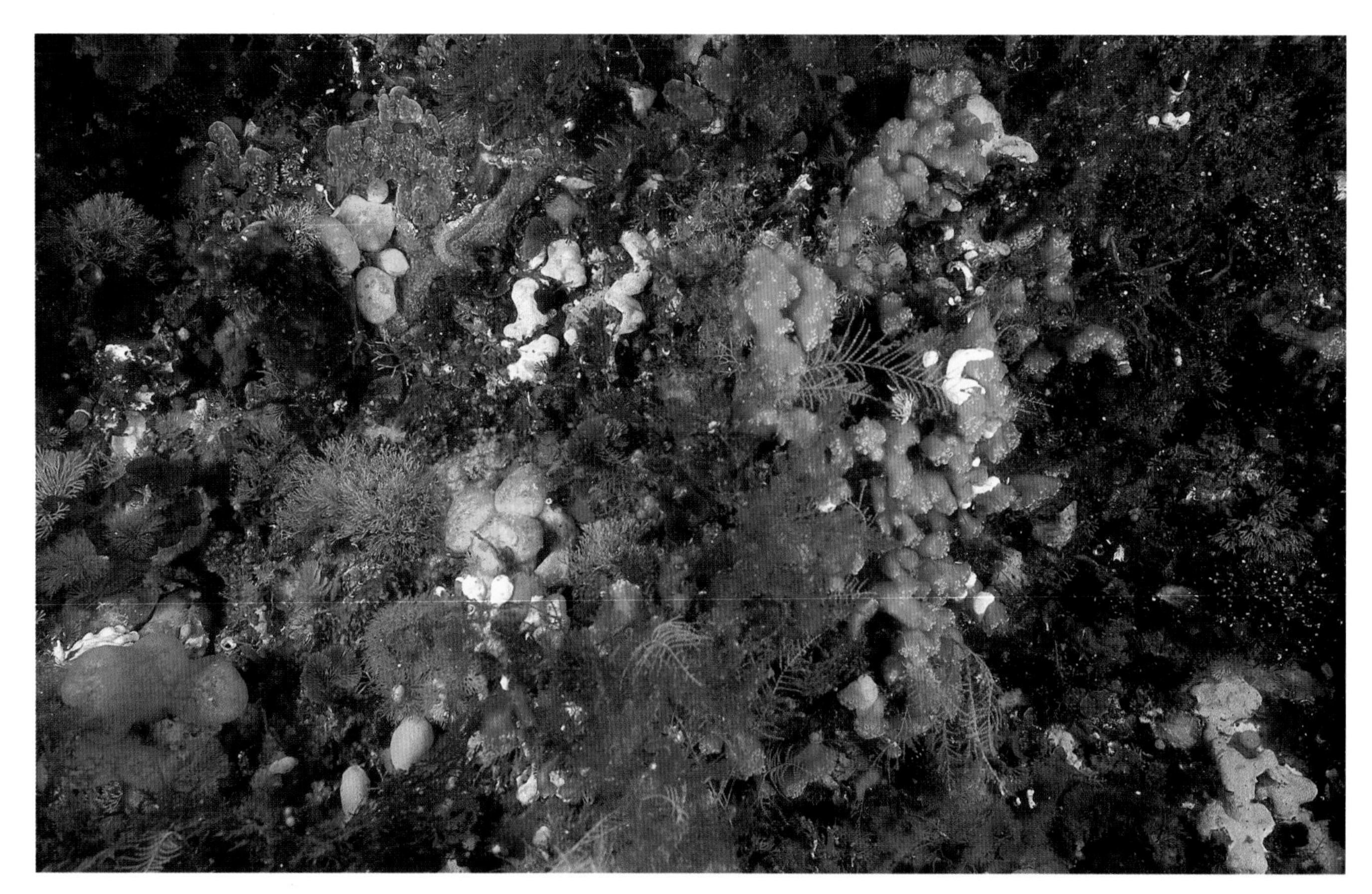

Fig. 44. Encrusting rock-wall community at 12 m in Doubtful Sound, Fiordland. *B.G. Stewart*

species diversity interspersed with 'barren' patches of low diversity. The size of these patches is of the order of hundreds of metres. The high diversity patches (Fig. 44) possess numerous erect bryozoans (e.g. *Galeopsis, Caberea, Cellaria, Hastingsia*), encrusting bryozoans (e.g. *Beania*), colonial ascidians *Didemnum*, solitary ascidians *Pyura* and *Cnemidocarpa*, brachiopods *Notosaria nigricans*, sponges (e.g. *Clathrina* and *Eurypton*) and red coral

15 BOTTLENOSE DOLPHINS OF DOUBTFUL SOUND

A group of about 60 bottlenose dolphins *Tursiops truncatus* resides in the Doubtful-Thompson-Bradshaw Sounds system. They leave this system rarely and usually only for a few hours. Most of the individuals have been identified photographically, using unique patterns of nicks and scars on the dorsal fin. This has allowed their ecology and social lives to be studied in great detail. Movement patterns within the fiord system change seasonally, the dolphins avoiding the colder surface waters of the inner arms of the fiords in winter. Individuals within the group form stable 'cliques'. The social structure of the group centres on mature females, who become more socially 'important' when pregnant or nursing a calf.

Philip Mladenov

Bottlenose dolphins in Doubtful Sound. *S. Dawson*

Errina novaezelandiae, as well as mobile invertebrates such as sea stars (*Patiriella regularis, Coscinasterias muricata, Sclerasterias mollis* and *Pentagonaster pulchellus*), sea urchins (*Evechinus chloroticus* and *Pseudechinus*) and gastropods (*Cookia sulcata* and *Astraea heliotropium*). Black coral *Antipathes fiordensis* is often the visually dominant species in these high diversity areas (Box 14). The barren areas are dominated primarily by encrusting calcareous algae (e.g. *Lithothamnion*), bare rock and sediments. Some of the barren areas may result from over-grazing by sea urchins. Others may be the result of disturbance by large-scale landslips, which occur commonly in the fiords.

A deep-water muddy bottom fauna has been sampled from the basins of the fiords. Conspicuous members of the fauna include heart urchins (*Echinocardium cordatum* and *Brissopsis oldhami*), burrowing brittle stars *Amphiura rosea*, nut shells (*Saccella bellula* and *Neilo australis*) and tusk shells (scaphopods).

Vertebrate fauna

Over 150 species of fish have been recorded from the fiords. In general, the fish fauna is similar to that found elsewhere in southern New Zealand waters, although there is one species known only from the fiords, the fiord brotula *Fiordichthys slartibartfasti*. The fiord populations of some species have unique biological and ecological features. For example, wavy line perch *Lepidoperca tasmanica* in the fiords live at depths much shallower than those outside. Also, some species are far more abundant in the fiords than on the open coast and there appear to be some variants which are distinctive in their colour, but otherwise morphologically similar to those outside the fiord.

Other vertebrates of the fiords include bottlenose dolphins (Box 15), fur seals and an unusual lizard, the Fiordland skink (Box 16).

Human impacts

The pristine, forest-clad catchments of the fiords comprise a national park and World Heritage area. The fiords themselves, however, remain largely unprotected, with the exception of the Piopiotahi Marine Reserve in Milford Sound and the small Te Awaatu Channel (The Gut) Marine Reserve near the entrance to Doubtful Sound. Recreational and commercial fishing take place in the fiords, and commercial operators run scenic cruises and dive charters, particularly in Milford and Doubtful Sounds.

The Manapouri Power Station, which is about 180 metres underground, draws freshwater from Lake Manapouri and discharges it via two 10-km long tailraces into Deep Cove, Doubtful Sound. The rate of power station discharge fluctuates from about 100–510 cumecs (cubic metres per second) due to the electricity generation requirements of the power station and the need to maintain the level of Lake Manapouri within a prescribed range. The median rate of discharge is 392 cumecs. Median inflow into the Doubtful-Thompson-Bradshaw Sounds system from rainfall is about 100 cumecs. Median power station discharge is thus nearly four times the median discharge from natural inflow. Elizabeth Batham surveyed the shores of Doubtful Sound in the late 1950s and early 1960s, before the power station was commissioned in 1969. Another survey in 1995 showed that the intertidal biota had changed substantially since the early 1960s in terms of the distribution of communities and their species composition, most likely as a result of increased freshwater influence (Fig. 43). In general, the full range of marine intertidal species has become restricted to more seaward locations and brackish-water species have extended their range. Other possible effects of power station discharge on the marine environment of Doubtful Sound are difficult to establish because of a lack of baseline data for the fiord prior to the commissioning of the power station.

16 A LIZARD OF THE LITTORAL ZONE

Rock platforms on the often storm-lashed coast of Fiordland provide a most unusual habitat for a lizard. *A. H. Whitaker*

The Fiordland skink *Oligosoma acrinasum*. *A.H. Whitaker*

Coastal habitats of southern New Zealand house a variety of lizard species, with densities being especially high on offshore islands lacking introduced predators. Most of these species prefer life on dry land above high tide, as do the majority of lizards elsewhere.

A spectacular exception is found on the coastline of southern Fiordland. This small, near-black lizard, the Fiordland skink *Oligosoma acrinasum*, is known from about 40 islands and one mainland beach between Nancy Sound and Dusky Sound. It inhabits bare rock platforms and boulder beaches in the littoral zone, sometimes below high tide. When violent storms lash the southern coastline (as they frequently do) the lizards seek shelter from waves and rain in deep rock fissures or between boulders. When the sun shines the skinks quickly emerge to sun-bask, sometimes piling up on each other at communal basking sites. The species clearly tolerates salt water, as individuals have been observed to dive into rock pools when disturbed, and some rock stacks inhabited by the species are so small that waves break over them in storms.

Fiordland skinks often co-occur with fur seals, but populations appear depleted where rats have been introduced. Two islands that offered suitable habitat but that had Norway rats present (*Rattus norvegicus*, probably introduced by sealers) have recently been the sites of re-establishment of skink populations. In the case of Hawea Island (9 ha), translocation involved direct assistance from humans. Rats were eradicated by poisoning in 1986, and 40 Fiordland skinks were shifted there in 1988 from a nearby island. By 1992, at least 200 Fiordland skinks were estimated to be present and their distribution had increased, leading researchers to conclude that translocation was successful. On Breaksea Island (170 ha), rats were eradicated in 1988, but a plan to translocate Fiordland skinks there was put on hold when it was discovered that the skinks had got there first! Natural dispersal from a rock stack 50 m offshore is considered probable.

These observations indicate an unusual tolerance (for a lizard) of salt water and of such extreme coastal conditions. The success of the rat eradication and translocation programme for Fiordland skinks also attests to the resourcefulness and vision of biologists working in this inhospitable region.

Alison Cree

BIRDS OF THE COAST

Hamish Spencer and John Darby

Since all birds breed on land, it may seem somewhat arbitrary to attempt a distinction between coastal and oceanic birds. There is, however, a reasonably clear division. Coastal birds are those seen from the shore or that mostly feed on shore or in inshore waters. Oceanic birds, on the other hand, feed exclusively out at sea and are seen only rarely from the shore.

Gulls and terns

Commonest of the coastal birds are the red-billed gull *Larus novaehollandiae scopulinus* and the southern black-backed gull *L. dominicanus.* They feed both on land and at sea, in estuaries, coastal mudflats, lagoons and rubbish dumps, and are incorrigible scavengers. The red-billed is the smaller of the two and the adults have an almost completely white plumage and brilliant red bill, legs and feet. In contrast, the bill, legs and feet of juveniles are dark grey-brown. The juveniles can be confused by the inexpert eye with the black-billed gull *L. bulleri* for both are the same size, but in the latter the bill is black and the feet are usually black with occasionally a dark crimson showing through. Whereas the red-billed gull usually breeds on or near the coast, the black-billed is an inland breeder, wintering on the coast. The black-backed gull is found in greatest abundance at refuse tip sites or when nesting in their breeding colonies. It is a large gull, and does not reach adult plumage until its fourth year. Although considered coastal, it is also found well inland in cultivated and grassland areas to at least 1000 metres.

The region's commonest tern must be the white-fronted tern *Sterna striata* (Fig. 45) which breeds in colonies on the ground, from steep cliffs to flat expanses of shingle, often near those of the red-billed gull. It fishes by hovering and then diving from a low height, and is often pursued by skuas (usually the Arctic skua, *Stercorarius parasiticus*) for its catch. Also seen on the coast after its breeding season is the black-fronted tern *Sterna albostriata,* much more of an inland waterway breeder. The Antarctic tern *S. vittata* is occasionally seen on Stewart Island, although the few that breed in New Zealand tend to do so on islets off Stewart Island and in the New Zealand subantarctic. The largest New Zealand tern is the Caspian tern *S. caspia* with a wing span of 1.4 m and massive 7-cm orange-red bill. A study of a breeding colony near Invercargill has shown that they can live for well over 20 years. Southern birds move north after breeding, as far as Auckland's Manukau Harbour.

Fig. 45. The white-fronted tern *Sterna striata* in flight. *J. Darby.*

Fig. 46. The variable oystercatcher (below), which in southern regions is a deep, rich, overall black, is a true coastal species. The eggs are well camouflaged. *J. Darby.*

Oystercatchers

There can be little confusion between the two species of oystercatchers found in Otago and Southland. The South Island pied oystercatcher (SIPO) *Haematopus ostralegus finschi* mostly breeds inland east of the Main Divide, although occasionally also on Stewart Island. Birds that breed in the region tend to winter on the coast, whereas birds north of Canterbury frequently migrate to the North Island. The crisp, clean, black and white plumage of the SIPO contrasts with the deep, rich, all-over black of the variable oystercatcher *H. unicolor* (Fig. 46). Confined almost entirely to the coastline where it breeds, the variable oystercatcher may be found on Otago and Southland coasts, including Stewart Island and Fiordland. There is ample room for confusion on the naming of this species for, in its North Island stronghold, its plumage is very much like that of the SIPO. However, in southern regions it is often referred to as the black oystercatcher, since rarely does one see any white feathers on the belly or underwing.

The pied stilt *Himantopus himantopus leucocephalus* is a delightful element of the coastal fauna, although it is found inland during the breeding season as well as on the coast.

After breeding, birds migrate to coastal lagoons and wetlands, where their yapping call adds lustre to the tranquillity of such areas.

Waders

Many migrant waders visit New Zealand's food-rich mudflats and coastal lagoons. Commonest is the eastern bar-tailed godwit, or kuaka *Limosa lapponica baueri*, which breeds in north-eastern Siberia and north-western America. It can be seen in large flocks south to Paterson Inlet, most individuals 'wintering' in the austral summer, from October to March. Similar species occasionally found among these flocks include two other godwits, the Hudsonian *L. haemastica* and the black-tailed *L. limosa melanuroides*; and two species with long, downwardly curved beaks, the far-eastern curlew *Numenius madagascariensis* – the largest of the migrant waders found in New Zealand – and the whimbrel *N. phaeopus*.

Almost always with the larger godwits, the lesser knot *Calidris canutus* is a regular visitor to Southland and, less often, Otago lagoons. The smallest annual summer visitor is the red-necked stint *C. ruficollis*, usually seen in Southland, particularly at the Waituna Lagoon. Among the rarer sandpipers that occur, often with pied stilts, in fresh or brackish water coastal habitats during summer, are the sharp-tailed sandpiper *C. acuminata*, its close relative with which it often associates, the pectoral sandpiper *C. melanotus*, the needle-billed marsh sandpiper *Tringa stagnatilis* and the similar but larger greenshank *T. nebularia*. Associating with pied stilt, the Siberian or grey-tailed tattler *T. brevipes* has been seen on sandflats such as at Aramoana at the mouth of Otago Harbour and on Southland shores. That most restless of waders, the terek sandpiper *T. terek* occurs most years in Southland.

The dumpy ruddy turnstone *Arenaria interpres* occurs in small but regular numbers every year, both in estuaries and on shingle ocean beaches, where it busily fossicks for invertebate prey with its stout bill. The Pacific golden plover *Pluvialis fulva* is found in small flocks in the less salty parts of estuaries and coastal wet pastures. At least in Southland lagoons, it occurs in most years. The southern New Zealand dotterel *Charadrius obscurus obscurus* is a regular but rare and threatened breeder on the mountains of Stewart Island and occasionally strays to Southland beaches and at times farther inland. The black-fronted dotterel *C. melanops* was not known in New Zealand until 1954, when it migrated from Australia. A small group is known to breed close to the Manuherikia River in Otago. The banded dotterel *C. bicinctus* is the most numerous of the small plovers to be seen, both on the coast and inland. Its breeding range is extra-ordinarily wide, from unforested mountains and the gentler slopes of Central Otago, to dry shingle beds, inland lakes, rivers and coastal dunes. Many southern birds may migrate after breeding to Australia and northern New Zealand coasts. Several plover species have been recorded as rare vagrants, mostly on the larger Southland coastal lagoons.

Two heron species are found in coastal areas. By far the most common, the white-faced heron *Ardea novaehollandiae* occurs in almost all aquatic habitats. Rarely seen is the reef heron *Egretta sacra*, which has declined since the arrival of the former species. It has been suggested that the decline is due to competition between the two, but there is no real evidence for this. The reef heron prefers to feed on rocky shores, although it may be seen on mudflats. The royal spoonbill *Platalea regia* is a relatively recent arrival in Otago (Fig. 47). Whereas herons stalk and probe for their prey, spoonbills have a leisurely side-to-side scything motion as they wade stealthily through shallow water. First sighted breeding in Otago on Maukiekie Island off Moeraki in the early 1980s, this species is now an established breeder on Green Island, off the Kaikorai Estuary mouth near Dunedin and on Omaui Island near Invercargill.

Cormorants and shags

Breeding colonies of shags and cormorants are abundant in Otago and Southland. Probably the most prominent species is the smallest, the little pied cormorant *Phalacrocorax melanoleucos brevirostris*, which inhabitats estuaries and harbours as well as

Fig. 47. Spoonbills have been known to breed at Okarito on the Westland coast since 1949, but it was not until the early 1980s that they were found to be breeding in Otago, on Maukiekie Island off Moeraki. This chick is almost fully grown. *J. Darby*

Fig. 48. There are at least five species of shag or cormorant in southern New Zealand. The smallest and one of the most prominent is the little pied cormorant found in harbours, estuaries and sheltered waters. *J. Darby*

Fig. 49 (below). While the little pied cormorant is extremely variable, the Stewart Island shag has only two forms, absolute pied and absolute bronze. *J. Darby*

some inland waters (Fig. 48). Like the variable oystercatcher, it shows a cline in plumage from the north of New Zealand to the south. Most of the southern birds are the white-throated form (rather than the pied form, in which the whole lower part of the head and body is white). The larger pied cormorant *P. varius* is confined in the region to the coasts of Fiordland, western Southland and Stewart Island, where it nests in large trees (e.g. southern rata) near water. The largest species, the black cormorant *P. carbo novaehollandiae*, is more a freshwater species, although sometimes it lives on the coast. The Stewart Island shag *Leucocarbo chalconotus* also occurs in two colour phases, one pied and the other all dark bronze (Fig. 49). Confined to southern New Zealand, it has substantial breeding numbers on the Otago Peninsula, in North Otago and on Stewart Island. Its chimney-pot nests are a feature of the breeding colony at Taiaroa Head. One of the most attractive shag species, especially in its breeding plumage, is the spotted shag *Stictocarbo punctatus*. It is a cliff-nester, abundant on Otago and Southland coasts, including Stewart Island, where the form once known as the blue shag is common. Both Stewart Island and spotted shags feed on small fish and invertebrates, caught during dives, sometimes up to 15 kilometres from land.

COASTAL FISHERIES

Tony Brett

Given the diversity of coastal forms and habitats around southern New Zealand, it is not surprising that the region supports a range of coastal fisheries. A feature of the eastern and southern coasts that strongly influences fisheries is the sediment delivered by large rivers (Fig. 4). This has created long, sweeping beaches punctuated by peninsulas and short stretches of rocky reef. The scattered areas of foul ground and rocky reef are small both in number and size, while the beaches are numerous and large. The areas of rocky outcrop, especially the peninsulas, support the best inshore fishing grounds along this coast, as they offer shelter from the prevailing weather and provide habitat for the most important fishing species – paua, rock lobster and blue cod – as well as various lesser fished species, including mussels, kina, moki and butterfish, or greenbone. Beyond the shore, the distribution of seabed sediment on the continental shelf also strongly influences the type and location of fisheries environments.

By contrast, on the southwest coast, mountain building and glacial action have carved the steep rocky shores of Fiordland. Although within the fiords there is only a limited development of beaches and shore platforms, they support healthy commercial and recreational fisheries for rock lobster, paua and blue cod, as well as recreational fisheries for scallops, kina and groper *Polyprion oxygeneios*.

Fisheries have an inherent human component, being defined by the species we target. Early southern Maori relied heavily on moas and seals, but as these declined they turned increasingly to fish and shellfish (see Chapter 7). Barracouta *Thyrsites atun* was one of the most important species, but they also took red cod *Pseudophycis bachus*, ling *Genypterus blacodes*, blue cod and spotty, and collected shellfish such as cockle and pipi. Fishing using small single-hulled canoes, sometimes with a sail, was undertaken by Ngai Tahu in the coastal waters of Otago and Southland. Using traditional preservation techniques, fish caught during summer and autumn months could be stored over winter. That some of the more distant offshore fishing areas were known and named by Ngai Tahu indicates that they fished them at least periodically.

Customary fishing patterns have been altered by the accessibility of stocks close inshore, the effects of habitat modification and pollution, and the change in lifestyles of many contemporary Maori. With Maori reviving the cultural base of their communities, use requirements are expanding and traditional management measures are being given greater emphasis in sustaining local fisheries. The aspirations of tangata whenua are to restore access to and abundance of the resource in their areas, to an extent that allows future generations to provide fully for the non-commercial sustenance and cultural requirements of their communities.

Nowadays, the most popular species with all groups in the south are blue cod, rock lobster and paua. These all inhabit hard shore, rocky reef environments. As a result, Otago and Southland have characteristic fishing communities around the coast situated in the lee of rocky peninsulas or nearby harbours and river mouths. Examples are Moeraki (Box 17), Karitane, Taieri Mouth, Waikawa, Riverton and Oban on Stewart Island. While these towns contribute much to the character of the region, they underline the importance of these fishery areas, which are not only vital to the commercial fishery, but are popular and safe recreational fishing areas, have a long association with customary/Maori fishing and, further, are popular sites with sea birds and marine mammals.

The coastal fisheries of Otago and Southland are diverse and abundant, and fall into two categories: shellfish (including rock

17 MOERAKI FISHING VILLAGE

Moeraki village. *P.K. Probert*

Small fishing villages are characteristic of the Otago and Southland coast. Forty kilometres south of Oamaru, and just south of the Moeraki Boulders, lies the fishing village of Moeraki, a quiet, close-knit community geared mainly for the fishing industry, fishers and family. Its port is an open harbour and subject to the north-easterly swell. Twenty small (6–13 m long) local fishing vessels anchor offshore in the bay and dinghies are rowed to and from the water's edge. Dinghies, catch and fishers are winched ashore to two processing sheds on the cliff edge. Day fishing takes place on local grounds that extend northwards for 15 km and south to Karitane. Potting for rock lobsters is the major fishing activity, with the main season between June and December. Lining for groper and blue cod and jigging for squid are traditional off-season pursuits for most fishers. Some fishers also net spiny dogfish and rig.

Tony Brett

18 SOUTHERN AQUACULTURE

Southern New Zealand has a number of marine localities suitable for aquaculture. Marine farming in the south started in the late 1970s in Big Glory Bay, Stewart Island, and this area remains the centre of development for the region. Production from Big Glory Bay is currently 1500 tonnes of salmon *Onchorynchus tshawytscha* and 3000 tonnes of mussels *Perna canaliculus.* However, a rapid expansion of mussel production is anticipated. Species farmed in other localities, most notably Bluff Harbour, include oysters *Tiostrea chilensis,* paua and paua pearls *(Haliotis iris* and *H. australis*) and seaweed *Porphyra columbina.* Major farming methods include cages for salmon and buoyed longlines for bivalves, while paua are mostly farmed in shore-based facilities using pumped seawater. Concerns about the localised effect on seabed communities under farms is offset by the requirement of marine farming to maintain the highest water quality. Big Glory Bay salmon farming is uniquely managed by working within a predetermined level of nitrogen in the water column. Future developments will no doubt take advantage of improved fisheries production achievable from seabed enhancement and artificial reefs.

Tony Brett

Salmon farming in Big Glory Bay, Stewart Island.
Courtesy of the Ministry of Fisheries

lobster) and finfish, with a combined annual commercial value that exceeds NZ$70 million. Shellfish species are the region's most valuable inshore stocks, with rock lobster returning annual revenues of more than NZ$40 million, whilst Foveaux Strait oysters and paua both earn more than NZ$10 million each year. Other smaller shellfisheries include those for queen scallops *Zygochlamys delicatula,* kina and cockles. Finfish landings are dominated in financial terms by blue cod, flatfish, eels and stargazer *Kathetostoma giganteum.* Worth more than NZ$4 million annually, blue cod is the most valuable. Around 1300 tonnes of blue cod are caught commercially each year in the region, mostly in Foveaux Strait with smaller amounts at Otago ports. Flatfish are caught predominantly in Otago waters, and to a lesser extent in Foveaux Strait. School shark *Galeorhinus galeus* is an important fishery in Southland, whilst smaller fisheries in the region include those for red cod, squid (mostly in Otago), skate, groper (mostly in Southland) and tarakihi *Nemadactylus macropterus.*

Aquaculture – the farming of animals and plants in freshwater and marine environments – is a relatively new industry in southern New Zealand. The farmed production of fish has potential to augment the wild fishery harvest. With aquaculture currently providing in excess of 15 per cent of global fisheries production, this 'managed fishery' holds a special position for fisheries in the future (Box 18).

Hard shores

Southern rocky reefs and headlands provide suitable habitat for many species important to fisheries. Red, or spiny, rock lobster *Jasus edwardsii* are normally found in groups in crevices or other shelter (Fig. 50). Crepuscular opportunists, they find food primarily by smell, eating almost anything, including shellfish,

Fig. 50. Rock lobster *Jasus edwardsii* are important opportunist feeders of rocky areas, and are fished commercially from Moeraki to Fiordland. *M.A. McArthur*

crabs, fish and sometimes seaweed. Average size can vary greatly, and there are some localities (e.g. Karitane) where small rock lobsters always seem to have been characteristic. Rock lobster are commercially fished from Moeraki around to Fiordland, and are also an important recreational species. However, a substantial area between Nugget Point and Long Point, 25 km to the southwest, is permanently closed to commercial fishing.

The rock lobster of the south undergoes a spectacular migration. Juvenile animals congregate in the North Otago area until they begin to move south. The lobsters walk against the current around the south of the South Island and northwards along the Fiordland coast. Where they go from this point is unknown, as fishermen report their disappearance somewhere north of Bruce Bay (south Westland). It is thought that the migration is a mechanism to counter the effect of larval drift by the prevailing current.

Abalone, known locally as paua, are gastropods of rocky intertidal and subtidal habitat. The blackfoot paua *Haliotis iris*, is the largest of the three endemic species, reaching about 200 mm shell-length (Fig. 16). The yellowfoot paua *Haliotis australis* grows to a shell-length of about 110 mm. Paua come out at dawn and dusk to browse on seaweeds. They are a traditional Maori food and an important recreational species. Management of the paua fishery has been improved by the creation of substantially smaller Quota Management Areas for Otago and Southland.

Blue cod *Parapercis colias* are bottom-dwelling carnivores found on reef edges and on shingle/gravel or sandy bottoms close to rocky outcrops (Fig. 51). They are not a true cod, but a member of the weaver family (Pinguipedidae). Blue cod can grow to 60 cm in length and weigh up to 4 kg. Large male blue cod are territorial. Although the evidence is not yet conclusive, it seems likely that blue cod change sex from female to male as they grow. The commercial blue cod fishery is exclusively a cod pot fishery, supporting vessels from Moeraki, Karitane, Port Chalmers, Taieri Mouth, Waikawa, Bluff, Riverton and Stewart Island. They are also a prized recreational fish and a traditional Maori catch. Recreational fishers account for almost two-thirds of the annual blue cod catch in the Otago area.

The green-lipped mussel *Perna canaliculus* and the blue mussel *Mytilus galloprovincialis* (Fig. 8) are filter-feeding bivalves, widely distributed along the coastline. The green-lipped mussel (so named for the distinctive green band on the mantle fringe), has a low tide to subtidal habitat and can grow to lengths exceeding 230 mm. The blue mussel occurs higher on the shore, and grows to about 100 mm. Most mussels are taken by recreational fishers, hand gathering from the shore. They have always been an important food for Maori.

The common sea urchin or kina *Evechinus chloroticus* inhabits shallow water generally less than 12 m deep, and can form dense aggregations on sub-tidal reefs (Fig. 17). They are nocturnal grazers, using their tube feet to move around and forage for kelp or, in the absence of kelp, other algae and encrusting organisms. Kina support a small commercial dive fishery in the Stewart Island area. They are also a traditional Maori food. Customary and recreational divers and shore gatherers harvest kina along the length of the coast.

Soft shores

Southern New Zealand has a variety of harbours, inlets and semi-enclosed bays where soft sediment flats are covered and uncovered by the tide. Two of the most common inhabitants of these areas are flatfish and cockles. Consequently these accessible places are popular for gathering shellfish and netting or spearing flatfish. A general term applied to a number of species, flatfish includes sand flounder *Rhombosolea plebeia*, yellow-belly flounder

Fig. 51. Blue cod *Parapercis colias*. This coastal species characteristic of southern waters supports a valuable cod-pot fishery and is prized by recreational fishers. *M. Francis*

Fig. 52. Toheroa *Paphies ventricosa*. Only beaches on Southland's south coast support a recreational fishery for toheroa. *Courtesy of the Ministry of Fisheries*

R. leporina, greenback flounder *R. tapirina*, black flounder *Rhombosolea retiaria* and common sole *Peltorhamphus novaezeelandiae*. Adult flatfish move out from the channels and over the tidal flats with the rising tide, seeking food such as shellfish, polychaete worms and organic detritus. For many species of flatfish these waters also act as nursery areas, offering sheltered conditions and a rich variety of food for the young fish. Flatfish tend to be short-lived species (2–3 years), but with high fecundity (up to a million eggs per fish). Recruitment can be highly variable from year to year, depending on conditions during the breeding season.

One of the most abundant animals of inlets and bays is the cockle *Austrovenus stutchburyi* (Fig. 38). Stocks of this bivalve in Blueskin Bay Inlet (north of Dunedin) and Papanui Inlet (Otago Peninsula) are large enough to support a valuable 500-tonne per annum commercial harvest. Other highly prized shellfish species such as the pipi *Paphies australis* and the flat or dredge oyster *Tiostrea chilensis* are found in lesser abundance in these inlets and bays.

Local land-based management practices can threaten these important fisheries areas and shellfish habitat through sedimentation, non-point eutrophication and reclamation. Species such as the pipi may be especially vulnerable, and there is anecdotal evidence that the number of locations where it can be found has declined during this century.

Toheroa *Paphies ventricosa* is the large 'cousin' of the pipi and the tuatua. Beds of toheroa can be found throughout New Zealand in the intertidal zone of stable, fine sand on surf beaches.

Fig. 53. Fiordland fisheries. The fishing vessel *Trojan* in a stiff south-westerly off George Sound, Fiordland, but still working. *P. Peychers*

However, in recent times the only toheroa beds large enough to sustain some recreational access have been at Oreti Beach and Te Waewae Bay on Southland's south coast (Fig. 52). One-day recreational seasons at Oreti Beach have become very popular, that of 1993 attracting some 20,000 fishers. Unfortunately, 1995 and 1996 saw a number of large 'strandings' of toheroa at Oreti Beach following unfavourable winters and a summer bloom of toxic algae, prompting a postponement of recreational access to allow the beds to recover. With the recent drop in numbers of adult toheroa at Oreti, there is interest in enhancement and customary management of the beds.

Fiordland fisheries

Fiordland fisheries have a special status for New Zealanders. The image of rock lobster fishers battling the elements in this remote and wild area, contrasted with the romance and adventure of the recreational fishery, has contributed to the area's mystique (Fig. 53). Surprisingly, however, there has been commercial fishing in the area for nearly 100 years. This was initially based on blue cod with a fishing village in Chalky Inlet. Fresh Southland blue cod was available in Melbourne, Australia, in the 1930s. This fishery effectively closed during World War II, but was revitalised with improved boat and refrigeration technology and the opening of the American rock lobster market in the early 1950s.

Fig. 54. Fiordland's commercial fishery is based around rock lobster, with 20–25 per cent of the total New Zealand catch coming from the area. *P. Peychers*

19 PATERSON INLET

Paterson Inlet, Stewart Island, is an area of natural beauty, clean, clear waters and rich marine life. The Inlet supports a diversity of aquatic communities and healthy fish populations. Blue cod is the major finfish harvested, followed by trumpeter, flatfish and sea perch (or Jock Stewart). Other species in the Inlet are blue moki, butterfish (greenbone), tarakihi, rig *Mustelus lenticulatus*, red cod, gurnard *Chelidonichthys kumu*, spotties and other parrotfish, barracouta, dogfish and other shark species.

Paterson Inlet is renowned for its large scallops *Pecten novaezelandiae*, and is virtually the only area in southern New Zealand that supports an accessible recreational scallop fishery. Paua is the other frequently harvested shellfish, although mussels, rock lobster, kina and cockles are also taken. The Inlet has a healthy fishery prized by the local community and recreational fishers from around New Zealand and further afield. In recognition of the values of the Inlet for recreational fishing, the commercial sector volunteered to withdraw from the Inlet, and since 1994 it has been classified as a non-commercial area. *Tony Brett, Photo P.K. Probert*

Today, the commercial fishery is still based around rock lobster, with 20 to 25 per cent of the total New Zealand catch coming from the area (Fig. 54). Commercial paua fishing is an important fishery with a 142-tonne total allowable catch. Groper, blue cod and school shark, among others, are also fished. Potential fisheries have also been identified for kina and sea cucumber. Outside the fiords, the continental shelf is narrow and the continental slope descends steeply to depths of more than 4000 m. Species caught outside the fiords include albacore tuna *Thunnus alalunga*, school shark, groper and bluenose *Hyperoglyphe antarctica*.

The area's remoteness has meant that recreational fishing has become a feature of Fiordland only over the last 20 years. Opening the Wilmot Pass between Lake Manapouri and Doubtful Sound has facilitated a growing charter industry, and Fiordland wilderness fishing is now much sought after by recreational fishers from throughout New Zealand and overseas. Blue cod, rock lobster and groper are the most popular species taken, although for many the 'Fiordland experience' itself remains the major drawcard. Mindful that the steep-sided fiords result in less fisheries habitat than immediate impressions might suggest, and also of the implications of increasing use, the Guardians of Fiordland's Fisheries, a stakeholder group with members representing Ngai Tahu, recreational, charter boat and commercial sectors, has been established to ensure that Fiordland continues to make a special contribution to fisheries.

HUMAN IMPACTS ON THE COAST

Keith Probert and Barrie Peake

Humans have exploited marine resources of southern New Zealand for at least 800 years (Chapter 7). Early Maori of the area hunted fur seals, fished and collected shellfish. In the case of fur seals, colonies were reduced by the fourteenth century. Pressure on the region's marine resources increased enormously following the arrival of Europeans and today a number of the region's fish and shellfish species remain heavily exploited. Other important human impacts on the region's marine environment concern the physical alteration of the coastline and the input of pollutants.

The main human pressure on the region's marine resources continues, however, to be the exploitation of fish and shellfish. Heavy fishing not only affects target species. With the large-scale removal of a key species by fishing, there are likely to be changes in the abundance of non-target species in the community as they take up the slack. Some fishing methods, such as trawling and the use of set nets, can be unselective so that large quantities of unwanted species, or bycatch, are also taken. Trawling disturbs the seabed, crushes benthic organisms and may alter benthic communities.

Aquaculture

With wild-caught stocks coming under increasing pressure worldwide, there are incentives to use aquaculture methods. An important marine farming activity in southern New Zealand has been the sea-cage rearing of quinnat salmon *Onchorynchus tshawytscha* in Big Glory Bay, Stewart Island (Box 18). But salmon farming is not without its environmental costs. The deposition of uneaten food pellets and fish faeces results in an impoverished benthic community under salmon cages, similar to the impact caused by other organic effluents. There are concerns too that increased levels of nutrients may trigger abnormal growth of phytoplankton populations. Significant salmon losses from a major bloom of a toxic phytoplankton species in Big Glory Bay in 1989, combined with lower returns from salmon exports, has led to a decline in the quantity of salmon farmed in this area. On the other hand, the farming of green-lipped mussels *Perna canaliculus* in this area has gained considerably in importance. This is less labour intensive and is predicted to have smaller impacts on the marine environment.

Physical alteration of the coast

Intensive aquaculture may be deemed aesthetically undesirable in such scenic areas. Visually, however, much of the region's east coast has already been transformed by the clearance of native coastal vegetation to make way for farmland and coastal towns and settlements. One of the coastal areas that has been most physically altered would be Otago Harbour, where urban and industrial growth and port activity has meant extensive reclamation, shoreline development and dredging. Since the 1860s about eight per cent of the original area of Otago Harbour has been infilled. This has occurred mainly in the upper harbour and has entailed extensive loss of sheltered coastal habitat. Lake Logan at the head of the harbour, reclaimed in 1913–25 and now Logan Park, would, for instance, have been a sizeable

Fig. 55. It is estimated that some eight per cent of the original area of Otago Harbour has been reclaimed. In recent years reclamation has occurred mainly at Port Chalmers to accommodate expansion of the container terminal and wood chip and log wharves. *P.K. Probert*

wetland area. In recent years reclamation has occurred mainly at Port Chalmers to provide for port development (Fig. 55). The shoreline of Otago Harbour has also been extensively altered by the construction of some 37 km of rock wall, and there are now relatively few stretches of unmodified rocky shore left in the Harbour. Those that are left are mainly around the Portobello peninsula, Roseneath, and the islands. The shipping channel has to be regularly dredged to maintain navigable water depths. Sand from the Clutha River moves northwards along the south Otago coast and some is swept into the harbour by strong flood tides. To counteract this, some 240,000 cubic metres of sediment are dredged from the harbour each year and dumped at disposal sites outside its mouth.

Runoff and effluents

Sediment runoff into Otago Harbour, and into other inlets and estuaries of the region, has probably increased markedly since European settlement where loss of the catchment's natural cover of vegetation has promoted soil erosion.

Runoff is also the main source of pollutants affecting coastal habitats of southern New Zealand. In particular, the predominantly agricultural catchments contribute nutrients (principally nitrate) and organic inputs from fertiliser and animal wastes. Nutrient enrichment can promote the growth of algae and, if the input to an estuary or inlet is excessive in relation to its rate of water exchange, the algae can form dense mats (Fig. 57). In extreme situations other marine life is smothered and oxygen depletion can occur when the algae decompose. A seaweed that responds readily to nutrient inputs is the sea lettuce, *Ulva lactuca*, which can form a bright green sward on tidal flats.

Wastewaters, such as those from sewage and stormwater outfalls, can also be important sources of pollutants, but as point-source discharges their adverse effects are potentially easier to address than diffuse inputs from runoff. Population growth has occurred largely along the open coast or around estuaries and inlets, and sewage has often been discharged directly into adjacent waters with minimal, if any, treatment. In recent years, increased awareness of this potential environmental hazard, together with requirements of the Resource Management Act, have led many local authorities to implement schemes for upgrading the treatment of sewage before it is discharged into coastal waters. The quality of North Otago coastal water has, for instance, improved significantly with the commissioning in 1994 of a plant involving oxidation ponds and a wetland to treat domestic and industrial waste water from the greater Oamaru area. Similarly, until recently, Otago Harbour received sewage discharges from a number of outfalls, but these have now all been routed to Dunedin's wastewater treatment plant for discharge to the open coast at Lawyers Head, and to further reduce environmental impact the treatment plant is to be upgraded and the ocean outfall extended.

Effluents from urban and industrial areas are likely to contribute a cocktail of contaminants that includes heavy metals, petroleum hydrocarbons and debris. Industrial wastewaters have been discharged into the region's coastal waters with some significant environmental consequences. Tannery effluent was for many years discharged untreated into Sawyers Bay in Otago Harbour, which has led to high levels of chromium in the sediment. Uncontrolled runoff from a timber preservation plant on the Oamaru foreshore gave rise to high levels of chromium, copper and arsenic in nearby beach sands, which have required remedial measures. Sewage-like wastes from freezing works and food processing plants in the region contain large amounts of organic matter (Fig. 56). The bacterial activity required to break down a large input of organic matter can create a heavy demand for dissolved oxygen. In sheltered waters with limited turnover this may seriously reduce the amount of oxygen available for other marine life. Facilities to treat industrial wastes before they are discharged are gradually being improved, particularly in order to meet the requirements of the Resource Management Act.

A major concern worldwide is the amount of debris entering the oceans, particularly plastics such as fishing nets, packaging,

Fig. 56. Waste organic matter from an abattoir discharging on a north Otago beach. Such outfalls are, however, becoming a thing of the past. *P.K. Probert*

Fig. 57. Runoff from agricultural land can increase the nutrient concentration of estuarine waters and trigger a luxuriant growth of opportunistic algae. *P.K. Probert*

plastic bags and bottles. These can pose a serious threat to marine wildlife as a result of entanglement and ingestion. A survey of the debris collected on five beaches in the Dunedin area in 1994 showed that 86 per cent of the items were plastic, mostly packaging materials from land-based sources.

Introduced species

Much of New Zealand's land environment has been completely transformed by introduced species. Marine habitats tend to be less vulnerable in this regard. Nevertheless, harbours and other coastal habitats worldwide have acquired many exotic species that have been accidentally or intentionally introduced. Over the centuries, numerous fouling and boring organisms have been carried around the world on ships' hulls and are now more widely distributed. Of increasing concern are species that may be carried from port to port in the ballast tanks of ships. They may be taken up in ballast water from one harbour, survive a trans-oceanic journey and be discharged in a foreign harbour when the vessel empties its ballast tanks to take on cargo. In some cases the adults of small species are transported, in others it may be a larval stage or resting cyst. If conditions such as water temperature and salinity are not too different, then the discharged species may be able to survive and reproduce. Several algae and marine invertebrates are thought to have been introduced into New Zealand in ballast water (Box 20).

Efforts to develop new fisheries have been responsible for many introductions of marine species. A fish hatchery was established at Portobello in Otago Harbour in 1904 specifically to rear and release European species of fish and shellfish but, despite enormous effort, none of the imported species became established in local waters. However, in recent years a recreational fishery has developed based on introduced quinnat or chinook salmon *Oncorhynchus tshawytscha*. At sea, the adult salmon feed mainly on sprat and *Munida*. Their diet is similar to that of other pelagic fish and blue penguin, but it is not known if this exotic species is having any significant impact on native marine life.

Humans have clearly made their mark on the coastal zone of southern New Zealand, particularly through their exploitation of its living resources and modifying coastal habitats. However, impacts on the marine environment have been far less profound than those affecting the land. Marine ecosystems of much of southern New Zealand have probably not been greatly disturbed by human activity, with the more remote areas still ranking among the world's most pristine marine environments.

20 AN INTRODUCED ASIAN SEAWEED

The kelp *Undaria pinnatifida* is a large brown seaweed native to Japan and Korea where it is an important commercial species ('wakame'). In 1987 *Undaria* was discovered in Wellington Harbour and since then has been found at several southern locations including Oamaru, Moeraki, Otago Harbour, Bluff and Big Glory Bay in Stewart Island.

The kelp has a microscopic (gametophyte) phase in its life cycle and it is this that is likely to have been brought to New Zealand in ballast water. Subsequently, the large plants (sporophyte phase) appear to have been spread by coastal shipping. The kelp colonises the lower shore from about mean low water to a few metres depth. How it will affect the native community is unknown at this stage, but it is likely to compete with smaller brown seaweeds such as *Sargassum*, *Cystophora* and *Carpophyllum*.

Keith Probert

The kelp *Undaria pinnatifida*, here growing at Moeraki. *C. Hay*

CHAPTER 12

THE OPEN SEA

TONY BRETT, LIONEL CARTER, ROBERT M. CARTER, JOHN DARBY, LLOYD DAVIS, STEPHEN DAWSON, JOHN JILLETT, KEITH PROBERT, HAMISH SPENCER

Dusky dolphins in clear winter water, far offshore. *S. Dawson*

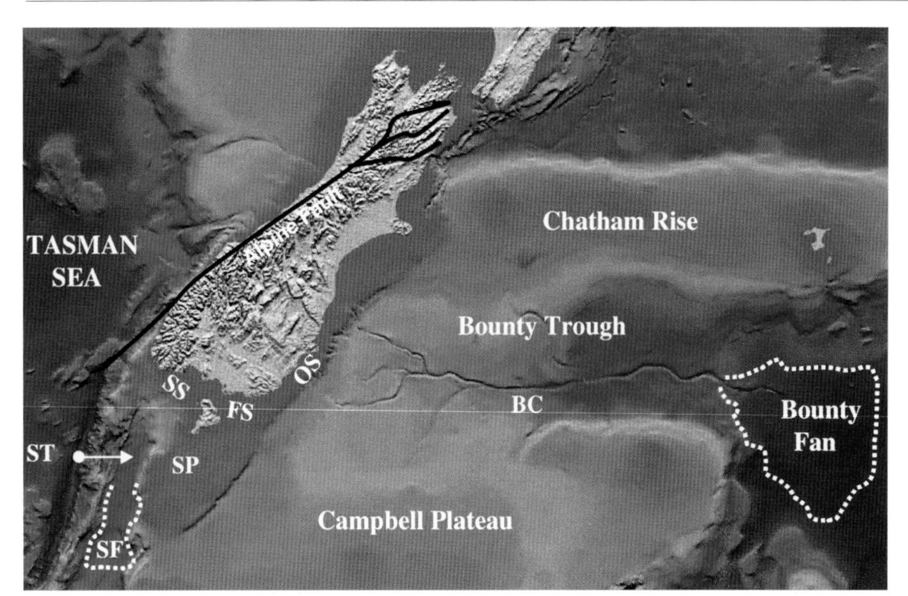

Fig. 1. The ocean floor off southern New Zealand from the continental shelf to the abyssal Southwest Pacific Ocean housing the vast Bounty Fan. Annotated features include the Southland shelf (SS), Foveaux Strait (FS), Otago shelf (OS), Bounty Channel (BC), Snares platform (SP), Solander Trough (ST) and Solander Fan (SF) which is fed by the Solander Channel (too small for chart scale) extending along the trough from the Southland shelf.

Fig. 2 (below). The Otago shelf in motion with the inner shelf wedge of modern sand from the Clutha River, shifting along the shelf and covering middle shelf relict gravel that formed at a time of lower sea level. Winnowing of the gravel by modern currents helped produce a belt of sand along the outer shelf which also includes a zone of shell-derived sediment.

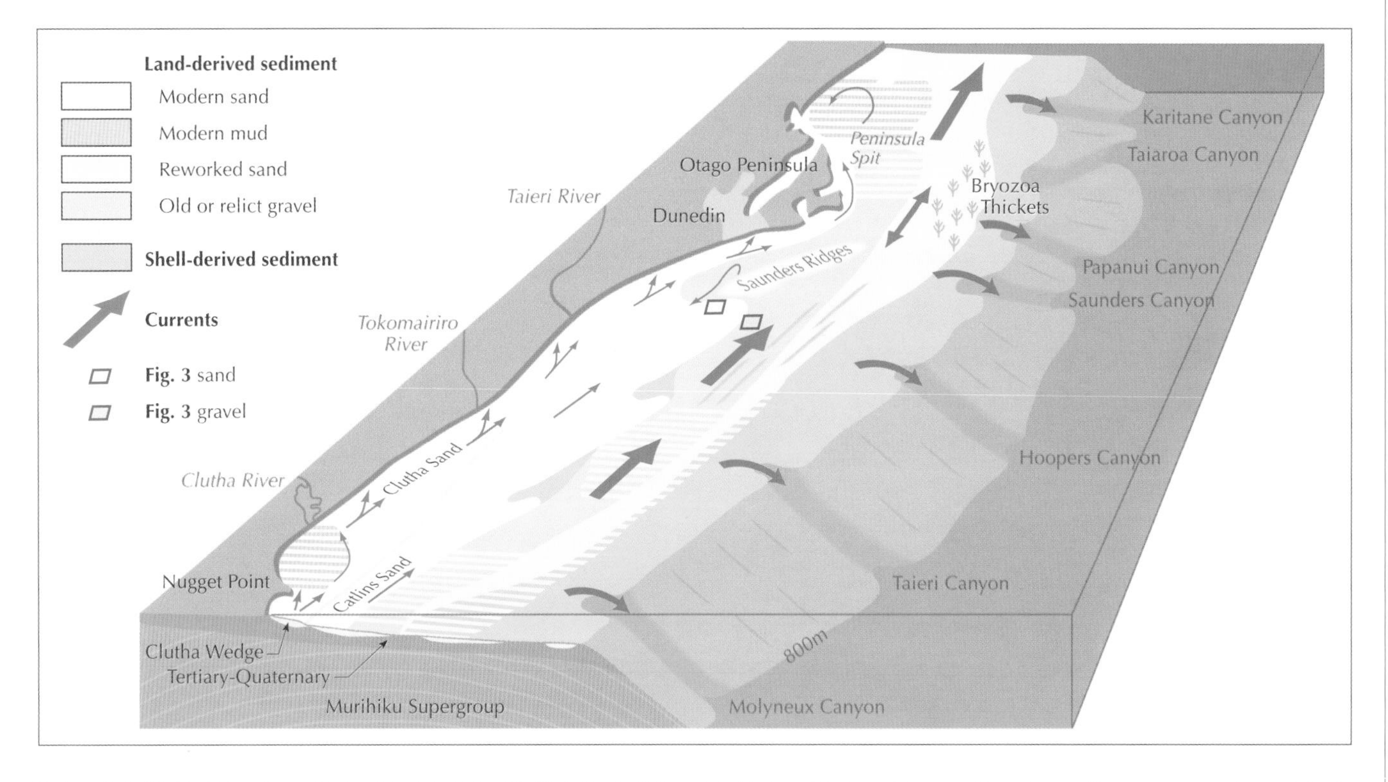

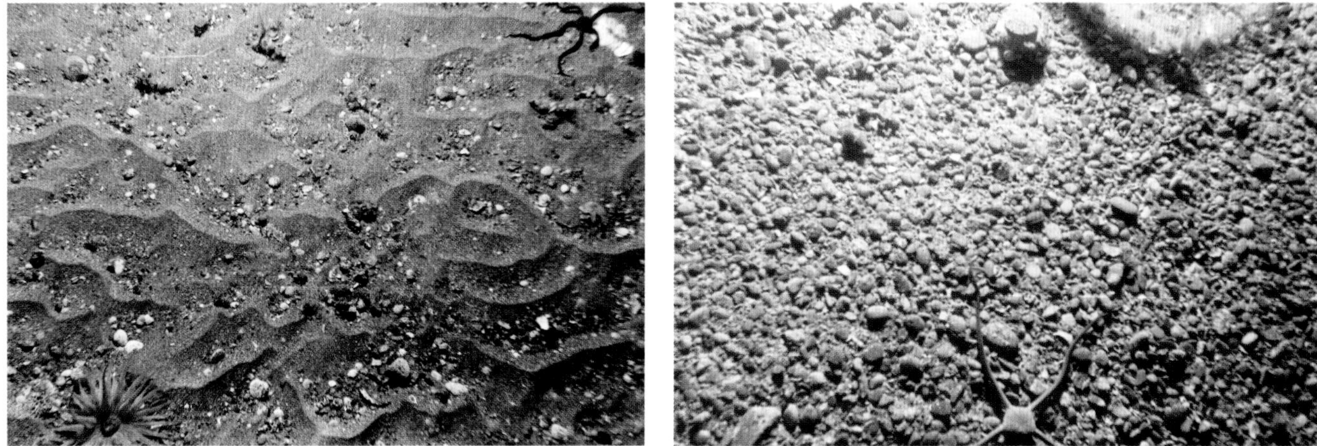

Fig. 3. Modern sand (left) and old gravel (right) deposited about 12,000 years ago when the sea level was 56 metres lower than today. *Courtesy of NIWA*

THE OPEN SEA

BENEATH THE SEAS – THE OTAGO/SOUTHLAND CONTINENTAL MARGIN AND DEEP OCEAN

Lionel Carter and Robert M. Carter

Setting the limits

'An amazing underwater landscape' is an apt description for the continental margin off southernmost New Zealand. From the shore, the margin begins as the continental shelf, which descends gradually seaward to the shelf edge at 125–150 m water depth. Off western Fiordland the shelf is a mere 1–5 km wide, the result of dissection by the offshore extension of the great Alpine Fault. By comparison, the southern Fiordland/Southland shelf is considerably wider, exceeding 90 km in the eastern approaches to Foveaux Strait and more than 270 km wide if the adjacent Snares platform is included. For much of its extent the southern shelf is a subdued plain, its monotony broken by scattered pinnacles, reefs, sand ridges and the occasional basin near Stewart Island. Between the two extremes is the Otago shelf with an average width of around 30 km. It also has a subdued relief with a few low terraces, ridges and pinnacles.

The other margin component, the continental slope, plunges from the shelf edge to 1300 m depth off Otago and to over 3000 m depth to the west of Fiordland. The last locality is a vivid testament to the impact of the Alpine Fault, which has carved a precipitous slope bearing many landslide scars, fractures and other deformational structures. This submarine tectonic landscape is complicated further by numerous erosional gullies and canyons, the larger features traversing the entire slope before opening on to the abyssal ocean floor. By comparison, the continental slope off Foveaux Strait appears to be a gently inclined plain which, west of Stewart Island, leads down to Solander Trough at 2750 m depth. However, at least two submarine canyons traverse the slope and, although survey coverage is incomplete, it is likely they merge to form Solander Channel. Larger than any New Zealand river, this channel passes more than 450 km along the floor of Solander Trough to eventually discharge to Solander Fan (Fig. 1).

The Otago continental slope also helps define the head of another major depression, the 300 km-wide Bounty Trough, which runs 900 km eastward to link with the abyssal Pacific Ocean. Like the Fiordland slope, the Otago counterpart is incised by submarine canyons. Between the Waitaki and Clutha Rivers, eleven major and eight minor canyons produce a topography that rivals any valley system onshore. Commencing near the shelf edge, the Otago canyons descend as narrow gorges with floors to 600 m below the surrounding seabed. They converge downslope to form three submarine channels that are wider and less incised than the canyons. Near 1700 m depth, the channels unite into one of the world's major ocean pathways, the 900 km-long Bounty Channel. Incised along the axis of Bounty Trough, this 1–2 kilometre-wide channel drains eastward to empty onto the vast, delta-shaped Bounty Fan – the submarine counterpart of the Canterbury Plains.

The face of the shelf

Sediments on the continental shelf reflect past and present changes in water depth, ocean currents, and the supply of detritus from rivers and coasts. Southerly swell is nearly always present and, together with the Southland Current, tides and storm-forced currents, can shift sand at all shelf depths. By themselves, tidal currents are too weak to transport sand on the open shelf. However, where the tides are constricted, as in Foveaux Strait and off Otago Peninsula, they have sufficient power to move sand on a daily basis. Overall, the swell and currents are too vigorous to permit much mud to settle. Consequently, the shelf is mantled mainly by sand and gravel deposits. The exception is embayments like Blueskin Bay where the Otago Peninsula offers protection from marauding southerly storms, thus allowing mud to accumulate.

At present, sand is delivered to the shelf by rivers and coastal erosion and it settles mainly on the inner continental shelf between the beach and 30–50 m water depth. Some sediment, however, fails to reach the shelf, especially at the Fiordland coast where it is effectively trapped within the 150–400 m-deep fiords. Back on the inner shelf, currents and swell frequently move sand to produce a prevailing anticlockwise transport around the bottom of the South Island to Otago. There, sand moves north-east along the shelf, as revealed by the distribution of Clutha River sand (Fig. 2). This distinctive deposit of sediment has advanced at least 150 km from the river mouth over the past 6500 years. This equates to an average rate of advance of 23 m per year.

Inner shelf sand usually gives way to finer grained material farther offshore. But around southernmost New Zealand, it is replaced by gravel deposits on the middle shelf in water depths of 30–80 m (Fig. 3). Today, gravel usually accumulates in shallow, high energy environments near river mouths and beaches. Thus, the presence of similar deposits at middle shelf depths suggests they formed when sea level was lower than now, but were later inundated as the sea returned to its present level (as discussed later). Originally, these ancient gravels contained sand, but modern currents have winnowed out the fine material and deposited it in a narrow belt along the outer shelf in water depths down to 100 m. Near the shelf edge, sediments are dominated by bryozoan and molluscan shell debris from recently live and long-dead organisms that dwelt there at different times of rising sea level. This shelly carpet reaches its peak on the Snares platform, where sediments contain 70–100 per cent shell debris. Such abundance reflects a low supply of terrestrial detritus, a vigorous current and wave regime to sweep mud from the area and, in the case of living organisms such as the brilliantly coloured queen scallop *Zygochlamys delicatula*, a healthy food supply.

Floods of sediment

As a source of sediment for the continental margin, New Zealand has the dubious distinction of being one of the world's major suppliers, accounting for nearly one per cent of the global input of sand and mud to the oceans. For southern New Zealand, the sediment input from Fiordland is over one million tonnes of fine sand and mud delivered annually. But most fails to reach the continental shelf because of entrapment within the fiords. Elsewhere, sediment is poured directly on to the shelf, mainly by the major rivers. Together, Southland and Otago rivers deliver over three million tonnes annually. But by past standards, this quantity is small. Construction of hydro-electric dams has diminished the input of some major rivers. Before completion of the Roxburgh Dam in 1956, the Clutha River annually discharged 2.3 million tonnes of sand and mud to the coast; now it discharges 0.4 million tonnes. River loads were even higher in the ice ages. Then, lakes such as Wanaka, Hawea and those of the McKenzie Basin were occupied by glaciers which delivered their debris directly to rivers below the lakes. Following retreat of the glaciers about 12,000 years ago, the restored lakes became giant settling ponds that captured most of the glacial and post-glacial debris to starve the rivers downstream.

Climate change and sea level

The impact of fluctuating sea level on the continental margin has been profound. During the great ice ages, the polar ice caps expanded and robbed the oceans of water, causing sea level to fall. At the peak of the last major ice age, some 20,000 years ago, sea level was about 120 m lower than at present. Consequently, most of the continental shelf was exposed. Cook and Foveaux Straits were blocked by the emergent land, causing major changes in coastal current systems. The shelf was a wind-swept plain, traversed by rivers that discharged at a shoreline located near the shelf edge. Strong glacial winds and powerful southerly swells pounded the ancient coast to disperse sediment to the deep ocean. As global temperatures climbed and the polar ice caps melted, sea level rose causing the shoreline to migrate landward in a series of rapid shifts that were punctuated by pauses lasting several centuries or more. One prominent pause occurred 12,000 years ago, when the shoreline was 56 m lower than now. Gravel beach ridges and muddy estuaries formed near Otago Peninsula, while other coastal deposits developed off the Waitaki River. Later, when sea level resumed its rapid ascent, these coastal environments were drowned and 'fossilised'. Even today, the ridges and estuaries can be recognised on echo-sounding and seismic records. Several pauses later about 6500 years ago, sea level reached its present position and, apart from a temporary rise of 0.9 m about 3500 years ago, has remained there ever since. With shoreline stability came the modern, current-driven 'along-shelf' sediment transport system: a marked change from the previous ice age, when river-guided 'across-shelf' transport was the norm.

To the deep ocean

Some of the sediment reaching the shelf escapes to the abyssal ocean at 4000–5000 m depth. One mode of escape is via submarine canyons. Sediment that is mobilised near the shelf edge is likely to be captured by canyons, especially off Fiordland

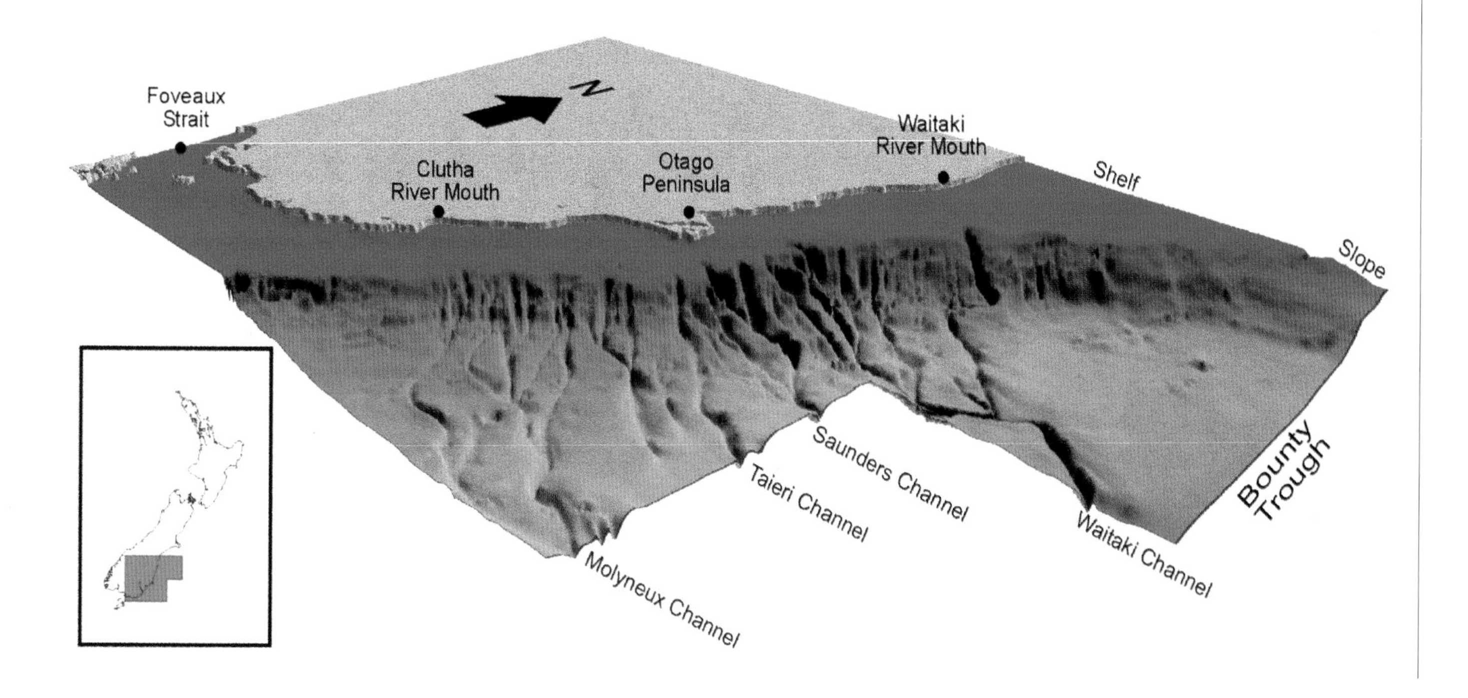

and Otago (Fig. 4). Once in the canyon, the only way is down. Such transport may be instigated by strong canyon currents, as witnessed by the presence of fresh sand ripples on canyon floors at depths to 850 m (Fig. 5). Transport is also facilitated by density or turbidity currents. These sediment-laden flows can develop from river flood discharges, seabed slumping or by resuspension of sediment under intense storms. The resultant slurries of water, mud and sand are denser than the clearer surrounding waters and, therefore, flow rapidly down-canyon, often scouring the seabed en route. On the Fiordland margin, turbidity currents either die out in 3000–4000 metre-deep basins at the base of the continental slope or continue further into the Tasman Sea by way of the Haast Channel. In western Foveaux Strait, canyon-guided flows enter Solander Channel, whereas the Otago canyons feed turbidity flows into Bounty Channel.

Under modern conditions of high sea level, turbidity currents are small and infrequent. Bounty Trough and Channel, for example, presently receive little mud from the land. Instead, the trough is mantled by calcium carbonate ooze derived from the shells of plankton that have settled out from overlying waters. This situation was very different at the height of the last ice age, when sea level was at its low point. In Fiordland, fiords were occupied by glaciers whose debris-laden meltwater discharged across the exposed continental shelf directly into the 3000–4000 m deep Tasman Sea. Elsewhere, the ancestral Clutha, Waitaki, Oreti and other rivers meandered across the emergent shelf and dumped their loads near or into the submarine canyons and troughs. Ice ages were also times when rivers carried large amounts of detritus, which favoured the frequent formation of turbidity currents. In just a few days, these turbid flows swept the full length of the Solander and Bounty channels. Periodically, flows spilled out of the channels to deposit mud ridges along the channel edges, much like the levees bordering the Mississippi River. But those flows contained by the channels eventually reached and fed the Solander and Bounty Fans.

Beyond the submarine channels

Although residing at abyssal depths, Solander and Bounty Fans may not be the final repose for the sediment borne by turbidity currents. Both fans are swept by the planet's largest deep ocean current – the Pacific Deep Western Boundary Current (DWBC), so-called because it moves along the western side of the Pacific Ocean in water depths exceeding 2000 m. In addition, the flow across Solander Fan is reinforced by the powerful Antarctic Circumpolar Current (ACC). As the combined ACC - DWBC passes north from the Southern Ocean, it scours sediment from Solander Fan and transports it along the base of Campbell Plateau to Bounty Trough. There, the ACC splits off to continue its eastward journey around Antarctica, leaving the DWBC to flow north into the Pacific. En route, the DWBC erodes the Bounty Fan and carries more mud around eastern Chatham Rise towards East Cape. Where the current is sluggish, mud settles out to form extensive deposits termed drifts, which underlie much of the current's path to East Cape. But even the drifts are only temporary havens for the much-travelled mud. Near East Cape, the drifts are carried into the Earth's mantle by the Pacific crustal plate as it descends beneath the Australian Plate along a line marked by the 7000 metre-deep Kermadec Trench. In the mantle, drift sediment is heated, mixed with other molten materials, and returned to the surface via volcanoes. Thus, sediment that was generated in the Southern Alps, transferred to the deep ocean by submarine channels, transported 5000 km around New Zealand by abyssal currents, and cycled through the mantle, was eventually returned to New Zealand by way of the volcanoes on Kermadec Ridge (Fig. 6). This is truly a remarkable recycling system.

Fig. 4 (left). The dissected Otago/Southland continental margin, with canyons that transfer sand and mud from the shelf to submarine channels. These small conduits merge further east into Bounty Channel.

Fig. 5. Active bottom currents sweep along the axis of Karitane Canyon to generate fresh sand ripples at 850 m depth. *Courtesy of NIWA*

Fig. 6 (right). What goes round comes round – the New Zealand recycling system where sediment from land is transported along submarine channels (SC = Solander, BC = Bounty, HC = Hikurangi channels) into the Deep Western Boundary Current (DWBC) which carries it back to land via the Kermadec subduction zone off East Cape. The Antarctic Circumpolar Current (ACC) reinforces the boundary current, south of Bounty Trough.

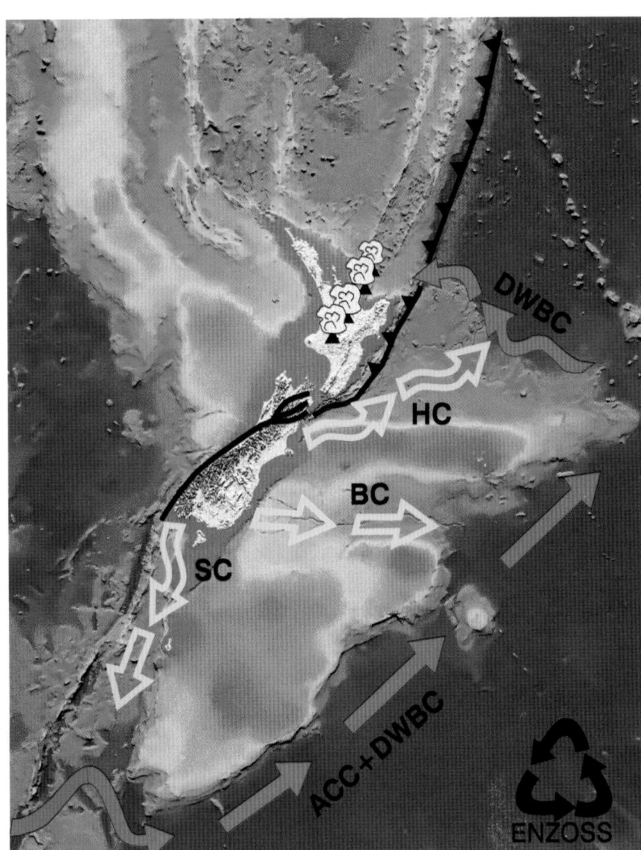

PHYSICAL OCEANOGRAPHIC SETTING AND WATER MASSES

John Jillett

Contrary to popular belief, the ocean waters that bathe New Zealand's southern coasts have subtropical origins. Though their subtropical character has been modified in the course of their passage from equatorial sources, these waters are still significantly warmer and saltier than the global average for such latitudes (45–48°S). On an annual basis there is a net loss of heat from ocean to atmosphere throughout the region. Ocean currents circulate very slowly, but have a long 'memory' compared with the overlying atmosphere.

The coastal surface waters of southern New Zealand originate in the southern Tasman Sea, which is filled with modified subtropical water derived from two different sources. The main source is from the tropical Pacific Ocean by way of the East Australian Current, most of which is deflected eastwards at about the latitude of Sydney, to be discharged to the open Pacific Ocean in a clockwise flow around the north of North Island. In the vicinity of this eastward deflection, however, some subtropical water is shed off, mostly as a succession of warm-cored eddies, into the southern Tasman Sea. A secondary subtropical source is the Indian Ocean, from which a monsoonally variable water flow is fed south of Australia and eventually also into the southern Tasman Sea.

The ocean currents that deliver these subtropical waters are wind-driven, so their speed varies with differences in the strength and patterns of atmospheric pressure systems on a hemisphere-wide scale. During El Niño events, the oceanic circulation becomes slower, as normal wind-fields break down, resulting in a greater loss of heat to the atmosphere during the prolonged southward flow. Hence El Niño events throughout New Zealand are characterised by cooler than average sea surface temperatures, the opposite of the situation in the eastern Pacific Ocean at such times (Box 1).

On their southern boundary these warm, saline waters of subtropical origin converge with cooler less saline surface water of the subantarctic West Wind Drift. Where the different water masses converge, a distinctive boundary feature, the Subtropical Front (STF), is formed (Fig. 7). The STF is a major feature of the Southern Ocean. It encircles the globe at about 40–45°S and is disrupted only by the South American continent. The mean location of the STF, more than any other factor, defines the northern biogeographic boundary of the subantarctic Southern Ocean.

Oceanic subtropical surface waters from the southern Tasman Sea approach the South Island from the west to separate into northward and southward currents in the vicinity of Jackson Head, in south Westland. The southward component becomes the distinctive Southland Current, which flows down the Fiordland coast, around the south of the South Island and Stewart Island, and then in a north-easterly direction along the Otago coast into the Canterbury Bight. As these subtropical waters of the Southland Current squeeze their way anticlockwise around the south of the New Zealand landmass, they push the cooler and less saline subantarctic water away from the coast. The southern boundary between these subtropical and subantarctic waters, the STF of the Tasman Sea, becomes much more sharply defined to the south and east of the South Island, where it is known as the Southland Front (Fig. 7). Along the Otago coast this Southland Front lies over the upper continental slope, coinciding with the 500 m depth contour (Box 2). Further to the north-east, the Southland Front becomes less intensely defined seawards off Banks Peninsula and reverts in name to the STF.

In its anticlockwise flow around southern New Zealand, the subtropical Southland Current is confined to the shallow continental shelf. Along the Otago coast, the Southland Current flows at about 18–20 cm per second, or 16–17 km per day, and continues to lose heat to the atmosphere. As a result, sea temperatures are warmer in an up-current direction. Stewart Island waters are warmer than those of the Catlins coast, which in turn are warmer than those of North Otago. The greatest difference is between sea temperatures on opposite sides of the South Island, with waters off northern Fiordland being some 3–

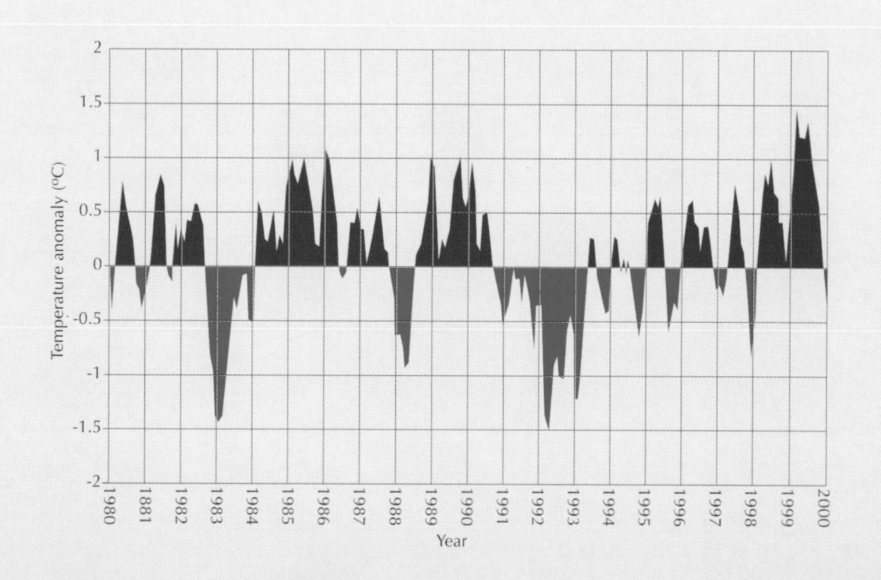

1 SEA SURFACE TEMPERATURES AND EL NIÑO

This record from Portobello in Otago Harbour shows the departure of sea surface temperature (3-month running means) from the long-term (45-year) average for 1980 to 2000. During this period, the 3-month mean sea temperatures have varied from the long-term average by a total 3°C (+1.5°C to –1.5°C). Little can be discerned of any long-term trend in the record, but three distinct El Niño events are marked by periods of cooler than average sea temperatures. The 1982–83 El Niño, regarded by meteorologists as perhaps the most intense this century, is rivalled in this record by the more recent 1991–93 event in both duration and intensity. A lesser El Niño occurred in the summer of 1987–88.

4°C warmer throughout the year than those off Dunedin. More than any other factor, the subtropical influence of the coastal seas plays a significant role in maintaining the temperate biogeographic character of the South Island and Stewart Island, by comparison with the distinctly subantarctic nature of the more southerly offshore islands.

While the Southland Front is the offshore boundary of the Southland Current, its inshore boundary is the coastline itself. However, the inshore 'neritic' water is distinctly modified by dilution, nutrient enrichment from land runoff and by seasonal thermal change, with greater warming in summer and cooling in winter. Coastal dilution occurs along the whole course of the Southland Current, from Fiordland around to the Canterbury Bight. The legendary high annual rainfall of Fiordland is rather greater in the north than the south, though this difference is compensated by the southern fiords having larger catchments than those in the north. Fiordland outflow has been amplified by the diversion of flow for hydro-electric purposes, from Lakes Te Anau and Manapouri, to Doubtful Sound instead of to Te Waewae Bay, along with other Southland rivers which discharge directly to Foveaux Strait. The Clutha River, the largest in the country with an average flow of around 500 cumecs and common peak flows in excess of 1100 cumecs, discharges into the coastal sea near Balclutha, some 70 km south of the Otago Peninsula. As the Clutha River is fed mainly from the headwater catchments of Lakes Hawea, Wanaka and Wakatipu, it has a different flow pattern from the rivers to the north, such as the Taieri and Shag Rivers.

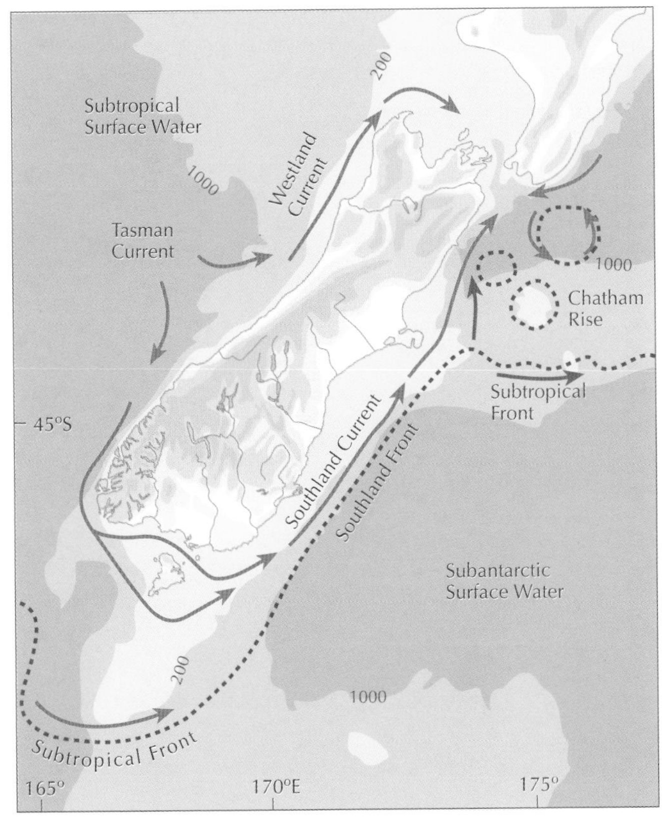

Fig. 7. Major surface water masses and currents.

All these freshwater outflows progressively dilute the coastal waters and modify their temperature on a seasonal basis. Their temperature tends to follow air temperature more closely than that of the coastal sea, so the outflows are cooler in winter and warmer in summer than the receiving coastal sea. At all seasons the freshwater runoff is less dense than the salty sea, so the outflows invariably form shallow buoyant plumes that persist until mixed by wind and tidal action. After a flood, the Clutha River plume can be a spectacular feature, plainly visible from vantage points as far north as on the Otago Peninsula, and up to 100 km downstream from the air and in satellite images (Box 3).

On the continental shelf the waters have considerable seasonal variation in temperature which becomes less with increasing distance from shore. In sheltered harbours and inlets the extreme seasonal range of water temperature is some 14°C, from less than 5°C in winter to over 19°C in summer, including diurnal variations. Mean monthly ranges for Otago Harbour are from 6.9°C in July to 16.1°C in January, with a year-round mean of 11.63°C. At depths greater than 200 m, beyond the edge of the continental shelf, temperatures do not vary seasonally but are much the same year round.

2 WATER MASSES OFF THE OTAGO PENINSULA

Neritic water hugs the shore as a low salinity wedge of surface water, thinning seawards and over-riding the higher salinity mainstream water of the Southland Current. Beyond the edge of the continental shelf, cool Subantarctic Surface Water of lower salinity is encountered. So at the surface, these three water masses are encountered seawards in succession. At a depth of about 800 m, beyond the edge of the contintental shelf, a salinity minimum marks the core of a fourth water mass, Antarctic Intermediate Water. The physical characteristics of these four water masses carry the signature of their origins.

John Jillett

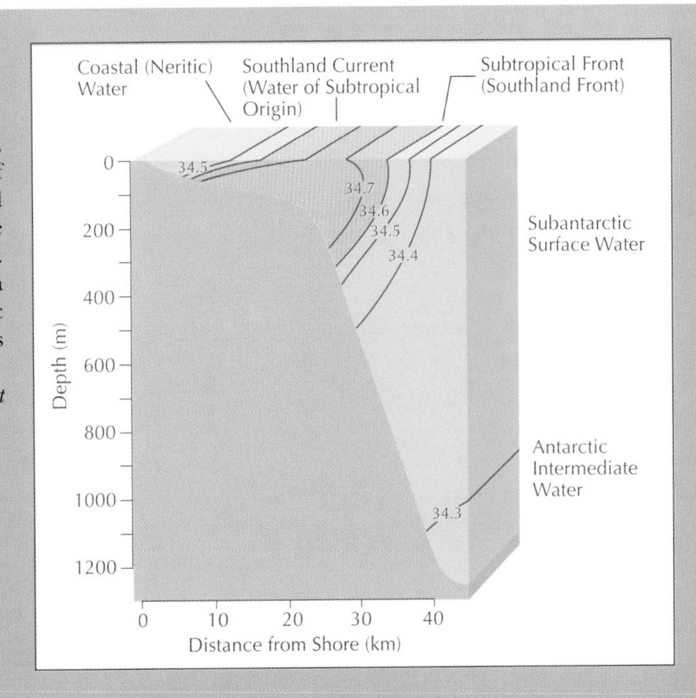

Cross-section of the typical distribution of water masses across the Otago continental shelf and upper slope. Contours are of salinity.

From a study by J.B. Jillett

3 CLUTHA RIVER FLOOD WATERS OFF OTAGO PENINSULA

J.B. Jillett

Flood waters form a discoloured buoyant plume extending from the river mouths into the coastal sea. Here the outer edge of the Clutha River plume is marked by a double crescent of foam-lines as it rounds Cape Saunders on the Otago Peninsula, about three days after entering the coastal sea some 85 km to the south.

At Cape Saunders the northward current flow is usually deflected offshore, generating a southward counter-current on its inshore boundary. The opposite directions of the offshore and inshore currents tend to form an eddy system in the downstream lee of Otago Peninsula. This eddy is a persistent feature of the coastal sea, though it does break down under some weather conditions. Similar eddies form in Molyneux Bay off the mouth of the Clutha River, in the downstream lee of Nugget Point, and off Shag Point and Moeraki. These eddies can strongly influence the way planktonic larvae are recruited into populations of benthic invertebrates and coastal fishes.

John Jillett

PLANKTON AND PELAGIC PRODUCTIVITY

John Jillett

Just as Otago and Southland seas are diverse in their physical characteristics and origins, each water mass has a distinctly different biological community that reflects its past history and origins. There are communities which are characteristic of each of the neritic, subtropical, subantarctic and Antarctic water masses found off southern South Island.

The term 'plankton' refers collectively to marine organisms that are drifted about by tidal and ocean currents. Many are microscopic and few can be seen with the naked eye. Diversely different sorts of organisms are included in the plankton, ranging from viruses and bacteria, through microscopic algae and protozoans, to worms, molluscs, crustaceans and even some fish. The largest members of the plankton are jellyfish, which can be half a metre in diameter. Some groups of organisms are typically planktonic throughout their lives, for example, flagellated microalgae, arrow-worms, salps and many groups of crustaceans. Other marine groups have planktonic forms at some stage of their life history, usually as larvae. Most fish, including nearly all those of commercial importance, have planktonic eggs and larvae. Often the breeding migrations of adult fish and larger crustaceans, for example flounder and rock lobster, are made against the current to compensate for the drift of their developing planktonic larvae (Box 4).

Plankton, then, is defined by lifestyle. Even though planktonic organisms are drifted about, many are very lively when seen under the microscope; it is only their small size that limits their ability to swim against a current. Many planktonic organisms, especially the crustaceans, are capable of impressive vertical migrations that can transfer them between different stratified water masses. Zooplankton can also become densely concentrated at hydrodynamic convergences, where a light water mass overrides a heavier one, and the zooplankton swim up towards the light along the line of downwelling.

Microscopic plant plankton is the basis of marine food chains in the open sea. Beyond shallow nearshore waters, light does not penetrate sufficiently to support the large fixed seaweeds which are so visually characteristic of rocky shores and shallow estuaries. Since there is sufficient light for photosynthetic production only in the upper water layers, all marine life in deeper layers depends either on primary production in the near-surface layers, or on organic food transported from elsewhere.

4 ROCK LOBSTER PLANKTONIC LARVAE

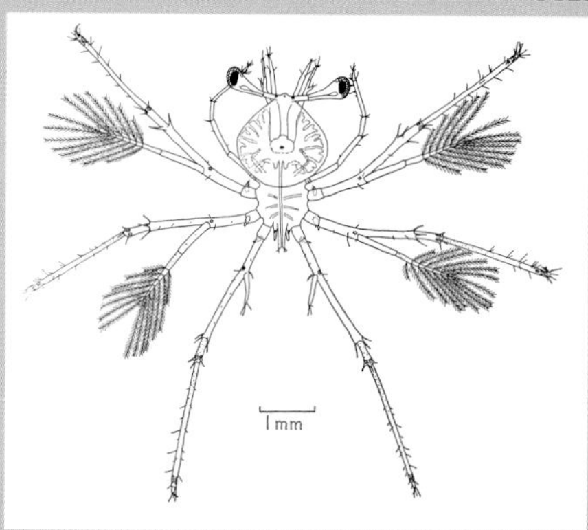

Phyllosome larva of the rock lobster.
After E.J. Batham

The eggs of the rock lobster *Jasus edwardsii* remain attached to the underside of the female's abdomen until they hatch as free-living, leaf-like larvae, which live in the plankton for 12 months or more. During planktonic development, these so-called phyllosome larvae drift with the currents far from their hatching sites. This drift is compensated by the adult lobsters undertaking long migrations, clockwise around southernmost New Zealand, against the prevailing currents (Fig. 7). Since the larval stage is so prolonged, it has been suggested that some larvae that settle in Fiordland could have drifted across the Tasman Sea. In any case, south-eastern Australian and New Zealand rock lobsters are now regarded as the same species.

John Jillett

5 PELAGIC FEEDING IN ACTION

A krill-induced feeding frenzy. *R.L. O'Driscoll*

Surface swarms of zooplankton, aggregated by the hydrodynamic convergence of surface waters of differing density, attract schools of planktivorous pelagic fish and are often associated with flocks of seabirds.

In this example, jack mackerel *Trachurus murphyi* are targeting the euphausiid krill *Nyctiphanes australis*, and are associated with red-billed gulls *Larus novaehollandiae*. Other zooplankton targeted by fish and birds in Otago waters include the pelagic juveniles of the squat lobster *Munida gregaria*, and pelagic larvae of the burrowing mantis shrimp *Heterosquilla tricarinata*. Fish species which form active plankton-feeding schools include other species of jack mackerel *Trachurus novaezelandiae* and *T. declivus*, barracouta *Thyrsites atun*, two species of sprat *Sprattus antipodum* and *S. muelleri*, and slender tuna *Allothunnus fallai*.

John Jillett

Microalgae are only visible to the naked eye when dense enough to discolour the water as 'blooms' in spring or summer. Blooms of the diatom *Heterosigma* sp. often visibly discolour upper Otago Harbour waters a reddish brown in summer. In recent years localised blooms of toxic planktonic algae have accumulated in filter-feeding clams, resulting in a continuing monitoring programme to check on the suitability of shellfish for human consumption. A ciliated protozoan *Mesodinium rubrum* also often discolours open coast waters a diffuse red in late summer and early autumn.

Marine food webs tend to be more complex than those on land, a reflection of the very small body size of the planktonic microalgae, compared with their terrestrial counterparts. Large herbivores cannot cope effectively with such small plants, so more steps are found in most marine food chains. Thus plankton communities encompass successive trophic levels from herbivores to predators.

While most zooplankton is microscopic and often transparent, a few larger species are found by beach walkers on the strandline of surf beaches, especially in late summer and autumn. These commonly include the blue bladders of the Portuguese man-of-war *Physalia physalis*, another siphonophoran *Physophora hydrostatica*, the purple amphipod *Themisto gaudichaudi* and closely related barrel shrimp *Phronima sedentaria*, and a variety of glassy-bodied salps, especially the firmly gelatinous cylindrical tests of *Iasis zonaria*. The last two species, the barrel shrimp and salp have an unusual interaction. Female *Phronima* actively capture *Iasis* tests which they then strip of salp tissue and fashion into open-ended barrels in which to brood their egg masses. Each female is spread-eagled inside its barrel which it propels through the water by the beating of abdominal swimmerets.

Perhaps the most distinctive and spectacular planktonic animal of southern coasts in summer months is the bright red pelagic juvenile stage of the squat lobster *Munida gregaria*, also known locally as whale-feed or, incorrectly, as red krill (Fig. 8). Krill is a term properly applied to euphausiid shrimps, of which *Nyctiphanes australis*, sometimes termed white krill, is a local

Fig. 8. Pelagic juveniles of the squat lobster *Munida gregaria* often form huge shoals in summer, forming red patches at sea. *P.K. Probert*

Fig. 9. The krill *Nyctiphanes australis*. *P.B. Batson*

example (Fig. 9). Both white krill and *Munida* strand in large quantities on southern New Zealand beaches from time to time during summer (page 281). Both species are important items in the diets of pelagic fishes and seabirds (Box 5), but while *Nyctiphanes* is planktonic throughout life, adult *Munida gregaria* are typically abundant on the sea floor of continental shelves of the circumpolar subantarctic zone and only the juvenile is free-swimming. The normal *Munida* swarming season is from November to February, though it can be very brief or sometimes extend into winter.

Nearshore zooplankton is dominated in spring and summer by the larval life-history stages of benthic and intertidal animals, which are present in almost overwhelming diversity and abundance. The advantage of planktonic larvae is thought to lie in the opportunities they offer for the dispersal of species into the widest possible habitats, rather like wind-blown seeds on land. In the open sea this advantage is countered by the possibility that the larvae will be carried to completely unfavourable sites. Along the Otago coast the self-contained eddies that form in the downstream lee of prominent headlands act to trap and retain higher densities of planktonic larvae than are found in the adjacent open sea.

Beyond the edge of the Otago continental shelf are found abundant populations of a globally important copepod, *Neocalanus tonsus*, one of the most significant and abundant marine animals of the Southern Ocean. Adult *N. tonsus* are each about the size of a dried rice grain, but they occur in huge numbers. Researchers have reported swarm densities in excess of 10,000 per cubic metre and have found these copepods to be the most abundant planktonic herbivores over nearly ten degrees of latitude in the circumpolar subantarctic zone. Most of what we know about *N. tonsus* on a global scale is derived from studies off Otago Peninsula. Intense feeding takes place over four or five months in spring and early summer, after which the population descends to over-wintering depths in the last but one life-history stage. In water from a few hundred metres to over 2000 m deep, fasting takes place for seven to eight months, the population eventually moulting to adults. Using fat reserves stored during the previous spring, the new brood is produced, and completes early development, rising to the surface in time to take advantage of the new spring cycle of primary production.

Fig. 10 (top). The wheel shell *Zethalia zelandica* is often very abundant in clean inshore sands, and can wash up in huge numbers on some beaches. Whereas most topshells are grazers, this species appears to be a filter feeder. *P.K. Probert*

Fig. 11. Southern New Zealand has a diverse seastar fauna. A common species of the continental shelf is the jewel star *Pentagonaster pulchellus*, individuals of which can vary considerably in their pattern of markings. The largest individual here is about 70 mm across. *E.J. Batham*

SEABED ANIMALS

Keith Probert

Around southern New Zealand the nature and topography of the seafloor and properties of the overlying water masses vary considerably, as we have just seen. These factors strongly influence the types of organisms that inhabit the seafloor – the benthos – and their distribution and abundance.

At depths greater than a few tens of metres there is insufficient light for algae to live so that the conspicuous organisms of the deeper-water benthos are all animals. Even so, the deeper benthos still depends ultimately on the fallout of organic particles from the productive surface waters. In general, as water depth and distance from land increase, so the quantity and quality of the organic input to the seabed decreases, and so does the amount of seabed life that can be supported. Around southern New Zealand, the biomass, or weight per unit area, of the benthos decreases – following the typical pattern – from some tens of grams wet weight per square metre on the continental shelf to only a few grams on the slope. However, there are probably marked variations to this pattern. Enhanced surface productivity at the Southland Front would, for instance, be expected to provide a greater supply of food to the underlying seabed.

Differences in seabed sediments also strongly influence the structure of benthic communities and the role these organisms play in marine food webs. The seabed off the Otago Peninsula has been extensively surveyed to determine the distribution of invertebrates. As we have seen, sediments of this area consist of a nearshore zone of sand to water depths of 30–50 m depth, whilst shelly, gravelly sediments characterise much of the middle and outer shelf. The Southland Current bathes the middle to outer shelf, whilst inshore there is the more variable neritic water.

The inshore sand zone provides few opportunities for animals that need a firm substratum for attachment and many of the animals of this zone are burrowers. One of the most distinctive animals is the wheel shell *Zethalia zelandica*, an attractive iridescent topshell (Fig. 10). Its empty shells often wash up on beaches in large numbers. In slightly deeper water the sand becomes siltier and *Z. zelandica* is replaced by the smaller topshell *Antisolarium egenum*. Other characteristic animals of this inshore zone are the swimming crab *Ovalipes catharus*, policeman crab *Ommatocarcinus macgillivrayi*, sand shrimp *Pontophilus australis*, midget cuttlefish *Sepioloidea pacifica* and the cushion star *Patiriella regularis*.

Particularly intriguing among the shelf sand fauna are bryozoans of the genus *Otionellina*. They resemble brown buttons up to about 1 cm in diameter and, unlike most bryozoans, are not attached to the seabed. Setae on the upper surface enable the colonies to free themselves of sediment particles and regain the surface if they become buried. Very few bryozoans have managed to adapt to life in such a dynamic sediment habitat.

6 BRYOZOAN THICKETS ON THE OTAGO SHELF

Although the individual zooids of bryozoans are minute, together they can form large almost coral-like colonies, such as the species *Cinctipora elegans*, *Hippomenella vellicata*, and *Celleporaria agglutinans*. The bryozoan patches form small mounds or thickets that represent oases where there is increased diversity of seabed life. Many other invertebrates associate with these bryozoan areas, either growing on the colonies or nestling in the crevices. It has been observed elsewhere that young fish of a number of valuable commercial species tend to associate with bryozoan thickets. Such bryozoan areas are, however, easily damaged by trawling.

Keith Probert

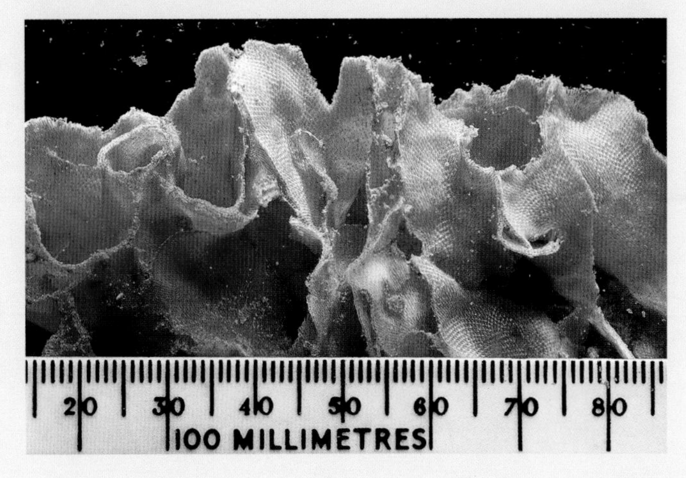

One of the bryozoans from the southern New Zealand continental shelf that forms large colonies. The individual zooids are just visible. *E.J. Batham*

Inhabiting the coarse sediments of the middle and outer shelf are many species of polychaete worms, molluscs and small crustaceans. Among the more conspicuous species on the surface of the seabed are the apricot anemone *Bunodactis chrysobathys*, circular saw shell *Astraea heliotropium*, tritons or trumpet shells *Argobuccinum tumidum* and *Fusitriton laudandus*, southern queen scallop *Zygochlamys delicatula* (the basis of a small fishery), common sea cucumber *Stichopus mollis*, snake star *Ophiopsammus maculata*, jewel star *Pentagonaster pulchellus* (Fig. 11) and apricot star *Sclerasterias mollis*. Common crabs include the red swimming crab *Nectocarcinus antarcticus*, triangle crab *Eurynolambrus australis* and the spider crab *Leptomithrax longipes*. This last often carries sea anemones, which may protect it from predation by octopus.

These more gravelly sediments of the middle and outer shelf provide opportunities for animals that need a stable surface on which to attach, and a feature of this zone is an abundance of sponges, hydroids, anemones, soft corals, tube worms, bryozoans and sea squirts. Many of these animals feed on waterborne particles of organic material which they filter from the overlying water. In this regard, a striking feature of the middle and outer shelf fauna off the Otago Peninsula is the abundance of bryozoans, filter-feeding animals, some of which build large calcareous colonies (Box 6). Their abundance here is probably supported by the Southland Current flowing more swiftly as it is squeezed between the out-jutting Peninsula and the offshore subantarctic water. This keeps the mid-shelf bottom sediments pebbly – good for colony attachment – and ensures a copious supply of waterborne food.

Other invertebrate groups are also well represented here. The seafloor supports, for instance, a rich echinoderm fauna, including many species of seastar – some of which are mentioned above – and sea urchins such as *Goniocidaris umbraculum* and species of *Pseudechinus*. Also conspicuous on the Otago shelf and upper slope are hermit crabs. The soft abdomen is modified to fit into a shell, usually that of a gastropod snail, which the hermit crab carries as a protective home. At least 17 species are known from this area, representing a third of the known New Zealand fauna. Bryozoans feature here, too, since several of these hermit crab species have symbiotic associations with bryozoans, whose colonies form tubes that the hermits inhabit (Fig. 12).

Fig. 12. Hermit crab inhabiting a tube made by a bryozoan. *E.J. Batham*

Related to hermit crabs are the squat lobsters or galatheids. One of these, *Munida gregaria*, is a very conspicuous element of the southern New Zealand fauna, and the dramatic seasonal appearance of the pelagic juveniles has already been mentioned (Fig. 8). The adults, however, are benthic and probably important as predators and scavengers. In fact, much of the benthic production of these coarse sediments appears to be consumed by invertebrate predators, including gastropods, decapod crustaceans and seastars.

Nevertheless, many species of fish also feed on benthic invertebrates. Among those common on the southern New Zealand shelf are rig *Mustelus lenticulatus*, carpet shark *Cepahaloscyllium isabella*, rough skate *Raja nasuta*, elephant fish *Callorhynchus milii*, red gurnard *Chelidonichthys kumu*, southern pigfish *Congiopodus leucopaecilus*, tarakihi *Nemadactylus macropterus*, blue moki *Latridopsis ciliaris*, common warehou *Seriolella brama*, witch *Arnoglossus scapha*, sand flounder *Rhombosolea plebeia*, greenback flounder *Rhombosolea tapirina*, brill *Colistium guntheri*, lemon sole *Pelotretis flavilatus* and common sole *Peltorhamphus novaezeelandiae*. A number of species are taken in the mixed-trawl fishery of the region.

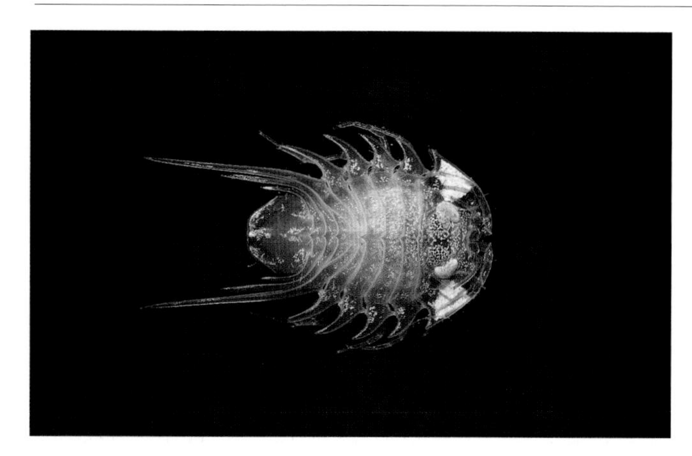

Fig. 13. The deep-sea isopod *Acutiserolis bromleyana*, one of the commoner and distinctive invertebrates of the continental slope off southern New Zealand. Individuals are typically up to about 30 mm in length. *E.J. Batham*

Benthic communities similar to those off the Otago Peninsula are widely distributed around southern New Zealand. Related faunas from coarse sediments cover extensive shelf areas off Otago and Southland, and include such conspicuous molluscs as the dog cockle *Glycymeris laticostata*, purple cockle *Venericardia purpurata*, morning star shell *Tawera spissa*, queen scallop and trumpet shell *Fusitrition laudandus*. A similar shelly gravel fauna occurs in Foveaux Strait, although in this case the best known member of the benthos is the Bluff oyster *Tiostrea chilensis* (page 326).

At the shelf edge (about125–150 m depth) the seabed gives way to the steeper gradient of the continental slope. A feature of the slope off Otago is a series of large canyons, tributaries of the Bounty Trough to the east, that cut deeply into the outer shelf-upper slope area (Fig. 4). There is a change in the benthos too and by about 450 m water depth we meet a distinctive and diverse slope fauna bathed by subantarctic water. Unlike the generally coarse sediments of the shelf, these slope sediments are predominantly muddy and many members of the seabed community are small burrowing species. Conspicuous animals on the surface of the sediment include a small stalked sponge *Stylocordyla borealis*, the quill worm *Hyalinoecia tubicola*, hermit crabs and the large isopod *Acutiserolis bromleyana* (Fig. 13). Among the molluscs are an iridescent deep-sea topshell *Otukaia blacki*, a pale pink whelk *Aeneator recens*, a miniature translucent scallop *Parvamussium maorium*, a trumpet shell *Cymatona kampyla* and a golden volute *Iredalina mirabilis*. Rocky outcrops on the canyon walls provide opportunities for attached animals including a deep-water cup coral *Flabellum knoxi* and a perch for the yellow feather star *Florometra austini*. However, knowledge of New Zealand's continental slope benthos is still sketchy. Although few slope areas have been studied in detail, it appears that many upper slope species are widely distributed in the region.

Fig. 15 (opposite). The blue shark *Prionace glauca*, a widely distributed species of the open ocean that frequents New Zealand waters. *M. Francis*

FISH AND FISHERIES

Tony Brett

The composition of a region's fish fauna – like that of other groups of marine organisms – is the product of several factors, in particular evolutionary history and range of environmental conditions and habitats available. Southern New Zealand is influenced by water masses with different characteristics, in terms of temperature, salinity and nutrients levels, which in turn affect productivity and food-web structure. Thus a key factor for a number of southern pelagic species is the seasonal occurrence of prey species such as the krill *Nyctiphanes australis*, juvenile *Munida gregaria* and arrow squid (*Nototodarus* spp.). For bottom-living species, water depth, bottom topography and sediment type are also important. Depth and substrate type influence not only species' distributions but also the method of fishing that may be employed.

More than a thousand species of fish are known from New Zealand waters. Although many are widely distributed, one can recognise marked regional differences in the country's fish fauna, including species that have a distinctly southern distribution or are more abundant in the south. For instance, predominantly reef and coastal species that are largely southern include the southern pigfish (Fig. 14), girdled wrasse *Notolabrus cinctus*, thornfish *Bovichtus variegatus*, blue cod *Parapercis colias* (page 306), common warehou and Maori chief *Paranotothenia angustata*. The latter belongs to a family (Nototheniidae) of mainly Antarctic and subantarctic fishes for which southern New Zealand is a northern limit. The latrids, a group of fishes of cold temperate waters, are well represented around southern New Zealand by blue moki, copper moki *Latridopsis aerosa*, trumpeter *Latris lineata* and telescope fish *Mendosoma lineatum*. The flatfish include such mainly southern species as greenback flounder, brill, lemon sole, and common sole. Predominantly deeper water species of the outer shelf and upper continental slope that are mainly southern include red cod *Pseudophycis bachus*, sea perch or Jock Stewart *Helicolenus papillosus*, ling *Genypterus blacodes*, hake *Merluccius australis*, hoki *Macruronus novaezelandiae* and dark ghost shark *Hydrolagus novaezelandiae*.

South-east mixed fishery

The shelf waters of Otago and Southland are characterised by fisheries that take a range of species, with the principal groupings defined largely by depth. Inshore groupings consist of flatfish, skate and gurnard with lesser numbers of red cod, elephant fish, rig and blue cod. Further out on to the shelf, spiny dogfish *Squalus acanthias*, groper *Polyprion oxygeneios*, tarakihi, school shark *Galeorhinus galeus*, stargazer *Kathetostoma giganteum* and arrow squid are taken. Red cod occurs at all depths. At the shelf edge and on the upper continental slope, hoki and rattails (Macrouridae) become important.

The mid to outer shelf (50–200 m) is fished by trawlers, 18–30 m in length, operating from Timaru, Dunedin and Bluff. These vessels mainly fish for barracouta *Thyrsites atun*, ling, squid, red cod, tarakihi, stargazer and ghost shark, but may also take warehou, gemfish *Rexea solandri* and rattails. The inner shelf, to

Fig. 14. The southern pigfish *Congiopodus leucopaecilus*, a species characteristic of southern coastal waters. *M. Francis*

depths of about 50 m, is mainly fished by smaller local trawlers, setnetters and line fishers operating from local coastal towns and ports. This fishery has a seasonal component, usually as an alternative to rock lobster fishing. Principal species of the Otago inshore fishery are flounder and sole, red cod, gurnard and tarakihi, but barracouta, spiny dogfish and other cartilaginous fishes are also caught. A small line fishery for groper exists off Oamaru and Taieri Mouth.

Off the Southland coast, barracouta, stargazer and flatfish are the major species. Flatfish, in particular, are trawled from inshore soft sediment areas such as Te Waewae Bay, and farther north off Nugget Point and Blueskin Bay. West of Stewart Island the shelf is deeply cut by the Solander Trough, where lining for ling becomes important. The shelf is very narrow off the Fiordland coast and the predominant fisheries are netting and lining for school shark and groper, with a bycatch of bluenose *Hyperoglyphe antarctica*.

Sharks

A group of fish that arouses much public interest in the region is sharks, well known as carnivores or scavengers. A number of species frequent southern New Zealand waters and several contribute to valuable fisheries, including school shark or tope, spiny dogfish, rig and black shark, or seal shark, *Dalatias licha*. Species targeted by game fishers include the blue shark *Prionace glauca* (Fig. 15) and mako *Isurus oxyrinchus* on account of its fighting ability. However, the species most renowned in Southland and Otago waters is the great white shark, or white pointer *Carcharodon carcharias* (Fig. 16). This magnificent predator can attain a large size (sometimes at least 6 m in length) and is certainly dangerous, most attacks on people being attributable to this species (Box 7). In these southern waters it often haunts seal colonies during the breeding season, seals being reported as important prey. There is also evidence of mature great white sharks having a seasonal beat as they travel between hunting areas.

Fig. 16. This great white shark *Carcharodon carcharias* was found in a fishing net at the entrance to Otago Harbour. It was a female measuring 5 m in length and had in its stomach a skate and various molluscs and crabs. *P.K. Probert*

Sharks and rays, together with the small group that includes the ghost sharks and elephant fish, comprise the cartilaginous fishes. They differ markedly from bony fishes, most obviously by their cartilaginous skeleton and other morphological features, and by their reproductive biology. Fertilisation is internal. In some species each large egg is encased in a capsule and left on the seabed to hatch, but in many cartilaginous fish, the eggs develop inside the female and the young are born at an advanced stage of development. Whether egg- or live-bearing, the female can afford to produce only a small number of young, in stark contrast to most bony fish. In addition to having low fecundity, most species of cartilaginous fish are also slow growing, late to reach sexual maturity and long lived. These life-history characteristics render them very vulnerable if fishing pressure is not managed. In some countries certain species, such as the great white shark, are protected by legislation.

Dredge oyster

There are two important open-sea shellfisheries of southern New Zealand – those for dredge oysters *Tiostrea chilensis* and queen scallops *Zygochlamys delicatula.* The Bluff oyster fishery is one of the few remaining wild dredge or flat oyster fisheries in the world. It is also one of the oldest commercial fisheries in New Zealand. Oysters were first collected for sale from Port Pegasus, Stewart Island in the 1830s and later from Port Adventure in the 1860s. In 1888, large beds were found in the eastern and central regions of Foveaux Strait, and these formed the basis of the fishery for the next 80 years. As a result of more information on oyster density, together with declining catch rates, fishing progressively intensified in the western and central areas of the Strait from 1960onwards, reaching a maximum harvest of more than 164,000 sacks (about 13,000 tonnes) in 1967.

7 DUNEDIN SHARK ATTACKS

Dunedin has gained a certain notoriety among swimmers and surfers as a hotspot for shark attacks. In the 1960s there were three fatal attacks and one non-fatal attack off Dunedin beaches, probably all by great white sharks *Carcharodon carcharias.*

Following this spate, shark nets were installed in 1969 off Dunedin's St Clair and St Kilda beaches, and in 1976 Brighton beach, just south of Dunedin, was included in the programme. Two nets are laid off each of these three beaches over December to February and checked at least twice a week.

Each net is some 110 m long, 5.5 m high and has a mesh size of 30 cm. Since the nets were installed there has been one non-fatal attack at St Clair in 1971, but this occurred after the nets had been lifted for the season. Since then there have been no attacks at these netted beaches, though sightings of sharks in Dunedin waters are quite often reported.

This certainly gives the impression that netting provides protection. However, accurate records of catches from the netting programme have been kept only since 1986. From then until the 1997–98 season, the total number of sharks taken from the three netted beaches has averaged nineteen per season, with about half taken from Brighton. About 90 per cent of sharks caught over this period were broadnose sevengill sharks *Notorynchus cepedianus,* school sharks *Galeorhinus galeus* and thresher sharks *Alopias vulpinus.* No great white sharks were taken, though newspaper reports from the early 1970s (before accurate catch records were kept) suggested that between two and seven great white sharks were then being caught each summer.

It is not entirely clear, therefore, to what extent netting provides effective protection or if other factors may be at work, such as a widespread decline in the great white, a species known to attack humans. There is, however, a suggestion that waters of the Dunedin region have more than their fair share of sharks because of a temperature cul-de-sac. The current of subtropical origin that flows anticlockwise around southernmost New Zealand (Fig. 7) has cooled significantly by the time it reaches Dunedin, and this may inhibit some shark species from continuing up the coast.

Data from the Dunedin City Council's Shark Net Policy Review, March 1998

In recent years the fishery has had to be managed in response to an outbreak of the parasite *Bonamia*. This protozoan has been responsible for the demise of most dredge oyster fisheries around the world. Measures were undertaken in 1991 and 1992 to reduce the impact of the parasite on the beds. Unfortunately these generally proved unsuccessful and the oyster population dropped to below 20 per cent of the 1975 level. Consequently, all commercial dredging was prohibited between 1992 and 1995. By 1996 the beds had recovered sufficiently to allow a small commercial season (approximately 15,000 sacks). With careful management and fishing practices the fishery should continue to rebuild, although it is not known if *Bonamia* will recur, or if all beds in the Strait will recover to their pre-*Bonamia* size. To assist recovery, the industry has initiated an oyster reseeding programme in suitable areas of the Strait.

Queen scallops

A small fishery for the deep-water queen scallop has been operating since the mid 1980s off the Otago coast. The fishery, the only one of its type in New Zealand, has a sustainable catch level of 750 tonnes, although landings have always fallen well short of this. Queen scallops (Fig. 17) are smaller (60–70 mm) than the common scallop *Pecten novaezelandiae* and occur in considerably deeper water (80–200 m). They also have two convex shells rather than one convex and one flat shell. But, like the common scallop, they have a series of eyes around the mantle edge and can swim actively to avoid predators.

Squid

The New Zealand arrow squid fishery is based on two closely related species, *Nototodarus gouldi* and *N. sloanii*, and yields about 100,000 tonnes per annum, primarily by bottom trawling between December and June. Squid fishing in Otago and Southland occurs mainly off Taieri Mouth, where jigging vessels take around 200 tonnes per annum. Both species live for about one year, spawn and then die. Information suggests that there is a spring and autumn spawning, but details and location are unknown. Growth is rapid, up to 4–5 cm per month, and tagging experiments indicate that arrow squid are capable of daily movements of up to 5–6 km.

Recreational gamefish

Despite its cooler waters, the lower South Island has thriving recreational game fisheries. Between late November and March, fishers target blue, mako and porbeagle *Lamna nasus* sharks off the Otago coast, and catch the occasional thresher *Alopias vulpinus* and white pointer shark. Off Dunedin, slender tuna *Allothunnus fallai*, skipjack *Katsuwonus pelamis* and albacore *Thunnus alalunga* are targeted between March and June. From January to late May fishers are attracted to waters off Fiordland by the presence of tuna; butterfly tuna *Gasterochisma melampus*, southern bluefin tuna *Thunnus maccoyii* and albacore are all available during this time. Although more coastal, another significant recreational fishery of southern New Zealand is that for quinnat salmon *Oncorhynchus tshawytscha* (Box 8).

Fig. 17. A haul of queen scallop *Zygochlamys delicatula* dredged from the Otago shelf. The seastar is *Sclerasterias mollis*. *R. J. Street*

8 SALMON FISHERIES

Southern New Zealand supports important recreational salmon fisheries. In the traditional open coast fishery, anglers target adult salmon that are returning to the headwaters of rivers to spawn. The most successful fishing, for chinook or quinnat salmon *Oncorhynchus tshawytscha*, occurs at river mouths, especially the Waitaki and Clutha Rivers.

A recent innovation is the Otago Harbour salmon fishery. This was established in 1986 by the Otago Branch of the New Zealand Salmon Anglers' Association and has developed into a highly successful fishery. Each year contributors and sponsors fund smolt releases into the harbour. Adult salmon return after two to four years' growth at sea, highly prized by hopeful anglers fishing from boats and wharves. Salmon smolt are now also being released into Bluff Harbour, and this fishery may develop into a popular recreational opportunity for the people of Southland.

Tony Brett

Salmon taken in Otago Harbour. *K. Robinson*

SEABIRDS

Hamish Spencer and John Darby

The waters around southern New Zealand abound with seabirds. In fact, this region has more species of seabirds than of songbirds. Many are seen only from boats, although on occasions truly oceanic species can be seen feeding close to shore. Shags, gulls, terns and skuas can be found feeding up to 15 km out to sea (and sometimes further), but they are really coastal species (see Chapter 11). Many of the seabirds seen in this region nest in New Zealand's subantarctic region, although there are some notable exceptions. Species that breed elsewhere may be present in the local seas for only a short time each year, often during migration, although sometimes in very large numbers.

Albatrosses

The most prominent of the local breeders is probably the northern royal albatross *Diomedea epomophora sanfordi*, since the colony near Dunedin is a major tourist attaction. This bird holds a special place in southern biology by having established a small breeding colony at Taiaroa Head on the Otago Peninsula, giving it the distinction of being the only albatross to nest on the mainland (Box 9). The main population breeds at the Chatham Islands. Its close cousin, the southern royal albatross *D. e. epomophora* can also be seen off the Otago and Southland coasts, but breeds at the Campbell and Auckland Islands. The southern and northern royals are the largest albatrosses, feeding on squid, other pelagic invertebrates and fish caught at the sea surface usually at night. They can be distinguished on the wing by their respective wing and back colours. The northern subspecies has a solid brownish black upper wing and a contrasting white back (although juveniles do have some brownish black on their lower backs). The southern subspecies starts life similarly, but the wing progressively whitens from the front as the bird ages. The other large albatross, the wandering albatross *D. exulans*, is less common in this region. Its diet is similar to that of the royals, although unfortunately it will also take the baited hooks behind long-line tuna fishing boats, and recent declines in some populations have been linked to this mortality.

New Zealand is home to a greater variety of smaller albatrosses, or mollymawks, than any other country. The species most likely to be seen from shore is Buller's mollymawk *D. bulleri*, a handsome species with a bright yellow top and bottom to its bill, and a thick black front to the underwing. The shy mollymawk *D. cauta* is the largest of the mollymawks, with two subspecies (possibly species) seen regularly in the waters off Otago and Southland, namely the New Zealand white-capped mollymawk *D. c. steadi*, which breeds mainly at the Auckland Islands, and the darker headed Salvin's mollymawk *D. c. salvini*, which nests on the Bounty Islands and Snares Island. Both the New Zealand black-browed mollymawk *D. melanophrys impavida*, which breeds only on the northern part of Campbell Island, and the nominate southern subspecies *D. m. melanophrys*, which breeds in small numbers on several of the subantarctic islands, are also found off Otago and in Foveaux Strait. They feed on crustaceans, jellyfish and salps, usually caught at the surface during the day. Most mollymawks will also follow squid and tuna fishing boats, scavenging for offal or chasing bait. Significant numbers are accidentally killed each year by flying into or getting caught in fishing gear.

Petrels, shearwaters and prions

Albatrosses reported dead on local beaches almost invariably turn out to be one of the giant petrels, also called stinkers or nellies. Giant petrels have a wingspan of almost 2 m, which is greater than that of some of the small mollymawks. The southern giant petrel *Macronectes giganteus* comes in two colour phases, the more usual one dark grey-brown and the other white with small brown flecks, but both morphs have a greenish tip to the yellow bill, and dark eyes. The northern giant petrel *M. halli* is all but identical to the dark phase of the southern, but the bill has a reddish tip and the iris is pale. Graceful fliers, they are both aggressive scavengers on dead birds, whales and seals, also preying on a wide variety of fish and pelagic invertebrates, mostly during the day.

The sooty shearwater *Puffinus griseus* (Fig. 18), better known as the muttonbird or titi to most New Zealanders, once nested in large numbers on the Otago and Southland coasts. There

Fig. 18. Sooty shearwater *Puffinus griseus* on the ground (left) and in flight. *Jamie Newman*

9 THE ROYAL ALBATROSS OF TAIAROA HEAD

In the beginning ....

The first pair of royal albatross laid at Taiaroa Head apparently in November 1920: the egg was promptly confiscated by a nearby resident. From 1920 to 1934, at least one egg appears to have been laid annually, but was always removed. In 1935, however, for some strange reason, nobody wanted the egg, and a baby albatross was actually allowed to hatch, but unfortunately was killed soon after. In November 1936, I paid my first visit to Taiaroa Head, and there on a grassy path, before my astonished gaze, sat a male albatross incubating a large white egg. That egg, too, was removed. In 1937, the birds laid again, so I made up my mind to do all possible to prevent a repetition of previous losses ...

L.E. Richdale, 1950

Conservation of the royal albatross at Taiaroa Head on the Otago Peninsula has proved to be most successful, with the breeding population increasing from some eight birds in 1937 to more than 100 by 2003.

J. Darby

This was the beginning of what has become one of Otago's conservation successes. From a resident population of about eight adults in 1937, the albatross colony grew to over 100 by 2003. Most of these birds fledged from Taiaroa Head, although some have immigrated from the main population (6000–7000 breeding pairs) on The Sisters and Forty-Fours Islands in the Chatham Islands group. More than 200 chicks have fledged at Taiaroa Head, the first on 22 September 1938.

The presence of the northern race of the royal albatross *Diomedea epomophora sanfordi* at Taiaroa Head since the late 1800s is undoubtedly due to the clearance of the scrub and stunted bush that once graced the steep slopes of this mainland 'island'. Much of the present breeding area was fenced off from the public at the instigation of Lance Richdale in 1937, with the assistance of the Otago Branch of the Royal Society and the Otago Harbour Board. World War II interrupted further studies and conservation efforts with the refortification of Taiaroa Head. In 1950, using funds raised by Dunedin Rotary Clubs, and with the co-operation of the Department of Internal Affairs and the Otago Harbour Board, additional fencing was added and a full-time ranger appointed to protect the colony. The area was gazetted as a fauna and flora reserve in 1964, with control vested in the Minister of Internal Affairs. During 1968 the Otago Peninsula Trust initiated discussions with the Wildlife Service, with a view to opening the area for public viewing. This was achieved in 1972. Over 60,000 people now visit the colony annually.

The average age of first breeding is between nine and ten years, although young birds frequently appear at the colony at aged five years and onwards. A single white egg is laid in mid-November and incubation is shared between both sexes, with spells of 4–10 days. Chicks hatch some 79 days later in late January or early February. Chicks are guarded by at least one of the parents continuously for the first 36 days, and they fledge between 219–265 days after hatching, usually about September of the following year.

Because of this prolonged breeding season, royals only breed every second year. Pairs mate for life, rarely divorcing. The oldest known bird, fondly known as 'Grandma', died at sea in 1989 at the grand old age of at least 61 years. She had been banded by Richdale in December 1937, one of the first albatross ever studied in detail anywhere in the world.

Young birds spend their first five years at sea, first flying eastwards across the Pacific. They winter in the south Atlantic off the coast of South America, mainly in the Argentine region. Their long journey back to their breeding grounds consists of flying around the subantarctic, across the south Indian Ocean and southern Australian seas, and back to New Zealand.

Visiting the albatross centre at Taiaroa Head is a unique experience: an opportunity to see the graceful flight of the largest of the world's albatross, birds at the nest and the gams formed during the courtship. But this windswept headland also harbours a constant threat to the colony's survival. In most years, introduced mammals, usually stoats, ferrets, and the occasional cat, make their way through fences and quickly dispatch guarded or unguarded chicks. For the Department of Conservation rangers who patrol this area, predator control is a constant battle, a battle that from present trends will never cease.

John Darby and Hamish G. Spencer

remain a few remnant mainland breeding colonies at Cape Wanbrow, the Otago Peninsula, Nugget Point and further south, but most of these sites have been devastated by ferrets and stoats. Flocks of titi in their millions can still be seen, however, from southern headlands as they move southwards to their subantarctic breeding islands during September–October and on their return in late March to their wintering grounds in the North Pacific. They prey on small fish and crustaceans caught by swimming underwater, often to surprising depths, 67 m once being recorded.

Of the smaller petrels and shearwaters, the Snares cape pigeon *Daption capense australe* is a frequent companion of fishing vessels, sometimes even entering local harbours. They feed on surface crustaceans, usually by sitting on the water and pecking, pigeon-like (hence their name), but sometimes diving after food, especially offal. Up to a third of their food is caught at night. Unlike most seabirds, they are noticeably vocal at sea, constantly making a churring chatter.

Several prion species can be seen off the coast, sometimes in their thousands, or found blown far inland, and there are remnant colonies around the Otago coast. Probably all six species

10 YELLOW-EYED PENGUIN

The yellow-eyed penguin has a special place in the natural history of Otago and Southland. It was the subject of study by the late Dr Lance Richdale, the first person to individually mark penguins. He followed their lives and loves from 1936 to 1954 in one of the most detailed population studies of its day. But sadly, by the 1970s it had become the forgotten bird, possibly because Richdale's study had been so intensive that few felt there was much more that could be done. The Otago Museum, however, began a further study in 1979 and found a serious depletion of habitat and a marked reduction in population from those of earlier years. Severe predation by ferrets, stoats and feral cats was taking a very heavy toll of chicks. Small breeding areas were losing all chicks year after year, and larger breeding areas were experiencing chick losses averaging 66 per cent in contrast to the 10–12 per cent recorded by Richdale.

The Yellow-eyed Penguin Trust was formed in 1987 and, together with the Department of Conservation, members of the farming community and sponsorship from the Dunedin company Mainland Cheese, set about purchasing habitat for the species. The gradual decline in numbers was not arrested until 1991, when the mainland population increased from 130 pairs in 1990 to a high of 600 pairs in 1996. The number of breeding pairs fluctuates from year to year and between 1996 and 2002 the number ranged from 400 to 600 pairs. There has also been a marked shift in population distribution. Whereas in the 1940s and 1950s most breeding pairs were in the Catlins, two-thirds of the mainland population are now on the Otago Peninsula.

Although Richdale left few records of bird numbers, he estimated the number of breeding birds at Penguin Bay and Hinahina Cove in the Catlins in 1942. At the former site he estimated 200 birds and at Hinahina Cove 100 birds. By February 1982 the forest at Penguin Bay had all but disappeared, apart from a patch of about 0.4 ha which contained just three chicks and their parents, whilst at Hinahina more than half the forest had been destroyed and some eight pairs were left. One of the implications of habitat loss is that birds will not breed successfully if the nest is within sight of another pair. Birds occasionally eyeball each other from their nests, but invariably fail to breed in such situations.

The yellow-eyed penguin is found in scattered groups from Banks Peninsula in Canterbury to Slope Point in Southland, with their greatest concentration on the Otago Peninsula. They are also found on Stewart Island in low numbers and on the Auckland and Campbell Islands. *J. Darby*

Adults are resident the year round, although in poor food years, chicks and some adults may move as far north as Cook Strait. Breeding areas on the mainland are found from Cape Wanbrow at Oamaru, to Slope Point in Southland. Up to 400 pairs may be found on Stewart Island and its outliers, but numbers have been known to go below 250 pairs in recent years. Nevertheless, with a total population of 1500–1700 pairs the yellow-eyed is the world's rarest penguin.

About half the females begin breeding as two-year-olds (remarkably young for penguins) and 95 per cent of all females will have attempted breeding by the time they are three. Most males start breeding at 4–6 years. Courtship begins as early as July–August and birds appear to defend large territories throughout the year. Rarely do birds use the same nest from one year to another, although nests are usually sited within the overall territory of a pair. Territories can be as large as one hectare, or small ones less than 10 square metres. Two eggs are laid between mid-September and mid-October, although females breeding for the first time usually lay a single egg. Incubation takes six weeks, with the pair sharing equally the responsibility of incubation and chick rearing. Chicks hatch during the second to third week of November and fledge in February-March, some 106–120 days after hatching. At fledging they are usually heavier then their parents and may weigh up to 7 kg.

The population can fluctuate widely. Food shortages at sea appear to be the usual reason for declines. However, the loss of more than 50 per cent of all breeding birds in 1989–90 and almost 20 per cent in 1996–97 still has scientists puzzled, as most of the birds found dead at nest sites or washed ashore looked to be in good condition. Nevertheless, that the population can recover rapidly from such setbacks demonstrates the ability of this species to respond to management.

Management has included the creation of public and private reserves, extensive revegetation of breeding areas, and a commitment to constant trapping of predators, particularly ferrets and stoats. The very high loss of chicks to predation in the early 1980s – up to 86 per cent in large breeding sites – has declined to less than 12 per cent in recent years.

Conservation of the yellow-eyed penguin on the mainland has been a success story. The species now features as a major international tourist attraction, especially in the Dunedin area. The bird's image has appeared on a commemorative coin, the $5 banknote, tens of thousands of T-shirts and millions of cheese wrappers. However, these successes have placed an enormous load on the various agencies that took up the cause. If the species is to remain on the mainland in good numbers it will have to be managed in perpetuity, an expensive and time-consuming cause, but one surely worthwhile.

John Darby

occur, but they are very difficult to distinguish at sea. The commonest are undoubtedly the fairy prion *Pachyptila turtur* and the broad-billed prion *P. vittata*, both of which have healthy colonies on rat- and mustelid-free islands in Foveaux Strait and off Stewart Island. The latter species has an upper bill edged with comb-like lamellae that it uses to sieve small crustaceans from the water, but it also eats small squid caught at the surface. About a third of its feeding takes place at night. The fairy prion takes a wider variety of pelagic fish, molluscs and arthropods, almost exclusively during the day.

Several shearwaters occur around southern New Zealand. Hutton's shearwater *Puffinus huttoni* now breeds only on the Seaward Kaikouras in north-eastern South Island, but is frequently seen off the Otago coast. The very similar fluttering shearwater *P. gavia*, which breeds mostly on small islands around the North Island, also occurs off Otago. It is distinguished in good views by having whiter underparts. Both species can sometimes be seen in large flocks sitting on the water, with their heads under the surface looking for small fish and crustaceans, which they then chase by 'flying' under the water, down to about 10 m. The occasional Buller's shearwater *P. bulleri* may also be seen off the Otago coast in summer, well away from its breeding grounds, Northland's Poor Knights Islands. It usually feeds by sitting on the surface with open wings, lunging at small fish and crustaceans. Non-breeding flesh-footed shearwaters *P. carneipes* are found off the east coast as far south as Foveaux Strait in the summer. The subantarctic little shearwater *P. assimilis elegans* breeds mostly on the Antipodes Islands and probably occurs around Stewart Island and Foveaux Strait. In some years, large numbers of the Australian version of the sooty shearwater, the short-tailed shearwater *P. tenuirostris*, fly through Foveaux Strait on their way to or from breeding.

Two species of very similar large, dark, brown-black petrels occur in different parts of the region. Fairly common off Otago and Southland is the so-called white-chinned petrel *Procellaria aequinoctialis*, the New Zealand race of which has little if any noticeable white on the chin. It breeds during the summer on the Campbell, Auckland and Antipodes Islands. It seizes squid and crustaceans off the sea surface at all times of the day and night, and will also steal food from larger birds, such as giant petrels. The Westland petrel *P. westlandica* differs in having a dark tip to the bill (and having no white at all on its chin), but is a winter breeder, in just one place, coastal forest south of the Punakaiki River in Westland. It may be seen off the Fiordland coast and occasionally off North Otago. In recent years this species has more than doubled its population to over 15,000 birds, mainly as a result of the large-scale fishing off the South Island west coast. Fish waste has become the dietary mainstay of chicks.

Of the smaller petrel species, the mottled petrel *Pterodroma inexpectata* is mostly seen near its breeding grounds in Fiordland, Stewart, Codfish and Snares Islands, although in late summer large flocks may be encountered off the Otago coast on their post-breeding northward migration. They once bred extensively in the hills of Otago, in small burrows. This and many of the related species seem to feed exclusively at night. Cook's petrel *P. cooki* has a small colony on Codfish Island, and is sometimes seen foraging in Foveaux Strait. Predation by weka *Gallirallus australis* reduced this colony from 20,000 pairs to just 100 in about 80 years, but wekas have now been eradicated from the island. Sometimes other 'gadfly petrels', such as the soft-plumaged petrel *P. mollis*, the black-winged petrel *P. nigripennis*, Gould's petrel *P. leucoptera caledonica*, the white-headed petrel *P. lessonii* and the grey-faced petrel *P. macroptera gouldi* may also be seen in migration. Rarely seen in New Zealand waters before 1970, the gull-like Antarctic fulmar *Fulmarus glacialoides* and the Antarctic petrel *Thalassoica antarctica* seem to be increasing in numbers, with small flocks seen flying in Foveaux Strait and off Fiordland, and dead birds washing up regularly on Southland beaches.

Diving petrels and storm petrels

Two almost indistinguishable species of diving petrel, the common *Pelecanoides urinatrix* and the South Georgian *P. georgicus*, breed on Codfish Island, the former much more commonly and also on Stewart and Solander Islands. They are small stocky birds that fly 'like bumblebees', their wings constantly and rapidly beating as they skim over the top of, or even through, the waves. The smallest seabirds are the storm petrels, that appear to walk on the water as they dabble for small food items. Several species, in particular the grey-backed *Oceanites nereis*, the white-faced *Pelagodroma marina* and the black-bellied *Fregetta tropica*, can be expected at sea, but they are difficult to see and rarely reported. All three of these species breed on several of the subantarctic islands.

Gannet

The Australasian gannet *Sula serrator* is an enigmatic bird in southern New Zealand. Although frequently seen diving for fish off the coasts of Otago and Southland and known to breed at Nugget Point, the number of nesting pairs known in the south is less than a dozen and one can but wonder where these coastal flybys come from.

Penguins

Penguins are a distinctive feature of southern seas. Prominent on the east coast is the yellow-eyed penguin or hoiho *Megadyptes antipodes* (Box 10). But more abundant is the blue penguin *Eudyptula minor* (Fig. 19). It usually nests in groups in small burrows or under buildings, and is very noisy after coming ashore soon after dark. Between these two species in size, the Fiordland crested penguin *Eudyptes pachyrhynchus* breeds in good numbers – about 2200 pairs – on the west coast, from Fiordland to Stewart and Codfish Islands, and is a frequent visitor to the south and east coasts in late summer after breeding. Both it and the yellow-eyed penguin are forest nesters, but the latter's nests are usually out of sight of others, whereas those of the Fiordland crested are usually clumped in small colonies. All three species feed on a range of food that may include squid, octopus, a wide variety of fish species and occasionally krill. Yellow-eyed penguins have been recorded as deep as 160 m.

Otago and Southland coasts also boast a greater variety of visiting penguin species than any other coast in the South Pacific – at least eight species have been recorded as occasional visitors.

Fig. 19. The blue penguin *Eudyptula minor*, one of the two penguins colonising the east coast of Otago and Southland (see also Box 10). Visiting penguins include all three species of pygoscelid penguins, the Adelie, chinstrap and gentoo, four species of crested penguin, the rockhopper, royal, snares and erect crested penguins and the king penguin. *Darren Scott*

SEALS AND SEA LIONS

Lloyd Davis

New Zealand fur seal

Cape Saunders is one of the most dramatic places on the Otago coast. Sitting on the south-eastern tip of the Otago Peninsula, it is exposed to big winds and even bigger seas – a wild place of desolate beauty. Huge basalt cliffs form a wall around the perimeter of its half-moon bay. Swells sweep across the rocks at the base of the cliffs, whipping the fronds of bull kelp this way and that. And there, in this zone of contact between volcanic cliffs and angry seas, you will see fur seals: New Zealand fur seals *Arctocephalus forsteri* (Fig. 20). At first they are hard to discern from the rocks themselves, but as you continue to look, more and more are revealed. Large adult males, petite females and, perhaps if you are lucky, you will see a few pups.

It has not always been this way. Hunted initially by the Maori and from the late eighteenth century by European sealers, the fur seal was all but exterminated from New Zealand waters (see Chapter 7). Protected since 1916, fur seal numbers remained low, but in recent times they have begun to increase exponentially. As late as the 1970s the Otago Peninsula was used only by male seals as a non-breeding haul-out site, but by 1993 there were over 1100 pups born there and pup production is increasing at an average rate of 25 per cent per year.

Fur seals and sea lions (family Otariidae) are distinguished from true seals (family Phocidae) by having prominent ears, large fore-flippers and hind-flippers that when rotated forwards under the body assist with locomotion on land. The coat can be various shades of brown – depending upon the season, whether it is wet or dry, and the age of the animal – and is often darker on the back. The face is not unlike that of a large dog, with long whiskers that can be up to half a metre in length in adult males. There is a marked sexual dimorphism, with adult males being up to 185 kg and 2.5 m in length, while females on the Otago Peninsula average only 40 kg and 1.05 m.

Most of the breeding sites for the New Zealand fur seal are located around the South Island or on its offshore islands. The main sites in Otago and Southland are shown in Fig. 21. Seals are descended from terrestrial bear-like ancestors. While having evolved many adaptations for a marine lifestyle, they must still breed on land. This need to live a life in two worlds presents very different requirements that must be met before seals can successfully live and breed within an area.

Breeding and haul-out sites of New Zealand fur seals tend to be steep rocky beaches with large boulders that have plenty of crevices and tidal pools. These are especially important for thermoregulation, affording the seals and their pups the opportunity to get out of the sun's rays and to cool off during the heat of summer. The layer of blubber with a thick fur coat that enables seals to retain their body temperature while in the sea is such an effective insulator that overheating can be a big problem for them when on land.

To date, there has been relatively little research carried out on the marine requirements of the New Zealand fur seal, but new technologies that enable us to follow their movements at sea and record their behaviour (Box 11), together with new techniques for analysing spatial patterns, suggest that proximity to deep water trenches associated with feeding (see below) may be an important aspect of any breeding site.

Adult males arrive at the breeding sites in late October to early November, where they set up breeding territories that they defend vigorously until mid-January. During that period, territorial bulls remain ashore and do not feed. Females arrive at the breeding sites from late November onwards, ready to give birth to a pup that was conceived the previous year. The pups are born from late November to early January, with the mid-point in births being around Christmas Day. The mother then stays ashore with her pup for a period of about 12 days, the perinatal period, during which time she suckles the pup. Male pups are born significantly heavier (4.81 kg) than female pups (4.25 kg). The adult females come into oestrous about eight

Fig. 20. New Zealand fur seal *Arctocephalus forsteri*. Decimated by over-hunting, recovery of fur seal populations has been striking. Well over 1300 pups are born each year on the Otago Peninsula alone, and further south to Stewart Island numbers are increasing rapidly. *L.S. Davis*

Fig. 21 (opposite top). Map of breeding and non-breeding colonies of the New Zealand fur seal around southern New Zealand.

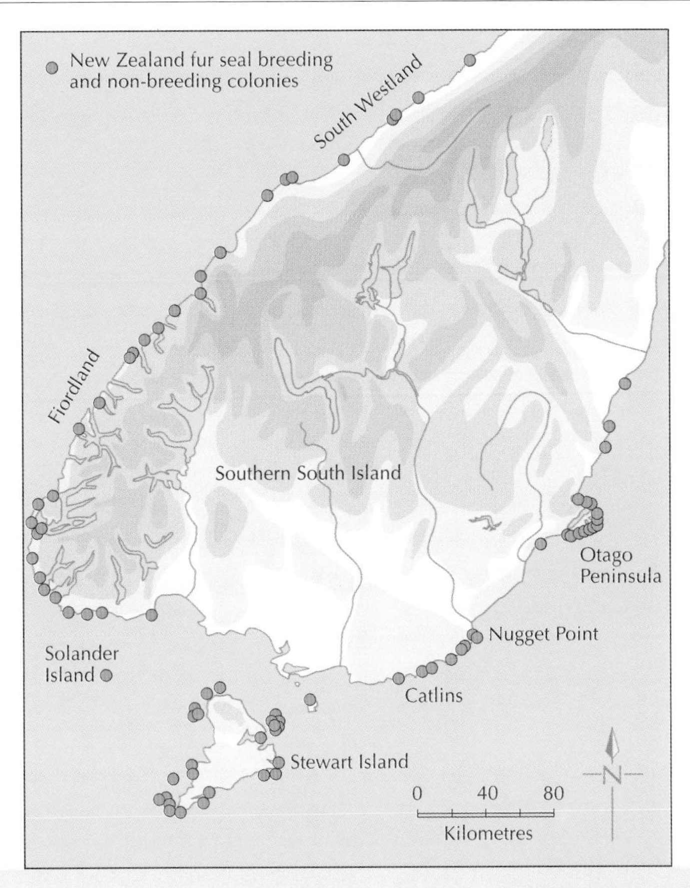

days after giving birth and will usually mate with the bull in whose territory they reside. The fertilised egg remains in a state of suspended animation until about April–May, when it implants in the uterine wall and development of the embryo continues.

Following the perinatal period, mothers must go to sea to catch food to sustain the milk supply needed for the growth of the pups. The adult males, by contrast, leave the breeding site by the end of January, their only contribution to the development of their progeny being the sperm they supplied the year before. Initially, the mothers go to sea for periods of 3–4 days, with an almost equal amount of time spent in between, suckling and caring for the pups. However, as the pups grow and their food demands increase, the females must spend progressively longer periods and a greater proportion of their time at sea. By autumn, their foraging trips last about 8 days with only a couple of days ashore, and by winter they are foraging for 10 days with only a day-and-a-half ashore between trips. Male and female pups both gain on average about 70 g per day (although body masses actually decline once the pups reach about 5.5 months old), but the growth rate of the larger male pups is more variable and more affected by the degree of maternal effort: male pups grow fastest when their mothers spend a longer time ashore with them during the perinatal period and take relatively shorter foraging trips to sea. Because sons place a greater demand on their

11 TECHNOLOGY ALL AT SEA

Until recently, what seals did at sea was largely a mystery; it was simply too difficult to track them. But recent technological advances now enable us to monitor where they go and what they do in the water.

Small transmitters attached to the heads or backs of fur seals breeding on the Otago Peninsula send signals to satellites that compute the location of the seals to the nearest kilometre. These have revealed that female fur seals forage relatively close inshore and not beyond the continental shelf during the summer months, when the pups are very young. Come autumn and winter, they range much further from the breeding site, up to 200 km or more and into deep water.

Other devices attached to the seals, known as time-depth recorders, record the depth the seal is at every 5–10 seconds. From these data, a profile of the swimming and diving behaviour of the seals can be built up. They dive for food mainly at night. During the summer, few dives exceed 100 m and those that do occur around dusk and dawn corresponding to the vertical migration of their prey (squid). However, in autumn and winter dives in excess of 100 m occur much more frequently and throughout the night, presumably as the seals go after bioluminescent prey such as lanternfish. The maximum depth measured for a female fur seal diving off the Otago Peninsula is 382 m and the maximum dive duration recorded is nearly seven minutes.

Lloyd Davis

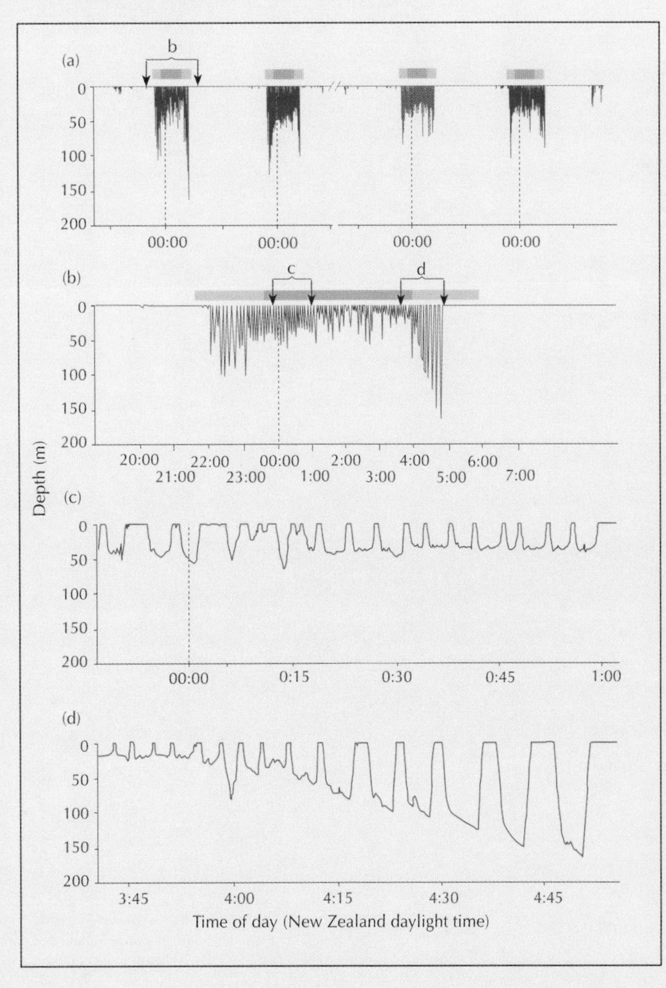

These dive traces are for an individual lactating female fur seal foraging off Otago Peninsula in December–January. In (a), showing a series of foraging trips, the marked nightly pattern is clear. The horizontal dark green bars indicate hours of darkness and the light green twilight. The record for a single night's foraging, shown in (b), displays the lack of feeding bout breaks during the entire night. The periods (c) and (d) are expanded in the two traces below, (c) displaying the repeated shallow U-shaped dives from the middle of the bout, and (d) the deep V-shaped foraging dives that occurred near dusk and dawn.

mothers, their survival is more susceptible to changes in food supply, with relatively more perishing in poor years. Pups are weaned in July or August and, thereafter, the adult females go to sea to prepare for the next breeding season.

Fur seals breeding on the Otago Peninsula rely heavily on arrow squid *Nototodarus sloanii* and to a lesser extent upon lanternfish *Symbolophorus barnardi* and *Lampanyctodes hectoris* during summer, when foraging trips are short, and autumn, when foraging trips are starting to increase. By winter, however, squid are largely unavailable and they switch their attention to fish such as ahuru *Auchenoceros punctatus* and lanternfish. As well as a change in diet, longer foraging trips and a fall off in the time spent ashore, there is other evidence to suggest that foraging mothers must work harder as their pups get older. Distances travelled from the breeding site increase, as do dive depths (Box 11).

Hooker's sea lion

Just around from Cape Saunders on the Otago Peninsula lies Papanui Beach. Fur seals breed here, too, on the rocky outcrops at the northern and southern ends of the beach. But on the wide expanse of sand that forms the crescent of the beach, there can be found another otariid seal: Hooker's sea lion *Phocarctos hookeri* (Fig. 22). Upwards of twenty male sea lions haul out here, especially during winter, and more haul out along the Catlins coast. Many of these are known to have travelled from the main breeding grounds for the species on the Auckland Islands. Hooker's sea lions are much larger than fur seals, and have a blunter nose. Adult males are dark brown, with a mane of thick hair about the neck and shoulders. They can weigh up to 410 kg and be over 3 m in length. Females are much lighter coloured, being creamy to yellow, and are much smaller too: under two metres and reaching a maximum of only 230 kg.

While the Otago/Southland coastline has typically been regarded as providing nothing more than haul-out sites for sea lions, in recent years there have been a few females pupping on the Otago Peninsula and in the Catlins. Pups are born in December, and the breeding sequence is similar to that for fur seals. Little is known of the past breeding distribution of Hooker's sea lions, but it was probably more extensive than today as sealers exploited sea lions in the early nineteenth century in the face of falling catches of fur seals.

Sea lions feed especially on squid, but they also take fish, krill, penguins and fur seal pups, and their large size permits them to dive to even greater depths than fur seals. In contrast to fur seals which are found on rocky, exposed coasts and seldom go more than a few metres from the water's edge, Hooker's sea lions frequent sandy beaches, have a gait similar to that of terrestrial mammals and frequently undertake walks of several hundred metres. Like fur seals, sea lions adjust their body postures, especially the exposure of the flippers (which act as heat radiators due to their large surface-to-volume ratios and lighter insulation), to help prevent overheating while on land. In addition, if temperatures rise above 14°C, sea lions will use their fore flippers to cover themselves with sand, instead of seeking shade like fur seals.

Elephant seal

If Cape Saunders is one of the most dramatic parts of the Otago coastline, then surely, Nugget Point must be among the most beautiful (Fig. 23). Fur seals also breed here, on the rock stacks at the tip of this small peninsula and along the boulder-strewn beaches on its northern side. The occasional sea lion may haul out here, too. But by far the most awesome sight is that of a southern elephant seal *Mirounga leonina* moving like a submarine between the rock stacks. They are the largest seals in the world: males can weigh up to 3700 kg and be 5 m in length (Fig. 24), whereas the much more svelte females are only half the length of a male and would struggle to tip the scales at a mere 400 kg. They are a brownish grey and the males have an inflatable proboscis, from which they get their name. Elephant seals are true seals (phocids) and, therefore, have no external ears and their hind limbs trail behind their bodies. They move on land by undulating their bodies, caterpillar-fashion.

The southern elephant seal has a circumpolar distribution, breeding mainly on subantarctic islands between 40°S and 62°S. The main population within New Zealand waters is on the Antipodes Islands and Campbell Island. However, in the caves at Nugget Point, a female elephant seal will often give birth to a pup. Pups have been recorded frequently between Oamaru and Nugget Point since 1965, and breeding has been known to occur as far north as Kaikoura. Pups are usually born in September–October. Like all the seals in the New Zealand region, they were hunted extensively in the nineteenth century.

Fig. 23. Nugget Point in South Otago is an important site for marine wildlife, not least because fur seals, Hooker's sea lion and southern elephant seal all occur here. *P.K. Probert*

Leopard seal

Leopard seals *Hydrurga leptonyx* are creatures of the Antarctic and subantarctic, although vagrants show up each year around the coast of southern New Zealand and may occasionally be encountered on beaches. It is as well to give them a wide berth; if you approach them too closely the thing you are most likely to notice is their enormous mouth (Fig. 25). They are fearsome predators, capable of killing a fur seal as easily as the fish or penguins they usually eat. They have a dark olive green back and creamy undersides unevenly covered with distinctive dark spots (from which they derive their name) and a large head. Leopard seals can be more than 3 m in length and weigh anything from 300–500 kg.

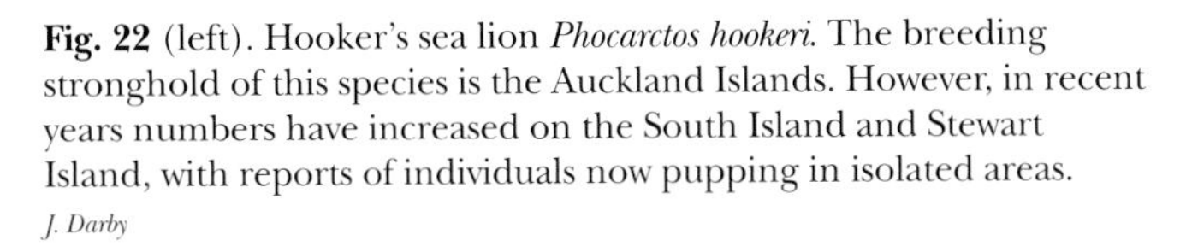

Fig. 22 (left). Hooker's sea lion *Phocarctos hookeri*. The breeding stronghold of this species is the Auckland Islands. However, in recent years numbers have increased on the South Island and Stewart Island, with reports of individuals now pupping in isolated areas. *J. Darby*

Fig. 24. The southern elephant seal *Mirounga leonina* is the world's largest seal, and the males have an inflatable proboscis, hence their name. *L.S. Davis*

Fig. 25 (bottom right). Leopard seal *Hydrurga leptonyx*. A vagrant to New Zealand shores, leopard seals most frequently fetch up on New Zealand shores in August-September, usually emaciated and ill. This does not, however, reduce their aggression towards those who may approach too closely. *J. Darby*

WHALES AND DOLPHINS

Stephen Dawson

Even though southern New Zealand is today well blessed with marine mammals, 200 years ago it was positively littered with them. Then, any coastal voyage in winter would have found southern right whales *Eubalaena australis* by the dozen. Slow and cumbersome, right whales prefer shallow waters, and are usually seen swimming very slowly at the surface. They were seen regularly in Otago Harbour and bred in large numbers off southern shores. They are odd-looking animals, the lack of a dorsal fin accentuating their rotundness. On their heads are conspicuous callosities – white, horny areas of thickened skin. Right whales are baleen whales, named from the long curtains of baleen plates hanging from each side of the upper jaw. Made of keratin (like fingernails), the fraying inside edge of the plates acts as a sieve to trap planktonic crustaceans. The baleen plates are more closely spaced in right whales than in any other whale species, enabling them to concentrate on one of the smallest of planktonic crustaceans, the copepods.

Humpback whales *Megaptera novaeangliae* were once common too, migrating past New Zealand on their way from Antarctic summer feeding grounds to winter breeding sites in the tropics. Humpbacks sing long, highly structured songs of extraordinary complexity. Each 'breeding ground' has its own characteristic song, which is embellished throughout the season as individuals add their improvisations. Only males are known to sing, and while most singing occurs during the breeding season, Royal New Zealand Navy recordings made during the 1960s and 1970s showed that some 'tune up' on the long swim north. On a quiet calm night 200 years ago, during the peak of the migration, a lucky coastal sailor listening through the wooden hull might easily have heard humpbacks practising.

Certainly Hector's dolphins (*Cephalorhynchus hectori*) would have been very common 200 years ago. These small, elegant dolphins are found only in New Zealand. They are the smallest and the most coastal of our dolphins (Fig. 26). Population modeling suggests that in the early 1970s, before gillnetting became widespread, population size on the east coast was roughly double what it is today. Historical records show that they were regularly seen in places where they are now absent. Current distribution is very patchy and some populations are genetically isolated, supporting the idea that they were once more widespread. From an 1866 book *Guide to Brighton and its Environs Containing Every Information for Visitors to this Otago Watering Place,* we learn that:

> There is a certain favourite Brighton sport, which we recommend to the notice of Otago riflemen, as combining utility with amusement .... Let the marksman sit at either side of the boat harbour on a fine day, at high tide, and he will soon see a shoal of porpoises, at which if he is quick, he will be able to get a fair shot. The bodies of these tenants of the deep, when hit can be recovered at low tide, and a considerable quantity of oil taken from them.

These porpoises were probably Hector's dolphins. They are not found in this area now, but thankfully attitudes have changed.

The regulars

Three cetacean species can still be seen daily around southern coasts, namely Hector's, bottlenose and dusky dolphins. And seven others are seen occasionally, roughly in order of frequency: right whales, killer whales, sperm whales, humpback whales, long-finned pilot whales, minke whales and common dolphins.

A small bunch of Hector's dolphins habitually cruise outside the breakers at Moeraki. In summer, the calm waters of Porpoise Bay in the Catlins shelter a bunch of twenty or so. They spend most of their time just outside the breakers, or surfing in them,

usually at the southern end of the bay. This makes them easy to find and easy for a swimmer to get close, but we have no idea why they spend so much time in so small an area.

Hector's dolphins have been at Porpoise Bay for as long as anyone can remember, and the bay was 'misnamed' after them. Despite looking vaguely porpoise-like (small, and blunt headed), Hector's are true dolphins (members of the family Delphinidae). They lack the characteristically flattened, spade-shaped teeth of porpoises, having instead conical teeth like other true dolphins.

Te Waewae Bay is the major southern hotspot for Hector's dolphins. Ninety were found in small groups there in a 1990 survey. Apart from two brief surveys, this population is unstudied. Though more samples are needed to be sure, genetic testing of a few dolphins found dead on the beach showed that this population is probably separated from the major west and east coast populations. These studies (of mitochondrial DNA) show that South Island east and west coast populations are definitely separate.

Hector's dolphins are better known than any other of New Zealand's dolphins and whales, thanks to detailed study for more than a decade. Population size is about 7400 individuals, about 98 per cent of which are 'South Islanders'. Though there is much detail still to be learnt, few aspects of their biology and behaviour remain unstudied. For example, females bear their first calf between seven and nine years, and normally give birth every 2–3 years. Maximum age is around 20 years. They are not picky eaters; just about any type of fish or squid of the right size is fair game. Hector's dolphins spend most time feeding near the bottom on fish like ahuru, red cod, stargazer and small flatfish. In mid-water, and near the surface, small arrow squid, hake and hoki are taken when available. Sometimes they can be seen feeding right at the surface, chasing yellow-eyed mullet or sprats.

Bottlenose dolphins *Tursiops truncatus* are seen daily in parts of Fiordland (Fig. 27), and occasionally elsewhere. About 60 reside in Doubtful Sound. Since most are identifiable from marks and scars, we know that the group is very stable and that individuals within the group form alliances with one or a few others, with whom they associate very closely. Being able to find dolphins daily, and identify individuals, also means that their day-to-day behaviour can be studied in rich detail. Doubtful Sound's dolphins leave the fiord only rarely, and usually for just a few hours. Movement patterns within the fiord change seasonally, the dolphins avoiding the cold water of the inner fiords in winter.

Dusky dolphins *Lagenorhynchus obscurus* (Fig. 28) are seen year-round, and often in groups numbering several hundred. They are the most acrobatic of New Zealand's dolphins. It is rare to see a large group that does not have several individuals jumping at once. They have been little studied in southern waters, but it is known that those off Kaikoura feed mainly at night beyond the edge of the continental shelf. There, dense aggregations of lantern fish, squid, salps, jellyfish and crustaceans migrate from

Fig. 26 (opposite). Hector's dolphins *Cephalorhynchus hectori* off Karitane beach. *S. Dawson*

Fig. 27. Bottlenose dolphins (*Tursiops truncatus*) in Fiordland. *S. Dawson*

100–200 m during the day to near the surface at night. This so-called 'deep scattering layer' (for the tell-tale trace it leaves on an echo-sounder) is very important for duskies (and for fur seals). The edge of the continental shelf, and hence the deep scattering layer, is reasonably close to Otago's coast, and it is probably this that brings these two marine mammals here in abundance.

The visitors

Sadly, the right whales that were once so common along southern coasts are now rare. They were easy targets and, like humpback whales, their New Zealand population was almost exterminated by whalers. Sightings can be guaranteed only much further south, at the Campbell and Auckland Islands, where perhaps as many as a hundred go to breed. Sightings in our waters are, however, increasing, with several recent sightings in Te Waewae Bay, off the Catlins and Otago Peninsula.

Humpbacks are seen less frequently than right whales, though they too were once very common. A marking programme during whaling days revealed a migration route from summer feeding grounds in Antarctica past New Zealand to the tropics, particularly Tonga. Such routes are probably learned, and passed on from generation to generation. Sightings of humpbacks in New Zealand waters have virtually ceased since the 1960s, yet sightings in Australian waters have increased. It is certainly possible that the New Zealand migration route has been forgotten, and our whales have followed their Australian cousins.

The past for sperm whales *Physeter macrocephalus* has been less catastrophic. They were among the first species to be whaled, but their populations were not hit as hard as the others. Currently sperm whales are second in abundance only to minke whales. Sperm whales are ultra-deep divers. Large males are apparently capable of diving to 1000 m routinely. Though giant squid have been recorded in sperm whale stomachs, most of the diet is more mundane, comprising mid-sized squid and bottom-living fish. Generally, sperm whales are found where water depths are 1000 m or more, and are concentrated where the depth changes rapidly. This explains why they favour Kaikoura waters and why they are seen offshore from the Otago Peninsula and off the western corner of Fiordland. These are places where depths of 1000 m and more are relatively close to the coast.

Killer whales *Orcinus orca* are occasional visitors to southern New Zealand, and always an impressive sight. They are the largest dolphins; mature males reach around 10 m, and feature a narrow, extremely tall dorsal fin. Females are about a third smaller, and have curved, more dolphin-like dorsal fins. Killer whales are among the longest lived of all mammals; females may live as long as 100 years. One 1994 sighting of killer whales on the Otago coast was especially interesting. A fishing boat in Blueskin Bay encountered four killer whales, together with several very large pieces of oily liver (one weighing at least 10 kg) floating on the surface. Most probably their prey had been a basking shark, a species which grows to over 6 m.

Minke whales *Balaenoptera acutorostrata* have two distinctions. Apparently they are delicious, and they are certainly abundant. The former makes some nations push for renewed whaling of them, and the latter makes it difficult for non-whaling nations to argue that such whaling would pose a significant threat. Despite their abundance, minke whales are seldom seen along southern shores. They are less obvious than most other whales, being small, fast moving, with indistinct blows. They are easy to miss.

Common dolphins *Delphinus delphis* and long-finned pilot whales *Globicephala melas* are infrequently seen, and usually only far offshore. Common dolphins feed in the deep scattering layer like duskies do, but unlike duskies do not spend the daytime inshore. Pilot whales are animals of deep water and open ocean. They are squid-eaters, often found in the same sorts of places as sperm whales.

Strandings

Strandings are reported relatively rarely on southern coasts, although a number of species have been recorded. For the Otago coast over the period 1988–97, some 15 cetacean species were recorded from standings. This includes only two mass strandings, both of long-finned pilot whales. Strandings are often the only view we get of the more obscure species, such as the beaked whales, a group of species hardly ever seen at sea. Beaked whales recorded from strandings on southern coasts include Shepherd's beaked whale *Tasmacetus shepherdi*, Scamperdown whale *Mesoplodon grayi*, Andrews' beaked whale *M. bowdoini*, strap-toothed whale *M. layardi*, Hector's beaked whale *M hectori*, Cuvier's beaked whale *Ziphius cavirostris* and large beaked whale *Berardius arnouxi*. Even on a beach, identification of beaked whales is not easy; really you need to look at the diagnostically shaped teeth. Just to make things difficult, these teeth (only two in most species) barely erupt above the gumline in females. Conclusive identification can often be made only from a clean skull. The stranding in September 1995 of a young spectacled porpoise *Australophocoena dioptrica*, the only true porpoise recorded from New Zealand waters, was truly one for the record books. The only other New Zealand record was of a skull found at the Auckland Islands two decades before.

The future

The outlook for whales and dolphins in southern New Zealand is bright. Recreational and commercial gillnetting is a potential threat to Hector's dolphins, which breed slowly so that even the accidental take of very small numbers is worrisome. But if fishing impacts are minimised, and standards for discharge of pollutants continue to be tightened, they should not become rarer. Given enough time, the large baleen whales might return in numbers. This hope is most realistic for right whales, which are recovering steadily elsewhere. Eventually they may once again be seen routinely in southern New Zealand's coastal waters.

Fig. 28. Dusky dolphins *Lagenorhynchus obscurus* in clear winter water, far offshore. S. Dawson

CHAPTER 13

NATURE CONSERVATION IN SOUTHERN NEW ZEALAND

JEFF CONNELL

Sutton Salt Lake near Middlemarch, here less than half full. Protected since 1991 within a 140 ha reserve, it is New Zealand's only inland saline lake. *Neville Peat*

Fig. 1. Catlins Coastal Rainforest Park northern edge, looking southwest. The Ajax Plateau is upper centre, and the coastline stretches away to Foveaux Strait behind. *Robin Thomas/DOC*

Fig. 2. Fiordland crested penguin. *Lou Sanson/DOC*

The views in this chapter are those of the writer, and not of the Department of Conservation.

NATURE CONSERVATION IN SOUTHERN NEW ZEALAND

This chapter is about the conservation of nature in southern New Zealand. Nationwide awareness of the need to protect natural resources against development first arose in the region with the Save Manapouri campaign (1969–72). Between Save Manapouri and the current prominence of 'green' politics on the national and international stage, a number of campaigns and issues have been significant, sometimes pivotal. They have marked a transition from concern about scenery preservation to concern for the comprehensive preservation of biological diversity. They have provoked significant changes in environmental administration and have helped to raise the profile of conservation issues in the general community. A brief account is given of these campaigns and issues by way of introduction. Then, the state of nature conservation in southern New Zealand is outlined, while the current issues and challenges are identified and discussed. Finally, some observations are made and goals suggested for the future.

A useful starting point is to outline the major areas now set aside for nature conservation purposes in the region. Protected under acts of parliament, these areas are administered by an agency set up to be accountable to the public of New Zealand through the government of the day for their proper conservation management. New Zealand's Department of Conservation is thought to be unique in the world, as a single purpose taxpayer-funded nature conservation agency.

Fiordland National Park

Pre-eminent among our national parks, Fiordland at 1,260,200 ha is the largest in New Zealand and one of the largest in the world. With its core having been set aside as early as 1904, the park was formally constituted in 1952. The area had been significant for tourism since the 1880s, due largely to the spectacular scenery of the world-famous Milford Track. It became more significant from the 1950s, when the Homer Tunnel opened up Milford Sound/Piopiotahi for mass tourism. In nature conservation terms, the park protects numerous special features. It contains the largest area of continuous indigenous forest left in New Zealand. This is cool temperate rainforest, with its characteristic diversity of mosses, liverworts and ferns covering every stable surface. While the forest is mainly beech, the alpine plant communities, wetlands and coastline are very diverse. The park boundary stops at the high tide line although two small marine reserves protect samples of the fiords' special marine ecosystems. The park provides habitat for threatened indigenous species, such as takahe (rediscovered there as recently as 1948), kakapo, whio (blue duck), mohua (yellowhead) and long-tailed bat. Endemic to Fiordland National Park (but not threatened) are the Fiordland crested penguin (Fig. 2), a suite of large-bodied invertebrates and a range of plants.

Mount Aspiring National Park

Originally conceived as an Otago national park, Aspiring now straddles the Main Divide of the Southern Alps and extends to within 20 km of the sea in South Westland. A mountaineers' park, it was established in 1964 from surrendered pastoral leasehold and other Crown land, as a result of public support kindled by a group of Dunedin enthusiasts. Now at 355,543 ha, the park has doubled in size since its inception. The most significant addition, that of the Red Hills, is discussed later. The national park lies immediately to the north of Fiordland, and is New Zealand's third largest. Like Fiordland, it has been heavily glaciated, but because of its higher average elevation it has many more extant glaciers and snowfields. The principal ecosystems protected in the park are beech forests (of three species, and irregularly distributed) and alpine plant communities. The park features areas of ultramafic rock with their peculiar sequences of dwarfed vegetation.

Te Wahi Pounamu (South West New Zealand) World Heritage Area

Both Fiordland and Mount Aspiring National Parks, along with adjacent protected areas, are now part of the Te Wahi Pounamu (South West New Zealand) World Heritage Area. While not altering the management requirements of the legislation under which they are administered, this status confers both an international profile and a degree of additional protection. As with the other 400 or so sites recognised globally under the World Heritage Convention (UNESCO) for their 'outstanding universal value', the New Zealand government is accountable internationally for the management of such areas, so that they retain their world heritage qualities. In the case of Te Wahi Pounamu, the quality recognised is that of wilderness.

Catlins Coastal Rainforest Park

The Catlins State Forest Park (Fig. 1) was gazetted in 1976 by the New Zealand Forest Service. The Forest Service was a multiple-use agency: in other words, its goals were set by the government to include exotic and indigenous forest timber production, the protection of indigenous forests for soil and water and ecosystem conservation purposes, and to include the development and promotion of public recreation in its forest settings. The Catlins Forest Park was typical in this respect of a number set up by the Forest Service throughout New Zealand. It included exotic forest plantings, proposed ecological areas, amenity areas such as campgrounds and walking tracks, and tracts of indigenous forest managed passively to limit soil erosion damage. In the government's environmental restructuring of the 1980s, the Catlins Forest Park was carved up. The great majority of its 60,000 ha were allocated to the Department of Conservation for nature conservation purposes. Within the orbit of the original park, the Department of Lands and Survey had managed approximately 4700 ha as scenic reserves and these are also now the responsibility of the Department of Conservation. They are now managed along with the former state forest areas as the Catlins Coastal Rainforest Park. No timber production will take place in the park.

The Catlins Coastal Rainforest Park is the largest area of indigenous forest remaining on the east coast of the South Island. The Catlins, the Waitutu area and Stewart Island contain the most significant lowland indigenous podocarp-hardwood forests left in southern New Zealand. The values of the Catlins include a diversity of forest types, from southern rata-fringed beaches, through podocarp dune forests, rimu-dominated podocarp and mixed podocarp-hardwood hill forests to silver beech and hardwood forests, and cedar and peatland communities on the highest hill crests. The coastline is readily accessible from many points and this makes the area very attractive to visitors. The forests contain a wide range of bush birds, with mohua (yellowhead) a particular feature. The coastal wildlife, however, is special. Visitors will see fur seal, New Zealand sealion and yellow-eyed and little blue penguins, and they can sometimes encounter elephant seal and Hector's dolphin.

Stewart Island/Rakiura

After early settlement and sawmilling failures, and limited interest in its mineral resources, the government had by 1907 set aside most of the island's crown land for scenery and flora and fauna preservation purposes. Today, 97 per cent of the main island (or 164,000 ha) is administered by the Department of Conservation for nature conservation purposes, and these lands have recently become New Zealand's latest national park.

The island contains many of the forest and wetland ecosystems of the adjoining mainland, although beech is absent. Its special features include shrubland (Fig. 3) and subalpine communities not found on the mainland, an absence of mice and mustelids and a general absence of introduced fish in its waterways. Its coastline is long and little modified. Even more significant, however, are the approximately 170 smaller islands in Crown and Maori ownership that fringe the main island. These collectively represent a resource of international conservation importance. Many are pest- and predator-free, and more are in the process of being rid of pests. They contain major seabird colonies and some are already key refuges for critically threatened species such as kakapo and South Island saddleback.

Fig. 3. Tin Range, Stewart Island/Rakiura. *Lou Sanson/DOC*

Fig. 4. Lake Manapouri from The Monument. *Alan Mark*

NATURE CONSERVATION EMERGES AS A POLITICAL ISSUE

Saving Lake Manapouri

In 1972 throughout New Zealand, 'Save Manapouri' bumper stickers began appearing on private motor vehicles. Many of the vehicle-owners had not seen the lake, were not members of a conservation organisation, and did not know the details of the issue. The widespread appearance of the 'Save Manapouri' bumper sticker was proof that ordinary people were concerned about nature, even if they were not directly affected. The appearance of the stickers had been preceded by a record 264,507-signature petition to parliament, and was part of a campaign that started in Southland but spread to engage the whole country. This mass movement was an echo of the great protest movements of the sixties against the Vietnam War and the exclusion of Maori from All Black teams to South Africa. For the first time, however, an environmental issue was at stake.

Fiordland National Park, including the largely pristine Lake Manapouri (Fig. 4), was protected under the National Parks Act 1952. It should have been safe, but in 1960 the government signed an agreement with an Australian company to allow it to raise and lower lake levels to enhance the value of a large-scale hydro-electric generating station in West Arm/Manapouri. Lake-level changes implied extensive damage to vulnerable and special lakeshore vegetation, from short turf communities to mature indigenous forest, and scarring of the hillsides adjoining the lake. Shortly afterwards, the government brushed aside the protection afforded by the National Parks Act, through the Manapouri-Te Anau Development Act 1960. Construction of the huge tailrace tunnel between the lake and the sound began in 1964.

In the early 1970s, the campaigners were up against powerful forces. The government was committed contractually, legislation bypassing the National Parks Act was in place and very little information was obtainable about the scheme, its costs and its environmental effects. The campaigners succeeded in protecting the lake, however, because they were able eventually to exert political pressure through the democratic process.

The political impact of the campaign resulted in a favourable recommendation from the Select Committee on the petition, and protecting the lake became an issue in the 1972 general election. The government lost the election, and openly blamed the loss of stronghold southern electorates on the mishandling of the Manapouri affair.

It is no accident that New Zealand's first major nationwide nature conservation campaign arose from an issue in the south. It was followed closely in the 1970s by an equally big campaign to protect native beech forest on the South Island West Coast, and then by an unsuccessful campaign to prevent the building of a second large-scale hydro dam on the Clutha River/Mata-Au. The South Island has vast natural resources that are not as developed as those in the north, and in many more places than in the north those resources coincide with high nature conservation values which are treasured by growing numbers of people.

Fig. 5. Lower Nardoo, April 1982. *Alan Mark*

The Manapouri campaign raised awareness of environmental issues generally, but it had another specific spin-off effect in the region. A parallel but smaller campaign was run to ensure that the level of Lake Wanaka was not raised, and in 1973 the Lake Wanaka Preservation Act was passed into law. This provides that the natural level of the lake and its outlet the Clutha River/ Mata-Au, may not be interfered with above the confluence with the Cardrona River.

FROM SCENERY PRESERVATION TO ISSUES OF ECOSYSTEM DIVERSITY

Tussock grassland of the Nardoo

The Department of Lands and Survey purchased the 15,600 ha Waipori Station (in the Lake Mahinerangi area west of Dunedin) in 1974, intending to develop it in order to settle civilian farmers on the land. In doing so, it bought a fight which contributed to the carving up of the department in 1986.

In the 1970s, tussock grasslands at lower altitudes were New Zealand's invisible disappearing ecosystem. While high-profile public campaigns were being fought to protect indigenous forests, with the New Zealand Forest Service usually cast in the role of villain, vast areas of the South Island's tussock grasslands were being irrevocably changed. Through processes started in the nineteenth century but accelerated in the twentieth by government subsidies, the grasslands were being converted (by burning, oversowing, topdressing, ploughing and intensified grazing) into pasture dominated by higher-producing exotic grasses and legumes. This process was invisible, because it was a gradual process occurring mostly on privately-owned or occupied land. The more amenable lower altitude lands were always the first to be converted.

But the purchase of Waipori Station by the Crown occurred during an era of growth in public awareness of environmental and conservation issues, largely sparked off by the Manapouri debate. The Department of Lands and Survey could no longer behave like a private landowner or occupier and carry out development in private. It had to commission and publish an Environment Impact Report (1977), and almost immediately the issue of adequate tussock grassland reserves arose.

In 1978 less than 13 per cent of the reserves and other protected areas in Otago were in tussock grassland, while it had once been the dominant vegetation cover over most of the region. These reserves amounted to a paltry 1112 ha, nearly all of which was montane or alpine snow tussock grassland. An even smaller proportion of tussock grassland was protected in Canterbury, northern Southland and Marlborough. In 1977 the Reserves Act (administered by the Department of Lands and Survey) called for 'the preservation of representative samples of all classes of natural ecosystems and landscape which in the aggregate originally gave New Zealand its own recognisable character'.

The Waipori Farm Settlement debate was about proposals to reserve for scientific purposes virtually the entire catchment of Nardoo Stream flowing into Lake Mahinerangi, representative of the tussock grassland ecosystems and landscape of the area. The Nardoo catchment ranged in altitude from 400 m to 900 m and contained a little-modified sequence of tussock grassland, as well as beech forest and shrubland remnants. The debate soon focused on the lower 400 ha of the Nardoo, an area of mixed fescue and snow tussock grassland at an altitude already developed in most other places. The issue was sharply defined – whether or not the number of farms to emerge from the settlement programme should be reduced by one (out of 23 possible farms) in order to allow for the reservation of the lower 400 ha. Those pushing for reservation (led by a group of scientists) called in the Ombudsman, who recommended to the Department's Land Settlement Board that the lower 400 ha should be left undeveloped for 15 years, at which time its reserve potential could be reassessed. The board rejected this recommendation, arguing that the reservation of the upper catchment was sufficient to result in an appropriate balance between farming and conservation. The board clearly missed the point about the lower area's significance as a representative sample of a fast-disappearing indigenous ecosystem.

Fig. 6. Lower Nardoo after ploughing, October 1987. *Alan Mark*

The debate remained heated, because the scientists arguing for the lower area's protection believed (with some justification) that the Dunedin Commissioner of Crown Lands had tried at two different stages of the saga to pre-empt the outcome by proceeding with development while the decision was subject to official review (Figs 5–6). The scientists were also frustrated because although the legislation (Reserves Act 1977) and policy statements (the Land Settlement Board's High Country Policy of 1980 provided for the protection of at-risk natural values) were on their side of the argument, the institutions were not. The Land Settlement Board had no scientific, conservation or recreation representatives on it at the time (although this was to change after the Nardoo saga), and the National Parks and Reserves Authority was brought into the discussion too late.

Unlike the Manapouri issue, there was no groundswell of public support for saving the lower altitude Nardoo tussock grasslands. The arguments were specialised, dealing with relatively esoteric ideas like unbroken sequences and representative samples of vegetation communities that even today not many people appreciate. Yet the Nardoo struggle did raise the profile of tussock grasslands in the conservation movement. It also demonstrated to those concerned with good environmental policy-making, that the Department of Lands and Survey and its institutions could not be expected to produce credible resolutions of development versus conservation issues. The same concern also applied to the New Zealand Forest Service. That realisation (coupled with a Treasury led drive for government organisations with commercial purposes to be set up under a different model) led eventually to the government's environmental restructuring of 1985–86 and the formation of the Department of Conservation on 1 April 1987. The new department became responsible for the part of the Nardoo catchment that was reserved, and in the land allocation process actually ended up with a small part of the lower 400 ha that had by then not been ploughed.

The Nardoo story did not end there. A survey of the Waipori Ecological District was published by the Department of Conservation in 1994. From the implementation of the recommendations of that survey and by a variety of negotiated means, the department has greatly increased the amount of protected tussock grassland and associated ecosystems in the area. In late 1997, the expanse of grassland to be protected adjoining the Nardoo had risen to over 20,000 ha and the Otago Conservation Management Strategy foreshadows the creation of a Te Papanui Conservation Park from that large public landholding. The problem remains, however, of adequate representation in the protected areas system of lower altitude tussock grasslands. In this respect, the opportunity to achieve protection of a full range of representative samples has probably gone forever. That issue is discussed later in this chapter.

The Red Hills

There is a published cartoon in which a mischievous Creator is shown chuckling over the placement of mineral deposits beneath areas of spectacular natural grandeur of national park quality. The Red Hills are a case in point, where asbestos and chromite deposits are present in an area recognised for its impressive and scientifically important geological features and its unique patterns of vegetation resulting from the variable mineral content in the soils of the area. The name Red Hills comes from the colour of its ultramafic rock, which is rich in iron and magnesium and is closer in composition to parts of the earth's core than to its surface. The sparse and often stunted vegetation of the mineral belt contrasts with the normal vegetation growing on adjoining schist and volcanic materials (Fig. 7). At lower altitudes, the normal vegetation is mixed beech-podocarp forest, whereas in the mineral belt there is a dwarf mixed forest of mountain beech, yellow-silver pine, stunted rata, manuka, celery pine and kamahi. At higher altitudes there are marked differences in the montane zone and in the treeline shrub belt. With the different vegetation zones on ultramafic rock extending over an altitude range of 1370 m, the Red Hills contains the greatest such undisturbed sequence in New Zealand.

The drive to protect the Red Hills by adding them to Mount

Aspiring National Park did not, however, come solely from scientists. It received a crucial push from mountaineers and trampers aiming to preserve the wild and undeveloped qualities of the region as a place where humans could experience nature entirely on its own terms.

The Red Hills were excluded from the park on its formation in 1964 because of the mineral potential of the area known from the 1890s. A period of mineral exploration by multi-national mining companies began in 1968 and culminated in 1973, with the driving of a bulldozer track from Jacksons Bay on the West Coast around to the Pyke River flats. By 1977, all exploration had ceased and no serious mining proposals had emerged. Local authority interests on the West Coast, however, continued to oppose the inclusion of the Red Hills in the park, correctly (as it turned out) supposing that such a step would be likely to prevent mining in the future. The Red Hills area, at 27,000 ha, was eventually included in Mount Aspiring National Park in 1990, long after demand for asbestos had ceased. In early 1997 the Olivine Wilderness (including the Red Hills) was formally gazetted, and this was followed in November of that year by an amendment to the Crown Minerals Act that banned all mining in national parks.

As far as the Red Hills are concerned, the cartoon will need to be redrawn.

Fig. 7. Simonin Pass, Red Hills, ultramafic rocks on the right, beech forest on the left. *Brian Ahern*

Saltmarsh or smelter?

At Aramoana, near the entrance to the Otago Harbour, a long environmental campaign was waged to protect a rare ecosystem and the lifestyle of people living near it. The saltmarsh and its associated dryland buffer zone, however, now have the protected status of an Ecological Area under the Conservation Act 1987.

This outcome would have been unthinkable for the Chairman of the Otago Harbour Board who in 1960 was trying to attract heavy industry to the Board's endowment land at Aramoana. 'The bigger the better', he said, announcing that the Board 'had 200 acres of land immediately available and another 300 acres could be reclaimed in short order'. He was talking about the last intact saltmarsh and associated tidal eelgrass flats and landward buffer zone left on the east coast of the South Island. Saltmarshes elsewhere had been reclaimed, overrun with weeds or used as municipal refuse dumps. But they are biologically significant. The Aramoana saltmarsh is the most important wader habitat in Otago, based on the number and variety of species that use it; it is a major fish nursery and its kai moana resources are treasured by Ngai Tahu.

After early success, the campaign to save the saltmarsh entered a new stage in 1979, when the Muldoon government began offering cut-rate surplus power from its controversial second Clutha hydro-electric dam, for heavy industry. The campaigners against the smelter proposal that then emerged had to overcome the power of the state, which was being wielded through the National Development Act and legislation introduced to authorise the reclamation of the saltmarsh. For one of the first times in New Zealand, environmental campaigners successfully used the courts to thwart the government, and by 1981 the proposal had been withdrawn.

The final protection of the saltmarsh was achieved in a series of stages. In the first of these the Dunedin City Council cooperated with the Department of Conservation to secure the buffer zone, then the department moved to secure the saltmarsh and tidal eelgrass flats.

By April 1996 the complex, totalling 359 ha, had been gazetted as a Conservation Area. Its gazettal as an Ecological Area confers upon it one of the highest protected area classifications available under New Zealand's conservation legislation.

Saving Waitutu forest

In western Southland, an impressive flight of 13 mudstone marine terraces rises from the sea to Hump Ridge at 1,000 m. Clothed in indigenous forest, the terraces represent one million years of tectonic uplift and are internationally recognised as a special feature in their entirety. The dissected upper terraces are administered by the Department of Conservation; in 1999 they were added to Fiordland National Park, which lies immediately to the west and north. The terraces have not been subject to glaciation, so they complement the heavily glaciated terrain that characterises the rest of the park. Above 300 m the vegetation is beech forest and much like the adjoining national park, but below that altitude (which takes in over half of the total area of the terraces) it is lowland forest, featuring rimu and silver beech with kahikatea locally prominent. Beech is entirely absent from the youngest (lowest) terraces, which are little dissected and present a major navigational challenge to their few human visitors.

The youngest terraces of all, supporting the richest lowland forest with a high proportion of rimu, are on some 2000 ha of Maori land owned by the Waitutu Incorporation.

The entire lowland forest is an outstanding wildlife habitat, containing probably the largest remaining South Island kaka population. The threatened mohua (yellowhead) is also present, along with South Island robin, yellow-crowned parakeet, falcon and the more common indigenous forest birds. The streams contain short-jawed and giant kokopu. The invertebrate biological diversity of the area is still to be fully surveyed.

The Waitutu Incorporation lands had been awarded to the original owners under the South Island Landless Natives Act 1906. Described recently by the Waitangi Tribunal as a 'cruel hoax', the 1906 legislation was intended to provide land for Maori who had none following the land sales of the previous century, but in fact the land awarded was then of no practical use. The Waitutu land was inaccessible (it is still 32 km from the nearest useable public road), clothed in heavy bush and is subject to a harsh climate.

As most of the accessible rimu forest in New Zealand was worked out through the twentieth century, the Waitutu rimu became progressively more valuable and of interest to logging companies. Ultimately, the only thing the owners could realistically do with the land in order to obtain a worthwhile income was to offer to sell the cutting rights. The prospect that these would be sold to logging interests brought them into conflict with those who wanted the habitat preserved for wildlife and to have the entire marine terrace sequence included in the national park.

In the controversy that followed, there was general sympathy for the Maori owners, while the heat was on the logging company and the government. In December 1993, the Waitutu Incorporation sold the cutting rights to logging company Paynter Group Limited on terms that have never been fully disclosed. By then, it was well known that the government had offered compensation to the owners not to log the forest, and that the land was exempt from the normal controls requiring indigenous forest to be sustainably managed (see page 353). It was widely speculated that the Paynter Group had acquired the cutting rights not with the intention of clearfelling, but with the aim of participating in the inevitable compensation agreement. The ante was high on both sides of the argument. Conservation groups such as the Royal Forest and Bird Protection Society were lobbying the government furiously, in a letter writing and publicity campaign that also targeted Paynter Group shareholders and the company's 1994 Annual General Meeting. To save Waitutu, Forest and Bird were able to call on the support of the internationally recognised conservationist David Bellamy (known from his 'Botanic Man' UK television series). They were also able to cite the 1992 Forest Accord, in which a number of large forestry companies had pledged to leave their indigenous forest holdings untouched. On the other hand the Waitutu Incorporation had lodged a claim with the Waitangi Tribunal alleging that controls on the exploitation of their indigenous forests were in breach of the Treaty of Waitangi.

With the support of Paynter's, the Incorporation tried to up the ante still further in 1995 by applying to the Southland District Council for a certificate stating that it could commence indigenous production without requiring a resource consent under the Southland District Plan. While the lawyers had a field day on questions such as whether 'farming' included clearing the trees first, and the resulting court case ground its way up to the High Court, the government's negotiator beavered away in the background.

An agreement was eventually reached, in which the Incorporation agreed to allow the forest to be managed as if it had been added to Fiordland National Park while remaining in its ownership. In return, the government agreed to pay the Incorporation $13.5 million in cash, plus cutting rights to 12,000 ha of Crown beech forest earmarked for a sustainable management regime in the nearby Longwood hills. The Incorporation also retained certain traditional use rights in the area, and the right to occupy a small part of it for overnight accommodation. The agreement is now enshrined in legislation which came into force in December 1997.

The Crown paid a high price to save a small proportion of the total Waitutu forest, but at the time the area owned by the Incorporation was ranked as the most important area of indigenous forest in private hands under threat in the country.

Fig. 8. The end of an era? Earnscleugh station merino wethers at 1500 m elevation, Old Man Range, summer. *Kate Wardle*

The value of the cutting rights is not known, so the total value of the settlement cannot be stated. It is clear, however, that the Waitutu deal has set a high benchmark price for future attempts to protect similar privately-owned forest habitat under threat elsewhere.

CURRENT PROTECTION ISSUES

Nature conservation in the pastoral high country

Establishing protected areas such as national parks in the unoccupied and rugged high-rainfall mountain lands of southern New Zealand was a laborious process. Such an undertaking, however, was always going to be more difficult in the rainshadow high country tussock grasslands and associated mountainlands of Otago and northern Southland. Occupied since the 1850s by pastoral farmers, these lands have become progressively more modified from the indigenous ecosystems that existed at the time. There are few natural areas left below 500 m in the high country. However, extensive natural areas remain at higher altitudes, although these are changing in places as a result of ploughing for exotic pasture establishment, burning the grasslands without replacing lost nutrients, and weed invasion. Although mostly owned by the Crown, the lands are privately occupied under pastoral leases (a form of perpetually renewable tenure administered under the Land Act 1948 that gave the runholders rights to exclude trespassers).

The need to preserve representative samples of rainshadow grasslands was slow to be recognised. Even after this need became official policy (Reserves Act 1977 and Land Settlement Board High Country Policy 1980), implementation did not proceed smoothly, as the Nardoo story shows. The Department of Lands and Survey on the recommendation of the National Parks and Reserves Authority developed a Protected Natural Areas Programme (PNAP) in 1983, in which the survey of the pastoral high country was given priority in the South Island. The first surveys commenced in early 1984. Among these were surveys of the high country between Otago and Southland in the Old Man, Umbrella and Nokomai Ecological Districts. Almost as soon as the results were published, the programme became controversial. Runholders felt threatened by the possibility of having key portions of their runs excised for reserves, particularly those farming small to medium-sized properties. Even if areas recommended for protection remained in the leases, runholders faced restrictions on their ability to burn, oversow and bulldoze firebreaks and farm tracks on the land. Surveys continued, but increasing difficulties were encountered, such as runholder resistance in the form of refusal of access. The Department of Lands and Survey failed to negotiate deals, and as the number of unresolved recommended areas for protection piled up, the tension increased.

By the time the Department of Conservation was established in 1987, the protection of a few token areas had been achieved, mainly by purchase. In addition, the Crown land allocation processes of 1986 and 1988, which occurred as a result of the government's restructuring of its environmental administration, resulted in several significant Crown grassland areas being protected (Mavora Lakes and Eyre Creek Conservation Areas in northern Southland). These successes, however, did nothing to ease the tension between the new department and the runholders. The department was given the official role of advocate for nature conservation values on pastoral leasehold land, but only blunt tools to do the job.

The Department of Conservation tried to ease the tension in 1990 by initiating the lifting of the Land Settlement Board's moratorium that restricted runholder activity on areas recommended for protection, but this achieved little. In 1991 the Commissioner of Crown Lands received legal advice that nature conservation was not one of the objects of the Land Act 1948. This meant that the moratorium was invalid in any case, and that the department had no legal basis upon which to influence the management of the pastoral leases under which most of the high country was occupied. Departmental staff continued to try to implement the Protected Natural Areas Programme. One key purchase was made (Lauder Basin

Conservation Area in the northern Dunstan Mountains) but funds for acquisitions were scarce. One or two covenants were negotiated, based on the continuation of status quo farming, and a small number of areas were protected as a result of surrenders resulting from earlier subsidisation agreements. Relations with runholders took a dive again when in 1993 the department used the designation process under the Resource Management Act to try to prevent the burning of a copper tussock grassland area recommended for protection on Little Valley Station in the Manorburn Ecological District. They recovered briefly once the Little Valley issue was settled by negotiation, but took a dive again as a result of attempts by the department to protect areas through District Plans.

In the meantime, falling primary produce prices and the rising costs of pest control were forcing some pastoral farmers to consider alternative land uses. The pastoral leasehold tenure, however, was designed to facilitate only the extensive grazing of sheep and cattle on land that continued to be owned by the Crown. It is a tenure not suited for other purposes, especially those of a capital intensive nature (where it is perceived to provide lenders with less security). The exact nature of the tenure is subject to debate, and being partly derived from the Land Act, is subject to change by parliament. Pastoral leases contain no right to freehold (the normal form of land ownership in New Zealand), so if freehold tenure was desired it needed to be negotiated from the Crown. Aware of these circumstances, the Crown's agent (Commissioner of Crown Land) and staff of the Department of Conservation invented a scheme for the exchange of property rights that could simultaneously deliver conservation gains and freehold title. In essence, the scheme was that the runholders could surrender their lessee interest over an area of nature conservation importance (which was then returned to full Crown ownership and allocated to the department for protection) in return for the Crown transferring its lessor interest in the balance of the property to the runholders, enabling freehold title to be issued. The scheme had the added advantage of lowering the freeholding cost to the lessee, in that only the part of the property of greatest farming value had to be 'purchased'. The scheme became known as a pastoral lease tenure review, and it soon began to produce 'win-win' outcomes that were widely hailed. By 1998 the scheme had become sufficiently successful to be given its own legislation (Crown Pastoral Lands Act 1998).

Many areas that were recommended for protection in PNAP surveys are now becoming conservation areas as a result of tenure review negotiations. Even where PNA survey information is not available, the opportunity to assess the conservation values on a property and negotiate an outcome is producing credible results. In some localities, tenure review is resulting in the formation of large contiguous blocks of conservation land earmarked for future conservation park status (Figs 8–10).

Tenure review has some limitations. Its possible adverse landscape effects will be discussed later, but its other main limitation is its voluntary nature. The Crown cannot compel a

Fig. 9. Copper tussock grassland, Manorburn Conservation Area adjacent to Little Valley Station. *Neville Peat/DOC*

runholder to agree to a tenure review. To do so would be a breach of the pastoral leasehold contract which entitles the runholder to renew the lease in perpetuity. The voluntary nature of tenure review means that some runholders will elect to stay with their leases. This could mean that a less than coherent approach is possible throughout the high country. That is, unless parliament shifts the goal posts.

So far, it has been assumed that nature conservation is in general best carried out by the appropriate Crown agency (i.e. the Department of Conservation) on land set aside for the purpose. Many high country farmers have disagreed with this. They say that they are good conservationists, and point to the remaining natural areas in the pastoral leasehold high country as evidence. They say that burning tussock grassland not only allows stock access and spreads grazing pressure but it also renews tussock vigour and increases localised species diversity. They have cast aspersions on the Crown as a land manager, pointing to weeds and animal pests on Crown land. They assert that the runholders on the property are better placed to 'look after' the land than an over-stretched government department with small groups of staff located in a few rural towns. On the other hand, it is probable that most natural areas remaining in the high country are in that condition in spite of, not because of, being farmed as part of pastoral leases. In addition, some aspects of ecosystem and species management are specialised and are fundamentally different from farming the land. There is insufficient space here to debate these issues fully. There are certainly risks for the Crown in taking responsibility for more high country land than it can properly manage. Notwithstanding, the general public seems to favour a larger role for the Department of Conservation in the high country. The public have access as of right to land administered by the department, whereas farmers can and some do refuse access. The southern New Zealand high country has a number of the most valued settings for tramping, cross country skiing and mountain biking in the country.

It might be thought that the Resource Management Act 1991 could deliver some conservation gains in the high country. The act states that the protection of areas of significant indigenous vegetation, habitats of indigenous fauna and outstanding natural features and landscapes are of national importance. Yet the early signs are that it will fail to achieve worthwhile results. Many district and regional councils, who have the main responsibility for making the act work, are reluctant to interfere with established land-use practices in the high country, preferring instead to take a low key educative approach, if any. Longstanding land-use practices may amount to existing use rights that ought not to be taken away without compensation. Suggestions that councils should apply protective policies through district plans to significant natural sites on privately-occupied land are being fiercely resisted by many landowners and occupiers, who see this as taking away their property rights. It would seem that one-to-one negotiation is the only way to achieve lasting results and that, at the moment, pastoral lease tenure review is the tool that is working best to achieve nature conservation outcomes in the high country.

Concern for the sustainability of high country natural resources is shared by conservationists, scientists, local and central government officials and farmers alike. Over the years there have been many attempts at developing solutions that allow sustainable land uses and the conservation and enjoyment of valued areas to coexist. None of these has been entirely satisfactory. Many commentators, including the Parliamentary Commissioner for the Environment, are now calling on the government to develop a National Policy Statement for the high country under the Resource Management Act. If approved, this would provide regional and district councils with clear direction for their policy statements and plans for the high country, and should provide the consistency that is currently lacking in attempts to tackle these issues.

Fig. 10. Taieri River headwaters, Te Papanui Conservation Area.
Neville Peat/DOC

Lowland wetlands

The lowland wetlands of southern New Zealand were once extensive. The greater part of the 20,000 ha Taieri plain south of Dunedin, for example, was described by Hocken in 1898 as 'an immense grass tree swamp, through which canals of black sluggish water wind in various directions, and interspersed with stagnant lagoons ...'. What he described, in ignorance, in such discouraging terms must have been a productive habitat for birds, invertebrates and freshwater fish and a rich source of mahinga kai and fibre for the Ngai Tahu inhabitants of the day.

Today, less than one tenth of the Taieri plain wetland remains and this typifies the depletion of New Zealand's lowland wetlands from their former extent. The remaining wetland of approximately 2000 ha between and including parts of Lakes Waipori and Waihola at the southern end of the plain, is still subject to a number of threats. Parts of it are privately owned. Key areas, totalling 140 ha, have been proposed by the owners for stopbanking and drainage for farming purposes. Other parts of the wetland are being invaded by willow species. The bed of Lake Waipori is infilling from sediments brought in from the catchments of the Taieri and Waipori Rivers. The whole system is being affected by non-point source pollution from a variety of sources.

The Waipori/Waihola wetland is the largest lowland freshwater wetland in the South Island (there are larger brackish and saline wetlands and much larger coastal lagoons). It is the most significant water-bird habitat in Otago. It regularly supports 12 species of freshwater fish and is a significant fish breeding habitat. The indigenous wetland vegetation is largely undisturbed, in particular on the river delta islands and levees. Its importance was recognised by the former New Zealand Wildlife Service of the Department of Internal Affairs and in 1975 discussions were initiated with landowners with a view to purchase for reserve purposes. No worthwhile progress was made, while a number of landowners continued to drain and otherwise modify parts of the wetland. So in 1980 the Minister of Works, at the request of the Wildlife Service, designated the key remaining parts of the wetland. The designation was bitterly opposed by the landowners, but was confirmed by the then Planning Tribunal in 1981. This provided interim protection, but the Wildlife Service was still obliged to try to negotiate purchase or to initiate compulsory purchase action. It had made no progress by the time its responsibilities were inherited by the Department of Conservation in 1987. More recently, the department has produced a Management Statement to indicate to landowners how it would manage the wetland if it was reserved, and has acquired the key wader habitat in private ownership.

The remaining wetland is now under no immediate threat of conversion to pasture. The Sinclair Wetlands component is protected under a Queen Elizabeth II National Trust covenant. It was purchased by the Crown and has been awarded to Ngai Tahu as part of the settlement of their claim under the Treaty of Waitangi. The Otago Regional Council administers the bed of Lake Waipori, its associated wetlands and most of the Waipori delta extending into Lake Waihola, and has indicated its general sympathy with the management aims of the Department of Conservation.

Now that the risk of modification has been averted, there remains the challenge of managing the wetland to preserve and enhance its wetland qualities. Willow control, riparian strip vegetation restoration and water level management are important, but the complex task of influencing the management of the Taieri and Waipori catchments to reduce adverse inflow-borne effects on the wetland will be even more demanding. This is a long-term commitment, involving participation in Resource Management Act processes. Ultimately, the success of these efforts depends on a significant degree of community support and understanding in relation to wetland conservation.

Similar management issues arise with regard to southern New Zealand's other major lowland wetlands, the Waituna Lagoon, Seaward Moss and Toetoes complex of coastal Southland. Unlike the Waipori/Waihola wetland, however, nearly the entire complex is under the single administration of the Department of Conservation. The Waituna Scientific Reserve has been designated a Wetland of International Importance under the RAMSAR Convention* and it is proposed to extend this status to the other two major wetlands. These wetlands are the most important bird habitat in Southland and are likewise nationally important for freshwater fish. Unique low-stature vegetation communities are also associated with the lagoons.

The Otago and Southland wetlands both experience the adverse effects of non-point source pollution. Where excessive nutrients run unimpeded off pasture into the catchments, they cause algal blooms and other adverse effects of excessive enrichment. Ammonia discharges, which occur particularly from land used for dairying (a growing land use in Otago and Southland) have direct toxic effects on aquatic invertebrates and fish. In heavy rain, when fine sediments are able to run directly into waterways, they can accumulate in the wetlands, producing unsuitable habitat conditions. The sediments store nutrients which are then resuspended as a result of wave action across shallow waters, recreating algal bloom conditions. Elevated water temperatures caused by an absence of shading vegetation exacerbates these effects.

In Southland, the existence of mineral resources in association with the wetlands (gravels and lignite deposits) is an additional source of potential conflict. Decisions on mining on land administered by the Department are taken by the Minister of Conservation under the Crown Minerals Act 1991. The legislation requires the Minister to focus on the natural values of the land, the requirements of the legislation and management plans or strategies under which it is administered in making the decision. Court rulings on this legislation would suggest that it will be effective in protecting biologically important areas from adverse mining effects.

Less easy to deal with is the threat to wetlands posed by illegal sphagnum moss harvesting. All of the major Southland wetlands and some of the Otago ones have been subjected to incidents of this sort. Commercial moss harvesting on departmental land in Otago and Southland is prohibited as a matter of policy on the grounds that it is not sustainable, but law enforcement is difficult in remote country, when the trade is sufficiently lucrative to warrant the use of helicopters. The damage is physical and may take a long time to heal. In severe cases the damage may cause permanent changes in a wetland's hydrology. As with all law enforcement activity, community support and good industry networks are crucial.

The Southland lowlands also feature a considerable number of smaller peat-based wetlands, some of which are protected. The nature conservation resources of the Southland plains are still in the process of being surveyed, in order to determine priorities for further protection work.

* Named after the location in Iran where it was negotiated, the RAMSAR Convention establishes criteria and management prescriptions for wetlands ranked as internationally significant.

Fig. 11. Upper Taieri Scroll Plain wetland, Styx Basin section. Neville Peat/DOC

Mid-altitude wetlands

Accessible mid-altitude wetlands in New Zealand have disappeared at more or less the same rate as lowland ones, although in the Te Anau basin the picture is a little different. The basin was developed as recently as the late 1970s. The Crown was the principal development agency and, as a result of the growing environmental awareness of the period, some 30 separate wetlands were excluded from farm settlements and are now administered by the Department of Conservation. The most significant are the Kepler Mire, incorporating an extensive patterned string-bog system unusual at lower altitudes, Dome Mire and Island tarn. Many other wetlands remain on private land in the basin. Together, they represent a resource of international significance as special landform/ecosystems, and as habitats for indigenous plants, birds, freshwater fish and invertebrates requiring wetland conditions.

The most significant mid-altitude wetland in Otago is on the Upper Taieri Scroll Plain (Fig. 11). The Taieri has its source high in the Lammerlaw mountains. It drops out of the mountains at Taieri Rapids and flows first of all through the Styx Basin and then across the Maniototo plain to a gorge at Kokonga. There is so little fall in the basin and plain that the river has meandered across both, forming a classic scroll plain wetland in each case, showing a high degree of curvature and all stages of ox-bow formation. Not only is this New Zealand's most fully developed scroll plain landform, but the wetland is rated as nationally important for wildlife and internationally important as a wader habitat. Unfortunately, the indigenous vegetation is modified in most places and in the past there have been modifications along some reaches of the river to make it flow in a confined channel rather than replenish the adjoining wetlands during flooding. These modifications were part of the Otago Catchment Board's River Channel Improvement and Drainage Scheme, which was completed in 1985. The most significant factor in preserving the remaining wetland is that less than 15 per cent of its 3000 ha has any legal protection. Virtually all of it is grazed, and although grazing of certain stock types at certain times of the year may not be harmful, only those parts administered by the Department of Conservation are grazed according to wetland management requirements. There is a growing awareness of the value of the wetland landform, and Resource Management Act processes may result in some protective policies being developed for it by the Central Otago District Council and the Otago Regional Council. It may, however, be beyond the resources of the Department of Conservation to involve itself in the direct management of much more of the wetland than it is responsible for at present.

High-altitude wetlands

All of the mountain ranges of southern New Zealand support wetlands, mostly in the form of bogs or mires of various types. These are usually small and are perched in hollows on spurs and ridges. Numerous examples are protected from disturbance within areas already mentioned, such as Fiordland National Park. All high-altitude wetlands are fragile, involving little-understood relationships between landform, substrate, precipitation and the biomass of their plants and invertebrates. The low relief block mountain ranges of Otago and northern Southland, however, support distinctive bogs which cover extensive areas, often blanketing large portions of the range crest. As well as their biological importance and landform significance, these bogs act as reservoirs feeding the streams that supply water-short areas below during summer droughts. Examples of these blanket bogs are protected within conservation areas on the Old Man and Old Woman Ranges, but some of them are accessible to vehicles, the misuse of which can cause extensive and irreparable damage. Many more exist unprotected on pastoral leasehold land. Many of these will eventually be protected through the pastoral lease tenure review programme.

New Zealand's largest and most complex string bog straddles the divide between the Nevis catchment of Otago and the Dome burn headwaters in northern Southland. One of very few such examples in the Southern Hemisphere, the Nokomai Patterned

Mire (as it is known) lies on pastoral leasehold land. It is partially protected by the Kawarau Water Conservation Order, which prevents any resource consent being granted that would disturb the portion that lies inside the Nevis catchment. The Nokomai Patterned Mire is not under any present threat, but other high-altitude wetlands have been obliterated in the past. The Great Moss Swamp, originally several thousand hectares in extent, used to lie in a hollow between the Rock and Pillar and Lammermoor ranges of Otago. It has been replaced by the Loganburn Reservoir. Copper tussock wetlands in the headwaters of the Teviot River in Otago have been replaced by Lake Onslow. Remnants of the wetlands lie in the tributary creeks flowing into these man-made reservoirs, but they are subject to periodic inundation not reflecting any natural cycle.

Lowland indigenous forests

Samples of the various types of lowland indigenous forests now surviving in southern New Zealand are already represented in the protected lands administered by the Department of Conservation. The challenges involved in the conservation management of these forests will be discussed in a later section, but because these challenges do exist, it is important to consider the extent to which that representation is adequate. Beech forests of all types are well represented, and will not be discussed further here, although some of them are important habitat for threatened species such as mohua (yellowhead). Forests dominated by hardwoods (mainly kamahi and southern rata, Fig. 12) are also well represented in the protected areas system.

Less well represented are the forest types dominated by podocarp trees, such as rimu, matai and kahikatea. Not surprisingly, these are the most valuable timber trees. Such forests support the greatest diversity of wildlife. They have largely been removed from the southern New Zealand landscape, and any sizeable remaining example is important for nature conservation.

A complex national debate about the protection of indigenous forest versus the private property rights of owners is part of the context for any discussion of this issue in the southern region. The values of indigenous forests for scenery and as wildlife habitat have long been generally recognised in this country. More recently, their intrinsic value has been recognised as well. At the time of the environmental restructuring of 1986, the government of the day adopted the policy that all Crown-owned indigenous forests (with the exception of certain forests on the West Coast and in Southland) would be protected. Recently (2000) this policy has been taken a step further, with the government deciding to protect most of the West Coast forests that were previously available for exploitation. Attempts to constrain the removal of indigenous forest on private land began under the Town and Country Planning Act 1977, and the Clutha County (now District) Council was one of the first to try to do so. The council recognised that the privately-owned indigenous forests of the Catlins were fast disappearing as a result of clearfelling for woodchip export, and that the district's once extensive indigenous forests were becoming fragmented. In 1986, relying on the provisions of the act allowing for matters of national importance to be addressed, the council proposed controls on bush clearance through its draft District Scheme. Landowners reacted strongly to what they saw as a threat to their rights, and in some cases indulged in 'panic' felling. The council backed off. Pressure was then applied on the government by conservation organisations and as a result controls on the woodchip export trade were tightened successively, in an attempt to achieve indirect protection for the forests. The government found itself in the awkward position of having to interfere in existing contractual arrangements between landowners and those running the export trade. Compensation was provided to those affected by the cancellation of contracts, but surprisingly the government failed to secure the protection of the forest at the time it paid compensation. This meant that when the government's Forest Heritage Fund (set up in 1992) was subsequently used to buy a number of key privately-owned indigenous forest areas in the Catlins, the government had effectively paid twice. The government did, however, succeed in halting the woodchip export trade and in 1993 it enacted the

Fig. 12. Southern rata in flower, Poison Bay, Fiordland National Park. *Lou Sanson/DOC*

Forests Amendment Act requiring most indigenous forestry on private land to be conducted on sustainable principles. This achieved little in the Catlins, since it did not control forest clearance for farm development or firewood extraction, and did not apply to key areas held under the Landless Natives Act 1906. The Resource Management Act 1991 replaced the Town and Country Planning Act, and gave more explicit recognition to the national importance of significant indigenous vegetation (such as lowland indigenous forest). So the Clutha District Council is still trying to meet its environmental responsibilities. In its latest District Plan, the indigenous forests in a Coastal Protection Zone are given a high degree of protection. Unfortunately the practice of felling native trees in the Catlins now seems to be so long established in some places that it has taken on the character of an existing use. The Resource Management Act provides that certain existing land uses cannot be constrained by the provisions of District Plans, so long as they do not increase in scale or intensity. Only those forests administered by the Department of Conservation or subject to protection covenants in perpetuity seem to be safe from logging or clearance. While protection negotiations will be attempted from time to time, the Waitutu outcome (see above) has raised expectations of a very high market value for podocarp timber in particular. This makes further lowland forest protection a slow and expensive business. Nor (except in the case of Waitutu) has the issue of logging in forests held under the Landless Natives Act 1906 yet been resolved. Key indigenous forest areas in the Catlins are owned by a group of trusts, some of which are selectively logging without scrutiny in reliance on the exemption for 1906-Act land, and some of which are applying pressure for government negotiators to make higher offers for protection. The priority 1906-Act forests on Stewart Island have recently been protected, but Waitutu-scale compensation was paid by the Crown. In the developed landscapes of other parts of the country, the remaining privately owned indigenous forest patches are usually small. They are normally treasured by their owners, who often voluntarily enter into arrangements such as covenants under the Queen Elizabeth II National Trust legislation without expecting compensation. Such owners will usually have other sources of income. Many of the owners of 1906-Act land in the Catlins and on Stewart Island, on the other hand, do not have other sources of income. Although most of them are Ngai Tahu, the Treaty claim settlement of November 1997 will not financially benefit them directly. In any case, the settlement was intended to redress historic grievances rather than address contemporary social problems.

It would seem that unless there is some determination on the part of the country to negotiate protection deals with generous compensation, the remaining indigenous forests of southern New Zealand that are in private and Maori ownership will become increasingly chequered ... like their history.

Fig. 13. *Myosotis uniflora*, Pisa Flats. *Neill Simpson/DOC*

Other lowland ecosytems

Southern New Zealand contains other special lowland ecosystems. The saline soil remnants found in Central Otago are found elsewhere in New Zealand only in the MacKenzie basin. To call them soils may be too generous, for they are silty and pebbly and have a very low biomass. The Central Otago remnants comprise scattered small (none is larger than 100 ha) examples of salty ground, together with their associated halophytic (salt tolerant) native plants. The plants in turn support particular native invertebrates whose larvae are dependent on them for food. The saltiness of the ground derives from minerals leached from schist rocks, and deposited out under conditions of strong evaporation. These remnants are part of what were once extensive saline ecosystems, now reduced as a result of agricultural development. The remaining sites are mostly known, and a number of them have formal legal protection, although the full range of different types is not yet protected. These sites are the habitat of a number of species that are found nowhere else. Their scientific value belies their small size, as all salt-tolerant genetic material is valuable on a global scale. This is because vast tracts of the world's agricultural lands are becoming increasingly saline as repeated irrigation brings mineral-laden groundwater to the surface. The main threat to these small ecosystems continues to be agricultural development. In recent years, a number of small sites have been lost to the plough, damaged by road widening, affected by stock concentrations or altered by changed irrigation practices.

As well as the saline sites, Central Otago contains New Zealand's lowest rainfall areas. These drylands are the home of a small number of plants, invertebrates and lizards found only here (Fig. 13). Little is known about the subadult life cycle of invertebrates such as the Cromwell and Alexandra chafer beetles, characteristic of the drylands, and the lizard species are still being described. Representative examples of the drylands have been protected at places like Flat Top Hill, the Cromwell Chafer Beetle Nature Reserve (the world's first reserve established to protect the habitat of an insect, namely inland low dunes topped by silver tussock) and Aldinga's Conservation Area, where two species of chafers overlap. The challenges of managing such areas for nature conservation are many. Rabbit plagues are a threat to the plants. Predators attracted by the presence of rabbits can turn to eat the native invertebrates and lizards. Introduced grasses, herbs and woody weeds can outcompete the indigenous plants, most of which are tiny. There are many unknowns about what will happen when the grazing pressure of sheep or rabbits is suddenly lifted from such areas.

Intact short tussock grasslands (i.e. those that have retained their associated native plants and invertebrates) have nearly disappeared from southern New Zealand. Such grasslands were once extensive, following the removal of native forest by fire from the eastern South Island, but were among the first to be utilised for grazing last century. They have since been ploughed, burnt repeatedly, overgrazed, oversown with exotic pasture species and invaded by herbaceous and woody weeds. Even the valley floors in Mount Aspiring and Fiordland National Parks have been grazed and modified, although in recent years the Department of Conservation has made strenuous efforts to negotiate the cessation of grazing. Probably the best remaining substantially intact short tussock grasslands left in the southern region are in parts of the Mararoa (Fig. 14), Greenstone and Nevis Valleys, while there are more modified examples in the upper parts of the Manuherikia and Dunstan Creek Valleys in Central Otago. None of these is protected as yet, although the Mararoa and Greenstone are eventually to be managed for conservation purposes as a result of the Ngai Tahu Treaty claim settlement. Both are subject to grazing at present. Pastoral lease tenure review could result in parts of the other valleys being protected. All are subject to invasion by hawkweed (*Hieracium* species) which can inhibit the recovery of native plants.

The final lowland ecosystem types that should be mentioned as requiring further protection are non-forest coastal communities. Dunelands featuring the native pingao are a feature of the western Southland coast and the coast of Stewart Island, and small examples are found in the Catlins (Fig. 15). Many of these examples are protected. The main threat to these dunelands is the aggressive introduced marram grass. Coastal turf communities comprising ground-hugging native herbs and small grasses are found in southern New Zealand to a greater extent than elsewhere in the country. The main threat to these is the impact of grazing animals. Inventory work on the coastal turfs is currently under way to determine conservation priorities.

Fig. 14. Cushion bog and tussock grassland on a bench above the upper Mararoa Valley, looking north. *Peter Johnson/DOC*

Fig. 15. Pingao near Preservation Inlet. *Lou Sanson/DOC*

Marine ecosystem protection

The protection and management of marine areas (i.e. below mean high water spring tides) for nature conservation purposes lags far behind terrestrial area protection in New Zealand. This is particularly true in the southern region where at the time of writing only two small marine reserves have been created.

Southern New Zealand encompasses a range of characteristic marine ecosystems. The east coast of Otago features nearshore communities dominated by bull kelp (*Durvillea*) or bladder kelp (*Macrocystis*) species depending on the degree of wave exposure. These communities provide habitat for a southern fauna including sea tulip, various species of brachiopods and fish such as southern pigfish and Jock Stewart. Such communities are also found on the Southland coast and the coast of Stewart Island/ Rakiura, but the latter contains its own special features such as *Lenormandia* seaweeds, and black corals in some places. The seaweeds of Stewart Island are especially rich and diverse for New Zealand. The marine ecosystems of the south-west coast fiords are internationally recognised for their subtidal communities of rare deep-water species found at shallow depths. These ecosystems have limited algal (seaweed) growth due to low light levels, but include brachiopods, black and red corals, sponges and deep-sea fishes.

The existing marine reserves protect examples of some of the many fiord marine ecosystems. Te Awaatu Channel (93 ha) in Doubtful Sound contains a deep reef system including species such as sea pens, zoanthid anemones and red hydrocorals. Piopiotahi (690 ha) along the northern side of Milford Sound (Fig. 16) contains typical fiord rock wall communities of black and red corals and brachiopods. Elsewhere in southern New Zealand there are places where voluntary protection is sought by local divers and fishers, but these cannot be regarded as secure because they are unenforceable and compliance is uneven. The Ministry of Fisheries manages harvestable marine species through its quota systems in order to try to ensure that catches are sustainable, but their effort is species-focused and does not address the protection of marine ecosystems. Ngai Tahu resource management (see later) is also focused on species of cultural significance, not on ecosystems.

The threats facing the natural integrity of marine ecosystems in southern New Zealand include the over-exploitation or illegal exploitation of harvested species. Apart from the impact on the species themselves, there are impacts up and down the food chain and physical impacts on habitats from trawling and dredging. Marine farming (which occurs on Stewart Island/Rakiura) can substantially modify natural ecosystems. Other threats include oil spills and the introduction of unwanted organisms by shipping, such as toxic dinoflagellates. Big Glory Bay and Paterson Inlet on Stewart Island are being monitored for the latter. The aggressive colonising exotic seaweed *Undaria* has already gained a foothold on most of the Otago coast and the Department of Conservation is attempting to control it within

Paterson Inlet/Whaka a Te Wera on Stewart Island.

The Department of Conservation with the support of various conservation organisations is attempting to improve the extent to which marine protected areas in southern New Zealand are representative of the full range of southern marine ecosystems. The department has chosen Nugget Point/Tokata as its preferred site for the first marine reserve for Otago. The proposed reserve (1500 ha) will encompass both sheltered bladder kelp and exposed shore bull kelp communities, as well as a special community of encrusting organisms on the vertical undersea walls of the Nuggets themselves. The reserve is undergoing a long gestation period, mainly due to concerns raised by commercial fishers and some locals who perceive that harvesting opportunities will be lost, and concerns raised by Ngai Tahu that their traditional uses of the marine environment are being excluded by reservation proposals. The department is also promoting a marine reserve in the eastern side of Paterson Inlet/Whaka a Te Wera between Ulva Island and the main island. This proposal has community support but has yet to be concurred with by the Minister of Fisheries.

A particular conservation challenge is posed by the situation in the fiords. In marine ecosystem conservation terms they are a high priority. There are many variations in marine ecosystems in the fiords, both in terms of latitude, aspect, exposure and freshwater influence. There are also particular threats, including those posed by intensive tourist and fishing vessel traffic, various forms of onshore and offshore development, and proposals for large ocean-going vessels to enter the fiords to take on fresh water for export. Above mean high water spring tides the fiords are national park land, recognised internationally as a World Heritage Area renowned for its wilderness values and given strong legal protection through a single-purpose nature conservation agency. Below mean high water springs, with the exception of two tiny marine reserves, the fiords have no ongoing legal protection and a multitude of bodies (Ministry of Fisheries, Southland Regional Council, Southland District Council, Maritime Safety Authority, etc.) will need to co-ordinate their activities for protection outcomes to be achieved. A strong case can be made that the entire marine/terrestrial interface in a special place like the fiords should be able to be protected under the National Parks Act, not just the part above the waterline.

Overshadowing many of these difficulties is the fact that relatively little is known about the functioning of marine ecosystems generally, and those of southern New Zealand in particular. Therefore it is impossible to be certain about whether or not any given form of disturbance (whether natural or caused by humans) will be absorbed with little adverse effect or whether it will set off a chain reaction producing an ecological disaster years later. For example, despite strenuous attempts to identify the cause, the reason for the die-off of royal albatross and yellow-eyed penguin breeding adults in the 1989–90 summer is still essentially unknown. A marine perturbation resulted in the latter's breeding population crashing to a critical 130 pairs. The rapid recovery of the population to pre-1989–90 levels in seven years was helped greatly by human intervention through

Fig. 16. Milford Sound/Piopiotahi from above the Homer Tunnel.
Lou Sanson/DOC

predator trapping at breeding areas on shore. There has been speculation that the die-off was linked to abnormally high sea surface temperatures the preceding year, which affected continental shelf food resources. The die-off coincided with other disturbances involving a massive bonamia infection of the Foveaux Strait oyster beds and lower than usual wetfish catches. The fact that very large-scale but difficult to detect perturbations can occur in the marine environment underlines the importance of marine protected areas for baseline studies to increase our knowledge, as well as to protect marine ecosystems for their intrinsic values.

CONSERVATION MANAGEMENT CHALLENGES

So far, the emphasis of this chapter has been on issues to do with the prevention of damage to important natural areas caused by direct impacts such as the logging of forests, conversion of native grasslands to exotic pasture, drainage of wetlands, industrial development and deliberate exploitation. Such damage is normally prevented by applying a protective legal regime to the area. This will vary from administration by the Department of Conservation (i.e. direct conservation management by the government), to the covenanting of private land with basic conservation management being carried out by the landowner. Once protected, though, all such areas will require some degree of ongoing conservation management to ward off or limit the adverse effects of other threatening factors.

At the simplest level, all terrestrial protected areas need a fence. This keeps out unwanted browsing animals and helps to delineate the boundary. The latter is especially important if development is continuing to occur next door. Fencing might seem to be the simplest aspect of conservation management, but it is often the most neglected. In southern New Zealand, fences in the mountain lands are susceptible to collapse under heavy snow as well as the usual problems of falling trees, land subsidence and rockfall. Special fencing techniques (siting and fence design) are required in snow-prone country, and often the protected area boundary has to be dictated by the practical fenceline as much as by the nature conservation values.

Probably the single most threatening factor in conservation management is fire. A hot wildfire can wipe out many decades of native forest growth in hours, it can open up grasslands to wind erosion or weed invasion and it can kill precious wildlife. Southern New Zealand might be thought not to be especially fire-prone but fire can be a threat to nature anywhere if the conditions are right. The Catlins are among the wettest parts of southern New Zealand, but in 1994 an escaped neighbouring burn-off entered the yellow-eyed penguin reserve at Te Rere and killed 52 birds, almost half of the breeding adults there (Fig. 17). Most were burnt on the nest, but some died in agony trying to walk through the hot ashes, having never experienced fire before.

A fire started accidentally alongside State Highway 6 between Lakes Wanaka and Hawea in 1995 destroyed hundreds of hectares of tussock grassland and 30 years' worth of woodland regeneration, and would have gone further but for a favourable wind shift. Fires along the heavily used tourist highway to Milford not uncommonly threaten parts of Fiordland National Park. The Department of Conservation in southern New Zealand maintains a capacity to respond to vegetation fires that threaten the areas it administers, but prevention is always the preferred approach.

Fire risk management for conservation purposes is a troublesome issue because many private landowners and occupiers still use fire as a regular management tool. Thus, every late winter and early spring, fires are lit throughout the pastoral high country of Otago and Southland to control growth that is not wanted (heavy tussock or woody growth of little grazing value). The use of fire in this way has occurred for generations and because the practice is regarded as beneficial, some high country farmers have a casual attitude towards it. Farmer-lit fires that escape from the area intended to be burnt are common, and these sometimes threaten or enter areas of conservation importance. A tolerant farming neighbour might not be concerned, but if the area into which the fire escapes is administered by the Department of Conservation, the department will follow its current policy and try to limit the damage and bring the fire under control. The department is empowered to recover its firefighting costs from the person who lights the fire in the first place, and although all farmers carry insurance, there is an understandable reluctance to have the department on the scene.

On the other side of the argument, the farmers see ungrazed and unburnt conservation areas as serious fire risks in themselves. Tussock grasslands and shrublands can accumulate a surprising amount of fuel, and fire (especially under dry and windy conditions) can travel fast and far and threaten life (including the lives of firefighters) and property. Such fires cannot be controlled by conventional urban methods. Risky (backburning) and expensive (helicopters and monsoon buckets) techniques may have to be used. Then there are those who argue persuasively that burning in the high country followed by grazing is an unsustainable practice. Even some forms of fire risk management are problematic. One way to limit the number of escaped fires is to maintain firebreaks, either in the form of a strip bulldozed down to mineral soil or as a strip of vegetation (for example, clover) that does not carry fire. But such firebreaks can sound a discordant note in the landscape, and be visible for great distances. So the use of fire and fire risk management will continue to divide conservationists and the farming community for some time to come. All aspects of the debate need further illumination by research. Many questions surround the

Fig. 17. Yellow-eyed penguins, casualties of the Te Rere fire. *J. Darby*

Fig. 18. *Phyllocladus-hoheria* woodland, Top Forks, upper Shotover. Neill Simpson/DOC

sustainability issue, which has yet to be explored by the Environment Court. Little is known about the effects of fire on grassland managed for conservation purposes and the Department of Conservation and associated agencies have begun a long-term experiment in Otago to assess the effects of various fire regimes on ungrazed tussock grasslands. This might eventually yield information that would provide a basis for fuel reduction burning (i.e. a light burn to reduce available fuels but not permanently damage the ecosystem) in conserv-ation areas as occurs (albeit in more fire-tolerant plant communities) in Australia, or burning for native grassland maintenance, as occurs in the United States of America.

Animal pests and predators as a threatening factor have already been mentioned in this book. A comprehensive overall assessment of animal pests is beyond the scope of this chapter, but some observations should be made. Different pests are significant in different ecosystems, and different predators threaten different species at different times of the year, so an overall ranking seems unhelpful. But there are some pests and predators that are particularly damaging.

Possums have now reached all parts of southern New Zealand except for south-western Fiordland and some lake and offshore islands. They are progressively altering the composition of forests and shrublands, where they prefer certain food species. They are a serious threat to southern rata and mistletoe and can locally eliminate preferred species such as fuchsia, kamahi, totara and members of the genus *Pseudopanax*. They are implicated in the removal of chicks and eggs from the nests of native birds. They also carry bovine tuberculosis. The extent of the possum infestation is such that the authorities have to be selective in their control attempts. The only prospect for controlling possums generally is some form of biological control, which may be decades away and may be controversial if it involves genetically modified organisms.

Feral goats are widespread in the region, except for Stewart Island/Rakiura where they are present as farmed animals but have not yet escaped. They occupy forest and shrubland habitats and can move into alpine areas under pressure. In all habitats their impacts are confined to the understorey, but they are a serious pest because they can prevent regeneration and locally eliminate palatable herbs and shrubs. The Department of Conservation has successfully used 'Judas' goats to keep feral populations at very low levels in some Otago shrubland and open forest ecosystems. After the main populations are knocked down by conventional methods, selected animals are captured and released with radio collars. Because of their herding instinct, the radio-collared 'Judas' animals soon join up with others and hunters in helicopters home in on them. After one or two such raids, the 'Judas' animal soon learns to duck under a rock when it hears the noise of a helicopter, and waits while its companions are destroyed. These operations are brutally efficient and mean that feral goat numbers can be kept very low over large areas at a moderate cost. The department has virtually eliminated feral goats from the upper Shotover catchment (Fig. 18), which among other values, contains the largest remaining population of the threatened shrub *Hebe cupressoides*.

The 'Judas' goat technology has now been adapted to the control of Himalayan thar, a pest of high alpine areas of national parks in the South Island. Thar tend to camp in and eat out the herbs and grasses of alpine basins if they are not controlled. Under a controversial national thar control strategy, they are being retained in limited numbers as a hunting resource in Canterbury and Central Westland, while officially they are to be excluded from southern New Zealand. They migrate in ones and twos, however, south along the Southern Alps/Ka Tiritiri o te Moana to a bottleneck around the upper reaches of the Hunter catchment in Otago. There, regular control with the assistance of 'Judas' animals placed to intercept the migrants might have achieved the exclusion objective. Thar, however, continue to 'appear' in the wild much further south, and the inference has to be drawn that they are being put there illegally by humans, as a hunting resource. The national control plan commits the Department of Conservation to 'Judas' and search-and-destroy operations in perpetuity at both extremities of the thar feral range, whereas a policy of complete eradication would remove all ambiguities and opportunities for illegal transfers. There are plans to stop chamois (also a pest of high alpine areas) from spreading south of Milford Sound/Piopiotahi in Fiordland National Park, using similar methods. The lessons of history have shown that pests are better eradicated while they are controllable, not after they have got out of control.

Deer (red deer in particular) were once a seriously damaging pest in the mountainlands and forests of the southern region, such that whole forests were threatened with the type of collapse that is now imminent in the dry beech forests of the North Island. In the late 1970s, southern deer were brought under control (and largely remain so) by the aerial deer recovery industry, which burgeoned after demand for deer farming stock and venison skyrocketed. Localised problems with deer remain, however, as recreational hunters exert little pressure and selective browse impacts are occurring. On Stewart Island/Rakiura, white-tailed deer are widespread, numerous and favour subcanopy hardwoods as food. This is altering forest composition across the main island. In the mainland recreational hunting areas (Blue Mountains and upper Wakatipu beech forests) the ecosystem impacts of sizeable herds of fallow deer are being tolerated in the interests of providing a particular kind of recreational opportunity. These arrangements are being questioned within and outside the Department of Conservation, which is responsible for these areas, and are the subject of a current debate as the department implements its Deer Plan.

Rabbits are a serious agricultural pest in the rainshadow areas of southern New Zealand. Their role as a pest in nature conservation terms is not so easy to assess. They most often inhabit areas long modified by farming activity and it is difficult to separate their impacts from other factors. Rabbits undoubtedly browse or disturb native herbs, grasses and seedlings of shrubs and trees, and reduce plant biomass (which in turn reduces soil moisture and fertility and contributes to soil erosion). But they also suppress the weeds and the seedlings of the woody weeds that compete with the native plants, and their scratchings may provide seedbed opportunities for those plants. They may accordingly be instrumental in maintaining the open habitat needed by small stature native grasses and herbs, for example those that are unique to the drylands of Central Otago. Rabbits also provide prey for the predators (to be discussed shortly) of some of the native animals of those areas. Whether the role of rabbits is positive overall in terms of indigenous biological diversity is an interesting academic exercise, but they are here to stay for the present and the real issues relate to how to manage them. Traditional forms of management have either been labour-intensive (with many men having been employed in the past as rabbiters) or unpopular (such as the repeated aerial spreading of 1080 poison baits). The effectiveness of any one method is always likely to diminish in the long term. Poison-bait shyness has occurred, and it has even been suggested that rabbits in some areas have developed an aversion to any new food and will not take unpoisoned baits put out to encourage them to take the poisoned ones later.

In 1997 farmers introduced RHD (Rabbit Haemorrhagic Disease, formerly known as RCD Rabbit Calicivirus) after an application for its official release had been turned down by the Ministry of Agriculture. The disease was widely spread and it killed many rabbits. The chief nature conservation concern raised during the application process was that starving rabbit predators would switch to native animals, such as ground-nesting birds, reptiles and invertebrates, perhaps pushing some of them over the edge to extinction. Localised prey-switching effects had previously been noted following large-scale aerial 1080 operations, so the government provided additional resources to the Department of Conservation for the threatened black stilt, the highest priority for protection against the prey-switching risk thought to be associated with the release of RHD. There was scientific evidence, that prey-switching to native species did occur in the MacKenzie basin of Canterbury, and anecdotes of increased predation of some seabird colonies along the Otago coastline coinciding with the arrival of RHD. There were no known extinctions, and now that the disease is becoming less effective (see page 144), this concern is subsiding. During the window of low rabbit numbers, the accelerated spread of woody and herbaceous weeds is just as great an issue for the conservation of plant species that require open habitats. This is probably balanced overall by other native plants benefiting from improved growing conditions. During the forthcoming uncertainties the Department of Conservation will be challenged to understand what is going on and make the correct choices about applying its limited resources in protection of the species that may be vulnerable to prey-switching or weed competition.

RHD does not transmit to hares. These are widespread in the high country grasslands of southern New Zealand, and their numbers seem to be increasing in some places. They chew on the native herbs and grasses of the alpine zone but as yet have not caused any extensive damage.

The animal pests discussed above are causing impacts mainly at the ecosystem level. There are other animal pests that impact on nature mainly at the species level. These are the introduced predators such as ferrets, stoats, wild cats and rats. Special southern species such as yellow-eyed penguin, albatross and other seabirds, and the native forest and open country birds, all evolved in the absence of this suite of predators, and their young have no defence mechanisms against them. On Big South Cape Island, off Stewart Island/Rakiura, one of New Zealand's worst ecological disasters occurred as recently as 1964. Ship rats overran the island and caused the extinction of two unique southern bird species, Stead's bush wren and the Stewart Island snipe. The then Wildlife Service only just managed to rescue sufficient pairs of the South Island saddleback to start new colonies on rat-free islands nearby, before the rats finished that species off for good. In an echo of what had already happened on the mainland, the rats on Big South Cape ate their way through colonies of the native greater short-tailed bat, the island's skinks and geckos, several species of large invertebrates and many bush birds including robin, parakeet and morepork. Nowadays, critically important wildlife breeding areas are intensively managed to reduce predation risks. Takahe eggs are taken from the wild and reared past the vulnerable stage before the young birds are reintroduced to the Murchison Mountains of Fiordland National Park where the habitat is specially protected for their benefit. Burrows containing kakapo chicks on Codfish Island may be monitored using video technology round the clock. Predator trapping is carried out in and around key yellow-eyed penguin colonies at critical times of the year. The Taiaroa Head royal albatross colony (unique in the world as it is a mainland site) is ringed with traps and other devices year round. Where possible, threatened species recovery programmes are carried out on predator-free islands. From time to time the Department of Conservation carries out eradication programmes to rid suitable islands of rats, possums or mustelids, and the predator-free status of key islands is assiduously maintained. For example, in recent years Ulva Island in Paterson Inlet/Waka a Te Wera has been cleared of possums and rats. Breaksea Island in southern Fiordland has been cleared of rats, allowing South Island saddleback and knobbed weevil to be introduced there. Mohua (yellowhead) have been transferred to predator-free islands, such as Pigeon Island/Wawahi Waka in Lake Wakatipu and Centre Island in Lake Te Anau. The department leads the world in this work.

In a few mainland locations, the department is attempting to replicate the successes of its island refuge work. The Eglinton Valley of Fiordland National Park is one of the most important habitats for forest birds in southern New Zealand. It contains a major population of the threatened mohua, and is also important for yellow-crowned parakeet, South Island kaka, robin and long-tailed bat. In addition, it holds colonies of two red mistletoe species favoured by possums. The department is attempting to manage the valley on an ecosystem basis through integrated pest control and scientific research aimed at effectively and efficiently controlling pests such as deer, possum and stoat.

Fig. 19. Wilding pine control, Remarkables west face. *Neville Peat/DOC*

It is currently not feasible to attempt this sort of work in more than one or two places in southern New Zealand, because of limits on the necessary resources. Intensive ecosystem and species recovery work in a wider range of mainland locations will have to await the development of new efficient and effective pest and predator control tools. These tools are only likely to be developed in the long term.

Plant pests that threaten specific low stature species by out-competing them have already been mentioned. There are other plant pests that are having larger effects in southern New Zealand. For example, wilding exotic conifers are spreading in places and have already locally displaced indigenous grasslands and shrublands. Wilding Douglas fir are even invading some beech forest areas. This threat to indigenous ecosystems could in the longer term cause some plant species of the grasslands and shrublands to become extinct. On the way though, the distinctive open landscapes of the southern New Zealand high country would be forever changed in character on a scale too vast to be presently imagined. The more conspicuous wilding exotic conifer infestations can be seen spreading around Queenstown (Fig. 19), on Mid Dome in northern Southland, on the Blue Mountains and around Naseby Forest in the Maniototo. More insidious though, are the beginnings of infestations hidden from view deep in the Eyre and Kakanui Mountains, through the northern Southland tussock grasslands east of Mid Dome, on Maungatua west of Dunedin and part way up the Remarkables west face.

The Department of Conservation, with the support of volunteer groups such as the Royal Forest and Bird Protection Society, is endeavouring to control some of these infestations before it is too late to do so, and at the same time has had to overcome a degree of community ignorance and indifference. Exotic conifer plantings are on the increase in southern New Zealand, particularly of Douglas fir, so the sources of wildings could continue to grow. Control is possible through an increasing variety of techniques, but the department cannot succeed on its own. Broad community awareness and action is necessary for the large-scale threat to be averted, and support from district and regional councils is needed. Owners of wilding sources will have to take responsibility for management to limit their spread, and restrictions on plantings near 'seed takeoff' points should be included in District Plans. Such plans should also try to manage the adverse effects of exotic conifer plantings in areas that are outstanding or sensitive in landscape terms. It seems that New Zealand is at the cutting edge here as well, as there is virtually no known overseas experience to draw on for wilding tree control. The Department of Conservation has recently developed special techniques for dealing efficiently with wildings in isolated mountain locations, and is trialling various combinations of toxins. New technologies are urgently required, particularly for dealing with difficult species such as Douglas fir.

There are numerous other plant pests that threaten nature conservation values in Southern New Zealand. *Undaria* has already been mentioned in the marine context. The reclaiming grass *Spartina* is present in the New River, Catlins and Waikouaiti Estuaries, but is being controlled by the department. The smothering lakeweed *Lagarosiphon* sp. is being controlled in Lake Wanaka, but is growing rapidly in the shallower parts of Lake Dunstan. The other major southern lakes are free of *Lagarosiphon* at present but vigilance is being maintained by the authorities lest the special values of these lakes be lost. Willow growth is displacing native plant communities in some wetlands, and is inhibiting water movement. This can result in silt deposition that eventually leads to a loss of wetland characteristics. Woody weeds such as gorse and broom are spreading into native grasslands and along waterways. The flatweed *Hieracium pilosella* has taken over many areas in the drier depleted tussock grasslands. It inhibits the recovery of native grasses and herbs and may out-compete them for moisture. The more upright *Hieracium* species (tussock hawkweed *H. lepidulum* and king devil hawkweed *H. praeltum*) can dominate the intertussock community or form a dense sward in the forest or shrubland understorey. The Hieracium Control Trust has tested and released a number of biological control organisms for *H. pilosella* and there is some evidence that in moderate to high rainfall areas, an unburnt and ungrazed snow tussock cover will keep *Hieracium* spp. out. Many farmers reduce *Hieracium* seed production by grazing the flower heads but this is not a complete answer as the plant reproduces vegetatively.

In the long term, new technologies such as biological controls must be developed for the purposes of animal and plant pest management or eradication in mainland areas. Investments in scientific research must continue to be made for these purposes. Everyone involved in nature conservation should also be encouraged to continue to develop and share innovative forms of conservation management while 'on the job'. Techniques such as island eradications and 'Judas' goats should be developed and built upon, backed by sound scientific evaluation.

NGAI TAHU AND NATURE CONSERVATION

Ngai Tahu (or Kai Tahu in the southern Maori dialect) are the indigenous people of southern New Zealand. Ngai Tahu whanui (the wider family of Ngai Tahu) is a term which encompasses and recognises the earlier tribal groupings of Waitaha and Ngati Mamoe, who are now linked by whakapapa to the Ngai Tahu cultural overlay. Successively these peoples have occupied the southern landscape for hundreds of years, and in doing so formed a close association with the natural world in which they lived, especially their mahinga kai (those resources used for their enjoyment and survival). In their terms, they are the manawhenua, the people whose mana is closely associated with the southern lands.

The signing of the Treaty of Waitangi in 1840 marked a period of decline in their fortunes. Land sale agreements reserving certain resources to Ngai Tahu were not honoured. Their population was decimated by new diseases. The natural resources they traditionally relied upon were taken over and in many cases destroyed or polluted by the European settlers or the communities that grew up near them.

By 1997, the strength of the Ngai Tahu iwi had recovered, growing to comprise some 20,000 people of Ngai Tahu descent who have been registered on the tribal record. The cultural life of the iwi is at the present time centred on a number of papatipu runanga (local tribal groupings) and their marae (or traditional meeting places). The overall government of iwi affairs is carried out by a tribal parliament known as Te Runanga o Ngai Tahu.

Ngai Tahu wish to renew their traditional associations with the natural world, as an essential aspect of the maintenance and revival of their culture. Historically, the indigenous resources of southern New Zealand provided food, shelter, fibre, stone tools, items of adornment, treasured objects and trade items. A variety of different food resources were sought in different places during all seasons. Ngai Tahu people explored the landscape looking for useable resources. These included pounamu, a highly treasured stone (mainly nephrite, semi-nephrite and Bowenite) with the toughness and edge-holding characteristics of metal. The search for pounamu in particular resulted in major feats of exploration in inhospitable mountain lands without the equipment and clothing available to modern trampers and mountaineers.

Over the years of their occupation of the landscape, Ngai Tahu whanui became familiar with the slower rhythms of the colder southern lands, compared with the warmer north. They became familiar with the different southern fauna and flora, and learned that the food resources in particular were not inexhaustible. They developed social structures and concepts to manage the risks involved in relying on limited natural resources for survival. Central to this system was the concept of kaitiakitanga. Resources generally were a tribal responsibility, but specific resources became the responsibility of groups of families (hapu) or wider families (whanau), the leading members of which were identified as kaitiaki or guardians of that resource for the collective good. Individuals might also be recognised as holding special traditional knowledge about resources. These structures went further than what we know of today as structures for sustainable management. They also involved belief in the supernatural qualities of nature, which was used both to explain the workings of the natural world and to reinforce appropriate behaviours. In practice, restrictions on the use of natural resources decreed by the kaitiaki were declared ceremonially as rahui. Compliance with rahui was achieved by social cohesion and the exercise of rangatiratanga (chiefly authority), reinforced where necessary by a supernatural component or tapu.

Ngai Tahu involvement in the management of natural resources, however, had been marginalised by successive colonial governments and this situation has persisted until recently. A significant exception is the regime for the utilisation for food of titi or muttonbird on a group of islands around Stewart Island/Rakiura. Under the deed of cession of Stewart Island/Rakiura (1864), local Ngai Tahu and their descendants formally reserved the right to harvest titi from nominated islands, and this right has been carried forward into present-day law. Rakiura Maori determine the timing and extent of the titi harvest in order to sustainably manage the resource, with the Department of Conservation in an advisory role. Recently, the titi kaitiaki have begun to supplement their traditional knowledge by applying modern scientific research techniques.

The Department of Conservation is required to give effect to the principles of the Treaty of Waitangi by its charter, the Conservation Act 1987. But the full recognition of kaitiakitanga and rangatiratanga has been held up by a lack of certainty about what the principles of the Treaty really are and how they relate to legislation that seems to provide for the absolute protection of indigenous animals and plants. The resolution of these issues has, however, progressed as a result of the settlement of the Ngai Tahu claim. The claim was based on grievances arising from dishonoured land sales agreements and loss of access to and involvement with natural resource decision-making. The settlement is in three parts, an apology aimed at restoring the honour of the Crown, a package of assets aimed at providing an improved economic base for Ngai Tahu, and a set of measures aimed at reconnecting Ngai Tahu with the remaining natural resources in the southern landscape. It is the latter set of measures that has significant implications for nature conservation, both in terms of their scope and the fact that many of them have now been legislated for. The measures provide for Ngai Tahu involvement in processes rather than provide for specific outcomes. They are, however, comprehensive and far-reaching. Among the measures are formal recognition for resource management and conservation purposes of Ngai Tahu's interest in listed lakes, rivers and wetlands which are the remnants of their once-vast mahinga kai. They include reconnecting Ngai Tahu with the management of taonga (treasured) species, hitherto managed solely by the Crown. The traditional importance to Ngai Tahu of coastal food sources and freshwater fish is to be recognised through a number of measures including mahinga mataitai areas (sections of the coast where Ngai Tahu will manage inshore resources for traditional use purposes) and closer involvement in marine and freshwater fisheries decision-making and regulation.

Ngai Tahu understand that most of the natural resources of southern New Zealand are seriously depleted, and that resources that can sustain regular harvest, such as titi, are rare exceptions. They agree that with respect to many species, the practical goal for the foreseeable future is the avoidance of local or total extinction. They are supportive of habitat management for conservation purposes, and are already supporting habitat restoration projects. An example of this is the work of the Taieri Moturata Whanau on Moturata (off the mouth of the Taieri river) where rabbits are being controlled and manuka slash is being laid to halt the deterioration of the island's vegetation. Ngai Tahu (in conjunction with the Department of Conservation) have also carried out animal pest eradication programmes on islands off Stewart Island/Rakiura and have established programmes for the cultivation of pingao (a threatened plant of sand dunes, much valued for weaving) near numerous marae on the mainland. These examples suggest that the greater involvement of Ngai Tahu in the management of the region's natural resources should be welcomed.

THE FUTURE OF NATURE CONSERVATION IN SOUTHERN NEW ZEALAND

A quick stocktake of the state of nature conservation in the southern region shows a mixed picture. Of the natural (i.e. indigenous) ecosystems remaining on land, representative examples of most are safe from development, but their management for conservation purposes is fraught with challenges. Marine ecosystems are essentially unprotected and little understood. At the species level, recovery programmes are in place for most threatened birds, plants and reptiles. The conservation of freshwater fish and invertebrates is in its infancy. The protection of large areas of natural habitat is, however, probably achieving a great deal by way of conservation in relation to invertebrates and the less threatened vertebrate species. The conservation of natural (i.e. indigenous) landscapes is only occuring in the large protected areas, which are mainly at high altitude.

What is likely to happen in the future? Let us assume that human intervention in support of nature remains at present levels of effectiveness and community commitment for, say, the next 50 years. The most threatened species such as kakapo and South Island saddleback will have had to remain on offshore islands. The range of some threatened species on the mainland, such as mohua, South Island kaka and Otago's giant skinks, will have contracted back to places where intensive predator control is carried out. Whether or not we still have royal albatross and yellow-eyed penguin nesting on the southern mainland will depend to a large extent on what happens in the marine environment, but they will still have contracted back to nesting areas where protection from nest predators is carried out. This implies some reductions in their overall populations. Possum and feral goats will have continued to spread and alter vegetation communities, and will have been controlled in a small number of areas so that some examples of natural ecosystems can remain intact . Wilding conifers will have begun to alter the appearance of vast areas of the high country, and will have achieved a closed canopy in some places, thus eliminating the natural communities beneath. There will be 'showplace' areas where pest and weed control has allowed indigenous communities to survive, but outside these, native species will be succumbing to exotic invaders. The national parks and other large protected areas will still contain core areas that are largely intact, but these will not represent the full biological diversity of southern New Zealand that presently exists.

Is this acceptable? Is it avoidable? These are separate questions. The first is a philosophical question, and the second is a largely practical one. They are interwoven, however, because community insistence that a decline in indigenous biodiversity is unacceptable can influence the amount of effort applied to arresting the decline.

With present technology and resourcing committed to nature conservation, there are unlikely to be significant extinctions of southern New Zealand species in the future. Many of these species, however, will be inaccessible to the general public because they are on island refuges. That can satisfy the legitimate demand that the species be preserved for their intrinsic worth, but is this enough? It seems logical that community support for retaining indigenous biodiversity will be greater if that biodiversity can be appreciated more directly and preferably within an appropriate ecosystem context. The goal then becomes one of ensuring there is an opportunity for such appreciation, i.e. that as many threatened and non-threatened species as possible that naturally inhabited the southern New Zealand mainland, should be present on it in suitable places, in the appropriate ecosystem context so that people can experience them directly. Clearly, this should not be done if it puts the entire species at risk, so there may need to be 'insurance' populations in refuges. Most people, however, would prefer relatively easy access to living parts of their natural heritage, and would value the opportunity highly. More so, if the opportunity is authentic, such as a chance to see and hear blue duck on a swift, clear mountain stream instead of on a muddy pond, or a chance to see and hear kaka, kakariki, mohua, bellbird and brown creeper together in a rich beech forest rather than separately in aviary enclosures. These are the goals that should be pursued for the long term.

There is a significant public demand, however, for more than just a glimpse of some native species. New Zealanders in general and southerners in particular also value highly their opportunity to harvest some native species. The whitebait catch is made up of native migratory galaxiids, and there would be widespread support for a community conservation goal that improved the management of southern rivers and creeks for an increased harvest. Some attitude change will be required. Most whitebaiters complain about dwindling catches, but put nothing back into the management of the resource. There is a lot they could do, from supporting the regulations more conscientiously, to voluntary work on riparian fencing and planting programmes. The point to be made, though, is that if the notion of a sustainable harvest of whitebait is broadly acceptable, then surely the social conditions exist for an eventual acceptance that other indigenous species can be sustainably harvested as a way of reinforcing broad support for their protection.

The philosophical and social issues are more complex when it comes to dealing with ecosystems and landscapes. The 'protection of representative natural areas' policy goal of the Reserves and National Parks Acts may seem straightforward in principle, but there is no widely accepted classification of nature beyond the species level, and the legislation does not say how deep the subdivisions should be taken. For example, is it enough to secure a representative area dominated by the common snow tussock species? Or is a suite of areas required in which snow tussock is growing on a full range of soil types, aspects and under all possible variations in precipitation? Clearly the subdivisions can be taken to absurd lengths, and there is currently no way of knowing when the goal of representativeness has been attained. This difficulty is not insurmountable. It ought to be possible to nominate the main ecosystem types and give their protection and management first priority, and give a lower priority to subsets of those main types. Community support for this goal is not as apparent as in the case of preserving biodiversity at the species level.

Fig. 20. Camping in the Upper Long Burn, Eyre Mountains. *John Barkla/DOC*

There is also more room for argument. For example, there is a view that the high country tussock grasslands below the historic treeline are not natural, having been induced by fires lit by the first inhabitants in the landscape, and that the only truly natural areas in these places are the tiny 'museum piece' remnants of native trees or shrubs that survived the fires. This view, however, should be dismissed as too narrow. Pollen analysis shows that the rainshadow high country of southern New Zealand has been covered extensively with grasslands and shrublands thoughout post-glacial times and that accordingly present-day grassland ecosystems are no less natural than the indigenous forests that they have largely displaced. Even during times when forests predominated, there were grasslands in boggy hollows, on frost flats and intermontane basins and above the treeline. There is another argument about whether or not to allow natural succession to operate in indigenous communities. There is a strong case to be made that natural succession processes are an essential aspect of biological diversity. There is a competing argument that says succession should be interrupted in order to prevent extinctions of species. Even that is debatable, however, because extinctions can themselves be seen as natural events. Bearing in mind that people have so disturbed the natural world that no extinction from now on can be safely regarded as natural, the better view is probably that natural succession should be allowed to occur everywhere, unless a risk of extinction of a species or loss of a special ecosystem is evident.

Representative examples of the main southern ecosystem types are now under conservation management, but the task of improving them continues. Representative landscapes occur on a much larger scale. Some of the larger protected areas, both the existing ones and those that are emerging from processes like tenure review, will include representative landscapes. Others, such as the Upper Taieri Scroll plain wetland, may receive recognition under the Resource Management Act. The Environment Court has recently made a distinction between 'natural' and 'working' landscapes, with a higher degree of modification being indicated for the latter. The representative natural landscapes of southern New Zealand will be found in the former category.

But it can hardly be said that there is a broad community consensus on these matters, nor any consensus about what the ecosystem and landscape preservation goals for southern New Zealand should be. Some leadership is required, and concurrently the public should be informed and engaged in these issues. This book contributes to that process.

The goals of the future should be to retain the indigenous species, ecosystem and landscape diversity and natural processes of the region in a way and in places that are accessible to people. Extensive protected areas and altitudinal sequences of ecosystems may be necessary to accommodate the dynamic processes that appear to be an integral component, as well as a predictable effect, of global warming. This chapter will conclude with some reflections on the achievability of these goals, and what is required to attain them.

The principal practical obstacles to the achievement of these goals are mainly to do with the efficiency and effectiveness of predator and ecosystem pest control (both plants and animals) and issues of reserve design. Long-term government-sponsored research has been under way for some years to underpin and eventually find a biological control tool for possums. Experiments have begun on the Otago coast to see if the predators of yellow-eyed penguin chicks can be controlled by secondary poisoning, i.e. through the poisoning of their prey species. The Department of Conservation is fine-tuning its 'Judas' goat and thar techniques and planning to extend them to other animal pests, in order to attain higher degrees of control efficiency. A growing range of slow-acting biological control agents is being trialled on plant pests. These include weevils, rusts, mites and the larvae of various other invertebrates. All of these developments have occurred in the last ten years and it is not unrealistic to suppose that control agents and improved efficiencies yet undreamt of will be developed in the next 50, provided the commitment to research is maintained. It is probable that new pests will emerge during that period also, but provided there is vigilance and early control, these should not become as unmanageable as the ones the country is currently faced with. The current emphasis on altitudinal sequences, connectance and buffering in reserve design should provide increased security for dealing with the impacts of global warming on nature conservation.

Education and community support for conservation is also crucial to the achievability of these goals. For example, there has been a marked shift in landowner recognition of the threat posed by wilding exotic conifers in southern New Zealand's open indigenous landscapes. Many landowners have turned this into action, and now swing the grubber at conifer seedlings as well as thistles. This shift has in turn influenced regional councils to include conifer species in their pest control plans. With a further consolidation of that shift, councils might be motivated to place conditions on forestry land-use consents to avoid the planting of trees at seed take-off points or in sensitive positions in the landscape. Plantation owners may take greater responsibility for wildings. Community support can also manifest itself in greater voluntary involvement in conservation work, such as wilding conifer control and habitat restoration planting. Voluntary involvement can be highly effective, for example the Yellow-eyed Penguin Trust with the support of a commercial sponsor has acquired and is restoring a number of significant nesting habitat areas on the Otago coast. Commercial support for kakapo conservation is enabling increased activity. Forest and Bird branches in the south run numerous conservation projects on the ground. Ngai Tahu from their cultural perspective, and as an increasingly influential and well-educated and resourced section of the southern New Zealand community, are likely to be involved in more and more projects and issues that have conservation benefits.

Southern New Zealand is well placed for community support for conservation to increase and to produce tangible results. So much of the southern landscape is natural (i.e. with indigenous vegetation and striking natural landforms still dominant, see page 367) and awareness and appreciation of this is growing, along with the increasing awareness in society generally that nature is there to be nurtured, not 'tamed'. Awareness of the processes that are working for and against these qualities in the southern landscape is growing also. Natural southern landscapes and features are valued as adding to and sustaining economic activity generated by overseas tourists and domestic visitors. These landscapes are appreciated by more and more of those who work, live or play in them (Fig. 20). Fewer people are seeing protected natural areas as 'locked up', and it is important that conservation managers and the owners of natural features ensure that access is available to the public in all but the most clearly justified situations.

Southern New Zealanders are blessed with a unique and internationally significant opportunity to ensure that their natural heritage is retained for its intrinsic value and for the appreciation and enjoyment of others now and in the future. This is a responsibility as well as a blessing.

TENURE REVIEW OF HIGH COUNTRY PASTORAL LEASEHOLD LAND

Most of the 2.6 million ha of tussock grasslands and associated high mountain lands, east of the Southern Alps in central and eastern South Island, generally referred to as the 'high country' (some ten per cent of New Zealand's land area) is Crown land which is leased from the government for pastoral farming. Of the 340 high country 'runs', some 182 are in Otago-Southland, occupying about 1.2 million ha. The average size of a run is about 7700 ha, but some are much bigger and may occupy whole mountain range and valley systems, as with Glenavy Station (about 54,000 ha) in northern Southland.

The lessees have certain rights: occupation of the property, use of its pasturage or grazing, determination of who enters the property (that is, trespass rights) and renewal of the lease on the original terms every 33 years. In addition, a lessee can apply to undertake activities associated with farm development, such as cultivation, vegetation clearance, fencing, burning, exotic forestry, etcetera. These approvals must be obtained from the department which oversees the administration and management of the leases: formerly Lands and Survey, then Landcorp, and now Land Information New Zealand (LINZ). None of this vast area was formally protected for conservation or baseline research until the late 1960s.

There has been serious land degradation (soil erosion and nutrient loss, animal pests, weed invasion and so on) over much of the South Island high country and at lower altitudes there has been extensive development, with aerial oversowing and topdressing on the more responsive land. Clearance of indigenous vegetation (tussock grasslands, native shrublands and wetlands) has occurred through drainage and/or cultivation, with replacement, usually by exotic pasture.

Among runholders there has been increasing interest in diversifying away from pastoral farming to more intensive land use, including subdivision and particularly various forms of tourism. Public interests in conservation and recreation, on the other hand, have been keen to see some high country retired from farming and managed to protect these particular values and activities.

The government recognised these diverse interests and, wishing to cease being a landlord, initiated the process of tenure review in the early 1990s.This is a formal process under special legislation (Crown Pastoral Land Act 1998), whereby a lessee applies to LINZ to have the lease reviewed. The outcome of this review may allow the runholder to freehold the more productive land while areas valuable for conservation (with 'significant inherent values'), as well as for landscape and recreation values, would revert to the Crown, to be managed in the public interest by the Department of Conservation.

The government is promoting tenure review because it presents a major opportunity for permanent re-allocation of much of the South Island high country. It is a one-off chance to correct a major imbalance in New Zealand's protected natural areas system by including in our national parks and reserves network adequately representative areas of tussock grasslands, shrublands, wetlands and associated mountain lands in diverse environments, in the rainshadow region of the South Island. It should also secure adequate and permanent access for public use and enjoyment of these areas.

The Department of Conservation has an important role in identifying the land needed for conservation/recreation management and also ensuring adequate public access to it. Preliminary proposals, following negotiation between a lessee and LINZ, with inputs from the department, are released for public submissions at an advanced stage of the process. Outcomes that are satisfactory to both the lessee and the government are then finalised. However, a lessee can withdraw from the process at any stage and a few have done so. Reviews completed to date have provided some important additions to the public conservation lands in southern New Zealand. The upper slopes of the Old Man Range now comprise the Kopuwai Conservation Area, through the review of Earnscleugh Station. The Te Papanui Conservation Park on the Lammerlaw-Lammermoor Ranges was officially opened by the Minister of Conservation in March 2003, and resulted largely from tenure reviews of Rocklands (above) and Halwyn Stations. The Remarkables Conservation Park, Pisa Conservation Park and Oteake Conservation Park (on the St Bathans, Hawkdun and Ida Ranges), outlined in the Department of Conservation's Otago Conservation Management Strategy of 1998, are also starting to take shape through tenure review of several runs.

The outcome of tenure review, when completed, is likely to result in a much more representative network of protected natural areas than seemed possible even a decade ago. Considerable challenges remain, however. Some tenure reviews have failed to deliver the conservation benefits that were assumed from the legislation. Considerable areas with high conservation and/or recreation values, including some high mountain areas, as well as most low- to mid-altitude areas, have been free-holded, sometimes with covenants attached, but with little or no restrictions on grazing. Even though they often have high invertebrate diversity and habitat values, low- to mid-altitude ecosystems are inadequately protected through tenure review because they are usually more modified and have higher production potential. Access provisions in several cases are also considered to be inadequate. As well, there have been problems with the large number of properties being reviewed at one time, which can affect the quality of the response that is possible on some runs. Since this is a one-off opportunity, there is general agreement that the current 'carve-up' of the South Island high country must be carried out in a responsible and mutually satisfactory way.

Alan Mark and William Lee

Above: View of Te Papanui Conservation Park across the rolling landscape of the Lammermoor Range of the eastern Otago uplands. The dominant subalpine narrow-leaved snow tussock grassland shown here characterises much of this 20,000 ha park.
Gilbert van Reenen

FURTHER READING

1 Geology

Several excellent textbooks on general geology are available, most very well illustrated. Publications on New Zealand geology include maps and field guides from the Crown Research Institute, the Institute of Geological and Nuclear Sciences. Field guides are also produced by the Geological Society of New Zealand, whose Misc. Publication 82 *Earth Science Resource List* is a useful compilation of local titles.

2 Landforms

Bishop, D.G. and Forsyth, P.J. 1988. *Vanishing Ice: an introduction to glaciers based on the Dart Glacier.* John McIndoe & New Zealand Geological Survey, Dunedin. 56 pp.

Turnbull, I.M. (compiler) 2001. Wakatipu Sheet, 1: 250 000 Geological Map, Institute of Geological and Nuclear Sciences.

3 Fossils and the History of Life

Beu, A.G., Maxwell, P.A., and Brazier, R.C. 1990. Cenozoic Mollusca of New Zealand. *New Zealand Geological Survey paleontological bulletin* 58: 1–518.

Fleming, C.A. 1959. Oceania. Fascicule 4. New Zealand. (Lexique Stratigraphique International, 6) Centre National de la Recherce Scientifique, for Stratigraphic Commission of International Geological Congress, Paris. 527 pp.

Fordyce, R.E. 1991. A new look at the fossil vertebrate record of New Zealand. Pages 1191–1316 in P.V. Rich, J.M. Monaghan, R.F. Baird, and T.H. Rich (eds), *Vertebrate palaeontology of Australasia.* Melbourne: Pioneer Design Studio and Monash University. 1437 pp.

Gage, M. 1957. The geology of Waitaki subdivision. *New Zealand Geological Survey bulletin* n.s. 55: 1–135.

Hornibrook, N. de B., Brazier, R.C., and Strong, C.P. 1989. Manual of New Zealand Permian to Pleistocene foraminiferal biostratigraphy. *New Zealand Geological Survey paleontological bulletin* 56: 1–175.

Landis, C.A., Campbell, H.J., Aslund, T., Cawood, P.A., Douglas, A., Kimbrough, D.L., Pillai, D.D.L., Raine, J.I., and Willsman, A. 1999. Permian-Jurassic strata at Productus Creek, Southland, New Zealand: implications for terrane dynamics of the eastern Gondwanaland margin. *New Zealand Journal of Geology and Geophysics* 42 (2): 255–278.

Pole, M.S. 1993. Early Miocene flora of the Manuherikia Group, New Zealand. Parts 4-10. *Journal of the Royal Society of New Zealand* 23 (4): 283–426.

Speden, I.G. and Keyes, I.W. 1981. Illustrations of New Zealand fossils. *Department of Scientific and Industrial Research Information Series* 150: 1–109.

Suggate, R.P., Stevens, G.R., and Te Punga, M.T. (eds). 1978. *The geology of New Zealand* (2 vols). Wellington: Government Printer. 820 pp.

Thornton, J. 1985. *Field guide to New Zealand geology. An introduction to rocks, minerals and fossils.* Reed Methuen, Auckland. 226 pp.

4 Climate

Bishop, G. and Forsyth, P. J. 1988. *Vanishing Ice: an introduction to glaciers based on the Dart Glacier.* Dunedin: John McIndoe & New Zealand Geological Survey.

Brenstrum, E. 2001. *The New Weather Book.* Nelson: Craig Potton Publishing. 128pp.

Brown, M.L. 1959. Typical rainfall patterns over Central Otago. *New Zealand Journal of Geology and Geophysics,* 2, 84–94.

Coulter, J.D. 1975. The Climate. In: Kuschel, G. (ed.). *Biogeography and Ecology of New Zealand.* The Hague: Dr W. Junk B.V. Publishers.

Fitzharris, B.B. 1988. The effect of greenhouse gas warming on a region: a case study of the province of Otago. In: Ministry for the Environment. *Climate Change, the New Zealand Response.* Wellington. Chapter 13, 130–134.

Fitzharris, B.B. and Endlicher, W. 1996. Climatic conditions for wine grape growing, Central Europe and New Zealand. *New Zealand Geographer,* 52 (1), 1–11.

Garnier, B.J. 1958. *The Climate of New Zealand.* London.

Mark, A.F. 1965. Vegetation and mountain climate. In: Lister, R.G. and Hargreaves, R.P. (eds). *Central Otago,* Christchurch: New Zealand Geographical Society. 69–91.

Maunder, W.J. 1965. Climatic Character. In: Lister, R.G. and Hargreaves, R.P. (eds). *Central Otago,* Christchurch: New Zealand Geographical Society. 69–91

Maunder, W.J. 1971. Climatic Areas of New Zealand. In: J. Gentilli (ed.). *World Survey of Climatology, 13: Climates of Australia and New Zealand.* Amsterdam: Elsevier. 265–68.

Neale, A.A. and Thompson, G.H. 1977: Meterological conditions accompanying heavy snow falls in Southern New Zealand. *New Zealand Meterological Service Technical Information Circular No. 155.*

New Zealand Meteorological Service, 1983. *Climate Regions of New Zealand.* Misc. Publication No. 174, Wellington (+ map).

Sturman, A. and Tapper, N. 1996. *The Weather and Climate of Australia and New Zealand.* Melbourne: Oxford University Press. 476 pp.

5 Biogeography

Dugdale, J.S. 1994. Hepialidae (Insecta: Lepidoptera). *Fauna of New Zealand* 30. Lincoln: Landcare Research. 163 pp.

Emerson, B.C., Wallis, G.P., Patrick, B.H. 1997. Biogeographic area relationships in southern New Zealand: a cladistic analysis of Lepidoptera distributions. *Journal of Biogeography* 24: 89–99.

Heads, M. 1989. Integrating earth and life sciences in New Zealand natural history. *New Zealand Journal of Zoology* 16: 549–585.

Heads, M. 1990. A taxonomic revision of Kelleria and Drapetes (Thymelaeaceae). *Australian Systematic Botany* 3: 595–652.

Heads, M. 1994c. Biogeography and evolution in the Hebe complex: (Scrophulariaceae): Leonohebe and Chionohebe. *Candollea* 49: 81–119.

Heads, M. 1997. Regional patterns of biodiversity in New Zealand. *Journal of the Royal Society of New Zealand* 27: 337–354.

Patrick, B.H. 1990a. Panbiogeography and the general naturalist with special reference to conservation implications. *New Zealand Journal of Zoology* 16: 749–756.

Patrick, B.H. 1990b. Occurrence of an upland grassland moth in a coastal saltmarsh in Otago. *Journal of the Royal Society of New Zealand.* 20: 305–307.

Patrick, B.H. 1994a. *Lepidoptera of the Southern Plains and Coast of New Zealand.* Department of Conservation, Dunedin. Misc. Ser. 17. 44 pp.

Patrick, B.H. 1994b. *Valley floor Lepidoptera of Central Otago.* Department of Conservation, Dunedin. Misc. Ser. 19. 65 pp.

Patrick, B.H., Rance, B.G., Barratt, B.I.P., Tangney, R., 1987. *Entomological survey – Longwood range.* Department of Conservation, Dunedin. 86 pp.

Patrick, B.H., Rance, B.D., Barratt, B.I.P. 1992. *Alpine insects and plants of Stewart Island.* Department of Conservation, Dunedin. 66 pp.

6 Environmental Change

Heusser, L.E. and G. van der Geer, 1994. Direct correlation of terrestrial and marine paleoclimatic records from four glacial-interglacial cycles – DSDP site 594 southwest Pacific. *Quaternary Science Reviews,* 13:273–282.

McGlone, M.S., Anderson, A.J., Holdaway, R.N. 1994. An ecological approach to the Polynesian settlement of New Zealand. In: D.G. Sutton (ed.). *The origins of the first New Zealanders.* Auckland: Auckland University Press. pp. 136–163.

McGlone, M.S., Mark, A.F., Bell, D. 1995. Late Pleistocene and Holocene vegetation history, Central Otago, South Island, New Zealand. *Journal of the Royal Society of New Zealand* 25:1–22.

McGlone, M.S., Mildenhall, D.C., Pole, M.S. 1996. History and paleoecology of New Zealand Nothofagus forests. T.T. Veblen, R.S. Hill, and J. Read. (eds) *The Ecology and Biogeography of* Nothofagus *forest.* New Haven: Yale University Press. pp. 83–130.

McGlone, M.S., Wilmshurst, J.M. 1999. A Holocene record of climate, vegetation change and peat bog development, east Otago, South Island, New Zealand. *Journal of Quaternary Science* 14: 239–254.

McGlone, M.S., Wilson, H.D. 1996. Holocene vegetation and climate of Stewart Island, New Zealand. *New Zealand Journal of Botany* 34: 369–388.

McKinnon, M. (ed.) 1997. *New Zealand Historical Atlas.* Auckland: David Bateman.

Nelson, C.S., Hendy, C.H., Jarrett, G.R., Cuthbertson, A.M. 1985. Near-synchroneity of New Zealand alpine glaciations and Northern Hemisphere continental glaciations during the past 750 kyr. *Nature* 318: 361–363.

Stevens, G., McGlone, M.S., McCulloch, B. 1988. *Prehistoric New Zealand.* Auckland: Heinemann Reed.

Wardle, P. 1991. *Vegetation of New Zealand.* England: Cambridge University Press.

Worthy, T. H., 1998. Quaternary fossil faunas of Otago, South Island, New Zealand. *Journal of the Royal Society of New Zealand* 28(3): 421–521.

Worthy, T. H., 1998. The Quaternary fossil avifauna of Southland, South Island, New Zealand. *Journal of the Royal Society of New Zealand* 28(4): 537–589.

Worthy, T.H. 1999. The role of climate change versus human impacts – avian extinction on South Island, New Zealand. In: Olson, Storrs L. (ed.). Avian paleontology at the close of the 20th Century: Proceedings of the 4th international Meeting of the Society of Avian Paleontology and Evolution, Washington, D.C. 4–7 June 1996. *Smithsonian Contributions to Paleobiology* 89.

7 The Human Factor

Allen, R.B., Partridge, T.R., Lee, W.G. and Efford, M. 1992. Ecology of *Kunzea ericoides* (E.Rich) J Thompson (kanuka) in east Otago, New Zealand. *New Zealand Journal of Botany* 30: 135–150.

Anderson, A.J. 1981. A fourteenth-century fishing camp at Purakanui Inlet, Otago. *Journal of the Royal Society of New Zealand* 11: 201–221.

Anderson, A.J. 1981. Pre-European hunting dogs in the South Island, New Zealand. *New Zealand Journal of Archaeology* 3: 15–20.

Anderson, A.J. 1983. *When all the moa ovens grew cold.* Dunedin: Otago Heritage Books.

Anderson, A.J. 1989. *Prodigious Birds.* Cambridge University Press.

Anderson, A.J. 1991. The chronology of colonisation in New Zealand. *Antiquity* 65: 767–795.

Anderson, A.J., Allingham, B. and Smith, I. (eds). 1996. *Shag River Mouth The Archaeology of an Early Southern Maori Village.* Canberra: ANH Publications.

Anderson, A.J. and McGlone, M. 1992. Living on the edge – prehistoric land and people in New Zealand. In: Dodson, J. (ed.) *The Naive Lands Prehistory and Environmental Change in Australia and the Southwest Pacific.* pp.199–241. Melbourne: Longman Cheshire.

Anderson, A.J. and McGovern-Wilson, R. 1990. The pattern of prehistoric Polynesian colonisation in New Zealand. *Journal of the Royal Society of New Zealand* 20: 41–63.

Anderson, A.J. and McGovern-Wilson, R. 1991. Beech Forest Hunters. *New Zealand Archaeological Association Monograph 18.*

Anderson, A.J. and Smith, I.W.G. 1992. The Papatowai site: new evidence and interpretations. *Journal of the Polynesian Society* 101: 129–158.

Anderson, A J and Smith, I.W.G. 1996 The transient village in southern New Zealand. *World Archaeology* 27: 359–371.

Anderson, A.J., Worthy, T., and McGovern-Wilson, R. 1996. Moa remains and taphonomy. In: Anderson, A.J., Allingham, B. and Smith I. (eds), *Shag River Mouth The Archaeology of an Early Southern Maori Village.* Canberra: ANH Publications.

Atkinson, I.A.E. and Moller, H. 1987. Kiore, Polynesian rat. In: King, C.M. (ed.) *The Handbook of New Zealand Mammals.* Auckland: Oxford University Press.

Bourner, T.C., Glare, T.C., O'Callaghan, M. and Jackson, T.A. 1996. Towards greener pastures – pathogens and pasture pests. *New Zealand Journal of Ecology* 20: 101–107.

Boyd, B., McGlone, M., Anderson, A. and Wallace, R. 1996. Late Holocene vegetation history at Shag River Mouth. In: Anderson, A.J, Allingham, B. and Smith, I. (eds), *Shag River Mouth The Archaeology of an Early Southern Maori Village.* Canberra: ANH Publications.

Coutts, P. 1982. Fiordland. In: Prickett, N. (ed.) *The First Thousand Years* Palmerston North: Dunmore Press.

Davidson, Janet. 1984. *The Prehistory of New Zealand.* Auckland: Longman Paul, Ltd.

Department of Conservation.n.d. Conservation Management Strategy: Stewart Island. Unpublished report, Department of Conservation, Invercargill.

Fisher, M.E., Satchell, E. and Watkins J.M. 1970. *Gardening with New Zealand Plants, Shrubs and Trees.* Auckland: Collins.

Gibb, J.A. and Williams, J.M. 1987. European rabbit. In: King, C.M. (ed.) *The Handbook of New Zealand Mammals.* Auckland: Oxford University Press.

Hamel, G.E. 1978. Hawksburn Revisited: an ecological assessment. *New Zealand Archaeological Association Newsletter* 22:116–128.

Hamel, G.E. 1980. Pounawea, the last excavation. Unpublished report to the NZ Historic Places Trust.

Holdaway, R. 1989. New Zealand's pre-human avifauna and its vulner-ability. *New Zealand Journal of Ecology* 12 (Supplement): 11–25).

Holdaway, R. 1999. A spatio-temporal model for the invasion of the New Zealand archipelago by the Pacific rat *Rattus exulans. Journal of the Royal Society of New Zealand* 29: 91–105.

Howard, Basil. 1940. *Rakiura, a history of Stewart Island, New Zealand.* A.H. and A.W. Reed, Dunedin.

Johnson, Peter. 1986. *Wildflowers of Central Otago.* Dunedin: John McIndoe.

King, C.M. 1987. Stoat. In: King, C.M. (ed.) *The Handbook of New Zealand Mammals.* Auckland: Oxford University Press.

Leach, H.M. 1994. Native plants and national identity in New Zealand gardening: an historical review. *Horticulture in New Zealand* 5: 28–33.

Lockerbie, L. 1959. From Moa-hunter to Classic Maori in southern New Zealand. In: J.D. Freeman and W.R. Geddes (eds). *Anthropology in the South Seas,* pp 75–110. New Plymouth: Avery.

McGlone, M.S., Anderson, A.J., and Holdaway, R.N. 1994. An ecological approach to the settlement of New Zealand. In: Sutton, D.G. (ed.) *The origins of the first New Zealanders.* Auckland University Press.

McGovern-Wilson, R., Kirk, F., and Smith, I. W.G. 1996. Small bird remains. In: Anderson, A.J., Allingham, B. and Smith, I. (eds), *Shag River Mouth The Archaeology of an Early Southern Maori Village.* Canberra: ANH Publications.

McIlroy, J.C. 1987. Feral Pig. In: King, C.M. (ed.) *The Handbook of New Zealand Mammals.* Auckland: Oxford University Press.

Metcalf, L. J. 1972. *The Cultivation of New Zealand Trees and Shrubs.* Auckland: Reeds.

Roberts, W.H.S. 1895. *Southland in 1856.* Invercargill: Southland Times.

Smith, I.W.G. 1989. Maori impact on the marine megafauna: Pre-European distributions of New Zealand sea mammals. In: Sutton, D.G. (ed.) Saying so doesn't make it so. Papers in honour of B. Foss Leach. *New Zealand Archaeological Association Monograph* 17.

Teviotdale, D. 1932. The material culture of the moa hunters in Murihiku.*Journal of the Polynesian Society* 41: 81–120.

8 Forests and Shrubland

Daugherty, C.H., Patterson, G.B. and Hitchmough, R.A. 1994. Taxonomic and conservation review of the New Zealand herpetofauna. *N. Z. J. Zool.* 21: 317–323.

Forster, R.R. & L.M. 1973. *New Zealand Spiders. An Introduction.* Auckland: Collins.

Gill, B. and Whitaker, T. 1996. *New Zealand frogs and reptiles.* Auckland: David Bateman.

Heather, B.D. and Robertson, H.A. 1996. *The Field Guide to the Birds of New Zealand.* (Illustrated by Derek J. Onley). Auckland: Viking.

Miller, D. 1971. *Common Insects in New Zealand.* Wellington: Reed. 178 pp.

Wardle, P. 1991. *Vegetation of New Zealand.* Cambridge University Press. 672 pp.

Wilson, H.D. 1987. Vegetation of Stewart Island, New Zealand: a supplement to the New Zealand Journal of Botany. Wellington: DSIR Science Information Publishing Centre. 131 pp.

9 Tussock Grasslands

Barlow B.A. 1986. Flora & Fauna of Alpine Australasia. *Ages & Origins.* CSIRO.

Bigelow R.S. 1967. *The Grasshoppers of New Zealand.* Christchurch: University of Canterbury.

Cockayne, L. 1967. *New Zealand Plants and their Story.* 4th ed. Wellingotn: Government Printer.

Galloway, D.J. 1985. *Flora of New Zealand – Lichens.* Wellington: Government Printer.

Gibbs G.W. 1980. *New Zealand Butterflies,* Auckland: Collins.

Gill, B. and Whitaker, T. 1996. *New Zealand frogs and reptiles.* Auckland: David Bateman.

Mark, A.F. 1992. Indigenous grasslands of New Zealand. In: R.T. Coupland (ed). *Natural grasslands – Eastern Hemisphere: Ecosystems of the World* 8B. Amsterdam: Elsevier. pp. 361–410.

Mark, A.F., Dickinson, K.J.M. 1996. New Zealand alpine ecosystems. Ch. 14. In: F.E. Wielgolaski (ed.). *Polar and Alpine Tundra. Ecosystems of the World 3.* Amsterdam: Elsevier.

Patrick B.H., Rance B.D. & Barratt B.I.P. 1992. *Alpine Insects and Plants of Stewart Island.* Department of Conservation, Dunedin. 66 pp.

Wilson, H.D. 1987. Vegetation of Stewart Island, New Zealand. *N.Z. J. Bot. Supplement.* 131 pp.

10 Inland Waters and Wetlands

Chapman, M.A. and M.H. Lewis. 1976. *An Introduction to the Freshwater Crustacea of New Zealand.* Auckland: Collins.

Collier, K.J. and M.J. Winterbourn (eds). 2000. *New Zealand Stream Invertebrates: Ecology and implications for management.* Christchurch: New Zealand Limnological Society.

McDowall, R.M. 1990. *New Zealand Freshwater Fishes: a natural history and guide.* Auckland: Heinemann.

Rowe, R. 1987. *The Dragonflies of New Zealand.* Auckland University Press, 260 pp.

Winterbourn, M.J. 1973. A guide to the freshwater mollusca of New Zealand. *Tuatara* 20: 141–159.

Winterbourn, M.J., K.L.D. Gregson and C.H. Dolphin. 2000. Guide to the aquatic insects of New Zealand. *Bulletin of the Entomological Society of New Zealand* 13 (3rd edition).

11 The Coast

Adams, N.M. 1994. *Seaweeds of New Zealand: an Illustrated Guide.* Christchurch: Canterbury University Press.

Batham, E. J. 1956. Ecology of southern New Zealand sheltered rocky shore. *Transactions of the Royal Society of New Zealand* 84: 447–465.

Batham, E. J. 1958. Ecology of southern New Zealand exposed rocky shore at Little Papanui, Otago Peninsula. *Transactions of the Royal Society of New Zealand* 85: 647–658.

Bradstock, M. 1985. *Between the Tides: New Zealand Shore and Estuary Life.* Auckland: Reed Methuen.

Francis, M. 1996. *Coastal Fishes of New Zealand: an Identification Guide.* Auckland: Reed.

Healy, T.R. and Kirk, R.M. 1992. Coasts. In: J. M. Soons and M. J. Selby (eds.), *Landforms of New Zealand,* 2nd edition. pp. 161–186. Auckland: Longman Paul Ltd.

Heather, B. D. and Robertson, H. A. 1996. *The Field Guide to the Birds of New Zealand.* Auckland: Viking.

Jones, M.B. 1983. *Animals of the Estuary Shore: Illustrated Guide and Ecology.* Christchurch: University of Canterbury.

Morton, J., and Miller, M. 1973. *The New Zealand Sea Shore,* 2nd edition. London: Collins.

Paulin, C. and Roberts, C. 1992. *The Rockpool Fishes of New Zealand.* Wellington: Museum of New Zealand Te Papa Tongarewa.

Robertson, C.J.R. (ed.). 1985. *Reader's Digest Complete Book of New Zealand Birds..* Sydney: Reader's Digest.

Ryan, P. and Paulin, C. 1998. *Fiordland Underwater: New Zealand's Hidden Wilderness.* Auckland: Exisle Publishing Ltd.

12 The Open Sea

Ayling, T. and Cox, G.J. 1982. *Collins Guide to the Sea Fishes of New Zealand.* Auckland: Collins.

Baker, A.N. 1990. *Whales and Dolphins of New Zealand and Australia: an Identification Guide.* Wellington: Victoria University Press.

Carter, L., Carter, R. M., McCave, I. N. and Gamble, J. 1996. Regional sediment recycling in the abyssal Southwest Pacific Ocean. *Geology* 24: 735–738.

Carter, R. M., Carter, L., Williams, J. J. and Landis, C. A. 1985. Modern and relict sedimentation on the South Otago continental shelf, New Zealand. *New Zealand Oceanographic Institute Memoir* 93: 43 p.

Crawley, M..C. 1990. Suborder Pinnipedia. In: C.M. King (ed.), *The Handbook of New Zealand Mammals,* pp. 243-280. Auckland: Oxford University Press.

Dawson, S. and Slooten, E. 1996. *Downunder Dolphins: the Story of Hector's Dolphin.* Christchurch: Canterbury University Press.

Harcourt, R.G., Schulman, A.M., Davis, L.S. and Trillmich, F. 1995. Summer foraging by lactating New Zealand fur seals (*Arctocephalus forsteri*) off Otago Peninsula, New Zealand. *Canadian Journal of Zoology* 73: 678–690.

Harper, P.C. 1987. Feeding behaviour and other notes on 20 species of Procellariiformes at sea. *Notornis* 34 (3): 169–192.

Jillett, J.B. 1969. Seasonal hydrology of waters off the Otago Peninsula, southeastern New Zealand. *New Zealand Journal of Marine and Freshwater Research* 3: 349–375.

Jillett, J.B. 1976. Zooplankton associations off Otago Peninsula, southwestern New Zealand, related to different water masses. *New Zealand Journal of Marine and Freshwater Research* 10: 543–557.

Paul, L. 2000. *New Zealand Fishes: Identification, Natural History & Fisheries.* Auckland: Reed.

Westerskov, K. and Probert, K. 1981. *The Seas around New Zealand.* Wellington: Reed.

INDEX

*References in **bold** indicate items that appear in captions to illustrations. Check both caption and main text on these pages.*